Essential Mathematics for Engineering

Essential Mathematics for Engineering

W. Bolton

Butterworth-Heinemann
Linacre House, Jordan Hill, Oxford OX2 8DP
A division of Reed Educational and Professional Publishing Ltd

A member of the Reed Elsevier plc group

OXFORD BOSTON JOHANNESBURG
MELBOURNE NEW DELHI SINGAPORE

First published 1997

British Library Cataloguing in Publication Data
A catalogue record for this book is available from the British Library

ISBN 0 7506 3621 1

Printed and bound in Great Britain by Hartnolls Limited, Bodmin, Cornwall

Contents

Preface

Aims

- To provide an accessible, readable introduction to engineering mathematics for university and college students at degree and Higher National Diploma/Certificate levels.
- To smooth the transition from school and college mathematics at Advanced level, GNVQ Advanced level and BTEC National Diploma/Certificate, not making too many assumptions regarding competence and understanding of mathematics at these levels.
- To show how mathematics is relevant to engineering through the incorporation of specific chapters showing the applications of mathematics in engineering.

Audience

This book is aimed at engineering students starting BSc, BEng or Higher National Diploma/Certificate courses. It has not been assumed that the students are starting with a good grasp of mathematics.

Format

The book is divided into a number of parts, each containing chapters concerned with principles and an application chapter illustrating the relevance of those principles to some topics in engineering. Within each chapter, following a discussion of principles there are worked examples, followed by revision problems. At the end of each chapter there are further problems. Answers to all the problems are given at the end of the book.

Content

The range of material covered in this book is seen as appropriate for a starting course in mathematics for degree and Higher National Diploma/Certificate engineering students. While primarily designed for engineers, this book is also likely to be relevant to students of the physical sciences. Mathematics traditionally at this level has been included, together with topics such as the Fourier, Laplace and z-transforms which are becoming of increasing relevance to engineers.

Part 1 deals with functions, including linear, polynomial, trigonometric, exponential, logarithmic and hyperbolic functions, and sequences and sets. Most students are likely to be familiar with most of this part and thus will act as a review of basic principles. The application chapter shows how engineers represent signals by functions.

Part 2 is devoted to complex numbers, considering them in polar form, Cartesian form and exponential form. The application chapter considers their use in representing phasors in electrical circuit analysis.

Part 3 provides an introduction to vector algebra, considering the basic rules of addition and subtraction, representation in terms of unit vector components and scalar and vector products. The application chapter shows the use of such algebra in mechanics.

Part 4 is an introduction to discrete mathematics, covering such topics as set theory, Boolean algebra and logic gates. The application chapter shows the use of such techniques in digital systems.

Part 5 concerns linear algebra and includes Gaussian elimination, matrices, determinants, eigenvalues and eigenvectors, and iteration methods. The application chapter shows the use of such methods in the solution of problems of circuit analysis.

Part 6 is devoted to calculus, there being chapters on differentiation, numerical differentiation, partial differentiation, integration and numerical integration. The application chapter shows the use of calculus in the solutions of problems involving areas, volumes, moments, means and root-mean-square values.

Part 7 provides an introduction to ordinary differential equations, including chapters on modelling with such equations, the solution of problems involving first-order and second-order differential equations and numerical methods. The application chapter shows how such equations arise in studies of the dynamic response of engineering systems.

Part 8 is devoted to the Fourier series, considering the series for periodic and non-periodic functions and represented by both trigonometric and exponential forms of the series. An introduction is included showing the development of the Fourier transform from the Fourier series. A chapter on harmonic analysis is included, the application chapter showing the use of the Fourier series and transform in the determination of the responses of systems to various forms of input.

Part 9 provides and introduction to the Laplace and *z*-transforms, the application chapter showing how the transforms can be used in analysing the response of systems.

Part 10 is a consideration of probability and basic statistics, including considerations of means, standard deviations, standard errors, binomial, Poisson and normal distributions and the least squares method of fitting a straight line to a set of data points. The application chapter is a consideration of the determination of the overall errors in quantities resulting from errors in measurements used in their determination.

W. Bolton

Part 1
Functions

1 Introducing functions
2 Polynomials
3 More functions
4 Non-linear functions
5 Sequences and series
6 Application: Representing signals

The aims of this part are to enable the reader to:

- Use the concept of a function and its inverse, recognising continuous, discontinuous, periodic, odd and even forms of functions.
- Manipulate and evaluate linear, polynomial, trigonometric, exponential, logarithmic and hyperbolic expressions.
- Use partial fractions to simplify expressions.
- Apply graphical and iterative methods to determine the roots of non-linear equations.
- Use sequences and series, determining limits, sums and whether convergent or divergent.
- Use the Taylor polynomial to give approximations for functions.
- Determine the Maclaurin series for functions.
- Apply the function and sequence concepts to describing continuous and discrete signals such as sinusoidal, unit step and impulses and those composed of such basic forms.

Much of this part is concerned with developing/revising basic concepts and terminology that will be used in later parts of the book. For this part, dexterity with mainly algebra and graphs is assumed. For Chapter 3 some familiarity with the notation of differentiation is assumed and Chapter 5 requires, for Sections 5.4 and 5.5, the ability to differentiate simple functions. Such sections might be left until differentiation is encountered in Part 6. The application chapter, Chapter 6, provides the basis for describing signals that are used in many other later chapters in this book.

1 Introducing functions

1.1 Introduction

This chapter is about the mathematical concept of a function, how we can represent it and the basic terminology used. Chapter 2 considers some of the more commonly encountered functions and their characteristics.

It is assumed that the reader is familiar with graphs and, in particular, the straight line graph and its representation by an equation of the form $y = mx + c$. Chapter 2 revises these items.

1.2 Relationships

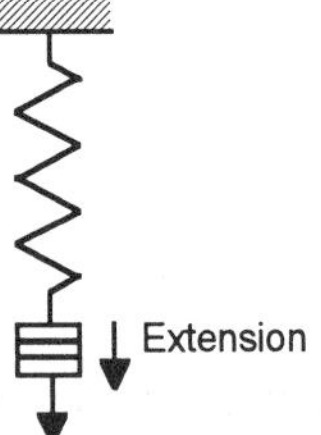

Figure 1.1 *Loading a spring*

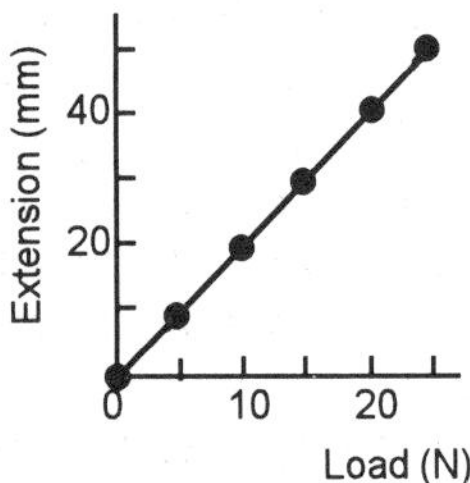

Figure 1.2 *Graph*

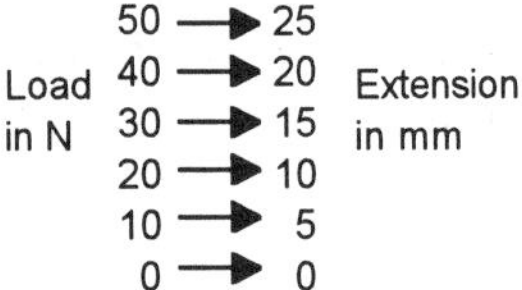

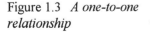

Figure 1.3 *A one-to-one relationship*

Suppose we carry out an experiment and measure the extension of a spring when subject to different loads (Figure 1.1). We might obtain the results:

Load 0 N, extension 0 mm
Load 10 N, extension 5 mm
Load 20 N, extension 10 mm
Load 30 N, extension 15 mm
Load 40 N, extension 20 mm
Load 50 N, extension 25 mm

The results are a set of paired values. The quantities that vary, i.e. the extension and the load, are called the *variables*. The convention is adopted of terming the chosen variable the *independent variable* and the variable that then changes as a consequence, the *dependent variable*. Because in the experiment we choose to vary the load, the load is the independent variable and the extension, which varies as a consequence of the load changes, is the dependent variable. With graphs the convention is used of plotting the dependent variable as the vertical axis (y) and the independent variable (x) as the horizontal axis. Thus for the spring, the vertical axis is the extension and the horizontal axis the load. When each paired set of values is plotted and a curve drawn through the points, the result is as shown in Figure 1.2.

The extension of a spring depends the load applied to stretch it. We can thus state that there is a *relationship* between the extension and the load. In a similar way we can have a relationship between the current in an electrical circuit and the resistance in the circuit, the temperature of a cooling hot object and the time for which it has been cooling, the distance fallen by a freely falling object and the time for which it has been falling. Relationships can be of the form shown above for the spring which is *one-to-one*. Each value of the extension is related to a single value of the load (Figure 1.3). Other forms of relationship are *many-to-one* when we have two or more values of the dependent variable related to one value of the independent variable, and *many-to-many* when two or more values of the dependent variable are related to two or more values of the independent variable. Figure 1.4 illustrates such relationships and the types of graphs produced.

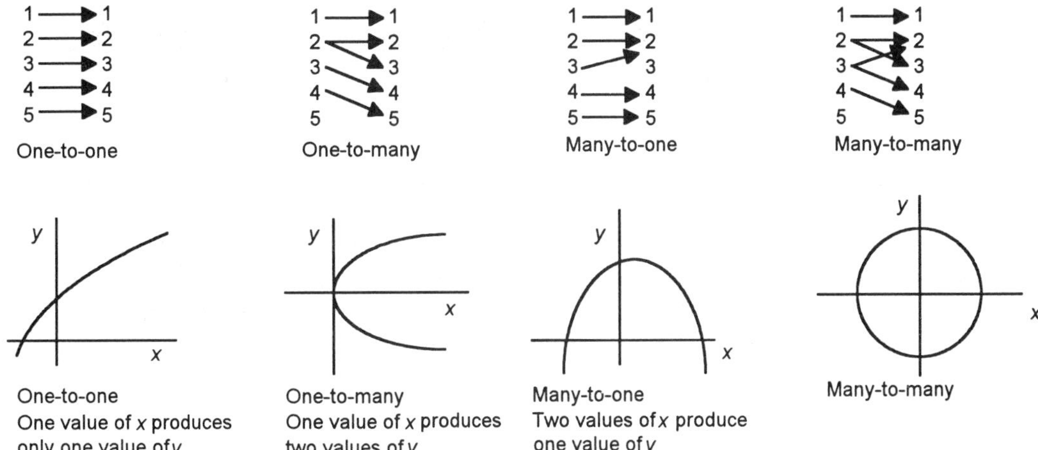

Figure 1.4 *Examples of relationships*

1.2.1 Functions

A *function* is a particular type of relationship. It is a relationship which has for each value of the independent variable a unique value of the dependent variable, i.e. for each value of x there is a unique value of y. Not all relationships are thus functions. With the spring, there is a unique value of the extension for each value of applied load (see Figure 1.3). Thus we can say that the extension is a function of the load.

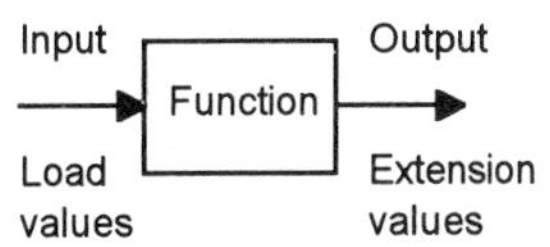

Figure 1.5 *The load-extension function*

In the same way that we can think of an amplifier as an electronic system which takes as an input a small electrical signal and produces as an output a larger version of the signal, then we can think of a function as a system which takes an input of one quantity and gives an output of another quantity. For each value of the input there is a unique value of the output. Thus the function describing the behaviour of the loaded spring has an input of load values and an output of extension values (Figure 1.5).

In general language the term function is sometimes loosely used to mean a relationship. In the mathematics sense of using the term function there cannot be a one-to-many-valued function, such a situation is just a relationship. A function is a special type of relationship since with a function there can only be one value of the dependent variable for each value of the independent variable. There is thus no ambiguity with a function regarding the value of the dependent variable produced when we have an input of the independent variable.

A function is a relationship between two variables with the restriction that no value of the independent variable can give rise to more than one value of the dependent variable.

1.2.2 Function notation

If y is a function of x then we might be able to express this as an equation involving y and x. However, the equation may not always be known or we may be concerned with a general relationship. For this reason we use the general notation $y = f(x)$ to represent y being a function of x. $f(x)$ does not mean f multiplied by x but merely is a label for 'function of x'. For the spring, with the extension e being a function of the load L we can write $e = f(L)$. We write $y = f(x = a)$, or more simply $y = f(a)$, to denote the value of y when the variable x is equal to a.

When we are dealing with a number of different functions it is usual to use different letters for the function label. For example, we might have $y = f(x)$ and $z = g(x)$.

1.2.3 Equations

Functions may be defined by equations. The equations give the instructions for calculating the dependent variable for values of the independent variable. For example, for a resistor the potential difference V across it is a function of the current I through it, i.e. $V = f(I)$. The equation defining the functional relationship (Ohm's law) is $V = RI$, where R is a constant. Thus, given a value for the current we can use the equation to obtain a value of the potential difference. Thus, when $R = 10\ \Omega$ we have for a current of 2 A, i.e. $V = f(2) = 20$ V.

For an object freely falling from rest, the distance fallen s is a function of the time t for which it has been falling, i.e. $s = f(t)$. The defining equation is $s = \frac{1}{2}at^2$, where a is a constant. Thus, given a value for the time we can use the equation to obtain a value for the distance fallen. When $a = 10$ m/s^2 we have for a time of 3 s, $s = f(3) = 45$ m.

A function may be defined by several equations, with each giving the instructions for calculating the dependent variable for different values of the independent variable. For example, for the voltage signal shown in Figure 1.6, a so-called *step voltage*, we have $v = f(t)$ and the relationship

$v = 0$ for t between 0 and 2 s, i.e. $0 \leq t < 2$
$v = 2$ V for t greater than 2 s, i.e. $2 \leq t$

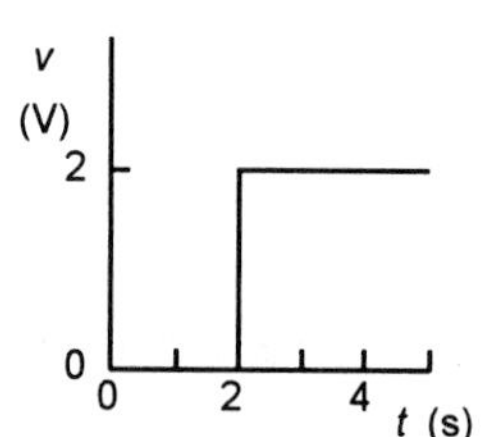

Figure 1.6 *Step voltage*

The value of $v = f(1)$ is thus 0 and of $v = f(3)$ is 2 V.

Example

If we have y as a function of x and described by the relationship $y = x^2$, what are the values of (a) $f(0)$, (b) $f(2)$?

(a) The function is described by $y = f(x) = x^2$. Thus $f(0)$ is the value of the function when $x = 0$ and so is 0.
(b) $f(2)$ is the value of the function when $x = 2$ and so is 4.

Example

Determine the values of (a) $f(2)$, (b) $f(4)$ if we have y as a function of x and defined by:

$$y = 1 \text{ for } 0 \leq x < 3, y = 2(x - 3) + 1 \text{ for } 3 \leq x$$

(a) The value of the function at $x = 2$ is given by the first relationship as 1.
(b) The value of the function at $x = 4$ is given by the second relationship as $y = 2(4 - 3) + 1 = 3$.

Revision

1 If we have y as a function of x and defined by the equation $y = 2x + 3$, what are the values of (a) $f(0)$, (b) $f(1)$?

2 If we have y as a function of x and defined by the equation $y = x^2 + x$, what are the values of (a) $f(0)$, (b) $f(2)$?

3 Determine the values of (a) $f(0.5)$, (b) $f(2)$ if we have y as a function of x and defined by:

$$y = 2 \text{ for } 0 \leq x < 1, y = 1 \text{ for } 1 \leq x$$

4 Determine the values of (a) $f(1)$, (b) $f(3)$ if we have y as a function of t and defined by:

$$y = 0 \text{ for } 0 \leq t < 2, y = 2(t - 2) \text{ for } 2 \leq t$$

5 The period of oscillation T of a simple pendulum is a function of the length L of the pendulum, being defined by the equation

$$T = 2\pi\sqrt{\frac{L}{g}}$$

where g is the acceleration due to gravity. What are the values of (a) $f(1)$, (b) $f(10)$ if g can be taken as 10 m/s^2?

6 The velocity v in metres per second of a moving object is a function of the time t in seconds, being defined by $v = 2 + 5t$. What are the values of (a) $f(0)$, (b) $f(1)$?

1.2.4 One-to-one, one-to-many and many-to-one relationships

A relationship is termed a function if a particular input always produces the same output and different inputs always produce different outputs. With a graph, if a vertical line only cuts the graph line once, then the graph represents a function. Thus for $V = RI$, the same current always results in the same potential difference and different currents give different potential differences.

But suppose we consider power $P = RI^2$, then we can obtain the same power by a current of +2 A as a current of −2 A. We can thus obtain a single output by two different inputs. This is an example of a *many-to-one* relationship. It is still a function because each value of I gives only one value of P. As will be indicated in Chapter 2, a trigonometric function, such as one described by $y = \sin x$, is a many-to-one relationship and a function because each value of the independent variable x gives rise to only one value of the dependent variable y. We can, however, have the same value of y from different values of x. For example, $1 = \sin \pi/2 = \sin 5\pi/2 = \sin 9\pi/2$.

For some relationships we might have a defining equation of the form $y = \pm\sqrt{x}$. Then for a single input of x there are two possible outputs. This is an example of a *one-to-many* relationship. Figure 1.7(a) illustrates this, showing a particular input value linked to more than one output value. Though we can draw a graph of the relationship it is not a function. This is because for a single value of x there are two values possible for y. However, if we restricted the relationship to just positive values of y, i.e. $y = +\sqrt{x}$, then we have a function (Figure 1.7(b)). Likewise if we restricted the relationship to just negative values of y, i.e. $y = -\sqrt{x}$, then we have a function. It has been assumed for both the figures that we are restricting the values of x to positive values only.

The strict mathematical definition of the term function is restricted to those relationships for which one input value gives rise to just one output value. We might thus have each input value giving rise to a unique output value or have a number of input values giving rise to the same output value, i.e. one-to-one or many-to-one relationships. With one-to-many relationships, if we are to have a relationship which is without ambiguity, i.e. a one-to-one or many-to-one function, then restrictions have to be placed on the values which are admissible. The set of values that the independent variable is allowed to take is termed its *domain* and the set of values that the dependent variable then has is termed its *range*.

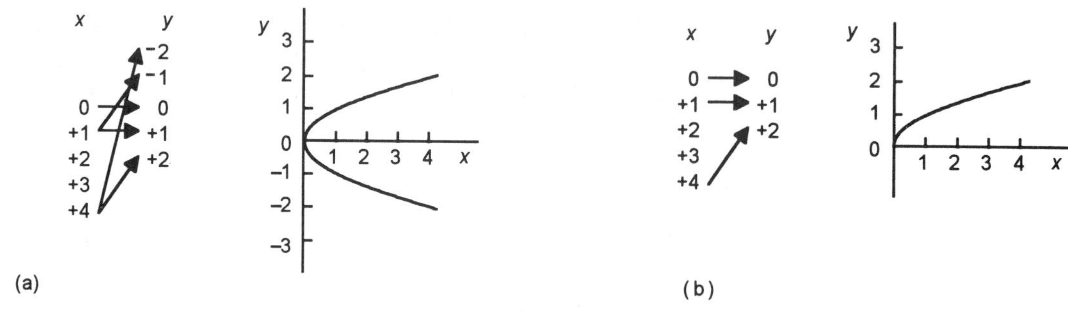

Figure 1.7 *(a)* $y = f(x) = \pm\sqrt{x}$, *(b)* $y = f(x) = +\sqrt{x}$

1.3 Continuous and discontinuous functions

Consider the function $y = f(x) = 2x$. Figure 1.8(a) shows the graph. For all values of x from $-\infty$ to $+\infty$ we have a continuous line on a graph. Such a function is said to be a continuous function. Now consider the function $y = f(x) = 1/x$. Figure 1.8(b) shows the graph. There is a break, or discontinuity, in the graph at $x = 0$. The function is said to be discontinuous at $x = 0$. If a graph of a function has no breaks then it is said to be a *continuous function* and if it contains a break it is said to be a *discontinuous function.*

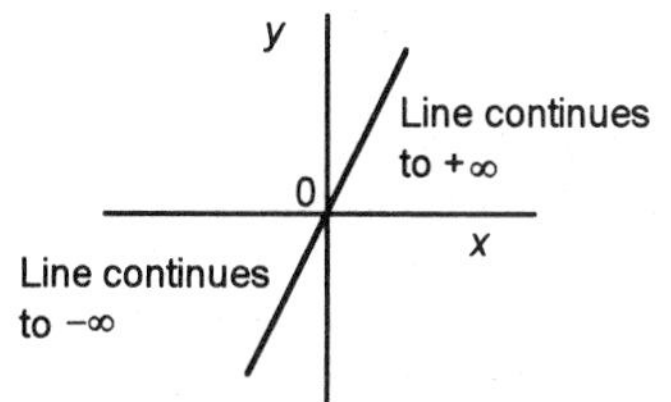

(a)

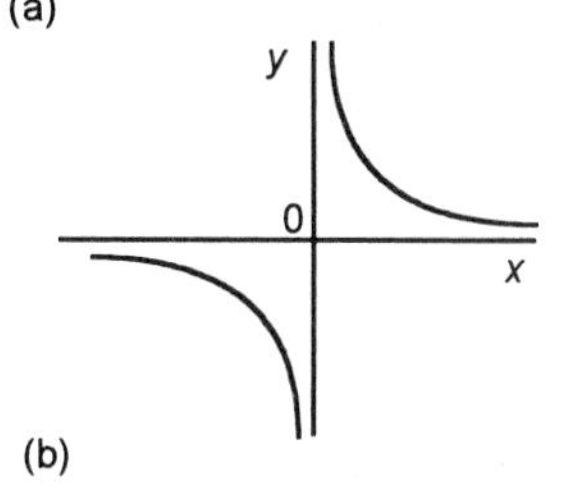

(b)

Figure 1.8 *(a) Continuous, (b) discontinuous*

Sometimes a function is defined by different equations for different parts of its graph, for example the step voltage function shown in Figure 1.5. Each part of the graph may be continuous but there is a discontinuity at the end points of each part. Such a function is said to be *piecewise continuous.*

> *A piecewise continuous function has a finite number of discontinuities in a given interval.*

Revision

7 Describe the discontinuous function shown in Figure 1.9 by equations.

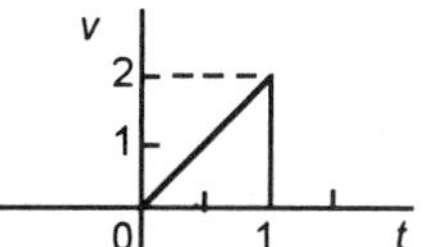

Figure 1.9 *Revision problem 7*

1.3.1 Periodic functions

A *periodic function* is a function which involves a relationship which is repeated at regular intervals. Thus we might have a current which is a function of time and has a waveform which is repeated every 3 s. We can represent this by describing the function which represents the first period of the waveform and state that to obtain the waveform in the second period we just add 3 s to the time values, for the next period we add 2 × 3 s, etc. What we have is: $f(t) = f(t + 3) = f(t + 6)$ = etc. In general:

$$f(t) = f(t + nT) \qquad [1]$$

where T is referred to as the *periodic time* and n = 0, 1, 2, 3, ..., etc. We thus have for the basic definition of a periodic function:

> *A periodic function is defined as one which, for all values of t within the range of the definition, f(t) = f(t + nT), where n is an integer and T is the smallest value for which the repetition occurs and is known as the period.*

Example

Plot the periodic function for which the first period is described by:

$y = f(t) = 4t/T$ for $0 \le t < T/4$
$y = f(t) = 2 - 4t/T$ for $T/4 \le t < 3T/4$
$y = f(t) = -4 + 4t/T$ for $3T/4 \le t < T$

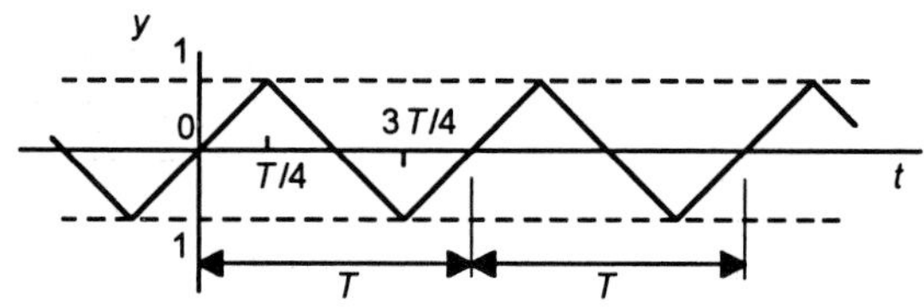

Figure 1.10 *Periodic triangular waveform*

Figure 1.10 shows the waveform. The equations define the straight line parts of the first period and this is repeated for further periods.

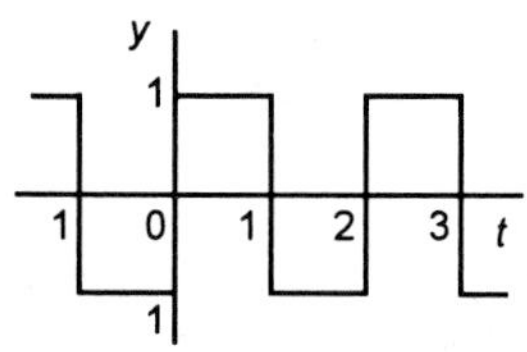

Figure 1.11 *Revision problem 9*

Revision

8 State the form of the periodic function described by the following equations:

$$y = f(t) = 1 \text{ for } 0 \leq t < 1,\ y = f(t) = 0 \text{ for } 1 \leq t < 2, \text{ period } T = 2$$

9 State the equations describing the waveform shown in Figure 1.11.

1.3.2 Even and odd functions

The terms *even* and *odd* function are used to describe the symmetry of a function $y = f(x)$ about the y-axis. An even function is one for which:

$$f(-x) = f(x) \qquad [2]$$

An odd function is one for which:

$$f(-x) = -f(x) \qquad [3]$$

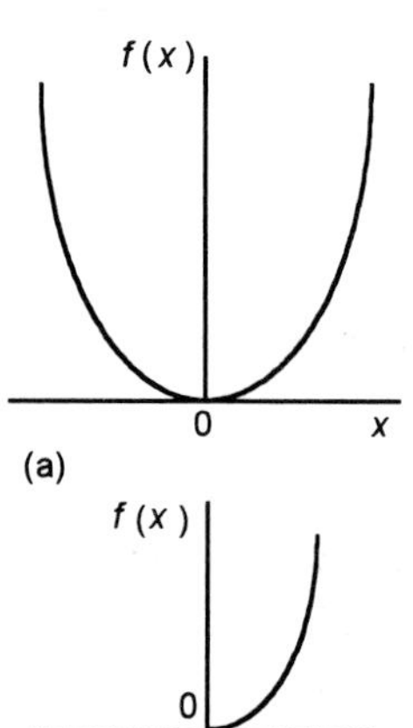

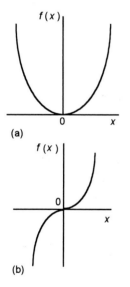

Figure 1.12 *(a) Even, (b) odd*

Figure 1.12(a) shows the graph of an example of an even function, (b) an example of an odd function. Not all functions are even or odd.

1.3.3 Combinations of functions

Many of the functions encountered in engineering and science can be considered to be combinations of other functions. Suppose we have the function $y = f(x) = x^2 + 2x$. We can think of the function $f(x)$ as resulting from the combination of two functions $g(x)$ and $h(x)$. One of the functions takes an input of x and gives an output of x^2 and the other takes an input of x and gives an output of $2x$. The two outputs are then added and we have $f(x) = g(x) + h(x)$. Figure 1.13 illustrates this.

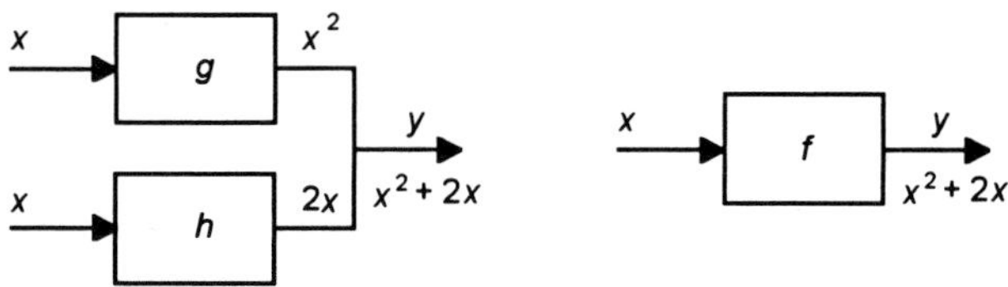

Figure 1.13 *Combination of functions to give the function f(x)*

Another way we can combine functions is by applying them in sequence. For example, if we have $h(x) = 2x$ and $g(x) = x^2$, then suppose we have the arrangement shown in Figure 1.14. The input of x to the g function box results in an output of x^2. The h function box takes its input and doubles it. Thus for an input of x^2 we have an output of $2x^2$, thus $f(x) = h\{g(x)\} = 2x^2$.

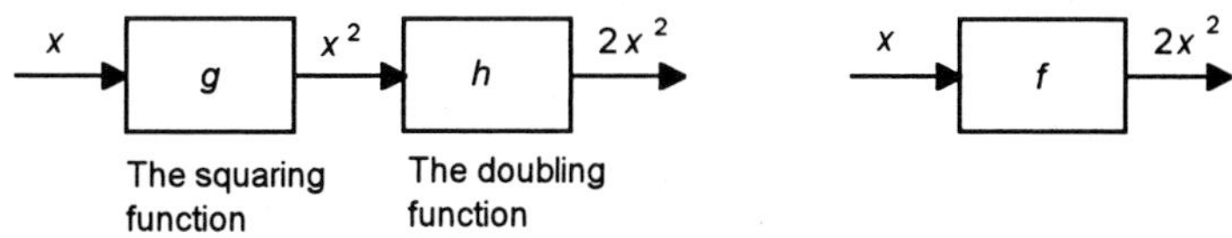

Figure 1.14 *Combination of functions to give the function f(x)*

Note that the order of the function boxes is important. If we had an input of x to the h box then the output would be $2x$. If we now applied this as an input to the g box then the output would be $(2x)^2$, thus $f(x) = g\{h(x)\} = 4x^2$. Figure 1.15 illustrates this.

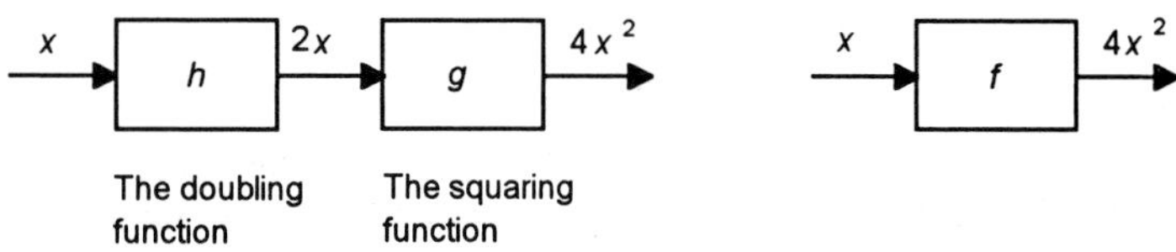

Figure 1.15 *Combination of functions to give the function f(x)*

Revision

10 If $g(x) = 2x$ and $h(x) = x + 1$, what are (a) $g(x) + h(x)$, (b) $g\{h(x)\}$, (c) $h\{g(x)\}$?

1.4 The inverse function

So far, in the treatment of a function we have started with a value of the independent variable x and used the function to find the corresponding value of the dependent variable y (Figure 1.16(a)). But suppose we are given a value for y and want to find x (Figure 1.16(b)). For example, we might have distance s as a function of time t, e.g. $s = 2t$. Given a value of the independent variable t we can use the function to determine s. However, suppose we are given a value of the dependent variable s and have to determine the corresponding t value? With the given equation we can rearrange it to give $t = s/2$. The function from t to s is $f(t)$, the function from s to t is a different function $g(s)$.

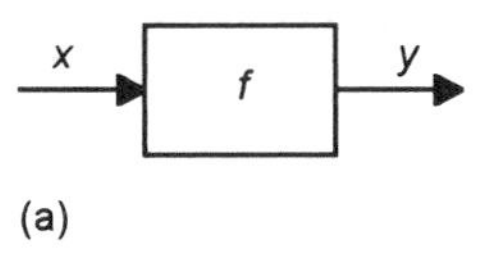

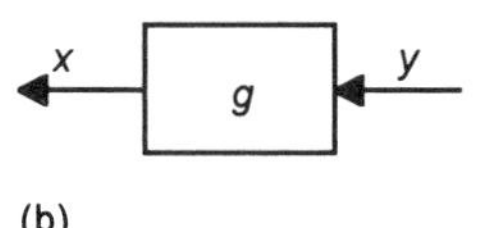

Figure 1.16 *(a) y = f(x), (b) x = g(y)*

Figure 1.17(a) shows some values for the $s = f(t)$ function described by the equation $s = 2t$. Figure 1.17(b) shows the function obtained by reversing the arrows, i.e. starting with time values deducing the corresponding distance values. This figure represents the *inverse* relationship.

There is a simple point which is of significance: if we use $s = 2t$ to calculate a value for s given a value of t and then use the inverse by taking that value of t to calculate a value of s, we end up back where we started with our original value of s. This leads to a method of specifying an inverse function. Consider the arrangement shown in Figure 1.18. The g function system box operates on the output from the f function box in order to undo the work of the f box. Because the g function is undoing the work of the f function it is said to be the *inverse* of f. We have:

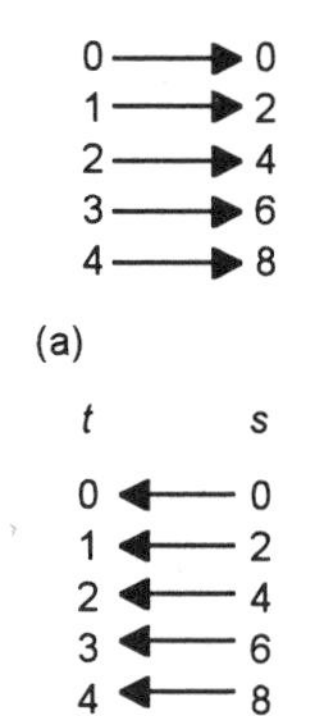

Figure 1.17 *(b) is inverse of (a)*

$$g\{f(x)\} = x \quad [4]$$

This equation is used to define an inverse function:

If f is a function of x then the function g which satisfies $g\{f(x)\} = x$ for all values of x in the domain of f is called the inverse of f.

The inverse of a function f of x is written as $f^{-1}(x)$. Note that $f^{-1}(x)$ does *not* mean $1/f(x)$, it is just a notation to indicate the inverse function, the -1 not indicating a power. $f^{-1}(x)$ takes an input which is some function of x and inverts it to give an output of x. Thus the above definition gives:

$$f^{-1}\{f(x)\} = x$$

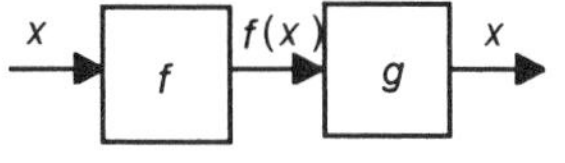

Figure 1.18 $x = g\{f(x)\}$

Consider a function f which adds 2, i.e. we have $f(x) = x + 2$ (Figure 1.19(a)). Then the inverse is a function that subtracts 2 in order to undo the action of the f function. Thus $f^{-1}(x) = x - 2$. Consider a function f which multiplies by 3, i.e. we have $f(x) = 3x$ (Figure 1.19(b)). Then the inverse is a function that divides by 3 in order to undo the action of the f function. Thus $f^{-1}(x) = x/3$. Consider a function which multiplies by 3 and then adds 2, i.e. $f(x) = 3x + 2$ (Figure 1.19(c)). Then the inverse must be a function that subtracts 2 and then divides by 3. Thus $f^{-1}(x) = x/3 - 2$. Note that you must undo things in the reverse order to which they were done with the function f.

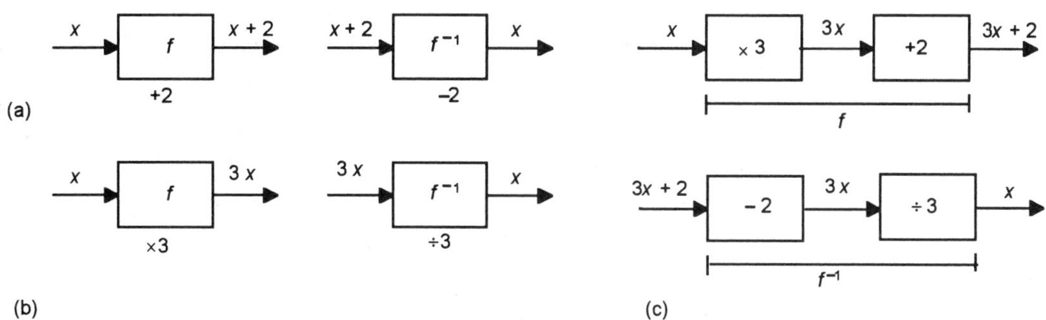

Figure 1.19 *Functions and their inverses*

A one-to-one relationship will give an inverse which is a one-to-one relationship. A one-to-many relationship will give an inverse which is a many-to-one relationship. A many-to-one relationship will give an inverse of a one-to-many relationship. Thus, if we do not impose any restrictions, only one-to-one relationships can be said to have inverse functions. This is considered in Chapter 2 in relation to the inverse trigonometric functions.

Example

If $f(x) = 2x$, what is the inverse function?

The inverse function is when we have $f^{-1}\{f(x)\} = x$. Since $f(x) = 2x$, what do we have to multiply it by to give x? The answer is $1/2x$ and so the inverse function is $f^{-1}(x) = 1/2x$.

Example

If $f(x) = 2x + 3$, what is the inverse function?

The inverse function is when we have $f^{-1}\{f(x)\} = x$. Since $f(x) = 2x + 3$, what do we have to multiply it by to give x? The function f involves doubling the input and then adding 3. The inverse is thus subtracting 3 from the input and then halving. Thus the inverse function is:

$$f^{-1}(x) = \frac{x-3}{2}$$

Revision

11 Determine the inverses of (a) $f(x) = 5x - 3$, (b) $f(x) = 4 + x$, (c) $f(x) = x^3$, (d) $f(x) = 2x^3 - 1$.

12 Determine the inverse of $f(x) = x^2$, restricting the domain so that there is an inverse function.

1.4.1 Graphs of f and f^{-1}

We can use the above rules for a function and its inverse to find the graph of an inverse function from a graph of the function. Consider the graph of $y = f(x)$ shown in Figure 1.20(a). This is the graph described by the equation $y = x^2$. What is the graph of the inverse function $f^{-1}(x)$? This will be the graph of $y = \sqrt{x}$ (Figure 1.20(b)) since the function $\sqrt{x}$ is what we need to apply to undo the function x^2.

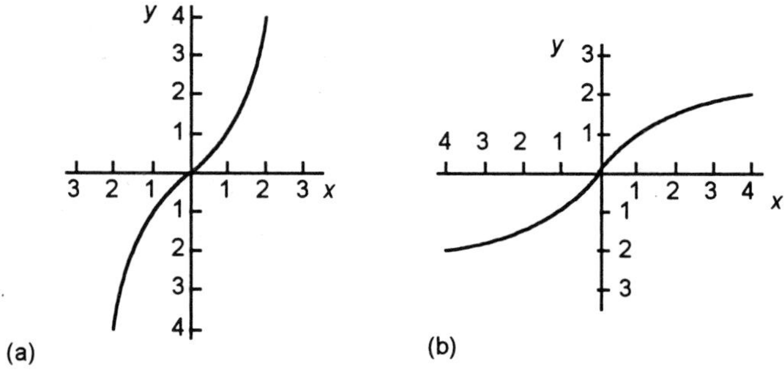

Figure 1.20 *(a)* $y = f(x)$, *(b)* $y = f^{-1}(x)$

If we examine the two graphs we find that the inverse f^{-1} is just the reflection of the graph of f in the line $y = x$ (Figure 1.21). This is true for any function when it possesses an inverse.

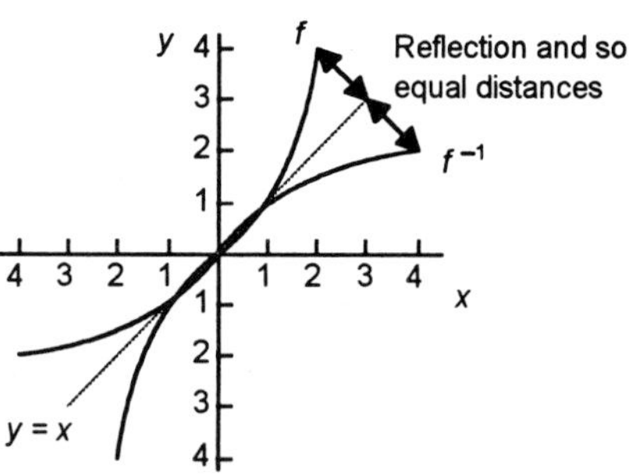

Figure 1.21 *The inverse as a reflection in* $y = x$

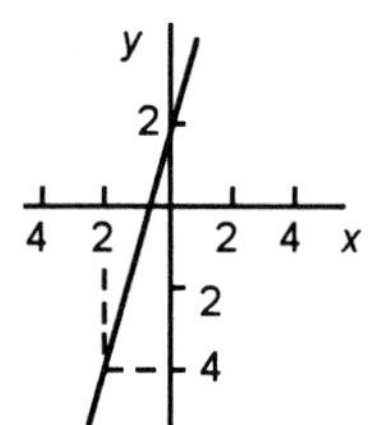

Figure 1.22 *Revision problem 13*

Revision

13 For the graph shown in Figure 1.22, plot its inverse.

Problems

1 If we have y as a function of x and defined by the following equations, what are the values of $f(0)$ and $f(1)$?

(a) $y = x^2 + 3$, (b) $y = x + 4$, (c) $y = (x + 1)^2 - 3$

2 If we have y as a function of x and defined by $y = 2$ for $0 \le x < 4$, $y = 0$ for $4 \le x$, what are the values of (a) $f(2)$, (b) $f(5)$?

3 If we have y as a function of x and defined by $y = 2x$ for $0 \le x < 2$, $y = 4$ for $2 \le x$, what are the values of (a) $f(1)$, (b) $f(4)$?

4 The cost c of manufacturing a product is a function of the number of units made, the cost being related to the number u units of a product by the equation $c = 500 + 3u$. What is the value of (a) $f(0)$, (b) $f(100)$?

5 The voltage in an electrical circuit is supplied by a constant voltage source of 10 V. If the voltage is switched on after a time $t = 2$ s, state the equations defining the step voltage.

6 Sketch the periodic waveform described by the following equations:

$y = f(t) = t$ for $0 \le t < 2$, $y = f(t) = 2 - t$ for $2 \le t < 4$, period 4

7 State which of the following equations describe functions:

(a) $y = x^2$, (b) $y^2 + x^2 = 1$, (c) $y = \pm\sqrt{x}$

8 If $f(x) = x^2 + 1$, $g(x) = 3x$ and $h(x) = 3x + 2$, determine:

(a) $f(x) + g(x)$, (b) $f\{g(x)\}$, (c) $g\{f(x)\}$, (d) $f(x) - h(x)$, (e) $f\{h(x)\}$

9 Determine the inverses of the following functions:

(a) $f(x) = 2x + 3$, (b) $f(x) = 4x$, (c) $f(x) = 3x - 4$, (d) $f(x) = \sqrt{x} + 2$, (e) $f(x) = (x + 1)^3 - 3$

10 Sketch the graph of the function $f(x) = \sqrt{x} + 2$ and on the same axes sketch the inverse.

11 Does the function $f(x) = x^2$ have an inverse for all real values of x?

12 For each of the following functions restrict the domain so that there is an inverse and then determine it.

(a) $f(x) = (x - 1)^2$, (b) $f(x) = (x + 1)^2 - 4$

2 Polynomials

2.1 Introduction

The e.m.f. E produced by a thermocouple is a function of the temperature T and is given by an equation which is typically of the form:

$$E = aT + bT^2$$

The deflection y of a point along a cantilever of length L as a result of a uniform load of w per unit length is a function of the distance x from the clamped end, being given by the equation:

$$y = \frac{w}{24EI}(6L^2x^2 - 4Lx^3 + x^4)$$

The above are examples of what are termed *polynomial functions*.

A *polynomial* function $p(x)$ is defined as having the general form:

$$p(x) = A + Bx + Cx^2 + Dx^3 + \ldots \qquad [1]$$

where A, B, C, D, etc. are constants termed the *coefficients*. The *degree* of a polynomial is the value of the highest power present. A polynomial with degree 0 is thus $p(x) = A$ and is just a *constant*. A polynomial with degree 1 is $p(x) = A + Bx$ and is termed a *linear* function since it gives a straight line graph. A polynomial with degree 2 is $p(x) = A + Bx + Cx^2$ and is known as a *quadratic*. A polynomial with degree 3 is $p(x) = A + Bx + Cx^2 + Dx^3$ and is known as a *cubic*.

This chapter is an introduction to polynomials, considering first the linear function and then higher order polynomials, methods of determining their roots and the use of partial fractions to simplify expressions.

It is assumed that the reader is familiar with graph plotting and algebraic manipulations. Section 2.3 assumes a knowledge of basic differentiation. Chapter 3 considers other functions, Chapters 4 and 5 extending the consideration of polynomials.

2.2 Linear functions

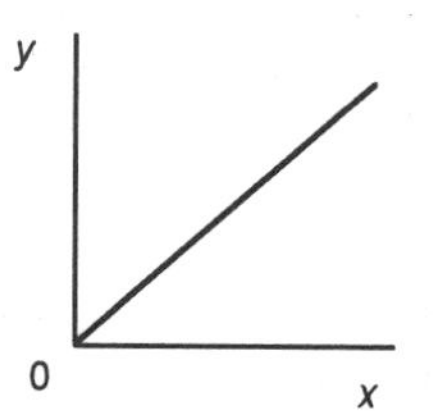

Figure 2.1 *Straight line graph*

The potential difference V across a resistor is a function of the current I through it. If the resistor obeys Ohm's law then $V = RI$, the potential difference is proportional to the current. If the current is doubled then the potential difference is doubled, if the current is trebled the potential difference is trebled. This means that a graph of V plotted against I is a straight line graph passing through the origin. *Gradient* is defined as the change in y value divided by the change in x value. Thus, for all straight line graphs passing through the origin (Figure 2.1), the gradient is constant and given by gradient $m = y/x$. Hence the equation of such a straight line is of the form:

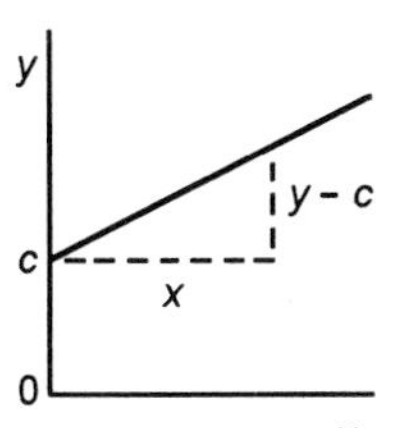

Figure 2.2 *Straight line graph*

$$y = mx \qquad [2]$$

where m is the gradient of the line. *Only* when we have such a relationship is y directly proportional to x.

Straight line graphs which do not pass through the origin (Figure 2.2) have a gradient, change in y value divided by change in x value, given by $m = (y - c)/x$, where c is the value of y when $x = 0$, i.e. the intercept of the straight line with the y-axis. Thus, such lines have the equation:

$$y = mx + c \qquad [3]$$

This is the equation which defines a straight line and is termed a *linear equation*. It is important to realise that with $c \neq 0$ that y is *not proportional* to x. However, by changing the position of the origin we can write the equation as:

$$(y - c) = mx$$

The term on the left is then proportional to x. Thus all such linear equations can be made to have the proportionality property.

Linear equations are ones in which each variable appears only with a power of one, and there is no product of variables.

Thus, for example, $y = 2x + 7$ is a linear equation but $y = 2x^2 + 3$ and $yx = 2$ are not.

Electrical components for which the output variable is proportional to the input variable are said to be *linear components*. A resistor which obeys Ohm's law is such a component. A spring is a linear spring if its output, the extension, is proportional to its input, the load. Systems that are linear have two basic properties:

1 An *additive* property.
If we have an input of x_1 giving an output of y_1, an input of x_2 giving an output of y_2, then an input of $(x_1 + x_2)$ will give an output of $(y_1 + y_2)$:

$$f(x_1 + x_2) = f(x_1) + f(x_2)$$

For example, for a linear resistor of 10 Ω, a current of 1 A will give a potential difference across the resistor of 10 V. A current of 2 A will give a potential difference of 20 V. Because it is linear, a current of (1 + 2) A will give a potential difference of (10 + 20) V.

2 A *homogeneous* property.
If the input is multiplied by some constant then the output is multiplied by the same constant. Thus for $y = f(x)$, when the input x is multiplied by a constant a:

$$f(ax) = af(x) = ay$$

Thus for a linear resistor, where the potential difference across the resistor is a function of the current through it, doubling the current through it doubles the potential difference across it.

Any function which does not satisfy both the above properties is termed *non-linear*. For example, the distance s fallen by a freely falling object is a function of the time t for which it has been falling. The defining equation is $s = 4.9t^2$. For a time of 1 s we have a distance fallen of 4.9 m. For a time of 2 s we have a distance fallen of 19.6 m. For a time of 3 s we have a distance fallen of 44.1 m. But this cannot be obtained by just adding 4.9 m and 19.6 m, $f(1) + f(2) \neq f(3)$. If we double the time we do not double the distance fallen, $f(2t) \neq 2f(1t)$, but the distance fallen is increased by a factor of 4.

Example

State the gradients and intercepts of the graphs of the following equations: (a) $y = 2x + 3$, (b) $y = 2 - x$, (c) $y = x - 2$.

(a) This has a gradient of +2 and an intercept with the y-axis of +3. A positive gradient means that y increases as x increases.
(b) This has a gradient of −1 and an intercept with the y-axis of +2. A negative gradient means that y decreases as x increases.
(c) This has a gradient of +1 and an intercept with the y-axis of −2.

Revision

1 State which of the following will give a straight line graph and, if so, whether it passes through the origin:
(a) A graph of the extension of a spring plotted against the applied load when the extension is proportional to the applied load.
(b) A graph of the resistance R of a length of resistance wire plotted against the temperature t when $R = R_0(1 + \alpha t)$, with R_0 and α being constants.
(c) A graph of the distance d travelled by a car plotted against time t when $d = 10 + 4t^2$.
(d) A graph is plotted of the pressure p of a gas against its volume v, the pressure being related to the volume by Boyle's law, i.e. pv = a constant.

2 Determine the straight line equations for the following data:
(a) The current i and time t over a period of time if at the beginning of the time we have $i = 2$ A and $t = 0$ s and at the end we have $i = 3$ A and $t = 2$ s.
(b) The extension e of a strip of material as a function of its length L when subject to constant stress, given that:

e in mm	0.60	0.72	0.84	0.96
L in m	0.5	0.6	0.7	0.8

2.2.1 Principle of superposition

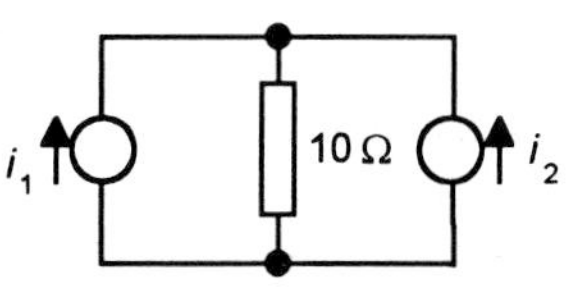

Figure 2.3 *Linear circuit*

In a linear electrical circuit, the additive property for linear systems, generally here termed the *principle of superposition*, means that the total current in any branch is the algebraic sum of the separate currents produced in the branch by the individual sources of e.m.f. considered one by one in isolation. An electrical circuit consisting of linear components, independent sources and linear dependent sources is said to be a *linear circuit*. For example, when we apply Kirchhoff's laws to the circuit in Figure 2.3, with two independent input current sources of i_1 and i_2, say 3 A and 2 A, then:

$$v = 10i_1 - 10i_2 = 30 - 20 = 10 \text{ V}$$

where v is the potential difference across the resistor. Suppose we now make $i_2 = 0$, then $v_1 = 10i_1 = 30$ V. If we had made $i_1 = 0$ we would have $v_2 = -10i_2 = -20$ V. Adding the two equations obtained by considering the responses when each source is considered to be acting alone gives the equation obtained by considering both the sources acting together, i.e.

$$v = v_1 + v_2$$

The principle of superposition can considerably simplify problems in electrical, electronic, mechanical, control, etc., engineering and thus, since may real-world systems are non-linear, engineers are likely to work with roughly equivalent linear systems rather than the non-linear ones, or impose restrictions on the non-linear systems so that they can work within a linear part of its characteristic.

Example

The deflection at the free end of a cantilever is 10 mm when the load applied at that end is 100 N. When the load is 200 N the deflection is 20 mm. What will be the deflection when the load is 300 N?

Assuming that the deflection is proportional to the load, i.e. the system is linear, then the deflection will be 30 mm. We can deduce this from writing the equation relating the deflection y and the load F as $y = kF$, where k is a constant with the value 100/10 N/mm. Alternatively we can use the principle of superposition. Thus a load of (100 + 200) N gives a deflection of (10 + 20) mm.

Revision

3 A linear electrical circuit gives a current of 2 mA when a voltage source of 1 V is applied to it. When a voltage source of 4 V is used the current is 8 mA in the same direction. What will be the current when both voltage sources are in the circuit?

2.2.2 Linear interpolation and extrapolation

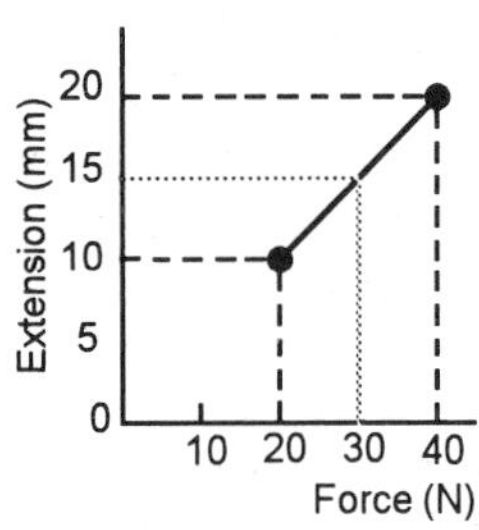

Figure 2.4 *Linear interpolation*

Suppose we have the values from an experiment of stretching a spring of:

Force 20 N, extension 10 mm; Force 40 N, extension 20 mm

If we assume a linear relationship between the force and extension then on a graph of extension plotted against force we can join the above points by a straight line (Figure 2.4). This enables us to predict the value of the extension for a force of 30 N, namely 15 mm. This method of prediction is called *linear interpolation*.

Thus if we have two sets of values, x_1 giving y_1 and x_2 giving y_2, then assuming a linear relationship, the value y_p of y predicted for some intermediate value x_p of x is given by:

$$\frac{y_p - y_1}{x_p - x_1} = \frac{y_2 - y_1}{x_2 - x_1}$$

Hence:

$$y_p = y_1 + \frac{x_p - x_1}{x_2 - x_1}(y_2 - y_1) \qquad [4]$$

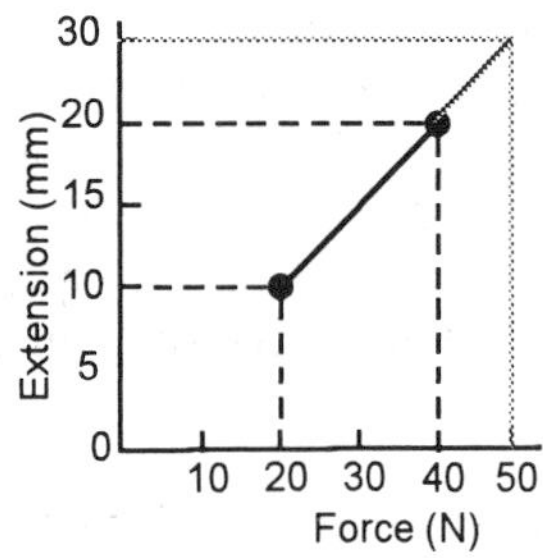

Figure 2.5 *Linear extrapolation*

Suppose we now, with our spring example, require the extension for a force of 50 N. This is a force which lies outside the data. If we assume that a linear relationship holds for the values we already have and that it extends to values outside that data, then we can obtain a value for the extension (Figure 2.5). This method is called *linear extrapolation*. Equation [4] can be used to make such a prediction. Extrapolation should be used with extreme caution as it can lead very easily to nonsensical results.

Revision

4 In an experiment the length of a piece of wire was measured with two different loads and the following results obtained:

Load 50 N, length 1.100 m; Load 90 N, length 1.200 m

(a) By linear interpolation determine the length for a load of 60 N and (b) by linear extrapolation determine the length for a load of 120 N.

2.3 Graphs of functions

The function described by the equation $y = mx + c$ is a straight line graph with a gradient of m and an intercept with the y axis of c. What about the types of graphs described by non-linear equations?

The following are briefly some points which can be useful describing the form, and plotting, of graphs of such functions. Given the equation describing a function $y = f(x)$ then its graph may be obtained by simply plotting a sufficient number of points obtained by calculating values of y for particular values of x. A smooth curve can then be drawn through the

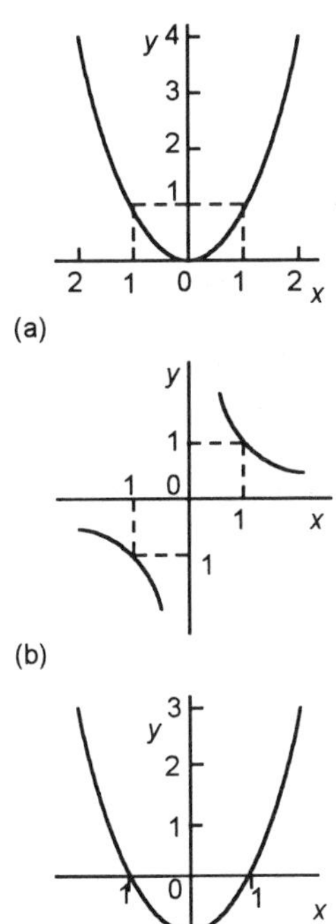

Figure 2.6 *Graphs*

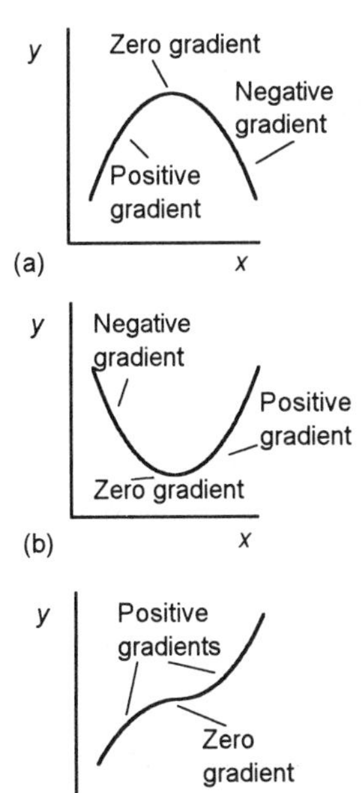

Figure 2.7 *(a) Maximum (b) minimum, (c) inflexion*

points. However, information about the general shape and limitations of the curve can often be obtained by an examination of the equation.

1 The graph will be symmetrical about the x-axis if it contains only even powers of y and symmetrical about the y-axis if it contains only even powers of x. For example, $y = x^2$ is symmetrical about the y-axis since the same values of y are obtained by $x = +1$ and $x = -1$ (Figure 2.6(a)).

2 The graph will be symmetrical about the origin if a change in sign of x causes a change in the sign of y without altering its numerical value. For example, $y = 1/x$ is symmetrical about the origin since changing x from $+1$ to -1 results in y changing sign (Figure 2.6(b)).

3 The points at which a curve cuts the axes can be determined by putting (a) $y = 0$, (b) $x = 0$. For example, $y = x^2 - 1$ cuts the x-axis at $x = +1$ and $x = -1$ and the y-axis at -1 (Figure 2.6(c)).

4 The graph will give an asymptote to the x-axis if there are values of y which make x infinite, likewise an asymptote to the y-axis if there are values of x which make y infinite. An *asymptote* is when the curve gets closer and closer to the axis but does not reach it until infinity. For example, with $y = 1/x$ if $y = 0$ then $x = \infty$ and so it gives an asymptote with the x-axis; with $x = 0$ then $y = \infty$ and so the curve also gives an asymptote with the y-axis (Figure 2.6(b)).

5 If the curve has any maximum or minimum values or points of inflexion (Figure 2.7) then they can be obtained by differentiating the equation and finding the values which make the derivative zero. Thus, for example, $y = x^2 - 1$ when differentiated gives $dy/dx = 2x$. For the derivative to be zero then $x = 0$. Hence there is a minimum at $x = 0$. We can determine that this is a minimum by considering that the derivative changes from being negative before the point to positive after it. A maximum has the derivative changing from being positive to negative. A point of inflexion has the derivative not changing in sign.

Example

Sketch the graphs of the functions: (a) $y = x^3$, (b) $y = x^4$.

(a) The function has both odd powers of y and x and is thus not symmetrical about either the x- or y-axes. A change in the sign of x causes a change in the sign of y without altering its numerical value. When $x = +1$ then $y = +1$, and when $x = -1$ then $y = -1$. Thus the graph is symmetrical about the origin. The graph passes through the origin since when $x = 0$ we have $y = 0$. The derivative of the function dy/dx is $3x^2$. For the derivative to be zero we must have $x = 0$. When $x = -1$ then the derivative is $+3$, when $x = +1$ then the derivative is $+3$. Since the derivative is zero at $x = 0$, and is positive before this point and positive after it, then there is a point of inflexion at $x = 0$. Inspection of the

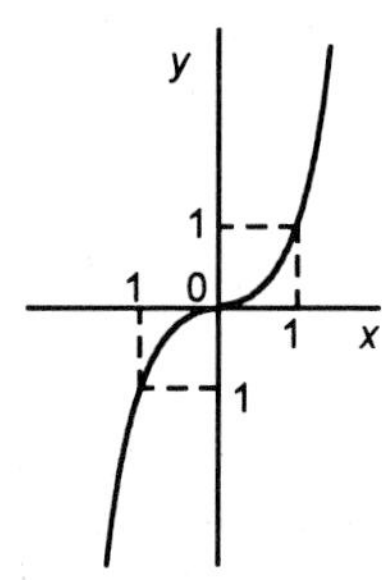

Figure 2.8 $y = x^3$

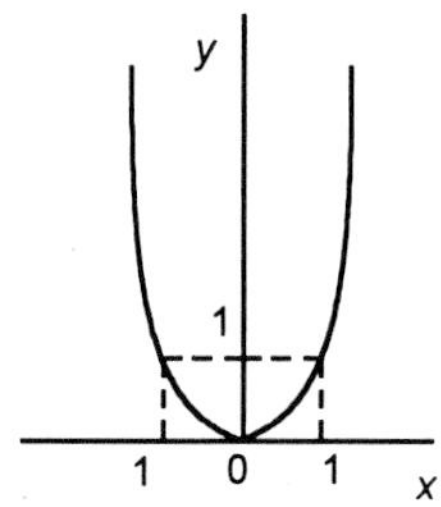

Figure 2.9 $y = x^4$

equation reveals that as x increases then y will increase much more rapidly, doubling x increasing y by a factor of eight. Figure 2.8 shows the graph.

(b) This has an even function of x, therefore the graph is symmetrical about the y-axis. When $x = +1$ then $y = +1$, and when $x = -1$ then $y = +1$. Whether x is positive or negative, y is always positive. Thus the graph will never cross the x-axis to give negative y values. The graph passes through the origin because when $x = 0$ we have $y = 0$. The derivative of the function is $dy/dx = 4x^3$. For the derivative to be zero we must have $x = 0$. When $x = -1$ the derivative is -4, when $x = +1$ the derivative is $+4$. Since the derivative is zero at $x = 0$, and is negative before this point and positive after it, then there is a minimum at $x = 0$. Inspection of the equation reveals that as x increases then y will increase much more rapidly, doubling x increasing y by a factor of sixteen. Figure 2.9 shows the graph.

Revision

5 Sketch the graphs of the following functions:

(a) $x^4 - 16$, (b) $\frac{1}{x} + 2$, (c) $x^2 + 2$

6 Sketch the graphs of the following functions:

(a) The bending moment M of a simply supported beam with a uniformly distributed load of w per unit length is a function of the distance x from one end support, being given by $M = \frac{1}{2}wLx - \frac{1}{2}wx^2$, where L is the distance between the two end supports and x can only have values between 0 and L.

(b) The reactance X of a capacitor is a function of the angular frequency ω, being given by $X = 1/\omega C$, where C is the capacitance and ω can only have values between 0 and infinity.

(c) The e.m.f. E generated by a thermocouple is a function of the temperature T, being given by $E = aT + bT^2$, where a and b are constants, only positive values of T being considered.

2.4 Properties of polynomials

A *polynomial function* is defined as having the general form:

$$p(x) = A + Bx + Cx^2 + Dx^3 + \dots$$

where A, B, C, D, etc. are constants termed the *coefficients*. There are two important properties for polynomials:

1 *If two polynomials are equal for all values of the independent variable then the corresponding coefficients of the powers of the variable are equal.*

2 *Any polynomial with real coefficients can be expressed as a product of linear and irreducible quadratic factors.*

A *factor* of a number is an exact divisor of that number, e.g. 2 is a factor of 6 since 6 can be divided by 2 and there is no remainder. The number 90 in terms of factors is $2 \times 3 \times 3 \times 5$. These numbers of 2, 3 and 5 cannot be reduced to smaller numbers.

> *To factorise a polynomial means writing it as a product of simple polynomials.*

A *linear factor* of a polynomial is a term of the form $(x + a)$ which is an exact divisor of the expression. An *irreducible quadratic factor* is of the form $ax^2 + bx + c$ and cannot be factored into the product of two linear terms with real coefficients.

We can illustrate property 1 by considering what the values of A, B and C should be for the two functions:

$$2x^2 + 3x + 1 \text{ and } A(x + 1) + B(x + 2)(x - 1) + C$$

to be equal for all values of x. Multiplying out the right-hand function gives:

$$Ax + A + Bx^2 + Bx - 2B + C$$

For this function to be equal for all values of x with $2x^2 + 3x + 1$, we must have the same coefficients for the x^2 terms, i.e. $2 = B$, the same for the x terms, i.e. $3 = A + B$, and the same for the constants, i.e. $1 = A - 2B + C$. Thus we must have $A = 1$, $B = 2$, and $C = 4$. Hence:

$$2x^2 + 3x + 1 = (x + 1) + 2(x + 2)(x - 1) + 4$$

This property is used in the discussion of partial fractions in Section 2.6.

We can illustrate property 2 by considering the polynomial function $x^2 + 2x - 3$. This quadratic can be factored to give two linear factors:

$$x^2 + 2x - 3 = (x + 3)(x - 1)$$

Consider the polynomial $x^3 - 3x^2 + 6x - 4$. We can extract a linear factor of $(x - 1)$ from the cubic function but no further linear factors. The term left after the extraction of the linear factor is an irreducible quadratic. The quadratic can be derived by long division.

$$\begin{array}{r|l} & x^2 - 2x + 4 \\ \hline x - 1 & x^3 - 3x^2 + 6x - 4 \\ & \underline{x^3 - x^2} \\ & \quad -2x^2 + 6x \\ & \quad \underline{-2x^2 + 2x} \\ & \qquad\qquad 4x - 4 \\ & \qquad\qquad \underline{4x - 4} \\ & \qquad\qquad\qquad 0 \end{array}$$

Thus $x^3 - 3x^2 + 6x - 4 = (x - 1)(x^2 - 2x + 4)$.

2.4.1 Finding factors

A factor of a polynomial is an exact divisor of that polynomial. Thus if we have a factor $(x - a)$ then:

polynomial function = $(x - a) \times$ other factors

So if we make $x = a$ the value of the function becomes zero. This is sometimes referred to as the *factor theorem*.

If $(x - a)$ is a factor of a polynomial, then putting $x = a$ makes the polynomial have a value of 0.

Thus, with the above example of $x^3 - 3x^2 + 6x - 4$ with a factor $(x - 1)$, this function has a zero value when $x = 1$. Then we have $1 - 3 + 6 - 4 = 0$. Thus we can find factors of a polynomial by determining the values of x that make the polynomial zero.

When $(x - a)$ is not a factor of a polynomial then there is a remainder after dividing the function by $(x - a)$ and putting $x = a$ does not give a zero value. The *remainder theorem* states that:

If a polynomial $f(x)$ is divided by $(x - a)$, the quotient will be a polynomial $g(x)$ of degree less than that of $f(x)$, together with a remainder R still to be divided by $(x - a)$, i.e.

$$\frac{f(x)}{x-a} = g(x) + \frac{R}{x-a}$$

Example

Determine the factors of the following polynomials:

(a) $x^2 - 9$, (b) $x^2 + 2x - 3$, (c) $x^2 + 2x$, (d) $x^3 + 2x^2 - 11x - 52$

(a) We can factorise $x^2 - 9$ as $(x - 3)(x + 3)$. When $x = +3$ and when $x = -3$ we have $x^2 - 9 = 0$.
(b) We can factorise $x^2 + 2x - 3$ as $(x + 3)(x - 1)$. When $x = -3$ and when $x = 1$ we have $x^2 + 2x - 3 = 0$.
(c) We can factorise $x^2 + 2x$ as $x(x + 2)$. When $x = 0$, the first term is effectively $(x - 0)$, and when $x = -2$ then $x^2 + 2x = 0$.
(d) We can obtain the factor $(x - 4)$, the remainder being irreducible. Thus we have $(x - 4)(x^2 + 6x + 13)$. When $x = 4$ then $x^3 + 2x^2 - 11x - 52 = 0$.

Revision

7 Determine the factors of the following polynomials:

(a) $x^2 - 16$, (b) $x^2 + 5x + 6$, (c) $x^2 + 2x + 1$, (d) $x^3 + 2x^2 + x$,

(e) $x^3 + 3x^2 + 6x + 4$, (f) $x^4 - 16$

8 Determine whether the following linear functions are factors of the given polynomials:

(a) $x^3 - 6x + 5$, $x - 1$, (b) $x^3 - 2x^2 + 3x - 6$, $x - 2$,

(c) $x^3 + x^2 - 3x + 2$, $x - 1$, (d) $x^4 - 2x^3 + 3x^2 - 4x + 2$, $x - 1$,

(e) $x^3 + 2x^2 + 3x + 2$, $x + 1$

2.5 Roots

The *roots* of a polynomial $y = p(x)$ are those values of x that make the polynomial have a zero value. They are thus the values of x at which a graph of the polynomial cuts the y-axis. Thus for the graph in Figure 2.6(a) the polynomial will have two roots, in Figure 2.10(b) one root and in Figure 2.10(c) no roots.

2.5.1 How many roots?

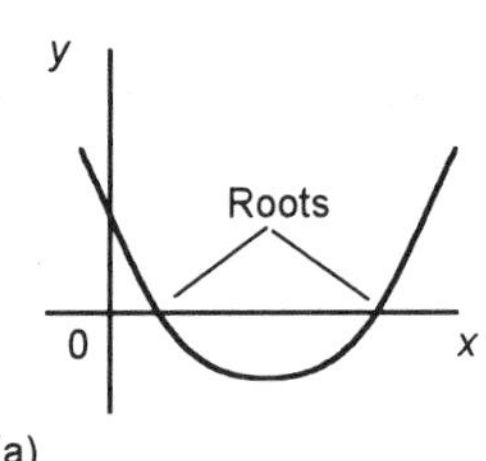

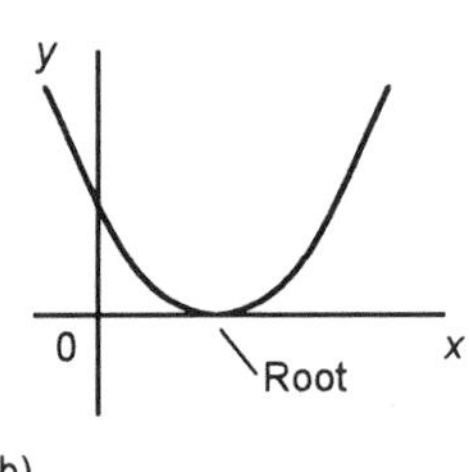

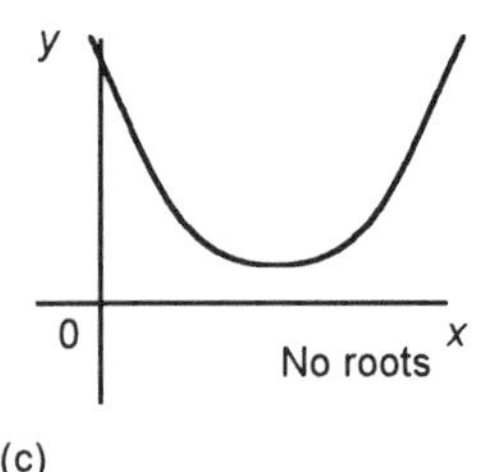

Figure 2.10 *Roots*

If $(x - a)$ is a factor of a polynomial then when $x = a$ the polynomial has a zero value. The maximum number of such factors we can have in a polynomial of degree n, i.e. a polynomial where the highest power of x is x^n, is n. Thus the maximum number of real roots (the term real is used to distinguish the roots from complex roots involving imaginary numbers, i.e. numbers involving the square root of -1) possible in a polynomial of degree n is n.

The actual number of real roots may, however, be less than n. A rule which can be used to determine the number of real roots with a polynomial is *Descartes' rule*.

The number of positive real roots is equal to the number of times for which the function p(x) has consecutive coefficients changing signs, or is less than that number by an even integer. The number of negative real roots is equal to the number of times for which the function p(–x) has consecutive coefficients changing signs, or is less than that number by an even integer.

To illustrate this, consider the polynomial:

$$p(x) = x^3 - 3x^2 - 5$$

This is cubic and may have three real roots. The coefficients are, in sequence, +1, –3, –5. There is only one sign change, so there is not more than one real positive root. Now consider $p(-x)$, i.e. the function when x is replaced by $-x$, i.e.

$$p(-x) = -x^3 - 3x^2 - 5$$

There are no sign changes, so there are no negative real roots.

Example

Determine the maximum number of real positive and real negative roots that can occur with the polynomial

$$p(x) = x^4 - 2x^3 + x^2 + 2x - 6$$

Using Descartes' rule, the coefficients for $p(x)$ have the signs +, − , +, +, − and so, since there are three sign changes, there will be a maximum of three real positive roots. For $p(-x)$ we have

$$p(-x) = x^4 + 2x^3 + x^2 - 2x - 6$$

The coefficients have the signs +, +, +, −, − and so, since there is one sign change there will be a maximum of one negative real root.

Revision

9 Determine the maximum number of real positive and real negative roots that can occur with the following polynomials:

(a) $x^3 + 2x^2 - 1$, (b) $x^3 - x^2 + 3$, (c) $x^5 + 2x^4 - 3x^3 - x^2 + 4x + 1$

10 Determine the number of real positive and real negative roots for the polynomials giving the graphs shown in Figure 2.11.

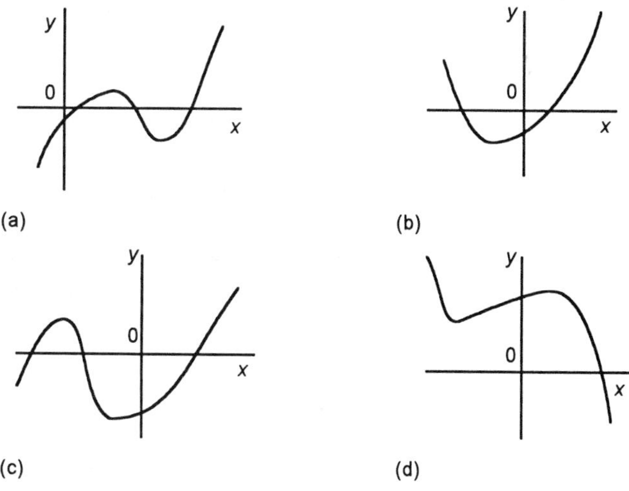

Figure 2.11 *Revision problem 10*

2.5.2 Graphical determination of roots

An approximation for a root can be found by graphical means. This involves plotting a few values of the function to find values which straddle the intercept of the function with the y-axis and so enable the rough location of the root to be determined. Consider the polynomial:

$$f(x) = x^2 - 5x - 7$$

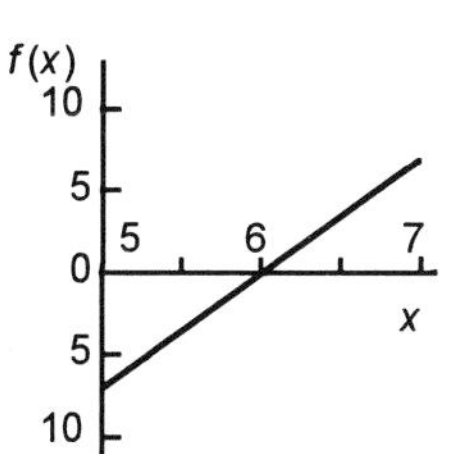

Figure 2.12 *Root estimate*

If we let $x = 5$ then we obtain $f(x) = -7$. With $x = 10$ we obtain $f(x) = +43$. Because there is a change in sign then there is a root between these two values. Since the value of the function with $x = 5$ is closer to $f(x) = 0$ value then it is likely that the root is nearer to 5 than 10. With $x = 6$ we obtain $f(x) = -1$. With $x = 7$ we obtain $f(x) = +7$. If we plot these values on a graph then we can make an estimate, using linear interpolation, of where the line cuts the x-axis. Figure 2.12 shows the graph. The root is thus about $x = 6.0$. Closer values either side of $x = 6.0$ lead to a better estimate of $x = 6.1$. The analysis could be repeated to also find another root at about $x = -1.1$. See Chapter 4 for a more formal discussion of this type of method as an iterative method.

There are a number of expressions which can give some clues as to the location of a root. One such expression, by Wilf, is:

$$|\text{root}| \leq 1 + \frac{\max(|a_0|, |a_1| \ldots |a_{n-1}|)}{|a_n|} \qquad [5]$$

where a_0 is the coefficient of the zero power term, a_1 that of the first degree terms, etc. in an n degree polynomial. The vertical lines either side of the coefficients means that the signs of the coefficients are disregarded and all are taken as positive. Thus, using this expression with the above example of $f(x) = x^2 - 5x - 7$ gives $|a_0| = 7$, $|a_1| = 5$, $|a_2| = 1$. The polynomial is a second degree polynomial, hence we have $|a_n| = 1$ and the |maximum value| of the remaining coefficients is 7. Thus $|\text{root}| \leq 1 + 7$ and the roots of the equation lie between -8 and $+8$.

It is possible to have two or more roots with the same value. For example, $x^2 + 2x + 2$ has the roots $x = 1$ and $x = 1$. If with a function we have two roots nearly equal then the graph cuts the axis in two points which are close together (Figure 2.13). If the roots are exactly equal then the two points coincide so that the curve just touches the axis at the value of the roots.

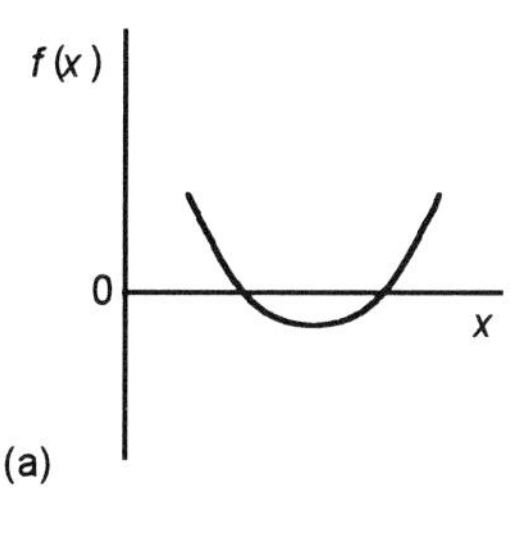

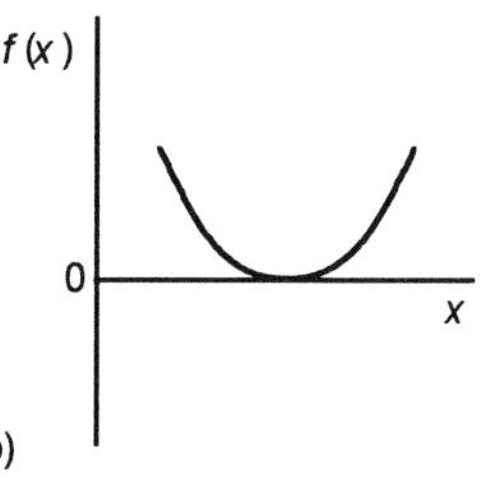

Figure 2.13 *(a) Two roots close together, (b) a double root*

Revision

11 Graphically, determine the roots of the following functions:

(a) $x^2 + x - 1$, (b) $x^2 + 2x - 1$, (c) $x^3 - 3x^2 - x + 9$,

(d) $x^3 + x^2 - 15x + 28$, (e) $x^3 + 4x^2 + 7x - 2$

2.5.3 Roots from factorisation by trial and error

We can find the roots of a polynomial if it factorises to give linear factors. The following are some commonly encountered forms of polynomials and factors:

$$(x + a)(x + b) = x^2 + (a + b)x + ab \qquad [6]$$

$$(x + a)^2 = x^2 + 2ax + a^2 \qquad [7]$$

$$(x - a)^2 = x^2 - 2ax + x^2 \qquad [8]$$

$$(x + a)(x - a) = x^2 - a^2 \qquad [9]$$

$$(x + a)^3 = x^3 + 3ax^2 + 3a^2x + a^3 \qquad [10]$$

$$(x - a)^3 = x^3 - 3ax^2 + 3a^2x - a^3 \qquad [11]$$

$$(x + a)(x^2 - ax + a^2) = x^3 + a^3 \qquad [12]$$

$$(x - a)(x^2 + ax + a^2) = x^3 - a^3 \qquad [13]$$

The general formula for positive integral powers of a sum, i.e. $(x + a)^n$, is given by the *binomial theorem* as:

$$(x + a)^n = a^n + na^{n-1}x + \frac{n(n-1)}{2!}a^{n-2}x^2 + \frac{n(n-1)(n-2)}{3!}a^{n-3}x^3 + \dots + x^n \qquad [14]$$

See Chapter 4 for a discussion of series and the binomial theorem.

Consider, for example, the polynomial $x^2 + 3x - 4$ and the determination of its roots. Suppose we think it can be represented by $(x + a)(x + b)$. This is of the form indicated by equation [6]. If this is to be the case then we must have $3 = a + b$ and $-4 = ab$. These equations can be satisfied if $a = 4$ and $b = -1$. Thus we can represent the polynomial by the linear factors $(x + 4)(x - 1)$. For the polynomial to be zero we must thus have $x + 4 = 0$, i.e. $x = -4$, or $x - 1 = 0$, i.e. $x = 1$. The roots are thus -4 and $+1$.

Revision

12 Using the equations given above for standard forms of polynomials, determine the roots of the following equations:

(a) $x^2 + 10x + 25$, (b) $x^2 - 16$, (c) $x^2 + 7x + 12$, (d) $x^2 + 4x - 12$,

(e) $x^2 + 8x + 16$

2.5.4 Completing the square

There is a technique involving *completing the square* that can be used with quadratics to enable roots to be obtained. It is based on getting equations into the form of a perfect square, i.e. as in equation [7] above:

$$(x + a)^2 = x^2 + 2ax + a^2$$

Consider, for example, the polynomial $x^2 + 3x - 4$. We can rewrite this as:

$$(x^2 + 3x) - 4 = 0$$

For the bracketed terms we make it into a square by adding 1.5^2 and to compensate subtract 1.5^2 outside the brackets. Squares are always obtained by adding the square of half the coefficient of the x term. Thus we have

$$(x^2 + 3x + 1.5^2) - 1.5^2 - 4 = (x + 1.5)^2 - 6.25$$

When x equals a root value then the function is zero. Thus, when we have a root value for x:

$$(x + 1.5)^2 - 6.25 = 0$$

Hence

$$(x + 1.5)^2 = 6.25$$

Taking square roots of both sides gives

$$x + 1.5 = \pm 2.5$$

Thus the roots are +1.0 or −4.0.

Example

Determine the roots of the function $x^2 + 2x - 3$.

We can rewrite this function as:

$$(x^2 + 2x) - 3$$

For the bracketed terms we make it into a square by adding 1 inside the brackets and to compensate we add a −1 outside the brackets. Thus we have:

$$(x^2 + 2x + 1) - 1 - 3 = (x + 1)^2 - 4$$

The function equals zero when x equals a root value. Thus we must have:

$$(x + 1)^2 - 4 = 0$$

Hence:

$$(x + 1)^2 = 4$$

Taking square roots of each side gives $x + 1 = \pm 2$ and hence the roots are $x = +1$ and -3.

Revision

13 Determine, by completing the squares, the roots of the following functions:

(a) $x^2 + 2x - 8$, (b) $x^2 + 4x - 8$, (c) $x^2 - 3x + 1$

2.5.5 The quadratic formula

The roots of a quadratic equation can be found by a simple equation. Consider the quadratic function $ax^2 + bx + c$, where a, b and c are constants and a does not equal zero. The roots are given by the values of x when the function equals zero, i.e.

$$ax^2 + bx + c = 0$$

Dividing throughout by a gives:

$$x^2 + \frac{b}{a}x + \frac{c}{a} = 0$$

We can rewrite this as

$$\left(x^2 + \frac{b}{a}x\right) + \frac{c}{a} = 0$$

Completing the square gives:

$$\left[x^2 + \frac{b}{a}x + \left(\frac{b}{2a}\right)^2\right] - \left(\frac{b}{2a}\right)^2 + \frac{c}{a} = 0$$

$$\left(x + \frac{b}{2a}\right)^2 = \left(\frac{b}{2a}\right)^2 - \frac{c}{a} = \frac{b^2 - 4ac}{4a^2}$$

Taking the square root of both sides of the equation gives:

$$x + \frac{b}{2a} = \pm\frac{\sqrt{b^2 - 4ac}}{2a}$$

Hence the roots are given by:

$$x = \frac{-b \pm \sqrt{b^2 - 4ac}}{2a} \qquad [15]$$

If $b^2 - 4ac$ is greater than 0, i.e. the term under the square root sign has a positive value, then the square root can be evaluated and we have two distinct roots. If $b^2 - 4ac$ equals 0 then we have $x = -b/2a$ and thus there are two equal roots. If $b^2 - 4ac$ is less than 0, i.e. has a negative value, then the function has no real roots since we cannot obtain a real number by taking the square root of a negative quantity.

Example

Determine, if they exist, the real roots of the following quadratic functions:

(a) $4x^2 - 7x + 3$, (b) $x^2 - 4x + 4$, (c) $x^2 + 2x + 4$

(a) The roots occur when $4x^2 - 7x + 3 = 0$. Using equation [15] we then have roots of:

$$x = \frac{+7 \pm \sqrt{49 - 48}}{8} = \frac{7 \pm 1}{8} = 1 \text{ or } 0.75$$

Hence we can represent the quadratic function by $(x - 1)(x - 0.75)$. This, when multiplied out gives $x^2 - 1.75x + 0.75$ and thus when multiplied by 4 gives the function $4x^2 - 7x + 3$.
(b) The roots occur when $x^2 - 4x + 4 = 0$. Using equation [15] we then have roots of:

$$x = \frac{+4 \pm \sqrt{16 - 16}}{2} = 2$$

Thus we have $x^2 - 4x + 4 = (x - 2)(x - 2)$.
(c) For $x^2 + 2x + 4 = 0$, equation [15] gives:

$$x = \frac{-2 \pm \sqrt{4 - 16}}{2}$$

Because we have the square root of a negative quantity, there are no real roots.

Example

A cantilever, length L, propped at its free end has a bending moment M which is a function of the distance x from the clamped end and is given by:

$$M = \tfrac{1}{2}wx^2 - \tfrac{5}{8}wLx + \tfrac{1}{8}wL^2$$

where w is the distributed load per unit length. Determine the points along the beam at which the bending moment is zero.

The problem involves the determination of the roots of the quadratic equation:

$$\tfrac{1}{2}wx^2 - \tfrac{5}{8}wLx + \tfrac{1}{8}wL^2 = 0$$

or, after simplification:

$$4x^2 - 5Lx + L^2 = 0$$

Using equation [15]:

$$x = \frac{5L \pm \sqrt{25L^2 - 16L^2}}{8}$$

Thus $x = L/4$ or L.

Revision

14 Determine, if they exist, the real roots of the following quadratic functions:

(a) $x^2 + 2x - 4$, (b) $x^2 + 3x + 1$, (c) $x^2 - 2x - 1$, (d) $x^2 + x + 2$

15 The e.m.f. E of a thermocouple is a function of the temperature T, being given by $E = -0.02T^2 + 6T$. The e.m.f. is in μV and the temperature in °C. Determine the temperatures at which the e.m.f. will be 200 μV.

16 When a ball is thrown vertically upwards with an initial velocity u from an initial height h_0, the height h of the ball is a function of the time t, being given by $h - h_0 = ut - 4.9t^2$. Determine the times for which the height is 1 m, if $u = 4$ m/s and $h_0 = 0.5$ m.

17 The deflection y of a simply supported beam of length L when subject to an impact load of mg dropped from a height h on its centre is obtained by equating the total energy released by the falling load with the strain energy acquired, i.e.

$$mgh + mgy = \frac{24EI}{L^3}$$

Hence obtain the value of the deflection y.

18 A right-angled triangle is to have right-angled sides of lengths L and $L + 5$. If the hypotenuse is to have a length of $L + 6$, what is x?

2.6 Partial fractions

This section is about a very useful mathematical tool, *partial fractions*, which is widely used to put complex expressions involving polynomial functions into simpler forms which can be more readily dealt with. In engineering and science, functions are often encountered which involve one polynomial function divided by another. Dealing with such functions is often easier when the functions are expressed as the sum of two or more simpler fractions. For example, many integrals involving fractions may be integrated by expressing the integral as the sum of two or more simpler fractions which can be integrated individually. In the analysis of electrical circuits or control systems by the means of the Laplace transform, it is frequently necessary to express a fraction as the sum of two or more simpler fractions in order that the inverse transform can be found. The simpler fractions resulting from the breaking down of a fraction are termed *partial fractions*. For example:

$$\frac{3x+4}{x^2+3x+2} = \frac{3x+4}{(x+1)(x+2)}$$

can be expressed as the partial fractions:

$$\frac{3x+4}{(x+1)(x+2)} = \frac{1}{x+1} + \frac{1}{x+2}$$

The type of function we are concerned with is known as a rational function. The term *rational function* is used for functions which have the general form:

$$f(x) = \frac{p(x)}{q(x)} \qquad [16]$$

where $p(x)$ and $q(x)$ are polynomials. If the degree of $p(x)$ is less than that of $q(x)$ the function $f(x)$ is said to be a *proper rational function*, otherwise it is *improper*. For example:

$$\frac{3x+4}{x^2+3x+2}$$

is a proper rational function. The top line of a fraction is called the *numerator* and the bottom line the *denominator*. The degree of either the numerator or the denominator function is the highest power of the variable in the expression. In the above fraction, the numerator function $3x + 4$ has the degree 1 while the denominator $x^2 + 3x + 2$ has the degree 2. The following is an example of an improper rational function:

$$\frac{x^3+4}{x^2+3x+2}$$

The numerator function has a degree of 3, while the denominator function has a degree of 2.

Any proper rational function can be expressed as a sum of simpler functions. An improper rational function can always be expressed as a polynomial plus a proper rational function by dividing the numerator by the denominator, i.e. obtaining the function in the form

$$f(x) = \frac{p(x)}{q(x)} = m(x) + \frac{r(x)}{q(x)} \qquad [17]$$

where $r(x)$ is the remainder after we have divided $p(x)$ by $q(x)$, $r(x)/q(x)$ being a proper rational function. Thus, for:

$$\frac{x^3+4}{x^2+3x+2}$$

we have:

$$\begin{array}{r|l} & \quad x - 3 \\ \hline x^2+3x+2 & x^3 \quad\quad + \quad 4 \\ & x^3 + 3x^2 + 2x \\ \hline & \quad -3x^2 - 2x + 4 \\ & \quad -3x^2 - 9x - 6 \\ \hline & \qquad\quad 7x + 10 \end{array}$$

and so:

$$\frac{x^3+4}{x^2+3x+2} = x - 3 + \frac{7x+10}{x^2+3x+2}$$

The rational function element can then be transformed into partial fractions. Hence:

$$\frac{x^3+4}{x^2+3x+2} = x - 3 + \frac{4}{x+2} + \frac{3}{x+1}$$

Revision

19 Transform the following improper rational functions into the sum of a polynomial and a proper rational function:

(a) $\dfrac{x^3}{x^2+1}$, (b) $\dfrac{x^3-x^2-5x+1}{x^2-3x+2}$, (c) $\dfrac{3x^2-x-2}{x+2}$

2.6.1 Obtaining partial fractions

When the degree of the denominator is greater than that of the numerator then an expression can be directly resolved into partial fractions. The form taken by the partial fractions depends on the type of denominator concerned.

1 If the denominator contains a *linear factor*, i.e. a factor of the form $(x + a)$, then for each such factor there will be a partial fraction of the form:

$$\frac{A}{(x+a)}$$

where A is some constant.

2 If the denominator contains *repeated linear factors*, i.e. a factor of the form $(x + a)^n$, then there will be partial fractions:

$$\frac{A}{(x+a)}+\frac{B}{(x+a)^2}+\ldots+\frac{C}{(x+a)^n}$$

with one partial fraction for each power of $(x + a)$.

3 If the denominator contains an *irreducible quadratic factor*, i.e. a factor of the form $ax^2 + bx + c$, then there will be a partial fraction of the form:

$$\frac{Ax+B}{ax^2+bx+c}$$

for each such factor.

4 If the denominator contains *repeated quadratic factors*, i.e. a factor of the form $(ax^2 + bx + c)^n$, there will be partial fractions of the form:

$$\frac{Ax+B}{ax^2+bx+c}+\frac{Cx+D}{(ax^2+bx+c)^2}+\ldots+\frac{Ex+F}{(ax^2+bx+c)^n}$$

with one for each power of the quadratic.

The values of the constants A, B, C, etc. can be found by either making use of the fact that the equality between the fraction and its partial fractions must be true for all values of the variable x or that the coefficients of x^n in the fraction must equal those of x^n when the partial fractions are multiplied out.

To illustrate this, consider the simplification of:

$$\frac{3x+4}{(x+1)(x+2)}$$

This has two linear factors in the denominator and so the partial fractions are of the form:

$$\frac{A}{x+1}+\frac{B}{x+2}$$

with one partial fraction for each linear term. Thus for the expressions to be equal we must have:

$$\frac{3x+4}{(x+1)(x+2)} = \frac{A}{x+1} + \frac{B}{x+2} = \frac{A(x+2)+B(x+1)}{(x+1)(x+2)}$$

Thus

$$3x + 4 = A(x + 2) + B(x + 1)$$

Consider the requirement that this relationship is true for all values of x. Then, when $x = -1$ we must have:

$$-3 + 4 = A(-1 + 2) + B(-1 + 1)$$

Hence $A = 1$. When $x = -2$ we must have:

$$-6 + 4 = A(-2 + 2) + B(-2 + 1)$$

Hence $B = 1$. Alternatively, we could have determined these constants by multiplying out the expression and considering the coefficients, i.e.

$$3x + 4 = A(x + 2) + B(x + 1) = Ax + 2A + Bx + B$$

Thus, for the coefficients of x to be equal we must have $3 = A + B$ and for the constants to be equal $4 = 2A + B$. These two simultaneous equations can be solved to give A and B.

When the degree of the denominator is equal to or less than that of the numerator, the denominator must be divided into the numerator until the result is the sum of terms with the remainder fraction term having a denominator which is of higher degree than its numerator. Consider, for example, the fraction:

$$\frac{x^3 - x^2 - 3x + 1}{x^2 - 3x + 2}$$

The numerator has a degree of 3 and the denominator a degree of 2. Thus, dividing has to be used.

$$\begin{array}{r|l} & x + 2 \\ \hline x^2 - 3x + 2 & x^3 - x^2 - 3x + 1 \\ & x^3 - 3x^2 + 2x \\ \hline & \quad 2x^2 - 5x + 1 \\ & \quad 2x^2 - 6x + 4 \\ \hline & \qquad\qquad x - 3 \end{array}$$

Thus

$$\frac{x^3 - x^2 - 3x + 1}{x^2 - 3x + 2} = x + 2 + \frac{x - 3}{x^2 - 3x + 2}$$

The fractional term can then be simplified using partial fractions.

$$\frac{x-3}{x^2-3x+2} = \frac{x-3}{(x-1)(x-2)} = \frac{A}{x-1} + \frac{B}{x-2}$$

Hence we must have:

$$x - 3 = A(x - 2) + B(x - 1)$$

When $x = 1$ we have $-2 = -A$ and so $A = 2$. When $x = 2$ we have $-1 = B$. Hence the partial fractions are $2/(x - 1)$ and $-1/(x - 2)$. Thus:

$$\frac{x^3-x^2-3x+1}{x^2-3x+2} = x+2+\frac{2}{x-1} - \frac{1}{x-2}$$

The procedure for obtaining partial fractions can thus be summarised as:

1 If the degree of the denominator is equal to, or less than, that of the numerator, divide the denominator into the numerator to obtain the sum of a polynomial plus a fraction which has the degree of the denominator greater than that of the numerator.

2 Write the denominator in the form of linear factors, i.e. of the form $(ax + b)$, or irreducible quadratic factors, i.e. of the form $(ax^2 + bx + c)$.

3 Write the fraction as a sum of partial fractions involving constants A, B, etc.

4 Determine the unknown constants which occur with the partial fractions by equating the fraction with the partial fractions and either solving the equation for specific values of x or equating the coefficients of equal powers of x.

5 Replace the constants in the partial fractions with their values.

Example

Use partial fractions to simplify the following expression:

$$\frac{2x-19}{(x-2)^2(x+3)}$$

The denominator is of higher degree than the numerator, so it can be transformed into partial fractions without first dividing. The denominator contains a repeated factor and so the partial fraction form is:

$$\frac{2x-19}{(x-2)^2(x+3)} = \frac{A}{(x-2)^2} + \frac{B}{(x-2)} + \frac{C}{(x+3)}$$

Thus we must have:

$$2x - 19 = A(x + 3) + B(x - 2)(x + 3) + C(x - 2)^2$$

When $x = -3$ then $-6 - 19 = 25C$ and so $C = -1$. When $x = 2$ then $4 - 19 = 5A$ and so $A = -3$. When $x = 0$ then $-19 = 3A - 2B + 4C$ and so $B = 1$. Thus

$$\frac{2x-19}{(x-2)^2(x+3)} = -\frac{3}{(x-2)^2} + \frac{1}{(x-2)} - \frac{1}{(x+3)}$$

Example

Use partial fractions to simplify the following expression:

$$\frac{3x}{(x^2-2x+5)(x+1)}$$

This expression contains a quadratic and a linear factor. Thus the partial fractions are of the form:

$$\frac{3x}{(x^2-2x+5)(x+1)} = \frac{A+Bx}{(x^2-2x+5)} + \frac{C}{(x+3)}$$

Hence we must have:

$$3x = (A + Bx)(x + 3) + C(x^2 - 2x + 5)$$

With $x = -3$ then we have $-9 = C(9 + 6 + 5)$ and so $C = -9/20$. With $x = 0$ we have $0 = 3A + 5C$ and so $A = 3/4$. With $x = 1$ we have $3 = 4(A + B) + 4C$ and so $B = 9/20$. Thus the partial fractions are:

$$\frac{3x}{(x^2-2x+5)(x+1)} = \frac{15+9x}{20(x^2-2x+5)} - \frac{9}{20(x+3)}$$

Example

Use partial fractions to simplify the following expression:

$$\frac{8(x+1)}{x(x^2-4)}$$

The denominator can be factorised to give:

$$\frac{8(x+1)}{x(x^2-4)} = \frac{8(x+1)}{x(x-2)(x+2)} = \frac{A}{x} + \frac{B}{x-2} + \frac{C}{x+2}$$

Hence:

$$8x + 8 = A(x-2)(x+2) + Bx(x+2) + Cx(x-2)$$

When $x = 2$ we have $24 = 8B$ and so $B = 3$. When $x = -2$ we have $-8 = 8C$ and so $C = -1$. When $x = 0$ we have $8 = -4A$ and so $A = -2$. Thus:

$$\frac{8(x+1)}{x(x^2-4)} = -\frac{2}{x} + \frac{3}{x-2} - \frac{1}{x+2}$$

Example

Use partial fractions to simplify the following expression:

$$\frac{x^2}{(x^2+1)^2}$$

We can write this fraction, which is a repeated quadratic, as:

$$\frac{x^2}{(x^2+1)^2} = \frac{Ax+B}{x^2+1} + \frac{Cx+D}{(x^2+1)^2}$$

Hence we must have:

$$x^2 = (Ax + B)(x^2 + 1) + Cx + D = Ax^3 + Bx^2 + Ax + B + Cx + D$$

Equating coefficients gives: for those of x^3 we have $A = 0$, for those of x^2 we have $B = 1$, for those of x we have $0 = A + C$ and so $C = 0$, for those of constants we have $0 = B + D$ and so $D = -1$. Hence we have:

$$\frac{x^2}{(x^2+1)^2} = \frac{1}{x^2+1} - \frac{1}{(x^2+1)^2}$$

Revision

20 Use partial fractions to simplify the following expressions involving linear factors:

(a) $\frac{x-6}{(x-1)(x-2)}$, (b) $\frac{x+5}{x^2+3x+2}$, (c) $\frac{x-13}{(x-1)(x+3)}$

21 Use partial fractions to simplify the following expressions involving repeated factors:

(a) $\frac{x^2}{(x-2)^2(x-1)}$, (b) $\frac{5x-1}{(x+1)^2(x-2)}$, (c) $\frac{2x-1}{(x+1)^2}$

22 Use partial fractions to simplify the following expressions involving quadratics:

(a) $\frac{x+5}{(x-1)(x^2+x+1)}$, (b) $\frac{3x+4}{(x-2)(x^2+2x+2)}$,

(c) $\frac{x-3}{(x-1)^2(x^2+2)}$

23 Use partial fractions to simplify the following expressions involving repeated quadratics:

$$\text{(a) } \frac{2x^2}{(x^2+1)^2}, \text{ (b) } \frac{2x^3}{(x^2+2)^2}, \text{ (c) } \frac{x^2+x+2}{(x^2+2)^2}$$

24 Use partial fractions to simplify the following improper rational functions:

$$\text{(a) } \frac{x^2+2}{(x+4)(x-2)}, \text{ (b) } \frac{3x^2-x-2}{x+2}, \text{ (c) } \frac{x^3+x+6}{(x-2)(x-4)}$$

Problems

1 Which of the following would give a straight line graph, and if so is it through the origin:
(a) A graph of the resistance R against the length L of resistance wire, the resistance being related to the length by the equation $R = \rho L/A$, with ρ and A being constants.
(b) A graph of the distance d fallen by a ball against the time t, the distance being related to the time by $d = 4.9t^2$.
(c) A graph of the length L of a rod against the temperature t, the length being related to the temperature by the equation $L = L_0(1 + at)$ with L_0 and a being constants.
(d) A graph of the circumference c of a circle against the radius r, the circumference being related to the radius by the equation $c = 2\pi r$.

2 Determine the straight line equations which fit the following situations:
(a) The potential difference V across a conductor is directly proportional to the current I through it. When the current is 0.5 A the potential difference is 4 V.
(b) The frictional force F between two surfaces is directly proportional to the normal reaction N. When the normal reaction is 20 N the frictional force is 8 N.
(c) The velocity v of a car as a function of the time t, the time being measured from when the car was at rest, if when accelerating in third gear the following results were obtained:

v in m/s	14.6	15.2	15.8	16.4
t in s	8.0	8.5	9.0	9.5

(d) The resistance R of a resistor as a function of temperature θ for the following results:

R in Ω	23.0	24.5	26.0	27.5
θ in °C	20	30	40	50

3 A linear electrical circuit has two independent voltage sources, one of 2 V and the other of 4 V. With the 2 V source alone the circuit has a

current of 3 mA, with the 4 V source alone it has a current of 1 mA. What will be the current when both sources are present?

4 Sketch the graphs of the following functions:

(a) $y = 2 - \frac{1}{x}$, (b) $y = 2x^2 - 1$, (c) $y = x^2 - 2x + 6$

5 Sketch the graphs of the following functions:
(a) The distance s travelled by an object as a function of time t and given by $s = \frac{1}{2}at^2$, where a is a constant.
(b) The volume V of a sphere as a function of its radius r, being given by $V = 4\pi r^2/3$ if only positive values of r can occur.

6 Determine the factors of the following polynomials:

(a) $x^2 - 1$, (b) $x^2 + 12x - 13$, (c) $x^2 + 4x$, (d) $x^3 + 2x^2 - x - 2$,

(e) $x^3 + 3x^2 + 3x + 2$

7 Graphically determine the roots of the following functions:

(a) $x^2 - 4x + 1$, (b) $x^2 + 3x - 2$, (c) $x^3 + x^2 - 3x - 3$,

(d) $2x^3 - 3x + 1$, (e) $x^3 - 4x + 3$

8 Determine, by factorisation, the roots of the following functions:

(a) $x^2 + 2x - 3$, (b) $x^2 - x - 2$, (c) $x^3 - 3x^2 + 2x$

9 Determine, if they exist, the real roots of the following quadratic functions:

(a) $x^2 - 2x - 5$, (b) $x^2 + 3x + 1$, (c) $x^2 - 2x + 5$, (d) $2x^2 + 4x + 2$

10 A rectangle has an area of 100 mm^2. If its width is 5 mm less than its length, what are the possible dimensions of the rectangle?

11 A ball is thrown vertically upwards. Its height h is a function of time t and given by $h = 20t - 5t^2$. Determine the times at which the height is 10 m.

12 When a bar of length L is subject to a longitudinal impact load of Mg, an equation relating the resulting deflection y with the load can be obtained by equating the energy supplied by the falling mass and the resulting strain energy, i.e.

$$Mg(h + y) = \frac{EAy^2}{2L}$$

Hence obtain a value for y.

13 The periodic time T of a rigid body about a horizontal axis a distance h from its centre of gravity is given by:

$$T = 2\pi\sqrt{\frac{h^2 + k^2}{gh}}$$

where k is the radius of gyration of the body about a parallel axis through the centre of gravity. Determine the values of h which will give rise to the same periodic time.

14 Use partial fractions to simplify the following expressions:

(a) $\dfrac{3x-1}{2x^2-x-1}$, (b) $\dfrac{3x+2}{x^2+3x-10}$, (c) $\dfrac{x+3}{(x-2)(x+4)}$,

(d) $\dfrac{1}{(x-2)(x+1)^3}$, (e) $\dfrac{x-1}{(x+1)(x-2)^2}$, (f) $\dfrac{2}{(x-1)(x^2+1)}$,

(g) $\dfrac{x^2+3}{x(x^2+2)}$, (h) $\dfrac{x^2+2x+5}{(x+1)(x^2+2x-2)}$, (i) $\dfrac{x^2-2}{x^2+2x-3}$,

(j) $\dfrac{x^3+3}{(x+1)(x-1)}$, (k) $\dfrac{x^3}{x^2-2x-3}$, (l) $\dfrac{x^3}{(x^2-4)^2}$,

(m) $\dfrac{5x-1}{(x-2)(x+1)^2}$, (n) $\dfrac{1}{x^2+x}$, (o) $\dfrac{x^2-x+1}{(x-2)(x^2+1)}$,

(p) $\dfrac{2x^3+x+6}{(x-1)(x+2)}$, (q) $\dfrac{x^3-x-1}{x^2-1}$, (r) $\dfrac{1}{(x-2)(x+1)^3}$,

(s) $\dfrac{x^2+x-1}{(x^2+1)^2}$, (t) $\dfrac{3-x}{(x+3)(x^2+3)}$, (u) $\dfrac{1}{x^2(x+2)}$,

(v) $\dfrac{x^3}{x^2+1}$, (w) $\dfrac{x^2+x-2}{(x+1)(x-2)^2}$, (x) $\dfrac{2x^3+3x^2-x-4}{x^2+x-1}$

3 More functions

3.1 Introduction

The function describing how the voltage across a capacitor changes with time when it discharges is an example of an exponential function. The signal ratio between the output power from an amplifier and its input is often expressed in terms of decibels, this being a logarithmic scale of units. The function describing how the current varies with time for the domestic alternating current supply is a sinusoidal function, one of a number of trigonometrical functions. Combinations of exponential functions often occur in engineering and science and are often expressed in terms of the hyperbolic functions.

This chapter is a consideration of exponential functions, logarithmic, trignometrical functions and hyperbolic functions. Algebraic manipulations involving such functions are reviewed, though some basic dexterity and differentiation notation are assumed.

3.2 Exponential function

The term *exponent* is just another name for a power or index and expressions involving exponents are termed *exponential expressions*. Thus a^x is an exponential expression with x being the exponent, a being termed the *base*.

> *An exponential function f(x) has the basic form* $f(x) = a^x$, *where a is a positive constant.*

Note that an exponential function is *not* a polynomial function. With a polynomial function the powers are constants, e.g. x^2, whereas with an exponential function the power is the independent variable x.

3.2.1 Laws of indices

Exponential expressions can be simplified and manipulated using the *laws of indices*. These can be summarised as:

$$a^m a^n = a^{m+n} \qquad [1]$$

$$\frac{a^m}{a^n} = a^{m-n} \qquad [2]$$

$$(a^m)^n = a^{mn} \qquad [3]$$

$$a^{-m} = \frac{1}{a^m} \qquad [4]$$

$$a^0 = 1 \qquad [5]$$

Example

Simplify (a) $\dfrac{a^{2x}a^{4x}}{a^{3x}}$, (b) $\dfrac{(a^{2x})^2}{a^x}$, (c) $a^{-2x}a^{4x}$.

(a) $\dfrac{a^{2x}a^{4x}}{a^{3x}} = \dfrac{a^{2x+4x}}{a^{3x}} = a^{6x-3x} = a^{3x}$

(b) $\dfrac{(a^{2x})^2}{a^x} = \dfrac{a^{2x\times 2}}{a^x} = a^{4x-x} = a^{3x}$

(c) $a^{-2x}a^{4x} = a^{-2x+4x} = a^{2x}$

Revision

1 Simplify (a) $a^{2x} \times a^{-3x} \times a^{5x}$, (b) $\dfrac{a^{3x}}{a^{2x}a^{5x}}$, (c) $\dfrac{(a^{3x})^4}{a^{5x}}$, (d) $\dfrac{a^{-3x}}{a^{-5x}}$.

3.2.2 The exponential function

An exponential function has the general form $f(x) = a^x$. Table 3.1 shows some values of the function when the bases have the values 1/3, 1/2, 2 and 3 and Figure 3.1 shows graphs of the functions. Note that $(1/3)^x = 3^{-x}$ and $(1/2)^x = 2^{-x}$. All the exponential functions have the value 1 when $x = 0$. As the graphs illustrate, when a is greater than 1, exponential functions grow very rapidly. When a is less than 1, the exponential function decreases very rapidly. In both cases there are never any negative values for the function.

The 2^x describes an event which doubles when x changes from 0 to 1, from 1 to 2, from 2 to 3, etc. The 3^x function describes an event which triples when x changes from 0 to 1, from 1 to 2, from 2 to 3, etc. The $(1/2)^x$ describes an event which halves when x changes from 0 to 1, from 1 to 2, from 2 to 3, etc. The $(1/3)^x$ describes an event which is reduced by a factor of 3 when x changes from 0 to 1, from 1 to 2, from 2 to 3, etc.

Table 3.1 *Exponentials*

x	$\left(\frac{1}{3}\right)^x$	$\left(\frac{1}{2}\right)^x$	2^x	3^x
−3	27	8	0.125	0.037
−2	9	4	0.25	0.111
−1	3	2	0.5	0.333
0	1	1	1	1
+1	0.333	0.5	2	3
+2	0.111	0.25	4	9
+3	0.037	0.125	8	27

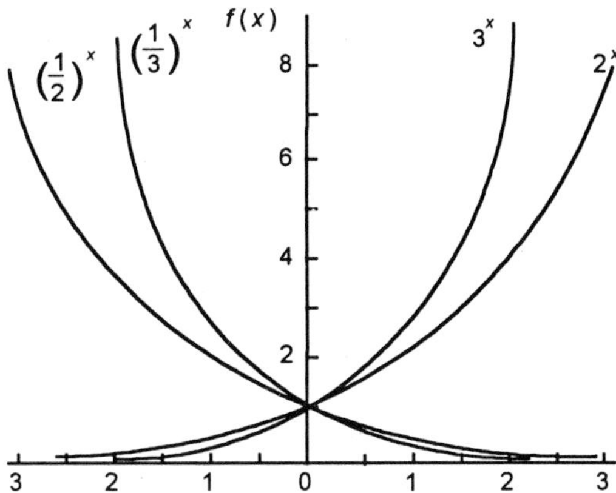

Figure 3.1 *Graphs of 0.5^x, 2^x and 3^x*

Consider the function of x described by $y = 2^x$. The change in y, Δy, when x changes by 1 depends on the value of x from which the change is made. We have:

x	0	1	2	3	4
y	1	2	4	8	16
Δy		1	2	4	8

Δy per change in x varies with x in the same way that y varies with x. We can thus write $\Delta y/\Delta x$ is proportional to y. For infinitesimally small changes in x we can write $\mathrm{d}y/\mathrm{d}x$ is proportional to y. If we repeat the above analysis for other exponential changes we find that for all exponential changes:

$$\frac{\mathrm{d}y}{\mathrm{d}x} \propto y \text{ and so } \propto a^x$$

Suppose we want the function that gives $\mathrm{d}y/\mathrm{d}x = y$. Consider the values obtained when $\Delta y/\Delta x = y$ and we take Δx values of 0.2. Starting with $y = 1$ at $x = 0$ we have:

x	0	0.2	0.4	0.6	0.8	1
y	1	1.2	1.44	1.73	2.08	2.50

At $x = 1$ we have $y = 2.50$. If smaller values of Δx are used then we tend to the value $y = 2.718$ at $x = 1$. Thus if we have $y = a^x$, then $y = a$ when $x = 1$. Thus the function described by $\mathrm{d}y/\mathrm{d}x = y$ is $y = \mathrm{e}^x$, where e represents the number 2.718. Exponentials of this form are often referred to as *the exponential*. Whenever an engineer or scientist refers to an exponential change, he or she is almost inevitably referring to an equation written in terms of e^x.

The fundamental property of the function e^x *is that* $\frac{d}{dx}e^x = e^x$.

While we can use any base to describe a change, it is generally more convenient to use the base e so that we have:

$$y = e^{kx} \tag{6}$$

and so:

$$\frac{dy}{dx} = ky \tag{7}$$

Table 3.2 shows values of the exponential functions $f(x) = e^x$ and $f(x) = e^{-x}$ for various values of x and Figure 3.2 the resulting graphs. As x increases, $f(x) = e^x$ becomes more positive with the value of the function increasing and tending to infinity as x tends to infinity. The function thus represents a growth. For $f(x) = e^{-x}$ as x tends to positive infinity then the value of the function tends to 0 and as x tends to $-\infty$ then the value of the function becomes infinite. The function thus represents a decay. For both functions we have $e^0 = 1$ at $x = 0$.

Table 3.2 *The exponential* e^x *and* e^{-x}

x	−3	−2	−1	0	+1	+2	+3
e^x	0.050	0.135	0.368	1	2.718	7.389	20.086
e^{-x}	20.086	7.389	2.718	1	0.368	0.135	0.050

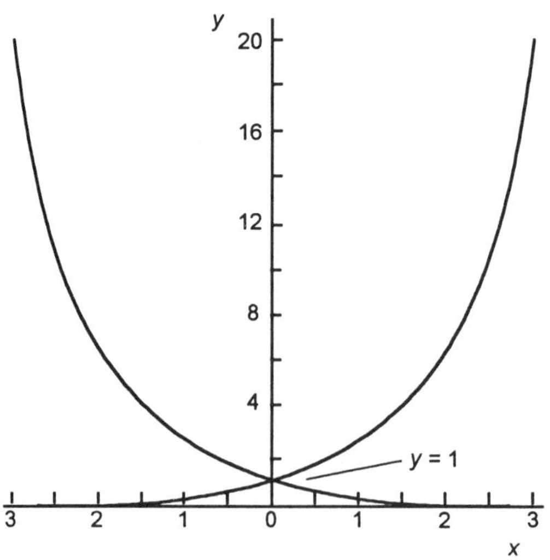

Figure 3.2 $y = e^x$ *and* e^{-x}

An example of exponential decay is that of the charge on a capacitor as it discharges through a resistor, the discharge equation being:

$$q = Q_0 \, e^{-t/CR}$$

where q is the charge on the capacitor at time t, Q_0 the charge at time $t = 0$, C the capacitance and R the resistance.

One form of equation involving exponentials that is fairly common in engineering and science is of the form:

$$y = A - A \, e^{-kx} \qquad [8]$$

When $x = 0$ then $e^0 = 1$ and so $y = 0$. When x tends to infinity then e^{-kx} tends to 0 and so y tends to the value A. Figure 3.3 shows the form of the graph. It shows a quantity that increases rapidly at first and then slows down to become eventually A. A graph of this form describes how the charge q on an initially uncharged capacitor changes with time t when it is charged, the equation being:

$$q = Q_0 - Q_0 \, e^{-t/CR}$$

When $t = 0$ then $e^0 = 1$ and so $q = 0$. When t tends to infinity then the exponential tends to 0, q tends to Q_0.

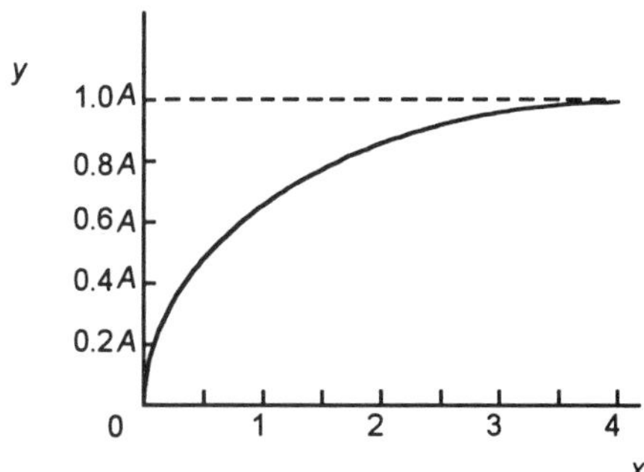

Figure 3.3 $y = A - A \, e^{-kx}$

Example

Determine, when (i) $x = 0$, (ii) $x =$ infinity, (c) $x = 1$, the values of (a) e^{2x}, (b) $4 \, e^{-3x}$, (c) $2 - 2 \, e^{-0.5x}$.

(a) (i) $e^{2x} = e^0 = 1$, (ii) $e^{2x} = e^{\infty} = 0$, (iii) $e^{2x} = e^2 = 0.693$.
(b) (i) $4 \, e^{-3x} = 4 \, e^0 = 4$, (ii) $4 \, e^{-3x} = 4 \, e^{\infty} = 0$, (iii) $4 \, e^{-3x} = 4 \, e^{-3} = 0.199$.
(c) (i) $2 - 2 \, e^{-0.5x} = 2 - 2 = 0$, (ii) $2 - 2 \, e^{-0.5x} = 2 - 0 = 2$, (iii) $2 - 2 \, e^{-0.5x} = 2 - 2 \, e^{-0.5} = 0.787$.

Example

For an electrical circuit containing inductance and resistance, the current in amperes is related to time t in seconds by $i = 3(1 - e^{-10t})$. Determine the current when (a) $t = 0$, (b) t = infinity, (c) $t = 0.1$ s.

(a) When $t = 0$ the exponential term has the value 1 and so $i = 0$.
(b) When $t = \infty$ the exponential term has the value 0 and so $i = 3$ A.
(c) When $t = 0.1$ s the exponential term is e^{-1} and has the value 0.368 and so $i = 1.896$ A.

Revision

2 Determine, using a calculator, to two decimal places the values for the function $f(x) = 5\ e^{x}$ when x is (a) −4, (b) −2, (c) 0, (d) +2, (e) +4.

3 Determine, using a calculator, to two decimal places the values for the function $f(x) = 5\ e^{-x}$ when x is (a) −4, (b) −2, (c) 0, (d) +2, (e) +4.

4 The voltage v, in volts, in an electrical circuit varies with time t, in seconds, according to the equation $v = 10\ e^{-2t}$. What will be the voltage when (a) $t = 0$, (b) $t = \frac{1}{2}$ s, (c) t tends to infinity?

5 A model used to predict the sales S of a product per month of a new product is $S = 4000 - 2000\ e^{-t}$, where t is the time on months. What will be the sales when (a) $t = 0$, (b) $t = 1$ month, (c) $t = 4$ months?

6 Simplify (a) e^{2x}/e^{3x}, (b) $e^{x}(e^{x} + 2)$, (c) $e^{-x}(1 + 2\ e^{x})$

3.2.3 The half value

In radioactivity the rate of decay of the number N of radioactive atoms with time is proportional to the time t, i.e.

$$\frac{dN}{dt} = kN \qquad [9]$$

where k is a constant called the *decay constant*. Hence:

$$N = N_0\ e^{-kt} \qquad [10]$$

The *half life* of a radioactive element is the time taken for half the number of atoms to decay. When $N = \frac{1}{2}N_0$ then $\frac{1}{2}N_0 = N_0\ e^{-kT}$, where T is the half life. Thus $0.5 = e^{-kT}$ and, taking logs to base e:

$$\ln 0.5 = \ln e^{-kT} = -kT$$

and so:

$$T = \frac{0.6931}{k} \qquad [11]$$

There are other situations described by an exponential function where a similar half value is used. For example, in considering the transmission of radiation through a material we can have a similar exponential relationship and refer to the half thickness, i.e. the thickness of material needed to reduce the intensity of the radiation by half.

Example

The half life of thoron is 52 s. How long will need to elapse for the amount of thoron in a sample to decay to (a) half the initial amount, (b) one-quarter the initial amount?

(a) The half life is the time taken for the amount to decay to half and so 52 s is required.
(b) If we have initially an amount A then after 52 s it has become $\frac{1}{2}A$. In the next 52 s this will again decay by half to give $\frac{1}{4}A$. Thus the time required is 104 s.

Revision

6 Strontium-90 has a decay constant of 0.028 day^{-1}. What is (a) the half life and (b) the time taken for the amount of strontium-90 to decay to one-quarter its initial amount?

3.2.4 e

Earlier in this section the reason given for choosing e as the base of exponentials was because the rate of change of a function $y = f(x) = e^x$ is equal to the same function e^x, i.e. $dy/dx = y$. If we consider finite increments then we can write for the exponential:

$$\frac{\Delta y}{\Delta x} = y$$

Consider what this implies for the value of y when it starts growing from $y = 1$ at $x = 0$. The value of y after one increment of Δx is $y + \Delta y$ where $\Delta y = y\ \Delta x = 1\ \Delta x$ and is thus $1 + \Delta x$. After a second increment we have $\Delta y = (1 + \Delta x)\ \Delta x$ and so the new value of y is $(1 + \Delta x)\ \Delta x + (1 + \Delta x) = (1 + \Delta x)^2$. After a third increment we have $\Delta y = (1 + \Delta x)^2\ \Delta x$ and so the new value of y is $(1 + \Delta x)^2\ \Delta x + (1 + \Delta x)^2 = (1 + \Delta x)^3$. After n increments the value of y will be $(1 + \Delta x)^n$.

Thus if we want to find the value of y at $x = 1$ and we divide the interval between $x = 0$ and $x = 1$ into ten increments, then $\Delta x = 1/10$ and so the value of y at $x = 1$ is $(1 + 1/10)^{10}$. This has the value 2.59. Suppose, however, we divided the interval into 100 increments, then $\Delta x = 1/100$ and the value of y at $x = 1$ is $(1 + 1/100)^{100}$. This has the value 2.70. If we use 1000 increments then y is 2.717. We can thus define e as:

e *is the value of* $(1 + 1/n)^n$ *as n tends to infinity.*

The above gives the value of the function when $x = 1$, i.e. it defines e^1. If we want the value of the function at x, i.e. the value of e^x, then instead of dividing the interval 1 into n segments we consider the interval x divided into n segments. Thus:

$$e^x = \left(1 + \frac{x}{n}\right)^n \quad [12]$$

3.3 Log functions

Consider the function $y = 2^x$. If we are given a value of x then we can determine the corresponding value of y. However, suppose we are given a value of y and asked to find the value of x that could have produced it. The inverse function is called the *logarithm function* and is defined, for $y = a^x$ and $a > 0$, as:

$$x = \log_a y \quad [13]$$

This is stated as 'log to base a of y equals x'.

If we take an input of x to a function $f(x) = a^x$ and then follow it by the inverse function $f^{-1}(x) = \log_a(x)$, as in figure 3.4, then because it is an inverse we obtain x. Thus:

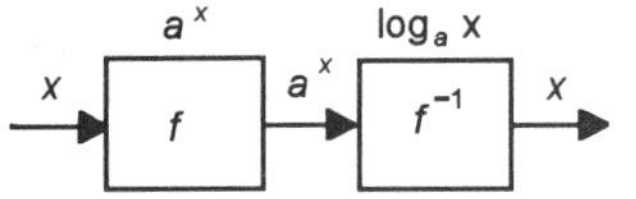

Figure 3.4 $f(x)f^{-1}(x) = x$

$$\log_a a^x = x \quad [14]$$

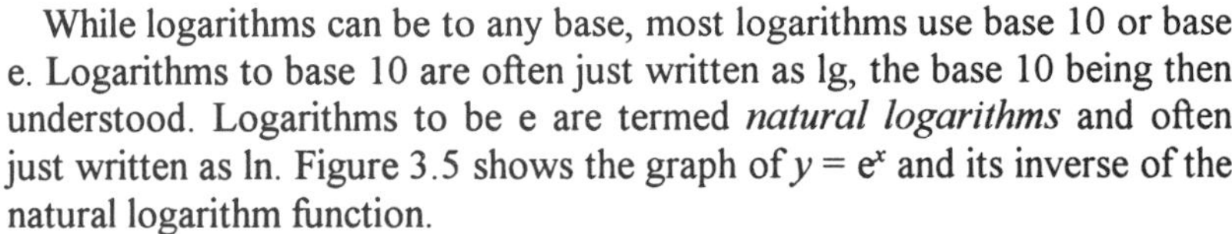

While logarithms can be to any base, most logarithms use base 10 or base e. Logarithms to base 10 are often just written as lg, the base 10 being then understood. Logarithms to be e are termed *natural logarithms* and often just written as ln. Figure 3.5 shows the graph of $y = e^x$ and its inverse of the natural logarithm function.

Since $a^{A+B} = a^A a^B$ then:

$$\log_a A + \log_a B = \log_a AB \quad [15]$$

$$n \log_a A = \log_a(A^n) \quad [16]$$

Since $a^{A-B} = \frac{a^A}{a^B}$ then:

$$\log_a A - \log_a B = \log_a \frac{A}{B} \quad [17]$$

Since $a^1 = a$ then:

$$\log_a a = 1 \quad [18]$$

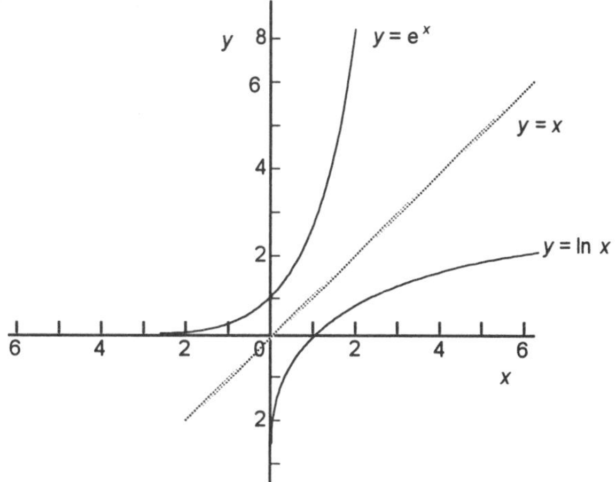

Figure 3.5 *The exponential and its inverse of the natural logarithm function*

Sometimes there is a need to change from one base to another, e.g. $\log_a x$ to $\log_b x$. Let $u = \log_b x$ then $b^u = x$ and so taking logarithms to base a of both sides gives $\log_a b^u = \log_a x$ and so $u \log_a b = \log_a x$. Since $u = \log_b x$ then $(\log_b x)(\log_a b) = \log_a x$ and so:

$$\log_b x = \frac{\log_a x}{\log_a b} \qquad [19]$$

Example

Write $\lg\left(\frac{\sqrt{a}}{bc^3}\right)$ in terms of $\lg a$, $\lg b$ and $\lg c$.

Using equation [17]:

$$\lg\left(\frac{\sqrt{a}}{bc^3}\right) = \lg\sqrt{a} - \lg(bc^3)$$

Hence, using equations [15] and [16]:

$$\lg\left(\frac{\sqrt{a}}{bc^3}\right) = \tfrac{1}{2}\lg a - \lg b - 3\lg c$$

Example

Simplify (a) $\lg x + \lg x^3$, (b) $3 \ln x + \ln(1/x)$.

(a) $\lg x + \lg x^3 = \lg(x \times x^3) = \lg x^4$

(b) $3 \ln x + \ln(1/x) = \ln x^3 + \ln(1/x) = \ln(x^3/x) = \ln x^2$

Example

Express $\log_{100} x$ in terms of $\log_{10} x$.

Using equation [19]:

$$\log_{100} x = \frac{\log_{10} x}{\log_{10} 100} = \frac{\log_{10} x}{\log_{10} 10^2} = \frac{\log_{10} x}{2}$$

Example

Solve for x the equation $2^{2x-1} = 12$.

Taking logarithms of both sides of the equation gives:

$$(2x - 1) \lg 2 = \lg 12$$

Hence:

$$2x - 1 = \frac{\lg 12}{\lg 2} = 3.58$$

Thus $x = 2.29$.

Revision

8 Simplify (a) $2 \lg x + \log x^2$, (b) $\ln 2x^3 - \ln(4/x^2)$.

9 Write the following in terms of $\lg a$, $\lg b$ and $\lg c$:

$$\text{(a) } \lg\left(\frac{b\sqrt{2}}{ac}\right), \text{ (b) } \lg\left(\frac{ab}{\sqrt{c}}\right)^3$$

10 Solve for x the equations: (a) $3^x = 300$, (b) $10^{2-3x} = 6000$, (c) $7^{2x+1} = 4^{3-x}$.

3.3.1 The decibel

The power gain of a system is the ratio of the output power to the input power. If we have, say, three systems in series (Figure 3.6) then the power gain of each system is given by:

$$G_1 = \frac{P_2}{P_1},\ G_2 = \frac{P_3}{P_2},\ G_3 = \frac{P_4}{P_3}$$

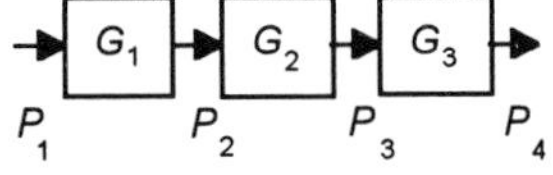

Figure 3.6 *Systems in series*

The overall power gain of the system is P_4/P_1 and is the product of the individual gains, i.e.

$$G = \frac{P_4}{P_1} = \frac{P_2}{P_1} \times \frac{P_3}{P_2} \times \frac{P_4}{P_3} = G_1 \times G_2 \times G_3$$

Taking logarithms gives:

$$\lg G = \lg G_1 + \lg G_2 + \lg G_3 \qquad [20]$$

We thus can add the log ratio of the powers. This log of the power ratio was said to be the power ratio in units of the *bel*, named in honour of Alexander Graham Bell:

$$\text{Power ratio in bels} = \lg \frac{\text{power out}}{\text{power in}} \qquad [21]$$

Thus the overall power gain in bels can be determined by simply adding together the power gains in bels of each of the series systems. The bel is an inconveniently large quantity and thus the *decibel* is used:

$$\text{Power ratio in decibels} = 10 \lg \frac{\text{power out}}{\text{power in}} \qquad [22]$$

A power gain of 3 dB is thus a power ratio of 2.0.

3.4 Trigonometric functions

Consider the angle θ in the right-angled triangle ABC shown in Figure 3.7. The trigonometric ratios sine, cosine and tangent of θ are defined as:

$$\sin\theta = \frac{\text{side opposite angle}}{\text{hypotenuse}} = \frac{\text{BC}}{\text{AC}} \qquad [23]$$

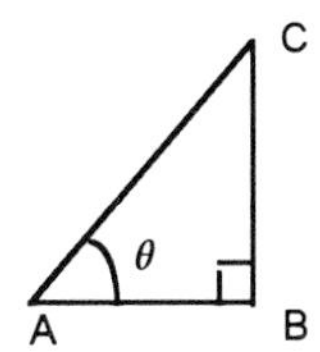

Figure 3.7 *Right-angled triangle*

$$\cos\theta = \frac{\text{side adjacent to angle}}{\text{hypotenuse}} = \frac{\text{AB}}{\text{AC}} \qquad [24]$$

$$\tan\theta = \frac{\text{side opposite angle}}{\text{side adjacent to angle}} = \frac{\text{BC}}{\text{AB}} \qquad [25]$$

We can also express the tangent as:

$$\tan\theta = \frac{\text{BC}}{\text{AB}} = \frac{\text{BC}}{\text{AC}} \times \frac{\text{AC}}{\text{AB}} = \frac{\sin\theta}{\cos\theta} \qquad [26]$$

For the triangle ABC (Figure 3.7), angle ACB = 90° − θ , or $\pi/2 - \theta$ when radian measure is used. But cos ACB = sin θ, thus:

$$\sin\theta = \cos\left(\frac{\pi}{2} - \theta\right) \qquad [27]$$

and sin ACB = cos θ, thus:

$$\cos\theta = \sin\left(\frac{\pi}{2} - \theta\right) \qquad [28]$$

The cosecant, secant and cotangent ratios are defined as the reciprocals of the sine, cosine and tangent:

$$\operatorname{cosec}\theta = \frac{1}{\sin\theta} \quad [29]$$

$$\sec\theta = \frac{1}{\cos\theta} \quad [30]$$

$$\cot\theta = \frac{1}{\tan\theta} \quad [31]$$

3.4.1 Pythagoras theorem

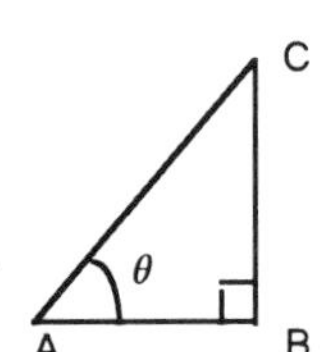

Figure 3.8 *Right-angled triangle*

For the right-angled triangle shown in Figure 3.8, the *Pythagoras theorem* gives $AB^2 + BC^2 = AC^2$. Dividing both sides of the equation by AC^2 gives:

$$\left(\frac{AB}{AC}\right)^2 + \left(\frac{BC}{AC}\right)^2 = 1$$

Hence:

$$\cos^2\theta + \sin^2\theta = 1 \quad [32]$$

Dividing this equation by $\cos^2\theta$ gives:

$$1 + \tan^2\theta = \sec^2\theta \quad [33]$$

and dividing equation [32] by $\sin^2\theta$ gives:

$$\cot^2\theta + 1 = \operatorname{cosec}^2\theta \quad [34]$$

Example

Simplify $\frac{\cos\theta}{1-\sin\theta} + \frac{\cos\theta}{1+\sin\theta}$.

$$\frac{\cos\theta}{1-\sin\theta} + \frac{\cos\theta}{1+\sin\theta} = \frac{\operatorname{cosec}\theta(1+\sin\theta) + \cos\theta(1-\sin\theta)}{(1-\sin\theta)(1+\sin\theta)}$$

$$= \frac{2\cos\theta}{1-\sin^2\theta} = \frac{2\cos\theta}{\cos^2\theta} = 2\sec\theta$$

Revision

11 Simplify the following: (a) $\frac{\sec\theta + \tan\theta}{1+\sin\theta}$, (b) $\frac{\cos\theta\cot\theta\sec^2\theta}{\operatorname{cosec}\theta}$

3.4.2 Trigonometric ratios of sums of angles

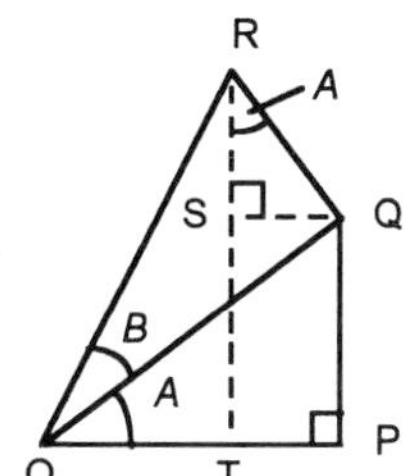

Figure 3.9 *Compound angle*

It is often useful to express the trigonometric ratios of angles such as $A + B$ or $A - B$ in terms of the trigonometric ratios of A and B. Consider the two right-angled triangles OPQ and OQR shown in Figure 3.9:

$$\sin(A+B) = \frac{\text{TR}}{\text{OR}} = \frac{\text{TS}+\text{SR}}{\text{OR}} = \frac{\text{PQ}+\text{SR}}{\text{OR}} = \frac{\text{PQ}}{\text{OQ}}\frac{\text{OQ}}{\text{OR}} + \frac{\text{SR}}{\text{QR}}\frac{\text{QR}}{\text{OR}}$$

Hence:

$$\sin(A + B) = \sin A \cos B + \cos A \sin B \qquad [35]$$

If we replace B by $-B$ we obtain:

$$\sin(A - B) = \sin A \cos B - \cos A \sin B \qquad [36]$$

If in equation [35] we replace A by $(\pi/2 - A)$ we obtain:

$$\cos(A + B) = \cos A \cos B - \sin A \sin B \qquad [37]$$

If in equation [37] we replace B by $-B$ we obtain:

$$\cos(A - B) = \cos A \cos B + \sin A \sin B \qquad [38]$$

We can obtain tan $(A + B)$ by dividing sin $(A + B)$ by cos $(A + B)$:

$$\tan(A+B) = \frac{\tan A + \tan B}{1 - \tan A \tan B} \qquad [39]$$

and likewise $\tan(A - B)$ by dividing $\sin(A - B)$ by $\cos(A - B)$:

$$\tan(A-B) = \frac{\tan A - \tan B}{1 + \tan A \tan B} \qquad [40]$$

By adding or subtracting equations from above we obtain:

$$2 \sin A \cos B = \sin(A + B) + \sin(A - B) \qquad [41]$$

$$2 \cos A \cos B = \cos(A + B) + \cos(A - B) \qquad [42]$$

$$2 \sin A \sin B = \cos(A - B) - \cos(A + B) \qquad [43]$$

$$2 \cos A \sin B = \sin(A + B) - \sin(A - B) \qquad [44]$$

3.4.3 Trigonometric ratios for double angles

If, for equations [35], [37] and [39], we let $B = A$ we obtain the double-angle equations:

$$\sin 2A = 2 \sin A \cos A \qquad [45]$$

$$\cos 2A = \cos^2 A - \sin^2 A = 1 - 2 \sin^2 A = 2 \cos^2 A - 1 \qquad [46]$$

$$\tan 2A = \frac{2 \tan A}{1 - \tan^2 A} \qquad [47]$$

Example

Solve the equation $\cos 2x + 3 \sin x = 2$.

Using equation [46] for $\cos 2x$ gives:

$$1 - 2 \sin^2 x + 3 \sin x = 2$$

This can be rearranged as:

$$2 \sin^2 x - 3 \sin x + 1 = 0$$

$$(2 \sin x - 1)(\sin x - 1) = 0$$

Hence $\sin x = ½$ or 1. For angles between 0° and 90°, $x = 30°$ or 90°.

Revision

12 Show that: (a) $\dfrac{1 - \cos 2A}{\sin 2A} = \tan A$, (b) $\sec 2A + \tan 2A = \dfrac{\cos A + \sin A}{\cos A - \sin A}$,

(c) $\dfrac{\sin(A + B)}{\cos A \cos B} = \tan A + \tan B$, (d) $\sin 3A = 3 \sin A - 4 \sin^3 A$

(Hint: write $3A$ as $2A + A$.)

13 Solve $2 \cos^2 x + 3 \sin x = 3$, for angles between 0° and 90°.

3.4.4 Trigonometric ratios for half angles

If we let $A = \theta/2$ then half-angle equations are obtained from equations [44], [45] and [46]:

$$\sin \theta = 2 \sin \frac{\theta}{2} \cos \frac{\theta}{2} \qquad [48]$$

$$\cos \theta = \cos^2 \frac{\theta}{2} - \sin^2 \frac{\theta}{2} = 1 - 2 \sin^2 \frac{\theta}{2} = 2 \cos^2 \frac{\theta}{2} \qquad [49]$$

$$\tan \theta = \frac{2 \tan \frac{\theta}{2}}{1 - 2 \tan^2 \frac{\theta}{2}} \qquad [50]$$

If we let $A + B = P$ and $A - B = Q$ then, with $A = \frac{1}{2}(P + Q)$ and $B = \frac{1}{2}(P - Q)$, equations [41], [42], [43] and [44] can be written as:

$$\sin P + \sin Q = 2 \sin \tfrac{1}{2}(P + Q) \cos \tfrac{1}{2}(P - Q) \qquad [51]$$

$$\sin P - \sin Q = 2 \cos \tfrac{1}{2}(P + Q) \sin \tfrac{1}{2}(P - Q) \qquad [52]$$

$$\cos P + \cos Q = 2 \cos \tfrac{1}{2}(P + Q) \cos \tfrac{1}{2}(P - Q) \qquad [53]$$

$$\cos P - \cos Q = -2 \sin \tfrac{1}{2}(P + Q) \sin \tfrac{1}{2}(P - Q) \qquad [54]$$

3.4.5 $a \cos \theta + b \sin \theta$

It is often very useful to reduce an equation of the form $a \cos \theta + b \sin \theta$ to a single term such as $r \cos(\theta - \alpha)$. Using equation [38] we must have:

$$r \cos(\theta - \alpha) = r(\cos \theta \cos \alpha + \sin \theta \sin \alpha)$$

Hence, in order to make the reduction, we require:

$$r(\cos \theta \cos \alpha + \sin \theta \sin \alpha) = a \cos \theta + b \sin \theta$$

and the coefficients of the $\cos \theta$ and $\sin \theta$ terms to be given by $r \cos \alpha = a$ and $r \sin \alpha = b$. Thus:

$$\tan \alpha = \frac{b}{a} \qquad [55]$$

This leads us to be able to describe the angle α by the right-angled triangle shown in Figure 3.10. Hence:

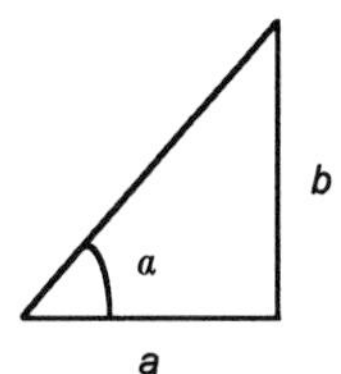

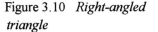

Figure 3.10 *Right-angled triangle*

$$r = \sqrt{a^2 + b^2} \qquad [56]$$

Thus:

$$a \cos \theta + b \sin \theta = r \cos(\theta - \alpha) \qquad [57]$$

where r is given by equation [56] and α by equation [55]. In a similar way we can obtain, for the same values of r and α:

$$a \cos \theta - b \sin \theta = r \cos(\theta + \alpha) \qquad [58]$$

$$a \sin \theta + b \cos \theta = r \sin(\theta + \alpha) \qquad [59]$$

$$a \sin \theta - b \cos \theta = r \sin(\theta - \alpha) \qquad [60]$$

Example

Express $3 \cos \theta + 4 \sin \theta$ in the form $r \cos(\theta - \alpha)$.

Using equation [56], $r = 5$. Using equation [55], $\tan a = 3/4$ and so $a = 36.9°$ or 0.64 rad. Thus the form required is $5\cos(\theta - 36.9°)$ or $5\cos(\theta - 0.64)$.

Revision

14 Express $3\sin\theta + 4\cos\theta$ in the form $r\sin(\theta + a)$.

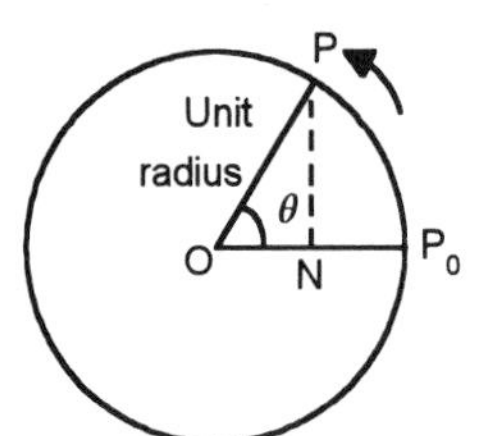

Figure 3.11 *Circular functions*

3.4.6 Circular functions

The trigonometric ratios for sine, cosine and tangent were defined earlier in Section 3.4 as ratios with a right-angled triangle. This limits the definition to angles between 0 and 90°. There is, however, another way of defining these ratios which allows us to define them for angles greater than 90°. This defines them as *circular functions*.

Consider the motion of a point P around a unit radius circle (Figure 3.11). P_0 is the initial position of the point and P the position to which it has rotated. The radial arm OP in moving from OP_0 has swept out an angle θ. The angle θ is measured between the radial arm and the OP_0 axis as a positive angle when the arm rotates in an anticlockwise direction. Since the circle has a unit radius, to obtain for angles up to 90° the same result as the trigonometric ratios defined in terms of the right-angled triangle, the perpendicular height NP defines the sine of the angle P_0OP and the horizontal distance ON defines the cosine of the angle P_0OP.

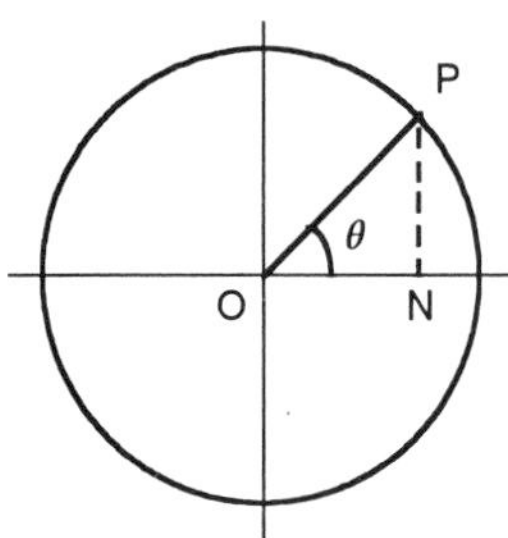

Figure 3.12 *First quadrant*

When NP is in an upward direction from OP_0 it is positive, when downward it is negative. When ON is measured to the right of O it is positive, to the left it is negative. Thus:

1 *Angles between 0 and 90°*
When the radial arm is in the first quadrant (Figure 3.12) with $0 \le \theta < \pi/2$, $0 \le \theta < 90°$, NP is positive and ON positive. Thus both the sine and the cosine of angle θ are positive. For example, $\sin 30° = +0.5$ and $\cos 30° = +0.87$.

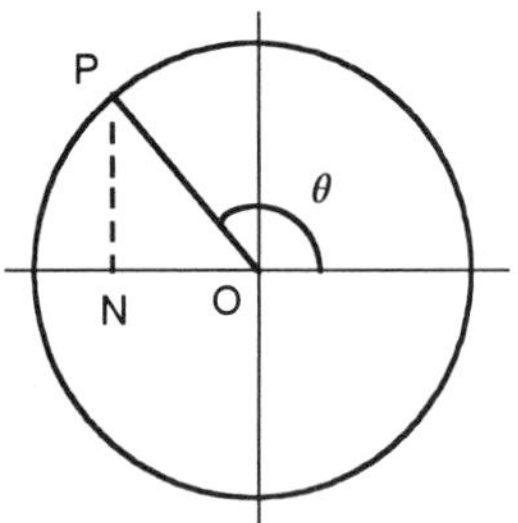

Figure 3.13 *Second quadrant*

2 *Angles between 90° and 180°*
When the radial arm is in the second quadrant (Figure 3.13) with $\pi/2 \le \theta < \pi$, $90° \le \theta < 180°$, NP is positive and ON negative. Thus the sine of angle θ is positive and the cosine negative. For example, $\sin 120° = 0.87$ and $\cos 120° = -0.5$.

3 *Angles between 180° and 270°*
When the radial arm is in the third quadrant (Figure 3.14) with $\pi \le \theta < 3\pi/2$, $180° \le \theta < 270°$, NP is negative and ON negative. Thus the sine of angle θ is negative and the cosine negative. For example, $\sin 210° = -0.5$ and $\cos 210° = -0.87$.

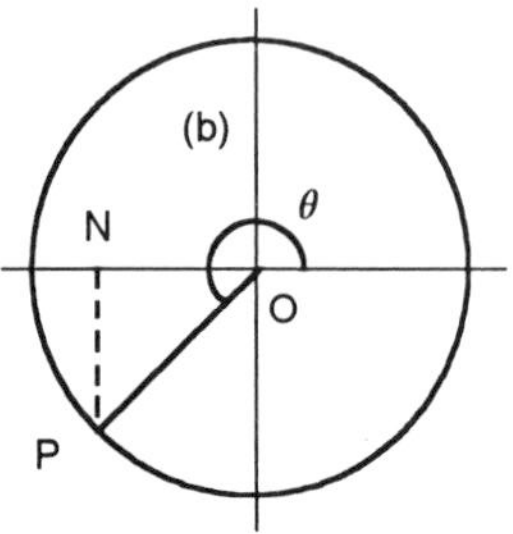

Figure 3.14 *Third quadrant*

4 *Angles between 270° and 360°*
When the radial arm is in the fourth quarter (Figure 3.15) with $3\pi/2 \leq \theta < 2\pi, 270° \leq \theta < 360°$, NP is negative and ON positive. Thus the sine of angle θ is negative and the cosine positive. For example, sin 300° = −0.87 and cos 300° = 0.5.

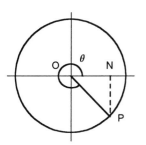

Figure 3.15 *Fourth quadrant*

For angles greater than 2π,360°, the radial arm simply rotates more than one revolution. Negative angles are interpreted as a clockwise movement of the radial arm from OP_0.

3.4.7 The sine, cosine and tangent functions

To obtain the graph of sin x we simply consider the rotation of a unit length radial arm OP in a circular path and read off the values of its vertical projection NP as the radial arm moves round the circle. Figure 3.16 shows the result. Since the graph describes a periodic function of periodic 2π, then:

$$\sin(x + 2\pi n) = \sin x \qquad [61]$$

where $n = 0, \pm 1, \pm 2$, etc.

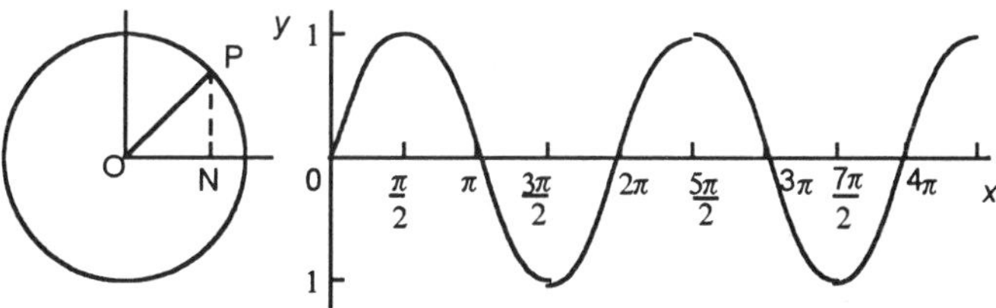

Figure 3.16 *Graph of y = sin x*

To obtain the graph of cos x we thus simply consider the rotation of a unit length radial arm OP in a circular path and read off the values of its horizontal projection ON as the radial arm moves round the circle. Figure 3.17 shows the result. Since the graph describes a periodic function of periodic 2π, then:

$$\cos(x + 2\pi n) = \cos x \qquad [62]$$

where $n = 0, \pm 1, \pm 2$, etc.

Note that the graph of $y = \sin x$ is the same as that of $y = \cos x$ moved $\frac{1}{2}\pi$ to the right, while that of $y = \cos x$ is the same as $y = \sin x$ moved $\frac{1}{2}\pi$ to the left, i.e. $\sin x = \cos(x - \frac{1}{2}\pi)$ and $\cos x = \sin(x + \frac{1}{2}\pi)$.

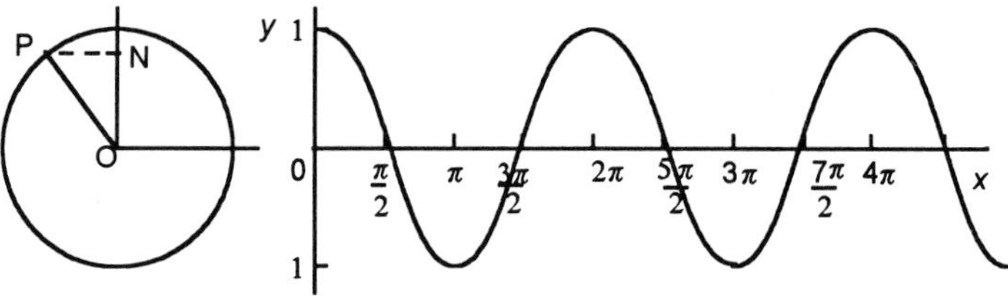

Figure 3.17 *Graph of y = cos x*

In engineering and science a function of the form $y = A\ \sin(\omega t + a)$ or $y = A\ \cos(\omega t + a)$ is often encountered. The A is the amplitude term and indicates that all the values of the sine or cosine are multiplied by A with the maximum and minimum values being $+A$ and $-A$. If the radial arm rotates with a constant angular velocity ω then the angle covered in a time t is ωt, thus the x-axis becomes a time axis. The time T for one complete rotation, i.e. one period, is $2\pi/\omega$. A graph of $y = \sin 2t$ would differ from a graph of $y = \sin 1t$ by having a frequency f of rotation ($f = 1/T = \omega/2\pi$) of the radius arm twice as much as that generating the sin $1t$ graph and so a period which is half that of the sin $1t$ graph. a is the angle by which the sine or cosine graph is moved to the left when positive and to the right when negative. In terms of time a/ω is the time by which the sine or cosine graph is moved to the left when positive and to the right when negative. Figure 3.18 illustrates the above points with graphs of (a) $y = 2 \sin t$, (b) $y = 2 \sin 2t$ and (c) $y = 2 \sin(t + \pi/3)$.

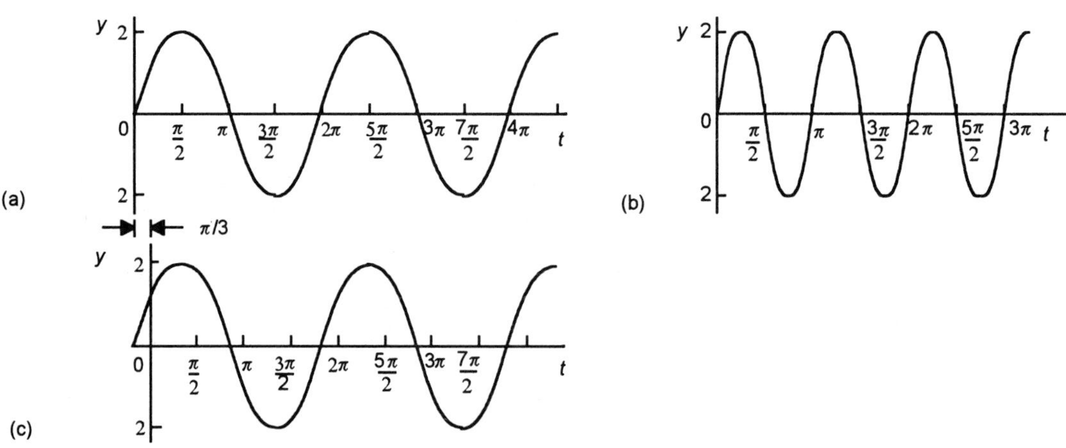

Figure 3.18 *(a) y = 2 sin t, (b) y = sin 2t, (c) y = 2 sin(t + π/3)*

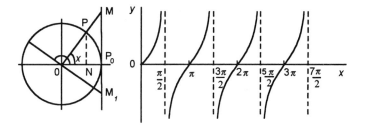

Figure 3.19 $y = \tan x$

Now consider the graph of the function $y = \tan x$. For the radial arm OP rotating in a circle, the tangent is PN/OP (Figure 3.19). But if we draw a tangent to the circle at P_0 then, for a unit radius circle the tangent of the angle is P_0M. When the radius arm has moved to an angle between 90° and 180° then the tangent is P_0M_1. The graph describes a periodic function which repeats itself every period of π. Thus:

$$\tan(x + \pi n) = \tan x \qquad [63]$$

for $n = 0, \pm1, \pm2$, etc.

Revision

15 Determine the general solutions of: (a) $\sin x = 0.25$, (b) $\tan x = 0.9$.

16 State the amplitude, period and phase angle for:

$$\text{(a) } 2\sin(5t + 1),\ \text{(b) } 6\cos 3t,\ \text{(c) } 5\cos\left(\frac{2t+1}{3}\right),\ \text{(d) } 2\cos(t - 0.6)$$

3.4.8 The inverse trigonometric functions

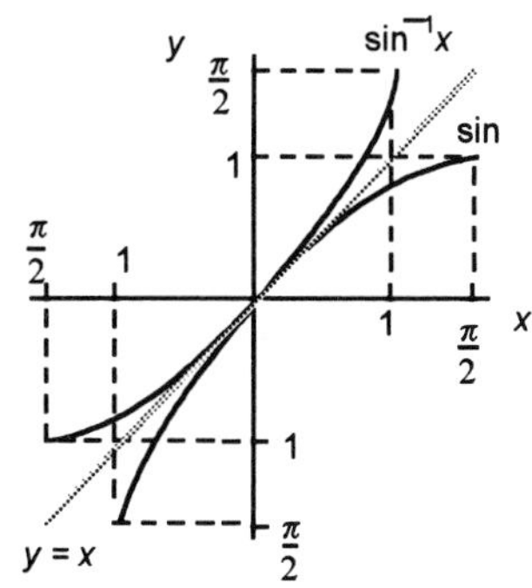

Figure 3.20 *sin x and its inverse*

If $\sin x = 0.8$ what is the value of x? This requires the inverse being obtained. There is an inverse if the function is one-to-one or restrictions imposed to give this state of affairs (see Section 1.5). The trigonometric function $y = \sin x$ is a many-to-one function, many values of x giving the same value of y. To obtain an inverse we have to restrict the domain of the function to $-\pi/2$ to $+\pi/2$. With that restriction $y = \sin x$ is a one-to-one function and so has an inverse. The inverse function is denoted as $\sin^{-1} x$ (sometimes also written as arcsin x). Note that the -1 is *not* a power here but purely notation to indicate the inverse. If $\sin x = 0.8$ then the value of x that gives this sine is the inverse and so $x = \sin^{-1} 0.8$, i.e. $x = 53°$. Figure 3.20 shows the graphs for $\sin x$ and its inverse function. In a similar way we can define inverses for cosines and tangents. Thus:

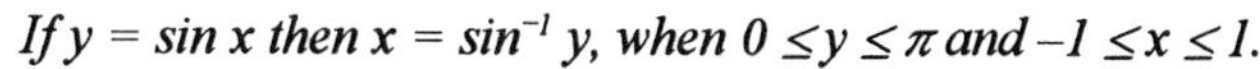

If $y = \sin x$ then $x = \sin^{-1} y$, when $0 \le y \le \pi$ and $-1 \le x \le 1$.

If $y = \cos x$ then $x = \sin^{-1} y$, when $0 \leq y \leq \pi$ and $-1 \leq x \leq 1$.
If $y = \tan x$ then $x = \tan^{-1} y$, when $-\tfrac{1}{2}\pi \leq y \leq +\tfrac{1}{2}\pi$ and x is any real number.

Example

Determine the values in radians of (a) $\sin^{-1} 0.34$, (b) $\cos^{-1} 0.30$, (c) $\tan^{-1} 0.20$, (d) $\sin^{-1}(-0.20)$.

Using a calculator, the values are (a) 0.35, (b) 1.27, (c) 0.20, (d) –0.20.

Revision

17 Determine the values in radians of (a) $\sin^{-1} 0.60$, (b) $\cos^{-1} 0.80$, (c) $\tan^{-1} 0.50$, (d) $\sin^{-1}(-0.60)$.

3.5 Hyperbolic functions

The sine, cosine and tangent are termed circular functions because their definition is associated with a circle. In a similar way, the sinh, cosh and tanh are *hyperbolic functions* associated with a hyperbola. Sinh is a contracted form of 'hyperbolic sine', cosh of 'hyperbolic cosine' and tanh of 'hyperbolic tangent'. Figure 3.21 shows the comparison of the circular and hyperbolic functions. The hyperbolic functions are defined as:

$$\sinh x = \tfrac{1}{2}(e^{x} - e^{-x}) \qquad [64]$$

$$\cosh x = \tfrac{1}{2}(e^{x} + e^{-x}) \qquad [65]$$

$$\tanh x = \frac{\sinh x}{\cosh x} = \frac{e^{x} - e^{-x}}{e^{x} + e^{-x}} \qquad [66]$$

Also we have sech $x = 1/\cosh x$, cosech $x = 1/\sinh x$ and coth $x = 1/\tanh x$.

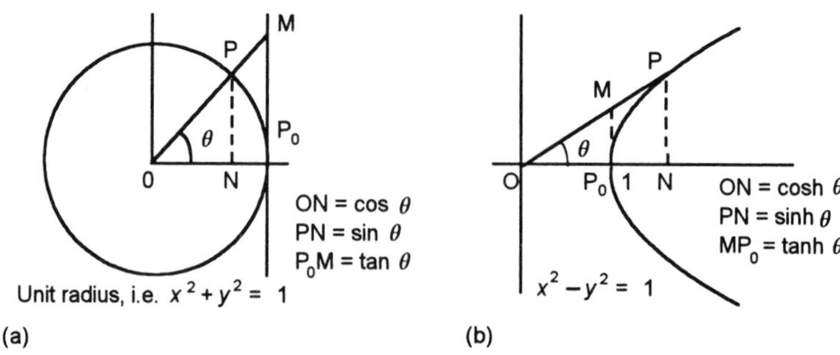

Figure 3.21 *(a) Circular functions, (b) hyperbolic functions*

Example

Determine, using a calculator, the values of (a) cosh 3, (b) sinh 3.

Some calculators have hyperbolic functions so that they can be evaluated by the simple pressing of a key, with others you will have to evaluate the exponentials.
(a) Evaluating the exponentials: $\cosh 3 = \frac{1}{2}(e^3 + e^{-3}) = 10.07$.
(b) Evaluating the exponentials: $\sinh 3 = \frac{1}{2}(e^3 - e^{-3}) = 10.02$.

Revision

18 Determine, using a calculator, the values of (a) sinh 2, (b) cosh 5, (c) tanh 2, (d) sinh(−2), (e) cosech 1.4, (f) sech 0.8.

3.5.1 Hyperbolic identities

Since $\cosh x = \frac{1}{2}(e^x + e^{-x})$ and $\sinh x = \frac{1}{2}(e^x - e^{-x})$, then:

$$\cosh x + \sinh x = e^x \qquad [67]$$

$$\cosh x - \sinh x = e^{-x} \qquad [68]$$

Thus:

$$(\cosh x + \sinh x)(\cosh x - \sinh x) = e^x e^{-x} = 1$$

and so:

$$\cosh^2 x - \sinh^2 x = 1 \qquad [69]$$

Dividing equation [67] by $\cosh^2 x$ gives:

$$1 - \tanh^2 x = \text{sech}^2 x \qquad [70]$$

Dividing equation [67] by $\sinh^2 x$ gives:

$$\coth^2 x - 1 = \text{cosech}^2 x \qquad [71]$$

Equation [67] gives $\cosh x + \sinh x = e^x$, squaring this gives:

$$\cosh^2 x + 2 \sinh x \cosh x + \sinh^2 x = e^{2x}$$

Equation [68] gives $\cosh x - \sinh x = e^{-x}$, squaring this gives:

$$\cosh^2 x - 2 \sinh x \cosh x + \sinh^2 x = e^{-2x}$$

Subtracting gives:

$$4 \sinh x \cosh x = e^{2x} - e^{-2x} = 2 \sinh 2x$$

and thus:

$$\sinh 2x = 2 \sinh x \cosh x \quad [72]$$

If we had added the two equations we would have obtained:

$$\cosh 2x = \cosh^2 x + \sinh^2 x = 1 + 2 \sinh^2 x = 2 \cosh^2 x - 1 \quad [73]$$

In a similar fashion we can show that:

$$\sinh(x + y) = \sinh x \cosh y + \cosh x \sinh y \quad [74]$$

$$\sinh(x - y) = \sinh x \cosh y - \cosh x \sinh y \quad [75]$$

$$\cosh(x + y) = \cosh x \cosh y + \sinh x \sinh y \quad [76]$$

$$\cosh(x - y) = \cosh x \cosh y - \sinh x \sinh y \quad [77]$$

If we compare the hyperbolic identities with the circular identities, there is great similarity. If we replace each circular function by the corresponding hyperbolic function and change the sign of every product or implied product of two sines, we obtain the hyperbolic identity. This is known as *Osborn's rule*.

Example

If $\sin 3x = 3 \sin x - 4 \sin^3 x$, use Osborn's rule to determine the corresponding hyperbolic identity.

$\sin^3 x$ implies $(\sin x \times \sin x) \times \sin x$ and so we have the implied product of two sines and need to change the sign when we transform the equation to a hyperbolic identity. Thus:

$$\sinh 3x = 3 \sinh x + 4 \sinh^3 x$$

Revision

19 Using Osborn's rule, write the hyperbolic identity corresponding to:

(a) $\sin x - \sin y = 2 \sin \frac{1}{2}(x - y) \cos \frac{1}{2}(x + y)$

(b) $\cos 3x = 4 \cos^3 x - 3 \cos x$

20 Simplify $\dfrac{1 + \sinh 2x + \cosh 2x}{1 - \sinh 2x - \cosh 2x}$.

3.5.2 Hyperbolic equations

The term *hyperbolic equation* is used for equations of the general form:

$$a \cosh x \pm b \sinh x = c \qquad [78]$$

where a, b and c are constants. Such equations can be solved by writing the hyperbolic functions in terms of exponentials and then manipulating the expression to obtain it in the form of a quadratic equation involving e^x. The equation can then have its roots determined in the usual way. An alternative is to express the equation in terms of cosh x. This involves rearranging the equation and squaring it so that the identity $\cosh^2 x - \sinh^2 x$ can be used to eliminate the $\sinh^2$ term. The resulting quadratic equation in cosh x can then have its roots determined in the usual way.

Example

Solve the equation $5 \cosh x + 3 \sinh x = 4$.

We can write this as:

$$\tfrac{5}{2}(e^x + e^{-x}) + \tfrac{3}{2}(e^x - e^{-x}) = 4$$

$$4\, e^x + e^{-x} = 4$$

Multiplying by e^x, and rearranging, gives:

$$4(e^x)^2 - 4\, e^x + 1 = 0$$

$$(2\, e^x - 1)(2\, e^x - 1) = 0$$

Hence $e^x = ½$ and $x = \ln 0.5 = -0.693$.

Alternatively, we can write the equation in the form:

$$5 \cosh x - 4 = -3 \sinh x$$

$$(5 \cosh x - 4)^2 = 9 \sinh^2 x = 9(\cosh^2 x - 1)$$

$$16 \cosh^2 x - 40 \cosh x + 25 = 0$$

$$(4 \cosh x - 5)(4 \cosh x - 5) = 0$$

Hence $\cosh x = 1.25$. This gives $x = -0.693$.

Revision

21 Solve the equations (a) $4 \cosh x = 5$, (b) $3 \cosh x + 2 \sinh x = 5$. Hint: chapter 2, equation [15] can be used to find the roots of a quadratic.

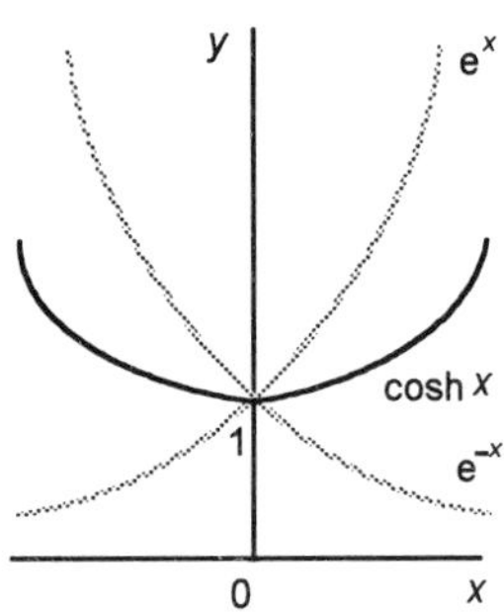

Figure 3.22 *cosh x*

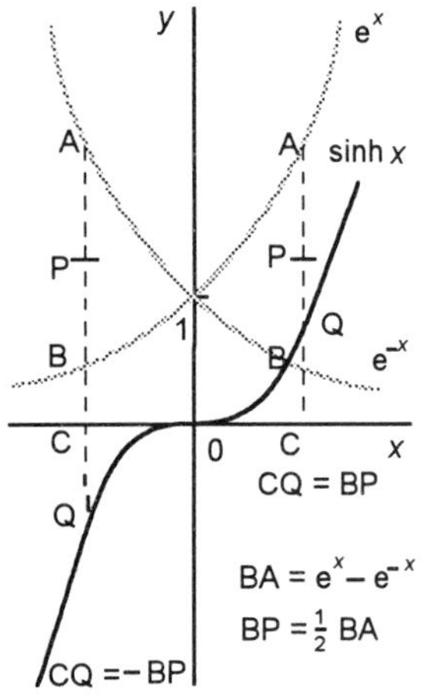

Figure 3.23 *sinh x*

3.5.3 Graphs of hyperbolic functions

Since cosh x is the average value of e^x and e^{-x} we can obtain a graph of cosh x as a function of x by plotting the e^x and e^{-x} graphs and taking the average value. Figure 3.22 illustrates this. Note that unlike cos x, cosh x is not a periodic function. At $x = 0$, cosh $x = 1$. The curve is symmetrical about the y-axis, i.e. $\cosh(-x) = \cosh x$ and is termed an even function.

To obtain the graph of sinh x from those of e^x and e^{-x}, at a particular value of x we subtract the second from the first and then take half the resulting value. Figure 3.23 illustrates this. Note that unlike sin x, sinh x is not a periodic function. When $x = 0$, sinh $x = 0$. The curve is symmetrical about the origin, i.e. $\sinh(-x) = -\sinh x$, and is said to be an odd function.

Figure 3.24 shows the graph of tanh x, obtained by taking values of e^x and e^{-x} and calculating values of tanh x for particular values of x. Unlike tan x, tanh x is not periodic. When $x = 0$, tanh $x = 0$. All the values of tanh x lie between -1 and $+1$. As x tends to infinity, tanh x tends to 1. As x tends to minus infinity, tanh x tends to -1. The curve is symmetrical about the origin, i.e. $\tanh(-x) = -\tanh x$, and is said to be an odd function.

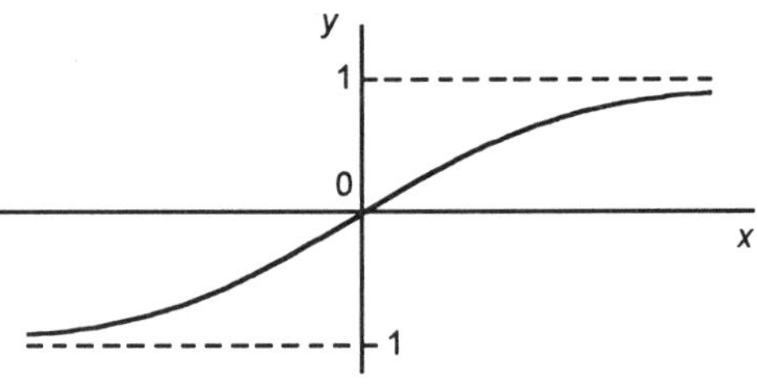

Figure 3.24 $y = \tanh x$

3.5.4 Inverse hyperbolic functions

If sinh x = 1.3, what is the value of x? This requires the inverse to be obtained, i.e. $\sinh^{-1} 1.3$. The functions sinh x and tanh x are one-to-one functions and so have an inverse, but cosh x when considered from $-\infty$ to $+\infty$ is a many-to-one function. We can, however, have an inverse if we restrict the domain to 0 to $+\infty$.

$x = \sinh^{-1} y$ implies $y = \sinh x$ and so:

$$\tfrac{1}{2}(e^x - e^x) = y$$

$$e^x - e^x = 2y$$

Multiplying by e^x gives:

$$(e^x)^2 - 1 = 2y\, e^x$$

$$(e^x)^2 - 2y\, e^x - 1 = 0$$

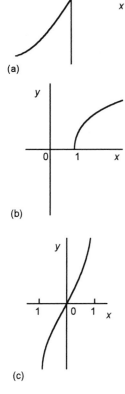

Thus, solving the quadratic equation gives:

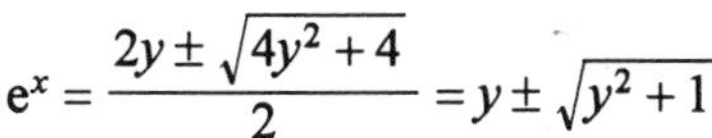

$$e^x = \frac{2y \pm \sqrt{4y^2 + 4}}{2} = y \pm \sqrt{y^2 + 1}$$

But e^x is always positive for real values of x, thus the only solution is:

$$e^x = y + \sqrt{y^2 + 1}$$

$$x = \sinh^{-1} y = \ln\left(y + \sqrt{y^2 + 1}\right) \qquad [79]$$

Similarly we can derive:

$$x = \cosh^{-1} y = \ln\left(y + \sqrt{y^2 - 1}\right) \text{ for } y \geq 1 \qquad [80]$$

$$x = \tanh^{-1} y = \tfrac{1}{2} \ln\left(\frac{1 + y}{1 - y}\right) \text{ for } -1 < y < 1 \qquad [81]$$

Figure 3.25 shows graphs of the inverse functions.

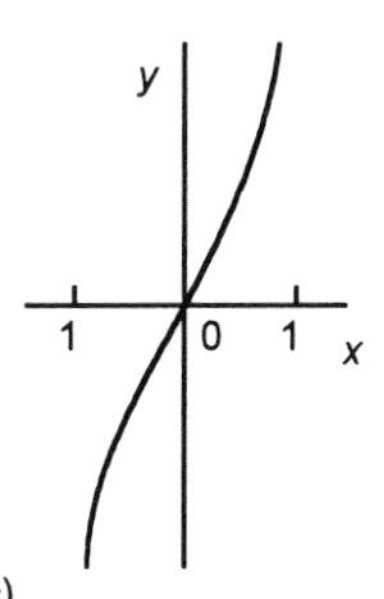

Figure 3.25 *(a)* $\sinh^{-1} x$, *(b)* $\cosh^{-1} x$, *(c)* $\tanh^{-1} x$

Example

Determinine x from $\sinh x = 1.3$.

While calculators with hyperbolic functions can be used to directly obtain inverse functions, writing it as exponentials can be used. We must have:

$$\tfrac{1}{2}(e^x - e^{-x}) = 1.3$$

and so:

$$e^x - e^{-x} = 2.6$$

Multiplying both sides of the equation by e^x gives:

$$(e^x)^2 - 1 = 2.6\, e^x$$

$$(e^x)^2 - 2.6\, e^x - 1 = 0$$

This quadratic equation has the roots (Chapter 2, equation [15]):

$$e^x = \frac{2.6 \pm \sqrt{2.6^2 + 4}}{2} = 1.3 \pm 1.640 = +2.940 \text{ or } -0.340$$

But e^x is always positive for real values of x, thus the only solution is $e^x = 2.940$. Hence $x = \ln 2.940 = 1.078$. Thus $\sinh^{-1} 1.3 = 1.078$.

Revision

20 Determine the values of:

(a) $\sinh^{-1} 0.5$, (b) $\tanh^{-1} 0.5$, (c) $\cosh^{-1} 2$, (d) $\sinh^{-1}(-2)$.

Problems

1 Simplify the following:

(a) $4^x 4^{2x}$, (b) $e^{2x} e^{4x}$, (c) $2^{3x} 2^{-5x}$, (d) $(1 + e^{2x})^2$, (e) $\dfrac{2\,e^{5x}}{8\,e^{4x}}$

2 Determine the values of y in the following equations when (i) $x = 0$, (b) x is infinite:

(a) $y = 2\,e^x$, (b) $y = 10\,e^{x/2}$, (c) $y = 2\,e^{-x}$, (d) $y = 2(1 - e^{-x})$

3 Determine, using a calculator, to two decimal places the values of y in the following equations when (i) $x = +1$, (ii) $x = +2$:

(a) $y = 10\,e^{0.2x}$, (b) $y = 10\,e^{-0.2x}$, (c) $y = 10(1 - e^{-0.2x})$

4 The charge q on a discharging capacitor is related to the time t by the equation $q = Q_0\,e^{-t/CR}$, where Q_0 is the charge at $t = 0$, R the circuit resistance and C the capacitance. Determine the charge on the capacitor after a time of 0.2 s if initially the capacitor had a charge of 1 μC, R is 1 MΩ and C is 4 μF.

5 The amount N of a radioactive isotope decays with time t in years according to the equation $N = N_0\,e^{-0.7t}$, where N_0 is the amount at time $t = 0$. What fraction of the isotope will be left after 5 years?

6 The current i, in amperes, in an electrical circuit varies with time t, in seconds, according to the equation $i = 2(1 - e^{-10t})$. What will be the current after (a) 0.1 s, (b) 0.2 s?

7 The current I through a junction diode when there is a voltage V across it is given by the equation $I = I_s(e^{qV/kT} - 1)$, where q is the charge on an electron (1.6×10^{-19} C), k is Boltzmann's constant (1.38×10^{-23} J/K), T is the temperature on the kelvin scale and I_s is the reverse saturation current.

(a) A germanium junction diode at 300 K gives a current of 1 mA when $V = +0.20$ V. What is the saturation current for the diode?
(b) A silicon junction diode at 300 K has a saturation current of 1.0×10^{-13} A. What will be the current through the diode when $V = +0.6$ V?

8 Express the following in terms of lg a, lg b and lg c:

(a) $\lg(abc^2)$, (b) $\lg(a\{bc\}^2)$, (c) $\lg\left(\frac{a}{bc}\right)$, (d) $\lg\left(\frac{\sqrt{ab}}{c^3}\right)$

9 The pH of a solution is defined as pH = $-\lg[H^+]$, where $[H^+]$ is the concentration of hydrogen ions in the solution. What is the pH of a solution with a hydrogen ion concentration of 3.25×10^{-7}?

10 Solve for x the equations:

(a) $2^x = 9$, (b) $3^x = 12$, (c) $5^{x+1} = 72$, (d) $3^{x+1} = 2^x$, (e) $5^{x-1} = 2^{3x-1}$,

(f) $e^x = 20$, (g) $2\,e^x = e^{3x}$, (h) $2\,e^{x/2} = 1$

11 Simplify the following:

(a) $\frac{\tan\theta+1}{\cot\theta+1}$, (b) $\frac{\sin^2\theta\cot\theta}{\cos\theta}$, (c) $\sin\theta\cos\theta\tan\theta$

12 Show that:

(a) $\tan(A+B) - \tan A = \frac{\sin B}{\cos A\cos(A+B)}$,

(b) $\tan A + \cot A = 2\ \mathrm{cosec}\ 2A$, (c) $\cot(A+B) = \frac{\cot A\cot B - 1}{\cot A + \cot B}$

(d) $\cos 4A = 8\cos^4 A - 8\cos^2 A + 1$

13 Solve the following equations for angles in the range 0° to 360°:

(a) $\cos 2x = \sin x$, (b) $\sin 2x - 1 = \cos 2x$, (c) $\tan x \tan 2x = 2$

14 Solve the following equations for angles in the range 0 to 2π:

(a) $\cos^2 x = 3/4$, (b) $2\sin^2 x - \sin x = 1$, (c) $\cos 2x - \cos x = 0$

(d) $\tan^2 x = 1$, (e) $\tan x + \sec x = 1$

15 Write $5\sin\theta + 4\cos\theta$ in the forms (a) $r\sin(\theta - \alpha)$, (b) $r\cos(\theta + \alpha)$.

16 Determine the values in radians of (a) $\sin^{-1} 0.74$, (b) $\cos^{-1} 0.10$, (c) $\tan^{-1} 0.80$, (d) $\sin^{-1}(-0.40)$.

17 State the amplitude, period and phase angle for:

(a) $6\sin(2t+1)$, (b) $2\cos 9t$, (c) $5\cos\left(\frac{2t-1}{5}\right)$, (d) $2\cos(t - 0.2)$,

(e) $5\sin\left(4t + \frac{\pi}{8}\right)$, (f) $\frac{1}{2}\sin\left(t - \frac{\pi}{6}\right)$

18 The potential difference across a component in an electrical circuit is 40 sin $40\pi t$, what are the maximum potential difference and the frequency?

19 What is the value of v when $t = 3 \times 10^{-5}$ s for an amplitude-modulated radio wave with a voltage v in volts which varies with time t in seconds according to $v = 50(1 + 0.02 \sin 2400\pi t) \sin (2 \times 10^5 \pi t)$.

20 Determine, using a calculator, the values of: (a) sinh 1.2, (b) cosh 2.1, (c) tanh 1.2, (d) sinh (−1.5), (e) sinh 0.38, (f) coth 0.38.

21 Determine the values of: (a) $\sinh^{-1} 3.7$, (b) $\tanh^{-1} 0.8$, (c) $\cosh^{-1} (-2)$, (d) $\sinh^{-1} 0.8$.

22 Express $3\,e^x - 2\,e^{-x}$ in terms of cosh x and sinh x.

23 Using Osborn's rule, write the hyperbolic identity corresponding to:

(a) $\cos (x + y) = \cos x \cos y - \sin x \sin y$, (b) $\tan 2x = \dfrac{2 \tan x}{1 - \tan^2 x}$

24 Solve the equations:

(a) $3 \cosh x - 5 \sinh x = 2$, (b) $5 \cosh x - 3 \sinh x = 5$,

(c) $2 \cosh x + 2 \sinh x = 5$

25 Show that: (a) $1 + \cosh x = 2\cosh^2 \frac{x}{2}$, (b) $2 \sinh^3 x = \sinh 3x - 3 \sinh x$.

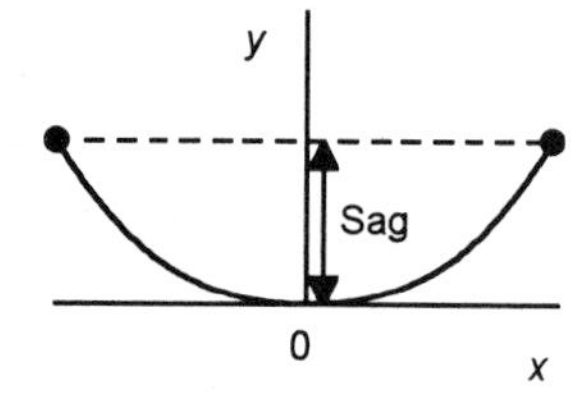

Figure 3.26 *Problem 26*

26 A flexible cable suspended between two horizontal points hangs in the form of a catenary (Figure 3.26), the equation of the curve being given by $y = c[\cosh(x/c) - 1]$, where y is the sag of the cable, x the horizontal distance from the midpoint to one end of the cable and c is a constant. Determine the sag of a cable when $c = 20$ and $2x = 16$ m.

27 The speed v of a surface wave on a liquid is given by:

$$v = \sqrt{\left[\left(\frac{g\lambda}{2\pi} + \frac{2\pi\gamma}{\rho\lambda}\right) \tanh \frac{2\pi h}{\lambda}\right]}$$

where g is the acceleration due to gravity, λ the wavelength of the waves, γ the surface tension, ρ the density and h the depth of the water. What will the speed approximately be for (a) shallow water waves when h/λ tends to zero, (b) deep water waves when h/λ tends to infinity?

4 Non-linear functions

4.1 Introduction

This chapter is concerned with the techniques that can be used to determine the roots of non-linear equations such as polynomials. Graphical and iterative techniques are discussed. An *iterative technique* is one where we take an approximation of a solution, then use that approximation to produce a better estimate. This estimate is then used to determine an even better estimate, and so on. The technique has been described as a method of successive approximations. Two iterative techniques are discussed in this chapter, the bisection method (sometimes referred to as interval halving) and Newton's method (sometimes referred to as the Newton-Raphson method). Newton's method is also discussed in Chapter 5 where it is derived from a consideration of Taylor's series; in this chapter it is from a consideration of graphs.

In engineering and science there are many situations where non-linear equations such as polynomial equations can arise and for which solutions are required. In real-world situations, the equations often cannot be solved by factorisation or the use of the quadratic formula (see Chapter 2) and an iterative method, generally Newton's method, is often used.

This chapter follows on from Chapter 2 and uses some of the functions given in Chapter 3. It assumes a basic knowledge of differentiation.

4.2 Graphical solutions

One way of determining approximate values for the roots of a function is to plot its graph and determine the values of the function when it crosses the x-axis (see Section 2.5.2 for examples with polynomials). With some functions, the plotting of the graph may present problems. The following is a graphical method that might make it easier to determine the roots.

Suppose we have two functions $f(x)$ and $g(x)$. If these functions have graphs that intersect then there must be values of x which give identical values of $f(x)$ and $g(x)$. Consider, for example, the functions $y = f(x) = \sin x$ and $y = g(x) = x/2$. Figure 4.1 shows graphs of the two functions. They intersect at two points, x_1 and x_2. An accurate drawing of the graphs would enable these points to be estimated as being about -1.9 and $+1.9$. At these points we have:

$$\sin x = \frac{x}{2}$$

which can be rewritten as:

$$\sin x - \frac{x}{2} = 0$$

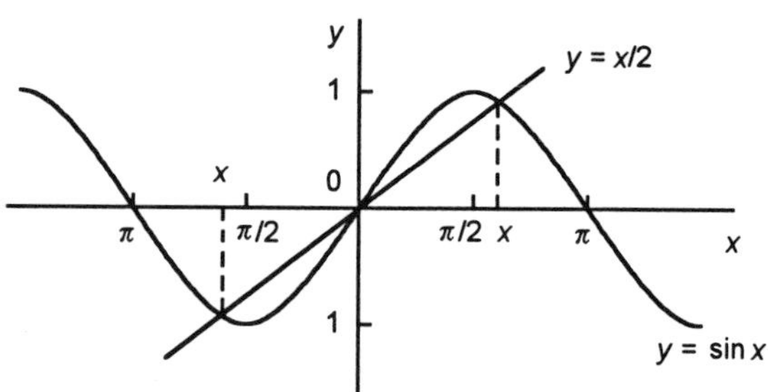

Figure 4.1 *The points of intersection*

This is the condition for x in the function $\sin x - x/2$ to give the roots of that function. Thus the splitting of the function $\sin x - x/2$ into two functions, $\sin x$ and $x/2$, has enabled the roots of the function to be determined. This is a technique that can often be used with complicated functions in order to make the graph plotting easier.

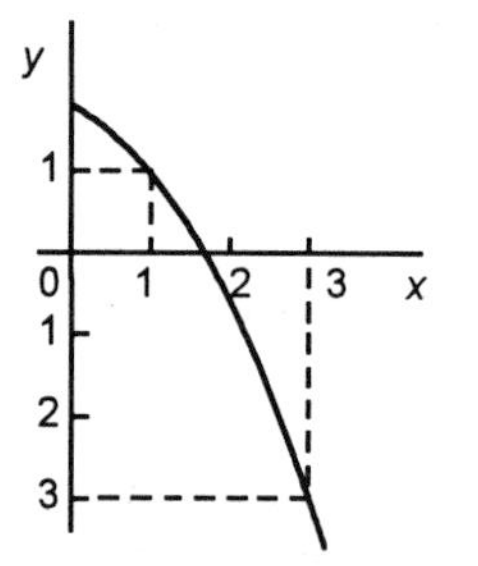

Figure 4.2 *Change of sign*

Another way of determining an approximate solution is to recognise that when a graph crosses the x-axis the value of the function changes sign. Figure 4.2 illustrates this, with the function having the value $y = +1$ at $x = 1$ and $y = -2$ at $x = 3$. Because there is a change of sign for the value of the function we can be confident that there is at least one root between $x = 1$ and $x = 2$. We can use this method of calculating values of the function to determine the interval in which a root lies and, by steadily reducing the size of the interval, end up with a value for the root. The following example illustrates this method. The next section, concerning the bisection method, details a more systematic way of using this technique to find roots.

Example

By splitting the function and plotting a graph, determine the roots of the function $x^3 - 6x^2 + 12$.

The roots are when the function has a zero value, i.e. when:

$$x^3 - 6x^2 + 12 = 0$$

This can be rearranged to give:

$$x^3 = 6x^2 - 12$$

The roots are thus the intersections of graphs of x^3 and $6x^2 - 12$. However, this involves plotting two graphs which are non-linear. The points of intersection of such graphs are more difficult to estimate than if one of the graphs is linear. Thus the function is better split in a different way. Thus we could have:

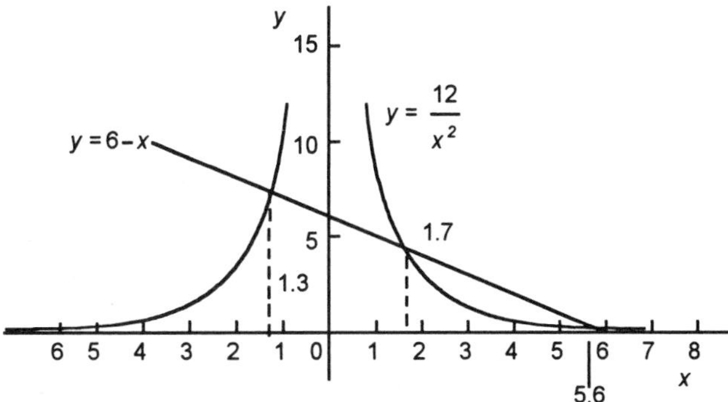

Figure 4.3 *Example*

$$6 - x = \frac{12}{x^3}$$

Figure 4.3 shows the graphs of $6 - x$ and $12/x^2$. The points of intersection are −1.3, +1.7 and +5.6.

Example

By calculation, determine the approximate value of the smallest positive root of the function $x^3 + x - 11$.

At $x = 0$ the function has the value −11. At $x = 1$ the function has the value −2, at $x = 2$ it is −1 and $x = 3$ it is +19. Thus there is a root between $x = 2$ and $x = 3$. If we try $x = 2.5$ then the value of the function is +7.1. Thus the root must lie between $x = 2$ and $x = 2.5$. Suppose we now try $x = 2.2$, then the value of the function is +1.8. Thus the root is between $x = 2$ and $x = 2.2$. If we now try $x = 2.1$ we obtain for the value of the function +0.4. Thus the root must be about $x = 2.1$. This may be accurate enough. We could, however, continue to obtain a closer estimate.

Revision

1 By splitting the functions and plotting graphs, determine the roots of the following:

(a) $x^3 - 3x + 1$, (b) $x - \tan x$, (c) $x^4 + x^3 - 1$, (d) $\cos x - x + 1$

Hint: for (c), consider rearranging the function as $x^3(x + 1) = 1$ and hence $x = (x + 1)^{1/3}$.

2 By calculation, determine the approximate value of the roots of the functions:

(a) $x^3 - 5x^2 - 4$, the smallest positive root, (b) $x^2 - 2$,

(c) $3x^2 - 4x + 5$, the negative root.

4.3 The bisection method

If we plot a graph of a continuous function $f(x)$ against x then the roots are where the graph line cuts the x-axis. Then, if we determine the value of the function at x_1 and x_2 and find that the values have different signs, there must be at least one root of $f(x)$ between x_1 and x_2. Figure 4.4(a) illustrates this for the case where there is one root. We now consider the value of the function at x_m, the value of x midway between x_1 and x_2, i.e.

$$x_m = \frac{x_1 + x_2}{2}$$

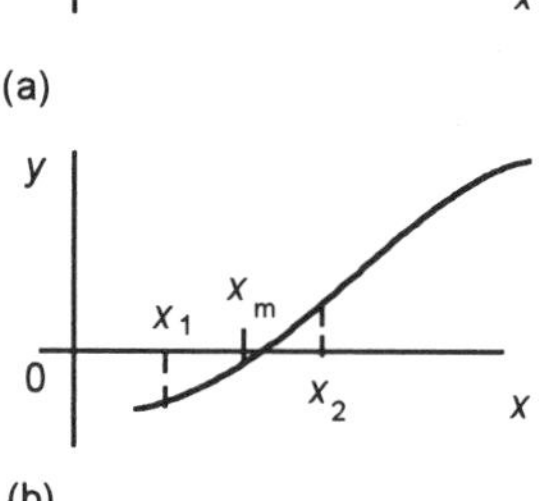

Figure 4.4 *Bisection method*

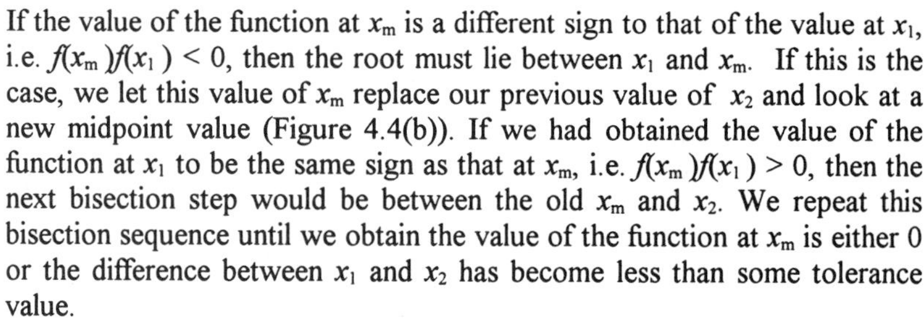

If the value of the function at x_m is a different sign to that of the value at x_1, i.e. $f(x_m)f(x_1) < 0$, then the root must lie between x_1 and x_m. If this is the case, we let this value of x_m replace our previous value of x_2 and look at a new midpoint value (Figure 4.4(b)). If we had obtained the value of the function at x_1 to be the same sign as that at x_m, i.e. $f(x_m)f(x_1) > 0$, then the next bisection step would be between the old x_m and x_2. We repeat this bisection sequence until we obtain the value of the function at x_m is either 0 or the difference between x_1 and x_2 has become less than some tolerance value.

The bisection method can be summarised as:

1. Start with two values of x of x_1 and x_2.
2. Let $x_m = (x_1 + x_2)/2$.
3. If $f(x_m)f(x_1) < 0$ then set $x_2 = x_m$, otherwise $x_1 = x_m$.
4. Repeat steps 2 and 3 until $f(x_m) = 0$ or $|x_2 - x_1| <$ tolerance value.

The following examples illustrate the use of this method, the first involving the solution of a polynomial and the second another form of non-linear function. An important point to realise is that in step 2 we only determine the signs of $f(x_m)$ and $f(x_1)$ in order to determine whether to set $x_2 = x_m$, or $x_1 = x_m$, the actual values do not matter.

With the bisection method, the root is located inside an interval of $x_2 - x_1$ and this interval is halved at each bisection. Thus after k bisections the root is located in an interval of $(x_2 - x_1)/2^k$. We require about 3 to 4 bisections to reduce the interval by a factor of about 10 ($2^3 = 8$, $2^4 = 16$). We thus require about 3 to 4 bisections to improve the approximation to a root by one decimal place. To improve it by two decimal places requires about 6 to 7 bisections. Since the root must lie within the interval $x_2 - x_1$ then the maximum error in taking the midpoint within that interval as the root is plus or minus half the interval. The error is thus halved at each bisection.

Example

Use the bisection method to determine the root of the function $x^3 + x^2 - 3x - 3$ which lies between $x = 1$ and $x = 2$.

First bisection

Let $x_1 = 1$ and $x_2 = 2$. Then $f(x_1) = -4$ and $f(x_2) = 3$. The change of sign indicates that we can expect a root between these two values. These values give $x_m = 1.5$ and $f(x_m) = -1.875$. Thus, these values do not give $f(x_m)f(x_1) < 0$. The root thus lies between x_m and x_2, i.e. 1.5 to 2.0. We thus replace x_1 by x_m.

Second bisection

Let $x_1 = 1.5$ and $x_2 = 2$. Then $f(x_1) = -1.875$ and $f(x_2) = 3$. The change of sign indicates that we can expect a root between these two values. These values give $x_m = 1.75$ and $f(x_m) = 0.172$. Thus, these values give $f(x_m)f(x_1) < 0$. The root thus lies between x_1 and x_m, i.e. 1.5 and 1.75. We thus replace x_2 by x_m.

Third bisection

Let $x_1 = 1.5$ and $x_2 = 1.75$. Then $f(x_1) = -1.875$ and $f(x_2) = 0.172$. The change of sign indicates that we can expect a root between these two values. These give $x_m = 1.625$ and $f(x_m) = -0.943$. Thus, these values do not give $f(x_m)f(x_1) < 0$. The root thus lies between x_m and x_2, i.e. 1.625 and 1.75. We thus replace x_1 by x_m.

Fourth bisection

Let $x_1 = 1.625$ and $x_2 = 1.75$. Then we have $f(x_1) = -0.943$ and $f(x_2) = 0.172$. The change of sign indicates that we can expect a root between these two values. These values give $x_m = 1.687\ 5$ and $f(x_m) = -0.409$. Thus, these values do not give $f(x_m)f(x_1) < 0$. The root thus lies between x_m and x_2, i.e. 1.687 5 and 1.75. We thus replace x_1 by x_m.

If the accuracy required is to one decimal place, i.e. the interval between the values is less than 0.1 and so the extremes are less than ±0.05 either side of the midpoint, then we can stop here. The root is thus 1.7.

Fifth bisection

Let $x_1 = 1.687\ 5$ and $x_2 = 1.75$. Then we have $f(x_1) = -0.409$ and $f(x_2) = 0.172$. The change of sign indicates that we can expect a root between these two values. These values give $x_m = 1.718\ 75$ and $f(x_m) = -0.125$. Thus, these values do not give $f(x_m)f(x_1) < 0$. The root thus lies between x_m and x_2, i.e. 1.718 75 and 1.75. We thus replace x_1 by x_m.

Sixth bisection

Let $x_1 = 1.718\ 75$ and $x_2 = 1.75$. Then we have $f(x_1) = -0.125$ and $f(x_2) = 0.172$. The change of sign indicates that we can expect a root between these two values. These values give $x_m = 1.734\ 37$ and $f(x_m) = 0.220$. Thus, these values give $f(x_m)f(x_1) < 0$. The root thus lies between x_1 and x_m, i.e. 1.718 75 and 1.734 37. We thus replace x_2 by x_m.

Seventh bisection
Let $x_1 = 1.718\ 75$ and $x_2 = 1.734\ 37$. Then we have $f(x_1) = -0.125$ and $f(x_2) = 0.220$. The change of sign indicates that we can expect a root between these two values. These values give $x_m = 1.726\ 56$ and $f(x_m) = -0.052$. Thus, these values do not give $f(x_m)f(x_1) < 0$. The root thus lies between x_m and x_2, i.e. 1.726 5 and 1.734 37. We thus replace x_2 by x_m.

If the accuracy required is to two decimal places, i.e. the interval between the values is less than 0.01 and so the extremes are less than ±0.005 either side of the midpoint, then we can stop here. The root is thus 1.73.

Example

Use the bisection method to determine a root of $f(x) = e^x - 3x$, given that the root lies between $x = 1$ and $x = 2$.

First bisection
Let $x_1 = 1$ and $x_2 = 2$. Then $f(x_1) = -0.28$ and $f(x_2) = 1.39$. The change of sign indicates that we can expect a root between these two values. These values give $x_m = 1.5$ and $f(x_m) = -0.018$. Thus, these values do not give $f(x_m)f(x_1) < 0$. The root thus lies between x_m and x_2, i.e. 1.5 to 2.0. We thus replace x_1 by x_m.

Second bisection
Let $x_1 = 1.5$ and $x_2 = 2$. Then $f(x_1) = -0.018$ and $f(x_2) = 1.39$. The change of sign indicates that we can expect a root between these two values. These values give $x_m = 1.75$ and $f(x_m) = 0.50$. Thus, these values give $f(x_m)f(x_1) < 0$. The root thus lies between x_m and x_1, i.e. 1.5 to 1.75. We thus replace x_2 by x_m.

Third bisection
Let $x_1 = 1.5$ and $x_2 = 1.75$. Then $f(x_1) = -0.018$ and $f(x_2) = 0.50$. The change of sign indicates that we can expect a root between these two values. These values give $x_m = 1.625$ and $f(x_m) = 0.20$. Thus, these values give $f(x_m)f(x_1) < 0$. The root thus lies between x_m and x_1, i.e. 1.5 to 1.625. We thus replace x_2 by x_m.

Fourth bisection
Let $x_1 = 1.5$ and $x_2 = 1.625$. Then $f(x_1) = -0.018$ and $f(x_2) = 0.20$. The change of sign indicates that we can expect a root between these two values. These values give $x_m = 1.562\ 5$ and $f(x_m) = 0.08$. Thus, these values give $f(x_m)f(x_1) < 0$. The root thus lies between x_m and x_1, i.e. 1.5 to 1.562 5. We thus replace x_2 by x_m.

At this point the difference between the values is less than 0.1. Thus if the root is only required to the accuracy of one decimal place we need proceed no further, the root then being 1.5.

Fifth bisection
Let $x_1 = 1.5$ and $x_2 = 1.562\ 5$. Then we have $f(x_1) = -0.018$ and $f(x_2) = 0.08$. The change of sign indicates that we can expect a root between these two values. These values give $x_m = 1.531\ 25$ and $f(x_m) = 0.03$. Thus, these values give $f(x_m)f(x_1) < 0$. The root thus lies between x_m and x_1, i.e. 1.5 to 1.531 25. We thus replace x_2 by x_m.

Sixth bisection
Let $x_1 = 1.5$ and $x_2 = 1.531\ 25$. Then we have $f(x_1) = -0.018$ and $f(x_2) = 0.03$. The change of sign indicates that we can expect a root between these two values. These values give $x_m = 1.515\ 625$ and $f(x_m) = 0.005$. Thus, these values give $f(x_m)f(x_1) < 0$. The root thus lies between x_m and x_1, i.e. 1.5 to 1.515 625. We thus replace x_2 by x_m.

Seventh bisection
Let $x_1 = 1.5$ and $x_2 = 1.515\ 625$. Then we have $f(x_1) = -0.018$ and $f(x_2) = 0.005$. The change of sign indicates that we can expect a root between these two values. These give $x_m = 1.507\ 812\ 5$ and $f(x_m) = -0.007$. Thus, these values do not give $f(x_m)f(x_1) < 0$. The root lies between x_m and x_2, i.e. 1.507 812 5 to 1.515 625. We thus replace x_1 by x_m.

The difference between the values is less than 0.01. Thus if the root is required to an accuracy of two decimal places then further bisections are unnecessary. The root to this accuracy is 1.51.

Eighth bisection
Let $x_1 = 1.507\ 812\ 5$ and $x_2 = 1.515\ 625$. Then $f(x_1) = -0.007$ and $f(x_2) = 0.005$. The change of sign indicates that we can expect a root between these two values. These give $x_m = 1.511\ 718\ 75$ and $f(x_m) = -0.006$. Thus, these values do not give $f(x_m)f(x_1) < 0$. The root lies between x_m and x_2, i.e. 1.511 718 75 to 1.515 625. To continue we replace x_1 by x_m. We can keep on taking bisections and so obtain yet higher accuracy for the root.

Revision

3 Use the bisection method to determine, to an accuracy of two decimal places, the specified roots of the following functions:

(a) $f(x) = x^3 - 5x - 3$, given the root lies between $x = 2$ and $x = 3$.

(b) $f(x) = x^3 - x + 3$, given the root lies between $x = -1$ and $x = -2$.

(c) $f(x) = x^3 + x^2 + 5x - 1$, given the root lies between $x = 0$ and $x = 1$.

(d) $f(x) = e^{-2x} - x$, given the root lies between $x = 0$ and $x = 1$.

(e) $f(x) = \tan x - x$, given the root lies between $x = 4$ and $x = 4.5$.

(f) $f(x) = \cos x - 0.6x$, given the root lies between $x = 0$ and $x = 1$.

4 Use the bisection method to determine, to within 0.01, the root of the function $f(x) = x^3 + 4x^2 - 10$, given that a root lies between $x = 1$ and $x = 2$.

5 Use the bisection method to determine, to an accuracy of two decimal places, the cube root of 20.
Hint: consider $x = 20^{1/3}$, hence $x^3 = 20$ and so the function of $x^3 - 20$.

4.4 Newton's method

A method that is widely used for the determination of roots is *Newton's method* or, as it is often called, the *Newton-Raphson method.* Consider the graph of the function $y = f(x)$ shown in Figure 4.5. If we have a first approximation to the root of x_1 then we draw a tangent to the curve at that point and take as the next approximation to the root the point where this tangent cuts the x-axis. This then gives us estimate x_2. We now draw a tangent to the curve at this point and take as the next approximation to the root the point where this new tangent cuts the x-axis. We can thus continue this process until successive values of the intercepts with the x-axis are close enough to give the required accuracy of the root.

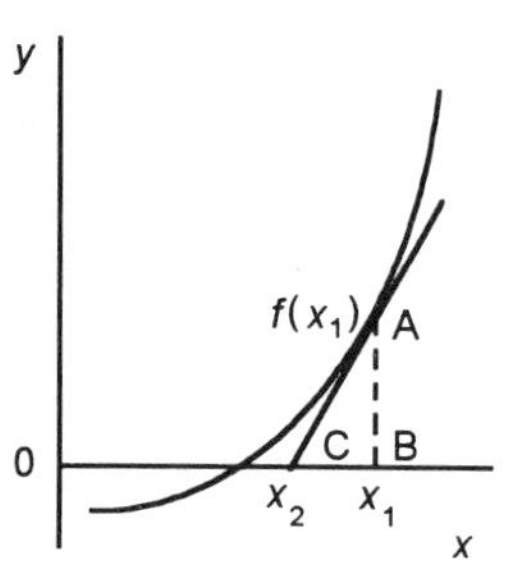

Figure 4.5 *Newton's method*

The gradient of the tangent in Figure 4.5 is:

$$\text{gradient} = \frac{\text{AB}}{\text{BC}} = \frac{f(x_1)}{x_1 - x_2}$$

But the gradient at x_1 for the function is the derivative (i.e. dy/dx) of the function at that value of x. Writing $f'(x_1)$ for this derivative, then:

$$f'(x_1) = \frac{f(x)}{x_1 - x_2}$$

Thus:

$$x_2 = x_1 - \frac{f(x_1)}{f'(x_1)} \qquad [1]$$

The procedure for using Newton's method can thus be summarised as:

1 Take initial estimate x_1.
2 Compute $f(x_1)$ and $f'(x_1)$.
3 Use equation [1] to determine x_2.
4 Set $x_1 = x_2$ and repeat operations 2 and 3 to obtain x_3.

The sequence is continued until the required accuracy is obtained for the root.

Thus if we had a function $x^3 - 6x^2 + 12$ and have estimated that there is a root at about $x = 2$, then taking this value as x_1 we have $f(x_1) = -4$ and, since $f'(x) = 3x^2 - 12x$, then $f'(x_1) = -12$. Thus, using equation [1], the improved root value x_2 is:

$$x_2 = 2 - \frac{(-4)}{(-12)} = 1.667$$

We can now repeat this procedure to obtain an improved estimate x_3. Since $f(x_2) = -0.041$ and $f'(x_2) = -11.667$, then equation [1] gives:

$$x_3 = 1.667 - \frac{(-0.041)}{(-11.667)} = 1.663$$

Thus, very rapidly, the root value is given to an accuracy of two decimal places as 1.66.

The Newton method has the great advantage over the bisection method in that the series of values generated at successive estimates more rapidly converges. The number of decimal places of accuracy nearly doubles at each iteration. However, in some cases, the series of values generated will not converge but diverge. This is when the tangent to the curve does not give a point closer to the root. Such situations can occur when $f'(x)$ is small or zero near the root. In some cases, divergence can occur if the initial approximation to the root is too far away from the root. Figure 4.6 illustrates such a case, the tangent to the curve at the first approximation not giving an intercept with the axis that is closer to the root. One of the examples that follow gives divergence.

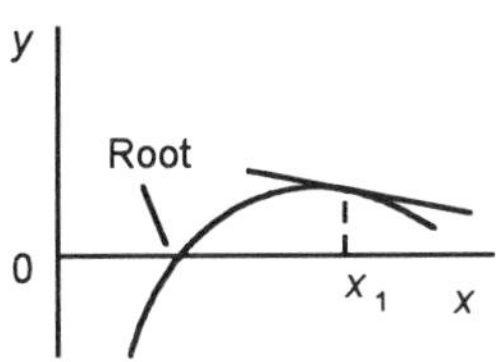

Figure 4.6 x_1 *is too far from the root*

Example

Use Newton's method to determine the root of $f(x) = x^3 + x^2 - 3x - 3$ which occurs about $x = 2$.

Let $x_1 = 2$, then $f(x_1) = 3$. Since $f'(x) = 3x^2 + 2x - 3$ then $f'(x) = 13$. Hence, using equation [1]:

$$x_2 = 2 - \frac{3}{13} = 1.769$$

Let $x_1 = 1.769$, then $f(x) = 0.358\ 2$. Since $f'(x) = 3x^2 + 2x - 3$ then $f'(x) = 9.9261$. Hence, using equation [1]:

$$x_3 = 1.769 - \frac{0.358\,2}{9.926\,1} = 1.732\,9$$

Thus after just two iterations the root is established to an accuracy of one decimal place. Let $x_1 = 1.732\ 9$, then $f(x) = 0.008\ 0$. With $f'(x) = 9.474\,6$ then equation [1] gives:

$$x_4 = 1.732\,9 - \frac{0.008\,0}{9.474\,6} = 1.732\,0$$

Thus after three iterations the root is established to an accuracy of three decimal places. Compare these results with the earlier bisection method example for the same function. Many more steps were needed there.

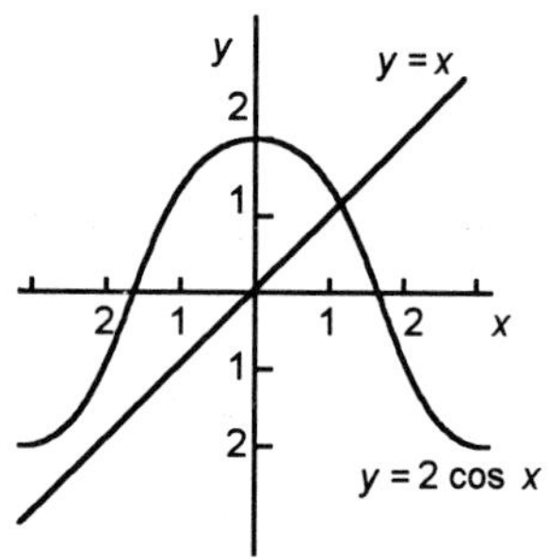

Figure 4.7 *Example*

Example

Using Newton's method, determine the value of x which is a solution of the equation $2 \cos x = x$.

This equation can be rearranged as

$$2 \cos x - x = 0$$

Thus the roots of the function $2 \cos x - x$ are the solutions. We can obtain approximate values for such roots by plotting a graph. Figure 4.7 shows sketch graphs of the functions $y = 2 \cos x$ and $y = x$. There is a point of intersection at about $x = 1$.

Thus if we let $x_1 = 1$, then $f(x_1) = 0.08$. Since $f'(x_1) = -2 \sin x - 1$ then $f'(x_1) = -2.68$. Thus, using equation [1]:

$$x_2 = 1 - \frac{0.08}{(-2.68)} = 1.03$$

Let $x_1 = 1.03$, then $f(x_1) = -0.001$. Since $f'(x_1) = -2.715$, then equation [1] gives:

$$x_3 = 1.03 - \frac{(-0.001)}{(-2.715)} = 1.0296$$

Thus, to two decimal places, the root is 1.03.

Example

Use Newton's method in an attempt to obtain the roots of the function $x^{1/3}$.

Since $f(x) = x^{1/3}$ then $f'(x) = \frac{1}{3}x^{-2/3}$. If we take an initial approximation for the root of $x_1 = 0.1$ then $f(x_1) = 0.464$ and $f'(x_1) = 1.548$. Hence, using equation [1], gives:

$$x_2 = 0.1 - \frac{0.464}{1.548} = -0.2$$

Let $x_1 = -0.2$, then $f(x_1) = -0.584$ and $f'(x_1) = 0.975$. Hence, using equation [1]:

$$x_3 = -0.2 - \frac{(-0.584)}{0.975} = 0.4$$

Let $x_1 = 0.4$, then $f(x_1) = 0.614$ and $f'(x_1) = 0.737$. Hence, using equation [1]:

$$x_4 = 0.4 - \frac{0.614}{0.737} = -0.8$$

The sequence of values is 0.1, −0.2, 0.4, −0.8, ..., the values oscillating from positive to negative values which become larger and larger and are not converging to any specific value. The actual root is 0.

Revision

6 Use Newton's method to determine, to four decimal places, the specified root of the following functions:

(a) $x - \cos x$, root at about 0.8, (b) $x^2 - 4x + 2$, root at about 0.6,

(c) $\tan x - x$, root at about 4.5, (d) $x^3 - 2x - 5$, root at about 2,

(e) $2x^3 + x^2 - x + 1$, root about −1.

7 Use Newton's method to determine, to four decimal places, the root or roots of the following functions:

(a) $\tan x - 2x$, (b) $e^x - 2x - 1$, (c) $x\,e^x - 3$, (d) $e^x - 3x$

Hint: use splitting of the function to enable you to sketch graphs and so find an initial approximation for the roots.

8 Use Newton's method to determine, to four decimal places, the cube root of 10.
Hint: this requires the solution of the equation $x = 10^{1/3}$ and so the root of the function $x^3 - 10$.

Problems

1 Draw the graph of $y = x^3$ between $x = -2$ and $x = +2$ and then, by drawing further graph lines, determine the roots of the functions: (a) $x^3 + 4x - 2$, (b) $x^3 - 1.5x + 0.5$.

2 By splitting the function, plot graphs and hence determine the roots of: (a) $x\,e^{3x} - 3$, (b) $e^x + x - 2$, (c) $2 \sin x - 2x + 1$, $0 \le x \le \pi$, (d) $e^x - 2 \cos x$, $-\pi/2 \le x \le \pi/2$.

3 Use the bisection method to determine, to an accuracy of two decimal places, the specified roots of the following functions:

(a) $f(x) = 2x^3 + 3x - 3$, given the root lies between $x = 0$ and $x = 1$,

(b) $f(x) = x^4 - 5x + 2$, given the root lies between $x = 0.3$ and $x = 0.5$,

(c) $f(x) = x^4 - 3x^2 - 3x + 1$, given the root lies between $x = 2$ and $x = 3$,

(d) $f(x) = \cos x + x$, given the root lies between $x = -1$ and $x = 0$,

(e) $f(x) = e^{-x} + x - 2$, given the root lies between $x = 1$ and $x = 2$,

(f) $f(x) = e^x - 2\cos x$, given the root lies between $x = 0$ and $x = 1$,

(g) $f(x) = 3\sin x - x$, given the root lies between $x = 2$ and $x = 2.5$,

(h) $f(x) = \cos x \cosh x - 1$, given the root lies between $x = 4.5$ and $x = 5$.

4 Use the bisection method to determine, to an accuracy of two decimal places, the cube root of 25.

5 Use Newton's method to determine, to four decimal places, the root or roots of the following functions:

(a) $x^3 + x^2 + 3x + 4$, root near -1.2, (b) $x^3 - 2x^2 - 5$, root near 2.5,

(c) $\sin x + 3x - e^x$, root near 0, (d) $e^x - 3x^2$, root near -0.5,

(e) $3\sin x - x$, (f) $3\sin x - 3x + 1$,

(g) $\tan x - e^x$, the least positive and the least negative roots.

6 Solve the following equations:

(a) $\sin x = x^2$, (b) $e^{-2x} = x$, (c) $x^2 = 2\,e^{-x} + 1$, (d) $x^2 = \ln(x + 1)$

7 A semicircle is bounded by its diameter with a line drawn from one end of this diameter across the semicircle to bisect the area. It can be shown that if the line makes an angle θ with the diameter that $2\theta + \sin 2\theta = \frac{1}{2}\pi$. Solve this equation and obtain θ to three decimal places.

8 The ratio of the voltage at the receiving end of a transmission line to the voltage at the sending end is a maximum when the length L of the line is given by $\beta \sin 2\beta L = \alpha \sinh 2\alpha L$. With $\alpha = 0.5 \times 10^{-3}$ km^{-1} and $\beta = 1.8 \times 10^{-3}$ km^{-1}, determine the value of the positive root to one decimal place. Hint: try an initial approximation of 1000 km.

9 A sphere of density ρ and radius r has a weight of $\frac{4}{3}\pi r^3 \rho g$. When floating in water the depth to which it sinks is h, the volume of the spherical segment below the surface then being $\frac{1}{3}\pi(3rh^2 - h^3)$. Archimedes' principle states that when an object floats the weight of fluid displaced must equal the weight of the object. Hence determine h if the density of the sphere is 400 kg/m^3 and that of the water 1000 kg/m^3.

5 Sequences and series

5.1 Introduction

This chapter is a look at sequences and series and the use of the Taylor polynomial to provide an approximation to a function. This approximation involves working from the value of the function and its derivatives at a particular point, i.e. for the function $y = f(x)$ we are given y, $\mathrm{d}y/\mathrm{d}x$, $\mathrm{d}^2y/\mathrm{d}x^2$, $\mathrm{d}^3y/\mathrm{d}x^3$, etc. at a particular value of x. From these values a polynomial approximation of the function is constructed. The result is what is termed a *Taylor polynomial*. Such polynomials can be used to transform functions such as $\sin x$ into polynomials. The substitution of a polynomial for a function can often make the solution of an engineering problem easier.

For the Taylor theorem it is assumed that the reader can differentiate functions. A point regarding notation used in this chapter, for the function $y = f(x)$ the derivative $\mathrm{d}y/\mathrm{d}x$ is more simply represented by $f'(x)$, with $\mathrm{d}^2y/\mathrm{d}x^2$ by $f''(x)$, $\mathrm{d}^3y/\mathrm{d}x^3$ by $f'''(x)$, etc. At $x = a$ the value of the function is represented by $f(a)$ and the derivatives by $f'(a)$, $f''(a)$, $f'''(a)$, etc.

5.2 Sequences

Consider the numbers, 1, 3, 5, 7, 9. Such a set of numbers is termed a *sequence* because the numbers are stated in a definite order, 1 followed by 3 followed by 5, etc. Another sequence might be 1, $\frac{1}{2}$, $\frac{1}{4}$, $\frac{1}{8}$, $\frac{1}{16}$. These sequences have a finite number of terms but often we can meet ones involving an infinite number of terms, e.g. 2, 4, 6, 8, 10, 12, ..., etc.

> *The term sequence is used for a set of quantities stated in a definite order.*

In general we can write a sequence as:

first value of variable, second value of variable, third value of variable, ..., etc.

or, if x is the variable:

$x[1]$, $x[2]$, $x[3]$, ..., etc.

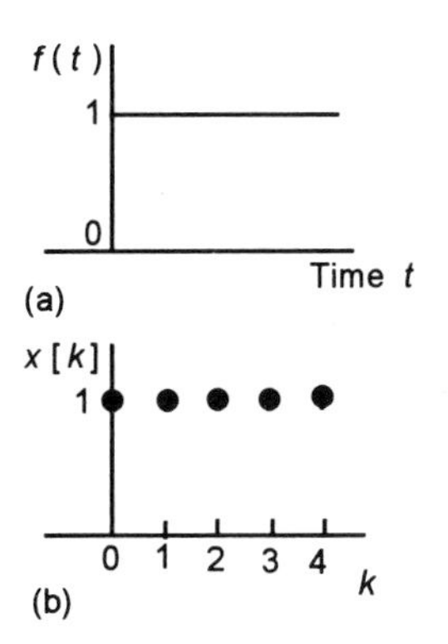

Figure 5.1 *(a) Unit step, (b) unit step sequence*

This is usually more compactly written as $x[k]$, where $k = 1, 2, 3, \ldots$, etc. Such a form of notation is commonly encountered in signal processing when perhaps an analogue signal is sampled at a number of sequential points and the resulting sequence of digital signal values processed. For example, if an analogue unit step signal is sampled the sampled data output might be expressed as $x[k] = 0$ for $k < 0$, $x[k] = 1$ for $k \geq 0$ with $k = 0, 1, 2, 3, 4$, etc. Figure 5.1 shows graphs of the unit step input and the sampled output. Since we have a sequence of discrete numbers, the graph takes the form of a collection of isolated points.

Sometimes it is possible to describe a sequence by giving a rule for the kth term, common forms being the arithmetic and geometric sequences.

An arithmetic sequence has each term formed from the previous term by simply adding on a constant value.

If a is the first term and d the common difference between successive terms, the terms are:

$$a, (a+d), (a+2d), (a+3d), ..., \text{etc.} \quad [1]$$

The kth term is $a + (k-1)d$, with $k = 1, 2, 3, 4, ...,$ etc. (note that if k has the values 0, 1, 2, etc. the kth term is $a + kd$). Thus for such a sequence we can write:

$$x[k] = a + (k-1)d \quad [2]$$

A geometric sequence has each term formed from the previous term by multiplying it by a constant factor, e.g. 3, 6, 12, 24, ...

If a is the first term and r the common ratio between successive terms, the terms are:

$$a, ar, ar^2, ar^3 + ..., \text{etc.} \quad [3]$$

The kth term is ar^{k-1}, with $k = 1, 2, 3, 4, ...,$ etc. Thus for such a sequence we can write:

$$x[k] = ar^{k-1} \quad [4]$$

The sequence $1, \frac{1}{2}, \frac{1}{3}, \frac{1}{4}, ...$ is termed the *harmonic sequence* and defined for $k = 1, 2, 3,$ etc. by:

$$x[k] = \frac{1}{k} \quad [5]$$

Sequences can be generated by other rules. For example, the sequence 1, 2, 5, 10, 17, ... is generated by $x[k] = 1 + (k-1)^2$, where $k = 1, 2, 3, ...$. This sequence is neither an arithmetic nor a geometric sequence.

Example

Write down the first five terms of the sequence $x[k]$ defined by $x[k] = \frac{1}{2}k^2 + k$ when $k \geq 0$.

When $k = 0$ we have $0 + 0$, when $k = 1$ we have $0.5 + 1$, when $k = 2$ we have $2 + 2$, and so on. The sequence is thus 0, 1.5, 4, 7.5, 12.

Revision

1 A sinusoidal signal $f(t) = \sin t$ is sampled every quarter period starting when $t = 0$. State the sequence of sampled values.

2 Write down the first five terms of the sequence $x[k]$ defined, for $k \geq 0$, by (a) $x[k] = k$, (b) $x[k] = e^{-k}$.

3 State the fifth term of (a) the arithmetic sequence given by 4, 7, 10, ..., (b) the geometric sequence given by 12, 6, 3, ...

4 Write an equation for the kth term, where k = 1, 2, 3, ..., for the following sequences (a) 1, −1, 1, −1, ..., (b) 5, 10, 15, 20, ..., (c) 2, 1, 5, 1, 0.5, ...

5 A car depreciates so that after each year its value is 60% of that at the beginning of the year. What is its value after k years if its initial value is £10 000?

5.2.1 Limits of sequences

The sequence 1, 0.1, 0.01, 0.001, 0.0001, ... , etc. gets smaller and smaller as k increases. In the limit as k tends to infinity, then $x[k]$ tends to zero. The sequences converges to 0:

$$\lim_{x \to \infty} x[k] = 0 \qquad [6]$$

The numbers in the sequence 1, 2, 3, 4, 5, ..., etc. go on increasing as k increases. In the limit as k approaches infinity then $x[k]$ tends to infinity. The series is said to diverge..

A sequence that converges to a finite limit is said to be convergent; one that does not converge to a finite limit being divergent.

The limit of an arithmetic sequence $x[k] = a + (k - 1)d$ (equation [2]) is infinite because as k tends to infinity so $a + (k - 1)d$ tends to infinity. Depending on whether d is positive or negative, so the limit is $+\infty$ or $-\infty$. With a geometric sequence $x[k] = ar^{k-1}$ (equation [4]), if $r = 1$ then the sequence is a, a, a, ..., etc. and the limit is clearly a. With $r > 1$, e.g. $r = 2$ giving a, a^2, a^3, ..., etc., then the limit is ∞. With $-1 < r < 1$, e.g. $r = -½$ giving $a, -a/2, a/4, -a/8$, ..., etc., then the limit is 0. With $r = -1$ we have $a, -a, a, -a$, ..., and there is no limit we can define. With $r < -1$, e.g. $r = -2$ giving $a, -2a, 4a, -8a$, ..., the magnitude of successive terms increases but the signs alternate and thus it is not possible to define a unique limit. With the harmonic series $x[k] = 1/k$ (equation [5]), as k increases so $1/k$ approaches zero.

Often we can regard a particular sequence as being formed from two or more other sequences. If the limit of $x[k]$ is X and the limit of $y[k]$ is Y then:

1 The limit of the sum $x[k] + y[k]$ is $X + Y$.
2 The limit of the difference $x[k] - y[k] = X - Y$.
3 The limit of the product $x[k]y[k]$ is XY.
4 The limit of the quotient $x[k]/y[k]$ is X/Y.

Example

Determine, if possible, the limits of the sequences (a) $x[k] = 2$, (b) $x[k] = 1 + 0.1^k$, (c) $x[k] = k^2$.

(a) The sequence is 2, 2, 2, 2, ..., etc. and so in the limit the sequence has the value 2.
(b) The sequence is 1.1, 1.01, 1.001, ..., etc. and so in the limit the sequence has the value 1.
(c) The sequence is 1, 4, 9, 16, ..., etc. and so the series is divergent with a limit of infinity.

Example

Determine the limit of the sequence given by $x[k] = \dfrac{4}{2 - \frac{1}{k^2}}$.

We can consider this as the limit of one sequence divided by the limit of another.

$$\lim_{k\to\infty} x[k] = \frac{\lim_{k\to\infty} 4}{\lim_{k\to\infty}\left(2 - \frac{1}{k^2}\right)} = \frac{4}{2-0} = 2$$

Revision

6 Determine, if possible, the limits of the sequences: (a) $x[k] = 1/k$, (b) $x[k] = 1 + k^2$, (c) $x[k] = 4 + 1/k$, (d) $x[k] = 4/k^2$, (e) $x[k] = 3 + 1/k^2$, (f) $x[k] = 2^k$.

5.3 Series

A *series* is formed by adding the terms of a sequence. Thus 1 + 3 + 5 + 7 + 9 + ..., etc. is a series.

A series is the sum of the terms of a sequence.

The sum of n terms of a series is written using *sigma notation* as:

$$S_n = \sum_{k=1}^{n} x[k] \qquad [7]$$

The first and the last values of k are shown below and above the sigma. For example, the series 1 + 3 + 5 + 7 + 9 would have the sum, over the five terms, written as:

$$S_5 = \sum_{k=1}^{5}(2k - 1)$$

5.3.1 Arithmetic and geometric series

An arithmetic series has each term formed from the previous term by simply adding on a constant value.

Such a series can be written in the general form as:

$$a + \{a + d\} + \{a + 2d\} + \{a + 3d\} + \ldots + \{a + (n - 1)d\} \qquad [8]$$

The sum to k terms is:

$$S_k = \{a\} + \{(a + d)\} + \{(a + 2d)\} + \{(a + 3d)\} + \ldots + \{a + (n - 1)d\}$$

If we write this back to front then:

$$S_k = \{a + (n - 1)d\} + \{a + (n - 2)d\} + \{a + (n - 3)d\} + \ldots \{a\}$$

Adding these two equations gives first term plus first term, second term plus second term, etc. and we obtain:

$$2S_k = \{2a + (n - 1)d\} + \{2a + (n - 1)d\} + \{2a + (n - 1)d\} + \ldots$$

for k terms. Thus $2S_k = n\{2a + (n - 1)d\}$ and so:

$$S_k = \tfrac{1}{2}n\{2a + (n - 1)d\} \qquad [9]$$

A geometric series has each term formed from the previous term by multiplying it by a constant factor.

Such a series can be written in the general form as:

$$a + ar + ar^2 + ar^3 + \ldots + ar^{n-1} \qquad [10]$$

The sum to the kth terms is:

$$S_k = a + ar + ar^2 + ar^3 + \ldots + ar^{n-1}$$

Multiplying by r gives:

$$rS_k = ar + ar^2 + ar^3 + ar^4 + \ldots + ar^n$$

Hence $S_k - rS_k = a - ar^n$, and so, provided $k \neq 1$:

$$S_k = \frac{a(1 - r^n)}{1 - r} \qquad [11]$$

Example

Determine the sum of the arithmetic series 1 + 5 + 9 + ... if it contains 10 terms.

Such a series has a first term a of 1 and a common difference d of 4. Thus, using equation [9]:

$$S_k = \tfrac{1}{2}\{2a + (k-1)d\} = \tfrac{1}{2} \times 10\{2 + 9 \times 4\} = 190$$

Example

Determine the sum of the geometric series 4 + 6 + 9 + ... if it contains 10 terms.

Such a series has a first term of 4 and a common ratio of 3/2. Thus, using equation [11]:

$$S_k = \frac{a(1-r^k)}{1-r} = \frac{4(1-1.5^{10})}{1-1.5} = 453.3$$

Revision

7 Determine the sums of the following arithmetic or geometric series if each contains 10 terms:

(a) 3 + 2.5 + 2.0 + ..., (b) 12 + 6 + 3 + ..., (c) 1 + 2 + 4 + 8 + ...

8 A contractor submits a quotation to sink a well at the rate of £0.50 for the first metre, £0.52 for the second metre, £0.54 for the third metre and an extra £0.02 for each successive metre. What will be the cost of a well of depth 80 m?

5.3.2 Series of powers of integers

The series 1 + 2 + 3 + 4 + ... + n is an arithmetic series and so its sum for the first k terms is given by equation [9] as:

$$S_k = \sum_{k=1}^{n} k = \tfrac{1}{2}n\{2a + (n-1)d\} = \tfrac{1}{2}n\{2 + (n-1)d\} = \tfrac{1}{2}n(n+1) \quad [12]$$

Now consider the series $1^2 + 2^2 + 3^2 + 4^2 + \dots n^2$. This is not an arithmetic or a geometric series. To determine the sum for the first k terms we make use of the identity:

$$(k+1)^3 = k^3 + 3k^2 + 3k + 1$$

We can rewrite this as:

$$(k+1)^3 - k^3 = 3k^2 + 3k + 1$$

Hence:

$$\sum_{k=1}^{n}\left[(k+1)^3 - k^3\right] = \sum_{k=1}^{n}(3k^2 + 3k + 1)$$

The left-hand side of the equation gives:

$$\{2^3 - 1^3\} + \{3^3 - 2^3\} + \{4^3 - 3^3\} + \ldots + \{(n+1)^3 - n^3\} = (n+1)^3 - 1$$

The right-hand side of the equation can be written as:

$$3\sum_{k=1}^{n} k^2 + 3\sum_{k=1}^{n} k + \sum_{k=1}^{n} 1$$

But the second term summation is just equation [12] and so $\frac{1}{2}n(n+1)$. The third term summation is the sum of the series $1 + 1 + 1 + \ldots$ and so is n. Hence:

$$3\sum_{k=1}^{n} k^2 + \tfrac{3}{2}n(n+1) + n = (n+1)^3 - 1$$

Thus:

$$\sum_{k=1}^{n} k^2 = \tfrac{1}{6}n(n+1)(2n+1) \qquad [13]$$

If we wanted to find the sum of the series $1^3 + 2^3 + 3^3 + \ldots n^3$ then we can proceed in a similar manner by developing the identity:

$$(k+1)^4 = k^4 + 4k^3 + 6k^2 + 4k + 1$$

to give $(k + 1)^4 - k^4$ and then carrying out the summation of the terms. The result is:

$$\sum_{k=1}^{n} k^3 = \left[\frac{n(n+1)}{2}\right]^2 \qquad [14]$$

Example

Determine the sum of the series $\sum_{k=1}^{n} k(3+2k)$.

We can write this as:

$$\sum_{k=1}^{n} k(3+2k) = 3\sum_{k=1}^{n} k + 2\sum_{k=1}^{n} k^2$$

Hence, using equations [12] and [13]:

$$\sum_{k=1}^{n} k(3+2k) = \tfrac{3}{2}n(n+1) + \frac{2n(n+1)(2n+1)}{6}$$

Revision

9 Determine the sums of the following series:

(a) $\sum_{k=1}^{4}(2+3k)$, (b) $\sum_{k=1}^{5} k(1+2k)$, (c) $\sum_{k=1}^{3}(k^2+1)$

5.3.3 Convergent and divergent series

So far we have considered the sums of series with a finite number of terms. What about the sum when we have a series with an infinite number of terms? For an infinite number of terms the sum will have a limiting value as the number of terms tends to infinity.

A series in which the sum of the series tends to a definite value as the number of terms tends to infinity is called a convergent series.

Consider an *arithmetic series* $a + (a + d) + (a + 2d) + \dots$ for an infinite number of terms. For k terms we have the sum (equation [9]) of:

$$S_k = \tfrac{1}{2}n\{2a + (n-1)d\}$$

As k tends to infinity then n tends to infinity and so the sum tends to infinity. The sum of an infinite arithmetic series is infinity. The series is said to be *divergent*.

Consider a *geometric series* $a + ar + ar^2 + \dots$ for an infinite number of terms. For k terms we have the sum (equation [11]) of:

$$S_k = \frac{a(1-r^n)}{1-r} = \frac{a}{1-r} - \frac{ar^n}{1-r} \qquad [15]$$

Suppose we have $-1 < r < 1$, as n tends to infinity then r^n tends to 0. Thus the second term converges to zero and we are left with just the first term. Thus such a series converges to the sum:

$$S_\infty = \frac{a}{1-r} \text{ for } -1 < r < 1 \qquad [16]$$

Thus the geometric series $x[k] = 3^{1/2}$ converges to the sum 6. However, if we had the geometric series $x[k] = 3^2$ then the sum is given by equation [15] as $-3 + 3 \times 2^n$ and thus as n tends to infinity the sum tends to infinity. For $|r| \geq 1$ the geometric series does not converge.

There are a number of ways that are used to determine whether a series will converge:

1 *Comparison test*
A series of positive terms is convergent if its terms are less than the corresponding terms of a positive series which is known to converge. Similarly, the series is divergent if its terms are greater than the corresponding terms of a series which is known to be divergent. As an example, consider the series:

$$1+\frac{1}{2^2}+\frac{1}{3^3}+\frac{1}{4^4}+\dots$$

Suppose we know that the series:

$$1+\frac{1}{2^2}+\frac{1}{2^3}+\frac{1}{2^4}+\dots$$

converges (it is a geometric series with $r = ½$), then if, after the first two terms, we compare terms we find that every term in our convergent series is greater than the one we are considering. Thus the series must also converge.

2 *D'Alembert's ratio test*
An infinite series is convergent if, as k tends to infinity, the ratio of each term u_{n+1} to the preceding term u_n is numerically less than 1 and divergent if greater than 1, i.e.

$$\lim_{n\to\infty}\left|\frac{u_{n+1}}{u_n}\right| < 1, \text{ the series converges,}$$

$$\lim_{n\to\infty}\left|\frac{u_{n+1}}{u_n}\right| > 1, \text{ the series diverges,}$$

$$\lim_{n\to\infty}\left|\frac{u_{n+1}}{u_n}\right| = 1, \text{ the series may converge or diverge.}$$

Consider the series:

$$1-\frac{1}{2!}+\frac{1}{3!}-\frac{1}{4!}+\dots$$

The nth term u_n is $|1/n!|$ and the $(n + 1)$th term u_{n+1} is $|1/(n + 1)!|$. Therefore:

$$\frac{u_{n+1}}{u_n}=\left|\frac{\dfrac{1}{(n+1)!}}{\dfrac{1}{n!}}\right|=\left|\frac{1}{n+1}\right|$$

As n tends to infinity then $\lim_{n\to\infty}\left|\frac{u_{n+1}}{u_n}\right| < 1$ and so the series converges.

Example

Find the sum to infinity of the series $4 + 2 + 1 + ½ + \dots$.

This is a geometric series with $a = 4$ and $r = ½$. Using equation [16]:

$$S_\infty = \frac{a}{1-r} = \frac{4}{1-\frac{1}{2}} = 8$$

Example

Determine, using the comparison test, whether the series $x[k] = 1/n^n$, i.e. $1 + 1/2^2 + 1/3^3 + 1/4^4 + \ldots$, is convergent.

If we exclude the first two terms we can compare it with the geometric series $1/2^3 + 1/2^4 + 1/2^5 + \ldots$ which is known to be convergent. Each term in this series being tested is smaller than the comparable term in the comparison series. Thus it must be convergent.

Example

Determine, using d'Alembert's ratio test, whether the series $1 + x + x^2/2! + x^3/3! + \ldots$ is convergent.

Using d'Alembert's ratio test, since $u_n = x^{n-1}/(n-1)!$ and $u_{n+1} = x^n/n!$:

$$\frac{u_{n+1}}{u_n} = \frac{\dfrac{x^{n-1}}{(n-1)!}}{\dfrac{x^n}{n!}} = \frac{x}{n}$$

In the limit as n tends to infinity then the ratio tends to 0. Thus the series is convergent.

Revision

10 Find the sum to infinity of the series:

(a) $6 + 3 + 1.5 + \ldots$, (b) $4 + 3 + 2.25 + \ldots$, (c) $12 + 3 + 0.75 + \ldots$

11 Using the comparison test, determine whether the following series are convergent or divergent:

(a) $x[k] = 1/3^n$ (compare with $1/2^n$), (b) $x[k] = 1.5^n$ (compare with 1^n)

12 Using d'Alembert's ratio test, determine whether the following series are convergent or divergent:

(a) $x - \frac{x^2}{2} + \frac{x^3}{3} + \ldots + \frac{(-1)^{n-1}x^n}{n} + \ldots$, (b) $3 + \frac{3^2}{2} + \frac{3^3}{3} + \ldots + \frac{3^n}{n} + \ldots$

5.3.4 Power series

A series of the type:

$$a_0 + a_1x + a_2x^2 + a_3x^3 + \dots + a_nx^n + \dots$$

is known as a *power series*. If we apply d'Alembert's ratio test then the series will be convergent when:

$$\lim_{n\to\infty} \left| \frac{a_{n+1}x^{n+1}}{a_nx^n} \right| < 1$$

This can be written as:

$$|x| \lim_{n\to\infty} \left| \frac{a_{n+1}}{a_n} \right| < 1$$

or:

$$|x| < \lim_{n\to\infty} \left| \frac{a_{n+1}}{a_n} \right| \qquad [17]$$

Thus there are conditions attached to the value of x if the series is to converge. Examples are given later in this chapter.

Example

For what values of x is the series $x[k] = x^n/n$ convergent?

Here $a_n = 1/n$ and $a_{n+1} = 1/(n + 1)$. Thus $|a_{n+1}/a_n| = (n + 1)/n = 1 + 1/n$ and so in the limit we have the value of 1 for the limit. Thus the condition for convergence is that $|x| < 1$ or $-1 < x < +1$.

Revision

13 Determine for what values of x the following series are convergent:

(a) $x[k] = nx^n$, (b) $x[k] = n^2x^n$

5.4 Approximating a function

Consider some function $y = f(x)$. It is often convenient to be able to represent this function by a polynomial $p(x)$, which is in the form of ascending powers of x:

$$p(x) = A + Bx + Cx^2 + Dx^3 + \dots \qquad [18]$$

A, B, C, D, etc. being constants which are selected to give the best fit. First, however, consider a number of simplified cases when the number of terms in the polynomial are limited. When we have only $p(x) = A$ then the polynomial is said to be of degree 0, when $p(x) = A + Bx$ of degree 1, when

$p(x) = A + Bx + Cx^2$ of degree 2, etc. The degree of the polynomial is the highest power of the x that it contains.

Degree 0 polynomial
Consider the matching of the polynomial with the function when we only consider the first term in the polynomial, i.e. $p(x) = A$. We must then have, for a match of the y values between the polynomial and the function at some point $x = a$:

$$A = f(a)$$

The polynomial is thus:

$$p(x) = f(a) \qquad [19]$$

This equation is referred to as the *zero degree Taylor polynomial.*

Degree 1 polynomial
Now consider matching when the polynomial has two terms, namely $p(x) = A + Bx$. To determine A and B we need two equations. At $x = a$, for a match of the y values we must have:

$$A + Ba = f(a) \qquad [20]$$

We can obtain a second equation if we also consider matching the first derivatives. Since $p'(x) = B$ then we must have:

$$B = f'(a)$$

Substituting this value into equation [20] gives:

$$A = f(a) - af'(a)$$

Thus the polynomial becomes:

$$p(x) = f(a) - af'(a) + f'(a)x = f(a) + (x - a)f'(a) \qquad [21]$$

This equation is referred to as the *first degree Taylor polynomial.*

Degree 2 polynomial
Now consider matching when the polynomial has three terms, i.e. $p(x) = A + Bx + Cx^2$. To determine A, B and C we need three equations. At $x = a$, for a match of the y values we must have:

$$A + Ba + Ca^2 = f(a) \qquad [22]$$

We can obtain two further equations by considering the first derivative and the second derivative. Since the first derivative of the polynomial is $p'(x) = B + 2Cx$, then for a match of first derivatives we must have:

$$B + 2Ca = f'(a) \qquad [23]$$

Since the second derivative of the polynomial is $p''(x) = 2C$, then for a match of second derivatives we must have:

$$2C = f''(a) \qquad [24]$$

Equation [24] gives the value of C. Substitution of this value into equation [23] gives:

$$B + af''(a) = f'(a)$$

Thus we have $B = f'(a) - af''(a)$ and so substitution of these values of B and C into equation [22] gives:

$$A + [f'(a) - af''(a)]a + \tfrac{1}{2}f''(a)a^2 = f(a)$$

Hence $A = f(a) - af'(a) + \frac{1}{2}a^2 f''(a)$. Thus the polynomial becomes:

$$p(x) = f(a) - af'(a) + \tfrac{1}{2}a^2 f''(a) + [f'(a) - af''(a)]x + \tfrac{1}{2}f''(a)x^2$$

$$= f(a) + (x - a)f'(a) + \frac{(x-a)^2}{2}f''(a) \qquad [25]$$

This equation is referred to as the *second degree Taylor polynomial.*

Degree n polynomial

We can go on increasing the number of terms in the polynomial and matching derivatives. The result of doing this is to develop the equation

$$p(x) = f(a) + \frac{(x-a)}{1!}f'(a) + \frac{(x-a)^2}{2!}f''(a) + \frac{(x-a)^3}{3!}f'''(a) + \dots$$

$$+ \frac{(x-a)^n}{n!}f^n(a) \qquad [26]$$

This equation is referred to as the *nth degree Taylor polynomial.*

Example

Determine the polynomial approximation to the function $f(x)$ if at the point $x = 0$ we have $f(0) = 2$, $f'(0) = 3$, and $f''(0) = -4$.

The number of terms given indicates that the approximating polynomial will be of degree 2, a higher degree cannot be developed because we only have terms up to the second derivative. Thus the approximating polynomial will be of the form:

$$p(x) = A + Bx + Cx^2$$

We can obtain the answer by working from first principles. For $p(0)$ to equal $f(0)$ we must have $A = 2$. Differentiating the polynomial gives:

$$p'(x) = B + 2Cx$$

Hence, for $p'(0) = f'(0)$ we must have $B = 3$. Differentiating again gives:

$$p''(x) = 2C$$

Hence for $p''(0) = f''(0)$ we must have $2C = -4$. Thus the approximating polynomial is:

$$p(x) = 2 + 3x - 2x^2$$

Alternatively we could have just used the second degree Taylor polynomial (equation [25]):

$$p(x) = f(a) + (x - a)f'(a) + \frac{(x-a)^2}{2} f''(a)$$

With $a = 0$ we have:

$$p(x) = 2 + (x - 0)3 + \frac{(x-0)^2}{2}(-4) = 2 + 3x - 2x^2$$

Example

Determine the third degree Taylor polynomial about $x = 0$ for the function $f(x) = \sin x$.

We require values at $x = 0$ for the 0 to 3rd degree derivatives for the function.

$$f(x) = \sin x \text{ thus } f(0) = 0$$

$$f'(x) = \cos x \text{ thus } f'(0) = 1$$

$$f''(x) = -\sin x \text{ thus } f''(0) = 0$$

$$f'''(x) = -\cos x \text{ thus } f'''(0) = -1$$

Hence the third degree Taylor polynomial (equation [26]) is:

$$p(x) = f(a) + (x - a)f'(a) + \frac{(x-a)^2}{2!} f''(a) + \frac{(x-a)^3}{3!} f'''(a)$$

$$= 0 + x + 0 - \tfrac{1}{3}x^3$$

Revision

14 Determine the polynomial approximation to the function $f(x)$ if:

(a) at the point $x = 0$ we have $f(0) = 1$, $f'(0) = 2$, $f''(0) = -8$, and $f'''(0) = 12$,
(b) at the point $x = 1$ we have $f(1) = 1$, $f'(1) = 1$ and $f''(1) = -2$,
(c) at the point $x = 2$ we have $f(2) = 0$, $f'(2) = 2$, $f''(2) = 4$, and $f'''(2) = 6$.

15 Determine the third degree Taylor polynomial approximation to the function: (a) $f(x) = 1/(1 + x)$ at $x = 0$, (b) $f(x) = \sin x$ at $x = 1$.

16 Determine the fourth degree Taylor polynomial approximation to the function $f(x) = \cos 2x$ at $x = 0$.

5.4.1 Linearisation

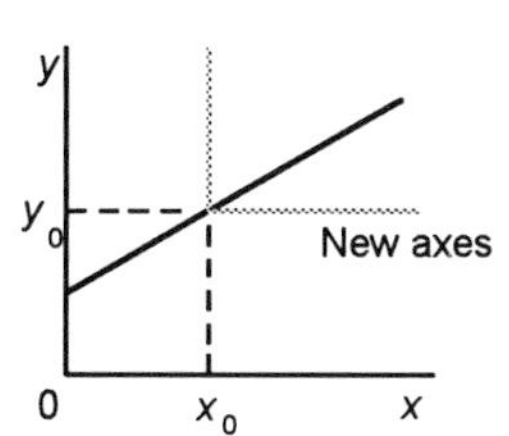

Figure 5.2 *Change of origin*

A function $y = f(x)$ for which we have the relationship of the form $y = mx$ is termed *linear*. For systems having such relationships the output is proportional to the input. However, a function represented by the relationship $y = mx + c$ does not have y proportional to x. Though it gives a straight line graph it does not pass through the origin. However, if the origin is shifted to some operating point (x_0, y_0) which is on the straight line (Figure 5.2) the $y = mx + c$ relationship can be turned into $(y - y_0) = m(x - x_0)$. Then the change in y from its operating value is proportional to the change in x from its operating value.

Many, though not all, mechanical and electrical elements can be considered to be linear over a reasonably large range of input values. However, thermal and fluid elements are frequently non-linear. If we represent the function by a first degree Taylor polynomial we can linearise non-linear curves for small changes about an operating point (Figure 5.3). For example, a linear equivalent about its operating point may be obtained for a transistor. The linear equivalent is generally referred to as the *small-signal equivalent circuit* for the device. In relation to control systems, linearisation is used for devices such as the flow of liquids through valves, since most design techniques are based on the assumption of linear elements.

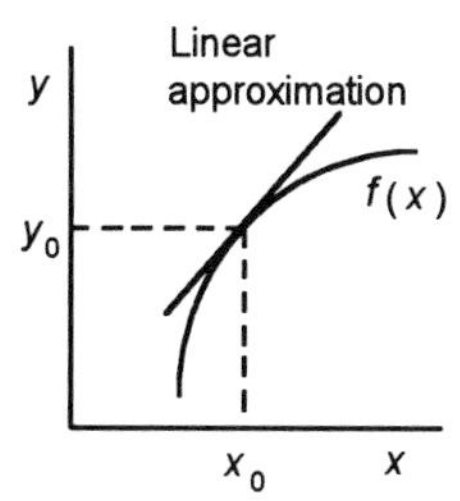

Figure 5.3 *Linear approximation*

Consider the linearisation of the function $y = f(x)$ about the operating point (x_0, y_0), as in Figure 5.3. The first degree Taylor polynomial is (equation [21]):

$$p(x) = f(a) + (x - a)f'(a)$$

Thus, at the operating point:

$$p(x) = f(x_0) + (x - x_0)f'(x_0)$$

If we now consider this polynomial to represent the function $f(x)$ at the operating point, i.e. $f(x) = p(x)$, then:

$$f(x) = f(x_0) + (x - x_0)f'(x_0)$$

We can rewrite this as:

$$f(x) - f(x_0) = (x - x_0)f'(x_0)$$

or:

$$y - y_0 = [f'(x_0)](x - x_0) \qquad [27]$$

Thus:

change in y from operating point $\propto$ change in x from operating point

This is a linear relationship, with the origin shifted to the operating point. The change in y from the operating value is proportional to the change in x from its operating value. The gradient is $f'(x_0)$. The range of values for which a linearised relationship is valid depends the accuracy required, the error being due to the higher degree Taylor polynomial terms that have not been used.

To illustrate linearisation, consider the obtaining of the small-signal equivalent circuit for a bipolar transistor. Figure 5.4 shows the typical form of relationship between the collector current I_C and the base-emitter voltage V_{BE}. With normal operating conditions we have the relationship:

$$I_C \approx I_{SE}\, e^{KV_{BE}}$$

where I_{SE} is the emitter junction saturation current and K is a constant. On the graph a line has been drawn which is the tangent to the curve at an operating point of $I_C = 2$ mA, $V_{BE} = 0.62$ V. This line gives the linearised relationship. We thus shift the operating point to 2 mA, 0.62 V. The first derivative is $I_{SE}K\, e^{0.62K}$. Thus, using the first degree Taylor polynomial, i.e. equation [27], we can obtain the equation for this line as:

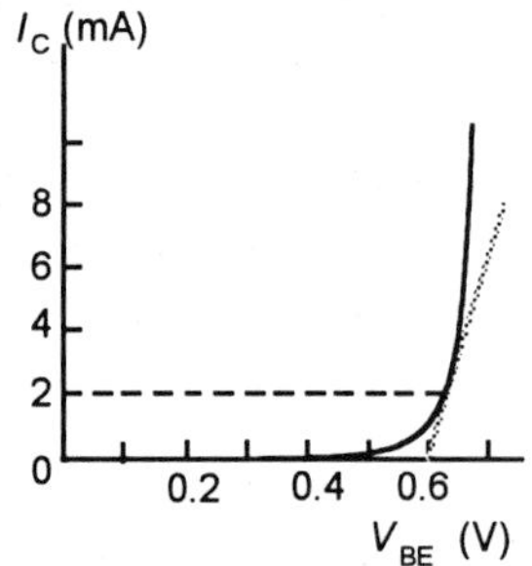

Figure 5.4 *Transistor characteristic*

$$I_C - 2 = (I_{SE}K\, e^{0.62K})(V_{BE} - 0.62)$$

$I_{SE}\, e^{0.65K}$ is the collector current at the operating point, i.e. 2 mA. Hence, since K is about 40 V^{-1}, then the linearised equation can be written as:

$$I_C - 2 = 80(V_{BE} - 0.62)$$

The gradient of the line, i.e. the 80 mA/V, is called the transconductance.

Example

The rate of flow q of a liquid through an orifice, e.g. a valve, is a function of the difference in pressure p across the orifice being given by:

$$q = C\sqrt{p}$$

where C is a constant. Linearise this equation about an operating point pressure difference of p_0.

If q_0 is the flow rate at the operating point pressure difference then, using equation [27]:

$$y - y_0 = [f'(x_0)](x - x_0)$$

$f'(x_0) = \mathrm{d}q/\mathrm{d}p$ at $p = p_0$. Since $\mathrm{d}q/\mathrm{d}p = \frac{1}{2}Cp^{-1/2}$, then:

$$q - q_0 = \left[\frac{C}{2\sqrt{p_0}}\right](p - p_0)$$

This is a linear relationship, the change in flow from the operating value being proportional to the change in pressure from the operating value. This equation can be used in place of the original equation for changes about the operating point.

Revision

17 The resistance R of a thermistor is a function of the temperature T, being described by $R = k\,\mathrm{e}^{-cT}$ where k and c are constants. Linearise this relationship about an operating temperature of T_0.

18 The e.m.f. E generated by a thermocouple is a function of the temperature T and described by the relationship $E = aT + bT^2$, where a and b are constants. Linearise this relationship about an operating temperature of T_0.

19 The power P dissipated in a resistor is a function of the current, being described by the relationship $P = RI^2$, where R is a constant. Linearise this relationship about an operating current of I_0. For $R = 10\ \Omega$ and an operating current of 1 A, determine by how much the power will change when the current increases by 0.1 A from this operating value. Compare the result with that calculated using the non-linear relationship.

20 Determine the small-signal (linearised) model, at an operating voltage for v_A of -2 V, for a three-terminal field-effect transistor for which $i_B = 16\left(1 + \frac{1}{4}v_A\right)^2$ mA.

5.4.2 Small increments

Consider y as a function of x and δy to be the change in y produced when x increases by δx. By definition of the derivative we have:

$$\frac{\mathrm{d}y}{\mathrm{d}x} = \lim_{\delta x \to 0} \frac{\delta y}{\delta x}$$

Hence when δx is very small:

$$\frac{\mathrm{d}y}{\mathrm{d}x} \approx \frac{\delta y}{\delta x}$$

and so:

$$\delta y \approx \frac{\mathrm{d}y}{\mathrm{d}x}\delta x \qquad [28]$$

The above equation only gives a reasonable estimate of δy when we are able to assume that δx is very small. A better estimate of δy is given by the use of the Taylor polynomial. For $y = f(x)$ we can write, when we increase the x value at $x = a$ by δx:

$$\delta y = f(a + \delta x) - f(a)$$

Taylor's polynomial is given by (equation [26]):

$$p(x) = f(a) + \frac{(x-a)}{1!}f'(a) + \frac{(x-a)^2}{2!}f''(a) + \ldots$$

Consider the polynomial at $x = a + \delta x$. Then we can write:

$$p(a + \delta x) = f(a) + \frac{\delta x}{1!}f'(a) + \frac{(\delta x)^2}{2!}f''(a) + \ldots$$

If the polynomial can be considered to be the same as the function at the point $f(a + \delta x)$ then:

$$\delta y = f(a + \delta x) - f(a) \approx \frac{\delta x}{1!}f'(a) + \frac{(\delta x)^2}{2!}f''(a) + \ldots \qquad [29]$$

The first term is what we obtained with equation [28]. Adding further terms improves the accuracy of the value given for δy.

Example

The range R of a projectile is a function of the angle of launch θ and is described by:

$$R = \frac{v^2}{g}\sin 2\theta$$

If v and g are constants, what will be the error in the range if there is an error in θ of $\delta\theta$ at an angle θ_0?

Using equation [29]:

$$\delta R \approx \frac{\delta\theta}{1!}f'(\theta_0) + \frac{(\delta\theta)^2}{2!}f''(\theta_0) + \ldots$$

$$\approx \frac{v^2}{2g}\left[2\delta\theta\cos 2\theta_0 - 4(\delta\theta)^2 \sin 2\theta_0 + \ldots\right]$$

Thus, for example, if we have a projectile with an initial velocity of 4 m/s and there was an error in the angle of 0.1 rad at an angle of projection of $\pi/6$ then the error would be:

$$\delta R \approx \frac{16}{9.8}[2 \times 0.1 \times 0.87 - 4 \times 0.01 \times 0.5 + \ldots] \approx 0.25 \text{ m}$$

Revision

21 The focal length F of a convex lens is given by:

$$\frac{1}{F} = \frac{1}{u} + \frac{1}{v}$$

where u is the distance of the object from the lens and v the image distance. The focal length is being determined from measurements of the object and images distances. What will be the effect on the accuracy of the focal length of an error in the image distance v of δv?

22 Use the Taylor polynomial to obtain the value of $\sin(\pi/6 + 0.01)$, given that $\sin \pi/6 = 1/2$ and $\cos \pi/6 = (\sqrt{3})/2$.

23 Show that if h is small:

$$\cos(a + h) = \cos a - h \sin a - \frac{h^2}{2!}\cos a + \ldots$$

5.4.3 Newton's method for roots

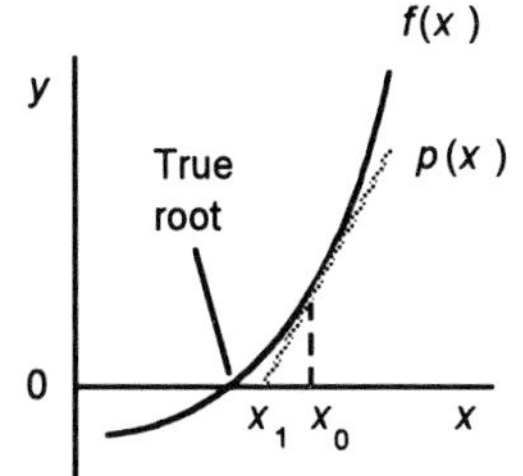

Figure 5.5 *Newton method*

Methods for the determination of the roots of functions were discussed in Chapter 4. One of the methods was Newton's method (Section 4.4), there being considered graphically. Here we derive Newton's method using the Taylor polynomial.

Consider the function $y = f(x)$. This will have roots when $f(x) = 0$ (Figure 5.5). We make a guess that there is a root near the point $x = x_0$. The first degree Taylor polynomial representation of the function at $x = x_0$ is:

$$p(x) = f(x_0) + (x - x_0)f'(x_0)$$

This polynomial will have a root when $p(x) = 0$, i.e. at $x = x_1$ when:

$$f(x_0) + (x_1 - x_0)f'(x_0) = 0$$

Hence:

$$x_1 = x_0 - \frac{f(x_0)}{f'(x_0)} \qquad [30]$$

This root x_1 is an improvement on the original estimate of the root x_0 of the function. If we now use x_1 as the estimate of the root, i.e. we take the root of the Taylor polynomial about x_1 instead of x_0, we can repeat the procedure to obtain a new estimate x_2.

$$x_2 = x_1 - \frac{f(x_1)}{f'(x_1)} \qquad [31]$$

This root x_2 is an improvement on the x_1 estimate. We can repeat this for yet further estimates. In general, equations [30], [31] and further ones can be written as:

$$x_{n+1} = x_n - \frac{f(x_n)}{f'(x_n)} \qquad [32]$$

This method of obtaining a root is called *Newton's method* or the *Newton-Raphson method.*

The following examples and problems illustrate the above method. See, however, Chapter 4 for more examples and problems.

Example

Use Newton's method to determine the square root of 5.

The square root of 5 is the positive root of the equation:

$$f(x) = x^2 - 5$$

Since $f'(x) = 2x$ then equation [32] can be written as:

$$x_{n+1} = x_n - \frac{f(x_n)}{f'(x_n)} = x_n - \frac{x_n^2 - 5}{2x_n} = \frac{1}{2}\left(x_n + \frac{5}{x_n}\right)$$

For a first estimate, let $x_0 = 3$. Then:

$$x_1 = \frac{1}{2}\left(3 + \frac{5}{3}\right) = 2.333$$

For a second estimate:

$$x_2 = \frac{1}{2}\left(2.333 + \frac{5}{2.333}\right) = 2.238$$

For a third estimate:

$$x_3 = \frac{1}{2}\left(2.238 + \frac{5}{2.238}\right) = 2.236$$

For a fourth estimate:

$$x_4 = \frac{1}{2}\left(2.236 + \frac{5}{2.236}\right) = 2.236$$

Thus to the accuracy of the figures quoted, the root of 5 is 2.236.

Example

Determine, to three decimal places, the root which is near to −1.3 of the function $f(x) = x^3 - 6x^2 + 12$.

There will be three roots to the equation but the one to be found is indicated as being near to −1.3. Since $f'(x) = 3x^2 - 12x$, then equation [32] can be written as:

$$x_{n+1} = x_n - \frac{f(x_n)}{f'(x_n)} = x_n - \frac{x_n^3 - 6x_n^2 + 12}{3x_n^2 - 12x_n}$$

Thus with the estimate of $x_0 = -1.3$ we obtain:

$$x_1 = -1.3 - \frac{-2.197 - 10.14 + 12}{5.07 + 15.6} = -1.284$$

For a second estimate:

$$x_2 = -1.284 - \frac{-2.117 - 9.892 + 12}{4.946 + 15.408} = -1.284$$

Thus, to an accuracy of three decimal places, the root is −1.284.

Revision

24 Use Newton's method to determine, to three decimal places, the square root of (a) 2, (b) 3, (c) 7.

25 Determine, to three decimal places, the root which is near to +2 of the function $f(x) = x^3 - 2x - 5$.

26 Determine, to three decimal places, the root which is near to + 0.25 of the function $f(x) = x^4 - 3x^2 - 3x + 1$.

5.5 Maclaurin series

Consider the Taylor polynomial (equation [26]). As more and more terms are included in the polynomial we obtain an infinite series which is termed the *Taylor series*:

$$p(x) = f(a) + \frac{(x-a)}{1!}f'(a) + \frac{(x-a)^2}{2!}f''(a) + \frac{(x-a)^3}{3!}f'''(a) + \ldots + \frac{(x-a)^n}{n!}f^n(a) \qquad [33]$$

A special, commonly used form of the Taylor series is obtained by setting $a = 0$, i.e. we are fitting the Taylor polynomial to the function at the point $x = 0$ rather than $x = a$. We then have:

$$p(x) = f(0) + \frac{x}{1!}f'(0) + \frac{x^2}{2!}f''(0) + \frac{x^3}{3!}f'''(a) + \ldots \qquad [34]$$

This is called the *Maclaurin series*.

The polynomial can represent a function if the series converges as the number of terms included tends to infinity. We can use the d'Alembert's ratio test to test for this.

Example

Determine the Maclaurin series for the function $y = e^x$.

For the function we have $f'(x) = e^x$. Likewise we have $f''(x) = e^x$, $f'''(x) = e^x$, etc. When $x = 0$ the $e^x = 1$, hence equation [34] becomes:

$$p(x) = 1 + \frac{x}{1!} + \frac{x^2}{2!} + \frac{x^3}{3!} + \ldots$$

Hence the Maclaurin series expansion of the function is:

$$e^x = 1 + \frac{x}{1!} + \frac{x^2}{2!} + \frac{x^3}{3!} + \ldots$$

We can test for convergence using the d'Alembert's ratio test. Thus:

$$\left|\frac{u_{n+1}}{u_n}\right| = \left|\frac{x^{n+1}/(n+1)!}{x^n/n!}\right| = \left|\frac{x}{n+1}\right|$$

As n tends to infinity then the ratio tends to zero. The series is thus convergent.

Example

Determine the Maclaurin series for the function $y = \sin x$.

For the function $f(x) = \sin x$ we have $f'(x) = \cos x$, $f''(x) = -\sin x$, $f'''(x) = -\cos x$, $f^{iv}(x) = \sin x$, $f^{v}(x) = \cos x$, etc. Thus at $x = 0$ we have $f(0) = 0, f'(0) = 1, f''(0) = 0, f'''(0) = -1, f^{iv}(0) = 0, f^{v}(0) = 1$, etc. The series is thus given by equation [34] as:

$$\sin x = 0 + x + 0 - \frac{x^3}{3!} + 0 + \frac{x^5}{5!} + \dots$$

$$\sin x = x - \frac{x^3}{3!} + \frac{x^5}{5!} + \dots$$

We can test for convergence using d'Alembert's ratio test:

$$|\text{ratio of successive terms}| = \left|\frac{x^{n+2}/(n+2)!}{x^n/n!}\right| = \left|\frac{x^2}{n+2}\right|$$

As n tends to infinity then the ratio tends to zero. The series converges.

Example

Determine the Maclaurin series for the function $y = \ln(1 + x)$.

For $f(x) = \ln(1 + x)$ we have $f'(x) = 1/(1+x)$, $f''(x) = -1/(1+x)^2$, $f'''(x) = 2/(1+x)^3$, etc. At $x = 0$ we thus have $f(0) = \ln 1 = 0, f'(0) = 1$, $f''(0) = -1, f'''(0) = 2$, etc. Thus the series is given by equation [34] as:

$$\ln(1 + x) = x - \frac{x^2}{2} + \frac{x^3}{3} + \dots$$

The terms can be represented as being:

$$(-1)^{n-1}\frac{x^n}{n}$$

Thus, when testing for convergence using d'Alembert's ratio test:

$$\left|\frac{u_{n+1}}{u_n}\right| = \left|\frac{x^{n+1}/(n+1)}{x^n/n}\right| = \left|\frac{nx}{n+1}\right| = \left|x\left(\frac{1}{1+1/n}\right)\right|$$

As n tends to infinity then the term in the brackets tends to 1 and so the ratio tends to $|x|$. The ratio will thus only have a value less than 1 if $|x| < 1$, i.e. $-1 < x < +1$. The representation of $\ln(1 + x)$ by this series is thus only valid with this condition.

Revision

27 Determine the Maclaurin series for the following functions:

(a) $y = \dfrac{1}{x+1}$, (b) $y = e^x \sin x$, (c) $y = (1 + x)^4$, (d) $y = \tan x$

28 Determine the Maclaurin series for $\tan(x + \pi/4)$ and hence use it to find the value for tan 46.5°, i.e. take x to be 1.5°.

29 Determine the Maclaurin series for the function:

$$\ln\left(\frac{1+x}{1-x}\right)$$

and hence use it to determine the value of ln 2, i.e. take x to be 1/3.

Note that if we attempted to determine the value of ln 2 by using the Maclaurin series for $\ln(1 + x)$, as determined in the example, then we have problems because of the lack of convergence.

5.5.1 Binomial series

For $(a + x)^2$ we can readily show that it can be written as $a^2 + 2ax + x^2$. If we multiply this by $(a + x)$ we obtain $a^3 + 3a^2x + 3ax^2 + x^3$. Multiplying by repeated factors of $(a + x)$ enables expansions of higher powers of $(a + x)$ to be generated. This is, however, rather cumbersone if, say, we wanted the expansion of $(a + x)^{10}$. There is, however, a pattern in the results:

$$\begin{aligned}
(a+x)^1 &= 1a + 1x \\
(a+x)^2 &= 1a^2 + 2ax + 1x^2 \\
(a+x)^3 &= 1a^3 + 3a^2x + 3ax^2 + 1x^3 \\
(a+x)^4 &= 1a^4 + 4a^3x + 6a^2x^2 + 4ax^3 + 1x^4
\end{aligned}$$

If we just write the coefficients the pattern is more readily discerned:

$$\begin{array}{c}
1 \quad 1 \\
1 \quad 2 \quad 1 \\
1 \quad 3 \quad 3 \quad 1 \\
1 \quad 4 \quad 6 \quad 4 \quad 1
\end{array}$$

Every coefficient is obtained by adding the two either side of it in the row above. Thus, for example, we have:

The above pattern is known as *Pascal's triangle*.

There is, however, another way we can arrive at the expansion. Consider the use of the Maclaurin series for the function $(1 + x)^n$. We have:

$$\begin{aligned}
f(x) &= (1+x)^n, \\
f'(x) &= n(1+x)^{n-1}, \\
f''(x) &= n(n-1)(1+x)^{n-2}, \\
f'''(x) &= n(n-1)(n-2)(1+x)^{n-3}, \text{ etc.}
\end{aligned}$$

Thus $f(0) = 1, f'(0) = n, f''(0) = n(n-1), f'''(0) = n(n-1)(n-2)$, etc. and the Maclaurin series is given by equation [34] as:

$$(1+x)^n = 1 + nx + \frac{n(n-1)}{2!}x^2 + \frac{n(n-1)(n-2)}{3!}x^3 + \dots \qquad [35]$$

This is known as the *binomial series*. This series converges when we have $-1 < x < 1$.

From the series of $(1+x)^n$ we can obtain a series for $(a+x)^n$.

$$(a+x)^n = \left[a\left(1+\tfrac{x}{a}\right)\right]^n = a^n\left(1+\tfrac{x}{a}\right)^n$$

$$= a^n\left[1 + n\left(\tfrac{x}{a}\right) + \frac{n(n-1)}{2!}\left(\tfrac{x}{a}\right)^2 + \frac{n(n-1)(n-2)}{3!}\left(\tfrac{x}{a}\right)^3 + \dots\right]$$

$$= a^n + nx\frac{n(n-1)}{2!}a^{n-2}x^2 + \frac{n(n-1)(n-2)}{3!}a^{n-3}x^3 + \dots \qquad [36]$$

Example

Expand by the binomial theorem $(1+x)^6$.

Using equation [35]:

$$(1+x)^6 = 1 + 6x + \frac{6\times 5}{2!}x^2 + \frac{6\times 5\times 4}{3!}x^3 + \frac{6\times 5\times 4\times 3}{4!}x^4$$

$$+ \frac{6\times 5\times 4\times 3\times 2}{5!}x^5 + \frac{6\times 5\times 4\times 3\times 2\times 1}{6!}x^6$$

$$= 1 + 6x + 15x^2 + 20x^3 + 15x^4 + 6x^5 + x^6$$

Example

Write the first four terms in the expansion of $(1+x)^{1/2}$.

$$(1+x)^{1/2} = 1 + \tfrac{1}{2}x + \frac{\tfrac{1}{2}\left(-\tfrac{1}{2}\right)}{2!}x^2 + \frac{\tfrac{1}{2}\left(-\tfrac{1}{2}\right)\left(-\tfrac{3}{2}\right)}{3!}x^3 + \dots$$

$$= 1 + \tfrac{1}{2}x - \tfrac{1}{8}x^2 + \tfrac{1}{16}x^3 + \dots$$

Example

Expand, using the binomial theorem, $(a-2b)^4$.

Using equation [36]:

$$(a-2b)^4 = a^4 + 4a^3(-2b) + \frac{4\times 3}{2!}a^2(-2b)^2 + \frac{4\times 3\times 2}{3!}a(-2b)^3$$
$$+ \frac{4\times 3\times 2\times 1}{4!}(-2b)^4$$

$$= a^4 - 8a^3b + 24a^2b^2 - 32ab^3 + 16b^4$$

Revision

30 Expand by the binomial theorem:

(a) $(1 + x)^4$, (b) $(1 + x)^{3/2}$, (c) $(1 - x)^{-5/2}$, (d) $(1 + 0.25)^{-1}$ for four terms,

(e) $(4 + x)^{1/2}$ for four terms.

5.5.2 Power series

Table 5.1 gives some commonly met functions and their Maclaurin series expansions. Generally we obtain the series for more complicated functions by manipulation of these series rather than direct calculation of the derivatives. The examples that follow illustrate this.

Table 5.1 *Power series*

	Function	Series	Validity
1	$\sin x$	$x - \frac{x^3}{3!} + \frac{x^5}{5!} - \frac{x^7}{7!} + \ldots$	For all x
2	$\cos x$	$x - \frac{x^2}{2!} + \frac{x^4}{4!} - \frac{x^6}{6!} + \ldots$	For all x
3	$\tan x$	$x + \frac{x^3}{3!} + \frac{2x^5}{15} + \frac{17x^7}{315} + \ldots$	$-\pi/2 < x < \pi/2$
4	e^x	$1 + x + \frac{x^2}{2!} + \frac{x^3}{3!} + \frac{x^4}{4!} + \ldots$	For all x
5	$\sinh x$	$x + \frac{x^3}{3!} + \frac{x^5}{5!} + \frac{x^7}{7!} + \ldots$	For all x
6	$\cosh x$	$1 + \frac{x^2}{2!} + \frac{x^4}{4!} + \frac{x^6}{6!} + \ldots$	For all x
7	$\ln(1 + x)$	$x - \frac{x^2}{2!} + \frac{x^3}{3!} - \frac{x^4}{4!} + \ldots$	$-1 < x < 1$
8	$(1 + x)^n$	$1 + nx + \frac{n(n-1)}{2!}x^2 + \frac{n(n-1)(n-2)}{3!}x^3 + \ldots$	$-1 < x < 1$

Example

Using series given in Table 5.1, determine the series expansion of the function $e^x \sin x$.

Entries 1 and 4 in Table 5.1 give:

$$\sin x = x - \frac{x^3}{3!} + \frac{x^5}{5!} + \ldots, \text{ valid for all values of } x$$

$$e^x = 1 + \frac{x}{1!} + \frac{x^2}{2!} + \frac{x^3}{3!} + \ldots, \text{ valid for all values of } x$$

We can multiply these two series to give

$$e^x \sin x = \left(1 + \frac{x}{1!} + \frac{x^2}{2!} + \frac{x^3}{3!} + \ldots\right)\left(x - \frac{x^3}{3!} + \frac{x^5}{5!} + \ldots\right)$$

$$= x + x^2 + \left(\tfrac{1}{2} - \tfrac{1}{6}\right)x^3 + \left(\tfrac{1}{6} - \tfrac{1}{6}\right)x^4$$

$$+ \left(\tfrac{1}{120} + \tfrac{1}{24} - \tfrac{1}{12}\right)x^5 + \ldots$$

$$= x + x^2 + \tfrac{1}{3}x^3 - \tfrac{1}{30}x^5 + \ldots$$

Example

Using Table 5.1, determine the series, as far as the x^3 term, for the function $y = e^{4x}$.

Item 4 in Table 5.1 gives:

$$e^x = 1 + \frac{x}{1!} + \frac{x^2}{2!} + \frac{x^3}{3!} + \ldots$$

If we substitute $4x$ for x then we obtain:

$$e^{4x} = 1 + \frac{4x}{1!} + \frac{16x^2}{2!} + \frac{64x^3}{3!} + \ldots = 1 + 4x + 8x^2 + \tfrac{32}{3}x^3 + \ldots$$

Revision

31 Using Table 5.1, determine the series for the following functions:

(a) $y = e^{2x}$, (b) $y = e^x \cos x$, (c) $y = (1 + x)^{-1/2}$,

(d) $y = e^x \ln(1 + x)$, (e) $y = \sec x$, (f) $y = \cos^2 x$

32 Show that, if x is small:

$$\frac{1}{1+x} - (1 - 2x)^{1/2} \approx \tfrac{3}{2}x^2$$

33 Determine the series expansion for $\sinh x$ using the relationship:

$$\sinh x = \tfrac{1}{2}(e^x - e^{-x})$$

34 For a continuous belt passing round two wheels, diameters d and D, with centres a distance x apart, the length L of belt required, if there is no sag, is:

$$L = 2x \cos a + \tfrac{1}{2}\pi(D + d) + (D - d)a$$

where $\sin a = (D - d)/2x$. Show that:

$$L \approx 2x + \tfrac{1}{2}\pi(D + d) + \frac{(D - d)^2}{4x}$$

35 The displacement x of the slider of a reciprocating mechanism depends on the crankshaft angle θ, being related by

$$x = r\cos\theta + L\sqrt{1 - \frac{r^2}{L^2}\sin^2\theta}$$

where r is the radius of the crankshaft and L the length of the connecting rod. Show, when r/L is considerably smaller than 1, that:

$$x \approx r\cos\theta + L - \frac{r^2}{2L}\sin^2\theta$$

Problems

1 A sinusoidal signal $f(t) = \cos t$ is sampled at $t = 0, 1, 2, 3, \ldots$, etc. State the first five terms of the sequence of sampled values.

2 Write down the first five terms of the sequence $x[k]$ defined, for $k \geq 0$, by (a) $x[k] = k^2$, (b) $x[k] = e^k$, (c) $x[k] = \tfrac{1}{2}k^2 + 2k$.

3 State the fifth term of the arithmetic progression given by 5, 7, 9, ...

4 State the fifth term of the geometric progression given by 8, 4, 2, ...

5 Write an equation for the kth term for the following sequences:

(a) $\frac{1}{4}, \frac{1}{16}, \frac{1}{64}, \ldots$, (b) $-2, +2, -2, \ldots$, (c) 3.1, 3.01, 3.001, ...

6 State the first three terms of the sequences given by:

(a) $(0.1)^k$, (b) $5 + (0.1)^k$, (c) $(-1)^k$

7 Determine, if possible, the limits of the sequences:

(a) $x[k] = 8 + 1/k$, (b) $x[k] = k$, (c) $x[k] = 1/(k + 1)$, (d) $x[k] = 5(-1)^k$,

(e) $x[k] = 3/(1 + 1/k)$, (f) $x[k] = 3^k$

8 A battery loses 10% of its charge every hour. Write an equation for the charge at the end of the kth hour if it starts with an initial charge C.

9 Determine the sums of the following series if each contains 12 terms:

(a) $2 + 5 + 8 + \ldots$, (b) $5 + \frac{5}{2} + \frac{5}{4} + \ldots$, (c) $4 + 3.6 + 3.24 + \ldots$

10 Determine the sums of the following series:

(a) $\sum_{k=1}^{5} 4k$, (b) $\sum_{k=1}^{5} 2k^2$, (c) $\sum_{k=1}^{4}(2 + k^3)$

11 Determine which of the following series is convergent and which divergent:

(a) $-1 + 1 - 1 + \ldots (-1)^n + \ldots$, (b) $1\,e^{-1} + 2\,e^{-2} + 3\,e^{-3} + \ldots n\,e^{-n} + \ldots$,

(c) $\dfrac{2\times1+1}{2} + \dfrac{2\times2+1}{2^2} + \dfrac{2\times3+1}{2^3} + \ldots + \dfrac{2n+1}{2^n} + \ldots$,

(d) $\sum_{k=0}^{\infty} \dfrac{10^n}{n!}$, (e) $\sum_{k=1}^{\infty} \dfrac{1}{n^2+2^2}$, (f) $\sum_{k=1}^{\infty} \dfrac{2^{n-1}}{10+(n-1)}$

12 Determine the polynomial approximation to the function $f(x)$ if at the point $x = 0$ we have $f(0) = 1, f'(0) = 1, f''(0) = -2$ and $f'''(0) = -2$.

13 Determine the polynomial approximation to the function $f(x)$ if at the point $x = 1$ we have $f(1) = 2, f'(1) = 2$ and $f''(1) = 6$.

14 Determine the third degree Taylor polynomial for the function $f(x) = e^x$ at $x = 0$.

15 For a semiconductor diode at room temperature the current I is a function of the voltage V and described by the equation

$$I = f(V) = I_s(e^{40V} - 1)$$

Derive (a) the first degree, (b) the second degree Taylor polynomials which approximate to this function at the operating voltage V_a.

16 Determine the third degree Taylor polynomial approximation to the function $f(x) = 1/x$ at $x = -1$.

17 The torque T applied to a simple pendulum is related to the angular deflection θ by $T = MgL \sin\theta$, where M, g and L are constants. Linearise this equation for the equilibrium angle $\theta_0 = 0$.

18 The extension x of a spring is a function of the applied stretching force F. If $F = kx^2$, where k is a constant, linearise the relationship about the operating extension of x_0.

19 For a MOSFET the relationship between I_D and V_{GS} is given by:

$$I_D = \tfrac{1}{2}\beta(V_{GS} - V_T)^2(1 + \lambda V_{DS})$$

Determine the small-signal (linearised) equivalent relationship for an operating point of $I_D = 2$ mA, $V_{GS} = 0.5$ V.

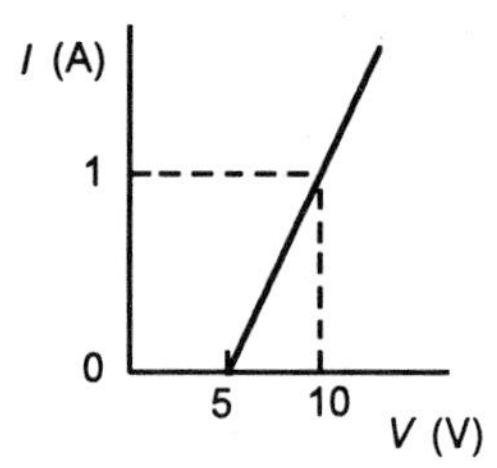

Figure 5.6 *Problem 20*

20 A diode has the non-linear relationship shown in Figure 5.6 of current i with the voltage v across it. Derive a linearised relationship about an operating voltage of 10 V.

21 Determine, by the use of the Taylor polynomial, the value of cos 31°, given that sin 30° = 1/2 and cos 30° = (√3)/2.

22 Show that if h is small:

$$\tan(a + h) = \tan a + h \sec^2 a + h^2 \sec^2 a \tan a + \ldots$$

23 An instrument gives a deflection θ which is related to the current i by $i = k \tan \theta$. Determine the error in the current reading resulting from an error in the reading of $\delta\theta$.

24 Use Newton's method to determine, to three decimal places, the square root of (a) 6, (b) 8, (c) 13.

25 Determine, to three decimal places, the root which is near to −2 of the function $f(x) = 3x^3 - 4x + 5$.

26 Determine, to three decimal places, the root which is between 0.3 and 0.4 of the function $f(x) = x^3 - 3x + 1$.

27 Solve the equation $f(x) = x - 2 \sin x = 0$ by means of Newton's method, given that there is a root at about $x = 2$.

28 Determine the Maclaurin series for the following functions:

(a) $y = \cos 2x$, (b) $y = \tan 3x$, (c) $y = e^{2x} \cos 3x$, (d) $y = x^2 + \sin x$

29 Using Table 5.1, determine the series for the following functions:

(a) $y = \tan 2x$, (b) $y = e^{x/2}$, (c) $y = (1 + x)^{1/2}$, (d) $y = \ln(1 + \sin x)$,

(e) $y = (1 - x)^{-3}$, (f) $y = e^x \sinh x$, (g) $x \cot x$, (h) $x \sin x + \cos x$

30 Show that, if x is small, $e^x \ln(1 + x) + \ln(1 - x) \approx x^4$.

31 The transverse deflection δ of a column of length L when subject to a vertical load F and a horizontal load H at the top is given by:

$$\delta = \frac{HL}{F}\left(\frac{\tan aL}{aL} - 1\right)$$

where $a^2 = F/EI$. Show that as F tends to a zero value that δ tends to $HL^3/3EI$.

32 Determine the series expansion for cosh x using $\cosh x = \frac{1}{2}(e^x + e^{-x})$.

33 Determine the series expansion for tan x using $\tan x = \frac{\sin x}{\cos x}$.

34 Use the binomial theorem to write the first four terms of:

(a) $(1 + x)^{12}$, (b) $(1 - 2x)^{-2}$, (c) $(3 - 2x)^{2/5}$, (d) $\frac{1}{1-x}$, (e) $(1 + 3x)^{-1/2}$,

(f) $\frac{1}{(1-x^3)^2}$

35 Use the binomial theorem to expand (a) $(2x + y)^4$, (b) $(2 - 3x)^5$.

36 By using the binomial theorem, determine the cube root of 1.04 to four decimal places. Hint: write 1.04 as 1 + 0.04.

6 Application: Representing signals

6.1 Introduction

This chapter is a brief consideration of how continuous and discrete signals can be represented mathematically. It illustrates the application of some of the principles discussed in earlier chapters in this part, in particular Chapters 1 and 5.

The discussion ranges over the representation of continuous signals that vary with time, e.g. the current waveform given by sinusoidal alternating current or the current variation with time in an electrical circuit where a capacitor is discharging through a resistance, and signals that can be considered to be a sequence of discrete impulses, e.g. the digital signals handled by a microprocessor.

6.2 Continuous and discrete signals

Continuous signals, because their waveforms are functions of time, can be represented as $f(t)$, existing for all values of the continuous-time variable t. This implies that there is the possibility of a signal value at every instant of time. An example of a continuous signal is the alternating current output from the mains supply; values of the current can be specified for any value of the time. Figure 6.1(a) shows an example of a continuous signal.

A *discrete signal* differs from a continuous signal in that such a signal is only available at discrete time intervals. The term *sampled-data signal* is often used since discrete signals are often obtained by taking samples of a continuous-time signal at various instants of time. Figure 6.1(b) shows an example of a discrete signal. The signal is shown as only existing at times of 0, $1T$, $2T$, $3T$, etc. Discrete signals can be considered to be sequences of impulses and so can be represented by the notation used for sequences, namely $x[k]$ where this term represents the size of the discrete signals for sequences values of perhaps $k = 0, 1, 2, 3, 4$, etc. In general, k can take on integer values between $-\infty$ and $+\infty$.

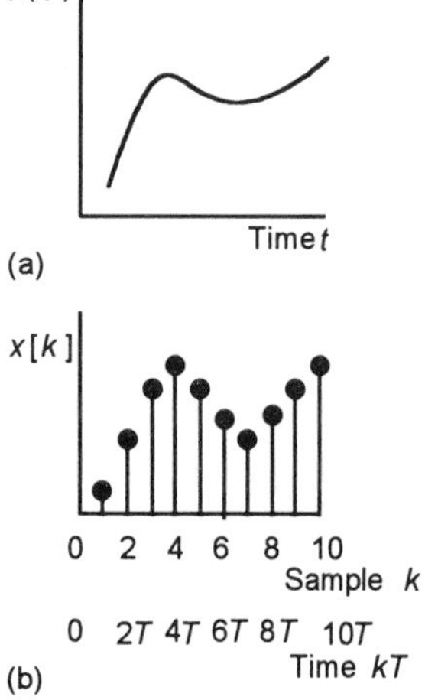

Figure 6.1 *(a) Continuous signal, (b) discrete signal*

Both the continuous-time and the discrete-time signals are termed *time-domain* functions since the independent variable is time. In Parts 2 and 9 other domains will be considered.

6.2.1 Even and odd signals

The terms *even* and *odd* are used to describe the relationship between how a function $y = f(x)$ varies for negative values of x compared with how it varies for positive values, i.e. its symmetry about the y-axis. If a graph has the property that $f(x) = f(-x)$ it is said to be an *even function*, if $-f(x) = f(-x)$ it is said to be an *odd function* (see Section 1.3.2). The graph of an even function is symmetrical about the y-axis while the odd function is asymmetrical about the y-axis. A continuous signal is said to be even if $f(t) = f(-t)$ and odd if $-f(t) = f(-t)$. Figure 6.2 shows examples of such functions. A discrete signal is even if $x[k] = -x[k]$ and odd if $-x[k] = x[-k]$. Figure 6.2 shows examples of such signals. Not all signals can be expressed as odd or even signals. However, signals which are not even or odd can be expressed as the sum of odd and even signals.

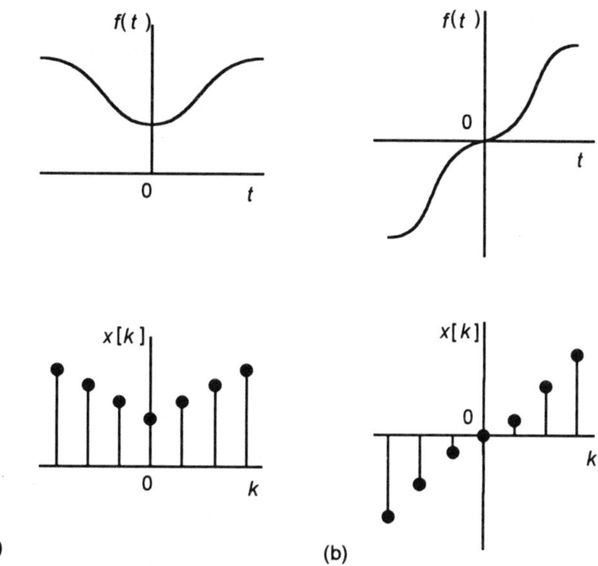

Figure 6.2 *Signals showing (a) even symmetry, (b) odd symmetry*

Revision

1 Which of the signals shown in Figure 6.3 are even and which are odd?

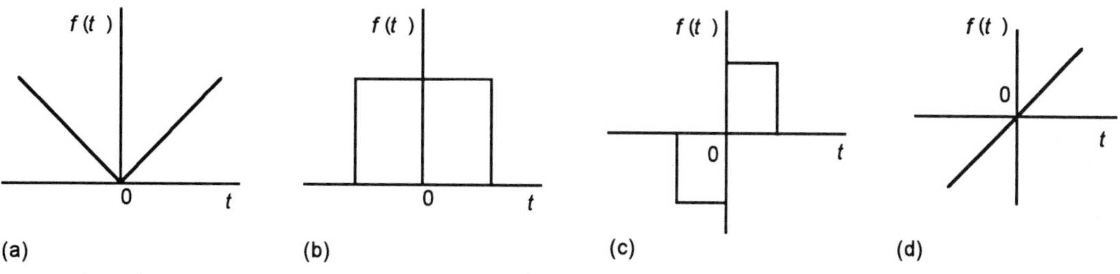

Figure 6.3 *Example*

6.2.2 Time shifting

The term *time shifting* is used when we consider one signal is just the same form as another but with its time shifted. Thus for a continuous signal described by $f(t)$, the shifted signal is $f(t + T)$, where T is the amount of time by which the signal has been shifted. For a discrete signal $x[k]$, the shifted signal is $x[k + K]$, where K is the amount by which the signal has been shifted. Figure 6.4 shows examples of shifted signals. For the continuous signal we have $f(t)$ in (a) and $f(t - T)$ in (b). For the discrete signal we have $x[k]$ in (a) and $x[k - K]$ in (b).

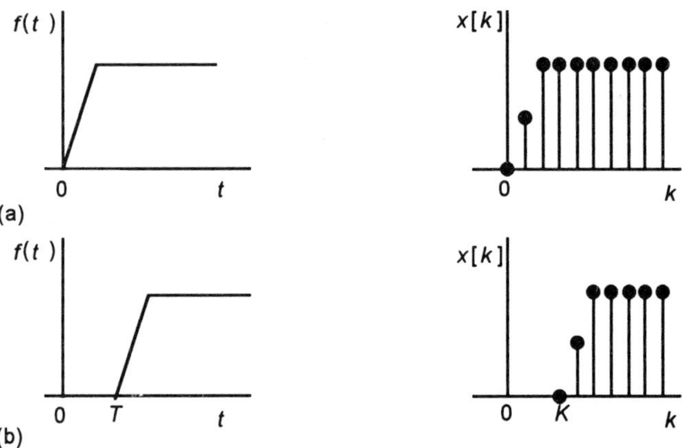

Figure 6.4 *(a) Original signals, (b) shifted signals*

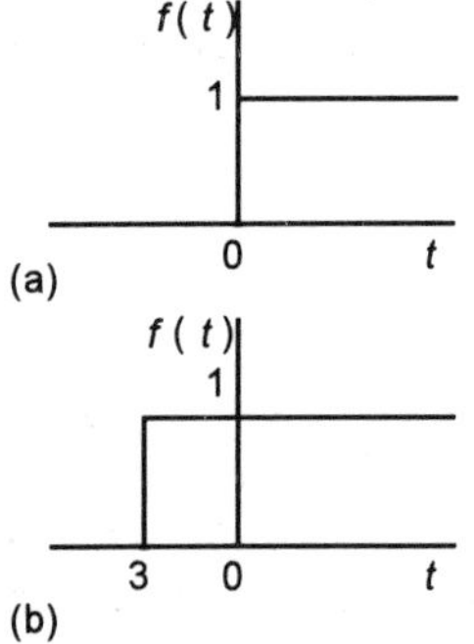

Figure 6.5 *Example*

Example

Sketch graphs of the signal $f(t) = 0$ for $t < 0$, $f(t) = 1$ for $t \geq 0$ and when it is time shifted by +3.

The signals are shown in Figure 6.5. For $f(t + 3) = 0$ we have $f(t) = 0$ for $t < -3$, $f(t) = 1$ for $t \geq -3$.

Revision

2 Sketch graphs of the signal $f(t) = 0$ for $t < 0$, $f(t) = 1$ for $t \geq 0$, when it is (a) delayed by $t = 2$, (b) starts earlier by $t = 2$.

3 Sketch a graph of $f(t - 2) = 1$ for $t < 2$ and $f(t - 2) = 0$ for $t \geq 2$.

6.2.3 Periodic signals

A periodic function is one which involves a relationship which is repeated at regular intervals and thus for which $f(t) = f(t + T) = f(t + 2T) = f(t + 3T)$,

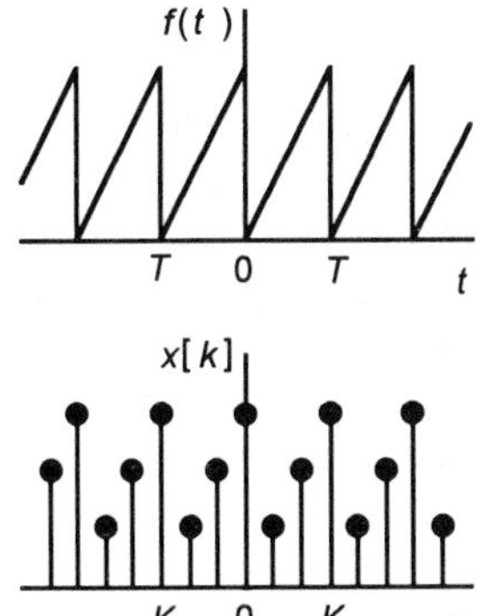

Figure 6.6 *Periodic signals*

etc., where T is the periodic time (see Section 1.3.1). Thus a continuous periodic signal is described by $f(t) = f(t + nT)$ and a discrete signal by $x[k] = x[k + nK]$, where K is the period of the waveform and n an integer. The signals are repeatedly time shifted. A periodic signal thus repeats itself indefinitely in the future and has repeated itself indefinitely in the past. Figure 6.6 shows an example of a continuous and a discrete periodic signal.

Example

Sketch the waveform of $f(t) = f(t + nT)$ if for a single period T of 2 we have the function describing the signal shown in Figure 6.7.

The signal repeats itself every $t = 2$ and thus the answer is as shown in Figure 6.8.

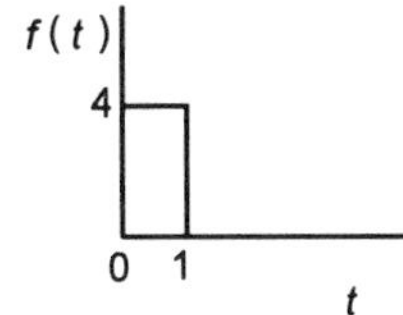

Figure 6.7

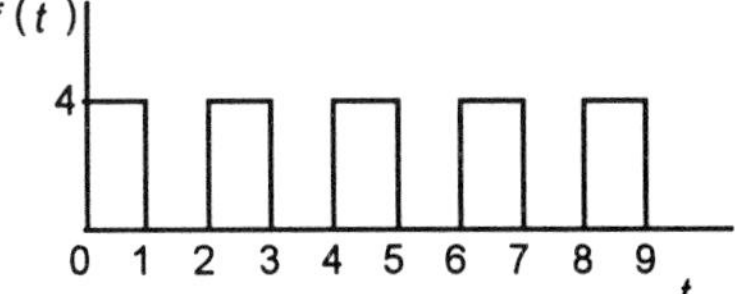

Figure 6.8

6.3 Basic signals

There are a number of basic signal waveforms that commonly occur and that are used to build up more complex waveforms. Thus many waveforms can be built up from the sinusoidal signal, Part 8 illustrating this with the Fourier series. The unit step and the impulse are other forms of basic signal waveforms.

6.3.1 Sinusoidal signal

A sinusoidal continuous signal can be described by $f(t) = A \sin(\omega t + \phi)$ or $f(t) = A \cos(\omega t + \phi)$, where A is the amplitude of the signal and ω, often referred to as the *angular frequency*, is equal to $2\pi f$, with f being the frequency. ϕ is the amount by which the sinusoidal wave is time shifted and is known as the *phase angle* (Figure 6.9).

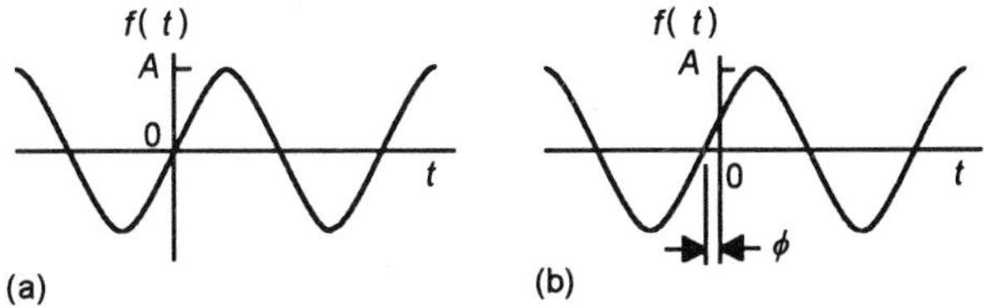

Figure 6.9 *(a)* $f(t) = A \sin \omega t$, *(b)* $f(t) = A \sin(\omega t + \phi)$

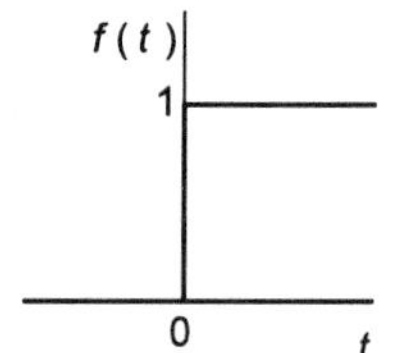

Figure 6.10 *Unit step*

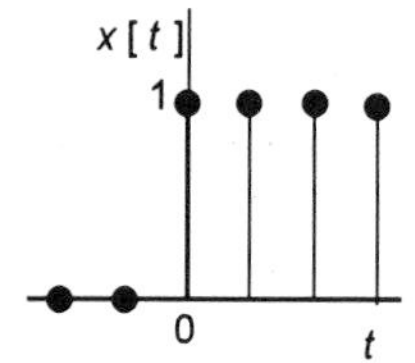

Figure 6.11 *Unit step*

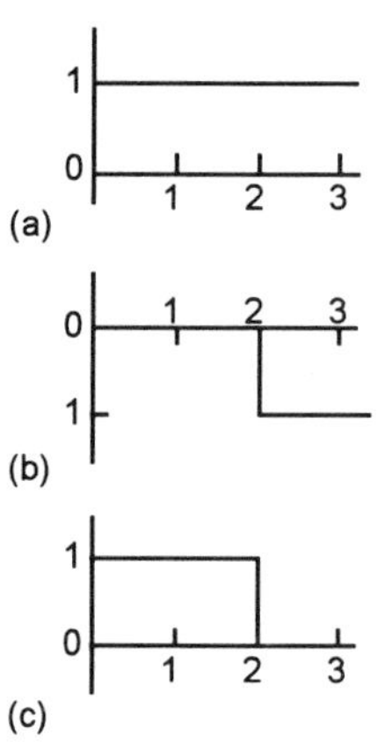

Figure 6.13 *(a) u(t), (b) −u(t − 2), (c) u(t) − u(t − 2)*

6.3.2 Step signal

Figure 6.10 shows a graph of the unit step signal. It is described by $f(t) = 0$ for all values of t less than 0 and $f(t) = 1$ for all values of t greater than 0. The step signal is one where there is an abrupt change in some quantity from 0 to a steady value 1 at the time $t = 0$. The unit step signal cannot be described by $f(t) = 1$ since this would imply a function that has the constant value of 1 for all values of t, both positive and negative. The convention is adopted of describing the unit step function by the symbol $u(t)$, thus the signal in Figure 6.10 is described by the function $u(t)$ and has the value 0 for $t < 0$ and the value 1 for $t \geq 0$. A discrete unit step signal is likewise described by $u[k]$ having the value 0 for $k < 0$ and 1 for $k \geq 0$ (Figure 6.11).

If the unit step signal is time shifted by T then we have the function $u(t - t_0)$ having the value 0 for $t < t_0$ and the value 1 for $t \geq t_0$. Figure 6.12 shows some time shifted signals.

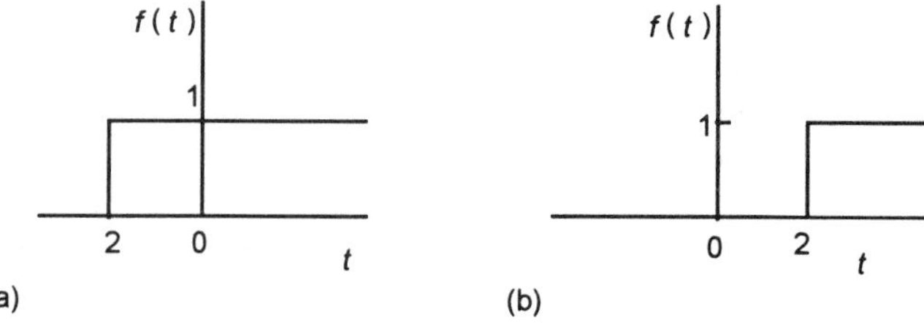

Figure 6.12 *(a) u(t + 2), (b) u(t − 2)*

The function $u(t) - u(t - 2)$ can be considered to involve the addition of two functions, namely $u(t)$ and $-u(t - 2)$. Figure 6.13 shows the result of such an addition. The function describes a unit step that starts at $t = 0$ and continues until it is cancelled at $t = 2$, i.e. it is a rectangular pulse of duration 2 time units.

Step functions can be used to define what might be termed a *window function*, i.e. a period of time in which some other function can occur. Thus if we have $g(t)\ u(t - 2)$, then we have the function $g(t)$ multiplied by $u(t - 2)$. But $u(t - 2)$ has the value 0 prior to $t = 2$ and the value 1 thereafter. Thus we have the function $g(t)$ only able to exist for t greater than 2 (Figure 6.14).

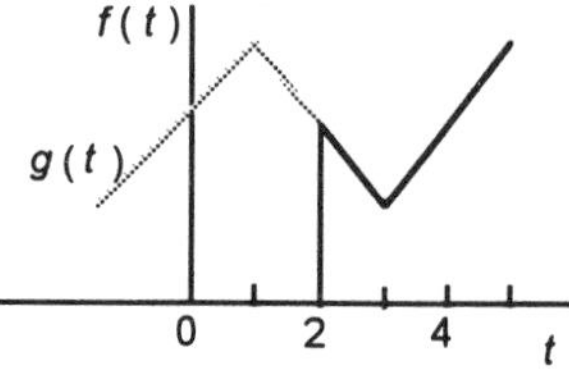

Figure 6.14 *g(t) u(t − 2)*

Revision

4 Sketch the waveforms: (a) $u(t-3)$, (b) $2u(t)$, (c) $2t\,u(t-2)$

6.3.3 Impulse signal

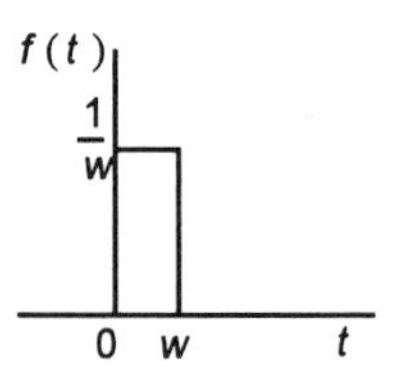

Figure 6.15 *Rectangular pulse*

Consider a rectangular pulse of size $1/w$ that occurs at time $t = 0$ and which has a pulse width w. The area of the pulse is thus 1. The pulse waveform is described by $f(t) = 1/w$ for $0 \le t < w$ and $f(t) = 0$ for $t > w$ and $t < 0$, Figure 6.15 showing the graph. If we maintain this constant pulse area of 1 and decrease the width of the pulse, then the height of the pulse increases. In the limit as w tends to infinitely small width then we end up with a vertical line at $t = 0$ with the top of the line going off to infinity. Such a line is used to represent an impulse. The impulse is said to be a unit impulse because the area enclosed by it is 1. The function is represented by $\delta(t)$, being termed the *unit impulse function* or the *Dirac-delta function*. The function $\delta(t - t_0)$ describes a unit impulse which has been time shifted and delayed by a time t_0. Figure 6.16 shows examples of impulse functions.

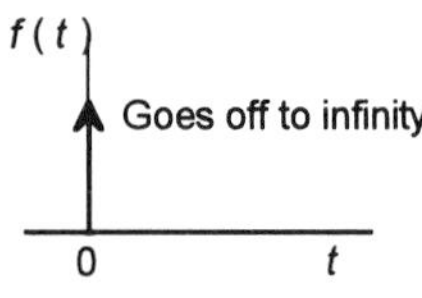

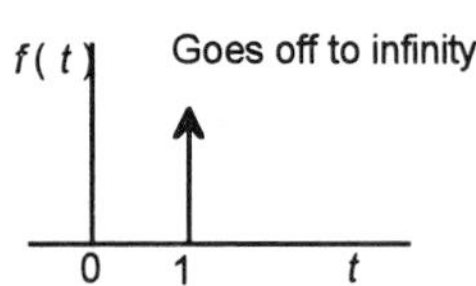

Figure 6.16 *(a) $\delta(t)$, (b) $\delta(t-1)$*

The discrete version of the unit impulse $\delta[k]$, or *unit* sample, is defined as $\delta[k] = 1$ for $k = 0$ and $\delta[k] = 0$ for $k \neq 0$. If we have $\delta[k-1]$ then the unit sample is delayed by 1, $\delta[k-2]$ a unit sample delayed by 2, etc. (Figure 6.17). A sequence of unit samples is thus:

$$\delta[k] + \delta[k-1] + \delta[k-2] + \delta[k-3] + \ldots$$

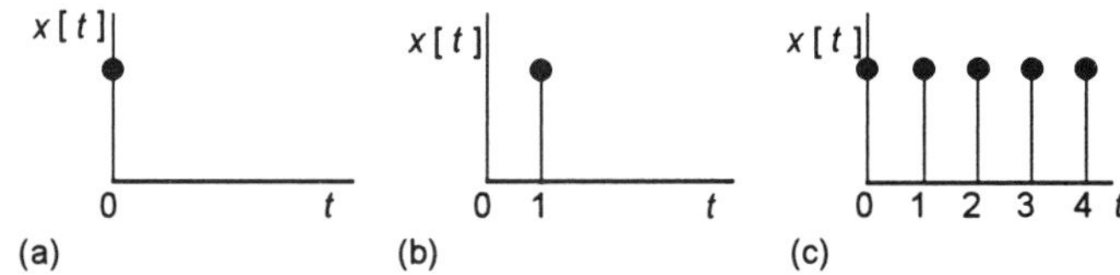

Figure 6.17 *Unit samples: (a) $\delta[k]$, (b) $\delta[k-1]$, (c) $\delta[k] + \delta[k-1] + \delta[k-2] + \delta[k-3] + \ldots$*

Revision

5 Write the mathematical expression for the functions shown in Figure 6.18.

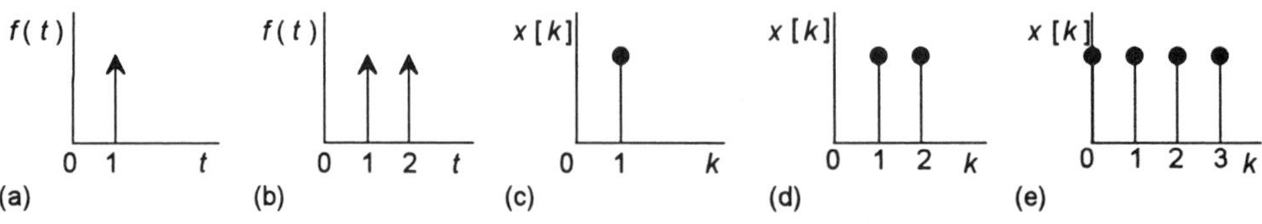

Figure 6.18 *Revision problem 5*

Problems

1 Which of the waveforms shown in Figure 6.19 are even and which are odd?

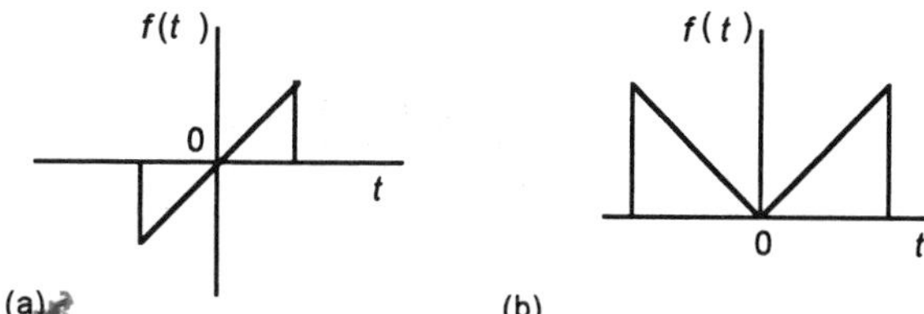

Figure 6.19 *Problem 1*

2 Sketch the waveforms:

(a) $f(t) = 2$ for $t < 0$ and $f(t) = 1$ for $t > 0$, (b) $u(t + 1)$,

(c) $u(t - 1) - u(t - 3)$, (d) $t\,u(t - 1)$, (e) $2(t - 1)\,u(t - 1)$,

(f) $\delta(t + 1)$, (g) $\delta(t) + \delta(t - 2)$

3 Write mathematical expressions for the functions shown in Figure 6.20.

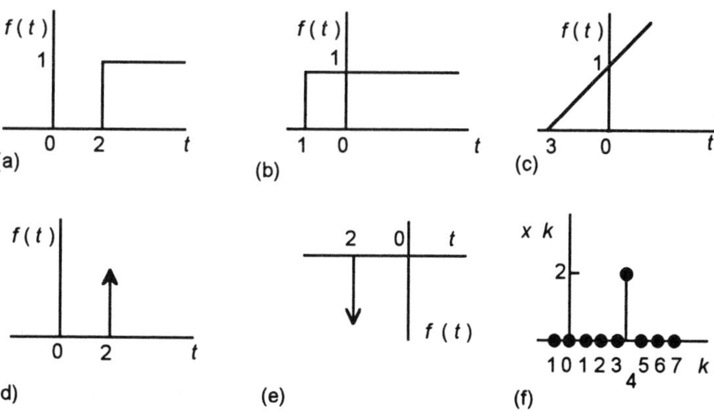

Figure 6.20 *Problem 3*

Part 2
Complex numbers

The aims of this part are to enable the reader to:

- Use complex numbers in polar form and Cartesian form and convert between the two.
- Draw Argand diagrams.
- Carry out addition, subtraction, multiplication and division of complex numbers.
- Use de Moivre's theorem for the powers of complex numbers, determining roots of numbers, cosines and sines of $n\theta$ and powers of cosines and sines.
- Use complex numbers in exponential form.
- Use Euler's formula, complex trigonometric functions, recognising their relationships with hyperbolic functions, and the complex logarithm function.
- Use complex numbers to represent phasors.

For this part, familiarity is assumed with basic algegra, trigonometric functions, hyperbolic functions, exponential and logarithmic functions. In Section 7.4 concerning applications of de Moivre's theorem, the binomial theorem is assumed. Chapter 8 assumes a knowledge of power series (as in Chapter 5).

Complex numbers are a powerful mathematical tool and are widely used in the description and analysis of many engineering systems, in particular with the alternating current circuits where they are used to describe phasors, these representing sinusoidal currents and voltages, and in doing so they considerably simplify the analysis of such circuits.

7 Complex numbers

7.1 Introduction

This chapter introduces complex numbers, the Argand diagram, arithmetic processes involving complex numbers and applications of de Moivre's theorem. Complex numbers are a powerful mathematical tool and figure in the description of many engineering and scientific systems. In particular they are used in the analysis of alternating current circuits where they are used to describe phasors, these representing sinusoidal currents and voltages (see Chapter 9 Phasors).

7.2 Real, imaginary and complex numbers

Real numbers can be represented as points along a line called the *number line* (Figure 7.1), each number representing a distance along the line from some origin. Each distance along the line is some multiple of a basic unit of length designated as 1.

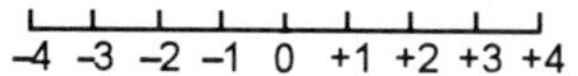

Figure 7.1 *The real number line*

The real number system has numbers which are multiples of 1.

7.2.1 Imaginary numbers

If we square the real number +2 we obtain +4, if we square the real number −2 we obtain +4. Thus the square root of +4 is ±2. But what is the square root of −4? To give an answer we need another form of number. If we invent a number $j = \sqrt{-1}$ (mathematicians often use i rather than j but engineers and scientists generally use j to avoid confusion with i used for current in electrical circuits), then we can write $\sqrt{-4} = \sqrt{-1} \times \sqrt{4} = \pm j2$. Thus the solution of the equation $x + 4 = 0$ is $x = \pm j2$.

A number system which has numbers which are multiples of j is termed imaginary.

$j = \sqrt{-1}$ and thus $j^2 = -1$. Since j^3 can be written as $j^2 \times j$ then $j^3 = -j$. Since j^4 can be written as $j^2 \times j^2$ then $j^4 = +1$.

$$j = \sqrt{-1}, \quad j^2 = -1, \quad j^3 = -j, \quad j^4 = +1 \qquad [1]$$

Revision

1 Simplify (a) j^5, (b) j^6.

2 Solve the equations: (a) $x^2 + 25 = 0$, (b) $x^2 + 81 = 0$.

7.2.2 Complex numbers

The solution of a quadratic equation of the form $ax^2 + bx + c = 0$ is given by (see Section 2.5.5):

$$x = \frac{-b \pm \sqrt{b^2 - 4ac}}{2a} \qquad [2]$$

Thus if we want to solve the quadratic equation $x^2 - 4x + 13 = 0$ then:

$$x = \frac{4 \pm \sqrt{16-52}}{2} = 2 \pm \sqrt{-9}$$

We can represent $\sqrt{-9}$ as $\sqrt{-1} \times \sqrt{+9} = \mathrm{j}3$. Thus the solution can be written as $2 \pm \mathrm{j}3$, a combination of a real and either plus or minus an imaginary number. Such a pair of roots is known as a *conjugate pair* (see Section 7.3.3).

> *The term complex number is used for a combination by addition or subtraction of a real number and a purely imaginary number.*

In general a complex number z can be written as $z = a + \mathrm{j}b$, where a is the real part of the complex number and b the imaginary part.

Example

Solve the equation $x^2 - 4x + 5 = 0$.

Using equation [2]:

$$x = \frac{4 \pm \sqrt{16-20}}{2} = 2 \pm \sqrt{-1} = 2 \pm \mathrm{j}$$

Revision

3 Solve the following equations:

(a) $x^2 - 4x + 13 = 0$, (b) $x^2 + 6x + 13 = 0$, (c) $x^2 - 2x + 5$

7.2.3 The Argand diagram

The effect of multiplying a real number by (–1) is to move the point on the number line from one side of the origin to the other. Figure 7.2 illustrates this for (+2) being multiplied by (–1). We can think of the positive number line radiating out from the origin being rotated through 180° to its new position after being multiplied by (–1). But $(-1) = \mathrm{j}^2$. Thus multiplication by j^2 is equivalent to a 180° rotation. Multiplication by j^4 is a multiplication by (+1) and so is equivalent to a rotation through 360°. On this basis it seems

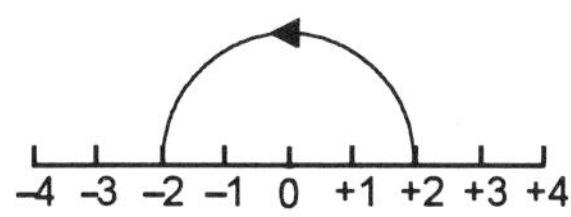

Figure 7.2 *(+2)* × *(–1)*

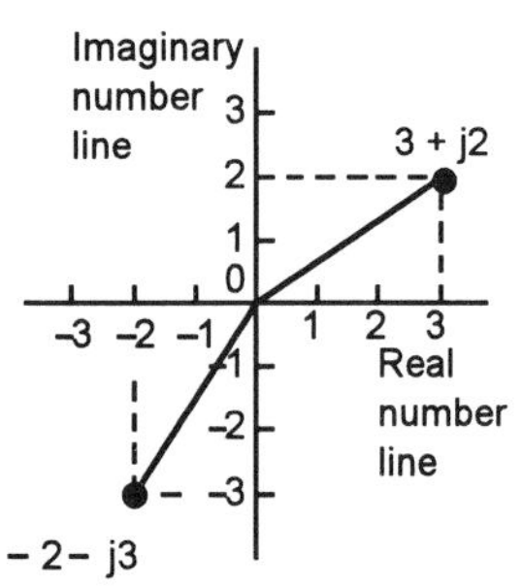

Figure 7.3 *Argand diagram*

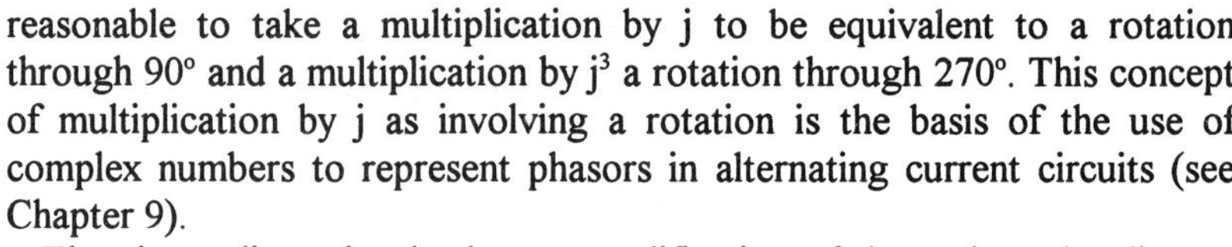

reasonable to take a multiplication by j to be equivalent to a rotation through 90° and a multiplication by j^3 a rotation through 270°. This concept of multiplication by j as involving a rotation is the basis of the use of complex numbers to represent phasors in alternating current circuits (see Chapter 9).

The above discussion leads to a modification of the real number line so that we can represent complex numbers. The modification is to consider a plane of points rather than just a line, with an imaginary number line being drawn at right angles to the real number line (Figure 7.3). Then rotation of the real number line by 90° gives the imaginary number line. The resulting diagram is called the *Argand diagram*. Figure 7.3 shows how we represent the complex numbers 3 + j2 and −2 − j3 on such a diagram. The line joining the number to the origin is taken as the graphical representation of the complex number.

Revision

4 Draw the following complex numbers on an Argand diagram:

(a) 2 − j1, (b) −1 + j2

7.2.4 Modulus and argument

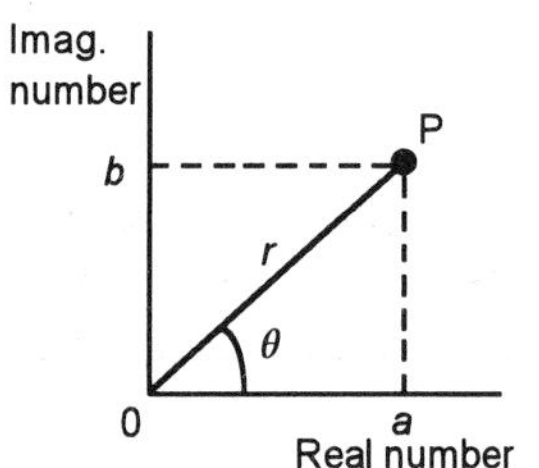

Figure 7.4 *Modulus and argument*

If the complex number $z = a + jb$ is represented on an Argand diagram by the line OP, as in Figure 7.4, then the length r of the line OP is called the *modulus* of the complex number and its inclination θ to the real number axis is termed the *argument* of the complex number. The length of the line is denoted by $|z|$ or modulus z and the argument by θ or arg z.

Using Pythagoras' theorem:

$$|z| = \sqrt{a^2 + b^2} \qquad [3]$$

and, since $\tan\theta = b/a$:

$$\arg z = \tan^{-1}\left(\frac{b}{a}\right) \qquad [4]$$

Since $a = r\cos\theta$ and $b = r\sin\theta$, we can write a complex number z as:

$$z = a + jb = r\cos\theta + jr\sin\theta = r(\cos\theta + j\sin\theta) \qquad [5]$$

Thus we can specify a complex number by either stating its location on an Argand diagram in terms of its *Cartesian co-ordinates* a and b or by specifying the modulus, $|z| = r$, and the argument θ. These are termed its *polar co-ordinates*. The specification in polar co-ordinates can be written as:

$$z = |z| \angle \arg z \text{ or } z = r\angle\theta \qquad [6]$$

Example

Determine the modulus and argument of the complex number 2 + j2.

Using equation [3]:

$$|z| = \sqrt{a^2 + b^2} = \sqrt{2^2 + 2^2} = 2.8$$

Using equation [4]:

$$\arg z = \tan^{-1}\left(\frac{b}{a}\right) = \tan^{-1}\left(\frac{2}{2}\right) = 45°$$

In polar form the complex number could be written as 2.8 ∠45°.

Example

Write the complex number –2 + j2 in polar form.

Using equation [3]:

$$|z| = \sqrt{a^2 + b^2} = \sqrt{(-2)^2 + 2^2} = 2.8$$

Using equation [4]:

$$\arg z = \tan^{-1}\left(\frac{b}{a}\right) = \tan^{-1}\left(\frac{2}{-2}\right)$$

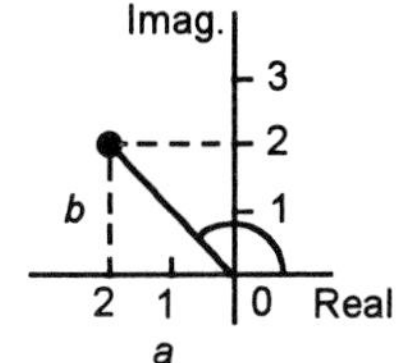

Figure 7.5 *Example*

If we sketch an Argand diagram (Figure 7.5) for this complex number we can see that the number is in the second quadrant. The argument is thus –45° + 180° = 135°. In polar form the complex number could be written as 2.8 ∠135°.

Example

Write the complex number 10 ∠60° in Cartesian form.

Using equation [5]:

$$z = r(\cos\theta + \mathrm{j}\sin\theta) = 10(\cos 60° + \mathrm{j}\sin 60°) = 5 + \mathrm{j}8.7$$

Revision

5 Express the following complex numbers in polar form: (a) 3 – j4, (b) 3, (c) 3 + j4, (d) –3 – j4, (e) –j4.

6 Express the following complex numbers in polar form: (a) 1 – j, (b) 5, (c) –2 + j2, (d) 2j, (e) –3 – j4.

7.3 Manipulation of complex numbers

Addition, subtraction, multiplication and division can be carried out on complex numbers in either the Cartesian form or the polar form. Addition and subtraction is easiest when they are in the Cartesian form and multiplication and division easiest when they are in the polar form.

For two complex numbers to be equal, their real parts must be equal and their imaginary parts equal. On an Argand diagram the two numbers then describe the same line. Thus 2 + j3 is *not* equal to 3 + j2 as Figure 7.6 shows.

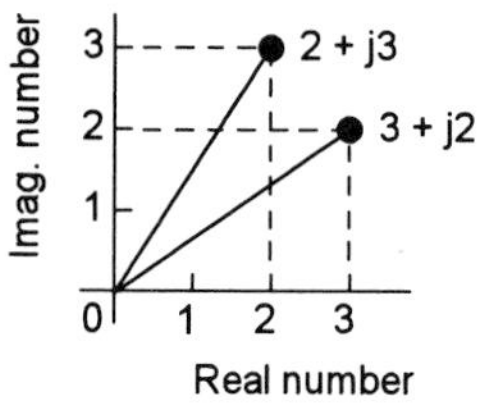

Figure 7.6 *2 + j3 and 3 + j2*

7.3.1 Addition and subtraction

To add complex numbers we add the real parts and add the imaginary parts:

$$(a + \mathrm{j}b) + (c + \mathrm{j}d) = (a + c) + \mathrm{j}(b + d) \qquad [7]$$

On an Argand diagram, this method of adding two complex numbers is the same as the vector addition of two vectors using the parallelogram of vectors, the line representing each complex number being treated as a vector (Figure 7.7).

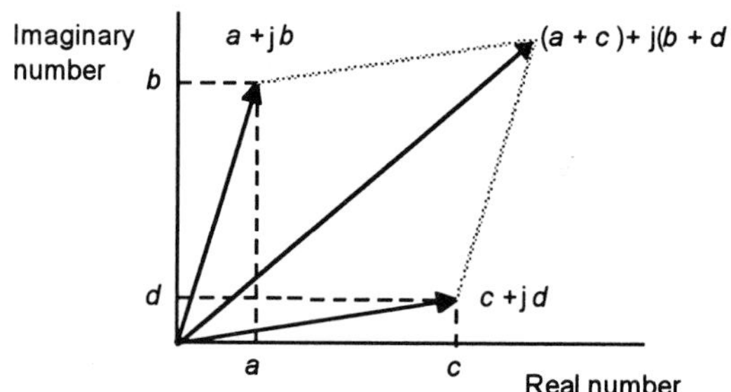

Figure 7.7 *Addition of complex numbers*

To subtract complex numbers we subtract the real parts and subtract the imaginary parts:

$$(a + \mathrm{j}b) - (c + \mathrm{j}d) = (a - c) + \mathrm{j}(b - d) \qquad [8]$$

On an Argand diagram, this method of subtracting two complex numbers is the same as the vector subtraction of two vectors. To subtract a vector quantity you reverse its direction and then add it using the parallelogram of vectors (Figure 7.8).

Example

If $z_1 = 4 + \mathrm{j}2$ and $z_2 = 3 + \mathrm{j}5$, determine (a) $z_1 + z_2$, (b) $z_1 - z_2$.

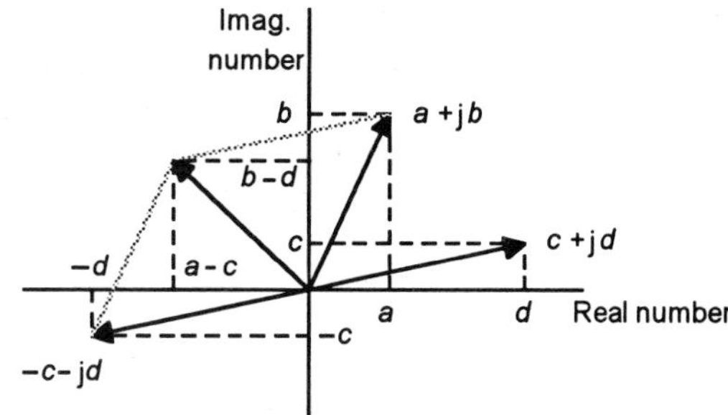

Figure 7.8 *Subtraction of complex numbers*

With $z_1 = 4 + j2$ and $z_2 = 3 + j5$, then:

$$z_1 + z_2 = (4 + 3) + j(2 + 5) = 7 + j7$$

$$z_1 - z_2 = (4 - 3) + j(2 - 5) = 1 - j3$$

Revision

7 If $z_1 = 2 + j5$, $z_2 = 1 + j3$ and $z_3 = 4 - j2$, determine (a) $z_1 + z_2$, (b) $z_2 + z_3$, (c) $z_1 + z_2 + z_3$, (d) $z_1 - z_2$, (e) $z_2 - z_3$.

7.3.2 Multiplication

Consider the multiplication of the two complex numbers in Cartesian form, $z_1 = a + jb$ and $z_2 = c + jd$. The product z is given by:

$$z = (a + jb)(c + jd) = ac + j(ad + bc) + j^2bd = ac + j(ad + bc) - bd \quad [9]$$

Now consider the multiplication of the two complex numbers in polar form, $z_1 = |z_1|\angle\theta_1$ and $z_2 = |z_2|\angle\theta_2$. Using equation [5] we can write:

$$z_1 = |z_1|(\cos\theta_1 + j\sin\theta_1) \text{ and } z_2 = |z_2|(\cos\theta_2 + j\sin\theta_2)$$

Thus the product z is given by:

$$z = |z_1|(\cos\theta_1 + j\sin\theta_1) \times |z_2|(\cos\theta_2 + j\sin\theta_2)$$

$$= |z_1z_2|\,[\cos\theta_1\cos\theta_2 + j(\sin\theta_1\cos\theta_2 + \cos\theta_1\sin\theta_2) + j^2\sin\theta_1\sin\theta_2]$$

$$= |z_1z_2|\,[(\cos\theta_1\cos\theta_2 - \sin\theta_1\sin\theta_2) + j(\sin\theta_1\cos\theta_2 + \cos\theta_1\sin\theta_2)]$$

Using equations [37] and [35] from Chapter 3:

$$z = |z_1 z_2|[\cos(\theta_1 + \theta_2) + \mathrm{j}\sin(\theta_1 + \theta_2)] \qquad [10]$$

Hence we can write for the complex numbers in polar form

$$z = |z_1 z_2|\angle(\theta_1 + \theta_2) \qquad [11]$$

The magnitude of the product is the product of the magnitudes of the two numbers and its argument is the sum of the arguments of the two numbers.

Example

Multiply the two complex numbers 2 – j3 and 4 + j1.

Operating in the way indicated by equation [9]:

$$(2 - \mathrm{j}3)(4 + \mathrm{j}1) = 8 + \mathrm{j}2 - \mathrm{j}12 - \mathrm{j}^2 3 = 8 + \mathrm{j}2 - \mathrm{j}12 + 3 = 11 - \mathrm{j}10$$

Example

Multiply the two complex numbers $3\angle 40°$ and $2\angle 70°$.

Using equation [11]:

$$3\angle 40° \times 2\angle 70° = (3 \times 2)\angle(40° + 70°) = 6\angle 110°$$

Revision

8 Determine the products of the following complex numbers:

(a) 3 – j5 and 4 + j2, (b) 1 – j2 and 3 + j4, (c) 2 – j1 and 3 + j4,

(d) $2\angle 10°$ and $5\angle 30°$, (e) $1\angle 50°$ and $2\angle(-10)°$, (f) $3\angle 110°$ and $1\angle 10°$

7.3.3 Complex conjugate

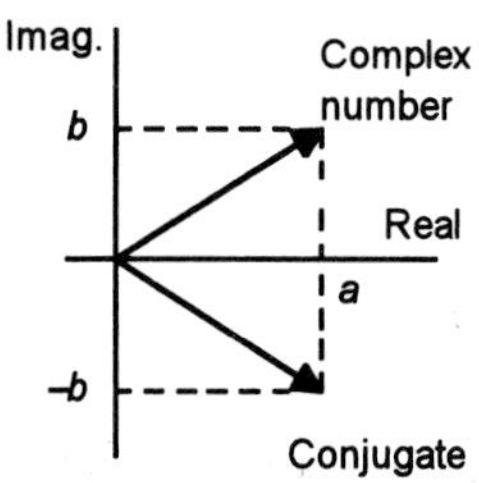

Figure 7.9 *A complex number and its conjugate*

If $z = a + \mathrm{j}b$ then the term *complex conjugate* is used for the complex number given by $z^* = a - \mathrm{j}b$. The imaginary part of the complex number changes sign to give the conjugate, conjugates being denoted as z^*. Figure 7.9 shows an Argand diagram with a complex number and its conjugate. The complex conjugate is the mirror image of the original complex number.

Consider now the product of a complex number and its conjugate:

$$zz^* = (a + \mathrm{j}b)(a - \mathrm{j}b) = a^2 - \mathrm{j}^2 b = a^2 + b^2 \qquad [12]$$

The product of a complex number and its conjugate is a real number.

Example

What is the conjugate of the complex number 2 + j4?

The complex conjugate is 2 – j4.

Revision

9 Determine the complex conjugates of the following complex numbers:

(a) 3 – j5, (b) –2 + j4, (c) –4 – j6

7.3.4 Division

Consider the division of $z_1 = a + jb$ by $z_2 = c + jd$, i.e.

$$z = \frac{z_1}{z_2} = \frac{a + jb}{c + jd}$$

To divide one complex number by another we have to convert the denominator into a real number. This can be done by multiplying it by its conjugate. Thus:

$$z = \frac{a + jb}{c + jd} \times \frac{c - jd}{c - jd} = \frac{(a + jb)(c - jd)}{c^2 + d^2} \qquad [13]$$

Now consider the division of the two complex numbers when in polar form, $z_1 = |z_1|\angle\theta_1$ and $z_2 = |z_2|\angle\theta_2$:

$$z = \frac{|z_1|(\cos\theta_1 + \text{j}\ \sin\theta_1)}{|z_2|(\cos\theta_2 + \text{j}\ \sin\theta_2)}$$

Making the denominator into a real number by multiplying it by its conjugate:

$$z = \frac{|z_1|(\cos\theta_1 + \text{j}\ \sin\theta_1)}{|z_2|(\cos\theta_2 + \text{j}\ \sin\theta_2)} \times \frac{|z_2|(\cos\theta_2 - \text{j}\ \sin\theta_2)}{|z_2|(\cos\theta_2 - \text{j}\ \sin\theta_2)}$$

$$= \frac{|z_1|}{|z_2|}\left[\frac{(\cos\theta_1 + \text{j}\ \sin\theta_1)(\cos\theta_2 - \text{j}\ \sin\theta_2)}{\cos^2\theta_2 + \sin^2\theta_2}\right]$$

But $\cos^2\theta_2 + \sin^2\theta_2 = 1$ (chapter 3, equation [32]) and so:

$$z = \frac{|z_1|}{|z_2|}[(\cos\theta_1\cos\theta_2 + \sin\theta_1\sin\theta_2) + \text{j}(\sin\theta_1\cos\theta_2 - \cos\theta_1\sin\theta_2)]$$

Using equations [38] and [36] from Chapter 3:

$$z = \frac{|z_1|}{|z_2|}[\cos(\theta_1 - \theta_2) + \mathrm{j}\sin(\theta_1 - \theta_2)] \qquad [14]$$

We can express this as:

$$z = \frac{|z_1|}{|z_2|}\angle(\theta_1 - \theta_2) \qquad [15]$$

Thus to divide two complex numbers in polar form, we divide their magnitudes and subtract their arguments.

Example

Divide 1 + j2 by 1 + j1.

Using the method indicated by equation [13]:

$$\frac{1+\mathrm{j}2}{1+\mathrm{j}1} = \frac{1+\mathrm{j}2}{1+\mathrm{j}1} \times \frac{1-\mathrm{j}1}{1-\mathrm{j}1} = \frac{1+\mathrm{j}1-\mathrm{j}^2 2}{1-\mathrm{j}^2} = \frac{3+\mathrm{j}1}{2} = 1.5 + \mathrm{j}0.5$$

Example

Divide $4\angle 40°$ by $2\angle 30°$.

$$\frac{4\angle 40°}{2\angle 30°} = \frac{4}{2}\angle(40° - 30°) = 2\angle 10°$$

Revision

10 Determine the values of the following:

$$\text{(a) } \frac{2+\mathrm{j}3}{3-\mathrm{j}4}, \text{ (b) } \frac{3+\mathrm{j}5}{1+\mathrm{j}1}, \text{ (c) } \frac{1}{1-\mathrm{j}1}, \text{ (d) } \frac{2+\mathrm{j}2}{5-\mathrm{j}2},$$

$$\text{(e) } \frac{4\angle 80°}{2\angle 30°}, \text{ (f) } \frac{5\angle 20°}{2\angle 70°}, \text{ (g) } \frac{1\angle 80°}{10\angle 70°}, \text{ (h) } \frac{2\angle 120°}{4\angle(-20°)}$$

7.4 Powers

The product of two complex numbers $z_1 = |z_1|\angle\theta_1$ and $z_2 = |z_2|\angle\theta_2$ is given by equation [11] as:

$$z = |z_1 z_2|\angle(\theta_1 + \theta_2)$$

This if we require z^2, i.e. we have $z_1 = z_2 = |z|\angle\theta$,then $z^2 = |z^2|\angle 2\theta$. If we require z^3 then $z^3 = |z^3|\angle 3\theta$. Hence if we require z^n:

$$z^n = |z^n|\angle n\theta \qquad [16]$$

To raise a complex number to a particular power, we raise the modulus to that power and multiply the argument by that power.

We can write equation [16] (see equation [5]) as:

$$z^n = |z^n|(\cos n\theta + \text{j} \sin n\theta) \qquad [17]$$

This equation is known as *de Moivre's theorem*. n can be either positive or negative, integer or fractional.

When n is fractional the equation is being used to determine all the distinct roots of a number. For example, when $n = 1/2$ we find the square root values and when $n = 1/3$ the cube root values. However, when we take the square root of a number there are two roots. When we take the cube root there are three roots. In order to obtain all these root values we must take account of the fact that the cosine and sine functions are periodic function with a period of 2π radians or 360°. The same value of a sine is given by $\sin \theta$, $\sin(\theta + 360°)$, $\sin(\theta + 2 \times 360°)$, $\sin(\theta + 3 \times 360°)$, and so on.

Example

Determine z^6 if $z = 2\angle\pi/6$.

Using de Moivre's theorem [17]:

$$z^6 = |z^6|(\cos 6\theta + \text{j} \sin 6\theta) = 2^6(\cos \pi + \text{j} \sin \pi) = 64(-1 + \text{j}0) = -64$$

Example

Solve the equation $z^5 = 1 + \text{j}1$.

What is required is the fifth root of the complex number. In polar form $1 + \text{j}1 = \sqrt{1^2 + 1^2}\angle(\tan^{-1} 1/1) = \sqrt{2}\angle 45°$. There are also other arguments which give the same value of $\tan^{-1} 1$, and so we also have $\sqrt{2}\angle(45° + 360°)$, $\sqrt{2}\angle(45° + 2 \times 360°)$, $\sqrt{2}\angle(45° + 3 \times 360°)$, $\sqrt{2}\angle(45° + 4 \times 360°)$, etc. Thus, using de Moivre's theorem [17] for the first of the possible polar forms:

$$z^{1/5} = |\sqrt[1/5]{2}|\left(\cos \tfrac{1}{5}45° + \sin \tfrac{1}{5}45°\right) = 1.07(\cos 9° + \sin 9°)$$

With the second polar form:

$$z^{1/5} = |\sqrt[1/5]{2}|\left(\cos \tfrac{1}{5}405° + \sin \tfrac{1}{5}405°\right) = 1.07(\cos 81° + \sin 81°)$$

With the third polar form:

$$z^{1/5} = |\sqrt[1/5]{2}|\left(\cos \tfrac{1}{5}765° + \sin \tfrac{1}{5}765°\right) = 1.07(\cos 153° + \sin 153°)$$

With the fourth polar form:

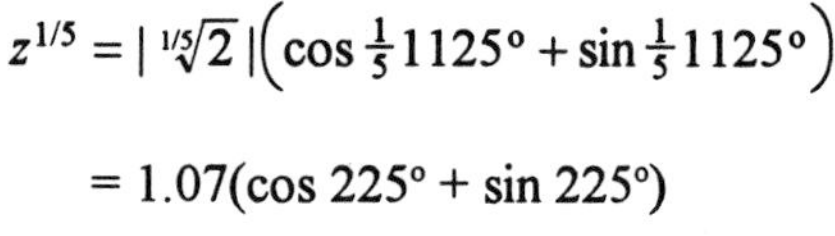

$$z^{1/5} = |\sqrt[1/5]{2}|\left(\cos \tfrac{1}{5}1125° + \sin \tfrac{1}{5}1125°\right)$$

$$= 1.07(\cos 225° + \sin 225°)$$

With the fifth polar form:

$$z^{1/5} = |\sqrt[1/5]{2}|\left(\cos \tfrac{1}{5}1485° + \sin \tfrac{1}{5}1485°\right)$$

$$= 1.7(\cos 297° + \sin 297°)$$

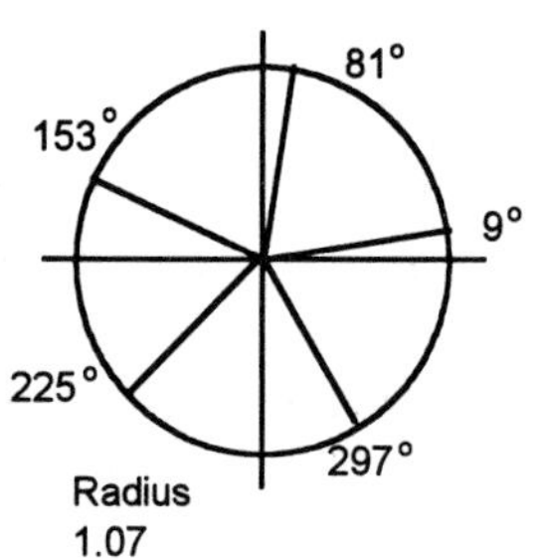

Figure 7.10 *Example*

With further values we just end up with the same values for the roots. Thus the roots are $1.07\angle 9°$, $1.07\angle 81°$, $1.07\angle 153°$, $1.07\angle 225°$, $1.07\angle 297°$. Figure 7.10 shows these roots plotted on an Argand diagram, the roots being the same length and equally spaced at 72° intervals. When converted to Cartesian form, these are 1.06 + j0.17, 0.17 + j1.06, –0.96 + j0.49, –0.76 – j0.76 and 0.49 – j0.96.

Example

Determine the cube roots of 5.

Putting z = 5 + j0, then in polar form we have $5\angle 0°$, $5\angle 360°$, $5\angle 720°$, etc. Thus using de Moivre's theorem [17] we have for $z^{1/3}$:

$$z^{1/3} = |5^{1/3}|(\cos \tfrac{1}{3}0° + \text{j} \sin \tfrac{1}{3}0°)$$

$$z^{1/3} = |5^{1/3}|(\cos \tfrac{1}{3}360° + \text{j} \sin \tfrac{1}{3}360°)$$

$$z^{1/3} = |5^{1/3}|(\cos \tfrac{1}{3}720° + \text{j} \sin \tfrac{1}{3}720°)$$

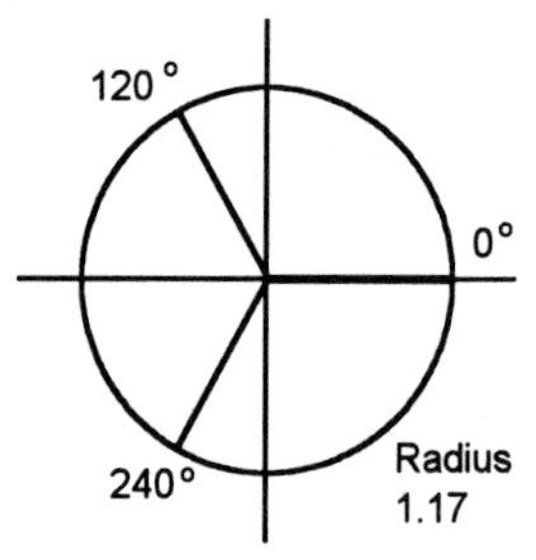

Figure 7.11 *Example*

The solutions are thus $1.17\angle 0°$, $1.17\angle 120°$ and $1.17\angle 240°$. Figure 7.11 shows the roots on an Argand diagram. In Cartesian form the roots are 1.17, –0.59 + j1.01 and –0.59 – j1.01.

Revision

11 Determine the three cube roots of $8\angle 120°$.

12 Determine the three cube roots of j8.

13 Determine the six sixth roots of 64.

14 Determine the four roots of $z^4 = 1\angle 2\pi/3$.

15 Determine $z^{1/3}$ if z = –0.5 + j0.5.

16 Determine (a) $(1 - \text{j}1)^5$, (b) $(-3 + \text{j}4)^3$.

7.4.1 Cosines and sines of $n\theta$

For a complex number $z = |z|(\cos\theta + \mathrm{j}\sin\theta)$ (equation [5]) and by de Moivre's theorem [17] we have $z^n = |z^n|(\cos n\theta + \mathrm{j}\sin n\theta)$. Thus we can write:

$$|z^n|(\cos n\theta + \mathrm{j}\sin n\theta) = [|z|(\cos\theta + \mathrm{j}\sin\theta)]^n$$

and so:

$$\cos n\theta + \mathrm{j}\sin n\theta = (\cos\theta + \mathrm{j}\sin\theta)^n$$

The right-hand side of the equation can be expanded by means of the binomial series (Chapter 5, equation [36]), being of the form $(a + b)^n$:

$$\begin{aligned}\cos n\theta + \mathrm{j}\sin n\theta = {} & \cos^n\theta + n\cos^{n-1}\theta \times \mathrm{j}\sin\theta \\ & + \frac{n(n-1)}{2!}\cos^{n-2}\theta \times \mathrm{j}^2\sin^2\theta \\ & + \frac{n(n-1)(n-2)}{3!}\cos^{n-3}\theta \times \mathrm{j}^3\sin^3\theta \\ & + \frac{n(n-1)(n-2)(n-3)}{4!}\cos^{n-4}\theta \times \mathrm{j}^4\sin^4\theta \\ & + \frac{n(n-1)(n-2)(n-3)(n-4)}{5!}\cos^{n-5}\theta \times \mathrm{j}^5\sin^5\theta \\ & + \dots\end{aligned}$$

$\mathrm{j}^2 = -1$, $\mathrm{j}^3 = -\mathrm{j}$, $\mathrm{j}^4 = 1$, $\mathrm{j}^5 = \mathrm{j}$, etc. Thus equating real terms gives:

$$\begin{aligned}\cos n\theta = {} & \cos^n\theta - \frac{n(n-1)}{2!}\cos^{n-2}\theta\sin^2\theta \\ & + \frac{n(n-1)(n-2)(n-3)}{4!}\cos^{n-4}\theta\sin^4\theta + \dots\end{aligned} \qquad [18]$$

Equating the imaginary terms:

$$\begin{aligned}\sin n\theta = {} & n\cos^{n-1}\theta\sin\theta - \frac{n(n-1)(n-2)}{3!}\cos^{n-3}\theta\sin^3\theta \\ & + \frac{n(n-1)(n-2)(n-3)(n-4)}{5!}\cos^{n-5}\theta\sin^5\theta + \dots\end{aligned} \qquad [19]$$

Example

Express $\cos 3\theta$ in terms of powers of cosines and sines of θ.

Using equation [18]:

$$\cos 3\theta = \cos^3 \theta - \frac{3(3-1)}{2!} \cos\theta \sin^2\theta$$

and so:

$$\cos 3\theta = \cos^3 \theta - 3 \cos \theta \sin^2 \theta$$

Revision

17 Express the following in terms of powers of cosines and sines:

(a) $\sin 3\theta$, (b) $\cos 4\theta$, (c) $\sin 5\theta$

7.4.2 Powers of cosines and sines

For a complex number $z = |z|\angle\theta$ we have $z = |z|(\cos \theta + \mathrm{j} \sin \theta)$ (equation [5]). If we have a number with magnitude 1 then we can write:

$$z = \cos \theta + \mathrm{j} \sin \theta$$

If we have $1/z$ then we have $1\angle 0/|z|\angle\theta = |1/z|\angle(-\theta)$. Thus for a number with magnitude 1 we can write:

$$\frac{1}{z} = \cos(-\theta) + \mathrm{j}\sin(-\theta) = \cos\theta - \mathrm{j}\sin\theta$$

$\cos(-\theta) = \cos \theta$ and $\sin(-\theta) = -\sin \theta$. Thus, adding the two equations:

$$z + \frac{1}{z} = 2\cos\theta \qquad [20]$$

and, subtracting:

$$z - \frac{1}{z} = \mathrm{j}2\sin\theta \qquad [21]$$

We can write equations [20] and [21] as:

$$\left(z + \frac{1}{z}\right)^n = 2^n \cos^n\theta \qquad [22]$$

$$\left(z - \frac{1}{z}\right)^n = \mathrm{j}^n 2^n \sin^n\theta \qquad [23]$$

We can derive equations for multiple angles by using de Moivre's theorem [17], $z^n = |z^n|(\cos n\theta + \mathrm{j} \sin n\theta)$. For unit magnitude we have:

$$z^n = \cos n\theta + \mathrm{j} \sin n\theta$$

If we have $1/z^n$ then we have $1\angle 0/|z^n|\angle n\theta = |1/z^n|\angle(-n\theta)$. Thus for a number with magnitude 1 we can write:

$$z^{-n} = \cos(-n\theta) + \mathrm{j}\sin(-n\theta) = \cos n\theta - \mathrm{j}\sin n\theta$$

Thus, adding the equations gives:

$$z^n + \frac{1}{z^n} = 2\cos n\theta \qquad [24]$$

and subtracting them gives:

$$z^n - \frac{1}{z^n} = \mathrm{j}2\sin n\theta \qquad [25]$$

Example

Express $\cos^3\theta$ in terms of cosines of multiples of θ.

Using equation [22]:

$$2^3\cos^3\theta = \left(z + \frac{1}{z}\right)^3 = z^3 + 3z^2\left(\frac{1}{z}\right) + 3z\left(\frac{1}{z^2}\right) + \frac{1}{z^3}$$

$$= z^3 + 3z + 3\frac{1}{z} + \frac{1}{z^3} = \left(z^3 + \frac{1}{z^3}\right) + 3\left(z + \frac{1}{z}\right)$$

Hence, using equations [20] and [24]:

$$2^3\cos^3\theta = 2\cos 3\theta + 3(2\cos\theta)$$

and so:

$$\cos^3\theta = \tfrac{1}{4}(\cos 3\theta + 3\cos\theta)$$

Revision

18 Express, in terms of multiples of θ, (a) $\sin^4\theta$, (b) $\cos^5\theta$.

Problems

1 Simplify (a) j^7, (b) j^8, (c) $\mathrm{j}^2 \times \mathrm{j}$, (d) $\mathrm{j}^5/\mathrm{j}^3$.

2 Solve the following equations:

(a) $x^2 + 16 = 0$, (b) $x^2 + 4x - 5 = 0$, (c) $2x^2 - 2x + 3 = 0$

3 Express the following complex numbers in polar form:

(a) $-4 + \mathrm{j}$, (b) $-3 - \mathrm{j}4$, (c) 3, (d) $-\mathrm{j}6$, (e) $1 + \mathrm{j}$, (f) $3 - \mathrm{j}2$

4 Express the following complex numbers in Cartesian form:

(a) $5\angle 120°$, (b) $10\angle 45°$, (c) $6\angle 180°$, (d) $2.8\angle 76°$,

(d) $2(\cos 30° + \text{j} \sin 30°)$, (e) $3(\cos 60° - \text{j} \sin 60°)$

5 If $z_1 = 3 + \text{j}2$ and $z_2 = -2 + \text{j}4$, determine the values of:

(a) $z_1 + z_2$, (b) $z_1 - z_2$, (c) $z_1 z_2$, (d) $\frac{1}{z_1}$, (e) $\frac{z_1}{z_2}$

6 Evaluate the following:

(a) $(2 + \text{j}3) + (3 - \text{j}5)$, (b) $(-4 - \text{j}6) + (2 + \text{j}5)$, (c) $(2 + \text{j}2) - (3 - \text{j}5)$,

(d) $(2 + \text{j}4) - (1 + \text{j}4)$, (e) $4(3 + \text{j}2)$, (f) $\text{j}2(3 + \text{j}5)$, (g) $(1 - \text{j}2)(3 + \text{j}4)$,

(h) $(2 + \text{j}2)(2 - \text{j}3)$, (i) $(1 + \text{j}2)(4 - \text{j}3)$, (j) $\frac{6 + \text{j}3}{4 - \text{j}2}$, (k) $\frac{1}{3 + \text{j}2}$,

(l) $\frac{1 + \text{j}1}{1 - \text{j}1}$, (m) $\frac{3 + \text{j}2}{1 - \text{j}3}$

7 If $z_1 = 10\angle 20°$, $z_2 = 2\angle 40°$ and $z_3 = 5\angle 60°$, evaluate the following:

(a) $z_1 z_2$, (b) $z_1 z_3$, (c) $\frac{1}{z_1}$, (d) $\frac{1}{z_2}$, (e) $\frac{z_1}{z_2}$, (f) $\frac{z_2}{z_3}$

8 Determine the roots of:

(a) $(-5 + \text{j}12)^{1/2}$, (b) $(16\angle 60°)^{1/2}$, (c) $(5 + \text{j}3)^{1/2}$, (d) $(1\angle 60°)^{1/4}$,

(e) $1^{1/3}$, (f) $(1 + \text{j}1)^{1/5}$

9 Determine the values of:

(a) $(2\angle 20°)^3$, (b) $(2 + \text{j}3)^3$, (c) $(6 + \text{j}5)^3$, (d) $(2 + \text{j}1)^6$, (e) $(-2 + \text{j}2)^{10}$

10 Express, in terms of powers of sines and cosines, (a) $\cos 5\theta$, (b) $\sin 7\theta$, (c) $\sin 9\theta$.

11 Express, in terms of multiples of θ, (a) $\cos^6 \theta$, (b) $\sin^8 \theta$.

12 Show that:

$$\sin^5\theta \cos^5\theta = \frac{1}{2^9}(\sin 10\theta - 5\sin 6\theta + 10\sin 2\theta)$$

13 Show that:

$$\sin^4\theta \cos^2\theta = \frac{1}{2^4}(\tfrac{1}{2}\cos 6\theta - \cos 4\theta - \tfrac{1}{2}\cos 2\theta + 1)$$

8 The exponential function

8.1 Introduction

The exponential function was introduced in Chapter 3 for real numbers; in this chapter the exponential function form of a complex number is considered. This chapter assumes a knowledge of the exponential, trigonometric, hyperbolic and logarithmic functions (Chapter 3), the power series for exponentials, sines and cosines (see Chapter 5 and Table 5.1) and follows on from the discussion of complex numbers in Chapter 7.

8.2 The exponential form of a complex number

A complex number z can be expressed as $z = |z|(\cos\theta + \mathrm{j}\sin\theta)$. Using the power series for sines and cosines (see Table 5.1) we can write this as:

$$z = |z|\left[\left(1 - \frac{\theta^2}{2!} + \frac{\theta^4}{4!} + \ldots\right) + \mathrm{j}\left(\theta - \frac{\theta^3}{3!} + \frac{\theta^5}{5!} + \ldots\right)\right]$$

$$= |z|\left(1 + \mathrm{j}\theta - \frac{\theta^2}{2!} - \mathrm{j}\frac{\theta^3}{3!} + \frac{\theta^4}{4!} + \mathrm{j}\frac{\theta^5}{5!} + \ldots\right)$$

Since $\mathrm{j}^2 = -1$, $\mathrm{j}^3 = -\mathrm{j}$, $\mathrm{j}^4 = 1$, $\mathrm{j}^5 = \mathrm{j}$, etc. we can write the equation as:

$$z = |z|\left(1 + \mathrm{j}\theta + \frac{\mathrm{j}^2\theta^2}{2!} + \frac{\mathrm{j}^3\theta^3}{3!} + \frac{\mathrm{j}^4\theta^4}{4!} + \frac{\mathrm{j}^5\theta^5}{5!} + \ldots\right)$$

But this is the power series for an exponential e^x (see Table 5.1) if we replace x by $\mathrm{j}\theta$. Thus:

$$z = |z|\,\mathrm{e}^{\mathrm{j}\theta} \qquad [1]$$

This is the exponential form of a complex number.

Since $z = |z|(\cos\theta + \mathrm{j}\sin\theta)$ then we must have:

$$\mathrm{e}^{\mathrm{j}\theta} = \cos\theta + \mathrm{j}\sin\theta \qquad [2]$$

This is known as *Euler's formula*. If we replace θ by $-\theta$ we obtain:

$$\mathrm{e}^{-\mathrm{j}\theta} = \cos\theta - \mathrm{j}\sin\theta \qquad [3]$$

Since $\cos 2\pi = 1$ and $\sin 2\pi = 0$, equation [2] gives $\mathrm{e}^{\mathrm{j}2\pi} = 1$. Likewise $\cos 2\pi\mathrm{n} = 1$ and $\sin 2\pi n = 0$, where n is an integer. Thus:

$$\mathrm{e}^{\mathrm{j}2\pi n} = 1 \qquad [4]$$

Similarly we can obtain $\mathrm{e}^{\mathrm{j}\pi} = \mathrm{e}^{-\mathrm{j}\pi} = -1$, $\mathrm{e}^{\mathrm{j}\pi/2} = \mathrm{j}$, $\mathrm{e}^{-\mathrm{j}\pi/2} = -\mathrm{j}$.

Since $e^{a+jb} = e^a e^{jb}$ then we can write:

$$e^{a+jb} = e^a(\cos b + j \sin b) \qquad [5]$$

Consider $e^{a+j2\pi} = e^a e^{j2\pi}$. Since $e^{j2\pi} = 1$, we have $e^{a+j2\pi} = e^a$.
We can also write:

$$e^{a-jb} = e^a(\cos b - j \sin b) \qquad [6]$$

Example

Determine the polar form of the complex number $2\ e^{j\pi/4}$.

This number is in the form of equation [1] and thus the modulus is 2 and the argument $\pi/4$. Hence $z = 2\angle\pi/4$.

Example

Determine the exponential form of the complex number 2 + j2.

In polar form we have a modulus of $\sqrt{2^2+2^2} = 2.8$ and an argument of $\tan^{-1}(2/2) = \pi/4$. Thus, using equation [1], the exponential form is $2.8\ e^{j\pi/4}$.

Example

Determine the Cartesian form of the complex number $e^{2-j\pi}$.

Using equation [6]:

$$e^{2-j\pi} = e^2(\cos \pi - j \sin \pi) = e^2\,(-1 - 0) = -e^2$$

Example

Evaluate $2\ e^{j\pi/4} \times 5\ e^{-j\pi/2}$.

We can simplify the expression to give:

$$2\ e^{j\pi/4} \times 5\ e^{-j\pi/2} = 10\ e^{j(\pi/4-\pi/2)} = 10\ e^{-j\pi/4}$$

Thus, using equation [3], we have:

$$10(\cos \pi/4 - j \sin \pi/4) = 7.07 - j7.07$$

Revision

1 Express the following complex numbers in exponential form:

(a) 1 + j1, (b) j3, (c) 1 – j1, (d) –1 + j1.

2 Express the following in terms of their real and imaginary parts:

(a) e^{j1}, (b) $e^{\pi-1}$, (c) $5\ e^{j\pi/2}$, (d) $e^{j3\pi/2}$, (e) $e^{j3\pi/4}$

3 Evaluate $2\ e^{j\pi/3} \times 4\ e^{-j2\pi/3}$.

8.3 Complex trigonometric and hyperbolic functions

Euler's formula (equation [2]) gives $e^{j\theta} = \cos\theta + j\sin\theta$, and if we replace θ by $-\theta$ (equation [3]) $e^{-j\theta} = \cos\theta - j\sin\theta$. Adding these equations gives:

$$\cos\theta = \tfrac{1}{2}(e^{j\theta} + e^{-j\theta}) \qquad [7]$$

Subtracting the equations gives:

$$\sin\theta = \frac{1}{2j}(e^{j\theta} - e^{-j\theta}) \qquad [8]$$

Example

Express sin j1 as exponentials.

Using equation [8]:

$$\sin j1 = \frac{1}{2j}(e^{j^2 1} - e^{-j^2 1}) = \frac{1}{2j}(e^{-1} - e^{1})$$

Example

By writing the trigonometric ratio in exponential form, express $\cos^3\theta$ in terms of cosines of multiple angles.

Using equation [7]:

$$\cos^3\theta = \frac{1}{2^3}(e^{j\theta} + e^{-j\theta})^3 = \tfrac{1}{8}(e^{j3\theta} + 3\ e^{j\theta} + 3\ e^{-j\theta} + e^{-j3\theta})$$

$$= \tfrac{1}{8}(e^{j3\theta} + e^{-j3\theta}) + \tfrac{3}{8}(e^{j\theta} + e^{-j\theta}) = \tfrac{1}{4}\cos 3\theta + \tfrac{3}{4}\cos\theta$$

Revision

4 By writing the trigonometric ratio in exponential form, express $\cos^4\theta$ in terms of cosines of multiple angles.

5 Verify that the identity $\cos^2\theta + \sin^2\theta = 1$ holds when the trigonometric ratios are expressed in exponential form.

6 Express cos 2 in exponential form.

8.3.1 Trigonometric and hyperbolic functions

Hyperbolic functions were defined in Section 3.5, equations [61], [62] and [63]. We have $\cosh x = \frac{1}{2}(e^x + e^{-x})$ and so if we put $x = j\theta$ then $\cosh j\theta = \frac{1}{2}(e^{j\theta} + e^{-j\theta})$. Hence, using equation [7] we have:

$$\cos\theta = \cosh j\theta \qquad [9]$$

$\sinh x = \frac{1}{2}(e^x - e^{-x})$ and so if we put $x = j\theta$ then $\sinh j\theta = \frac{1}{2}(e^{j\theta} + e^{-j\theta})$. Hence, using equation [8] we have:

$$j\sin\theta = \sinh j\theta \qquad [10]$$

If we replace θ by $j\theta$ in equation [9] we obtain:

$$\cos j\theta = \cosh(-\theta) = \cosh\theta \qquad [11]$$

If we replace θ by $j\theta$ in equations [10] we obtain:

$$j\sin j\theta = \sinh(-\theta) = -\sinh\theta = j^2\sinh\theta$$

and so:

$$\sin j\theta = j\sinh\theta \qquad [12]$$

Consider $\sin(a + jb)$. Using equation [8] we can replace θ by $a + jb$ and so obtain:

$$\sin(a+jb) = \frac{1}{2j}(e^{j(a+jb)} - e^{-j(a+jb)}) = \frac{1}{2j}(e^{-b}\,e^{ja} - e^{b}\,e^{-ja})$$

Using Euler's formula (equations [2] and [3]) we can replace the e^{ja} and e^{-ja} terms and obtain:

$$\begin{aligned}\sin(a+jb) &= \frac{1}{2j}\,e^{-b}(\cos a + j\sin a) - \frac{1}{2j}\,e^{b}(\cos a - j\sin a)\\ &= \sin a\left(\frac{e^b + e^{-b}}{2}\right) + j\cos a\left(\frac{e^b - e^{-b}}{2}\right)\end{aligned}$$

Thus:

$$\sin(a+jb) = \sin a\cosh b + j\cos a\sinh b \qquad [13]$$

There is an alternative way we could have obtained the above equation. Since $\sin(A + B) = \sin A\cos B + \cos A\sin B$ then:

$$\sin(a+jb) = \sin a\cos jb + \cos a\sin jb$$

and thus, using equations [11] and [12], we obtain:

$$\sin(a + \mathrm{j}b) = \sin a \cosh b + \mathrm{j} \cos a \sinh b$$

Similarly we can obtain the relationship:

$$\cos(a + \mathrm{j}b) = \cos a \cosh b - \mathrm{j} \sin a \sinh b \qquad [14]$$

Example

Express $\sin(1 + \mathrm{j}1)$ in the form $a + \mathrm{j}b$.

Using equation [13]:

$$\sin(1 + \mathrm{j}1) = \sin 1 \cosh 1 + \mathrm{j} \cos 1 \sinh 1$$

$\sin 1 = 0.841$, $\cos 1 = 0.540$, $\cosh 1 = \frac{1}{2}(\mathrm{e}^1 + \mathrm{e}^{-1}) = 1.543$ and $\sinh 1 = \frac{1}{2}(\mathrm{e}^1 - \mathrm{e}^{-1}) = 1.175$. Thus:

$$\sin(1 + \mathrm{j}1) = 1.298 + \mathrm{j}0.635$$

Revision

7 Express $\sin(2 + \mathrm{j}3)$ in the form $a + \mathrm{j}b$.

8 Write $\sinh(-\mathrm{j}4)$ as a sine.

9 Write $\sin \mathrm{j}$ as a sinh.

8.4 The complex logarithm function

Let w be the natural logarithm of a complex number z, i.e. $w = \ln z$. Then we must have $z = \mathrm{e}^w$. We can express z as $|z|\,\mathrm{e}^{\mathrm{j}\theta}$ and w as $u + \mathrm{j}v$. Thus:

$$|z|\,\mathrm{e}^{\mathrm{j}\theta} = \mathrm{e}^{u+\mathrm{j}v} = \mathrm{e}^u\,\mathrm{e}^{\mathrm{j}v}$$

Since the real parts on both sides of the equals sign must be equal we have $|z| = \mathrm{e}^u$ and so $\ln |z| = u$. For the imaginary parts to be equal we must have $\mathrm{e}^{\mathrm{j}\theta} = \mathrm{e}^{\mathrm{j}\pi}$. Thus, dividing both sides by $\mathrm{e}^{\mathrm{j}\pi}$:

$$\mathrm{e}^{\mathrm{j}(\theta - v)} = 1$$

We have $\mathrm{e}^{\mathrm{j}2\pi n} = \cos 2\pi n + \mathrm{j} \sin 2\pi n = 1$ (equation [4]) and thus:

$$\theta - v = 2\pi n$$

So $v = \theta + 2\pi n$ and:

$$w = u + \mathrm{j}v = \ln |z| + \mathrm{j}(\theta + 2\pi n)$$

Hence:

$$\ln z = \ln |z| + j(\theta + 2\pi n) \qquad [15]$$

Note that the complex logarithm is multi-valued, there being a value for each value of n.

Example

Evaluate ln(−1).

Since $-1 = 1\angle\pi$ or $1\angle 3\pi$ or $1\angle 5\pi$,etc., using equation [15]:

$$\ln(-1) = \ln 1 + j(\pi + 2\pi n)$$

Since $e^{j2\pi} = 1$ then ln 1 can be written as $j2\pi$. Thus:

$$\ln(-1) = j(3\pi + 2\pi n)$$

When $n = 0$ the value is $j3\pi$,when $n = -1$ it is $j\pi$,when $n = 1$ it is $j5\pi$, when $n = -2$ it is $-j\pi$,when $n = 2$ it is $j7\pi$ and so on. The values are thus $\pm j\pi$, $\pm j3\pi$, $\pm j5\pi$, $\pm j7\pi$, etc.

Example

Evaluate ln(2 + j1).

In polar form we have $\sqrt{2^2 + 1^2}\ \angle(\tan^{-1} 1/2)$, which, with the argument in radians, is $\sqrt{5}\ \angle 0.46$. Thus using equation [15]:

$$\ln(2 + j1) = \ln \sqrt{5} + j(0.46 + 2\pi n) = 2.24 + j(0.46 + 2\pi n)$$

Revision

10 Evaluate: (a) ln j2, (b) ln(3 + j1), (c) ln 1, (d) ln(1 + j1)

8.4.1 z^w

In general, if $y = a^x$ then $x = \log_a y$ and $\log_a y^n = n \log_a y$ (see Section 3.3). Thus if we have $\ln z^w = w \ln z$ then powers of z are defined by the equation:

$$z^w = e^{w \ln z} \qquad [16]$$

provided z is not zero.

Example

Determine the values of j^j.

Using equations [16] and [15]:

$$j^j = e^{j\ln j} = e^{j[j(\pi/2+2\pi n)]} = e^{-(\pi/2+2\pi n)}$$

Example

Determine the values of $(1+j)^j$.

Using equations [16] and [15]:

$$(1+j)^j = e^{j\ln(1+j)} = e^{j[\ln\sqrt{2}+j(\pi/4+2\pi n)]} = e^{-\pi/4-2\pi n+j\ln\sqrt{2}}$$

But $e^{j\ln\sqrt{2}} = \cos(\ln\sqrt{2}) + j\sin(\ln\sqrt{2})$. Thus:

$$(1+j)^j = e^{-\pi/4-2\pi n}[\cos(\ln\sqrt{2}) + j\sin(\ln\sqrt{2})]$$

Revision

11 Determine the values of: (a) j^{1+j}, (b) 2^{3-j}.

Problems

1 Express the following complex numbers in exponential form:

(a) $3 - j4$, (b) $5 + j3$, (c) $j3$

2 Express the following exponentials in terms of their real and imaginary parts:

(a) $3\,e^{2-j3}$, (b) $2\,e^{1+j4}$, (c) $2\,e^{j3}$, (d) e^{2+j3}, (e) $e^{2+j\pi/4}$

3 Evaluate $2\,e^{j\pi/2} \times 3\,e^{j7\pi/4}$.

4 By writing the trigonometric ratio in exponential form, express $\sin^5\theta$ in terms of cosines of multiple angles.

5 Verify that the identity $2\cos^2\theta = 1 + \cos 2\theta$ holds when the trigonometric ratios are expressed in exponential form.

6 Obtain $\tan\theta$ in exponential form.

7 Write $\sin(1 + j2)$ in the form $a + jb$.

8 Show that $\tanh j\theta = j\tan\theta$.

9 Evaluate: (a) $\ln(-9)$, (b) $\ln(1 - j2)$, (c) $\ln j1$, (d) $\ln(2 + j4)$, (e) $\ln j5$.

10 Determine the values of: (a) 2^j, (b) $(1 - j)^{1+j}$.

11 If $z_1 = 8 + j6$ and $z_2 = 3 - j4$, determine $\ln(z_1/z_2)^{1/2}$.

9 Application: Phasors

9.1 Introduction

We can represent a voltage v which varies sinusoidally with time t by the equation $v = V \sin \omega t$, where V is the maximum value of the voltage and ω the angular frequency (see Section 3.4.4). This equation is termed the *time-domain representation* of the voltage. We can imagine such a signal being produced by a radial line of length V rotating with a constant angular velocity ω (Figure 9.1). Thus instead of specifying the variation of the voltage with time by the above equation, we can specify it by the length of the line V and whether it starts at $t = 0$ at some angle, termed the *phase angle*, to the reference axis. The reference axis is usually taken as the horizontal axis. Such lines are termed *phasors* and the representation of an alternating voltage is said to be the *frequency-domain representation.*

This chapter illustrates how the complex number system can be used to describe phasors. It assumes mainly Chapter 7.

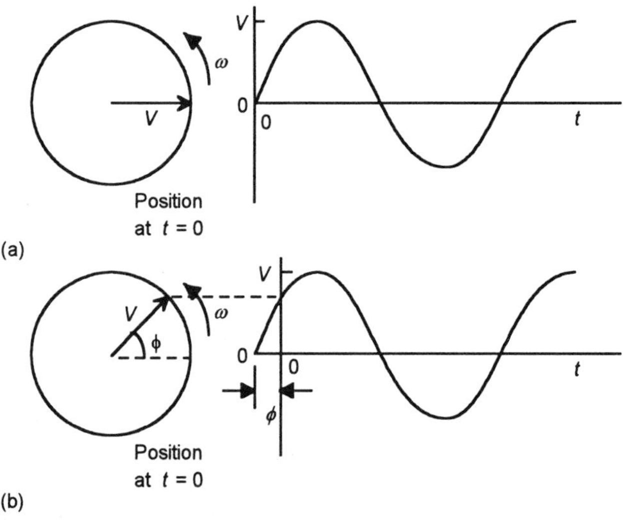

Figure 9.1 *(a)* $v = V \sin \omega t$, *(b)* $v = V \sin(\omega t + \phi)$

9.2 Describing phasors

A convenient way of specifying a phasor is by polar notation. Thus a phasor of length V and phase angle ϕ can be represented by $V\angle\phi$. Although the length of a phasor when described in the way shown in Figure 9.1 represents the maximum value of the quantity, it is more usual to specify the length as representing the root-mean-square value. The root-mean-square value is just the maximum value divided by $\sqrt{2}$ and so is just a scaled version of the one drawn using the maximum value.

When we are working in the time domain and want to find, say, the sum of two voltages at some instant of time we just add the voltages. Thus if we have a voltage across one component described by $v_1 = V_1 \sin(\omega t + \phi_1)$ and across a series component by $v_2 = V_2 \sin(\omega t + \phi_2)$, then the sum of the two voltages is:

$$v_1 + v_2 = V_1 \sin(\omega t + \phi_1) + V_2 \sin(\omega t + \phi_2)$$

This equation describes how the voltage sum varies with time.

When we are working with phasors and want to find the phasor representing the sum of two phasors we have to add the phasors in the way that vector quantities are added. Thus if we have a voltage across one component described by $V_1\angle\phi_1$ and across a series component by $V_2\angle\phi_2$, then the phasor representing the sum of the two voltages is that indicated in Figure 9.2. While we can draw such diagrams for simple situations and obtain the resultant phasor graphically, a more useful technique is to describe a phasor by a complex number and use the techniques for manipulating complex numbers.

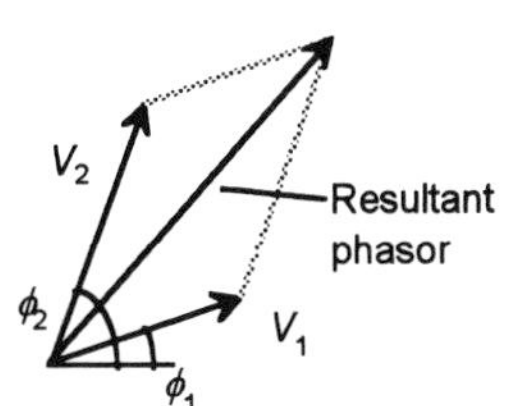

Figure 9.2 *Adding phasors*

Note that there is a difference between a phasor diagram and a vector diagram. A phasor diagram represents the phasors at one instant of time, a vector diagram represents the vectors without regard to time. Otherwise the mathematics of handling vectors is applicable to phasors.

9.2.1 Representing phasors by complex numbers

As indicated in Section 7.2.4 a complex number $z = a + jb$ can be represented on an Argand diagram by a line (Figure 9.3) of length $|z|$ at an angle θ. Thus we can describe a phasor used to represent, say, a sinusoidal voltage, by a complex number in this Cartesian form as:

$$\mathbf{V} = a + jb \qquad [1]$$

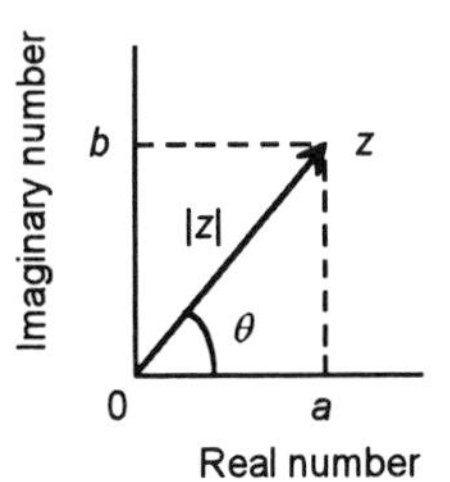

Figure 9.3 *Complex number*

Note that bold print is usually used to distinguish phasor quantities from other quantities which have only the attribute of their size to specify them with no angle specification and do not represent a rotating line. An alternative way of describing a complex number, and hence a phasor, is in polar notation, i.e. the length of the phasor and its angle to some reference axis. Thus we can describe it as:

$$\mathbf{V} = V\angle\theta \qquad [2]$$

where V is the magnitude of the phasor and θ its phase angle. The magnitude $|z|$ of a complex number z and its argument θ are given by (Chapter 7, equations [3] and [4]):

$$|z| = \sqrt{a^2 + b^2} \text{ and } \theta = \tan^{-1}\left(\frac{b}{a}\right) \qquad [3]$$

Since $a = |z| \cos\theta$ and $b = |z| \sin\theta$, then:

$$z = |z| \cos\theta + j|z| \sin\theta = |z|(\cos\theta + j\sin\theta) \qquad [4]$$

Thus if we have the voltage across one component described by $V\angle\phi$ then we can write this as:

$$\mathbf{V} = V(\cos\phi + j\sin\phi) \qquad [5]$$

Yet another way of describing a complex number, and hence a phasor, is in terms of an exponential (Chapter 8, equation [1]) $z = |z|\, e^{j\theta}$ and hence:

$$\mathbf{V} = V e^{j\phi} \qquad [6]$$

The exponential gives the same information as the polar form. We can convert this exponential form into Cartesian form by Euler's formula (Chapter 8, equation [3]) $e^{j\phi} = \cos\phi + j\sin\phi$.

Example

Describe the signal $v = 12 \sin(314t + \pi/4)$ V by a phasor.

The phasor has a magnitude, when expressed as the maximum value, of 12 and argument $\pi/4$. Thus we can describe it as $12\angle\pi/4$ V, or by using equation [4] as $12 \cos\pi/4 + j12 \sin\pi/4 = 8.49 + j8.49$ V, or by using equation [6] as $12\, e^{j\pi/4}$ V. If used root-mean-square values then we would have $8.49\angle\pi/4$ V, $3 + j3$ V, $8.49\, e^{j\pi/4}$.

Revision

1 Describe the following signals by phasors, expressed as the maximum values, in polar and Cartesian forms:

(a) $10 \sin 314t$, (b) $2 \sin(314t + \pi/3)$, (c) $5 \sin(6283t + \pi/2)$

9.2.2 Adding or subtracting phasors

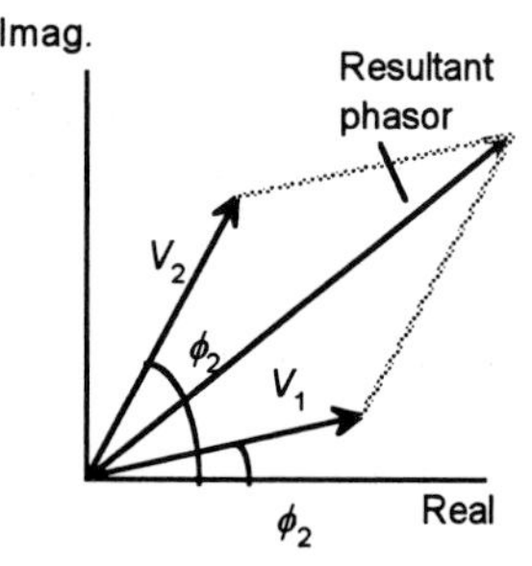

Figure 9.4 *Adding phasors*

If we have the voltage across one component described by $V_1\angle\phi_1$ then we can write: $\mathbf{V}_1 = V_1(\cos\phi_1 + j\sin\phi_1)$. If we have the voltage across a series component described by $V_2\angle\phi_2$, then: $\mathbf{V}_2 = V_2(\cos\phi_2 + j\sin\phi_2)$. The phasor for the sum of the two voltages is then obtained by adding the two complex numbers (Figure 9.4). Thus:

$$\mathbf{V} = \mathbf{V}_1 + \mathbf{V}_2 = V_1(\cos\phi_1 + j\sin\phi_1) + V_2(\cos\phi_2 + j\sin\phi_2)$$

$$= (V_1 \cos\phi_1 + V_2 \cos\phi_2) + j(V_1 \sin\phi_1 + V_2 \sin\phi_2) \qquad [7]$$

Subtraction is carried out in a similar manner. Since adding or subtracting complex numbers is easier when they are in Cartesian form rather than polar form, when phasors are to be added or subtracted they should be put in Cartesian form.

Example

A circuit has three components in series. If the voltages across each component are described by phasors 4 V, j2 V and 3 + j4 V, what is the voltage phasor describing the voltage across the three components?

Since the components are in series, the resultant phasor voltage is described by the phasor:

$$\mathbf{V} = 4 + j2 + 3 + j4 = 7 + j6$$

Example

A circuit has two components in series. If the voltages across each component are described by phasors $4\angle 60^\circ$ V and $2\angle 30^\circ$ V, what is the voltage phasor describing the voltage across the two components?

For adding complex numbers it is simplest to convert the phasors into Cartesian notation. Thus:

$$\mathbf{V} = (4 \cos 60^\circ + j4 \sin 60^\circ) + (2 \cos 30^\circ + j2 \sin 30^\circ)$$

$$= 2 + j3.46 + 1.73 + j1.73 = 3.73 + j5.19 \text{ V}$$

If we want this phasor in polar notation then, using equations [3]:

$$V = \sqrt{3.73^2 + 5.19^2} = 6.39$$

$$\phi = \tan^{-1}\frac{5.19}{3.73} = 54^\circ$$

Thus the phasor is $6.39\angle 54^\circ$ V.

Revision

2 Determine the phasor representing the sum of the voltages described by the following phasors, expressing the results in both Cartesian and polar notation:

(a) $2 + j3$ V and $1 - j5$ V, (b) 2 V and $-j5$ V, (c) $4\angle 0^\circ$ V and $3\angle 60^\circ$,

(d) $5\angle 0^\circ$ V and $10\angle 45^\circ$ V, (e) $1 + j3$ V, $3 + j1$ V and $4 - j3$ V,

(f) 10 V, j2 V and $-j6$ V

9.2.3 Multiplication or division of phasors

Multiplication or division of complex numbers can be carried out when they are in either Cartesian form or polar form, being easiest when they are in polar form. Thus, if we have a voltage across a component described by $\mathbf{V} = V\angle\phi$ and the current by $\mathbf{I} = I\angle\theta$ then the product of the two phasors is:

$$\mathbf{VI} = VI\angle(\phi + \theta) \qquad [8]$$

If the voltage and current were in Cartesian form, i.e. in the form $\mathbf{V} = a + \text{j}b$ and $\mathbf{I} = c + \text{j}d$ then the product is:

$$\mathbf{VI} = (a + \text{j}b)(c + \text{j}d) = (ac - bd) + \text{j}(bc + ad) \qquad [9]$$

For division, if we have a voltage across a component described by $\mathbf{V} = V\angle\phi$ and the current by $\mathbf{I} = I\angle\theta$ then:

$$\frac{\mathbf{V}}{\mathbf{I}} = \frac{V\angle\phi}{I\angle\theta} \qquad [10]$$

If the voltage and current were in Cartesian form, i.e. in the form $\mathbf{V} = a + \text{j}b$ and $\mathbf{I} = c + \text{j}d$ then:

$$\frac{\mathbf{V}}{\mathbf{I}} = \frac{a + \text{j}b}{c + \text{j}d} = \frac{a + \text{j}b}{c + \text{j}d} \times \frac{c - \text{j}d}{c - \text{j}d} = \frac{(a + \text{j}b)(c - \text{j}d)}{c^2 - d^2} \qquad [11]$$

Example

If phasor **A** is represented by $10\angle 30°$ and **B** by $2\angle 45°$, determine **AB** and **A/B**.

For the product the situation is similar to that in equation [8], thus:

$$\mathbf{AB} = (10 \times 2)\angle(30° + 45°) = 20\angle 75°$$

For the division the situation is similar to that in equation [10], thus:

$$\frac{\mathbf{A}}{\mathbf{B}} = \frac{10\angle 30°}{2\angle 45°} = 5\angle(-15°)$$

Example

If phasor **A** is represented by $1 - \text{j}2$ and **B** by $3 + \text{j}4$, determine **A/B**.

The situation is similar to that in equation [11], thus:

$$\frac{\mathbf{V}}{\mathbf{I}} = \frac{1 - \text{j}2}{3 + \text{j}4} = \frac{1 - \text{j}2}{3 + \text{j}4} \times \frac{3 - \text{j}4}{3 - \text{j}4} = \frac{3 - 8 - \text{j}10}{9 + 16} = -0.2 - \text{j}0.4$$

Revision

3 Determine **AB** if (a) phasor **A** is represented by 3 – j4 and **B** by 4 + j6, (b) phasor **A** is represented by $2\angle 30^\circ$ and **B** by $4\angle 60^\circ$, (c) phasor **A** is represented by $4\angle(-30^\circ)$ and **B** by $2\angle 60^\circ$.

4 Determine **A/B** if (a) phasor **A** is represented by 2 + j3 and **B** by 1 – j2, (b) phasor **A** is represented by $12\angle 80^\circ$ and **B** by $4\angle 60^\circ$.

9.3 Circuit analysis

Kirchhoff's laws apply to the voltages and currents in a circuit at any instant of time. Thus the voltage law that the sum of the voltages taken round a closed loop is zero means that, with alternating voltages having values of v_1, v_2, v_3, etc. at the same instant of time:

$$v_1 + v_2 + v_3 + \ldots = 0$$

and so, if these voltages are sinusoidal:

$$V_1 \sin(\omega t + \phi_1) + V_2 \sin(\omega t + \phi_2) + V_3 \sin(\omega t + \phi_3) + \ldots = 0$$

We can consider each of these sinusoidal voltages to be the vertical projection of the phasor describing it (as in Figure 9.1). Thus we must have:

$$\mathbf{V_1} + \mathbf{V_2} + \mathbf{V_3} + \ldots = 0$$

Kirchhoff's voltage law can thus be stated as:

The sum of the phasors of all the voltages around a closed loop is zero.

Kirchhoff's current law can be stated as the sum of all the currents at a node is zero, i.e. the current entering a junction equals the current leaving it. In a similar way we can state this law for sinusoidal currents as:

The sum of the phasors of the currents at a node is zero, i.e. the sum of the phasors for currents entering a junction equals that for those leaving it.

Example

A circuit has two components in parallel. If the currents through the components can be described by the phasors 2 + j4 A and 4 + j1 A, what is the phasor describing the current entering the junction?

Using Kirchhoff's current law we must have: the phasor for current entering junction = phasor sum for currents leaving the junction. Hence:

phasor for current entering = 2 + j4 + 4 + j1 = 6 + j5 A

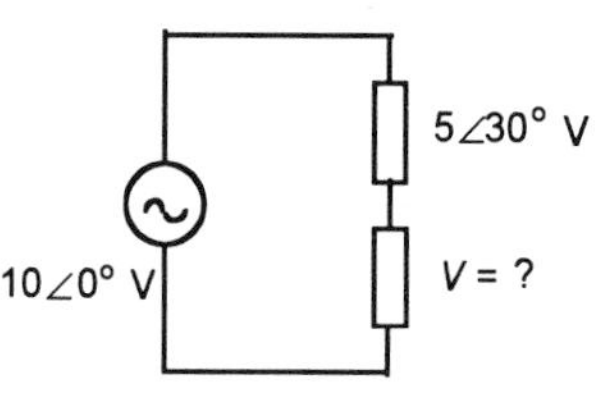

Figure 9.5 *Example*

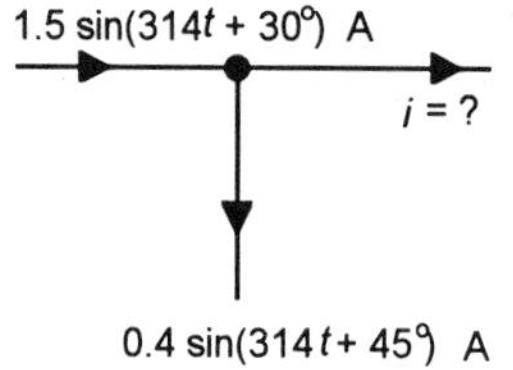

Figure 9.6 *Problem 5*

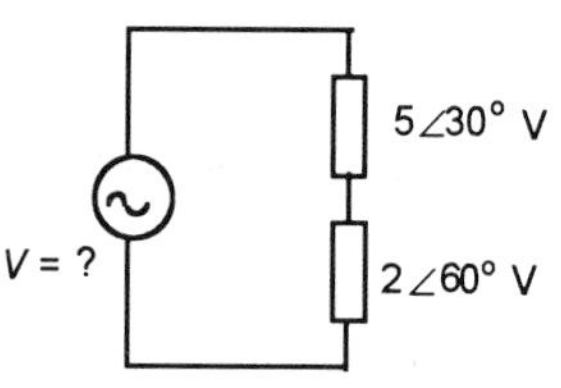

Figure 9.7 *Problem 6*

Example

For the a.c. circuit shown in Figure 9.5, determine the unknown voltage.

Using Kirchhoff's voltage law, and writing the phasors in Cartesian notation:

$$10 + j0 = (5 \cos 30^\circ + j5 \sin 30^\circ) + \mathbf{V}$$

Thus:

$$\mathbf{V} = 10 - 4.33 + j2.5 = 5.67 - j2.5 \text{ V}$$

or in polar notation:

$$\mathbf{V} = \sqrt{5.67^2 + 2.5^2} \angle \tan^{-1}(-2.5/5.67) = 6.2\angle 336.2^\circ \text{ V}$$

Revision

5 Determine the unknown alternating current for the circuit shown in Figure 9.6.

6 Determine the unknown alternating voltage for the circuit shown in Figure 9.7.

9.3.1 Impedance

The term *impedance* Z is defined as the ratio of the phasor voltage across a component to the phasor current through it:

$$Z = \frac{\mathbf{V}}{\mathbf{I}} \qquad [12]$$

Thus if we have $\mathbf{V} = V\angle\theta$ and $\mathbf{I} = I\angle\phi$ then:

$$Z = \frac{V\angle\theta}{I\angle\phi} = \frac{V}{I}\angle(\theta - \phi)$$

Impedance is a complex number but not a phasor since it does not describe a sinusoidally varying quantity. It describes a line on an Argand diagram but not one that rotates with an angular velocity. Hence I have not used bold print for it. In some textbooks, however, it is written in bold print because it is complex.

If we have impedances connected in series (Figure 9.8), then Kirchhoff's voltage law gives:

$$\mathbf{V} = \mathbf{V}_1 + \mathbf{V}_2 + \mathbf{V}_3$$

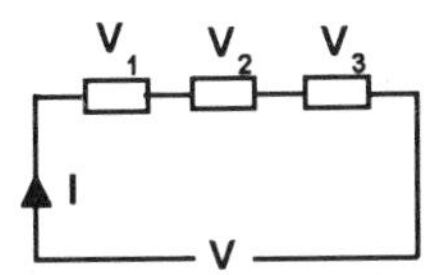

Figure 9.8 *Impedances in series*

Dividing by the phasor current, the current being the same through each:

$$\frac{\mathbf{V}}{\mathbf{I}} = \frac{\mathbf{V_1}}{\mathbf{I}} + \frac{\mathbf{V_2}}{\mathbf{I}} + \frac{\mathbf{V_3}}{\mathbf{I}}$$

Hence the total impedance Z is the sum of the impedances of the three impedances:

$$Z = Z_1 + Z_2 + Z_3 \qquad [13]$$

Consider the parallel connection of impedances (Figure 9.9). Kirchhoff's current law gives:

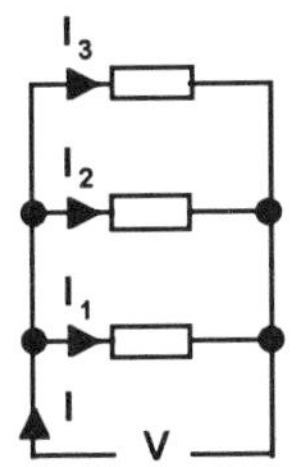

Figure 9.9 *Impedances in parallel*

$$\mathbf{I} = \mathbf{I_1} + \mathbf{I_2} + \mathbf{I_3}$$

Diving by the phasor voltage, the voltage being the same for each impedance:

$$\frac{\mathbf{I}}{\mathbf{V}} = \frac{\mathbf{I_1}}{\mathbf{V}} + \frac{\mathbf{I_2}}{\mathbf{V}} + \frac{\mathbf{I_3}}{\mathbf{V}}$$

Thus the total impedance Z is given by:

$$\frac{1}{Z} = \frac{1}{Z_1} + \frac{1}{Z_2} + \frac{1}{Z_3} \qquad [14]$$

Example

If the voltage across a component is 4 sin ωt V and the current through it 2 sin(ωt – 30°) A, what is its impedance?

Using equation [12] with the phasors put into polar notation:

$$Z = \frac{4\angle 0^\circ}{2\angle(-30^\circ)} = 2\angle 30^\circ\ \Omega$$

Thus, in Cartesian notation the impedance is 2 cos 30° + 2 sin 30° = 1.73 + j1 Ω.

Example

What is the total impedance of a circuit having impedances of 2 + j5 Ω, 1 – j3 Ω and 4 + j1 Ω in series?

The total impedance is given by equation [13] as:

$$Z = 2 + j5 + 1 - j3 + 4 + j1 = 7 + j3\ \Omega$$

Example

What is the total impedance of impedances $4\angle 30^\circ\ \Omega$ in parallel with $2\angle(-20^\circ)\ \Omega$.

Using equation [14]:

$$\frac{1}{Z} = \frac{1}{4\angle 30^\circ} + \frac{1}{2(-20^\circ)} = 0.25\angle(-30^\circ) + 0.5\angle 20^\circ$$

$$= 0.25 \cos(-30^\circ) + \text{j}0.25 \sin 30^\circ + 0.5 \cos 20^\circ + \text{j}0.5 \sin 20^\circ$$

$$= 0.686 + \text{j}0.296$$

$$= \sqrt{0.686^2 + 0.296^2}\ \angle \tan^{-1}(0.296/0.686) = 0.747\angle 23.3^\circ$$

Hence $Z = 1.339\angle(-23.3^\circ)\ \Omega$.

Revision

7 A circuit element has an impedance of $6\angle 60^\circ\ \Omega$. What will be the voltage across it when the current through it is $2\angle 25^\circ$ A?

8 If the voltage across a component is $2 \sin 314t$ V and the current through it $5 \sin(314t + 90^\circ)$ A, what is its impedance?

9 Determine, in Cartesian form, the total impedances of:

(a) $4\angle 20\ \Omega$ in series with $5\angle 60^\circ\ \Omega$,
(b) $2 + \text{j}3\ \Omega$ in series with $-1 + \text{j}2\ \Omega$,
(c) $10\ \Omega$ in parallel with $\text{j}10\ \Omega$,
(d) $2\angle 90^\circ\ \Omega$ in parallel with $5\angle 60^\circ\ \Omega$

9.3.2 Circuit elements

For a *pure resistor* the current through it is in phase with the voltage across it. Thus for a voltage phasor of $V\angle 0^\circ$ we must have a current phasor of $I\angle 0^\circ$ and so the impedance of the circuit element is:

$$Z = \frac{\mathbf{V}}{\mathbf{I}} = \frac{V\angle 0^\circ}{I\angle 0^\circ} = \frac{V}{I}\angle 0^\circ$$

The impedance is just the real number V/I which is the resistance R.

For a *pure capacitance* the current leads the voltage by 90°. Thus for a voltage phasor of $V\angle 0^\circ$ we must have a current phasor of $I\angle 90^\circ$ and so the impedance of the circuit element is:

$$Z = \frac{\mathbf{V}}{\mathbf{I}} = \frac{V\angle 0^\circ}{I\angle 90^\circ} = \frac{V}{I}\angle(-90^\circ)$$

The impedance is thus $-j(V/I)$ and is just an imaginary quantity. The term *capacitive reactance* X_C is used for the ratio of the maximum, or r.m.s., voltage and current and thus for a pure capacitance:

$$Z = -jX_C \quad [15]$$

For a *pure inductance* the current lags the voltage by 90°. Thus for a voltage phasor of $V\angle 0°$ we must have a current phasor of $I\angle(-90°)$ and so the impedance of the circuit element is:

$$Z = \frac{\mathbf{V}}{\mathbf{I}} = \frac{V\angle 0°}{I\angle(-90°)} = \frac{V}{I}\angle 90°$$

The impedance is thus $j(V/I)$ and is just an imaginary quantity. The term *inductive reactance* X_L is used for the ratio of the maximum, or r.m.s., voltage and current and thus for a pure inductance:

$$Z = jX_L \quad [16]$$

For a pure inductance in series with a pure resistance we have:

$$Z = R + jX_L \quad [17]$$

For a pure capacitance in series with a pure resistance we have:

$$Z = R - jX_C \quad [18]$$

For a pure inductance, pure capacitance and pure resistance in series we have:

$$Z = R + jX_L - jX_C = R + j(X_L - X_C) \quad [19]$$

For a pure inductance in parallel with a pure resistance we have:

$$\frac{1}{Z} = \frac{1}{R} + \frac{1}{jX_L} \quad [20]$$

For a pure capacitance in parallel with a pure resistance we have:

$$\frac{1}{Z} = \frac{1}{R} - \frac{1}{jX_C} \quad [21]$$

Example

Determine the impedance of a 100 Ω resistance in series with a capacitive reactance of 5 Ω.

Using equation [18]:

$$Z = 100 - j5\ \Omega$$

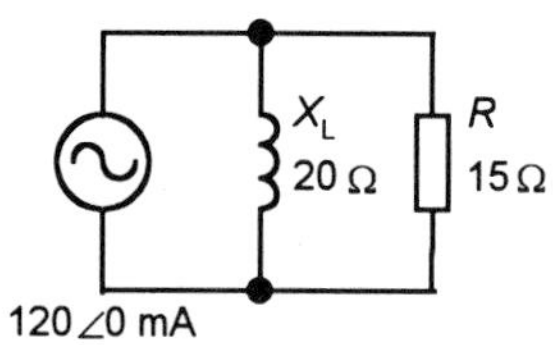

Figure 9.10 *Example*

Example

Determine the potential difference across the resistor for the circuit shown in Figure 9.10.

The circuit has an impedance due to a resistance in parallel with an inductance. Thus the impedance is given by equation [20] as:

$$\frac{1}{Z} = \frac{1}{15} + \frac{1}{j20} = \frac{15 + j20}{15 \times j20}$$

$$Z = \frac{j300}{15 + j20} \times \frac{15 - j20}{15 - j20} = \frac{6000 + j4500}{15^2 + 20^2} = 9.6 + j7.2\ \Omega$$

In polar notation this is $\sqrt{9.6^2 + 7.2^2} \angle \tan^{-1}(7.2/9.6) = 12\angle 36.9°\ \Omega$. The potential difference across this impedance, and hence across the resistance, is thus:

$$\mathbf{V} = Z\mathbf{I} = 12\angle 36.9° \times 120\angle 0° = 1440\angle 36.9°\text{ mV} = 1.44\angle 36.9°\text{ V}$$

Revision

10 Determine the impedance of the following:

(a) a resistance of 20 Ω in series with an inductive reactance 100 Ω,
(b) a resistance of 100 Ω in series with a capacitive reactance of 40 Ω,
(c) a resistance of 10 Ω in series with an inductive reactance of 20 Ω and a capacitive reactance of 5 Ω,
(d) a resistance of 20 Ω in parallel with an inductive reactance of 10 Ω,
(e) three components in parallel, a resistance of 200 Ω, an inductive reactance of 200 Ω and a capacitive reactance of 100 Ω

11 An alternating e.m.f. $30\angle 0°$ V is applied to a circuit consisting of a resistance of 200 Ω in series with an inductive reactance of 100 Ω. Determine, in polar notation, the phasor describing the circuit current.

Problems

1 Describe the following signals by phasors written in both polar and Cartesian forms, taking the magnitude to represent the maximum value:

(a) $10 \sin(2\pi 50t - \pi/6)$, (b) $10 \sin(314t + 150°)$,

(c) $22 \sin(628t + \pi/4)$

2 Determine, in both Cartesian and polar forms, the sum of the following phasors:

(a) $4\angle 0°$ and $3\angle 60°$, (b) $2 + j3$ and $-4 + j4$, (c) $4\angle \pi/3$ and $2\angle \pi/6$

3 If phasors **A**, **B** and **C** are represented by $\mathbf{A} = 10\angle 30°$, $\mathbf{B} = 2.5\angle 60°$ and $\mathbf{C} = 2\angle 45°$ determine: (a) **AB**, (b) **AC**, (c) **A**(**B** + **C**), (d) **A**/**B**, (e) **B**/**C**, (f) **C**/(**A** + **B**)

4 If $v_1 = 10 \sin \omega t$ and $v_2 = 20 \sin(\omega t + 60°)$, what is (a) the phasor describing the sum of the two voltages and (b) its time-domain equation?

5 If the voltage across a component is $5 \sin(314t + \pi/6)$ V and the current through it $0.2 \sin(314t + \pi/3)$ A, what is its impedance?

6 A voltage of 100 V is applied across a circuit of impedance $40 + j30\ \Omega$, what is, in polar notation, the current taken?

7 Determine, in Cartesian form, the total impedances of:

(a) 10 Ω in series with $2 - j5\ \Omega$,
(b) $100\angle 30°\ \Omega$ in series with $100\angle 60°\ \Omega$,
(c) $20\angle 30°\ \Omega$ in series with $15\angle(-10°)\ \Omega$,
(d) $20\angle 30°\ \Omega$ in parallel with $6\angle(-90°)\ \Omega$,
(e) 10 Ω in parallel with $-j2\ \Omega$,
(f) $j40\ \Omega$ in parallel with $j20\ \Omega$

8 Determine, in Cartesian form, the impedance of:

(a) a resistance of 5 Ω in series with an inductive reactance of 2 Ω,
(b) a resistance of 50 Ω in series with a capacitive reactance of 10 Ω,
(c) a resistance of 2 Ω in series with an inductive reactance of 5 Ω and a capacitive reactance of 4 Ω,
(d) three elements in parallel, a resistance of 2 Ω, an inductive reactance of 10 Ω and a capacitive reactance of 5 Ω,
(e) an inductive reactance of 500 Ω in parallel with a capacitive reactance of 100 Ω

9 An alternating voltage of $100 \sin 314t$ V is applied to a circuit consisting of three series elements, a resistance of 800 Ω, a capacitive reactance of 450 Ω and an inductive reactance of 1250 Ω. Determine the circuit current.

10 An alternating current of $240 \sin \omega t$ mA is applied to a circuit consisting of a resistance of 100 Ω in parallel with a capacitive reactance of 100 Ω. Determine the current through each element.

11 An alternating voltage of $20 \sin \omega t$ V is applied to a circuit consisting of a capacitive reactance of 10 Ω in series with a parallel arrangement of a resistance of 10 Ω and an inductive reactance of 10 Ω. Determine the circuit current.

12 An alternating voltage of 20 sin ωt V is applied to a circuit consisting of a resistance of 2 Ω in series with an inductive reactance of 4 Ω and a parallel arrangement of a resistance of 5 Ω and a capactive reactance of 10 Ω. Determine the phasors describing (a) the current drawn by the circuit, (b) the current through the capacitor.

13 A circuit consists of three components in series, a resistance of 5 Ω, a capacitive reactance of 6 Ω and an inductive reactance of 10 Ω. A root-mean-square voltage of $10\angle 0°$ V is applied to the circuit. Determine the voltage across each element of the circuit.

14 A circuit consists of three components in series, a resistance of 10 Ω, a capacitive reactance of 0.25 Ω and an inductive reactance of 10 Ω. If the voltage across the inductive reactance is $60\angle 40°$ V, what is the voltage applied to the circuit?

Part 3
Vector algebra

The aims of this part are to enable the reader to:

- Distinguish between scalar and vector quantities.
- Represent vectors by directed line segments and use the triangle rule, parallelogram rule and polygon rule with such directed line segments.
- Resolve vectors into components and represent such components in terms of unit vectors.
- Solve problems involving adding and subtracting position vectors described in terms of unit vector components.
- Solve problems involving scalar and vector products.
- Apply the vector algebra principles to mechanical systems, solving problems involving velocities, relative velocities, forces, work, moments and angular velocities.

This part introduces the concept of vectors and the rules needed for vector algebra. It assumes a knowledge of the Cartesian co-ordinate system. With some sections of chapters, equations have been presented in two formats, one being as determinants. While determinants need not be used, they do offer a concise way of representing the equations concerned and a knowledge of them would be useful. Determinants are covered in Chapter 18. Differentiation of vectors is not included in this part but is left to Part 6.

10 Vectors

10 Introduction

If we talk of, say, the mass of this book then we quote just a number, this being all that is needed to give a specification of its mass. However, if we quote a force then in order to fully describe the force we need to specify both its size and the direction in which it acts. Quantities which are fully specified by a statement of purely size are termed *scalars*. Quantities for which we need to specify both size and direction in order to give a full specification are termed *vectors*. Examples of scalar quantities are mass, distance, speed, work and energy. Examples of vector quantities are displacement, velocity, acceleration and force.

This chapter is an introduction to vectors, their addition and subtraction, and representation. Chapter 11 extends this consideration of vectors to products. A knowledge of the Cartesian co-ordinate system is assumed.

10.2 Representing vectors

To specify a vector we need to specify its magnitude and direction. Thus we can represent it by a line segment AB (Figure 10.1) with a length which represents the magnitude of the vector and a direction, indicated by the arrow on the segment, which represents the direction of the vector. We can denote this vector representation as

Figure 10.1 *Representing a vector*

$$\overrightarrow{AB}$$

the arrow indicating the direction of the line segment being from A to B. Note that:

$$\overrightarrow{AB} \neq \overrightarrow{BA}$$

One of the vectors is directed from A to B while the other is directed from B to A. An alternative notation is often used, lower case bold notation **a** being used in print, or underlining $\underline{a}$ in writing. With this notation, if we write **a** or $\underline{a}$ from the vector from A to B then the vector from B to A is represented as $-\mathbf{a}$ or $-\underline{a}$, the minus sign being used to indicate the vector is in the opposite direction.

The length of the line segment represents the *magnitude* of the vector. This is indicated by the notation:

$$|\overline{AB}| \quad \text{or} \quad |\mathbf{a}| \quad \text{or} \quad a$$

A vector which is defined as having a magnitude of 1 is termed a *unit vector*, such a vector often being denoted by the symbol $\hat{\mathbf{a}}$.

10.2.1 Like vectors

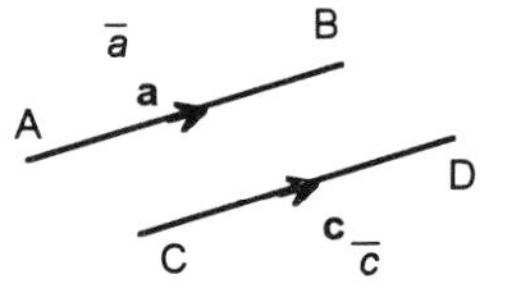

Figure 10.2 *Equal vectors*

Two vectors are equal if they have the same magnitude and direction. Thus the vectors in Figure 10.2 are equal, even if their locations differ. A vector is only defined in terms of its magnitude and direction, its location is not used in its specification. Thus, for Figure 10.2, we can write:

$$\overrightarrow{\mathrm{AB}} = \overrightarrow{\mathrm{CD}} \quad \text{or} \quad \mathbf{a} = \mathbf{c} \quad \text{or} \quad \underline{a} = \underline{c}$$

Vectors for which the location is not significant are termed *free vectors*. For example, the wind velocity might be stated as 15 km/h in a northerly direction. The velocity vector is not associated with any particular point or line of action. Unless stated otherwise it is customary to assume that vectors are free. There are exceptions to this property of the location of a vector having no significance. Force is a vector. If a force acts on a rigid body then the effect on that body depends on the line of action of the force. It thus depends on the location of the vector with the force considered as acting anywhere along its line of action. The term *sliding vector* is generally used for such a vector. The term *position vector* or *bound vector* is used for a vector that emanates from or is directed towards a particular point, e.g. when talking of the velocity at a point on a body. Distinctions between free, sliding and positions vectors do not give rise to any particular problems and indeed it is rarely necessary to do other than note the existence of the distinction.

10.2.2 Orthogonal vectors

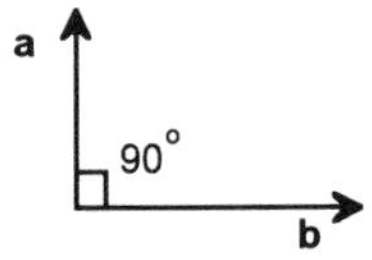

Figure 10.3 *Orthogonal vectors*

If the angle between two vectors **a** and **b** is 90° (Figure 10.3) then the vectors **a** and **b** are said to be *orthogonal*.

10.2.3 Multiplication of vectors by a number

If a vector is multiplied by a positive real number k then the result is another vector with the same direction but with a magnitude that is k times the original magnitude. This is multiplication of a vector by a scalar.

$$k \times \mathbf{a} = k\mathbf{a} \qquad [1]$$

We can consider a vector **a** with magnitude $|\mathbf{a}|$ as being a unit vector, i.e. a vector with a magnitude 1, multiplied by the magnitude $|\mathbf{a}|$, i.e.

$$\mathbf{a} = |\mathbf{a}|\hat{\mathbf{a}} \qquad [2]$$

Note that the magnitude $|\mathbf{a}|$ is a scalar.

10.3 Addition and subtraction of vectors

Consider the following situation involving displacement vectors. An aeroplane flies 100 km due west, then 60 km in a north-westerly direction. What is the resultant displacement of the aeroplane from its start point? If the initial displacement vector is **a** and the second displacement vector is **b**, then what is required is the vector sum **a** + **b**.

One way we can determine the sum involves the *triangle rule* and is shown in Figure 10.4(a).

*The triangle rule can be stated as: to add two vectors **a** and **b** we place the tail of the line segment representing one vector at the head of the line segment representing the other and the line that forms the third side of the triangle represents the vector sum of **a** and **b**.*

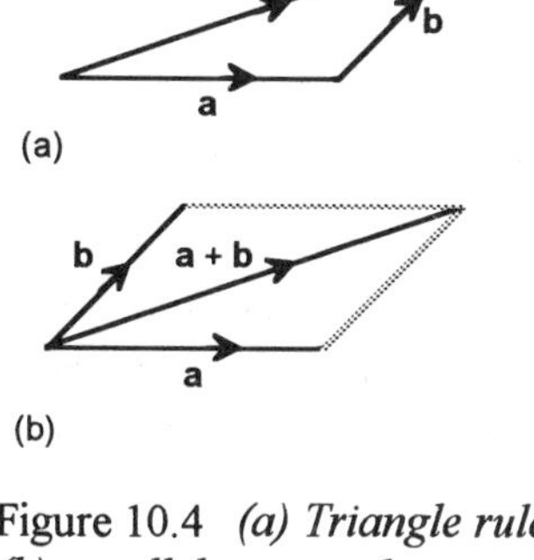

Figure 10.4 *(a) Triangle rule, (b) parallelogram rules*

Note that **a** and **b** have directions that go in one sense round the triangle and the sum **a** + **b** has a direction in the opposite sense. An alternative way of determining the sum involves the *parallelogram rule* and is shown in Figure 10.4(b).

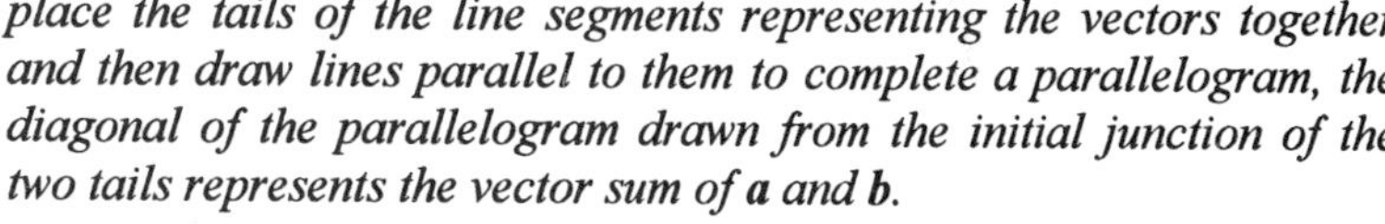

*The parallelogram rule can be stated as: to add two vectors **a** and **b** we place the tails of the line segments representing the vectors together and then draw lines parallel to them to complete a parallelogram, the diagonal of the parallelogram drawn from the initial junction of the two tails represents the vector sum of **a** and **b**.*

Subtraction of vector **b** from **a** is carried out by adding –**b** to **a**, i.e.

$$\mathbf{a} - \mathbf{b} = \mathbf{a} + (-\mathbf{b}) \qquad [3]$$

The addition of **a** and –**b** is carried out using the triangle (Figure 10.5(b)) or parallelogram rules (Figure 10.5(c)). Note that, whatever rule we use, the vector **a** – **b** can be represented by the vector from the end point of **b** to the end point of **a** (Figure 10.5(d)), the vector from the end point of **a** to the end point of **b** being **b** – **a** (Figure 10.5(e)).

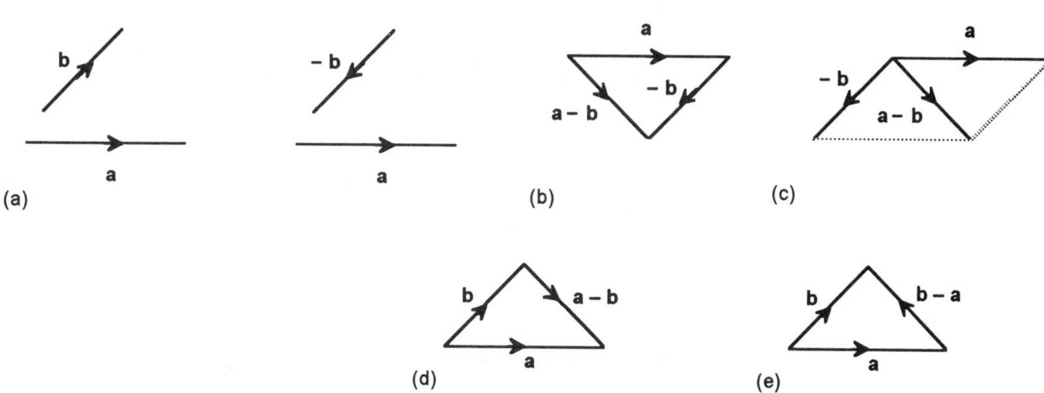

Figure 10.5 *(a) The vectors, (b) subtraction by the triangle rule, (c), subtraction by the parallelogram rule, (d)* **a** – **b**, *(e)* **b** – **a**

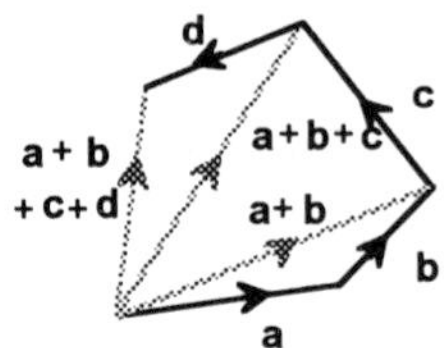

Figure 10.6 *Polygon of vectors*

The triangle rule for the addition of vectors can be extended to the addition of any number of vectors. If the vectors are represented in magnitude and direction by the sides of a *polygon* then their sum is represented in magnitude and direction by the line segment used to close the polygon (Figure 10.6). Essentially what we are doing is determining the sum of vector 1 and vector 2 using the triangle, then adding to this sum vector 3 by a further triangle and repeating this for all the vectors.

If we have a number of vectors and the vectors give a closed triangle or polygon, then, since the line segment needed to close the figure has zero length, the sum of the vectors must be a vector with no magnitude.

Example

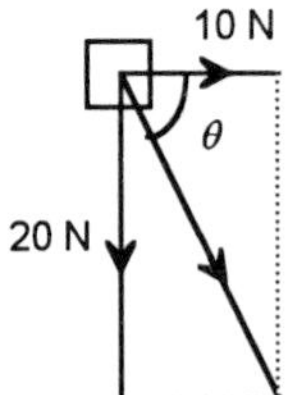

Figure 10.7 *Example*

An object is acted on by two forces, one of which is 10 N and acts horizontally and the other 20 N which acts vertically. Determine the resultant force.

Figure 10.7 shows the vectors and the use of the parallelogram rule to determine the sum. Using the Pythagoras theorem, the diagonal has a size of $\sqrt{20^2 + 10^2} = 22.4$ N. It is at an angle θ to the horizontal force, with $\theta = \tan^{-1}(20/10) = 63.4°$.

Example

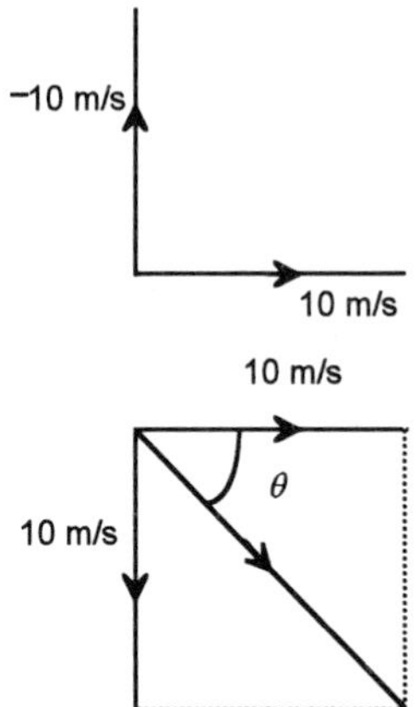

Figure 10.8 *Example*

Determine the resultant velocity if we have velocities of 10 m/s acting horizontally to the right and −10 m/s acting vertically upwards.

This problem requires the addition of two vectors, Figure 10.8 showing the vectors and the use of the parallelogram rule to determine the sum. The magnitude of the sum, i.e. the diagonal of the parallelogram, is given by the Pythagoras theorem as $\sqrt{(10^2 + 10^2)} = 14.1$ m/s and its at an angle below the horizontal of θ where $\theta = \tan^{-1}(10/10) = 45°$.

Example

For the triangle ABC (Figure 10.9) if **a** is the vector from A to B and **b** the vector from B to C, express the vector from C to A in terms of **a** and **b**.

Using the triangle rule, $\overrightarrow{AC} = \overrightarrow{BC} + \overrightarrow{AB}$ and since $\overrightarrow{CA} = -\overrightarrow{AC}$ then we have $\overrightarrow{CA} = -(\mathbf{a} + \mathbf{b})$.

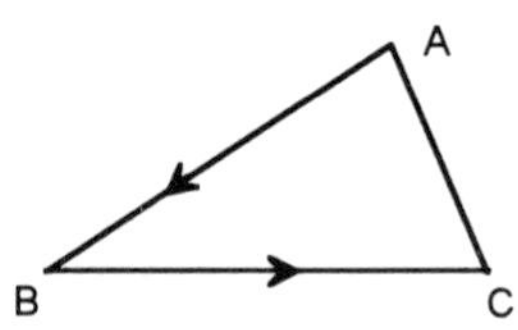

Figure 10.9 *Example*

Revision

1 Determine the sums of the following vectors:

(a) a velocity 30 km/h due east and a velocity of 40 km/h due north,
(b) a displacement of 10 m due north and 20 m due north-east,

(c) a force of 15 N acting at 45° to the right of the vertical and a force of 15 N acting on the same point at 45° to the left of the vertical

2 Determine the differences between the following vectors:

(a) a displacement of 30 m due east and 40 m due north,
(b) forces acting on a point of 50 N at 90° to the horizontal and 40 N at 180° to the horizontal

3 For the triangle ABC, D is the midpoint of AB. If $\overrightarrow{AB}$ represents the vector **a** and $\overrightarrow{BC}$ the vector **b**, express vector $\overrightarrow{DC}$ in terms of **a** and **b**.

4 Determine the vector sums of (hint: sketch possible figures):

(a) $\overrightarrow{AB}+\overrightarrow{BC}+\overrightarrow{CD}+\overrightarrow{DE}$, (b) $\overrightarrow{AB}+\overrightarrow{BC}+\overrightarrow{CD}+\overrightarrow{DA}$,

(c) $\overrightarrow{AB}+\overrightarrow{BC}+\overrightarrow{CD}+\overrightarrow{DB}$

10.3.1 Resolution into components

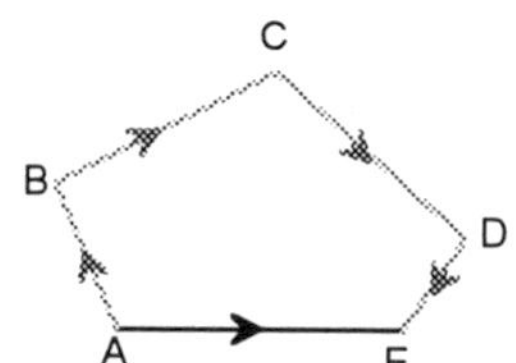

Figure 10.10 *Components*

We can add together any number of vectors using the triangle or polygon rules to give a sum vector. For example, $\overrightarrow{AB}+\overrightarrow{BC}+\overrightarrow{CD}+\overrightarrow{DE}$ has the sum $\overrightarrow{AE}$. This means that we can replace these vectors by the single vector from A to E. The converse is also possible, we can take a single vector $\overrightarrow{AE}$ and replace it by any number of component vectors so long as they form a closed figure which begins at A and ends at E (Figure 10.10).

In mechanics a common technique to aid in the solution of problems is to replace a single vector by two components which are at right angles to each other. Thus for the vector **a** in Figure 10.11 we have **h** and **v** as the horizontal and vertical components. Thus for the magnitudes we must have:

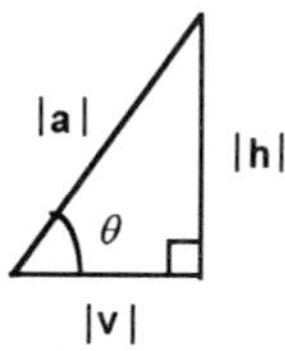

Figure 10.11 *Resolution*

$$|\mathbf{h}| = |\mathbf{a}| \cos \theta \quad [4]$$

$$|\mathbf{v}| = |\mathbf{a}| \sin \theta \quad [5]$$

Example

Express a force of 10 N at 40° to the horizontal in terms of horizontal and vertical components.

Using equations [4] and [5]:

horizontal component = 10 cos 40° = 7.7 N

vertical component = 10 sin 40° = 6.4 N

Revision

5 Determine the horizontal and vertical components of the vectors:

(a) an acceleration of 5 m/s^2 at 60° to the horizontal,
(b) a displacement of 200 mm at 30° to the horizontal,
(c) a force of 2 kN at 45° to the horizontal.

10.3.2 Zero vector

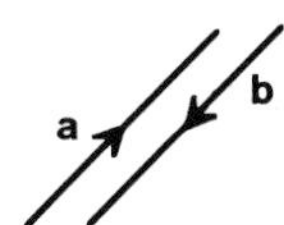

Figure 10.12 *Vector sum is zero*

If vector **b** has the same magnitude as vector **a** but is in the opposite direction (Figure 10.12) then the sum of the two vectors is zero:

$$\mathbf{a} + \mathbf{b} = \mathbf{0} \quad [6]$$

We say that the sum is a vector with zero length and this *zero vector* is denoted by **0**. Such a vector has an arbitrary direction, the direction not being defined.

10.3.3 Basic rules for vector algebra

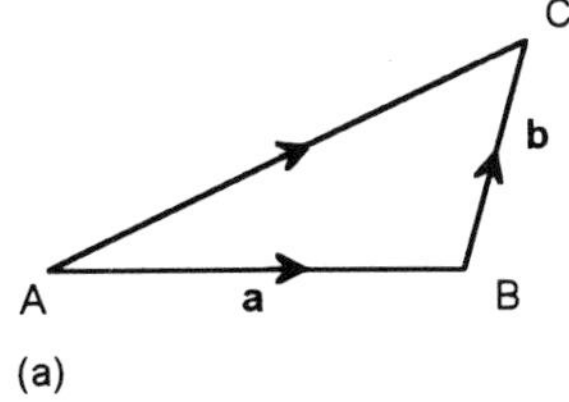

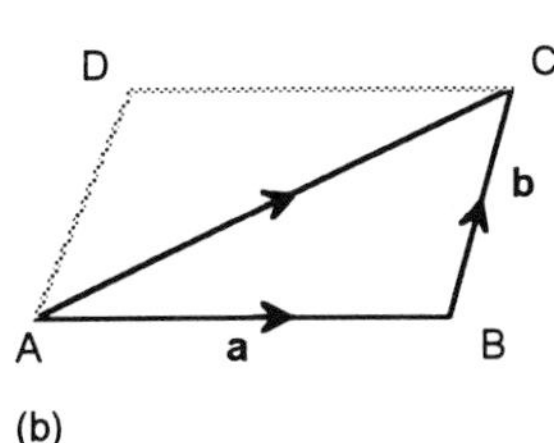

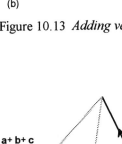

Figure 10.13 *Adding vectors*

Suppose we represent vectors **a** and **b** by AB and BC in Figure 10.13(a). Using the triangle rule of addition:

$$\mathbf{a} + \mathbf{b} = \overrightarrow{AC}$$

Now suppose we complete the parallelogram based on the triangle ABC (Figure 10.13(b)). If we now apply the triangle rule to triangle ADC:

$$\overrightarrow{AC} = \overrightarrow{AD} + \overrightarrow{DC} = \mathbf{b} + \mathbf{a}$$

Hence:

$$\mathbf{a} + \mathbf{b} = \mathbf{b} + \mathbf{a} \quad [7]$$

Vector addition is thus said to be *commutative* since the order in which vectors are added does not matter.

Consider the addition of vectors **a**, **b** and **c** (Figure 10.14). We can add, using the triangle rule, **a** and **b** to give (**a** + **b**) and then add **c** to give (**a** + **b**) + **c**. Or we might add **b** and **c** to give (**b** + **c**) and then add **a** to give **a** + (**b** + **c**). The result is the same in both cases, both being equal to **a** + **b** + **c**.

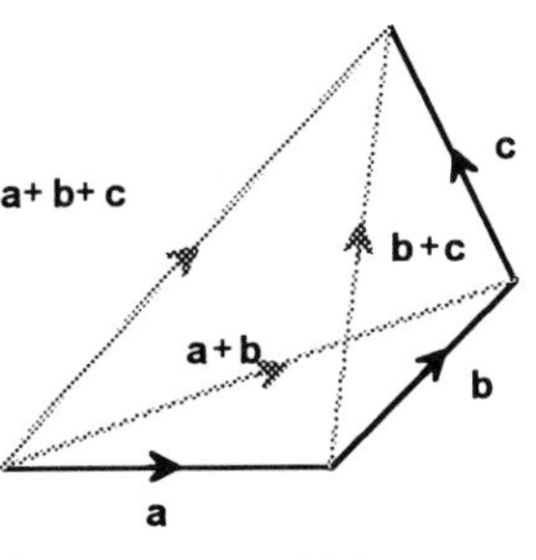

Figure 10.14 *Adding vectors*

$$(\mathbf{a} + \mathbf{b}) + \mathbf{c} = \mathbf{a} + (\mathbf{b} + \mathbf{c}) \quad [8]$$

Vector addition is thus said to obey the *associative law*.

Vector addition is also said to obey the *distributive law*. If we consider adding vectors **a** and **b** using the triangle rule to give **a** + **b**, then if we multiply the vectors by a number k, to give the addition of k**a** and k**b**, the sum is k(**a** + **b**). All we have done is change the scale of the triangle. This can be written as:

$$k(\mathbf{a} + \mathbf{b}) = k\mathbf{a} + k\mathbf{b} \qquad [9]$$

10.4 Components in terms of unit vectors

Consider the x-y plane shown in Figure 10.15. Point P has the co-ordinates (x, y) and is joined to the origin O by the line OP. This line from O to P can be considered to be a position vector **r**, it being a position rather than free vector because one end is located at a fixed point of the origin. The vector can be defined by its two components **a** and **b** along the x and y directions with:

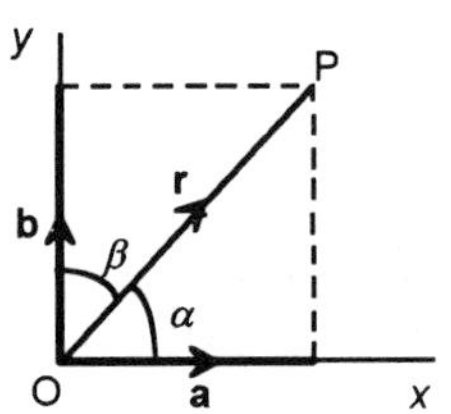

Figure 10.15 *Components*

$$\mathbf{r} = \mathbf{a} + \mathbf{b}$$

If we define **i** to be a unit vector along the x axis then **a** = a**i**, where a is the magnitude of the **a** vector (see Section 10.2.3). If we define **j** to be a unit vector along the y-axis then **b** = b**j**, where b is the magnitude of the **b** vector. Thus:

$$\mathbf{r} = a\mathbf{i} + b\mathbf{j}$$

But a is the x co-ordinate of P and b the y co-ordinate of P. Thus we can write:

$$\mathbf{r} = x\mathbf{i} + y\mathbf{j} \qquad [10]$$

For example, we might specify a position vector as 3**i** + 2**j**. This would mean a position vector from the origin to a point with the co-ordinates (3, 2).

The magnitude of the vector **r** is given by the Pythagoras theorem as:

$$|\mathbf{r}| = \sqrt{x^2 + y^2} \qquad [11]$$

If α and β are the angles the vector **r** makes with the x- and y-axes, then:

$$\cos\alpha = \frac{x}{|\mathbf{r}|} \quad \text{and} \quad \cos\beta = \frac{y}{|\mathbf{r}|} \qquad [12]$$

These are known as the *direction cosines* of **r**.

Example

If **r** = 4**i** + 7**j** determine |**r**| and the angle **r** makes with the x-axis.

Using equation [11]:

$$|r| = \sqrt{4^2 + 7^2} = 8.1$$

The angle with the x-axis is given by equation [12]:

$$\cos \alpha = \frac{4}{8.1}$$

Thus the angle is 60.4°.

Revision

6 For the following vectors determine their magnitudes and angles to the x-axis: (a) $\mathbf{r} = 3\mathbf{i} + 4\mathbf{j}$, (b) $\mathbf{r} = 2\mathbf{i} + 2\mathbf{j}$, (c) $\mathbf{r} = 7\mathbf{i} + 12\mathbf{j}$.

10.4.1 Addition and subtraction of vectors

Consider the addition of the two position vectors $\overrightarrow{OP}$ and $\overrightarrow{OQ}$ shown in Figure 10.16, P having the co-ordinates (x_1, y_1) and Q the co-ordinates (x_2, y_2). Thus:

$$\overrightarrow{OP} = x_1\mathbf{i} + y_1\mathbf{j}$$

$$\overrightarrow{OQ} = x_2\mathbf{i} + y_2\mathbf{j}$$

We can obtain the sum by the use of the parallelogram rule as $\overrightarrow{OR}$. R has the co-ordinates $(x_1 + x_2, y_1 + y_2)$. Thus:

$$\overrightarrow{OR} = (x_1 + x_2)\mathbf{i} + (y_1 + y2)\mathbf{j} \qquad [11]$$

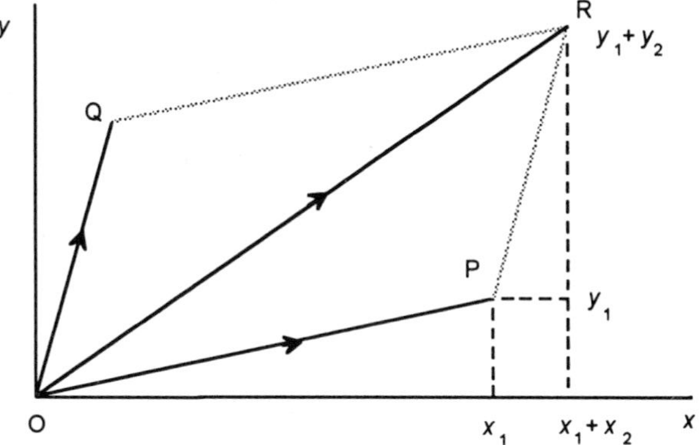

Figure 10.16 *Adding position vectors*

We thus end up with the basic rule for position vectors:

Adding or subtracting position vectors is achieved by adding or subtracting their respective co-ordinates.

Example

If $\mathbf{a} = 2\mathbf{i} + 4\mathbf{j}$ and $\mathbf{b} = 3\mathbf{i} + 5\mathbf{j}$, determine (a) $\mathbf{a} + \mathbf{b}$, (b) $\mathbf{a} - \mathbf{b}$, (c) $\mathbf{a} + 2\mathbf{b}$.

(a) Using the rule given above:

$$\mathbf{a} + \mathbf{b} = (2 + 3)\mathbf{i} + (4 + 5)\,\mathbf{j} = 5\mathbf{i} + 8\mathbf{j}$$

(b) Using the rule given above:

$$\mathbf{a} - \mathbf{b} = (2 - 3)\mathbf{i} + (4 - 5)\,\mathbf{j} = -1\mathbf{i} + -1\mathbf{j}$$

(c) Using the rule given above:

$$\mathbf{a} + 2\mathbf{b} = (2 + 6)\mathbf{i} + (4 + 10)\,\mathbf{j} = 8\mathbf{i} + 14\mathbf{j}$$

Revision

7 If $\mathbf{a} = 2\mathbf{i} + 3\mathbf{j}$ and $\mathbf{b} = -6\mathbf{i} + 4\mathbf{j}$, determine (a) $\mathbf{a} + \mathbf{b}$, (b) $\mathbf{a} - \mathbf{b}$, and (c) $\mathbf{a} + 2\mathbf{b}$.

8 If $\mathbf{a} = 3\mathbf{i} + 2\mathbf{j}$ and $\mathbf{b} = 4\mathbf{i} + 3\mathbf{j}$, determine (a) $\mathbf{a} + \mathbf{b}$, (b) $\mathbf{a} - \mathbf{b}$, and (c) $\mathbf{a} - 2\mathbf{b}$.

9 If $\mathbf{a} = 3\mathbf{i} + 3\mathbf{j}$, $\mathbf{b} = 2\mathbf{i} + 5\mathbf{j}$ and $\mathbf{c} = 2\mathbf{i} - 4\mathbf{j}$, determine (a) $\mathbf{a} + \mathbf{b} + \mathbf{c}$, (b) $\mathbf{a} - \mathbf{b} - \mathbf{c}$, (c) $\mathbf{a} + 2\mathbf{b} - 3\mathbf{c}$.

10.5 Vectors in space

In Section 10.4 the vectors were considered in terms of their components in two directions, i.e. on a plane. Here we extend the consideration to three dimensions. In Section 10.4 the vectors were specified in terms of their components on a pair of axes, x and y. For three dimensions we add a z-axis (Figure 10.17). A vector $\mathbf{r}$ from O to P, with co-ordinates (x, y, z), is then defined by its vector components in the directions x, y and z. If $\mathbf{i}$, $\mathbf{j}$ and $\mathbf{k}$ are the unit vectors in the directions x, y and z, then:

$$\mathbf{r} = x\mathbf{i} + y\mathbf{j} + z\mathbf{k} \qquad [12]$$

The magnitude of $\mathbf{r}$ is given by:

$$|\mathbf{r}| = \sqrt{x^2 + y^2 + z^2} \qquad [13]$$

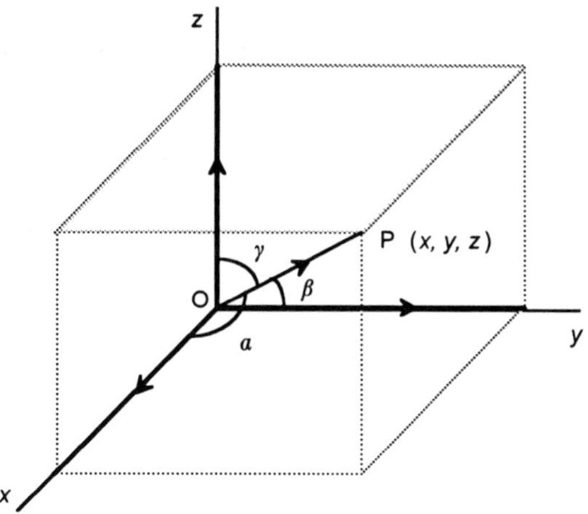

Figure 10.17 *Vector in space*

The direction of a vector in three dimensions is determined by the angles it makes with the three axes, x, y and z, i.e. the angles a, β and γ. With (x, y, z) the co-ordinates of the position vector:

$$\cos a = \frac{x}{|\mathbf{r}|}, \quad \cos\beta = \frac{y}{|\mathbf{r}|} \text{ and } \cos\gamma = \frac{z}{|\mathbf{r}|} \qquad [14]$$

These are termed the *direction cosines*.

Since $|\mathbf{r}| = \sqrt{x^2 + y^2 + z^2}$ (equation [13]) then:

$$|\mathbf{r}|^2 = |\mathbf{r}|^2 \cos^2 a + |\mathbf{r}|^2 \cos^2 \beta + |\mathbf{r}|^2 \cos^2 \gamma$$

and so:

$$\cos^2 a + \cos^2 \beta + \cos^2 \gamma = 1 \qquad [15]$$

If we let $l = \cos a$, $m = \cos \beta$ and $n = \cos \gamma$ then we can write the set of direction cosines for a vector as $[l, m, n]$ with equation [15] giving:

$$l^2 + m^2 + n^2 = 1 \qquad [16]$$

As with the two-dimensional case, we can develop the basic rule for position vectors:

Adding or subtracting position vectors is achieved by adding or subtracting their respective co-ordinates.

Example

Determine the magnitude and the direction cosines of the vector $\mathbf{r} = 2\mathbf{i} + 3\mathbf{j} + 6\mathbf{k}$.

Using equation [13]:

$$|\mathbf{r}| = \sqrt{2^2 + 3^2 + 6^2} = 7$$

The magnitude is thus 7. Using equations [14]:

$$l = \cos\alpha = \frac{2}{7},\quad m = \cos\beta = \frac{3}{7},\quad n = \cos\gamma = \frac{6}{7}$$

Thus the direction cosines are $[\frac{2}{7}, \frac{3}{7}, \frac{6}{7}]$.

Example

If $\mathbf{a} = 2\mathbf{i} + 3\mathbf{j} + 4\mathbf{k}$ and $\mathbf{b} = 3\mathbf{i} - 2\mathbf{j} + 1\mathbf{k}$, determine (a) $\mathbf{a} + \mathbf{b}$, (b) $\mathbf{a} - \mathbf{b}$, (c) $\mathbf{a} + 2\mathbf{b}$.

(a) Using the rule given above:

$$\mathbf{a} + \mathbf{b} = (2 + 3)\mathbf{i} + (3 - 2)\mathbf{j} + (4 + 1)\mathbf{k} = 5\mathbf{i} + 1\mathbf{j} + 5\mathbf{k}$$

(b) Using the rule given above:

$$\mathbf{a} - \mathbf{b} = (2 - 3)\mathbf{i} + (3 + 2)\mathbf{j} + (4 - 1)\mathbf{k} = -1\mathbf{i} + 5\mathbf{j} + 3\mathbf{k}$$

(c) Using the rule given above:

$$\mathbf{a} + 2\mathbf{b} = (2 + 6)\mathbf{i} + (3 - 4)\mathbf{j} + (4 + 2)\mathbf{k} = 8\mathbf{i} - 1\mathbf{j} + 6\mathbf{k}$$

Revision

10 If $\mathbf{a} = 7\mathbf{i} + 3\mathbf{j}$ and $\mathbf{b} = -2\mathbf{i} + 5\mathbf{j}$, determine (a) $\mathbf{a} + \mathbf{b}$, (b) $\mathbf{a} - \mathbf{b}$, (c) $2\mathbf{a} + \mathbf{b}$, (d) $2\mathbf{a} - 3\mathbf{b}$.

11 Determine the magnitude and the direction cosines of $\mathbf{r} = 2\mathbf{i} + 1\mathbf{j} + 2\mathbf{k}$.

12 Determine the magnitude and direction cosines of the sum of the vectors $2\mathbf{i} - 3\mathbf{j} + 4\mathbf{k}$ and $3\mathbf{i} - 7\mathbf{j} + 12\mathbf{k}$.

10.5.1 Angle between two vectors

Consider vectors $\mathbf{a}_1$ and $\mathbf{a}_2$ (Figure 10.18(a)) with $\mathbf{a}_1$ having direction cosines $[l_1, m_1, n_1]$ and $\mathbf{a}_2$ direction cosines $[l_2, m_2, n_2]$. Then:

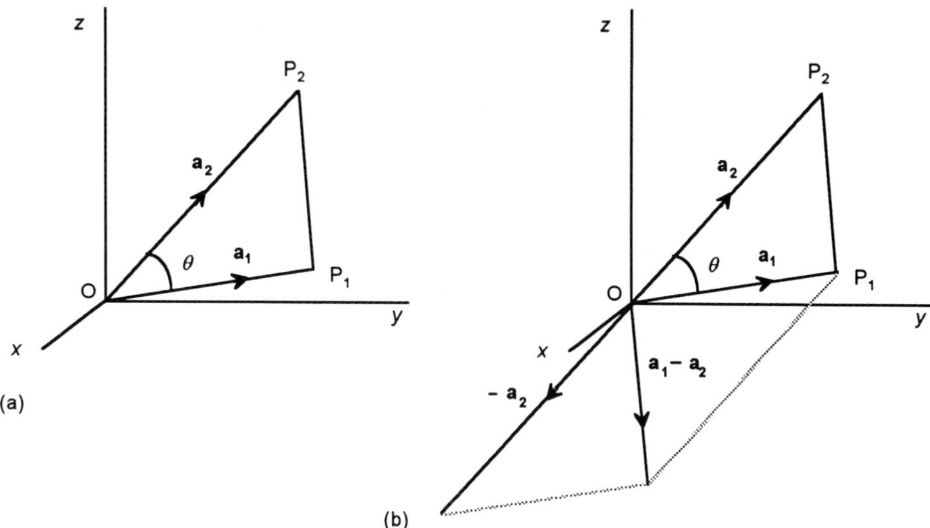

Figure 10.18 *Angle between two vectors*

$$\mathbf{a}_1 = x_1\mathbf{i} + y_1\mathbf{j} + z_1\mathbf{k} = |\mathbf{a}_1|(l_1\mathbf{i} + m_1\mathbf{j} + n_1\mathbf{k}) \qquad [17]$$

$$\mathbf{a}_2 = x_2\mathbf{i} + y_2\mathbf{j} + z_2\mathbf{k} = |\mathbf{a}_2|(l_2\mathbf{i} + m_2\mathbf{j} + n_2\mathbf{k}) \qquad [18]$$

As indicated by Figure 10.18(b), the line P_1P_2 represents $\mathbf{a}_1 - \mathbf{a}_2$ with:

$$\mathbf{a}_1 - \mathbf{a}_2 = (|\mathbf{a}_1|l_1 - |\mathbf{a}_2|l_2)\mathbf{i} + (|\mathbf{a}_1|m_1 - |\mathbf{a}_2|m_2)\mathbf{j} + (|\mathbf{a}_1|n_1 - |\mathbf{a}_2|n_2)\mathbf{k} \qquad [19]$$

Applying the cosine rule to the triangle OP_1P_2:

$$(P_1P_2)^2 = (OP_1)^2 + (OP_2)^2 - 2(OP_1)(OP_2)\cos\theta$$

OP_1 is the magnitude of $\mathbf{a}_1$ and is given by equation [17], OP_2 is the magnitude of $\mathbf{a}_2$ and is given by equation [18], P_1P_2 is the magnitude of $\mathbf{a}_1 - \mathbf{a}_2$ and is given by equation [19]. Thus:

$$\begin{aligned}(|\mathbf{a}_1|l_1 - |\mathbf{a}_2|l_2)^2 + (|\mathbf{a}_1|m_1 - |\mathbf{a}_2|m_2)^2 + (|\mathbf{a}_1|n_1 - |\mathbf{a}_2|n_2)^2 \\ = (|\mathbf{a}_1|l_1)^2 + (|\mathbf{a}_1|m_1)^2 + (|\mathbf{a}_1|n_1)^2 + (|\mathbf{a}_2|l_2)^2 + (|\mathbf{a}_2|m_2)^2 + (|\mathbf{a}_2|n_2)^2 \\ - 2[(|\mathbf{a}_1|l_1)^2 + (|\mathbf{a}_1|m_1)^2 + (|\mathbf{a}_1|n_1)^2][(|\mathbf{a}_2|l_2)^2 + (|\mathbf{a}_2|m_2)^2 + (|\mathbf{a}_2|n_2)^2]\cos\theta\end{aligned}$$

Therefore:

$$2(l_1l_2 + m_1m_2 + n_1n_2) = 2(l_1^2 + m_1^2 + n_1^2)(l_2^2 + m_2^2 + n_2^2)\cos\theta$$

Since $l_1^2 + m_1^2 + n_1^2 = l_2^2 + m_2^2 + n_2^2 = 1$, this can be simplified to give:

$$\cos\theta = l_1l_2 + m_1m_2 + n_1n_2 \qquad [20]$$

For parallel vectors $\theta = 0°$ and for perpendicular vectors $\theta = 90°$.

Example

Determine the angle between the position vectors **a** = 4**i** + 4**j** – 7**k** and **b** = 5**i** – 1**j** + 6**k**.

For **a** we have |**a**| = $\sqrt{(4^2 + 4^2 + 7^2)} = 9$ and thus the direction cosines are 4/9, 4/9, –7/9. For **b** we have |**b**| = $\sqrt{(5^2 + 1^2 + 6^2)} = \sqrt{62}$ and thus direction cosines of 5/√62, –1/√62, 6/√62. Thus, using equation [20]:

$$\cos\theta = \frac{(4\times 5)+(4\times -1)+(-7\times 6)}{9\sqrt{62}} = \frac{-26}{9\sqrt{62}} = -0.367$$

Hence $\theta = 111.5°$.

Revision

13 Determine the angle between the position vectors **a** = 4**i** – 4**j** + 7**k** and **b** = –1**i** + 4**j** + 8**k**.

14 Show that the vectors **a** = 4**i** – 1**j** + 5**k** and **b** = 9**i** + 6**j** – 6**k** are at right angles to each other.

Problems

1 If vector **a** is a velocity of 3 m/s in a north-westerly direction and **b** a velocity of 5 m/s in a westerly direction, determine (a) **a** + **b**, (b) **a** – **b**, (c) **a** – 2**b**.

2 If vector **a** is a displacement of 5 m in a northerly direction and **b** a displacement of 12 m in an easterly direction, determine (a) **a** + **b**, (b) **a** – **b**, (c) **b** – **a**, (d) **a** + 2**b**.

3 ABCD is a quadrilateral. Determine the single vector which is equivalent to (a) $\overrightarrow{AB}+\overrightarrow{BC}$, (b) $\overrightarrow{BC}+\overrightarrow{CD}$, (c) $\overrightarrow{AB}+\overrightarrow{DA}$.

4 If O, A, B, C and D are five points on a plane and $\overrightarrow{OA}$ represents the vector **a**, $\overrightarrow{OB}$ the vector **b**, $\overrightarrow{OC}$ the vector **a** + 2**b**, and $\overrightarrow{OD}$ the vector 2**a** – **b**, express (a) $\overrightarrow{AB}$, (b) $\overrightarrow{BC}$, (c) $\overrightarrow{CD}$, and (d) $\overrightarrow{AC}$ in terms of **a** and **b**.

5 ABCD is a square. A force of 6 N acts along AB, 5 N along BC, 7 N along DB and 9 N along CA. Determine the resultant force.

6 Determine the vector sums of:

(a) $\overrightarrow{AB}+\overrightarrow{BC}+\overrightarrow{CD}$, (b) $\overrightarrow{AB}-\overrightarrow{CB}+\overrightarrow{CD}+\overrightarrow{DE}$,

(c) $\overrightarrow{AB} + \overrightarrow{BC} - \overrightarrow{DC} - \overrightarrow{AD}$, (d) $\overrightarrow{AB} + \overrightarrow{BC} + \overrightarrow{CD} + \overrightarrow{DC}$

7 A point is acted on by two forces, a force of 6 N acting horizontally and a force of 4 N at 20° to the horizontal. Determine the resultant components of the forces in the vertical and horizontal directions.

8 For the following vectors determine their magnitudes and angles to the x-axis: (a) $\mathbf{r} = 2\mathbf{i} + 3\mathbf{j}$, (b) $\mathbf{r} = 5\mathbf{i} + 2\mathbf{j}$, (c) $\mathbf{r} = 3\mathbf{i} + 3\mathbf{j}$.

9 If $\mathbf{a} = -2\mathbf{i} + 3\mathbf{j}$ and $\mathbf{b} = 6\mathbf{i} + 3\mathbf{j}$, determine (a) $\mathbf{a} + \mathbf{b}$, (b) $\mathbf{a} - \mathbf{b}$, and (c) $\mathbf{a} + 2\mathbf{b}$.

10 If $\mathbf{a} = 5\mathbf{i} + 2\mathbf{j}$ and $\mathbf{b} = 2\mathbf{i} + 3\mathbf{j}$, determine (a) $\mathbf{a} + \mathbf{b}$, (b) $\mathbf{a} - \mathbf{b}$, and (c) $\mathbf{a} - 2\mathbf{b}$.

11 If $\mathbf{a} = 6\mathbf{i} + 3\mathbf{j}$, $\mathbf{b} = -2\mathbf{i} + 3\mathbf{j}$ and $\mathbf{c} = 5\mathbf{i} - 4\mathbf{j}$, determine (a) $\mathbf{a} + \mathbf{b} + \mathbf{c}$, (b) $\mathbf{a} - \mathbf{b} - \mathbf{c}$, (c) $\mathbf{a} + 2\mathbf{b} - 3\mathbf{c}$.

12 If $\mathbf{a} = 3\mathbf{i} - 2\mathbf{j} + 1\mathbf{k}$ and $\mathbf{b} = -1\mathbf{i} + 2\mathbf{j} - \mathbf{k}$, determine (a) $\mathbf{a} + \mathbf{b}$, (b) $\mathbf{a} - \mathbf{b}$, (c) $\mathbf{a} + 3\mathbf{b}$, (d) $2\mathbf{a} - \mathbf{b}$.

13 If $\mathbf{a} = 2\mathbf{i} + 2\mathbf{j} + 5\mathbf{k}$ and $\mathbf{b} = 3\mathbf{i} + 2\mathbf{j} - 2\mathbf{k}$, determine (a) $\mathbf{a} + \mathbf{b}$, (b) $\mathbf{a} - \mathbf{b}$, (c) $\mathbf{a} + 3\mathbf{b}$, (d) $2\mathbf{a} - \mathbf{b}$.

14 If $\mathbf{a} = 4\mathbf{i} + 4\mathbf{j} - 7\mathbf{k}$ and $\mathbf{b} = 5\mathbf{i} - 2\mathbf{j} + 6\mathbf{k}$, determine the direction cosines of $\mathbf{a}$, $\mathbf{b}$ and $(\mathbf{a} + \mathbf{b})$.

15 Determine the magnitude and the direction cosines of:

(a) $\mathbf{a} = 3\mathbf{i} + 7\mathbf{j} - 4\mathbf{k}$, (b) $\mathbf{a} = 2\mathbf{i} + 3\mathbf{j} + 5\mathbf{k}$, (c) $\mathbf{a} = -3\mathbf{i} + 5\mathbf{j} + 2\mathbf{k}$

16 Show that the vectors $\mathbf{a} = 6\mathbf{i} + 2\mathbf{j} + 4\mathbf{k}$ and $\mathbf{b} = 3\mathbf{i} + 1\mathbf{j} - 5\mathbf{k}$ are perpendicular to each other.

17 Determine the angle between the position vectors $\mathbf{a} = 2\mathbf{i} - 3\mathbf{j} - 4\mathbf{k}$ and $\mathbf{b} = 4\mathbf{i} + 3\mathbf{j} - 2\mathbf{k}$.

18 The position vectors of points P and Q are $2\mathbf{i} + 3\mathbf{j} - 5\mathbf{k}$ and $4\mathbf{i} - 2\mathbf{j} + 2\mathbf{k}$ respectively. Determine the length and direction cosines of the vector joining P and Q.

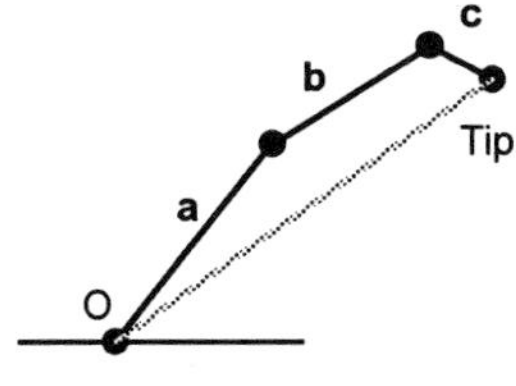

Figure 10.19 *Problem 19*

19 For a robot arm involving rigid links connected by flexible joints (Figure 10.19), the link vectors can be represented by $\mathbf{a} = 10\mathbf{i} + 12\mathbf{j} + 1\mathbf{k}$, $\mathbf{b} = 5\mathbf{i} - 2\mathbf{j} + 8\mathbf{k}$ and $\mathbf{c} = 2\mathbf{i} + 1\mathbf{j} - 4\mathbf{k}$. Determine the position vector of the tip of the robot from O and the length of each link.

11 Products

11.1 Introduction

This chapter is concerned with the algebra for a product of two vectors. The product of a vector and a number is another vector with its magnitude just scaled by the number (see Section 10.2.3). However, the product of two vectors is another matter. Force and displacement are vectors and work is the product of force and the displacement of the force along its line of action or the component of the force in a direction multiplied by the displacement in that direction (Figure 11.1(a)). Work is a scalar. Such a product is called a *scalar product*. The torque on a body is the product of the force and the length of its lever arm, the line of action of the force being perpendicular to the lever arm (Figure 11.1(b)). Torque is a vector. Such a product is called a *vector product*. In this chapter rules are defined for determining scalar products and vector products.

This chapter assumes Chapter 10 and thus a knowledge of the Cartesian system of co-ordinates. A knowledge of determinants is useful.

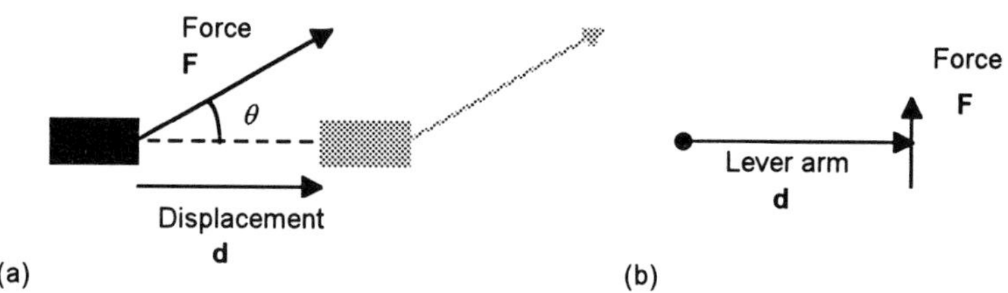

Figure 11.1 *(a) Work as a scalar product, (b) torque as a vector product*

11.2 Scalar product

The work done by a force is the product of the force and the displacement of the force along its line of action or the component of the force in a direction multiplied by the displacement in that direction. Thus for Figure 11.1(a), if we resolve the force into a component $\mathbf{F}\cos\theta$ in the direction of the displacement and $\mathbf{F}\sin\theta$ at right angles to it, the work done W by the force $\mathbf{F}$ moving its point of application through a displacement $\mathbf{d}$ is:

$$W = Fd\cos\theta$$

where F and d are the magnitudes of the force and the displacement. The product of the two vectors of force and displacement is thus a scalar. The product is termed the *scalar product* or *dot product* since we can write it with a dot between the two vectors as:

$$\mathbf{F}\cdot\mathbf{d} = Fd\cos\theta$$

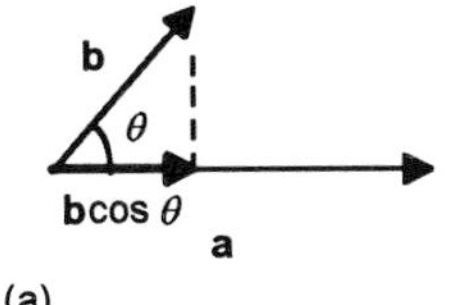

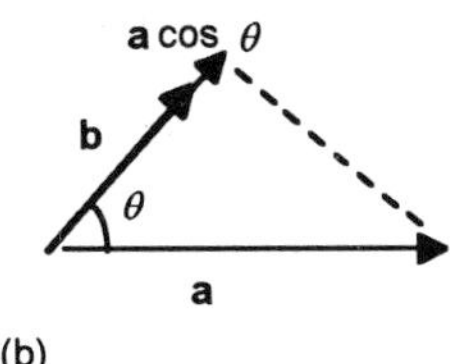

Figure 11.2 *Scalar product*

Thus we are led to define the scalar product of two vectors as:

The scalar product of two vectors is equal to the product of their magnitudes and the cosine of the angle between their directions.

In general for vectors **a** and **b** the scalar product is:

$$\mathbf{a}\cdot\mathbf{b} = ab\cos\theta \qquad [1]$$

We can interpret the scalar product as being equal to the product of the magnitude of vector **a** with the magnitude of the projection of **b** on **a** (Figure 11.2(a)) or the product of magnitude of vector **b** with the magnitude of the projection of **a** on **b** (Figure 11.2(b)).

$$\mathbf{a}\cdot\mathbf{b} = a(b\cos\theta) = b(a\cos\theta) \qquad [2]$$

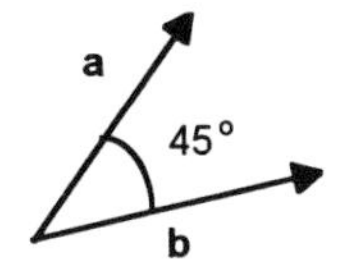

Figure 11.3 *Example*

Example

Determine the scalar product of the two vectors **a** and **b** shown in Figure 11.3. **a** has a magnitude 4 and **b** a magnitude 5.

The scalar product is given by equation [1] as:

$$\mathbf{a}\cdot\mathbf{b} = 4 \times 5\cos 45^\circ = 14.1$$

Revision

1 Determine the scalar product of the pairs of vectors shown in Figure 11.4.

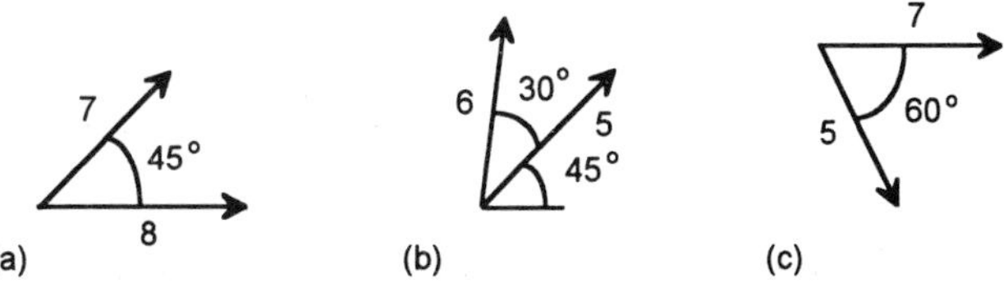

Figure 11.4 *Revision problem 1*

11.2.1 Properties of the scalar product

The following are some of the basic properties of the scalar product:

1 *Two perpendicular vectors*
If the vectors **a** and **b** are perpendicular to each other, then cos 90° = 0 and the scalar product:

$$\mathbf{a}\cdot\mathbf{b} = 0 \qquad [3]$$

Note that if a scalar product is zero, either $\mathbf{a} = 0$ or $\mathbf{b} = 0$ or $\mathbf{a}$ is perpendicular to $\mathbf{b}$.

2 *Two parallel vectors*
If the vectors $\mathbf{a}$ and $\mathbf{b}$ are parallel to each other, then $\cos 0° = 1$ and the scalar product:

$$\mathbf{a}\cdot\mathbf{b} = ab \qquad [4]$$

Note that the scalar product of a vector with itself, i.e. $\mathbf{a}\cdot\mathbf{a}$, is the square of its magnitude.

$$\mathbf{a}\cdot\mathbf{a} = a^2 \qquad [5]$$

3 *Commutative law*
Since (see Figure 11.2 and equation [2]) $\mathbf{a}\cdot\mathbf{b} = a(b \cos \theta) = b(a \cos \theta)$:

$$\mathbf{a}\cdot\mathbf{b} = \mathbf{b}\cdot\mathbf{a} \qquad [6]$$

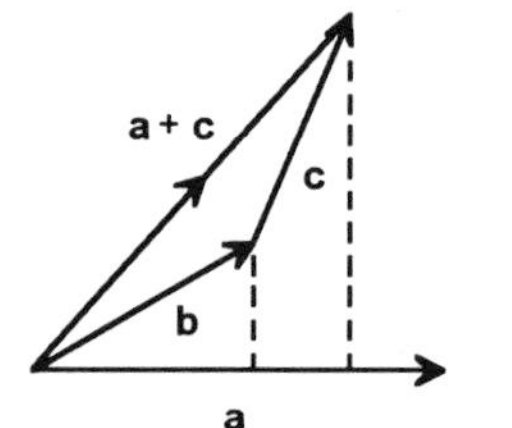

Figure 11.5 *Distributive law*

4 *Distributive law*
If we have three vectors $\mathbf{a}$, $\mathbf{b}$ and $\mathbf{c}$ then, as indicated in Figure 11.5, if we use the triangle of forces and replace $\mathbf{b}$ and $\mathbf{c}$ by $(\mathbf{b} + \mathbf{c})$, the projection of $(\mathbf{b} + \mathbf{c})$ on $\mathbf{a}$ is the same as that of $\mathbf{b}$ and $\mathbf{c}$ considered separately. Thus:

$$\mathbf{a}\cdot(\mathbf{b} + \mathbf{c}) = \mathbf{a}\cdot\mathbf{b} + \mathbf{b}\cdot\mathbf{c} \qquad [7]$$

Example

If $\mathbf{a} = 3\mathbf{b}$ simplify (a) $\mathbf{a}\cdot\mathbf{b}$, (b) $\mathbf{a}\cdot(\mathbf{a} + \mathbf{b})$.

(a) $\mathbf{a}\cdot\mathbf{b} = 3\mathbf{b}\cdot\mathbf{b}$ and thus the scalar product is $3b^2$.
(b) $3\mathbf{b}\cdot(3\mathbf{b} + \mathbf{b}) = 12\mathbf{b}\cdot\mathbf{b}$ and thus the scalar product is $12b^2$.

Revision

2 If $\mathbf{a} = \mathbf{b}$ simplify (a) $\mathbf{a}\cdot\mathbf{b}$, (b) $\mathbf{a}\cdot(\mathbf{a} + \mathbf{b})$, (c) $\mathbf{a}\cdot(\mathbf{a} + 2\mathbf{b})$.

11.2.2 Scalar product in terms of components of vectors

Now consider two vectors $\mathbf{a}$ and $\mathbf{b}$ expressed in terms of their components as $\mathbf{a} = a_1\mathbf{i} + a_2\mathbf{j} + a_3\mathbf{k}$ and $\mathbf{b} = b_1\mathbf{i} + b_2\mathbf{j} + b_3\mathbf{k}$. Then:

$$\mathbf{a}\cdot\mathbf{b} = (a_1\mathbf{i} + a_2\mathbf{j} + a_3\mathbf{k})\cdot(b_1\mathbf{i} + b_2\mathbf{j} + b_3\mathbf{k})$$

$$= a_1b_1\mathbf{i}\cdot\mathbf{i} + a_1b_1\mathbf{i}\cdot\mathbf{j} + a_1b_3\mathbf{i}\cdot\mathbf{k} + a_2b_1\mathbf{j}\cdot\mathbf{i} + a_2b_2\mathbf{j}\cdot\mathbf{j} + a_2b_3\mathbf{j}\cdot\mathbf{k} + a_3b_1\mathbf{k}\cdot\mathbf{i} + a_3b_2\mathbf{k}\cdot\mathbf{j} + a_3b_3\mathbf{k}\cdot\mathbf{k}$$

Since **i**·**i**, **j**·**j** and **k**·**k** are products of the same unit vector then, using equation [5], they have the value 1. Since **i**, **j** and **k** are three unit vectors at right angles to each other then, using equation [3], **i**·**j** = **i**·**k** = **j**·**i** = **j**·**k** = **k**·**i** = **k**·**j** = 0. Thus:

$$\mathbf{a}\cdot\mathbf{b} = a_1b_1 + a_2b_2 + a_3b_3 \qquad [8]$$

Note that this gives:

$$\mathbf{a}\cdot\mathbf{a} = a_1^2 + a_2^2 + a_3^2 \qquad [9]$$

Example

Determine the scalar products of: (a) 2**i** + 3**j** + 4**k** and 4**i** + 2**j** + 3**k**, and (b) 2**i** + 2**j** – 2**k** and 4**i** – 3**j** + 1**k**.

(a) Using equation [8]:

scalar product = (2 × 4) + (3 × 2) + (4 × 3) = 26

(b) Using equation [8]:

scalar product = (2 × 4) + (2 × –3) + (–2 × 1) = 0

The two vectors are at right angles to each other.

Revision

3 Determine the scalar products of:

(a) 3**i** + 2**j** – 8**k** and 2**i** + 5**j** + 2**k**, (b) 3**i** + 4**j** and 2**i** – 7**j**,

(c) 2**i** and 3**i** + 4**j** + 5**k**, (d) 1**i** – 4**j** + 5**k** and 2**i** – 3**j** + 2**k**

4 If **a** = 2**i** + 3**j** + 4**k**, determine the scalar product **a**·**a**.

11.2.3 Angle between vectors

From the definition of the scalar product (equation [1]):

$$\mathbf{a}\cdot\mathbf{b} = ab\cos\theta$$

thus the angle θ between vectors **a** and **b** is given by:

$$\cos\theta = \frac{\mathbf{a}\cdot\mathbf{b}}{ab} \qquad [10]$$

Example

Determine the angle between the vectors $3\mathbf{i} + 1\mathbf{j} + 2\mathbf{k}$ and $2\mathbf{i} + 1\mathbf{j} - 1\mathbf{k}$.

Using equation [10]:

$$\cos\theta = \frac{\mathbf{a}\cdot\mathbf{b}}{ab} = \frac{3\times2+1\times1+2\times(-1)}{\sqrt{3^2+1^1+2^2}\sqrt{2^2+1^2+(-1)^2}} = \frac{5}{\sqrt{84}}$$

Hence $\theta = 56.9°$.

Revision

4 Determine the angles between the following vectors:

(a) $1\mathbf{i} - 1\mathbf{j} - 1\mathbf{k}$ and $2\mathbf{i} + 2\mathbf{j} + 1\mathbf{k}$, (b) $2\mathbf{i} + 1\mathbf{j} + 1\mathbf{k}$ and $1\mathbf{i} + 3\mathbf{j} - 1\mathbf{k}$

11.3 Vector product

Consider the torque resulting from a force **F** applied to an object at a point P (Figure 11.6(a)). The torque is the product of the force and the lever arm vector **r** from the axis of rotation O, the directions of **F** and **r** being at right angles. The magnitude of the torque:

$$T = |\mathbf{r}||\mathbf{F}| = rF$$

If we consider the general case where the force **F** is not at right angles to the lever arm vector **r** but at some angle θ (Figure 11.6(b)), then we resolve the force into two components. The component **F** cos θ is along the same direction as **r** and exerts no torque on the body. The **F** sin θ is the component which exerts a torque. Thus the magnitude of the torque is given by:

$$T = rF \sin\theta$$

Torque is a vector quantity since it has both a direction and a magnitude.

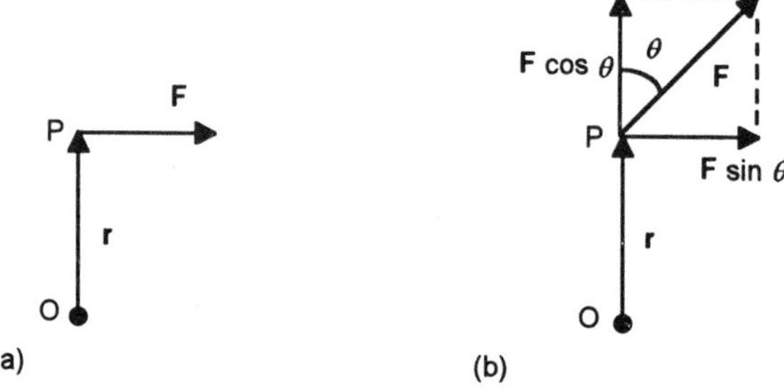

Figure 11.6 *Torque*

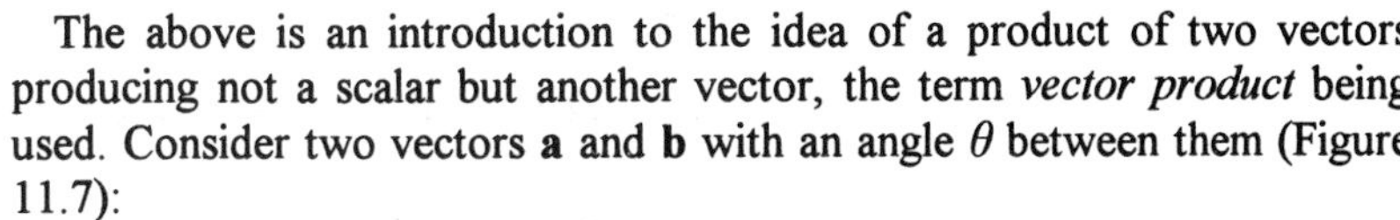
The above is an introduction to the idea of a product of two vectors producing not a scalar but another vector, the term *vector product* being used. Consider two vectors **a** and **b** with an angle θ between them (Figure 11.7):

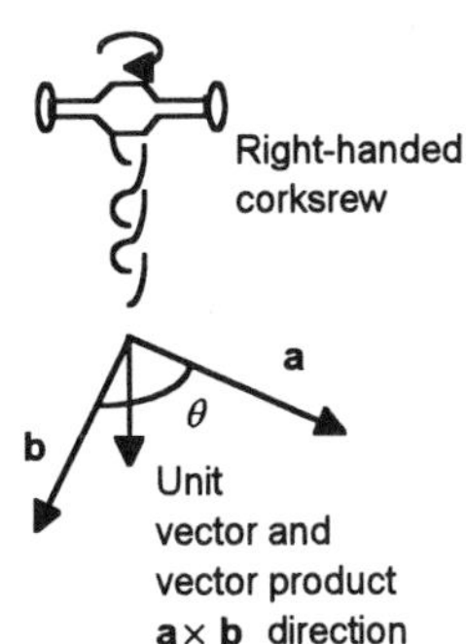

Figure 11.7 *Vector product*

The vector product of vectors **a** and **b** *is written as* $\mathbf{a} \times \mathbf{b}$, *this sometimes being referred to as the cross product, and is defined as being:*

$$\mathbf{a} \times \mathbf{b} = (ab \sin \theta)\mathbf{n} \qquad [11]$$

where **n** *is a unit vector at right angles to both* **a** *and* **b**. *The direction of the unit vector is given by the corkscrew rule: if a right-handed corkscrew rotates clockwise from* **a** to **b** *then it travels vertically in the direction of the unit vector.*

Note that the vector product $\mathbf{b} \times \mathbf{a}$ would be in the opposite direction, i.e. upwards in Figure 11.7. This is because the rotation is from **b** to **a**. Thus if we have $\mathbf{a} \times \mathbf{b} = (ab \sin \theta)\mathbf{n}$ then:

$$\mathbf{b} \times \mathbf{a} = (ab \sin \theta)(-\mathbf{n}) \qquad [12]$$

11.3.1 Vector area

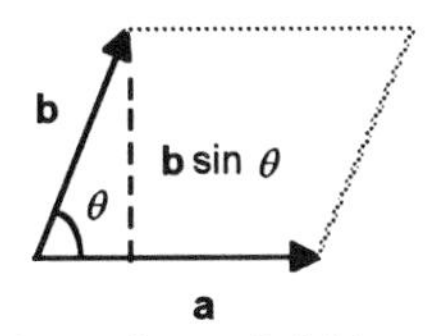

Figure 11.8 *Vector area*

The magnitude of the vector product $\mathbf{a} \times \mathbf{b}$ is $ab \sin \theta$. If we consider the parallelogram (Figure 11.8) formed by the vectors **a** and **b** then the area of the parallelogram is $ab \sin \theta$. The magnitude of the vector product is thus the area of the parallelogram that is defined by **a** and **b**.

11.3.2 Properties of the vector product

The following are some basic properties of the vector product:

1 **a** *and* **b** *parallel*
If **a** and **b** are parallel then $\theta = 0$ and $\sin \theta = 0$. Thus:

$$\mathbf{a} \times \mathbf{b} = 0 \qquad [13]$$

2 **a** *and* **b** *at right angles*
If **a** and **b** are at right angles then $\theta = 90°$ and $\sin \theta = 1$. Thus:

$$\mathbf{a} \times \mathbf{b} = (ab)\mathbf{n} \qquad [14]$$

3 $\mathbf{a} \times \mathbf{b} = -(\mathbf{b} \times \mathbf{a})$
This was discussed in Section 11.2 and arises from the definition. The vector product $\mathbf{b} \times \mathbf{a}$ is in the opposite direction, i.e. upwards in Figure

11.7, to $\mathbf{a} \times \mathbf{b}$ because the rotation is from **b** to **a**. Thus if we have $\mathbf{a} \times \mathbf{b} = (ab \sin \theta)\mathbf{n}$ then:

$$\mathbf{b} \times \mathbf{a} = (ab \sin \theta)(-\mathbf{n}) = -(\mathbf{b} \times \mathbf{a}) \qquad [15]$$

4 $(\mathbf{a} \times \mathbf{b}) \times \mathbf{c} \neq \mathbf{a} \times (\mathbf{b} \times \mathbf{c})$ *in general*
Only if $\mathbf{a} = \mathbf{0}$ or $\mathbf{b} = \mathbf{0}$ or $\mathbf{c} = \mathbf{0}$ is $(\mathbf{a} \times \mathbf{b}) \times \mathbf{c} = \mathbf{a} \times (\mathbf{b} \times \mathbf{c})$, each side then reducing to the zero vector. The vector $\mathbf{a} \times \mathbf{b}$ is perpendicular to both **a** and **b** and is thus at right angles to the plane containing **a** and **b**. The vector $\mathbf{b} \times \mathbf{c}$ is perpendicular to both **b** and **c** and is thus at right angles to the plane containing **b** and **c**. Hence, in general $(\mathbf{a} \times \mathbf{b}) \times \mathbf{c}$ and $\mathbf{a} \times (\mathbf{b} \times \mathbf{c})$ are different vectors.

5 $\mathbf{a} \times (\mathbf{b} + \mathbf{c}) = \mathbf{a} \times \mathbf{b} + \mathbf{a} \times \mathbf{c}$
The proof of this is rather protracted and is omitted here.

6 $\mathbf{a} \times (k\mathbf{b}) = k(\mathbf{a} \times \mathbf{b}) = (k\mathbf{a}) \times \mathbf{b}$
Because of the definition of the vector product (equation [11]) as $\mathbf{a} \times \mathbf{b} = (ab \sin \theta)\mathbf{n}$ then multiplying one of the vectors by a scalar k just multiplies $ab \sin \theta$ by k and it does not matter which of the vectors was multiplied by the scalar.

11.3.3 Vector product in terms of vectors in component form

Consider $\mathbf{a} = a_1\mathbf{i} + a_2\mathbf{j} + a_3\mathbf{k}$ and $\mathbf{b} = b_1\mathbf{i} + b_2\mathbf{j} + b_3\mathbf{k}$. Then, using property 6 given above:

$$\mathbf{a} \times \mathbf{b} = (a_1\mathbf{i} + a_2\mathbf{j} + a_3\mathbf{k}) \times (b_1\mathbf{i} + b_2\mathbf{j} + b_3\mathbf{k})$$

$$= a_1b_1\mathbf{i} \times \mathbf{i} + a_1b_2\mathbf{i} \times \mathbf{j} + a_1b_3\mathbf{i} \times \mathbf{k} + a_2b_1\mathbf{j} \times \mathbf{i} + a_2b_2\mathbf{j} \times \mathbf{j} + a_2b_3\mathbf{j} \times \mathbf{k} + a_3b_1\mathbf{k} \times \mathbf{i} + a_3b_2\mathbf{k} \times \mathbf{j} + a_3b_3\mathbf{k} \times \mathbf{k}$$

But, since for parallel vectors (equation [13]) the vector product is zero, then $\mathbf{i} \times \mathbf{i} = 0$, $\mathbf{j} \times \mathbf{j} = 0$ and $\mathbf{k} \times \mathbf{k} = 0$. For vectors at right angles (equation [14]) the vector product $\mathbf{i} \times \mathbf{j} = 1 \times 1 = 1$ in the direction of the **k** vector and so, since **k** is a unit vector, $\mathbf{i} \times \mathbf{j} = \mathbf{k}$. Similarly $\mathbf{j} \times \mathbf{k} = \mathbf{i}$ and $\mathbf{k} \times \mathbf{i} = \mathbf{j}$. Using equation [15], $\mathbf{i} \times \mathbf{j} = -(\mathbf{j} \times \mathbf{i})$, $\mathbf{j} \times \mathbf{k} = -(\mathbf{k} \times \mathbf{j})$ and $\mathbf{k} \times \mathbf{i} = -(\mathbf{i} \times \mathbf{k})$. Thus the above equation becomes:

$$\mathbf{a} \times \mathbf{b} = 0 + a_1b_2\mathbf{k} + a_1b_3(-\mathbf{j}) + a_2b_1(-\mathbf{k}) + 0 + a_2b_3\mathbf{i} + a_3b_1\mathbf{j} + a_3b_2(-\mathbf{i}) + 0$$

and so:

$$\mathbf{a} \times \mathbf{b} = (a_2b_3 - a_3b_2)\mathbf{i} + (a_3b_1 - a_1b_3)\mathbf{j} + (a_1b_2 - a_2b_1)\mathbf{k} \qquad [16]$$

This is generally more easily remembered when expressed in the determinant form (see Part 5) as:

$$\mathbf{a} \times \mathbf{b} = \begin{vmatrix} \mathbf{i} & \mathbf{j} & \mathbf{k} \\ a_1 & a_2 & a_3 \\ b_1 & b_2 & b_3 \end{vmatrix} = \mathbf{i}\begin{vmatrix} a_2 & a_3 \\ b_2 & b_3 \end{vmatrix} - \mathbf{j}\begin{vmatrix} a_1 & a_3 \\ b_1 & b_3 \end{vmatrix} + \mathbf{k}\begin{vmatrix} a_1 & a_2 \\ b_1 & b_2 \end{vmatrix} \qquad [17]$$

Example

Determine the $\mathbf{a} \times \mathbf{b}$ when $\mathbf{a} = 3\mathbf{i} - 2\mathbf{j} + 1\mathbf{k}$ and $\mathbf{b} = 2\mathbf{j} - 4\mathbf{k}$.

Using equation [16]:

$$\mathbf{a} \times \mathbf{b} = (a_2b_3 - a_3b_2)\mathbf{i} + (a_3b_1 - a_1b_3)\mathbf{j} + (a_1b_2 - a_2b_1)\mathbf{k}$$

$$= [(-2)(-4) - (1)(2)]\mathbf{i} + [(1)(0) - (3)(-4)]\mathbf{j} + [(3)(2) - (-2)(0)]\mathbf{k}$$

$$= 6\mathbf{i} + 12\mathbf{j} + 6\mathbf{k}$$

Alternatively, using equation [17]:

$$\mathbf{a} \times \mathbf{b} = \begin{vmatrix} \mathbf{i} & \mathbf{j} & \mathbf{k} \\ 3 & -2 & 1 \\ 0 & 2 & -4 \end{vmatrix} = \mathbf{i}\begin{vmatrix} -2 & 1 \\ 2 & -4 \end{vmatrix} - \mathbf{j}\begin{vmatrix} 3 & 1 \\ 0 & -4 \end{vmatrix} + \mathbf{k}\begin{vmatrix} 3 & -2 \\ 0 & 2 \end{vmatrix}$$

$$= [(-2)(-4) - (1)(2)]\mathbf{i} - [(3)(-4) - (0)(1)]\mathbf{j} + [(3)(2) - (-2)(0)]\mathbf{k}$$

$$= 6\mathbf{i} + 12\mathbf{j} + 6\mathbf{k}$$

Example

Determine the area of a triangle with two adjacent sides described by $\mathbf{a} = 2\mathbf{i} + 3\mathbf{j}$ and $\mathbf{b} = -2\mathbf{i} + 4\mathbf{j}$.

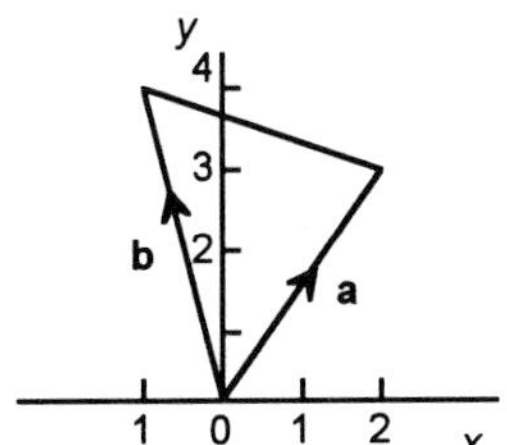

Figure 11.9 *Example*

Figure 11.9 shows the triangle. The area of the parallelogram with **a** and **b** as adjacent sides is given by the vector product $\mathbf{a} \times \mathbf{b}$. The area of the triangle is half this area.

$$\mathbf{a} \times \mathbf{b} = (2\mathbf{i} + 3\mathbf{j}) \times (-2\mathbf{i} + 4\mathbf{j}) = -4\mathbf{i} \times \mathbf{i} + 8\mathbf{i} \times \mathbf{j} - 6\mathbf{j} \times \mathbf{i} + 12\mathbf{j} \times \mathbf{j}$$

Since $\mathbf{i} \times \mathbf{i} = \mathbf{j} \times \mathbf{j} = 0$ and $\mathbf{j} \times \mathbf{i} = -\mathbf{i} \times \mathbf{j}$ then:

$$\mathbf{a} \times \mathbf{b} = 14\mathbf{i} \times \mathbf{j}$$

But $\mathbf{i} \times \mathbf{j} = \mathbf{k}$ and so $\mathbf{a} \times \mathbf{b} = 14\mathbf{k}$. The area of the parallelogram is 14 square units and so the area of the triangle is 7 square units.

Example

Determine a vector perpendicular to both $\mathbf{a} = 3\mathbf{i} + 1\mathbf{j} - 2\mathbf{k}$ and $\mathbf{b} = 2\mathbf{i} + 3\mathbf{j} - 1\mathbf{k}$.

This requires the determination of the vector product $\mathbf{a} \times \mathbf{b}$ since this is a vector perpendicular to $\mathbf{a}$ and $\mathbf{b}$.

$$\mathbf{a} \times \mathbf{b} = (3\mathbf{i} + 1\mathbf{j} - 2\mathbf{k}) \times (2\mathbf{i} + 3\mathbf{j} - 1\mathbf{k})$$

Using equation [16]:

$$\mathbf{a} \times \mathbf{b} = (a_2b_3 - a_3b_2)\mathbf{i} + (a_3b_1 - a_1b_3)\mathbf{j} + (a_1b_2 - a_2b_1)\mathbf{k}$$

$$= (-1 + 6)\mathbf{i} + (-4 + 3)\mathbf{j} + (9 - 2)\mathbf{k} = 5\mathbf{i} - 1\mathbf{j} + 7\mathbf{k}$$

Revision

6 Determine the $\mathbf{a} \times \mathbf{b}$ when:

(a) $\mathbf{a} = 2\mathbf{i} + 1\mathbf{j} - 5\mathbf{k}$ and $\mathbf{b} = 4\mathbf{i} - 2\mathbf{j} - 3\mathbf{k}$,

(b) $\mathbf{a} = -1\mathbf{i} - 1\mathbf{j} - 1\mathbf{k}$ and $\mathbf{b} = 4\mathbf{i} - 2\mathbf{j} + 3\mathbf{k}$,

(c) $\mathbf{a} = 2\mathbf{i} + 4\mathbf{j} + 3\mathbf{k}$ and $\mathbf{b} = 1\mathbf{i} + 3\mathbf{j} - 2\mathbf{k}$

7 Determine the area of the parallelogram with adjacent sides described by $\mathbf{a} = 3\mathbf{i} - 4\mathbf{j}$ and $\mathbf{b} = 4\mathbf{i} - 3\mathbf{j}$.

8 Determine a vector perpendicular to both $\mathbf{a} = 3\mathbf{i} - 1\mathbf{j} + 1\mathbf{k}$ and $\mathbf{b} = 1\mathbf{i} + 2\mathbf{j} + 2\mathbf{k}$.

9 Determine a unit vector perpendicular to both $\mathbf{a} = 2\mathbf{i} - 1\mathbf{j} + 3\mathbf{k}$ and $\mathbf{b} = 3\mathbf{i} + 2\mathbf{j} - 4\mathbf{k}$. (Hint: determine the vector and divide by its magnitude to obtain the unit vector.)

11.4 Product of three vectors

$(\mathbf{b} \times \mathbf{c})$ is a vector. Consider the product of that vector with a third vector $\mathbf{c}$. The product can take two basic forms:

1 $\mathbf{a}\cdot(\mathbf{b} \times \mathbf{c})$
This is known as the *scalar triple product*.

2 $\mathbf{a} \times (\mathbf{b} \times \mathbf{c})$
This is known as the *vector triple product*.

11.4.1 The scalar triple product

If $\mathbf{a} = a_1\mathbf{i} + a_2\mathbf{j} + a_3\mathbf{k}$, $\mathbf{b} = b_1\mathbf{i} + b_2\mathbf{j} + b_3\mathbf{k}$ and $\mathbf{c} = c_1\mathbf{i} + c_2\mathbf{j} + c_3\mathbf{k}$ then, using equation [16]:

$$\mathbf{b} \times \mathbf{c} = (b_2c_3 - b_3c_2)\mathbf{i} + (b_3c_1 - b_1c_3)\mathbf{j} + (b_1c_2 - b_2c_1)\mathbf{k}$$

Then, using equation [8]:

$$\mathbf{a}\cdot(\mathbf{b} \times \mathbf{c}) = a_1(\mathrm{b}_2\mathrm{c}_3 - b_3c_2) + a_2(b_3c_1 - b_1c_3) + a_3(b_1c_2 - b_2c_1) \qquad [18]$$

This may be written in determinant form as:

$$\mathbf{a}\cdot(\mathbf{b} \times \mathbf{c}) = \begin{vmatrix} a_1 & a_2 & a_3 \\ b_1 & b_2 & b_3 \\ c_1 & c_2 & c_3 \end{vmatrix} \qquad [19]$$

The geometrical interpretation of the scalar triple product is shown in Figure 11.10 with a parallelepiped with edges represented by the vectors **a**, **b** and **c**. The area of the base, a parallelogram, is $|\mathbf{a} \times \mathbf{b}| = |\mathbf{a}||\mathbf{b}| \sin \phi$ and thus $\mathbf{a} \times \mathbf{b} = |\mathbf{a}||\mathbf{b}| \sin \phi \ \hat{\mathbf{n}}$ (see Section 11.2.1), where $\hat{\mathbf{n}}$ is the unit vector. Thus:

$$(\mathbf{a} \times \mathbf{b})\cdot\mathbf{c} = |\mathbf{a}||\mathbf{b}| \sin \phi \ \hat{\mathbf{n}}\cdot\mathbf{c}$$

But $|\hat{\mathbf{n}}\cdot\mathbf{c}| = |\mathbf{c}| \cos \theta$ is the perpendicular height of the parallelepiped. Thus:

$$|(\mathbf{a} \times \mathbf{b})\cdot\mathbf{c}| = \text{(base area)(perpendicular height)}$$

$$= \text{volume of parallelepiped}$$

Thus:

The numerical value of the scalar triple product $(\mathbf{a} \times \mathbf{b})\cdot\mathbf{c}$ *gives the volume of a parallelepiped with sides represented by* **a**, **b** and **c**.

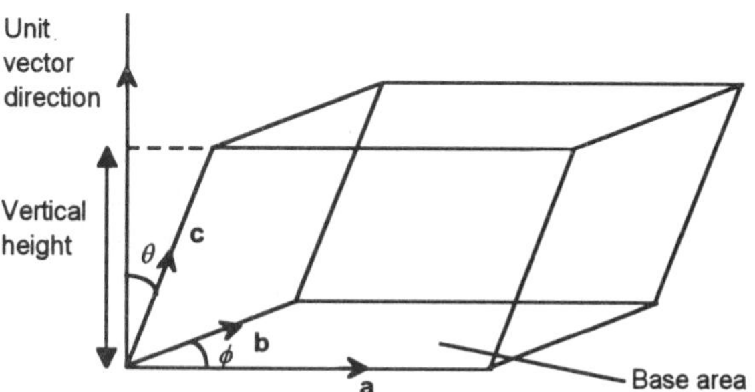

Figure 11.10 *Scalar triple product as the volume of a parallelepiped*

We can deduce a number of properties of the scalar triple product from considering it as representing the volume of a parallelepiped.

1 If two of the vectors **a**, **b** and **c** are parallel then $(\mathbf{a} \times \mathbf{b})\cdot\mathbf{c} = 0$. A consequence of parallel vectors is that the parallelepiped collapses to a plane and hence zero volume.

2 If the three vectors **a**, **b** and **c** are coplanar then $(\mathbf{a} \times \mathbf{b})\cdot\mathbf{c} = 0$. A consequence of the vectors all being in the same plane is that the parallelepiped collapses to a plane and hence zero volume.

3 If $(\mathbf{a} \times \mathbf{b})\cdot\mathbf{c} = 0$ then either $\mathbf{a} = 0$ or $\mathbf{b} = 0$ or $\mathbf{c} = 0$ or two of the vectors are parallel or the three vectors are coplanar.

4 $\mathbf{a}\cdot(\mathbf{b} \times \mathbf{c}) = \mathbf{b}\cdot(\mathbf{c} \times \mathbf{a}) = \mathbf{c}\cdot(\mathbf{a} \times \mathbf{b})$ since they all describe the same parallelepiped. Thus in the scalar triple product, the dot · and the cross × can be interchanged if the vectors are kept in the same cyclic order, i.e. **a**, **b**, **c**.

Example

Determine the scalar triple product of the three vectors $\mathbf{a} = 1\mathbf{i} + 2\mathbf{j}$, $\mathbf{b} = 1\mathbf{i} - 3\mathbf{j} - 1\mathbf{k}$ and $\mathbf{c} = 1\mathbf{i} + 2\mathbf{j} - 1\mathbf{k}$.

Using equation [18]:

$$\mathbf{a}\cdot(\mathbf{b} \times \mathbf{c}) = a_1(b_2c_3 - b_3c_2) + a_2(b_3c_1 - b_1c_3) + a_3(b_1c_2 - b_2c_1)]$$

$$= 1[(-3)(-1) - (-1)(2)] + 2[(-1)(1) - (1)(-1)] + 0$$

$$= 5$$

Alternatively, using equation [19]:

$$\mathbf{a}\cdot(\mathbf{b} \times \mathbf{c}) = \begin{vmatrix} a_1 & a_2 & a_3 \\ b_1 & b_2 & b_3 \\ c_1 & c_2 & c_3 \end{vmatrix} = \begin{vmatrix} 1 & 2 & 0 \\ 1 & -3 & -1 \\ 1 & 2 & -1 \end{vmatrix}$$

$$= 3 - 2 + 0 - 0 + 2 + 2 = 5$$

Example

Determine the value of λ so that the points A $(-\lambda, 3, 4)$, B $(\lambda, -1, 5)$, C $(-2, 1, -1)$ and D $(-\lambda, 2, 3)$ are coplanar.

We can draw a line to join A and B and represent this as a vector. Such a vector can be considered to be the position vectors from the origin to A subtracted from that from the origin to B. Thus, using the rule for adding vectors given in Section 10.5 the vector is $2\lambda\mathbf{i} - 4\mathbf{j} + 1\mathbf{k}$.

Likewise the vector from A to C is $(\lambda - 2)\mathbf{i} - 2\mathbf{j} - 5\mathbf{k}$ and that from A to D is $0 - 1\mathbf{j} - 1\mathbf{k}$. For the three vectors to be coplanar, the scalar triple product must be zero. Thus, using equation [19] (alternatively equation [18] could be used):

$$\mathbf{a}\cdot(\mathbf{b}\times\mathbf{c}) = \begin{vmatrix} a_1 & a_2 & a_3 \\ b_1 & b_2 & b_3 \\ c_1 & c_2 & c_3 \end{vmatrix} = \begin{vmatrix} 2\lambda & -4 & 1 \\ \lambda-2 & -2 & -5 \\ 0 & -1 & -1 \end{vmatrix} = 0$$

$$= 8\lambda + 0 - (\lambda - 2) - 0 - 10\lambda - 4(\lambda - 2) = -7\lambda + 10 = 0$$

Hence for the points to be coplanar λ must be 10/7.

Example

Using determinants, show that $\mathbf{a}\cdot(\mathbf{b}\times\mathbf{c}) = \mathbf{b}\cdot(\mathbf{c}\times\mathbf{a}) = \mathbf{c}\cdot(\mathbf{a}\times\mathbf{b})$.

Applying equation [18] gives:

$$\mathbf{a}\cdot(\mathbf{b}\times\mathbf{c}) = \begin{vmatrix} a_1 & a_2 & a_3 \\ b_1 & b_2 & b_3 \\ c_1 & c_2 & c_3 \end{vmatrix}$$

$$\mathbf{b}\cdot(\mathbf{c}\times\mathbf{a}) = \begin{vmatrix} b_1 & b_2 & b_3 \\ c_1 & c_2 & c_3 \\ a_1 & a_2 & a_3 \end{vmatrix}$$

$$\mathbf{c}\cdot(\mathbf{a}\times\mathbf{b}) = \begin{vmatrix} c_1 & c_2 & c_3 \\ a_1 & a_2 & a_3 \\ b_1 & b_2 & b_3 \end{vmatrix}$$

If we interchange the first and second rows and then the second and third rows of the determinant for $\mathbf{b}\cdot(\mathbf{c}\times\mathbf{a})$ we obtain:

$$\mathbf{b}\cdot(\mathbf{c}\times\mathbf{a}) = \begin{vmatrix} b_1 & b_2 & b_3 \\ c_1 & c_2 & c_3 \\ a_1 & a_2 & a_3 \end{vmatrix} = -\begin{vmatrix} c_1 & c_2 & c_3 \\ b_1 & b_2 & b_3 \\ a_1 & a_2 & a_3 \end{vmatrix} = +\begin{vmatrix} a_1 & a_2 & a_3 \\ b_1 & b_2 & b_3 \\ a_1 & a_2 & a_3 \end{vmatrix}$$

Thus $\mathbf{b}\cdot(\mathbf{c}\times\mathbf{a}) = \mathbf{a}\cdot(\mathbf{b}\times\mathbf{c})$. Similarly, if we interchange the second and third rows and then the first and second rows of the determinant for $\mathbf{b}\cdot(\mathbf{c}\times\mathbf{a})$ we obtain:

$$\mathbf{c}\cdot(\mathbf{a}\times\mathbf{b}) = \begin{vmatrix} c_1 & c_2 & c_3 \\ a_1 & a_2 & a_3 \\ b_1 & b_2 & b_3 \end{vmatrix} = -\begin{vmatrix} c_1 & c_2 & c_3 \\ b_1 & b_2 & b_3 \\ a_1 & a_2 & a_3 \end{vmatrix} = +\begin{vmatrix} a_1 & a_2 & a_3 \\ b_1 & b_2 & b_3 \\ a_1 & a_2 & a_3 \end{vmatrix}$$

Hence $\mathbf{c}\cdot(\mathbf{a}\times\mathbf{b}) = \mathbf{a}\cdot(\mathbf{b}\times\mathbf{c})$.

Revision

10 Determine the scalar triple product of the three vectors:

(a) $\mathbf{a} = 3\mathbf{i} + 2\mathbf{k}$, $\mathbf{b} = 3\mathbf{i} + 3\mathbf{j} - 4\mathbf{k}$ and $\mathbf{c} = 1\mathbf{i} - 2\mathbf{j} + 3\mathbf{k}$,

(b) $\mathbf{a} = 3\mathbf{i} + 2\mathbf{j}$, $\mathbf{b} = 2\mathbf{j} - 2\mathbf{k}$ and $\mathbf{c} = 2\mathbf{k}$,

(c) $\mathbf{a} = 1\mathbf{i} + 3\mathbf{j}$, $\mathbf{b} = -2\mathbf{i} + 3\mathbf{k}$, $\mathbf{c} = 4\mathbf{i} + 1\mathbf{k}$

11 Determine the volume of the parallelepiped having the following vectors as adjacent edges: $\mathbf{a} = 2\mathbf{i} + 1\mathbf{j}$, $\mathbf{b} = 1\mathbf{i} - 3\mathbf{j} + 2\mathbf{k}$ and $\mathbf{c} = -2\mathbf{j} + 3\mathbf{k}$.

12 Determine the value of λ that will make the following points coplanar: (1, 0, –3), (1, 1, –2), (2, 1, –1) and (λ, –1, 0).

13 If $\mathbf{a} = 2\mathbf{i} - 4\mathbf{j} + 3\mathbf{k}$, $\mathbf{b} = -3\mathbf{i} + 2\mathbf{k}$ and $\mathbf{c} = 3\mathbf{j} - 2\mathbf{k}$, determine the scalar triple products: (a) $\mathbf{a}\cdot(\mathbf{b}\times\mathbf{c})$, (b) $(\mathbf{b}\times\mathbf{a})\cdot\mathbf{c}$, (c) $\mathbf{c}\cdot(\mathbf{a}\times\mathbf{b})$ and $(\mathbf{a}\times\mathbf{c})\cdot\mathbf{b}$.

14 Show that $\mathbf{a}\cdot(\mathbf{b}\times\mathbf{c}) = (\mathbf{a}\times\mathbf{b})\cdot\mathbf{c}$.

11.4.2 The vector triple product

If **a**, **b** and **c** are three vectors then a *vector triple product* is $\mathbf{a}\times(\mathbf{b}\times\mathbf{c})$. In terms of the components (equation [16]):

$$\mathbf{b}\times\mathbf{c} = (b_2c_3 - b_3c_2)\mathbf{i} + (b_3c_1 - b_1c_3)\mathbf{j} + (b_1c_2 - b_2c_1)\mathbf{k}$$

Thus:

$$\begin{aligned}\mathbf{a}\times(\mathbf{b}\times\mathbf{c}) &= (a_1\mathbf{i} + a_2\mathbf{j} + a_3\mathbf{k}) \\ &\quad\times[(b_2c_3 - b_3c_2)\mathbf{i} + (b_3c_1 - b_1c_3)\mathbf{j} + (b_1c_2 - b_2c_1)\mathbf{k}] \\ &= (a_2b_1c_2 - a_2b_2c_1 - a_3b_3c_1 + a_3b_1c_3)\mathbf{i} \\ &\quad + (a_3b_2c_3 - a_3b_3c_2 - a_1b_1c_2 + a_1b_2c_1)\mathbf{j} \\ &\quad + (a_1b_3c_1 - a_1b_1c_3 - a_2b_2c_3 + a_2b_3c_2)\mathbf{k} \\ &= (a_2c_2 + a_3c_3)b_1\mathbf{i} + (a_1c_1 + a_3c_3)b_2\mathbf{j} + (a_1c_1 + a_2c_2)b_3\mathbf{k} \\ &\quad - (a_2b_2 + a_3b_3)c_1\mathbf{i} + (a_1b_1 + a_3b_3)c_2\mathbf{j} + (a_1b_1 + a_2b_2)c_3\mathbf{k}\end{aligned}$$

$$= (a_1c_1 + a_2c_2 + a_3c_3)(b_1\mathbf{i} + b_2\mathbf{i} + b_3\mathbf{j}) - (a_1b_1 + a_2b_2 + a_3b_3)(c_1\mathbf{i} + c_2\mathbf{i} + c_3\mathbf{j})$$

Thus:

$$\mathbf{a} \times (\mathbf{b} \times \mathbf{c}) = (\mathbf{a}\cdot\mathbf{c})\mathbf{b} - (\mathbf{a}\cdot\mathbf{b})\mathbf{c} \quad [20]$$

Similarly we can show:

$$(\mathbf{a} \times \mathbf{b}) \times \mathbf{c} = (\mathbf{a}\cdot\mathbf{c})\mathbf{b} - (\mathbf{b}\cdot\mathbf{c})\mathbf{a} \quad [21]$$

From this we can see that $\mathbf{a} \times (\mathbf{b} \times \mathbf{c}) \neq (\mathbf{a} \times \mathbf{b}) \times \mathbf{c}$. The position of the brackets in a triple product is thus important.

Other properties of the vector triple product are:

1 $k[\mathbf{a} \times (\mathbf{b} \times \mathbf{c})] = (k\mathbf{a}) \times (\mathbf{b} \times \mathbf{c}) = \mathbf{a} \times ((k\mathbf{b}) \times \mathbf{c}) = \mathbf{a} \times (\mathbf{b} \times (k\mathbf{c}))$
This can be proved in the following fashion. If we replace $(\mathbf{b} \times \mathbf{c})$ by vector $\mathbf{p}$ then $k[\mathbf{a} \times (\mathbf{b} \times \mathbf{c})] = k(\mathbf{a} \times \mathbf{p}) = k\mathbf{a} \times \mathbf{p} = (k\mathbf{a}) \times (\mathbf{b} \times \mathbf{c})$.

2 $(\mathbf{a} + \mathbf{b}) \times (\mathbf{c} \times \mathbf{d}) = \mathbf{a} \times (\mathbf{c} \times \mathbf{d}) + \mathbf{b} \times (\mathbf{c} \times \mathbf{d})$
This can be proved as follows:

$$(\mathbf{a} + \mathbf{b}) \times (\mathbf{c} \times \mathbf{d}) = ((\mathbf{a} + \mathbf{b})\cdot\mathbf{d})\mathbf{c} - ((\mathbf{a} + \mathbf{b})\cdot\mathbf{c})\mathbf{d}$$

$$= (\mathbf{a}\cdot\mathbf{d} + \mathbf{b}\cdot\mathbf{d})\mathbf{c} - (\mathbf{a}\cdot\mathbf{c} + \mathbf{b}\cdot\mathbf{c})\mathbf{d}$$

$$= (\mathbf{a}\cdot\mathbf{d})\mathbf{c} - (\mathbf{a}\cdot\mathbf{c})\mathbf{d} + (\mathbf{b}\cdot\mathbf{d})\mathbf{c} - (\mathbf{b}\cdot\mathbf{c})\mathbf{d}$$

$$= \mathbf{a} \times (\mathbf{c} \times \mathbf{d}) + \mathbf{b} \times (\mathbf{c} \times \mathbf{d})$$

Example

If $\mathbf{a} = 1\mathbf{i} + 2\mathbf{k}$, $\mathbf{b} = 2\mathbf{i} + 3\mathbf{j} - 3\mathbf{k}$ and $\mathbf{c} = -2\mathbf{j} + 5\mathbf{k}$, determine the vector triple products (a) $\mathbf{a} \times (\mathbf{b} \times \mathbf{c})$ and (b) $2\mathbf{a} \times (\mathbf{b} \times \mathbf{c})$.

(a) Using equation [20]:

$$\mathbf{a} \times (\mathbf{b} \times \mathbf{c}) = (\mathbf{a}\cdot\mathbf{c})\mathbf{b} - (\mathbf{a}\cdot\mathbf{b})\mathbf{c}$$

$$= (a_1c_1 + a_2c_2 + a_3c_3)(b_1\mathbf{i} + b_2\mathbf{i} + b_3\mathbf{j}) - (a_1b_1 + a_2b_2 + a_3b_3)(c_1\mathbf{i} + c_2\mathbf{i} + c_3\mathbf{j})$$

$$= (0 + 0 + 10)(2\mathbf{i} + 3\mathbf{j} - 3\mathbf{k}) - (2 + 0 - 6)(-2\mathbf{j} + 5\mathbf{k})$$

$$= 20\mathbf{i} + 38\mathbf{j} - 50\mathbf{k}$$

(b) Using equation [20] with $\mathbf{a}$ taken as $2\mathbf{i} + 4\mathbf{k}$:

$$\mathbf{a} \times (\mathbf{b} \times \mathbf{c}) = (\mathbf{a}\cdot\mathbf{c})\mathbf{b} - (\mathbf{a}\cdot\mathbf{b})\mathbf{c}$$

$$= (a_1c_1 + a_2c_2 + a_3c_3)(b_1\mathbf{i} + b_2\mathbf{i} + b_3\mathbf{j}) - (a_1b_1 + a_2b_2 + a_3b_3)(c_1\mathbf{i} + c_2\mathbf{i} + c_3\mathbf{j})$$

$$= (0 + 0 + 20)(2\mathbf{i} + 3\mathbf{j} - 3\mathbf{k}) - (4 + 0 - 12)(-2\mathbf{j} + 5\mathbf{k})$$

$$= 40\mathbf{i} + 76\mathbf{j} - 100\mathbf{k}$$

Alternatively we could have used property 1 listed above:

$$k[\mathbf{a} \times (\mathbf{b} \times \mathbf{c})] = (k\mathbf{a}) \times (\mathbf{b} \times \mathbf{c})$$

and thus the effect of the 2 is to double the answer given in (a), hence $40\mathbf{i} + 76\mathbf{j} - 100\mathbf{k}$.

Revision

14 If $\mathbf{a} = 2\mathbf{i} + 3\mathbf{j}$, $\mathbf{b} = 1\mathbf{i} + 2\mathbf{j} - 3\mathbf{k}$ and $\mathbf{c} = 1\mathbf{j} + 5\mathbf{k}$, determine the vector triple products (a) $\mathbf{a} \times (\mathbf{b} \times \mathbf{c})$, (b) $3\mathbf{a} \times (\mathbf{b} \times \mathbf{c})$, (c) $\mathbf{a} \times (2\mathbf{b} \times \mathbf{c})$, (d) $\mathbf{a} \times (2\mathbf{b} \times 2\mathbf{c})$.

Problems

1 Determine the scalar products of:

(a) $3\mathbf{i} + 1\mathbf{j} - 1\mathbf{k}$ and $2\mathbf{i} + 1\mathbf{j} + 2\mathbf{k}$, (b) $3\mathbf{i} + 2\mathbf{j} + 1\mathbf{k}$ and $2\mathbf{i} - 3\mathbf{j}$,

(c) $2\mathbf{j}$ and $3\mathbf{i} + 4\mathbf{j} + 1\mathbf{k}$, (d) $3\mathbf{i} - 5\mathbf{j} + 2\mathbf{k}$ and $2\mathbf{i} + 3\mathbf{j} + 1\mathbf{k}$

2 If $\mathbf{a} = 1\mathbf{i} + 2\mathbf{j} - 3\mathbf{k}$, determine the scalar product $\mathbf{a}\cdot\mathbf{a}$.

3 If $\mathbf{a} = 2\mathbf{b}$ simplify (a) $\mathbf{a}\cdot\mathbf{b}$, (b) $(\mathbf{a} + 3\mathbf{b})\cdot\mathbf{b}$.

4 Determine the angles between the following vectors:

(a) $1\mathbf{i} - 2\mathbf{j} - 1\mathbf{k}$ and $2\mathbf{i} + 1\mathbf{j} + 3\mathbf{k}$, (b) $2\mathbf{i} + 2\mathbf{j} + 3\mathbf{k}$ and $1\mathbf{i} + 2\mathbf{j} - 1\mathbf{k}$

5 If $\mathbf{a} = 3\mathbf{i} - 6\mathbf{j} + 3\mathbf{k}$ and $\mathbf{b} = 2\mathbf{i} + 2\mathbf{j} - \mathbf{k}$, determine (a) $\mathbf{a}\cdot\mathbf{b}$ and (b) the angle between the vectors.

6 Determine the $\mathbf{a} \times \mathbf{b}$ when:

(a) $\mathbf{a} = 2\mathbf{i} + 1\mathbf{j}$ and $\mathbf{b} = 2\mathbf{i} - 1\mathbf{j} + 1\mathbf{k}$,

(b) $\mathbf{a} = 2\mathbf{i} + 3\mathbf{j} + 1\mathbf{k}$ and $\mathbf{b} = -1\mathbf{i} + 2\mathbf{j} + 4\mathbf{k}$,

(c) $\mathbf{a} = 1\mathbf{i} - 2\mathbf{j} + 3\mathbf{k}$ and $\mathbf{b} = 2\mathbf{i} - 1\mathbf{j} - 1\mathbf{k}$.

7 Determine the area of a triangle with two adjacent sides described by $\mathbf{a} = -1\mathbf{i} + 4\mathbf{j}$ and $\mathbf{b} = 2\mathbf{i} + 6\mathbf{j}$.

8 Determine a vector perpendicular to both $\mathbf{a} = 2\mathbf{i} - 3\mathbf{j} + 1\mathbf{k}$ and $\mathbf{b} = 1\mathbf{i} + 2\mathbf{j} - 4\mathbf{k}$.

9 Determine a vector perpendicular to both $\mathbf{a} = 1\mathbf{i} + 1\mathbf{j}$ and $\mathbf{b} = 2\mathbf{i} + 1\mathbf{k}$.

10 If $\mathbf{a} \times \mathbf{b} = 2\mathbf{i} + 1\mathbf{j} - 3\mathbf{k}$, what is $\mathbf{b} \times \mathbf{a}$?

11 If $\mathbf{a} = 2\mathbf{i} + 3\mathbf{k}$, $\mathbf{b} = 1\mathbf{i} - 1\mathbf{j} + 2\mathbf{k}$ and $\mathbf{c} = 4\mathbf{i} + 3\mathbf{j} - 1\mathbf{k}$, determine:

(a) $\mathbf{a} \times (\mathbf{b} \times \mathbf{c})$, (b) $(\mathbf{a} \times \mathbf{b}) \times \mathbf{c}$, (c) $(\mathbf{a} \times \mathbf{b}) \times (\mathbf{a} \times \mathbf{c})$

12 Determine the scalar triple product of the three vectors:

(a) $\mathbf{a} = 1\mathbf{i} + 2\mathbf{k}$, $\mathbf{b} = 1\mathbf{i} + 3\mathbf{j} - 1\mathbf{k}$ and $\mathbf{c} = 2\mathbf{i} - 2\mathbf{j} + 3\mathbf{k}$,

(b) $\mathbf{a} = 2\mathbf{i} + 2\mathbf{j}$, $\mathbf{b} = 3\mathbf{j} - 1\mathbf{k}$ and $\mathbf{c} = 2\mathbf{j}$,

(c) $\mathbf{a} = 1\mathbf{i} + 2\mathbf{j}$, $\mathbf{b} = 2\mathbf{i} + 3\mathbf{j}$, $\mathbf{c} = 2\mathbf{i}$

13 Determine the value of λ that will make the vectors $\mathbf{a} = 1\mathbf{i} - 1\mathbf{j} + 1\mathbf{k}$, $\mathbf{b} = 1\mathbf{i} + 2\mathbf{k} - 2\mathbf{j}$ and $\mathbf{c} = 3\mathbf{i} + \lambda\mathbf{k} + 4\mathbf{j}$ coplanar.

14 Show that $\mathbf{a}\cdot(\mathbf{b} \times \mathbf{c}) = -\mathbf{c}\cdot(\mathbf{b} \times \mathbf{a})$.

15 If $\mathbf{a} = 1\mathbf{i} + 2\mathbf{k}$, $\mathbf{b} = 1\mathbf{i} + 2\mathbf{j} - 3\mathbf{k}$ and $\mathbf{c} = -2\mathbf{j} + 3\mathbf{k}$, determine the vector triple products (a) $\mathbf{a} \times (\mathbf{b} \times \mathbf{c})$ and (b) $\mathbf{a} \times (2\mathbf{b} \times \mathbf{c})$.

16 If $\mathbf{a} = 3\mathbf{i} + 4\mathbf{k}$, $\mathbf{b} = 2\mathbf{j} - 5\mathbf{k}$ and $\mathbf{c} = 1\mathbf{i} - 3\mathbf{j} + 6\mathbf{k}$, determine the vector triple products:

(a) $\mathbf{a} \times (\mathbf{b} \times \mathbf{c})$, (b) $\mathbf{a} \times (2\mathbf{b} \times \mathbf{c})$, (c) $3\mathbf{a} \times (\mathbf{b} \times \mathbf{c})$, (d) $3\mathbf{a} \times (2\mathbf{b} \times \mathbf{c})$

17 If $\mathbf{a} = 3\mathbf{i} - 2\mathbf{j} + 1\mathbf{k}$, $\mathbf{b} = -1\mathbf{i} + 3\mathbf{j} + 4\mathbf{k}$ and $\mathbf{c} = 2\mathbf{i} + 1\mathbf{j} - 3\mathbf{k}$, determine the vector triple products:

(a) $\mathbf{a} \times (\mathbf{b} \times \mathbf{c})$, (b) $\mathbf{a} \times (2\mathbf{b} \times \mathbf{c})$, (c) $3\mathbf{a} \times (\mathbf{b} \times \mathbf{c})$, (d) $3\mathbf{a} \times (2\mathbf{b} \times \mathbf{c})$

12 Application: Mechanics

12.1 Introduction

This chapter illustrates the use of vectors with mechanical systems, drawing on Chapters 10 and 11. Examples of vector quantities involved in such systems are displacement, velocity, momentum, force and acceleration. Such quantities are only fully specified when we know their magnitude and direction. Examples of scalar quantities involved in such systems are mass, work and energy. Such quantities are fully specified by just a statement of their magnitude.

12.2 Vector quantities in mechanics

A definition of a vector that is often used and which provides a convenient test as to whether some quantity is a vector is:

A vector is any quantity that has a magnitude and a direction in space and combines according to the triangle (or parallelogram) addition rule.

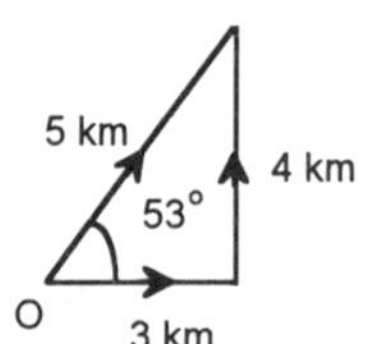

Figure 12.1 *Adding displacements*

We can use this test with displacements. Thus if we have a displacement of some object from a point O of 3 km in an easterly direction followed by 4 km in a northerly direction, then the resultant displacement of the object from O requires the use of the triangle rule to establish it (Figure 12.1). The sum of those two displacements is the same as if the object had received a single displacement of 5 km in a direction 53° north of east.

We can extend this to velocities. Consider the problem of a ship which is set to steam due east with a constant velocity relative to an observer on land of 3 km/h across a current which flows due north with a constant velocity relative to the same observer on land at 4 km/h. In one hour the ship would, in the absence of the current, of steamed 3 km in an easterly direction and the water surface would have been displaced 4 km in a northerly direction. The resultant displacement of the ship in one hour is thus the vector sum of the two displacements. In a time t the displacements due to the two velocities, ship $\mathbf{v}_s$ and water $\mathbf{v}_w$, would be $\mathbf{v}_s t$ and $\mathbf{v}_w t$ and the resultant displacement $\mathbf{v}t$ with $\mathbf{v}t$ being the vector sum of $\mathbf{v}_s t$ and $\mathbf{v}_w t$, i.e. $\mathbf{v}t = \mathbf{v}_s t + \mathbf{v}_w t$, and so:

$$\mathbf{v} = \mathbf{v}_s + \mathbf{v}_w$$

Velocity is a vector quantity.

We can use this triangle test with forces to establish them as vectors. If we apply three forces to act at the same point and adjust them so that they are in equilibrium with force $\mathbf{F}_3$ balancing the other two forces $\mathbf{F}_1$ and $\mathbf{F}_2$, perhaps using the apparatus shown in Figure 12.2, then we find that the directed line segments representing the forces form a triangle and so confirm forces as being vectors.

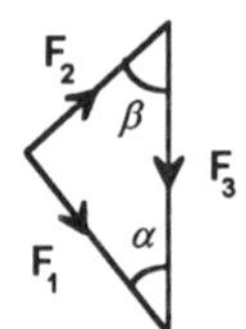

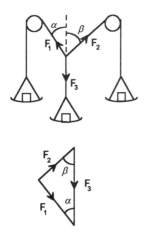

Figure 12.2 *Adding forces*

What we are doing with the triangle rule is representing forces as directed line segments and hence displacements on the sheet of paper. Forces have the same mathematical transformation properties as displacements. We can transform all vectors into displacements, i.e. represent them by directed line segments, and then add them using the triangle (or parallelogram) rule.

The definition of a vector is based on the properties of displacements and involves transforming vector quantities into displacements so they can be added using the triangle (or parallelogram) rule.

12.3 Velocity

The following are examples of some of the problems encountered in mechanics involving the velocities. As with other vectors, the velocity of an object can be resolved into components and represented in terms of unit vectors in the form:

$$\mathbf{v} = x\mathbf{i} + y\mathbf{j}$$

for two-dimensional problems and:

$$\mathbf{v} = x\mathbf{i} + y\mathbf{j} + z\mathbf{k}$$

for three-dimensional problems. **i**, **j** and **k** are the units vectors in the x, y and z directions of a Cartesian set of co-ordinates.

Example

Determine the magnitude of the velocity of an object moving with a velocity of $2\mathbf{i} - 1\mathbf{j} + 3\mathbf{k}$ m/s.

The magnitude is given by equation [13] in Chapter 11 as:

$$|\mathbf{v}| = \sqrt{2^2 + (-1)^2 + 3^2} = 3.7 \text{ m/s}$$

Example

An object has an initial position vector of $3\mathbf{i} + 2\mathbf{j} + 4\mathbf{k}$ m. If the object moves with a constant velocity of $4\mathbf{i} + 1\mathbf{j} - 2\mathbf{k}$ m/s, determine the position vector after 3 s.

The displacement occurring in the 3 s is $3(4\mathbf{i} + 1\mathbf{j} - 2\mathbf{k})$ and thus adding this displacement to the initial displacement gives a displacement of:

$$\text{displacement} = 3\mathbf{i} + 2\mathbf{j} + 4\mathbf{k} + 3(4\mathbf{i} + 1\mathbf{j} - 2\mathbf{k}) = 15\mathbf{i} + 5\mathbf{j} - 2\mathbf{k} \text{ m}$$

Example

A ship steams with a velocity of $4\mathbf{i} - 2\mathbf{j}$ m/s in still water but then steams where there is a water velocity of $-6\mathbf{i} + 4\mathbf{j}$ m/s. Determine the resultant velocity.

The resultant velocity is the sum of the two velocities and is thus:

$$\text{resultant velocity} = 4\mathbf{i} - 2\mathbf{j} - 6\mathbf{i} + 4\mathbf{j} = -2\mathbf{i} + 2\mathbf{j} \text{ m/s}$$

Example

Express a velocity of 10 km/h in a north-easterly direction in terms of unit vectors in the east (**i**) and north (**j**) directions.

The resolved component of the velocity in the easterly direction is $10 \cos 45° = 7.1$ km/h and the component in the northerly direction is $10 \sin 45° = 7.1$ km/h. Thus:

$$\mathbf{v} = 7.1\mathbf{i} + 7.1\mathbf{j} \text{ km/h}$$

Revision

1 Determine the magnitudes of the following velocities:

(a) $2\mathbf{i} + 3\mathbf{j}$ m/s, (b) $4\mathbf{i} - 5\mathbf{j}$ m/s, (c) $2\mathbf{i} + 3\mathbf{j} + 5\mathbf{k}$ m/s,

(d) $-3\mathbf{i} + 2\mathbf{j} - 6\mathbf{k}$ m/s

2 An object has an initial position vector of $2\mathbf{i} + 5\mathbf{j} + 2\mathbf{k}$ m. Determine the position vector of the object after 2 s if it has a constant velocity of $2\mathbf{i} + 3\mathbf{j}$ m/s.

3 A ship which has a course of due south is steaming across a current which is due west. The ship ends up with a velocity of 18 km/h in a direction 15° west of south. Determine the velocity set for the ship.

4 A ship has a velocity of 16 m/s in an easterly direction and is steaming in a sea which is running at a velocity of 2 m/s in a direction N 30° E. Determine the resultant velocity.

5 A pilot has to fly an aircraft from point A to point B where B is due east of A. There is a wind of 60 km/h blowing from the north-west. If the aircraft has a steady velocity of 360 km/h in still air, what course must the pilot set?

6 A person can row a boat in still water at 3.5 m/s. What course should the person set to directly cross a river if there is a current running at 2 m/s?

12.3.1 Relative velocity

Suppose objects A and B are moving with velocities of ${}_A\mathbf{v}_O$ and ${}_B\mathbf{v}_O$ relative to a fixed point O (Figure 12.3). The velocity of B relative to A can be obtained if we consider making A fixed by subtracting velocity ${}_A\mathbf{v}_O$ from it and then treating B in the same way and subtracting ${}_A\mathbf{v}_O$ from it. The velocity of B relative to A is thus:

$$ {}_B\mathbf{v}_A = {}_B\mathbf{v}_O - {}_A\mathbf{v}_O \qquad [1] $$

The velocity of B relative to A is thus the vector sum of the velocity of B and the negative velocity of A.

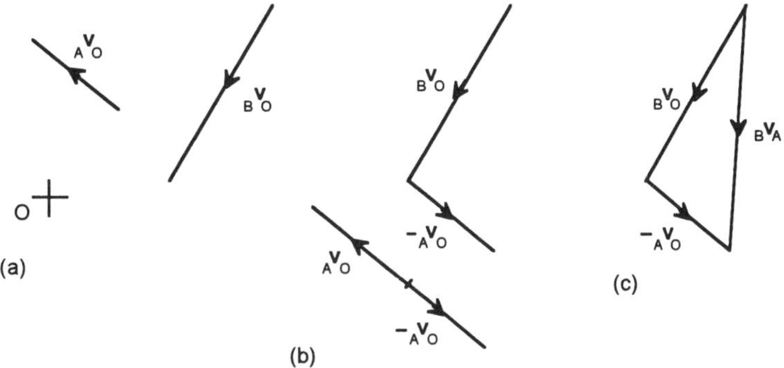

Figure 12.3 *(a) The velocities of A and B relative to O, (b) adding $-{}_A\mathbf{v}_B$ to both velocities, (c) the triangle rule to determine ${}_B\mathbf{v}_A$*

Example

A person on shore sees a ship moving with a velocity of 12**i** – 20**j** m/s and a small boat with a velocity of 8**i** + 4**j** m/s. What does the velocity of the boat appear to be to a passenger on the ship?

The velocity of the boat relative to the ship ${}_B\mathbf{v}_S$ is given in terms of the velocity of the boat relative to the shore ${}_B\mathbf{v}_O$ and the velocity of the ship relative to the shore ${}_S\mathbf{v}_O$ by equation [1] as:

$$ {}_B\mathbf{v}_S = {}_B\mathbf{v}_O - {}_S\mathbf{v}_O = 8\mathbf{i} + 4\mathbf{j} - (12\mathbf{i} - 20\mathbf{j}) = -4\mathbf{i} + 24\mathbf{j} \text{ m/s} $$

Example

The velocity of object A relative to object B is 3**i** – 5**j** m/s and the velocity of B relative to object C is 1**i** + 2**j** m/s. Determine the velocity of A relative to C.

The velocity of A relative to C is the sum of the velocities of A relative to B and B relative to C and is thus $3\mathbf{i} - 5\mathbf{j} + 1\mathbf{i} + 2\mathbf{j} = 4\mathbf{i} - 3\,\mathbf{j}$ m/s.

Revision

7 To a passenger in an aircraft flying with a velocity, relative to land, $120\mathbf{i} + 100\mathbf{j}$ km/h, the velocity of a liner appears to be $-90\mathbf{i} - 80\mathbf{j}$ km/h. What is the velocity of the liner relative to land?

8 The velocity of A relative to a fixed origin is $-2\mathbf{i} + 5\mathbf{j}$ m/s. The velocity of B relative to A is $4\mathbf{i} + 9\mathbf{j}$. Determine the velocity of B relative to the origin.

9 The velocity of object A relative to some origin is $2\mathbf{i} + 6\mathbf{j}$ m/s and the velocity of B relative to the same origin is $5\mathbf{i} + 2\mathbf{j}$ m/s. Determine the velocity of B relative to A.

12.4 Force

When there are several forces acting at a point then their net effect is the same as that of a single force, termed the *resultant*, which is equivalent to the vector sum of the forces and acting at the point. When there are two forces then the resultant can be found using the triangle or parallelogram rules. When there are more forces then the polygon rule can be used. The resultant can also be determined by expressing the forces in terms of their components in mutually perpendicular directions and then determining the algebraic sum of the components.

When the forces acting at a point are in equilibrium then the vector sum of the forces is zero. For three forces in equilibrium the three forces can be represented by the sides of a triangle, with more forces by the sides of a polygon. When expressed in terms of their components in mutually perpendicular directions, the algebraic sums of the components are zero.

Example

Determine the resultant force when forces of 5 N and 3 N act on an object and the angle between the forces is 60°.

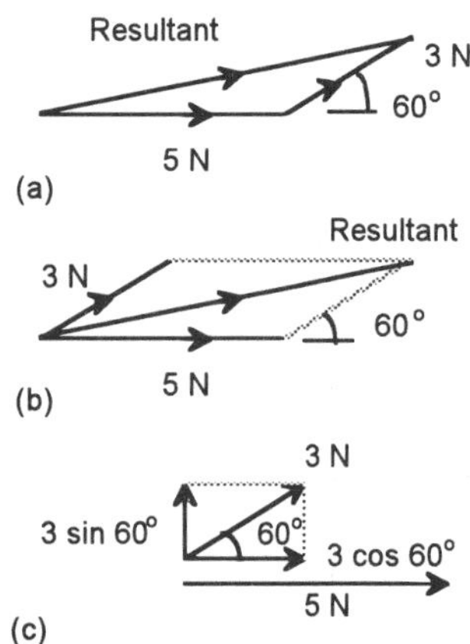

Figure 12.4 *Example*

Figure 12.4(a) shows the use of the triangle rule, and Figure 12.4(b) the parallelogram rule, to determine the resultant. Drawing to scale or calculation can be used to determine the resultant. Using the cosine rule ($a^2 = b^2 + c^2 - 2bc \cos A$) then:

$$(\text{resultant})^2 = 3^2 + 5^2 - 2 \times 3 \times 5 \cos 120°$$

Hence the resultant has a magnitude of 7.0 N. The angle θ between the resultant and the 5 N force can be found using the sine rule ($\sin A/a = \sin B/b$):

$$\frac{\sin\theta}{3} = \frac{\sin 120°}{7.0}$$

Hence $\theta = 21.8°$.

Alternatively we could resolve the forces into their components in two mutually perpendicular directions. Taking the directions to be in the direction of the 5 N force and at right angles to it (Figure 12.4(c)) then, when the components are expressed in terms of unit vectors, we have 5**i** N and (3 cos 60°)**i** + (3 sin 60°)**j**. Adding these gives 9.5**i** + 2.6**j** N. This resultant has a magnitude of $\sqrt{6.5^2 + 2.6^2}$ = 7.0 N and acts at an angle of $\tan^{-1}(2.6/6.5) = 21.8°$ to the 5 N force.

Example

Determine the magnitude and angle of the resultant force when forces of 7**i** + 3**j** N, –2**i** + 9**j** N and 1**i** + 3**j** N act at a point.

The resultant is:

resultant = 7**i** + 3**j** – 2**i** + 9**j** + 1**i** + 3**j** = 6**i** + 15**j** N

This has a magnitude of $\sqrt{6^2 + 15^2}$ = 16.2 N and is at an angle of $\tan^{-1}(15/6) = 68.2°$ to **i**.

Example

ABCD is a square. If forces of magnitudes 1 N, 2 N and 3 N act parallel to AB, BC and CD respectively, in the directions indicated by the order of the letters, determine the magnitude and direction of the resultant force.

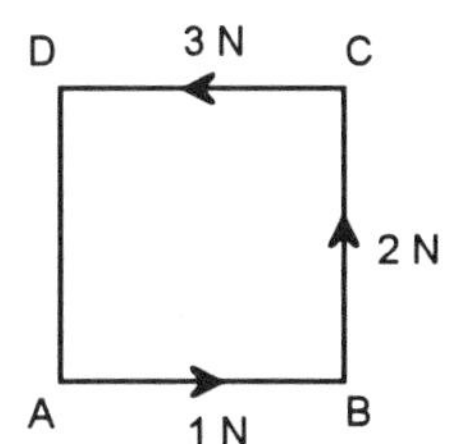

Figure 12.5 *Example*

Figure 12.5 shows the directions of the forces. Expressing the forces in terms of unit vector components then the force parallel to AB is 1**i**, parallel to BC is 2**j** and that parallel to CD is –3**i**. Thus the resultant is 1**i** + 2**j** – 3**i** = –2**i** + 2**j** N. This will have a magnitude $\sqrt{(-2)^2 + 2^2}$ = 2.8 N at an angle of $\tan^{-1}(2/-2) = 135°$ to AB.

Revision

10 Determine the resultants of the following forces acting at a point on an object:

(a) 7 N and 5 N with an angle of 30° between them,

(b) 5 N and 4 N with an angle of 65° between them,

(c) 7 N and 4 N with an angle of 25° between them.

11 Determine the resultant force when the following forces act at a point, expressing the results in terms of the unit vectors:

(a) $5\mathbf{i} + 3\mathbf{j}$ N, $2\mathbf{i} + 4\mathbf{j}$ N and $1\mathbf{i} + 2\mathbf{j}$ N,

(b) $-2\mathbf{i} + 3\mathbf{j}$ N, $-2\mathbf{i} + 4\mathbf{j}$ N and $1\mathbf{i} + 5\mathbf{j}$ N,

(c) $2\mathbf{i} + 3\mathbf{j}$ N, $-7\mathbf{i} + 2\mathbf{j}$ N and $1\mathbf{i} + 1\mathbf{j}$ N

12 The resultant of a force of $2\mathbf{i} + 9\mathbf{j}$ N and another force acting at a point is $5\mathbf{i} - 2\mathbf{j}$ N. What is the force?

13 ABC is an equilateral triangle. Forces of magnitudes 12 N, 10 N and 10 N act parallel to AB, BC and CA respectively, in the directions indicated by the order of the letters. Determine the magnitude and direction of the resultant force.

14 Forces of $2\mathbf{i} + 5\mathbf{j}$ N, $3\mathbf{i} + 7\mathbf{j}$ N and $-2\mathbf{i} - 4\mathbf{j}$ N act at a point. Determine the force needed to give equilibrium.

12.4.1 Force and acceleration

Newton's second law gives:

$$\mathbf{F} = m\mathbf{a} \qquad [2]$$

where the force **F** and the acceleration **a** are vectors and the mass m is a scalar. This equation indicates that the direction of the acceleration vector is the same as the direction of the force vector.

Example

Forces of $5\mathbf{i} - 5\mathbf{j}$ N and $-1\mathbf{i} + 3\mathbf{j}$ N act on a particle of mass 2 kg. Determine the resulting acceleration.

The resultant force is $5\mathbf{i} - 5\mathbf{j} - 1\mathbf{i} + 3\mathbf{j} = 4\mathbf{i} - 2\mathbf{j}$ N. Thus, using equation [2]:

$$4\mathbf{i} - 2\mathbf{j} = 2\mathbf{a}$$

Hence $\mathbf{a} = 2\mathbf{i} - 1\mathbf{j}$ m/s^2.

The acceleration has a magnitude of $\sqrt{2^2 + (-1)^2} = 2.2$ m/s^2 and is at an angle of $\tan^{-1}(-1/2) = -26.6°$ to the **i** direction.

Revision

15 Forces of $5\mathbf{i} - 12\mathbf{j}$ N and $4\mathbf{i} + 3\mathbf{j}$ N act on a particle of mass 3 kg. Determine the resulting acceleration.

16 Forces of $2\mathbf{i} + 7\mathbf{j}$ N and $-6\mathbf{i} + 3\mathbf{j}$ N act on a particle of mass 10 kg. Determine the resulting acceleration.

12.5 Work and energy

Consider the *work* done by the force **F** shown in Figure 12.6. The force is a vector at an angle θ to the horizontal plane. Thus the component of the force in the displacement direction, which is along the plane, is **F** cos θ. The work done in producing a displacement **d** is defined as the product of the magnitude of the force component in the direction of the displacement and the magnitude of the displacement, or the product of the magnitude of the force and resolved component of the displacement in the direction of the force, and is thus the scalar product (see Section 11.1) **F·d**:

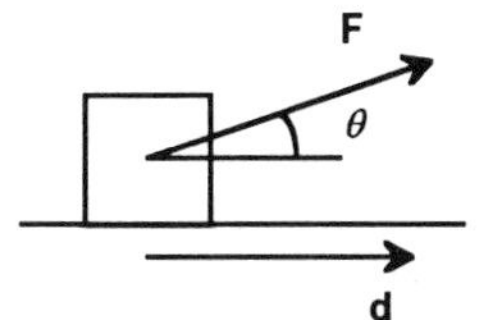

Figure 12.6 *Doing work*

$$\text{work done} = \mathbf{F}\cdot\mathbf{d} = Fd\cos\theta \qquad [3]$$

If $\mathbf{F} = X\mathbf{i} + Y\mathbf{j} + Z\mathbf{k}$ and $\mathbf{d} = x\mathbf{i} + y\mathbf{j} + z\mathbf{k}$ then (see Section 11.1.2 and equation [8]):

$$\text{work done} = \mathbf{F}\cdot\mathbf{d} = Xx + Yy + Zz \qquad [4]$$

Work is the energy change which is produced when a force moves its point of application through a distance. This can result in a change in kinetic energy, another scalar product:

$$\text{kinetic energy} = \tfrac{1}{2}m(\mathbf{v}\cdot\mathbf{v}) \qquad [5]$$

or perhaps gravitational potential energy, another scalar product:

$$\text{potential energy} = m(\mathbf{g}\cdot\mathbf{h}) \qquad [6]$$

Example

Determine the work done by a constant force of $2\mathbf{i} + 3\mathbf{j}$ N acting on a particle and moving it from point A to point B, with A having the position vector $1\mathbf{i} - 1\mathbf{j}$ m and B the position vector $2\mathbf{i} + 2\mathbf{j}$ m.

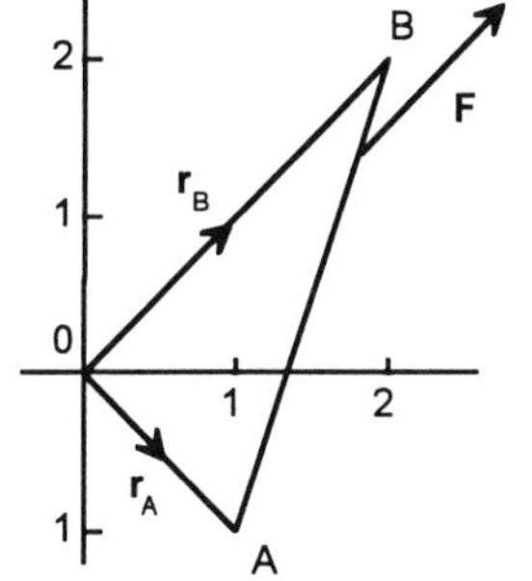

Figure 12.7 *Example*

Figure 12.7 shows the situation. The vector from A to B is $-(\mathbf{r}_A - \mathbf{r}_B)$ and is thus $-(1\mathbf{i} - 1\mathbf{j}) + (2\mathbf{i} + 2\mathbf{j}) = 1\mathbf{i} + 3\mathbf{j}$. Thus the work done is given by equation [4] as:

$$\text{work done} = \mathbf{F}\cdot\mathbf{d} = (2\mathbf{i} + 3\mathbf{j})\cdot(1\mathbf{i} + 3\mathbf{j}) = (2)(1) + (3)(3) = 11 \text{ J}$$

Revision

17 Determine the work done by the constant force **F** acting on a body and moving it from point A to point B in the following situations:

(a) A has the position vector $3\mathbf{i} - 5\mathbf{j}$ m, B the position vector $4\mathbf{i} - 2\mathbf{j}$ m and $\mathbf{F} = 1\mathbf{i} + 3\mathbf{j}$ N,

(b) A has the position vector $3\mathbf{i} + 1\mathbf{j}$ m, B the position vector $5\mathbf{i} + 3\mathbf{j}$ m and $\mathbf{F} = 2\mathbf{i} + 3\mathbf{j}$ N,

(c) A has the position vector $-2\mathbf{i} - 1\mathbf{j} + 1\mathbf{k}$ m, B the position vector $1\mathbf{i} + 2\mathbf{j} + 3\mathbf{k}$ m and $\mathbf{F} = 2\mathbf{i} + 3\mathbf{j} + 4\mathbf{k}$ N,

(d) A has the position vector $3\mathbf{i} + 5\mathbf{j} - 2\mathbf{k}$ m, B the position vector $-1\mathbf{i} + 3\mathbf{j} + 2\mathbf{k}$ m and $\mathbf{F} = 2\mathbf{i} + 3\mathbf{j} + 4\mathbf{k}$ N.

12.6 Moments

For calculations involving the equilibrium of two dimensional objects the *moment* of a force **F** about an axis is defined as having a magnitude *Fd* where *d* is the perpendicular distance from the axis to the line of action of the force (Figure 12.8). The moment defined in this way is a scalar quantity. Such a moment can result in a clockwise or anticlockwise rotation about an axis.

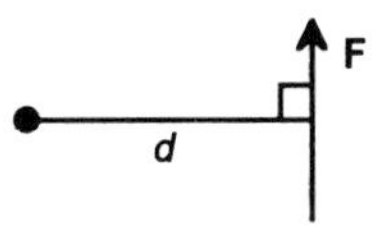

Figure 12.8 *Moment of a force*

Suppose, however, we have a rigid body free to rotate about some point O when acted on by a force **F** acting along some particular line of action, as in Figure 12.9. Consider any point P along that line of action, the position vector of P being **r**. Using the scalar definition of a moment given above, then the momentum has a magnitude |**M**| of *Fd*. Since $d = r \sin \theta$, then:

$$|\mathbf{M}| = |\mathbf{r}||\mathbf{F}| \sin \theta \qquad [7]$$

and thus the *vector moment* about the point O can be written as the vector product (see Section 11.2):

$$\mathbf{M} = \mathbf{r} \times \mathbf{F} \qquad [8]$$

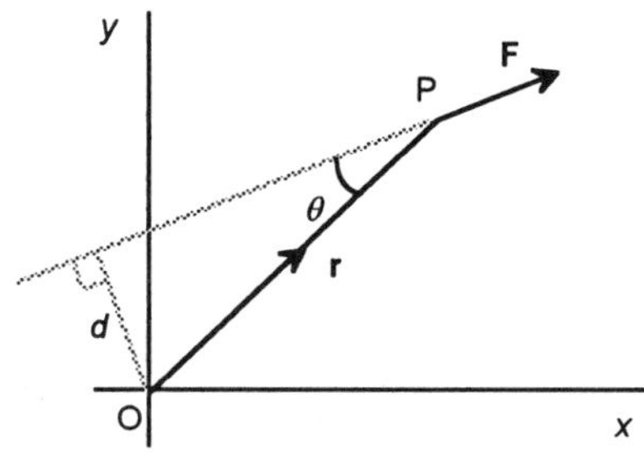

Figure 12.9 *Moment*

The moment is independent of the choice of the point P, as long as it is along the line of action of the force. The direction of the moment vector is taken as along the axis of the rotation in the direction a right-handed corkscrew would be driven by the rotation (Figure 12.10). All forces in the same plane will have vector moments perpendicular to the plane containing the forces, the senses of the moments being either into or out of the plane.

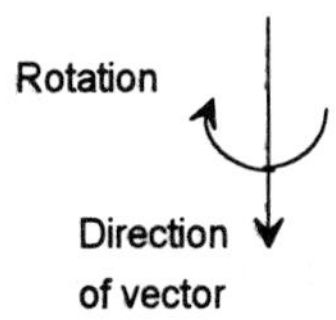

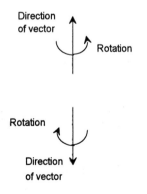

Figure 12.10 *Directions of moment vector*

12.6.1 Vector moment in terms of components

The vector moment about a point O is defined (equation [8]) as $\mathbf{M} = \mathbf{r} \times \mathbf{F}$. Drawing x-, y- and z-axes through the point O (Figure 12.11), we can write, for any point P along the line of action of the force, the position vector as $\mathbf{r} = x\mathbf{i} + y\mathbf{j} + z\mathbf{k}$ and the force as $\mathbf{F} = F_x\mathbf{i} + F_y\mathbf{j} + F_z\mathbf{k}$, (x, y, z) being the co-ordinates of point P and F_x, F_y and F_z the force components parallel to the x-, y- and z-axes. Then:

$$\mathbf{M} = \mathbf{r} \times \mathbf{F} = (x\mathbf{i} + y\mathbf{j} + z\mathbf{k}) \times (F_x\mathbf{i} + F_y\mathbf{j} + F_z\mathbf{k})$$

This gives (see equation [17] from Chapter 11):

$$\mathbf{M} = \begin{vmatrix} \mathbf{i} & \mathbf{j} & \mathbf{k} \\ x & y & z \\ F_x & F_y & F_z \end{vmatrix} = \mathbf{i}\begin{vmatrix} y & z \\ F_y & F_z \end{vmatrix} - \mathbf{j}\begin{vmatrix} x & z \\ F_x & F_z \end{vmatrix} + \mathbf{k}\begin{vmatrix} x & y \\ F_x & F_y \end{vmatrix}$$

$$= \mathbf{i}(F_z y - F_y z) + \mathbf{j}(F_x z - F_z x) + \mathbf{k}(F_y x - F_x y) \qquad [9]$$

We thus have components of the vector moment of $M_x = (F_z y - F_y z)$ in the x direction, $M_y = (F_x z - F_z x)$ in the y direction and $M_z = (F_y x - F_x y)$ in the z direction.

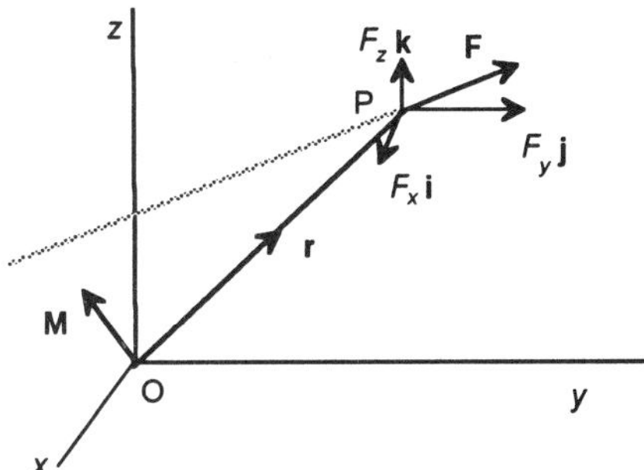

Figure 12.11 *Vector moment*

The scalar quantities M_x, M_y and M_z are the respective moments of the force **F** about the x-, y- and z-axes through O.

> *The moment of a force* **F** *about any straight line through a point O is defined as being the component along this line of the moment of* **F** *about O.*

Thus if we have some line OL then the component M_L along this line is:

$$M_L = \mathbf{M} \cdot \mathbf{L} = (\mathbf{r} \times \mathbf{F}) \cdot \mathbf{L} \qquad [10]$$

where **L** is the unit vector along the line OL. This is a triple scalar product (see Section 11.3.1).

Example

Determine the moment about the origin of a force of $2\mathbf{i} + 2\mathbf{j} - 3\mathbf{k}$ N at a point P which is defined by the position vector $1\mathbf{i} + 2\mathbf{j} + 3\mathbf{k}$ m.

Using equation [8]:

$$\mathbf{M} = \mathbf{r} \times \mathbf{F} = (1\mathbf{i} + 2\mathbf{j} + 3\mathbf{k}) \times (2\mathbf{i} + 2\mathbf{j} - 3\mathbf{k})$$

Using equation [9]:

$$\mathbf{M} = \mathbf{i}[(2)(-3) - (2)(3)] - \mathbf{j}[(1)(-3) - (2)(3)] + \mathbf{k}[(1)(2) - (2)(2)]$$

$$= -12\mathbf{i} + 9\mathbf{j} - 2\mathbf{k} \text{ N m}$$

The magnitude of the moment is $\sqrt{(-12)^2 + 9^2 + (-2)^2} = 15.1$ N m.

Example

A force of 100 N acts along a line between the points with the co-ordinates of (2, 1, 1) m and (5, 0, 4) m. Determine the moments of this force about the x-, y- and z-axes.

The force acts along the line joining the two points. This line is the difference between the position vectors from the origin to each point and is thus:

$$(5 - 2)\mathbf{i} + (0 - 1)\mathbf{j} + (4 - 1)\mathbf{k} = 3\mathbf{i} - 1\mathbf{j} + 3\mathbf{k}$$

This line has a magnitude of $\sqrt{3^2 + (-1)^2 + 3^2} = 4.26$ m. The direction cosines (see Section 10.5) are thus 3/4.26, –1/4.26 and 3/4.26. Hence the components of the force in the x, y and z directions are 100 N multiplied by the relevant direction cosines and hence 70.4 N, –23.5 N and 70.4 N. Thus we can write the force as:

$$\mathbf{F} = 70.4\mathbf{i} - 23.5\mathbf{j} + 70.4\mathbf{k} \text{ N}$$

We can determine the vector moment using any point on the line of action of the force. Thus using equation [9] with point (2, 1, 1), position vector $2\mathbf{i} + 1\mathbf{j} + 1\mathbf{k}$:

$$\mathbf{M} = \mathbf{r} \times \mathbf{F} = \mathbf{i}(F_z y - F_y z) + \mathbf{j}(F_x z - F_z x) + \mathbf{k}(F_y x - F_x y)$$

$$= \mathbf{i}[(70.4)(1) - (-23.5)(1)] + \mathbf{j}[(70.4)(1) - (70.4)(2)] + \mathbf{k}[(-23.5)(2) - (70.4)(1)]$$

$$= 93.9\mathbf{i} - 70.4\mathbf{j} - 117.5\mathbf{j}$$

The moments about the x-, y- and z-axes are thus 93.9 N m, −70.4 N m and −117.5 N m.

Revision

18 Determine the moment about the origin of a force of $1\mathbf{i} + 2\mathbf{j} - 3\mathbf{k}$ N at a point P which is defined by the position vector $1\mathbf{i} + 2\mathbf{j} + 3\mathbf{k}$ m.

19 Determine the moment about the origin of a force of $3\mathbf{i} + 2\mathbf{j} - 1\mathbf{k}$ N at a point P which is defined by the position vector $2\mathbf{i} + 2\mathbf{j} + 1\mathbf{k}$ m.

20 Determine the moment about the point O for the force shown in Figure 12.12. The force is acting on one corner of a rectangular plate with the line of action indicated.

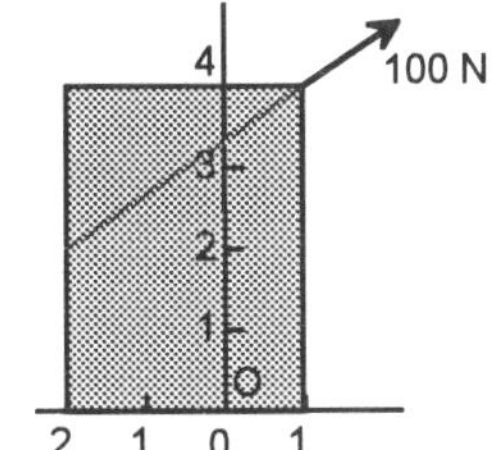

Figure 12.12 *Revision problem 20*

21 A force of 21 N acts along a line between the points with the co-ordinates of (2, 1, 1) m and (10, −3, 2) m. Determine the moments of this force about the x-, y- and z-axes.

12.7 Angular velocity

A point P rotating about a fixed axis OO' (Figure 12.13) with a constant *angular speed* ω rotates through equal angular displacements in equal intervals of time. Since one revolution is 2π radians and takes a time T called the period, then $\omega = 2\pi/T$. Since the point rotates in a circle of radius r it travels a distance of $2\pi r$ in the time T and thus its speed is $2\pi r/T$. Hence the speed is $r\omega$. Angular speed is a scalar quantity. The linear velocity $\mathbf{v}$ of P is tangential to the circular path and has a magnitude of the speed.

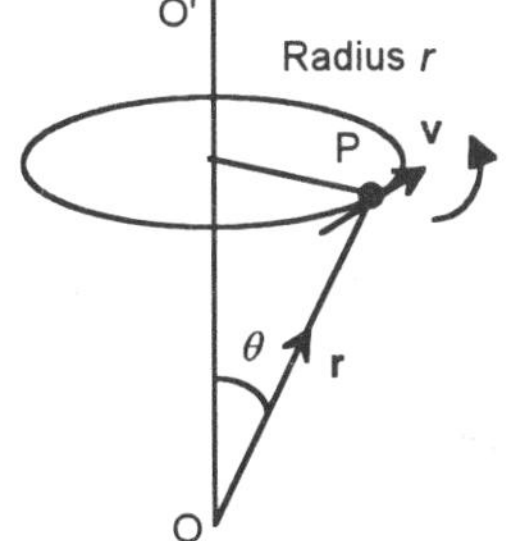

Figure 12.13 *Rotation*

Angular velocity $\boldsymbol{\omega}$ is a vector quantity. We specify it as being given by:

$$\boldsymbol{\omega} = \omega\mathbf{n} \qquad [11]$$

where ω is the magnitude of the angular velocity, i.e. the angular speed, and $\mathbf{n}$ is the unit vector along the direction of the axis of rotation. The angular velocity vector has a direction defined as being along the axis of the rotation and in the direction a right-handed corkscrew would be driven by the rotation.

If the position of the point P in Figure 12.13 is specified by the position vector $\mathbf{r}$ measured from the fixed reference point O, then $r = |\mathbf{r}| \sin \theta$. Thus $v = \omega|\mathbf{r}| \sin \theta$. The direction of the linear velocity $\mathbf{v}$ of P at any instant is tangential to the circular path and so at right angles to $\mathbf{r}$. It is also at right angles to the axis of rotation and hence, since the angular velocity vector is in that direction, at right angles to the angular velocity vector. Thus we can write the linear velocity as the vector product (see Section 11.2):

$$\mathbf{v} = \boldsymbol{\omega} \times \mathbf{r} \qquad [12]$$

Example

Determine the velocity vector for a point with the angular velocity vector $1\mathbf{i} - 1\mathbf{k}$ rad/s about an axis through the origin if the point has the position vector $2\mathbf{i} - 1\mathbf{j} - 2\mathbf{k}$ m.

Using equation [12]:

$$\mathbf{v} = (1\mathbf{i} - 1\mathbf{k}) \times (2\mathbf{i} - 1\mathbf{j} - 1\mathbf{k})$$

$$= \mathbf{i}[(0)(-2) - (-1)(-1)] - \mathbf{j}[(-1)(2) - (1)(-2)] + \mathbf{k}[(1)(-1) - (0)(2)]$$

$$= -1\mathbf{i} - 1\mathbf{k} \text{ m/s}$$

Revision

22 Determine the velocity vector for a point with the angular velocity vector $2\mathbf{i}$ rad/s about an axis through the origin if the point has the position vector $2\mathbf{i} - 3\mathbf{j}$ m.

23 Determine the velocity vector for a point with the angular velocity vector $5\mathbf{k}$ rad/s about an axis through the origin if the point has the position vector $3\mathbf{i} + 2\mathbf{k}$ m.

Problems

1 Determine the magnitudes of the following velocities:

(a) $4\mathbf{i} + 3\mathbf{j}$ m/s, (b) $2\mathbf{i} - 5\mathbf{j}$ m/s, (c) $2\mathbf{i} + 5\mathbf{j} + 5\mathbf{k}$ m/s,

(d) $-1\mathbf{i} + 2\mathbf{j} - 3\mathbf{k}$ m/s

2 An object has an initial position vector of $3\mathbf{i} + 2\mathbf{j} - 2\mathbf{k}$ m. Determine the position vector of the object after 2 s if it has a constant velocity of $2\mathbf{i} + 3\mathbf{j} + 1\mathbf{k}$ m/s.

3 Determine the resultants of the following sets of velocities:

(a) $2\mathbf{i} + 1\mathbf{j}$ m/s, $3\mathbf{i} - 5\mathbf{j}$ m/s, (b) $2\mathbf{i} + 4\mathbf{j}$ m/s, $-2\mathbf{i} + 5\mathbf{j}$ m/s,

(c) $12\mathbf{i} + 5\mathbf{j}$ m/s, $2\mathbf{i} + 3\mathbf{j}$ m/s, $-5\mathbf{i} + 12\mathbf{j}$ m/s

4 Determine the resultants of the following sets of velocities:

(a) 12 m/s due north, 9 m/s due west,

(b) 4 km/h in a north-westerly direction, 6 km/s due east,

(c) 10 m/s N 35° W, 15 m/s S 40° W.

5 A person can swim with a velocity of 5 km/h in still water. In what direction should the person swim if they wish to cross a river from one point on its bank to another point directly opposite and there is a water current of 3 km/h.

6 Object A has a velocity of 12**i** + 6**j** m/s relative to some origin and object B a velocity of 4**i** + 5**j** m/s relative to the same origin. Determine the velocity of A relative to B.

7 The velocity of object A relative to some origin is 3**i** + 5**j** m/s. Object B has a velocity of –2**i** + 3**j** m/s relative to A. Determine the velocity of B relative to the origin.

8 To a cyclist cycling due north at 16 km/h the wind appears to be blowing from the north-west at 10 km/h. Determine the actual wind velocity in the terms of the unit vectors **i** and **j**.

9 Determine the resultant force when the following forces act at a point, expressing the results in terms of the unit vectors:

(a) 1**i** + 2**j** N, 2**i** + 3**j** N and 3**i** + 2**j** N,

(b) –4**i** + 3**j** N, –5**i** + 2**j** N and 2**i** + 5**j** N,

(c) 2**i** – 3**j** N, –3**i** + 2**j** N and 1**i** + 7**j** N

10 ABCDEF is a regular hexagon. Forces of magnitudes 3 N, 4 N, 2 N and 6 N act in the directions AB, AC, EA and AF respectively, in the directions indicated by the order of the letters. Determine the magnitude and direction of the resultant force.

11 Forces of 1 N, 2 N and 3 N act at a point O with their directions along each of the diagonal faces of a cube with a corner at O. Determine the magnitude and the direction cosines of the resultant force.

12 ABC is an equilateral triangle. Forces of magnitude 10 N act in the direction parallel to AB, BC and AC, in the directions indicated by the order of the letters. Determine the magnitude and direction of the resultant force.

13 ABCDEFGH is a regular octagon. Forces of magnitude 2 N, 6 N, 8 N and 4 N act in the direction AB, AD, AE and AG, in the directions indicated by the order of the letters. Determine the magnitude and direction of the resultant force.

14 Forces of –2**i** – 5**j** N and –1**i** + 8**j** N act on a particle of mass 3 kg. Determine the resulting acceleration.

15 Determine the work done by the constant force **F** acting on a body and moving it from point A to point B in the following situations:

(a) A has the position vector 1**i** – 2**j** m, B the position vector 4**i** + 2**j** m and **F** = 5**i** + 3**j** N,

(b) A has the position vector –3**i** + 1**j** m, B the position vector 2**i** + 3**j** m and **F** = 2**i** + 1**j** N,

(c) A has the position vector 2**j** m, B the position vector 5**i** – 2**j** m and **F** = 2**i** + 1**j** N,

(d) A has the position vector –1**i** – 1**j** – 1**k** m, B the position vector 2**i** + 2**j** + 2**k** m and **F** = 2**i** + 1**j** + 1**k** N.

16 A particle is acted on by forces of 1**i** + 2**j** – 2**k** N and –4**i** – 2**j** + 4**k** N and move it from point A to point B where A has the position vector 3**i** + 2**j** – 2**k** m and B the position vector –1**i** + 5**j** + 2**k** m. Determine the work done.

17 A force of 3**i** + 4**j** N acts on a mass. What is the work done when the mass moves a distance of (a) 4 m in the **i** direction, (b) 4 m in the –**i** direction, (c) 4 m in the **j** direction?

18 Determine the moment about the origin of a force 1**i** + 2**j** + 3**k** N at a point P which is defined by the position vector 3**i** + 2**j** + 1**k** m.

19 A force of 50 N acts along a line between the points with the co-ordinates of (8, 2, 3) m and (2, –6, 5) m. Determine the moments of this force about the *x*-, *y*- and *z*-axes.

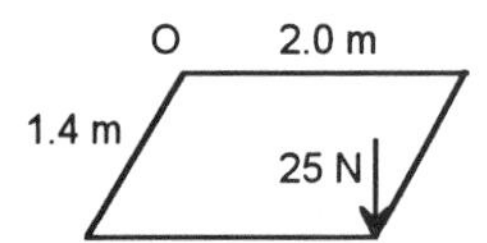

Figure 12.14 *Problem 20*

20 Figure 12.14 shows a horizontal plate of length 2.0 m and width 1.4 m which is subject to a vertical force of 25 N at one corner. Determine the resulting moment at the diagonally opposite corner O.

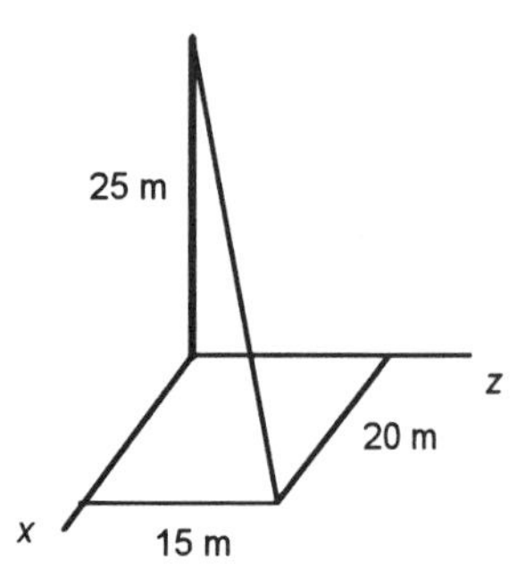

Figure 12.15 *Problem 21*

21 Figure 12.15 shows a TV mast which has a cable from the top of the mast to a tethering point on the ground. Determine the moment about the *z*-axis passing through the base of the mast as a result of a tension of 500 N being applied to the cable.

22 Determine the velocity vector for a point with the angular velocity vector 2**i** + 3**j** rad/s about an axis through the origin if the point has the position vector 2**i** m.

Part 4 Discrete mathematics

The aims of this part are to enable the reader to:

- Use the language of sets to describe situations.
- Carry out operations involving the algebra of sets.
- Describe switching circuits in terms of truth tables.
- Use the laws of Boolean algebra, in particular for switching circuit problems.
- Describe the AND, OR, INVERT, NAND, NOR and XOR logic gates in terms of their truth tables and Boolean functions.
- Combine gates to generate particular truth tables or Boolean functions and vice versa.
- Solve digital circuit problems involving logic gates and their Boolean algebra descriptions.
- Explain the denary, binary octal and hex number systems.
- Carry out binary addition and subtraction, using two's complement.
- Explain the basic forms of half-adder and full-adder circuits and their use in carrying out binary arithmetic.

Discrete mathematics is the term used to describe mathematics involving topics such as set theory, logic and Boolean algebra. This part assumes little in the way of previous knowledge. Chapter 15 assumes Chapter 14 has been covered.

13 Set theory

13.1 Introduction

Set theory provides a language that is used frequently for specifying mathematical work and is applicable to types of problem such as switching circuits and considerations of probability. *Set theory* is concerned with identifying one or more common characteristics among objects, a *set* being any collection of objects, things or states.

This chapter is an introduction to the basic principles of set theory. Switching circuits are considered in Chapters 14, 15 and 16 and probability in Part 10.

13.2 Definitions and notation

A set is any collection of objects, things or states.

The term *element* is used for a member of the set. The elements in a set may be defined by some rule or in a descriptive manner. Thus we might, for example, have a set numbers of perhaps of days of the week. A set may be *finite* and contain only a finite number of elements or *infinite* with an infinite number of elements. Thus the set of the days of the week will be finite but a set of all the integers will be infinite. We can describe a finite set by listing all the elements between braces { }. Thus for set A as the set of the days of the week (note that capital letters are used for sets):

A = {Sunday, Monday, Tuesday, Wednesday, Thursday, Friday, Saturday}

Another example is B as the set of possible states of a switch:

B = {off, on}

Another example might be the set of the binary digits:

$$C = \{1, 0\}$$

To state that an element belongs to a particular set we use the symbol $\in$, this symbol meaning 'is a member of'. Thus for set A above:

Sunday $\in A$

indicates that Sunday is a member of set A. The symbol $\notin$ is used to indicate that an element is not a member of a particular set. Thus for set C above:

$$2 \notin C$$

Another way of defining a set is to give a rule by which all its members can be found. Thus we might have:

$A = \{x : x \text{ is whole number and } 1 \leq x \leq 5)$

This reads as: A is the set of x values, where the value of x is a whole number and x is greater than or equal to 1 and less than or equal to 5. The colon (:) is used for the word 'where'. The set written out in full is {1, 2, 3, 4, 5}.

To aid in specifying rules, special symbols are often used:

- $\mathbb{N}$ The set of non-negative whole numbers, i.e. 0, 1, 2, 3, 4, 5, etc.
- $\mathbb{N}^+$ The set of positive whole numbers, i.e. 1, 2, 3, 4, 5, etc. Note that 0 is excluded.
- $\mathbb{R}$ The set of all real numbers.
- $\mathbb{R}^+$ The set of positive real numbers.
- $\mathbb{R}^-$ The set of negative real numbers.
- $\mathbb{Z}$ The set of whole numbers, positive, negative and zero.
- $\mathbb{Q}$ The set of numbers of the form p/q, p and q being integers, $q \neq 0$.

Thus we might have:

$A = \{x : x \in \mathbb{N} \text{ and } 1 \leq x \leq 5\}$

This reads as: A is the set of x values, where x is a member of the set of non-negative whole numbers and x is greater than or equal to 1 and less than or equal to 5. The set written out in full is {1, 2, 3, 4, 5}.

Revision

1 Use set notation to describe the set of x values where x is a member of the set of:

(a) real numbers that lie in the range 0 to 1, including 0 and 1,

(b) real numbers that lie in the range −3 to +3, including −3 and 3,

(c) whole numbers that lie in the range 0 to 3, including 0 and 3

13.2.1 Equal sets

Two sets A and B are equal if every element of each is also an element of the other. We can then write A = B.

For example, the sets $A = \{3, 4, 5\}$ and B = {5, 3, 4} are equal. The order in which the elements are written is immaterial. The sets $A = \{3, 4, 5\}$ and $B = \{3, 3, 4, 5\}$ are also equal. Repetition of elements is ignored.

The sets $A = \{3, 4\}$ and $B = \{x : x^2 - 7x + 12 = 0\}$ are also equal since the rule given for set B leads to the same elements as set A.

13.2.2 Subsets

If every element of set A is also an element of set B then A is said to be a *subset* of B. For example, $A = \{1, 2, 3\}$ is a subset of $B = \{1, 2, 3, 4, 5, 6\}$. The notation used to indicate that A is a subset of B is $A \subset B$. Note that this does *not* include the possibility that $A = B$, the term *proper subset* being used. If we want to include the possibility that $A = B$ then we use the symbol $A \subseteq B$.

13.2.3 Union and intersection

Suppose we have two sets: $A = \{1, 2, 3\}$ and $B = \{4, 5, 6\}$. The set which contains all the elements of A and B, i.e. $\{1, 2, 3, 4, 5, 6\}$ is said to be the *union* of A and B. It is denoted by $A \cup B$, this reading as 'A union B'. Thus:

$$A \cup B = \{x : x \in A \text{ or } x \in B \text{ or both}\} \qquad [1]$$

x is an element of A, or B, or both sets.

If we have two sets $A = \{1, 2, 3, 4\}$ and $B = \{3, 4, 5, 6\}$ then the set which contains elements common to both A and B, i.e. $\{3, 4\}$ is called the *intersection* of A and B. It is denoted by $A \cap B$, this reading as 'A intersection B'. Thus:

$$A \cap B = \{x : x \in A \text{ or } x \in B \text{ or both}\} \qquad [2]$$

If the sets A and B have no elements in common the sets A and B are said to be *disjoint* and we write:

$$A \cap B = \varnothing \qquad [3]$$

A set with no elements is called an *empty set* and denoted by $\varnothing$.

Example

Determine the union of the sets A and B when:

$$A = \{x : x \in \mathbb{R} \text{ and } 0 \le x \le 2\} \text{ and } B = \{x : x \in \mathbb{R} \text{ and } 1 \le x \le 3\}$$

Set A has the elements in the interval $0 \le x \le 2$ and B has the elements in the interval $1 \le x \le 3$. The interval containing all these numbers is $0 \le x \le 3$. Thus:

$$A \cup B = \{x : x \in \mathbb{R} \text{ and } 0 \le x \le 3\}$$

The union of sets A and B is the set of values of x where x is a real number and has values between, and including, 0 and 3.

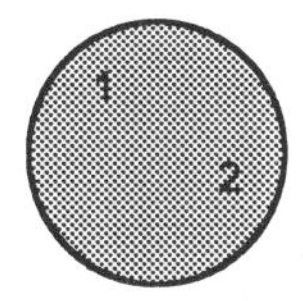

Figure 13.1 *Set {1, 2}*

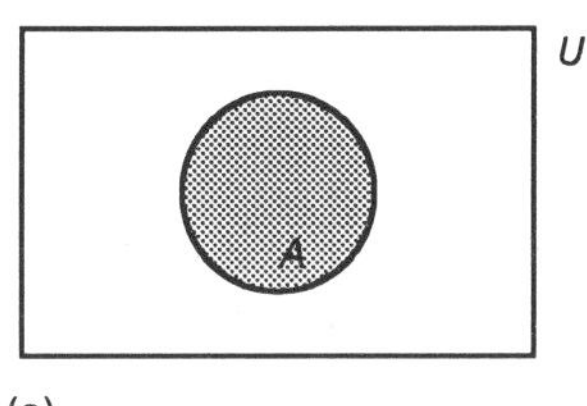

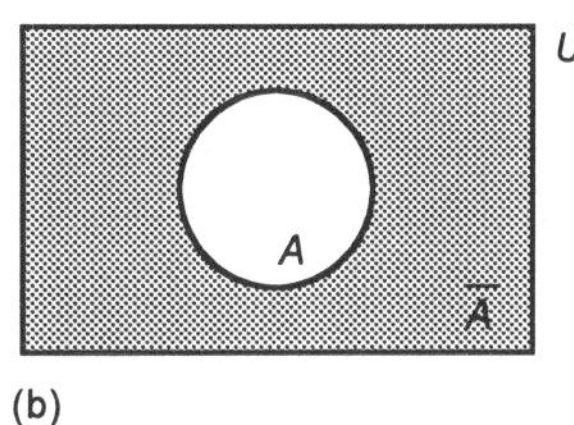

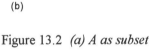

Figure 13.2 *(a) A as subset of U, (b) the complement of A*

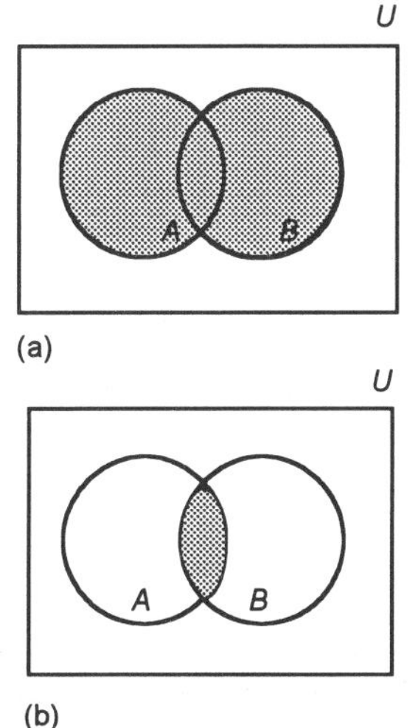

Figure 13.3 *(a)* $A \cup B$, *(b)* $A \cap B$

Example

Determine the intersection of the sets A and B when:

$$A = \{x : x \in \mathbb{R} \text{ and } 0 \le x \le 2\} \text{ and } B = \{x : x \in \mathbb{R} \text{ and } 1 \le x \le 3\}$$

The intersection set has to contain the elements that belong to both the sets. Set A has the elements in the interval $0 \le x \le 2$ and B has the elements in the interval $1 \le x \le 3$. The interval containing the numbers belonging to both sets is $1 \le x \le 2$. Thus:

$$A \cap B = \{x : x \in \mathbb{R} \text{ and } 1 \le x \le 2\}$$

Revision

2 For the following sets, determine $A \cup B$ and $A \cap B$:

(a) $A = \{1, 2, 3, 4\}$, $B = \{3, 4, 5, 6\}$,

(b) $A = \{x : x \in \mathbb{R} \text{ and } 0 \le x \le 5\}$, $B = \{x : x \in \mathbb{R} \text{ and } -1 \le x \le 3\}$

(c) $A = \{x : x \in \mathbb{R} \text{ and } 2 \le x \le 9\}$, $B = \{x : x \in \mathbb{R} \text{ and } 5 \le x \le 13\}$

13.2.4 Venn diagrams

A graphical representation of the relations between sets can be given by *Venn diagrams*. With such diagrams, a set is represented by a closed figure, usually a circle, which encloses all the elements of that set. Figure 13.1 illustrates this for a set containing the elements 1 and 2, i.e. {1, 2}.

We define a universal set U as one which contains all the items of interest. Then all the sets we are dealing with in a particular context become subsets of the universal set.

Figure 13.2(a) illustrates this with the rectangle containing all the elements of interest and so representing the universal set and the circle representing set A which contains some of the universal set elements. We can also define the *complement* of a set A as being the elements that remain in the universal set U when the elements of set A are removed. Figure 13.2(b) illustrates this with the shaded area representing the complement of set A. The complement of set A is denoted by $\bar{A}$.

Figure 13.3(a) shows how we can represent the union of two sets A and B, the total area (shaded) of the two circles indicating union set. Figure 13.3(b) shows how we can represent the intersection of two sets A and B, the overlap area (shaded) indicating the intersection set.

13.3 Algebra of sets

From the definitions of union, intersection and complement the following rules can be developed.

1 *Idempotent law*
The union of set A with set A is A:

$$A \cup A = A \qquad [4]$$

2 *Idempotent law*
The intersection of set A with set A is A:

$$A \cap A = A \qquad [5]$$

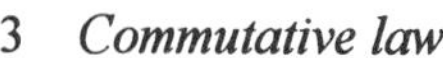

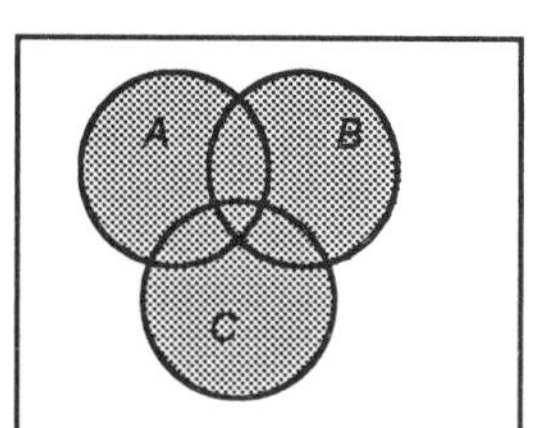

Figure 13.4 *Union of sets A, B and C*

3 *Commutative law*
The union of set A with set B is the same as the union of set B with set A:

$$A \cup B = B \cup A \qquad [6]$$

4 *Commutative law*
The intersection of set A with set B is the same as the intersection of set B with set A:

$$A \cap B = B \cap A \qquad [7]$$

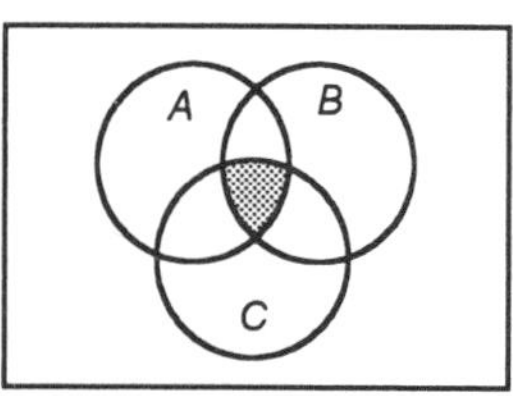

Figure 13.5 *Intersection of sets A, B and C*

5 *Associative law*
The union of set A with set B generates a new set $A \cup B$. When this set is combined to give a union with a third set C we then have $(A \cup B) \cup C$ (Figure 13.4). The result is a set which is the same as one that would have been produced if we had first taken the union of set B with C to give $B \cup C$ and then combined that with set A to give $(B \cup C) \cup A$.

$$(A \cup B) \cup C = A \cup (B \cup C) \qquad [8]$$

6 *Associative law*
The intersection of A with B gives a new set $A \cap B$. The intersection of this with set C gives $(A \cap B) \cap C$ (Figure 13.5). The result is a set which is the same as if we had first taken the intersection of B with C to give $B \cap C$ and then found the intersection of this with set A.

$$(A \cap B) \cap C = A \cap (B \cap C) \qquad [9]$$

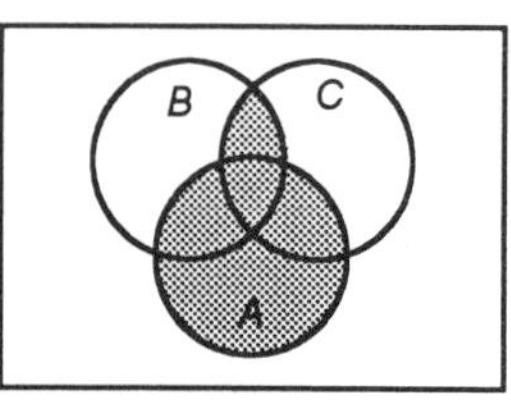

Figure 13.6 *Union of A with intersection of B and C*

7 *Distributive law*
The union of set A with the set formed by the intersection of set B with set C (Figure 13.6) is the same as the intersection of the set formed by the union of set A with set B and that formed with the union of set A with set C.

$$A \cup (B \cap C) = (A \cup B) \cap (A \cup C) \qquad [10]$$

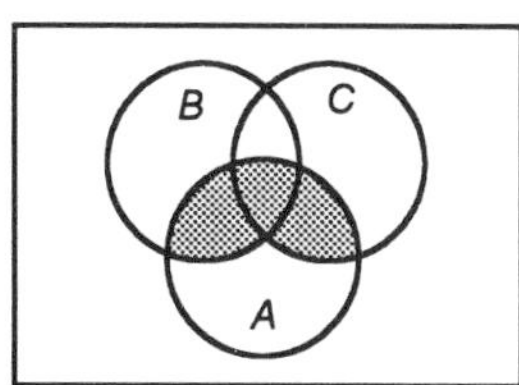

Figure 13.7 *Intersection of A with union of B and C*

8 *Distributive law*
The intersection of set A with the set formed by the union of set B with set C (Figure 13.7) is the same as the union of the set formed by the intersections of set A with B and that formed by the intersection of set A with set C.

$$A \cap (B \cup C) = (A \cap B) \cup (A \cup C) \qquad [11]$$

9 *Complementary law*
The union of set A with its complement is the universal set. This arises from the definition of the complement (see Figure 13.2).

$$A \cup \bar{A} = U \qquad [12]$$

10 *Complementary law*
The intersection of set A with its complement is the empty set, there thus being no intersection between the two (see Figure 13.2).

$$A \cap \bar{A} = \varnothing \qquad [13]$$

11 *Identity law*
The union of set A with an empty set is A:

$$A \cup \varnothing = A \qquad [14]$$

12 *Identity law*
The intersection of set A with the universal set is A:

$$A \cap U = A \qquad [15]$$

If you look at the above laws, they occur in pairs. Thus if in one pair you replace $\cup$ by $\cap$ and interchange $\varnothing$ and U, then you obtain the other law in the pair. This is referred to as the *principle of duality*.

In addition to the above laws there are two other laws, known as the *De Morgan laws*' which are very useful. These are:

1 The complement of the union of two sets is the intersection of the two complements (Figure 13.8(a)).

$$\overline{A \cup B} = \bar{A} \cap \bar{B} \qquad [16]$$

2 The complement of the intersection of two sets is the union of the two complements (Figure 13.8(b)).

$$\overline{A \cap B} = \bar{A} \cup \bar{B} \qquad [17]$$

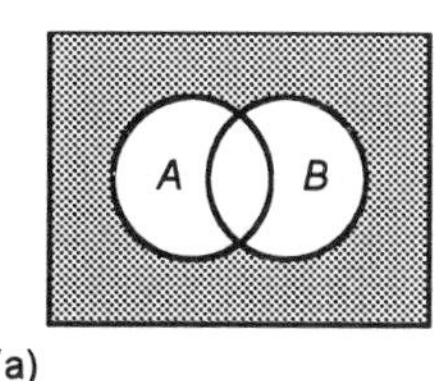

(a)

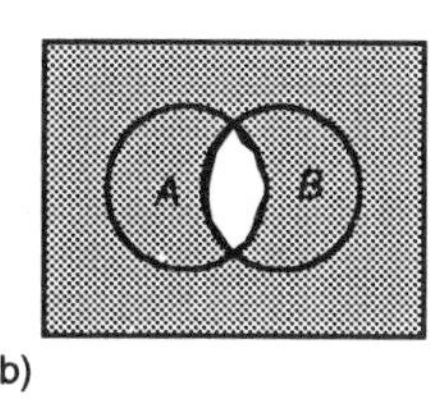

(b)

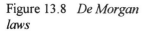

Figure 13.8 *De Morgan laws*

Example

On a Venn diagram indicate $\overline{A \cup B}$.

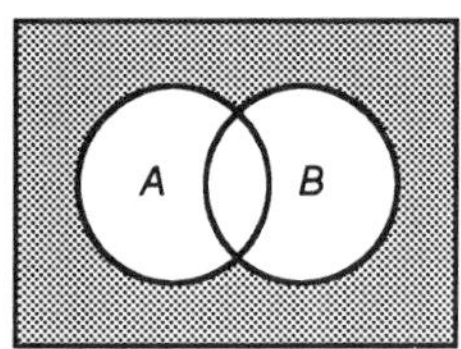

Figure 13.9 *Example*

This is the complement of the union of set A and set B and is thus as shown in the Venn diagram in Figure 13.9.

Example

Show that $(A \cap B) \cup (A \cap \bar{B}) = A$.

Using law [11], $A \cap (B \cup C) = (A \cap B) \cup (A \cup C)$, with $\bar{B}$ replacing C, and then using equation [12]:

$$(A \cap B) \cup (A \cup \bar{B}) = A \cap (B \cup \bar{B}) = A \cap U$$

The intersection of A with the universal set is just A.

Revision

3 Represent by Venn diagrams: (a) $\bar{A} \cap \bar{B}$, (b) $\overline{A \cap B}$, (c) $A \cap \bar{B}$.

4 Show that:

(a) $A \cap (A \cup B) = A$, (b) $(\bar{A} \cap \bar{B}) \cup (A \cup B) = U$,

(c) $(\overline{A \cap B}) \cup (\overline{\bar{A} \cap \bar{B} \cap C}) \cup A = U$

13.4 Sets and events

Suppose we have a situation where a number of products are subject to tests and a number, A, found to be defective in some way. We can consider the total number of products tested to be the universal set and the subset A within the universal set to be those that are defective (Figure 13.10).

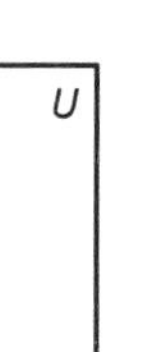

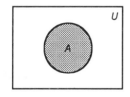

Figure 13.10 *Defectives as set A*

In such applications, the set of all the possible outcomes, i.e. the universal set, is called the *sample space* and denoted by S and an *event* is any subset of S. Thus in the case of the quality testing of products, the sample space is the total number of items tested and the event is failure of the tests and so the subset of such an event contains the defective items.

Suppose we roll a six-sided die. The set of all the possible outcomes is {1, 2, 3, 4, 5, 6}. If the event is the obtaining of the outcome of an even number then the outcome will be represented by the set {2, 4, 6}. If the die is rolled and yields a 2 then the outcome is within the event set and the event has occurred, i.e. $2 \in \{2, 4, 6\}$. If the die is rolled and yields a 3 then the outcome is not within the event set, i.e. $3 \notin \{2, 4, 6\}$, and the event has not occurred.

Two events are said to be *mutually exclusive* if they cannot occur together in a single trial. The two event sets are then disjoint (Figure

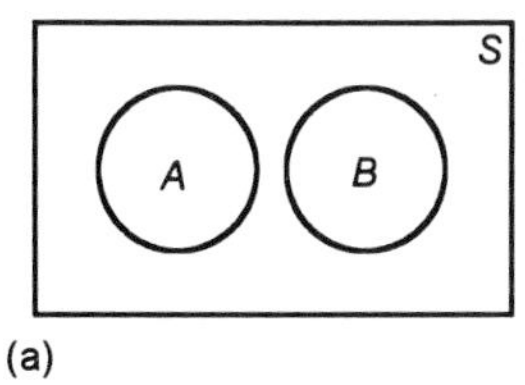

(a)

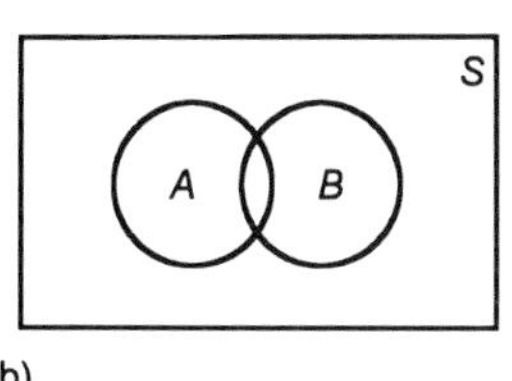

(b)

Figure 13.11 *(a) Mutually exclusive even, (b) not mutually exclusive event*

13.11(a)). If the two events are not mutually exclusive then the two sets intersect (Figure 13.11(b)).

Example

A, *B* and *C* are three events in the sample space. Determine the set that represents the event (a) *A* and *B* occur, (b) *A* occurs but *B* or *C* do not.

(a) The set for the two events to occur is $A \cap B$.

(b) The set for *B* or *C* not occurring is $\overline{B \cup C}$, thus the set for *A* occurring but *B* and *C* not is $A \cap (\overline{B \cup C})$.

Example

100 cases of material failure were investigated. In 30 of the cases the failure was due to just faulty heat treatment, in 40 cases due to just excessive loading, in 50 of the cases to a lack of fatigue resistance. If 7 of the failures were due to two types of failure mechanism, how many were due to all three?

Figure 13.12 shows the Venn diagram which describes the situation, *H* representing the heat treatment set, *L* the excessive load set and *F* the lack of fatigue resistance set. We have:

$$H \cup L \cup F = 100$$

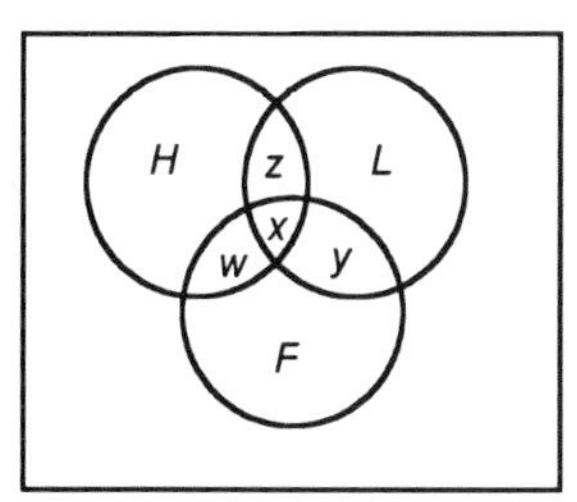

Figure 13.12 *Example*

and are required to determine $H \cap (L \cap F)$. Let this equal *x* and $L \cap F = y + x$, $L \cap H = z + x$ and $H \cap F = w + x$. Then the above equation gives:

$$30 - y - z - x + 40 - y - w - x + 50 - w - z - x = 100$$

$$20 = 2(y + z + w) + 3x$$

But $y + x + w = 7$. Thus $x = 2$.

Revision

5 Suppose that *A*, *B* and *C* are three events in the sample space. What is the set that represents (a) none of the events occurring, (b) all three events occur together, (c) only *A* occurs, (d) *B* does not occur, (e) *C* occurs but *A* does not.

6 A manufacturer carried out a survey of 100 customers to find which of products A, B and C were being used in a particular application. 30 of the customers were using just A, 22 just B, 18 just C, 8 using A and B, 9 using A and C, 7 B and C and 14 none of A, B or C. How many were using A and C but not B?

Problems

1 Use set notation to describe the set of x values where x is a member of the set of:

(a) real numbers that lie in the range 0 to 4, including 0 and 4,

(b) real numbers that lie in the range –3 to +5, including –3 and 5,

(c) whole numbers that lie in the range 0 to 6, including 0 and 6

2 For the following sets, determine $A \cup B$ and $A \cap B$:

(a) $A = \{b, c, d, e\}$, $B = \{d, e, f, g\}$,

(b) $A = \{x : x \in \mathbb{R} \text{ and } 0 \leq x \leq 3\}$, $B = \{x : x \in \mathbb{R} \text{ and } -1 \leq x \leq 1\}$,

(c) $A = \{x : x \in \mathbb{N} \text{ and } 1 \leq x \leq 9\}$, $B = \{x : x \in \mathbb{N} \text{ and } 5 \leq x \leq 12\}$

3 Represent by Venn diagrams: (a) $A \cup B$, (b) $\bar{A} \cup \bar{B}$, (c) $\overline{A \cup B}$.

4 Show that:

(a) $A \cup (A \cap B) = A$, (b) $(A \cup B) \cap (A \cup \bar{B}) = A$,

(c) $(\bar{A} \cup \bar{B}) \cap (A \cap B) = \varnothing$, (d) $(A \cup B) \cap (A \cup \bar{B}) = A$

5 A company makes products A, B and C from a number of basic components. A requires components a, b and c. B requires components b, d and e. C requires components a and f. Determine (a) $A \cup B$, and (b) $A \cup B \cup C$ and explain what is meant by these sets.

6 An analysis of 100 faults occurring on a production line showed that in 30 of the cases the fault was due to faulty material, 50 of the cases to incorrect use of a tool and 70 of the cases to poor workmanship. If there were 19 cases involving two of these faults, how many involved all three?

7 Quality control on a production line check each product against three tests. When checking 400 items it was found that 39 failed test A, 30 failed test B, and 28 failed test C. Of these 7 failed tests A and B, 6 failed tests B and C and 5 failed tests A and C. 3 failed all the tests. Determine the number of items that failed just one test.

14 Boolean algebra

14.1 Introduction

In digital circuits extensive use is made of logic circuits. A switch is either on or off with these states being denoted by the digits 1 or 0. A logic circuit can be considered as a collection of switching circuits. In this chapter the basic mathematics necessary to analyse and synthesise such circuits is introduced. The mathematics involved is named after George Boole (1815–64) who first developed the modern ideas of the mathematics concerned with the manipulation of logic statements. In this chapter, Boolean algebra is approached by means of the analysis of switching circuits, using the algebra of sets introduced in Chapter 13.

14.2 Switching circuits

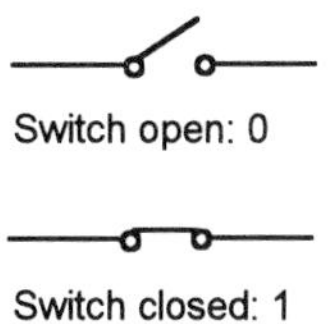

Figure 14.1 *The two states of a switch*

Consider a simple on-off switch (Figure 14.1). If we denote a closed contact by 1 and an open contact by 0 then the switch has just two possible states: 1 or 0. We can consider this to be the set $\{0, 1\}$. Note that the 0 and the 1 do not represent actual numbers but the state of the voltage or current in the circuit controlled by the switch. With electronic switches the 0 might be assigned to a voltage in the range 0 to nearly 1 V with the 1 any voltage in the range about 2 to 5 V, the switch operating between values in the two ranges. The term *logic level* is often used with the voltage being said to be at logic level 0 or logic level 1.

Suppose we have two switches a and b in series. Each switch has two possible states, 0 and 1. Figure 14.2 shows the various possibilities for switches. In (a) both switches are open, in (b) a is open and b is closed, in (c) a is closed and b is open and in (d) a and b are both closed. With (a) the effect of both switches being open is the same as would be obtained by a single open switch; (b) and (c) likewise are equivalent to a single open switch but (d) is equivalent to a single closed switch. Thus we can say that the two elements are equivalent to 0 for (a), (b) and (c) but 1 for (d). In tabular form we can represent the state of the circuit by Table 14.1:

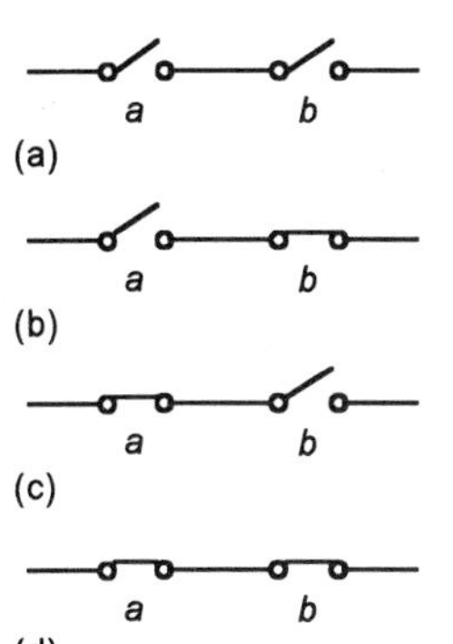

Figure 14.2 *Switches in series*

Table 14.1 a AND b

a	b	$a \cdot b$
0	0	0
0	1	0
1	0	0
1	1	1

Such a table is known as a *truth table*. If a AND b are 1 then the result is 1.

For series connections a is considered to be *multiplied* by b. The multiplication of two elements is denoted in a number of different ways in texts, the notations $\cdot$, $\times$, $*$, $\wedge$, and $\cap$ being used. Multiplication is here analogous to the intersection of sets. If a is 0 and b is 0 then the intersection is 0. If a is 0 and b is 1 there is no intersection. If a is 1 and b is 1 the intersection is 1. In this text · is used for the product. From the above table we have the rules:

$$0 \cdot 0 = 0, \quad 0 \cdot 1 = 0, \quad 1 \cdot 0 = 0, \quad 1 \cdot 1 = 1 \qquad [1]$$

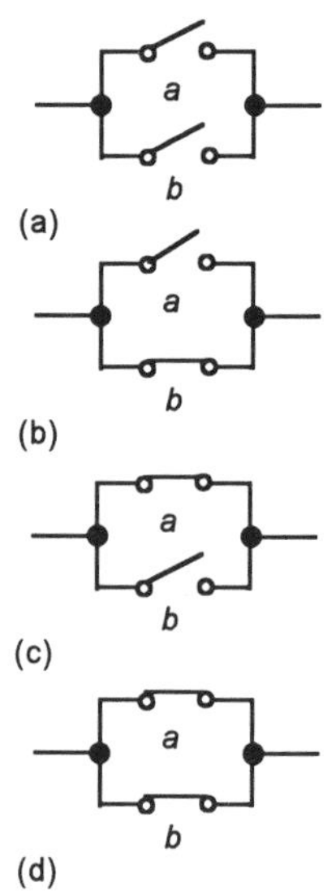

Figure 14.3 *Parallel switches*

Consider two switches in parallel. Figure 14.3 shows the various possibilities for switches. In (a) both switches are open, in (b) a is open and b is closed, in (c) a is closed and b is open and in (d) a and b are both closed. With (a) the effect of both switches being open is the same as would be obtained by a single open switch; (b), (c) and (d) are equivalent to a single closed switch. Thus we can say that the two elements are equivalent to 0 for (a), and 1 for (b), (c) and (d). In tabular form we can represent the state of the circuit by the truth table (Table 14.2):

Table 14.2 a OR b

a	b	$a+b$
0	0	0
0	1	1
1	0	1
1	1	1

If a OR b is 1 then the result is 1.

For parallel connections a is considered to be *added* to b. The sum of two elements is denoted in a number of different ways in text, the notations +, ⊕, ∨ and ∪ being used. Multiplication is analogous to union in sets. Thus the union of a and b when both are 0 is 0, when one of them is 0 and the other 1 is 1 and when both are 1 is 1. In this text + will be used for addition. From the above table we have the rules:

$$0 + 0 = 0, \quad 0 + 1 = 1, \quad 1 + 0 = 1, \quad 1 + 1 = 1 \qquad [2]$$

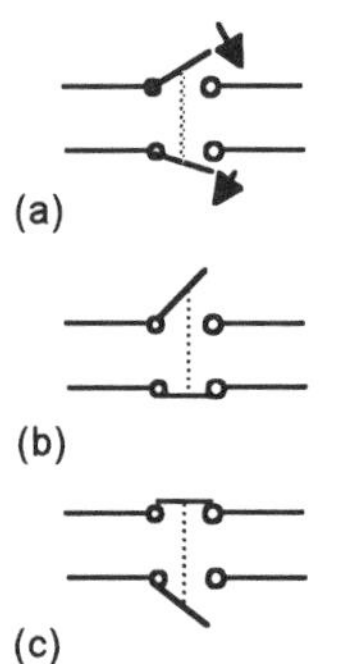

Figure 14.4 *Complement*

Another possible form of switch circuit is where two switches are connected together so that the closing of one switch results in the opening of the other. Figure 14.4(a) illustrates the switch action with (b) showing the upper switch open when the lower switch is closed and (c) the upper switch closed when the lower switch is open. The lower switch is said to give the *complement* of the upper switch. If the upper switch is denoted by a then the lower switch is denoted by $\bar{a}$. Table 14.3 is the truth table:

Table 14.3 NOT

a	$\bar{a}$
0	1
1	0

If one switch is 1 then the other switch is NOT 1. From the above table we have the rules:

$$\bar{0} = 1, \qquad \bar{1} = 0 \qquad [3]$$

Revision

1 Complete the following:

(a) $1 + 0 = ?$, (b) $1 \cdot 1 = ?$, (c) $\bar{1} = ?$

14.3 Laws of Boolean algebra

The binary digits 1 and 0 are the *Boolean variables* and, together with the operations ·, + and the complement, form what is known as *Boolean algebra*. By constructing the appropriate truth tables the following laws can be derived, being analogous to those given in Chapter 13 for sets:

Table 14.4

a	a	$a+a$
0	0	0
1	1	1

Table 14.5

a	a	$a \cdot a$
0	0	0
1	1	1

1 *Idempotent law*
See Table 14.4.

$$a + a = a \qquad [4]$$

2 *Idempotent law*
See Table 14.5.

$$a \cdot a = a \qquad [5]$$

3 *Commutative law*
This is illustrated by Table 14.2.

$$a + b = b + a \qquad [6]$$

4 *Commutative law*
This is illustrated by Table 14.1.

$$a \cdot b = b \cdot a \qquad [7]$$

5 *Distributive law*
See Table 14.6.

$$a + (b \cdot c) = (a + b) \cdot (a + c) \qquad [8]$$

6 *Distributive law*
See Table 14.7.

$$a \cdot (b + c) = a \cdot b + a \cdot c \qquad [9]$$

Table 14.6

a	b	c	$b \cdot c$	$a + b \cdot c$	$a + b$	$a + c$	$(a + b) \cdot (a + c)$
0	0	0	0	0	0	0	0
0	0	1	0	0	0	1	0
0	1	0	0	0	1	0	0
0	1	1	1	1	1	1	1
1	0	0	0	1	1	1	1
1	0	1	0	1	1	1	1
1	1	0	0	1	1	1	1
1	1	1	1	1	1	1	1

Table 14.7

a	b	c	$b + c$	$a \cdot (b + c)$	$a \cdot b$	$a \cdot c$	$a \cdot b + a \cdot c$
0	0	0	0	0	0	0	0
0	0	1	1	0	0	0	0
0	1	0	1	0	0	0	0
0	1	1	1	0	0	0	0
1	0	0	0	0	0	0	0
1	0	1	1	1	0	1	1
1	1	0	1	1	1	0	1
1	1	1	1	1	1	1	1

Table 14.8

a	$\bar{a}$	$a + \bar{a}$
0	1	1
1	0	1

Table 14.9

a	$\bar{a}$	$a \cdot \bar{a}$
0	1	0
1	0	0

Table 14.10

a	$a + 0$	$a + 1$
0	0	1
1	1	1

Table 14.11

a	$a \cdot 1$	$a \cdot 0$
0	0	0
1	1	0

7 *Complementary law*
See Table 14.8.

$$a + \bar{a} = 1 \quad [10]$$

8 *Complementary law*
See Table 14.9.

$$a \cdot \bar{a} = 0 \quad [11]$$

9 *Identity law*
See Table 14.10.

$$a + 0 = a, \quad a + 1 = 1 \quad [12]$$

10 *Identity law*
See Table 14.11.

$$a \cdot 1 = a, \quad a \cdot 0 = 0 \quad [13]$$

Table 14.12

a	b	$a+b$	$\overline{a+b}$	$\bar{a}$	$\bar{b}$	$\bar{a}\cdot\bar{b}$
0	0	0	1	1	1	1
0	1	1	0	1	0	0
1	0	1	0	0	1	0
1	1	1	0	0	0	0

Table 14.13

a	b	$a\cdot b$	$\overline{a\cdot b}$	$\bar{a}$	$\bar{b}$	$\bar{a}+\bar{b}$
0	0	0	1	1	1	1
0	1	0	1	1	0	1
1	0	0	1	0	1	1
1	1	1	0	0	0	0

In addition there are the *De Morgan* laws:

1 The complement of the outcome of switches a and b in parallel, i.e. an OR situation, is the same as when the complements of a and b are separately combined in series, i.e. the AND situation. Table 14.12 shows the validity of this.

$$\overline{a+b}=\bar{a}\cdot\bar{b} \quad [14]$$

2 The complement of the outcome of switches a and b in series, i.e. the AND situation, is the same as when the complements of a and b are separately considered in parallel, i.e. the OR situation. Table 14.13 shows the validity of this.

$$\overline{a\cdot b}=\bar{a}+\bar{b} \quad [15]$$

Using the rules given above, complicated switching circuits can be reduced to simpler equivalent circuits.

Example

Simplify the following Boolean function: $f= a\cdot c+(a+b)\cdot\bar{c}$.

Using equation [9]: gives $a\cdot(b+c)=a\cdot b+a\cdot c$

$$(a+b)\cdot\bar{c}=a\cdot\bar{c}+b\cdot\bar{c}$$

Hence we can write:

$$f=a\cdot c+a\cdot\bar{c}+b\cdot\bar{c}$$

Using equation [9] for the first two terms gives:

$$f = a \cdot (c + \bar{c}) + b \cdot \bar{c}$$

Then using equations [7] and [10] gives:

$$f = a \cdot 1 + b \cdot c = a + b \cdot \bar{c}$$

Example

Simplify the Boolean function: $f = a + a \cdot b \cdot c + \bar{a} \cdot \bar{c}$.

Using equation [13] we can replace a by $a \cdot 1$. The function can then be written as:

$$f = a \cdot 1 + a \cdot (b \cdot c) + \bar{a} \cdot \bar{c}$$

Then using equation [9]:

$$f = a \cdot (1 + (b \cdot c)) + a \cdot \bar{c}$$

Using the second of the equations in [12] gives $1 + (b \cdot c) = 1$ and so:

$$f = a \cdot 1 + \bar{a} \cdot \bar{c}$$

Since $a \cdot 1 = a$ (equation [10]), and applying equation [8]:

$$f = a + \bar{a} \cdot \bar{c} = (a + \bar{a}) \cdot (a + \bar{c})$$

But $a + \bar{a} = 1$ (equation [10]) and so, using equation [13]:

$$f = a + \bar{c}$$

Revision

2 Simplify the following Boolean functions:

(a) $a(\bar{a} + a \cdot b)$, (b) $a + b + c + \bar{a} \cdot b$, (c) $(a + b) \cdot (a + b)$,

(d) $a \cdot \bar{b} \cdot c + a \cdot \bar{b} \cdot \bar{c}$, (e) $a + \bar{a} \cdot \bar{b}$

14.3.1 Switching circuits

The operations ·, + and the complement can be used to write the Boolean functions for complex switching circuits, the states of such circuits being determined by developing the truth table to indicate all the various switching possibilities.

Example

Write, for the circuit shown in Figure 14.5, (a) the truth table and (b) the Boolean function to describe that truth table.

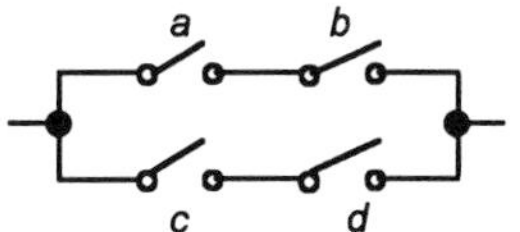

Figure 14.5 *Example*

(a) a and b are in series, and in parallel with the series arrangement of c and d. The result of using the switches is that only when either a and b are closed or c and d are closed will there be an output. Table 14.14 shows the truth table.

Table 14.14

a	b	c	d	Result
0	0	0	0	0
0	0	0	1	0
0	0	1	0	0
0	0	1	1	1
0	1	0	0	0
0	1	0	1	0
0	1	1	0	0
0	1	1	1	0
1	0	0	0	0
1	0	0	1	0
1	0	1	0	0
1	0	1	1	1
1	1	0	0	1
1	1	0	1	1
1	1	1	0	1
1	1	1	1	1

(b) The Boolean function for two switches in series is $a \cdot b$, the AND function, and thus, since the function for two items in parallel is OR, the function for the circuit as a whole is:

$$a \cdot b + c \cdot d$$

Example

Derive the Boolean function for the switching circuit shown in Figure 14.6.

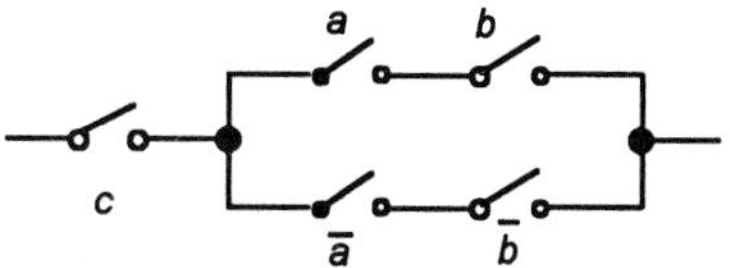

Figure 14.6 *Example*

In the upper parallel arm of the circuit, the switches a and b are in series and so have a Boolean expression of $a \cdot b$. In the lower arm the complements of a and b are in series. Thus the Boolean expression for that part of the circuit is $\bar{a} \cdot \bar{b}$. Because the two arms are in parallel the expression for that part of the circuit is $a \cdot b + \bar{a} \cdot \bar{b}$. In series with this is switch c. Thus the Boolean function for the circuit is:

$$c \cdot (a \cdot b + \bar{a} \cdot \bar{b})$$

Revision

3 State a Boolean function that can be used to represent each of the switching circuits shown in Figure 14.7.

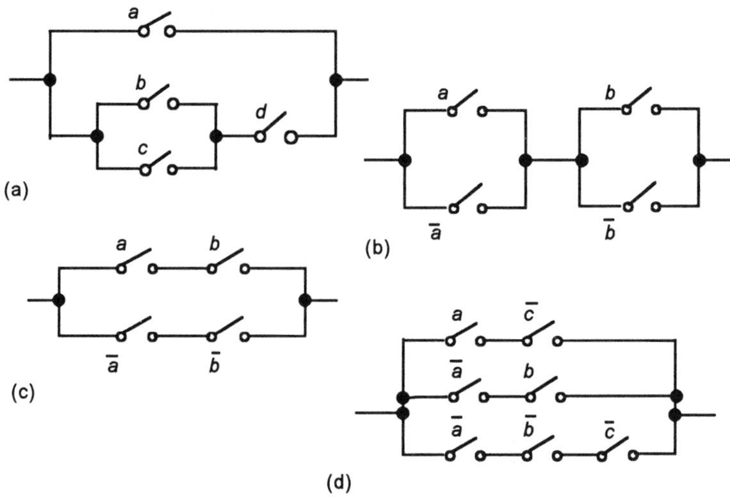

Figure 14.7 *Revision problem 3*

4 Give the truth table for the switching circuit corresponding to the Boolean function $(a \cdot \bar{b}) + (\bar{a} \cdot b)$.

5 Draw switching circuits to represent the following Boolean functions:

(a) $a \cdot (a + b)$, (b) $a \cdot (a \cdot b + c)$, (c) $\bar{a} \cdot (a + \bar{b} \cdot (a + c))$

14.4 Logic circuits

The Boolean variable involves the realisation of two states, 0 and 1. These are said to be the two *logical states*. With a mechanical switch these can represent the switch being open and closed. With electronic switches, 0 is taken to be a low voltage level and 1 a high voltage level for what is called *positive logic*, although the opposite convention (*negative logic*) can be used with 0 being represented by a high voltage level and 1 by a low voltage level. The basic building blocks of digital electronic circuits are called *logic gates*. A logic gate is an electronic block which has one or more inputs and an output. The output can be either high or low depending on the digital levels at the input terminals. The following sections take a look at the logic gates: AND, OR, INVERT/NOT, NAND, NOR and XOR.

14.4.1 AND gate

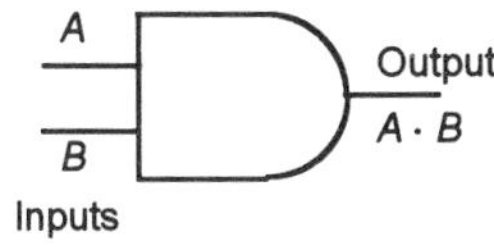

Figure 14.8 *AND gate*

The AND gate gives an output corresponding to the Boolean function $A \cdot B$, there being an output 1 when both input A and input B are 1. Different sets of standard circuit symbols have been developed in Britain, Europe and the United States, those of US origin being the most widely used and so the ones used in this text. Figure 14.8 shows the symbol, the associated truth table being given in Table 14.15.

Table 14.15 AND gate

A	B	$A \cdot B$
0	0	0
0	1	0
1	0	0
1	1	1

14.4.2 OR gate

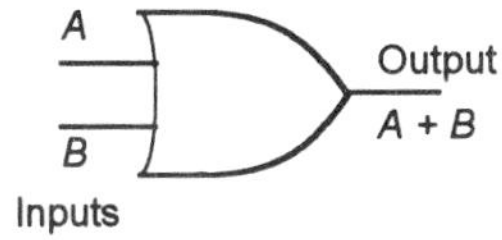

Figure 14.9 *OR gate*

The OR gate gives an output 1 when either input A or input B is 1 and corresponds to the Boolean function $A + B$. Figure 14.9 shows the gate symbol and Table 14.16 the truth table.

Table 14.16 OR gate

A	B	$A+B$
0	0	0
0	1	1
1	0	1
1	1	1

14.4.3 INVERT/NOT gate

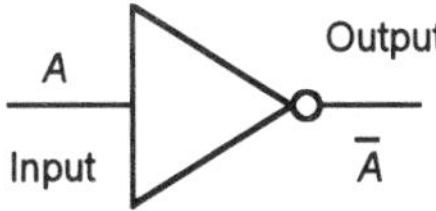

Figure 14.10 *NOT gate*

The INVERT or NOT gate has a single input and gives a 1 output when the input is 0, corresponding to the Boolean function $\bar{A}$. The gate inverts the input, giving a 1 when the input is 0 and a 0 when the input is 1. Figure 14.10 shows the gate symbol and Table 14.17 gives the truth table.

Table 14.17 NOT

A	$\bar{A}$
0	1
1	0

14.4.4 NAND gate

This gate is logically equivalent to a NOT gate in series with an AND gate (Figure 14.11(a)), NAND standing for NotAND. The symbol for the gate (Figure 14.11(b)) is the AND symbol followed by a small circle, the small circle being used to indicate negation. The gate represents the Boolean function $\overline{A \cdot B}$ and has the truth table shown in Table 14.18. There is a 1 output when A and B are both not 1, i.e. are both 0.

Note that by using a De Morgan law (equation [15]) we can also write the Boolean function for the NAND gate as $\overline{A \cdot B} = \bar{A} + \bar{B}$.

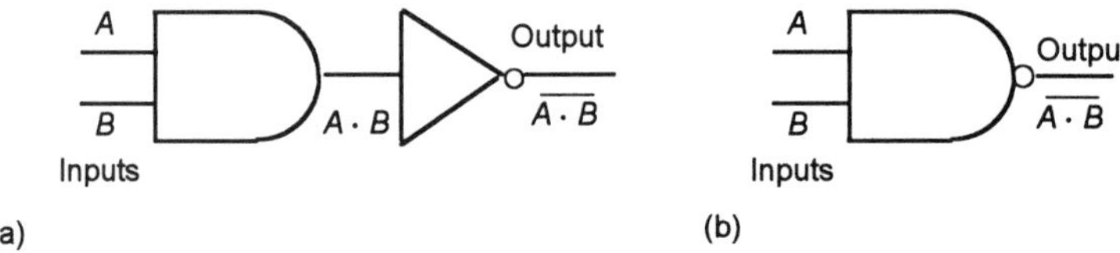

Figure 14.11 *NAND gate*

Table 14.17 NAND gate

A	B	$\overline{A \cdot B}$
0	0	1
0	1	1
1	0	1
1	1	0

14.4.5 NOR gate

This gate is logically equivalent to a NOT gate in series with an OR gate (Figure 14.12(a)). It is represented by the OR gate symbol followed by a small circle to indicate negation (Figure 14.12(b)). The gate represents the Boolean function $\overline{A + B}$. Note that using a De Morgan law (equation [14]) this can also be written as $\overline{A + B} = \bar{A} \cdot \bar{B}$. Table 14.19 gives the truth table, there being a 1 output when neither A nor B is 1.

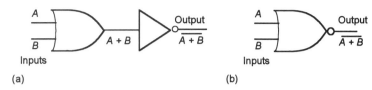

Figure 14.12 *NOR gate*

Table 14.19 NOR gate

A	B	$\overline{A+B}$
0	0	1
0	1	0
1	0	0
1	1	0

14.4.6 EXCLUSIVE OR (XOR) gate

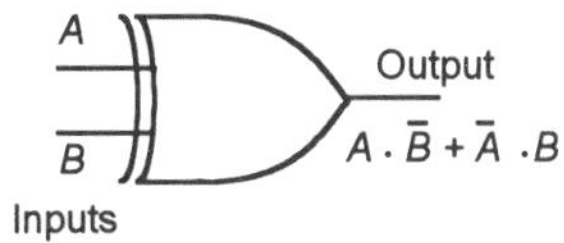

Figure 14.13 *XOR gate*

The OR gate gives an output 1 when either input A or input B is 1 or both A and B are 1. The EXCLUSIVE OR gate gives an output 1 when either input A or input B is 1 but not when both are. It is described by the Boolean function $A \cdot \bar{B} + \bar{A} \cdot B$. Figure 14.13 shows the gate symbol and Table 14.20 the truth table.

Table 14.20 XOR gate

A	B	$A \cdot \bar{B} + \bar{A} \cdot B$
0	0	0
0	1	1
1	0	1
1	1	0

14.4.7 Combining gates

By combining logic gates it is possible to represent other Boolean functions.

Example

Determine the Boolean function describing the relation between the output from the logic circuit shown in Figure 14.14. Hence, consider how the circuit could be simplified.

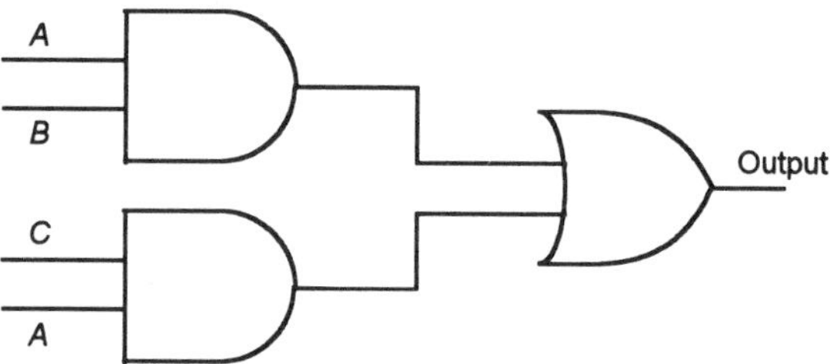

Figure 14.14 *Example*

This might be a circuit used with a car warning buzzer so that it sounds when the key is in the ignition (A) and a car door is opened (B) or the headlights are on (C) and a car door is opened (A). We have two AND gates and an OR gate. The output from the top AND gate is $A \cdot B$, and from the lower AND gate $C \cdot A$. These outputs are the inputs to the OR gate and thus the output is $A \cdot B + C \cdot A$. The circuit can be simplified by considering the Boolean algebra. Using equation [9] the Boolean function can be written as:

$$A \cdot B + C \cdot A = A \cdot (B + C)$$

We now have A and B or C. This function now describes a logic circuit with just two gates, an OR gate and an AND gate. Figure 14.15 shows the circuit.

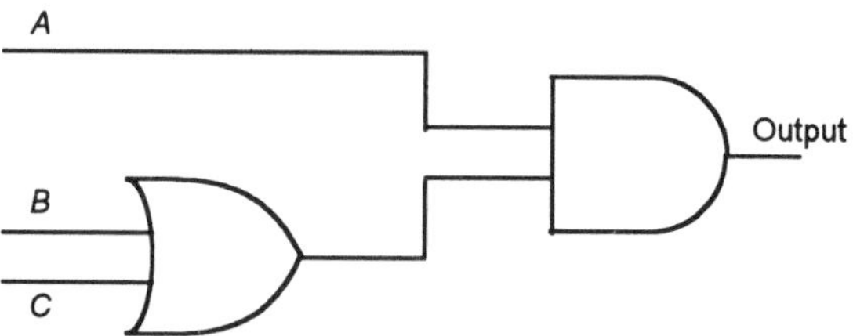

Figure 14.15 *Example*

Example

Devise a logic gate system to generate the Boolean function $A \cdot \bar{B} + C$.

$A \cdot B$ requires an AND gate, but as the B input has to be inverted we precede the input from B to the AND gate by a NOT gate. We then require an OR gate for the output from the AND gate and C. Figure 14.16 shows the gate system.

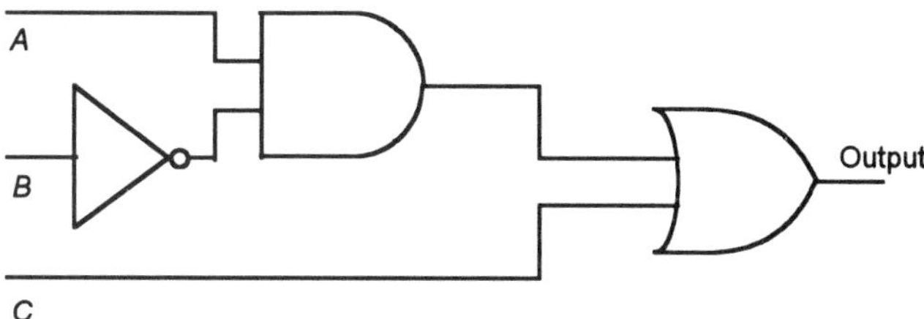

Figure 14.16 *Example*

Revision

6 Determine the Boolean functions that could generate the outputs in Figure 14.17.

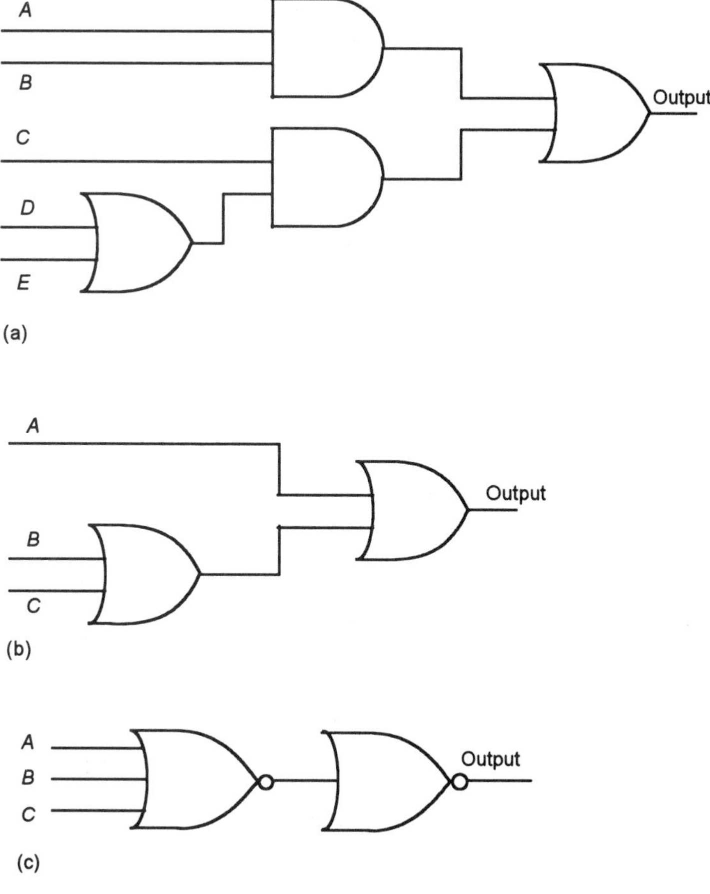

Figure 14.17 *Revision problem 6*

7 Determine the Boolean equations describing the logic circuits in Figure 14.18, then simplify the equations and hence obtain simplified logic circuits.

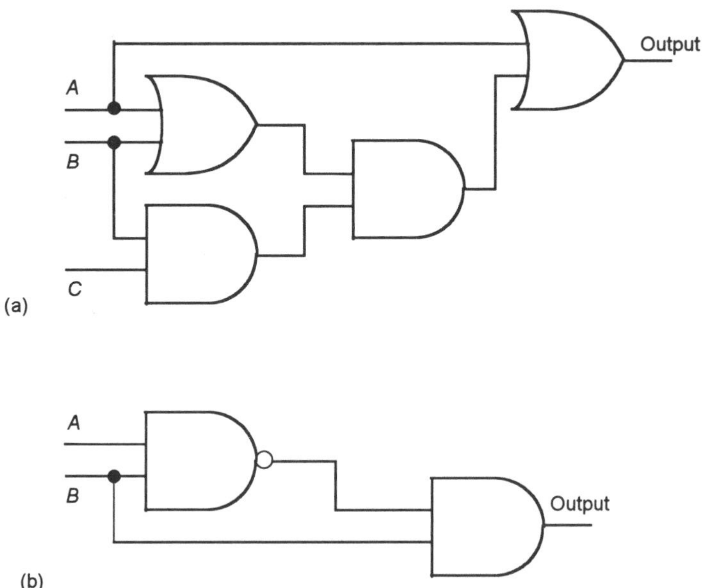

Figure 14.18 *Revision problem 7*

Problems

1 Simplify the following Boolean functions:

(a) $a\cdot b\cdot d+a\cdot b\cdot c\cdot d$, (b) $\overline{\bar{a}b\bar{c}}$, (c) $\bar{a}\cdot b\cdot\bar{c}+a\cdot b\cdot\bar{c}+b\cdot\bar{c}\cdot d$,

(d) $a+\bar{a}\cdot\bar{b}\cdot c$, (e) $\overline{(\overline{\bar{a}+b})(\overline{a\cdot b+\bar{c}})}$, (f) $(\overline{a+c})+\overline{a\cdot b}\cdot(b+c)$

2 State a Boolean function that can be used to represent each of the switching circuits shown in Figure 14.19.

3 Give the truth tables for the switching circuits represented by the Boolean functions (a) $(a+\bar{b})+(a+\bar{c})$, (b) $\bar{a}\cdot(a\cdot b+\bar{b})\cdot\bar{b}$.

4 Draw switching circuits to represent the Boolean functions (a) $a\cdot b$, (b) $a\cdot b+b$, (c) $c\cdot(a\cdot b+a\cdot\bar{b})$, (d) $a\cdot(a\cdot b\cdot\bar{c}+a\cdot(\bar{b}+c))$.

5 Derive the Boolean functions for the truth tables in Table 14.21(a) and (b).

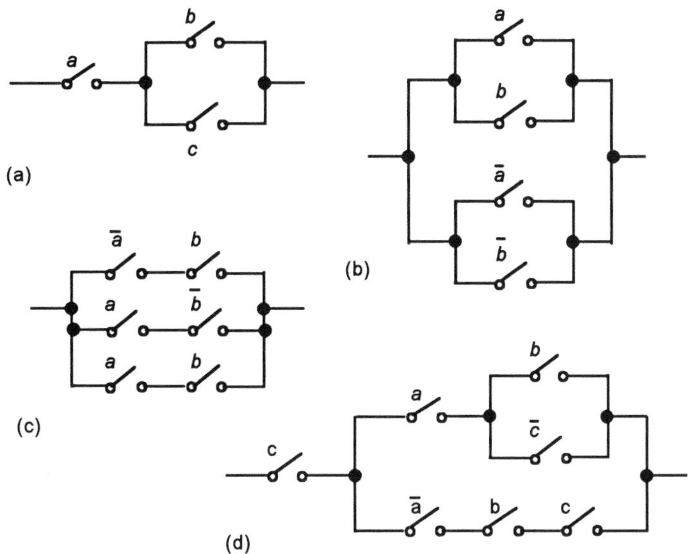

Figure 14.19 *Problem 2*

Table 14.21(a)

a	*b*	*c*	*Function*
0	0	0	0
0	0	1	1
0	1	0	0
0	1	1	0
1	0	0	0
1	0	1	0
1	1	0	1
1	1	1	0

Table 14.21(b)

a	*b*	*c*	*Function*
0	0	0	0
0	0	1	0
0	1	0	0
0	1	1	1
1	0	0	0
1	0	1	0
1	1	0	0
1	1	1	1

6 State the Boolean functions for the logic circuits shown in Figure 14.20.

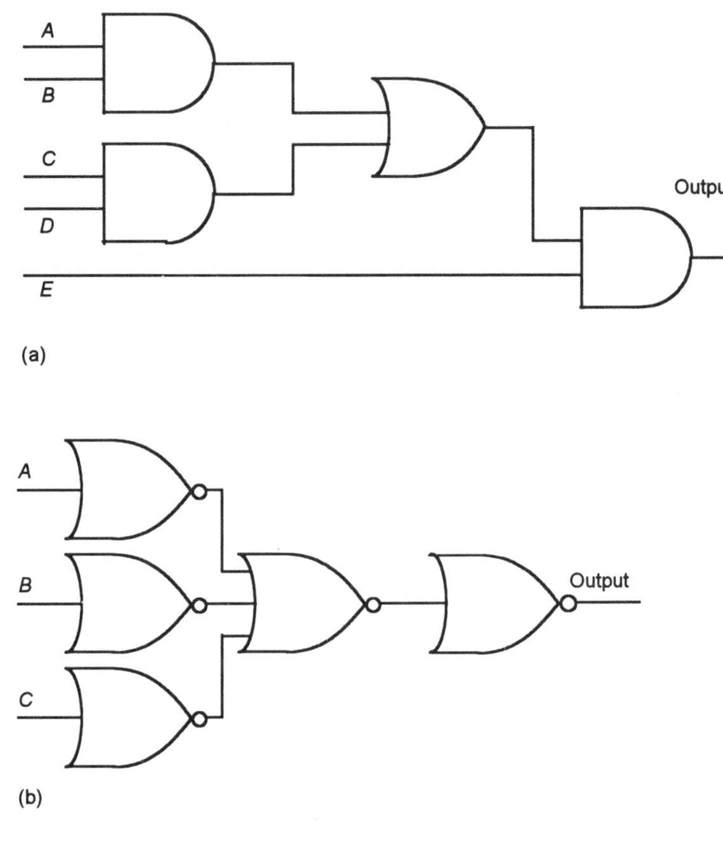

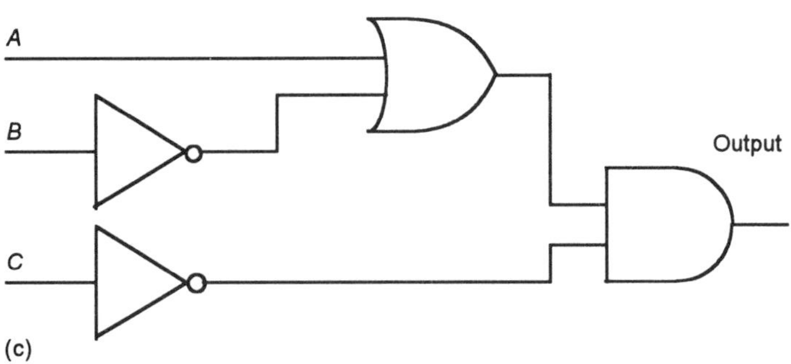

Figure 14.20 *Problem 6*

7 Devise logic gate systems to give the following Boolean functions:

(a) $\bar{A}+B+C$, (b) $A \cdot B+C$, (c) $A \cdot \bar{B}+B \cdot C$, (d) $\bar{D} \cdot (\bar{A}+\bar{B} \cdot \bar{C})$

8 Determine the Boolean equations describing the logic circuits in Figure 14.21, then simplify the equations and hence obtain simplified logic circuits

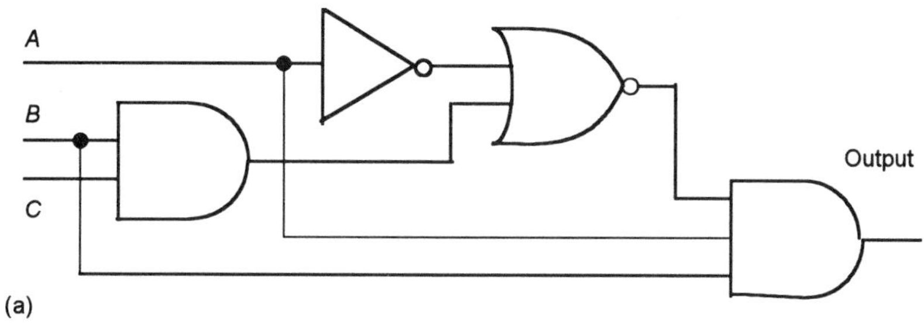

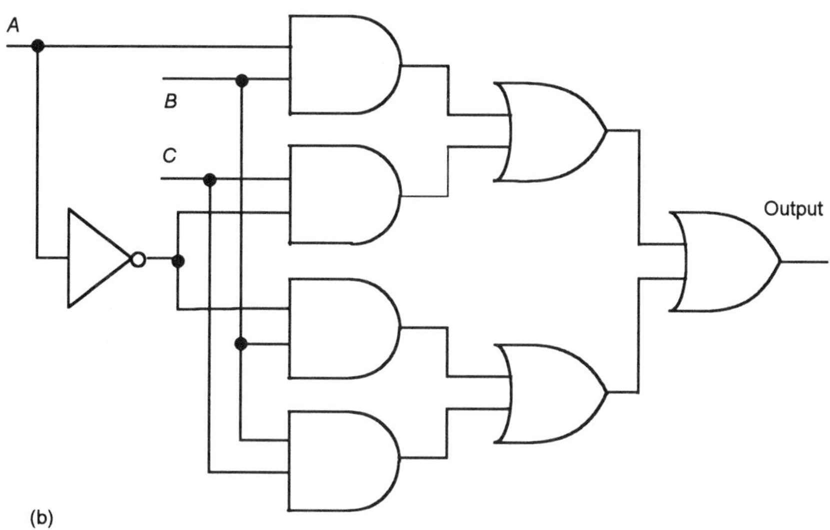

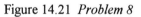

Figure 14.21 *Problem 8*

15 Application: Digital systems

15.1 Introduction

This chapter extends the discussion of logic gates given in Chapter 14 to a consideration of their use in digital electronic systems. In some applications a knowledge of arithmetic operations with binary numbers is required. An outline of such arithmetic operations is included.

Digital electronic gates are cheap and available as integrated circuits. They have a wide range of applications. For example, they might be used to determine whether an input signal will be allowed to lead to an output or not. They might be used to sound an alarm or initiate some action when any one of several sensors is activated. This might be because the temperature rises above some critical level or the pressure falls below some critical level. An important feature of digital systems, such as digital computers, is the execution of arithmetic operations. At the heart of such systems are digital electronic gates.

This chapter gives a brief indication of the ways in which logic gates are used.

15.2 Integrated circuit logic gates

Logic gates are available as integrated circuits. Figure 15.1 shows examples of such integrated circuits. Figure 15.1(a) shows the gate systems available in integrated circuit 7408, it having four two-input AND gates and is supplied in a 14-pin package. Power supply connections are made to pins 7 and 14, these supplying the operating voltage for all the four AND gates. In order to indicate at which end of the package pin 1 starts, a notch is cut between pins 1 and 14. Figure 15.1(b) shows the gate systems available in integrated circuit 7402. This has four two-input NOR gates in a 14 pin package, power connections being to pins 7 and 14.

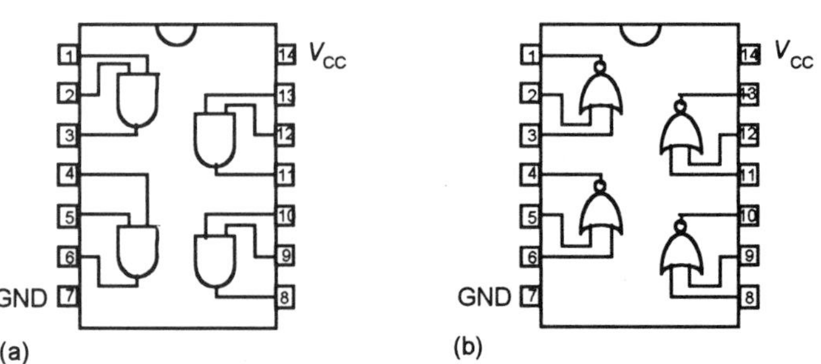

Figure 15.1 *Integrated circuits (a) 7408, (b) 7402*

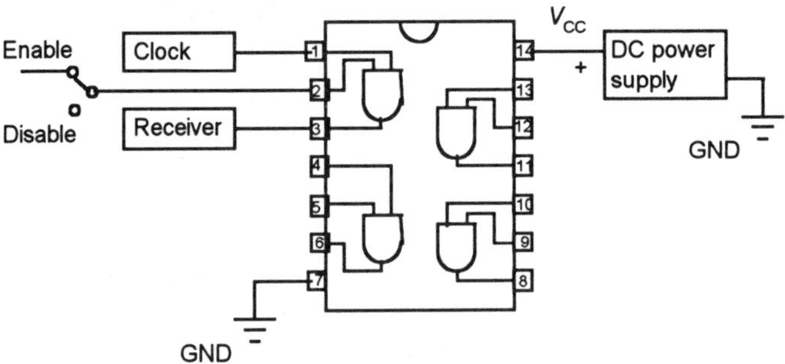

Figure 15.2 *Clock enable circuit*

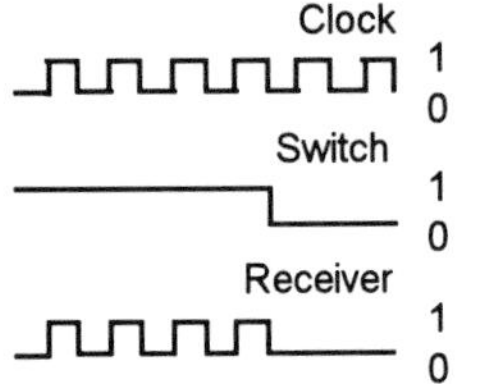

Figure 15.3 *Clock enable*

As a simple illustration of the use of the 7408 integrated circuit, Figure 15.2 shows its use to enable signals from a clock to be passed on to some receiving device when a switch is set to the enable position and not to be switched when set to the disable position. The clock emits a sequence of regular pulses, i.e. a sequence of 0 and 1 signals (Figure 15.3). The switch gives a 1 signal when in the enable position and a 0 signal when in the disable position. The AND gate will give an output when both the input signals are 1. Thus the output will be a sequence of pulses as long as the switch is in the enable position.

Revision

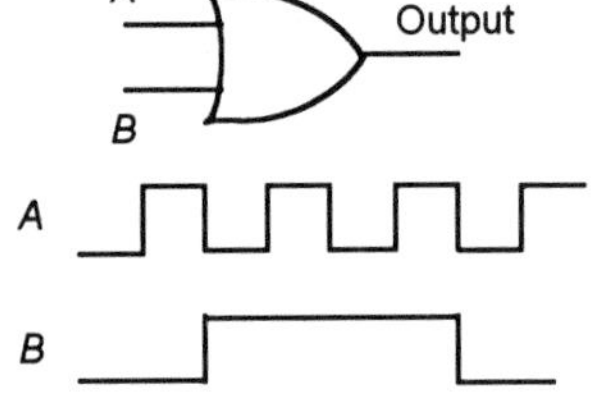

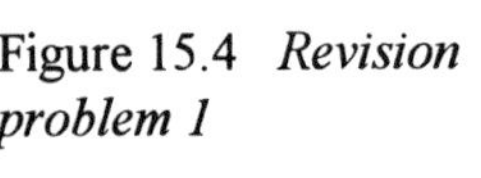

Figure 15.4 *Revision problem 1*

1 Sketch the output waveform for the two-input OR gate when the input waveforms are as shown in Figure 15.4.

2 Sketch the output waveform for the two-input AND gate when the input waveforms are as shown in Figure 15.5.

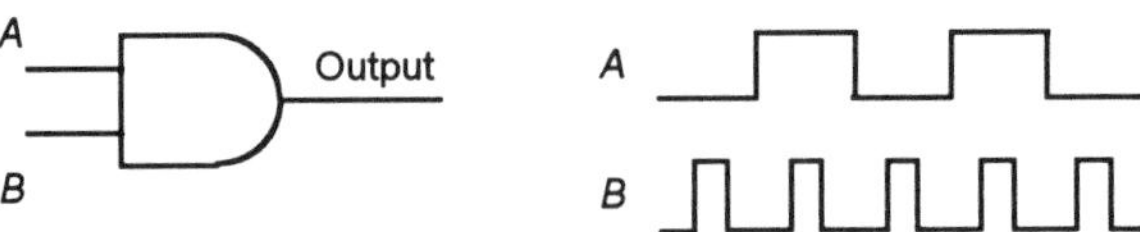

Figure 15.5 *Revision problem 2*

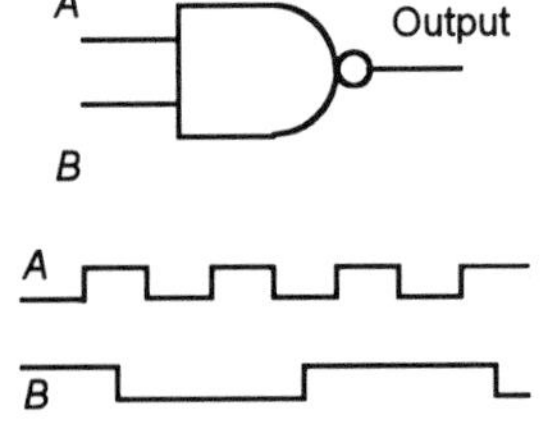

Figure 15.6 *Revision problem 3*

3 Sketch the output waveform for the two-input NAND gate when the input waveforms are as shown in Figure 15.6.

4 A three-input NAND gate has input waveforms A and B from sensors and input C as a control input. When the control input goes high, what is the condition that there is a high output?

15.2.1 NAND and NOR gates

By combining NAND gates in appropriate ways, all the other forms of gate can be produced. A NAND gate (Section 14.4.4) has the Boolean function:

$$\text{output} = \overline{A \cdot B}$$

The INVERT gate (Section 14.4.3) requires the Boolean function of output $= \bar{A}$ and can be realised by making $B = A$ to give:

$$\text{output} = \overline{A \cdot A}$$

But equation [5] of Chapter 15 gives $A \cdot A = A$, thus the INVERT gate can be realised from a NAND gate by Figure 15.7(a).

The OR gate (Section 14.4.2) requires the Boolean function of output = $A + B$. This can be obtained with just NAND gates (Figure 15.7(b)) by using the arrangement of a NAND gate to give INVERT and inverting each input. The result is then inputs of $\bar{A}$ and $\bar{B}$ to a third NAND gate and so:

$$\text{output} = \overline{\bar{A} \cdot \bar{B}}$$

The De Morgan law (equation [15] in Chapter 14) indicates that this is the same as $A + B$, since a double complement of A equals A and a double complement of B equals B.

The AND gate (Section 14.4.1) requires the Boolean function of output $= A \cdot B$. Inverting the output from a NAND gate (Figure 15.7(c)) gives:

$$\text{output} = \overline{\overline{A \cdot B}} = A \cdot B$$

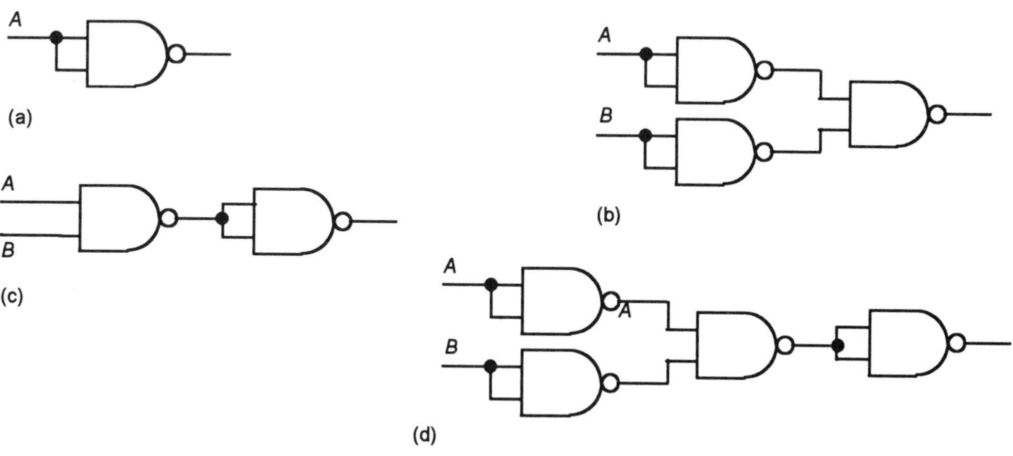

Figure 15.7 *Using NAND gates to produce (a) INVERT, (b) OR, (c) AND, (d) NOR gates*

The NOR gate (Section 14.4.5) has the Boolean function $\overline{A+B}$ and can be realised with NAND gates by taking the arrangement of them giving an OR gate and inverting the output (Figure 15.7(d)).

Likewise, by combining NOR gates in appropriate ways, all the other forms of gate can be produced. The NOR gate (Section 14.4.5) has the Boolean function:

$$\text{output} = \overline{A+B}$$

For INVERT (Figure 15.8(a)), we make $B = A$ and so, using equation [2] from Chapter 14:

$$\text{output} = \overline{A+A} = \bar{A}$$

For OR (Figure 15.8(b)), the output from a NOR gate is inverted. Then, recognising a double complement is no complement:

$$\text{output} = \overline{\overline{A+B}} = A+B$$

For AND (Figure 15.8(c)), the inputs to a NOR gate are both inverted. Thus, using De Morgan's law (equation [14], Chapter 14):

$$\text{output} = \overline{\bar{A}+\bar{B}} = A \cdot B$$

For NAND (Figure 15.8(d)), the AND form of the gate is followed by a NAND gate arranged as an inverter. Then:

$$\text{output} = \overline{A \cdot B}$$

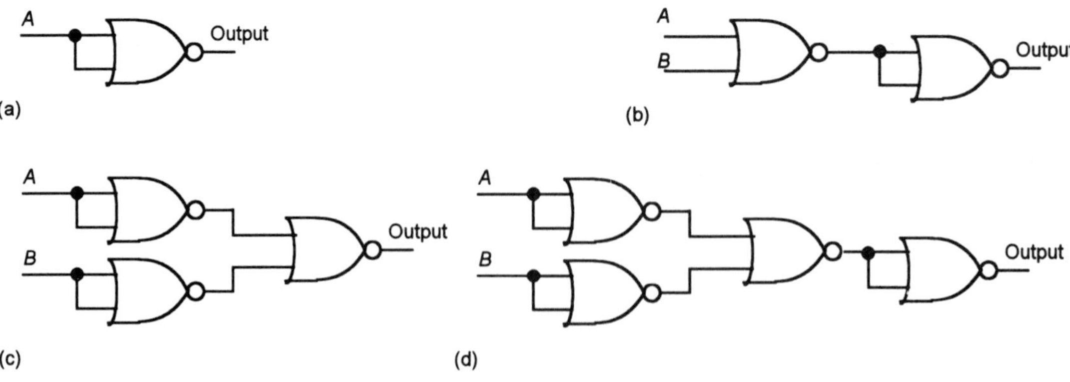

Figure 15.8 *Using NOR gates to produce (a) INVERT, (b) OR, (c) AND, (d) NOR gates*

Revision

5 Using NOR gates, obtain the Boolean function $\bar{A} + B$.

6 Using NAND gates, obtain the Boolean function $\bar{A} + \bar{B} \cdot \bar{C}$.

15.3 Arithmetic operations

An important function that is often required with digital circuits is the implementation of arithmetic operations. Before considering such operations, the following is a brief discussion of number systems.

15.3.1 Number systems

The number system used for everyday calculations is the *denary* or *decimal system*. This is based on the use of the 10 digits: 0, 1, 2, 3, 4, 5, 6, 7, 8, 9. With a number represented by this system, the digit position in the number indicates the weight attached to each digit, the weight increasing by a factor of 10 as we proceed from right to left. Hence we have:

...	10^3	10^2	10^1	10^0
	thousands	hundreds	tens	units

The *binary system* is based on just two digits: 0 and 1. These are termed *bi*nary digi*ts* or *bits*. When a number is represented by this system, the digit position in the number indicates the weight attached to each digit, the weight increasing by a factor of 2 as we proceed from right to left. Hence we have:

...	2^3	2^2	2^1	2^0
	bit 3	bit 2	bit 1	bit 0

The bit 0 is termed the *least significant bit* (LSB) and the highest bit the *most significant bit* (MSB). For example, with the binary number 1010, the least significant bit is the bit at the right-hand end of the number and so is 0. The most significant bit is the bit at the left-hand end of the number and so is 1. When converted to a denary number we have, for the 1010:

	2^3	2^2	2^1	2^0
	bit 3	bit 2	bit 1	bit 0
	MSB			LSB
Binary	1	0	1	0
Denary	$2^3 = 8$	0	$2^1 = 2$	0

Thus the denary equivalent is 10. The conversion of a binary number to a denary number thus involves the addition of the powers of 2 indicated by the number.

The conversion of a denary number to a binary number involves looking for the appropriate powers of 2. We can do this by successive divisions by 2, noting the remainders at each division. Thus if we have the denary number 31:

$31 \div 2 = 15$ remainder 1 This gives the LSB
$15 \div 2 = 7$ remainder 1
$7 \div 2 = 3$ remainder 1
$3 \div 2 = 1$ remainder 1 This gives the MSB

The binary number is thus 1111. The first division gives the least significant bit because we have just divided the 31 by 2, i.e. 2^1 and found 1 left over for the 2^0 digit. The last division gives the most significant bit because the 31 has then been divided by 2 four times, i.e. 2^4 and the remainder is 1.

Binary numbers are used in computers because the two states represented by 0 and 1 are easy to deal with in switching circuits where they can represent off and on. A problem with binary numbers is that comparatively small numbers require a large number of digits. For example, the denary number 9 which involves just a single digit requires four when written as the binary number 1001. The denary number 181, involving three digits, in binary form is 10110101 and requires eight digits. Because of this, octal or hexadecimal numbers are sometimes used to make numbers easier to handle.

The *octal system* is based on eight digits: 0, 1, 2, 3, 4, 5, 6, 7. When a number is represented by this system, the digit position in the number indicates the weight attached to each digit, the weighting increasing by a factor of 8 as we proceed from right to left. Thus we have:

$$\ldots \quad 8^3 \quad 8^2 \quad 8^1 \quad 8^0$$

For example, the denary number 15 is 17 in the octal system. To convert from binary into octal, the binary number is written in groups of three bits starting with the least significant bit. For example, the binary number 11010110 would be written as:

11 010 110

Each group is then replaced by the corresponding digit 0 to 7. The 110 binary number is 7, the 010 is 2 and the 11 is 3. Thus the octal number is 326. As another example, the binary number 100111010 is:

100	111	010	Binary
4	7	2	Octal

Octal to binary conversion involves converting each octal digit into its 3-bit equivalent.

The *hexadecimal system (hex)* is based on 16 digits/symbols: 0, 1, 2, 3, 4, 5, 6, 7, 8, 9, A, B, C, D, E, F. When a number is represented by this system, the digit position in the number indicates that the weight attached to each digit increases by a factor of 16 as we proceed from right to left. Thus we have:

$$\ldots \qquad 16^3 \qquad 16^2 \qquad 16^1 \qquad 16^0$$

For example, the decimal number 15 is F in the hexadecimal system. To convert binary numbers into hexadecimal numbers, group the binary numbers into fours starting from the least significant number. Thus, for the binary number 1110100110 we have:

11	1010	0110	Binary number
3	R	6	Hex number

For conversion from hex to binary, each hex number is converted to its 4-bit equivalent.

Because the external world tends to deal mainly with numbers in the denary system and computers with numbers in the binary system, there is always the problem of conversion. There is, however, no simple link between the position of digits in a denary number and the position of digits in a binary number. An alternative method that is often used is the ***binary*** *coded decimal system (BCD)*. With this system, each denary digit is coded separately in binary. For example, the denary number 15 has the 5 converted into the binary number 0101 and the 1 into 0001 to give in BCD the number 0001 0101.

Table 15.1 gives examples of numbers in the denary, binary, octal, hex and BCD systems.

Revision

7 Convert the following binary numbers to denary numbers:

(a) 011 011, (b) 001 100, (c) 111 100

8 Convert the following denary numbers to binary numbers:

(a) 100, (b) 111, (c) 42

9 Convert the following hex numbers to denary numbers:

(a) 9F, (b) 67E, (c) 1E

10 Convert the following hex numbers to binary numbers:

(a) 1D, (b) E, (c) 3B9

Table 15.1 *Number systems*

Denary	Binary	Octal	Hex	BCD
0	00000	0	0	0000 0000
1	00001	1	1	0000 0001
2	00010	2	2	0000 0010
3	00011	3	3	0000 0011
4	00100	4	4	0000 0100
5	00101	5	5	0000 0101
6	00110	6	6	0000 0110
7	00111	7	7	0000 0111
8	01000	10	8	0000 1000
9	01001	11	9	0000 1001
10	01010	12	A	0001 0000
11	01011	13	B	0001 0001
12	01100	14	C	0001 0010
13	01101	15	D	0001 0011
14	01110	16	E	0001 0100
15	01111	17	F	0001 0101
16	10000	20	11	0001 0110
17	10001	21	12	0001 0111

15.3.2 Binary arithmetic

Addition of binary numbers uses the following rules:

$$0 + 0 = 0$$

$$0 + 1 = 1 + 0 = 1$$

$$1 + 1 = 10$$

$$1 + 1 + 1 = 11$$

Consider the addition of the binary numbers 01 110 and 10 111:

	01 110
	10 111
Sum	100 001

For the LSB bit 1 in the sum, 0 + 1 = 1. For bit 1 in the sum, 1 + 1 = 10 and so we have 0 with 1 carried to the next column. For bit 3 in the sum, 1 + 0 + the carried 1 = 10. For bit 4 in the sum, 1 + 0 + the carried 1 = 10. We continue this through the various bits and end up with the 100 001.

Subtraction of binary numbers uses the following rules:

$0 - 0 = 0$

$1 - 0 = 1$

$1 - 1 = 0$

When evaluating 0 – 1, a 1 is borrowed from the next column on the left containing a 1. The following example illustrates this with the subtraction of 01110 from 11011:

	11 011
	01 110
Difference	01 101

For the LSB we have 1 – 0 = 1. For bit 1 we have 1 – 1 = 0. For bit 2 we have 0 – 1. We thus borrow 1 from the next column and so have 10 – 1 = 1. For bit 3 we have 0 – 1, remember we borrowed the 1. Again borrowing 1 from the next column, we then have 10 – 1 = 1. For bit 4 we have 0 – 0 = 0, remember we borrowed the 1.

Revision

11 Carry out the following arithmetic operations:

(a) 000 101 + 000 010, (b) 011 111 + 000 111,

(c) 010 010 011 + 001 001 011, (d) 011 011 – 001 010,

(e) 011 000 000 – 000 000 011

15.3.3 Signed numbers

The binary numbers considered so far contain no indication whether they are negative or positive and are said to be *unsigned*. Since there is generally a need to handle both positive and negative numbers there needs to be some way of distinguishing between them. This can be done by adding a sign bit. When a number is said to be *signed*, the most significant bit is used to indicate the sign of the number with a 0 being used if the number is positive and a 1 if it is negative. Thus for an 8-bit number we have:

XXXX XXXX
↑
Sign bit

When we have a positive number then we write it in the normal way with a 0 preceding it. Thus a positive binary number of 10 110 would be written as 010 110. A negative number of 10 110 would be written as 110 110.

Although the use of a sign bit can enable computers to handle negative binary numbers, a more commonly used method is the *two's complement method.* When negative numbers are transformed into their signed two's complement form they can be treated just like positive numbers. A binary number has two complements, known as the *one's complement* and the *two's complement.* When a binary number and its one's complement are added the result is always 1, when a binary number and its two's complement are added the result is always 10, i.e. denary 2. The one's complement of a binary number is obtained by subtracting each digit from 1; this is, however, the same as changing all the 1s in the number into 0s and the 0s into 1s. Thus if we have the binary number 101 101 then the one's complement of it is 010 010. The two's complement can be obtained from the one's complement by adding 1 to the least significant bit of the one's complement. Thus 010 010 becomes 010 011.

When we have a negative number then, to obtain the signed two's complement, we can obtain the two's complement of the unsigned number and then sign it with a 1. Consider the representation of the decimal number −6 as a signed two's complement number when the total number of bits is to be eight. We first write the unsigned binary number for 6, i.e. 0 000 110, then obtain the one's complement of 1 111 001, add 1 to give 1 111 010, and finally sign it with a 1 to indicate it is negative. The result is thus 11 111 010.

Unsigned binary number when sign ignored	0 000 110
One's complement	1 111 001
Add 1	1
Unsigned two's complement	1 111 010
Signed two's complement	11 111 010

An alternative way of obtaining the signed two's complement for a negative number is to take the signed binary number of the positive number, take the one's complement, then add 1 to give the signed two's complement which is then that of the negative number. Carrying out the two's complement operation on a signed number will change a positive number to a negative number.

Signed binary number for +6	00 000 110
One's complement	11 111 001
Add 1	1
Signed two's complement for −6	11 111 010

Another method of obtaining the two's complement is to start with the least significant bit, proceed along the number until a 1 is found. Then complement all the bits after the first 1. Thus for 00 000 110:

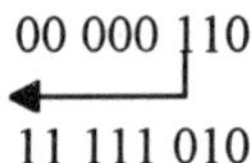

Table 15.2 lists some signed two's complements, given to 4 bits, for denary numbers. Note that the equivalent system with denary numbers, the ten's complement, enables us to write –1 as 999, –2 as 998, –3 as 997. Thus when we have 3 + (–3) we write 3 + 997 = 1000. Then ignoring the carry we have 000. All that the two's complement does is give rules by which we can write negative numbers as positive numbers which decrease as the negative number becomes more negative.

Table 15.2 *Signed two's complement*

Denary number	Signed two's complement
–5	1 011
–4	1 100
–3	1 101
–2	1 110
–1	1 111

Subtraction of a positive number from a positive number can be considered to be the addition of a negative number to a positive number. Thus we obtain the signed two's complement of the negative number and then add it to the signed positive number. Hence, for the subtraction of the denary number 6 from the denary number 4 we can consider the problem as being (+4) + (–6). Hence we add the signed positive number to the signed two's complement for the negative number.

Binary form of +4	00 000 100
(–6) as signed two's complement	11 111 010
Sum	11 111 110

The most significant bit, i.e. the sign, of the outcome is 1 and so the result is negative. This is the 8-bit signed two's complement for –2. Note that with two's complement arithmetic, *any carries-out of the most significant bit are meaningless and should be ignored.*

If we wanted to add two negative numbers then we would obtain the signed two's complement for each number and then add them. Whenever a number is negative we use the signed two's complement, when positive just the signed number.

Revision

12 Determine the two's complement of (a) 101 101, (b) 00 100 011.

13 State which of the following signed two's complement numbers are positive:

(a) 10 000 000, (b) 00 111 110, (c) 10 100 011, (d) 011 111 111

14 Represent, using a total of 8 bits which includes the sign, the following denary numbers in the signed two's complement form: (a) −5 , (b) −23, (c) −125.

15 Carry out the following arithmetic operations using signed two's complements: (a) 20 − 12, (b) 32 − 17, (c) 660 − 12, (d) 10 − 24.

15.3.4 Arithmetic circuits

Consider the addition of two 1-bit binary numbers when represented in the form of a truth table (Table 15.3):

Table 15.3 *Truth table for least significant bit*

First number	Second number	Sum	Carry-out
0	0	0	0
0	1	1	0
1	0	1	0
1	1	0	1

The sum is as a 1-bit number with the carry-out bit to be added to the next position. The above is the truth table for the least significant bit in a binary number. For a more significant bit we have Table 15.4:

Table 15.4 *Truth table for more significant bit*

First number	Second number	Carry-in	Sum	Carry-out
0	0	0	0	0
0	0	1	1	0
0	1	0	1	0
0	1	1	0	1
1	0	0	1	0
1	0	1	0	1
1	1	0	0	1
1	1	1	1	1

The above truth tables indicate what is required of circuits that are to be used for addition. Consider the logic circuit needed to realise Table 15.3. We require the sum to be 1 when the first number or the second number, but not both numbers, is 1. This is what is obtained with an XOR gate (see Section 14.4.6). The carry-out bit is to be 1 when the first number and the second number are both 1. This is obtained with an AND gate (see Section 14.4.1). The required circuit to realise Table 15.3 is termed a *half-adder* and is shown in Figure 15.8.

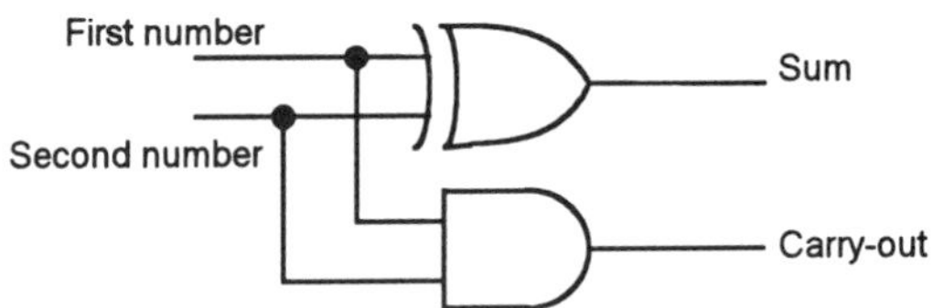

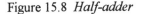

Figure 15.8 *Half-adder*

Now consider the logic circuit needed to realise Table 15.4, such a circuit being termed a *full-adder*. The sum output arises from three inputs, namely the first number, the second number and the carry-in number. This gives a 1 output when there is an odd number of 1 inputs. This can be realised with two XOR gates. The carry-out output is 1 when any two of the inputs are 1. This can be accomplished with three AND gates and an OR gate. Figure 15.9 shows the resulting logic circuit. There are other logic circuits which can be used to realise the same outcome.

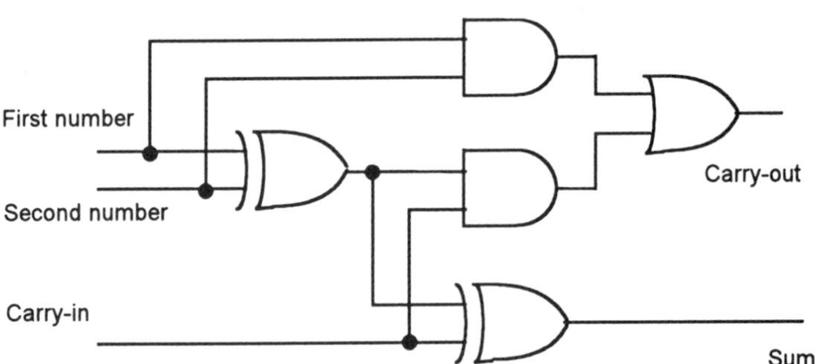

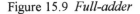

Figure 15.9 *Full-adder*

Now consider the realisation of a *multi-bit adder*. We can build up such an adder from a half-adder for the least significant bit and full adders for the other bits. Thus for a 4-bit adder we would have the arrangement shown in Figure 15.10. Rather than draw the details of the full-adders and the half-adder, they have been represented as blocks with the appropriate inputs and outputs.

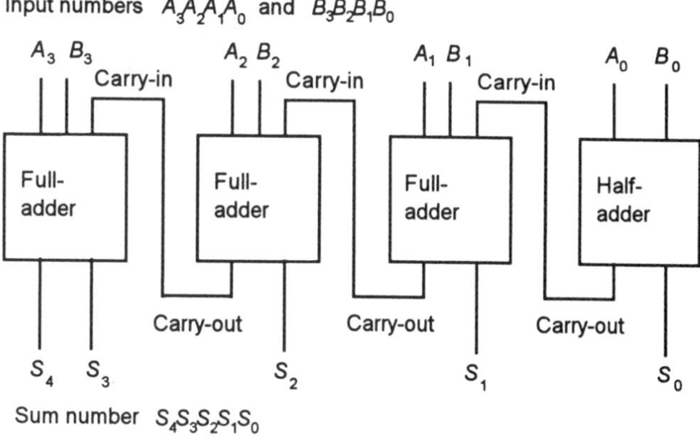

Figure 15.10 *A 4-bit adder*

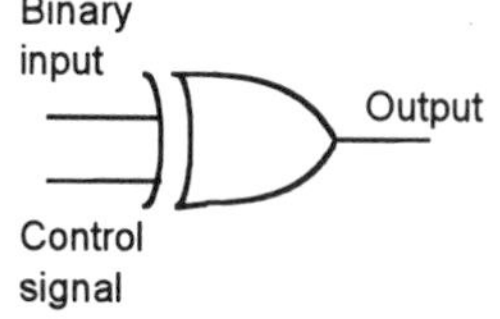

Input	Control	Output
0	0	0
0	1	1
1	0	1
1	1	0

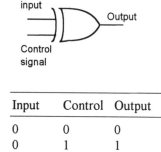

Figure 15.11 *XOR gate*

4-bit adders are available as integrated circuits. Two such circuits can be used to give an 8-bit adder if the carry-out from the first 4-bit adder is connected to the carry-in for the second 4-bit adder.

The above represents the system when positive numbers are added to positive numbers. When we have negative numbers or want to subtract numbers then the positive number in ordinary binary is added to the negative number when in signed two's complement. We thus need to add to the above a circuit which will, for negative numbers, convert ordinary binary numbers into signed two's complement numbers. The first operation in transforming a negative binary number to two's complement is to change all the 0s to 1s and all the 1s to 0s. Thus we need a circuit which will invert the input when given an input indicating a negative number, but if given an input indicating a positive number will not invert the input. An XOR gate (see Section 14.4.6) will do this, inverting if a control signal is 1 and not inverting if it is 0 (Figure 15.11). The second operation in transforming to two's complement is to add 1. This can be done if the control signal 1 is also connected so that it adds to the output. Figure 15.12 shows how a 4-bit adder can be connected so that it can handle number A having a positive or negative number B added to it or subtraction of a positive number B from a positive number A. If A is also to have the option of being positive or negative then a two's complement circuit for it can also be used.

Revision

16 Write the Boolean expression for (a) the half-adder circuit shown in Figure 15.8, (b) the full-adder circuit shown in Figure 15.9.

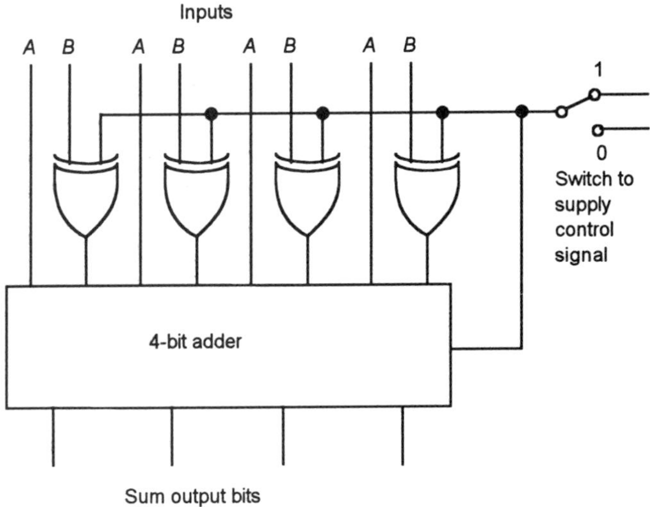

Figure 15.12 *4-bit two's complement adder/subtractor*

Problems

1 Sketch the waveform resulting from the application to a two-input OR gate of the signals shown in Figure 15.13.

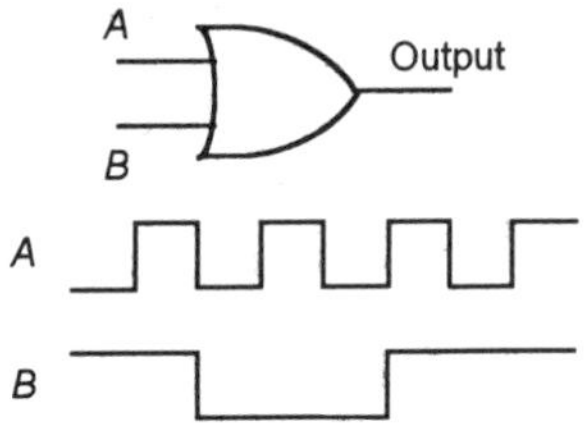

Figure 15.13 *Problem 1*

2 Sketch the output waveform for the two-input AND gate when the inputs are the waveforms shown in Figure 15.14.

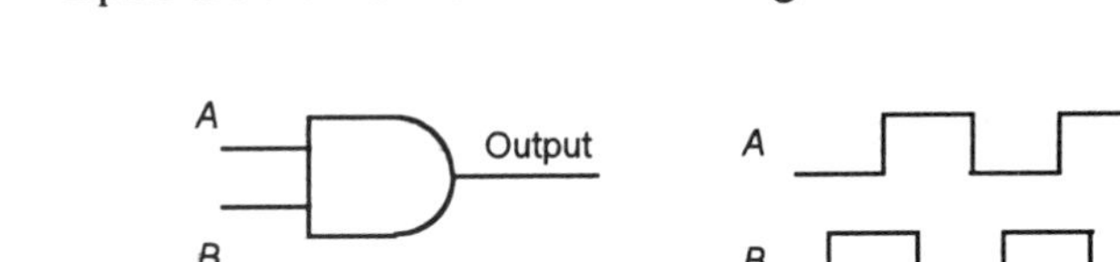

Figure 15.14 *Problem 2*

3 Sketch the output waveform for the two-input NOR gate when the inputs are the waveforms shown in Figure 15.15.

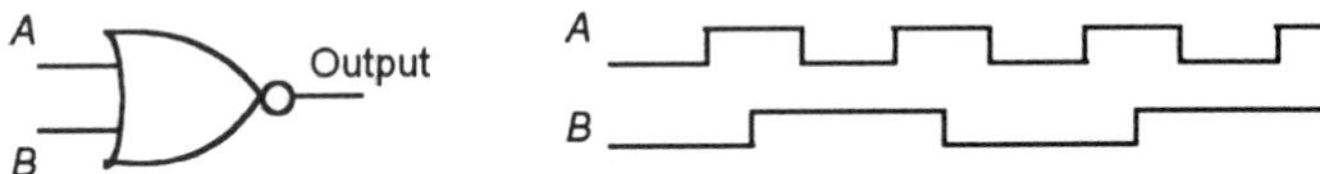

Figure 15.15 *Problem 3*

4 Using just NAND gates, devise a system to give the Boolean function $A + B \cdot C$.

5 Using just NAND gates, devise a system to give the Boolean function $A \cdot \bar{B} + B \cdot \bar{C}$.

6 Using just NOR gates, devise a system to give the Boolean function $(\bar{A} + B) \cdot (\bar{C} + D)$.

7 Convert the following binary numbers to denary numbers: (a) 000 110, (b) 010 101, (c) 111 110 001 111.

8 Convert the following denary numbers to binary numbers: (a) 29, (b) 57, (c) 63.

9 Convert the following hex numbers to denary numbers: (a) 1C (b) 9F, (c) ABCD.

10 Convert the following hex numbers to binary numbers: (a) 1C, (b) B, (c) 1F.

11 Carry out the following arithmetic operations:

(a) 011 011 + 010 110, (b) 110 110 + 101 010,

(c) 000 010 101 – 011 110 011, (d) 000 111 011 – 010 100 000

12 Determine the two's complement of (a) 01 100 010, (b) 00 100 011.

13 State which of the following signed two's complement numbers are positive:

(a) 10 110 000, (b) 10 111 110, (c) 00 100 011, (d) 011 111 100

14 Represent, using a total of 8 bits which includes the sign, the following denary numbers in the signed two's complement form: (a) –1, (b) –3, (c) –8.

15 Carry out the following arithmetic operations using signed two's complements: (a) 28 – 13, (b) 43 – 38, (b) 62 – 21, (d) 10 – 28.

16 Show that the logic circuit shown in Figure 15.16 operates as a half-adder.

17 Figure 15.17 shows a circuit involving two half-adders. Construct a truth table for it and hence show that it operates as a full-adder.

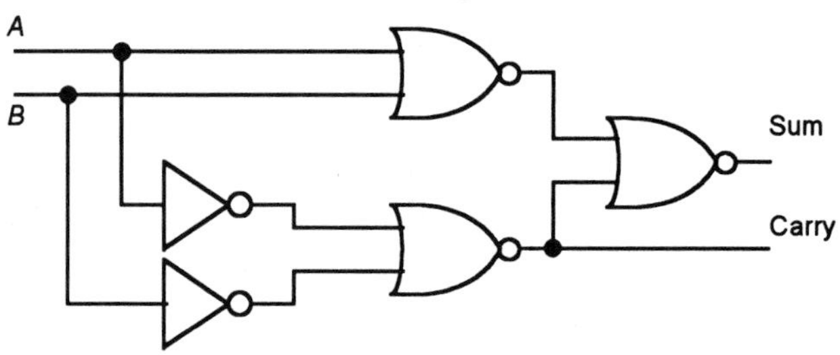

Figure 15.16 *Problem 16*

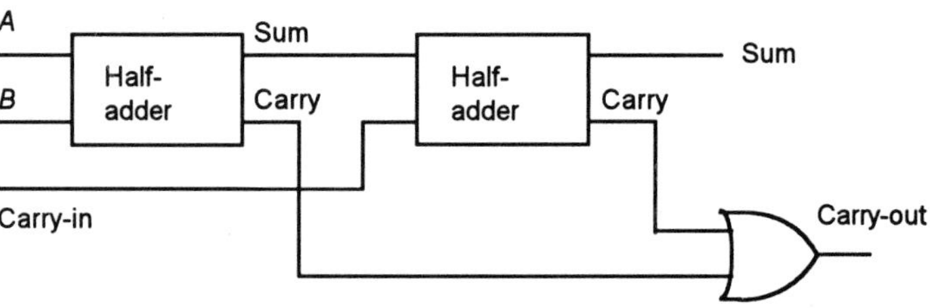

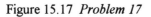

Figure 15.17 *Problem 17*

Part 5
Linear algebra

The aims of this part are to enable the reader:

- Solve systems of linear equations by Gaussian elimination, using augmented matrices.
- Carry out the arithmetic operations of addition, subtraction and multiplication involving matrices.
- Use the inverse matrix in the solution of sets of linear equations.
- Evaluate determinants by means of the diagonal rule and cofactors.
- Use row manipulations to simplify determinants.
- Solve systems of linear equations using Cramer's rule.
- Determine the inverse of a matrix by the use of cofactor and adjoint matrices.
- Explain the terms eigenvalues and eigenvectors and evaluate them in simple situations.
- Use Jacobi and Gauss-Seidel iteration methods for the evaluation of systems of linear equations.
- Develop and solve systems of linear equations by the use of node analysis or mesh analysis for electrical circuits.

The section requires algebraic dexterity with linear equations. Chapter 17 has been written assuming Chapter 16, though it could equally well have been written to precede it. Chapters 18 and 19 have been written to follow on from Chapter 17. In the main, Chapter 20 is independent of the rest of the part. Chapter 21 involves the developing and solving of systems of linear equations for electrical circuits. The methods used to develop the equations are node analysis and mesh analysis, these assuming a knowledge of, and ability to use, Kirchhoff's laws.

16 Systems of linear equations

16.1 Introduction

The equation $y = mx + c$ is the equation of a straight line graph with gradient m and intercept on the y-axis of c (see Section 2.2). Such an equation can be said to be a linear equation. In general we use the term *linear equation* for one in which all the unknowns all occur to the first power so that there are no squares or cubes of higher powers and no products of two or more unknowns. Such equations frequently arise in engineering and science. Indeed situations which are non-linear are often linearised to aid in their manipulation (see Section 5.4.1).

Often we have a *system of linear equations*, i.e. a set of simultaneous equations. For example, when Kirchhoff's laws are applied to a circuit containing more than one mesh we obtain a linear equation for each mesh. We then are required to solve the equations and determine the unknowns, e.g. the currents in each branch of the circuit (see Chapter 21).

This chapter introduces the Gaussian elimination method for solving systems of linear equations and uses matrices as a shorthand way of writing equations. An ability to manipulate algebraic expressions is assumed.

16.1.1 Definition of a linear equation

We can represent a straight line on a graph of the variables y against x by an equation of the form:

$$a_1x + a_2y = b$$

where a_1, a_2 and b are real constants. This is just another way of writing $y = mx + c$. Such an equation is termed a linear equation in the variables x and y.

In general we define a linear equation in the n variables x_1, x_2, x_3, ... x_n, where a_1, a_2, a_3, ... a_n are constants, to be of the form:

$$a_1x_1 + a_2x_2 + a_3x_3 + \ldots + a_nx_n = b \qquad [1]$$

A linear equation does not involve any products or roots of variables, all variables only occurring to the first power. It does not involve any products of variables or the variables occurring in trigonometric, logarithmic or exponential functions. Thus the equations $y - 2x = 5$ and $y + 2x + 3z = 5$ are linear but the equations $y - 2 \sin x = 0$, $x^2 + 3xy + 4 = 0$, $y = 2\,e^{3x}$ and $y = \ln x$ are not linear.

16.2 Solutions

The term *solution* of the linear equation $x - y = 1$ is used to describe the values of x and y that satisfy the equation when we substitute them for x and y. Thus we might have a solution of $x = 2$ and $y = 1$, or $x = 3$ and $y = 2$. These solutions specify points that lie on the straight line graph given by the equation (Figure 16.1). We thus have as solutions all the pairs of values that specify points lying on the straight line. Thus there is an infinite set of solutions. The set of all the solutions is referred to as the *general solution* of the equation.

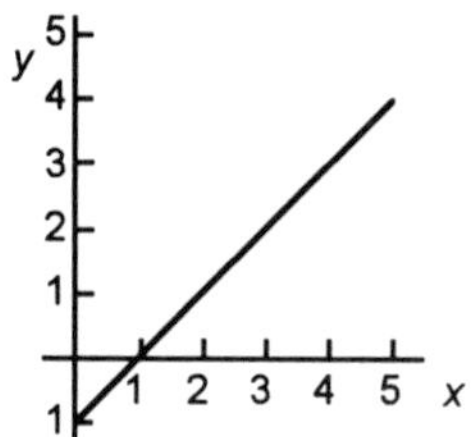

Figure 16.1 *General solution is the set of points lying on the line*

Now suppose we have two equations with the same variables: $x - y = 1$ and $2x - y = 4$. If the two equations *simultaneously* apply then the values of x and y that satisfy one equation must also be ones that satisfy the other equation. If we plot the graphs of the equations (Figure 16.2), then the values that satisfy both equations are where the graphs intersect. The only solution in this case is $x = 3$ and $y = 2$.

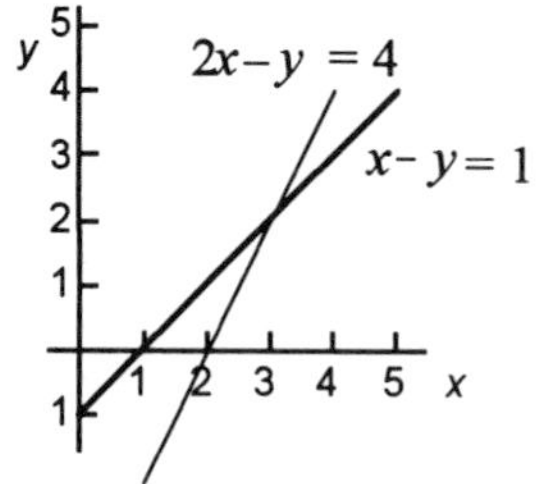

Figure 16.2 *One solution*

Suppose, however, we had the two equations $x - y = 1$ and $x - y = 2$. There are no values of x and y that simultaneously satisfy these equations. This is because the lines representing these equations on a graph do not intersect but are parallel (Figure 16.3).

Now suppose we have the two equations $x - y = 1$ and $2x - 2y = 2$. There are an infinite number of points of intersection and so an infinite number of values of x and y that simultaneously satisfy these equations. These describe lines on a graph that coincide (Figure 16.4).

In general:

> *Every system of linear equations has either no solution, one solution or an infinite number of solutions.*

A system of equations that has no solutions is said to be *inconsistent*, a *consistent* set being one that has at least one solution.

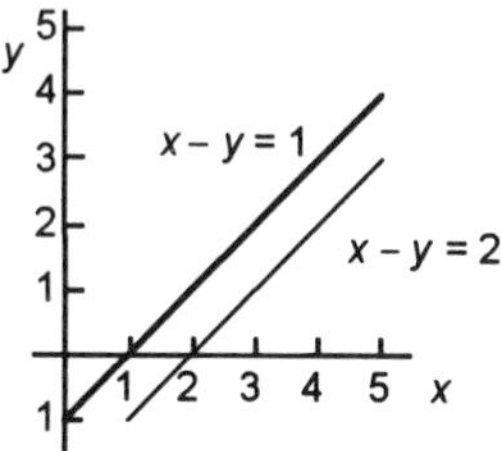

Figure 16.3 *No solution*

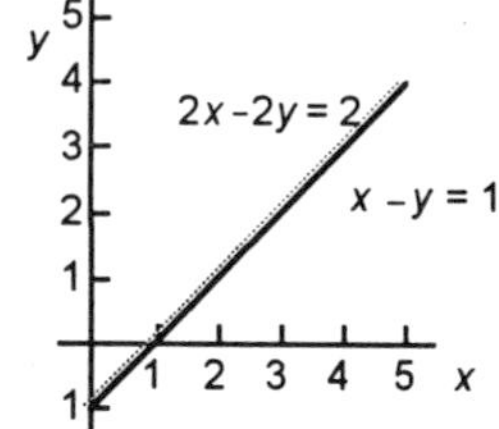

Figure 16.4 *Infinite number of solutions*

16.2.1 Two equations with two unknowns

When we have two linear equations with two unknowns, such a system of equations can be solved by *substitution*. With this method we rearrange one of the equations to isolate one of the variables. This variable is then substituted into the other equation. For example, if we have the equations

$x - y = 1$ and $2x - y = 4$, then we can rearrange the first equation to give $x = 1 + y$ and then substitute this value of x into the second equation, giving $2(1 + y) - y = 4$. Hence $y = 2$. Putting this value back into one of the equations gives $x = 3$. Thus the solution is $x = 3$, $y = 2$.

Alternatively we can solve the two linear equations by *elimination*. This procedure involves taking a pair of equations, then eliminating one of the variables by combining the equations in a suitable manner. Thus with $x - y = 1$ and $2x - y = 4$, if we want to eliminate x we can double the first equation and subtract it from the second equation:

$$\begin{array}{lr} & 2x - y = 4 \\ \text{minus} & \underline{2x - 2y = 2} \\ & y = 2 \end{array}$$

We can obtain the value of x by substituting the value of y into one of the equations. Alternatively we can obtain it by eliminating y between the equations. We can do this by subtracting the second equation from the first:

$$\begin{array}{lr} & x - y = 1 \\ \text{minus} & \underline{2x - y = 4} \\ & -x = -3 \end{array}$$

Thus $x = 3$.

Revision

1 Solve by the following pairs of equations by (a) substitution and (b) elimination:

(a) $x + y = 3$, $2x - y = 3$, (b) $4x + y = 7$, $5x - y = 2$,

(c) $x + 2y = 1$, $-x + 2y = 3$, (d) $2x - 3y = 7$, $x + y = 1$

16.2.2 Solving *n* equations with *n* variables

With two equations and two variables we can eliminate one of the variables by combining the two equations in a suitable manner. If we have three equations with three variables then we can first eliminate one of the variables by combining two of the equations, say equations one and two. Then we can similarly eliminate the same variable from a different pair of equations by combining the third equation with either equation one or two. We now have two equations with just two variables and so can solve them to obtain values for two of the variables. The third variable can be obtained by substituting these values into one of the original equations. The following example illustrates this method with the equations:

$$3x + 2y + 3z = 3 \qquad [A]$$

$$4x - 5y + 7z = 1 \qquad \text{[B]}$$

$$2x + 3y - 2z = 6 \qquad \text{[C]}$$

First, we eliminate x from equations [A] and [B]. We can do this by dividing equation [A] by the coefficient of x, giving:

$$x + \tfrac{2}{3}y + z = 1 \qquad \text{[D]}$$

then eliminating x from equation [B] by multiplying equation [D] by the coefficient of x in equation [B] and subtracting it from [B]:

$$\begin{array}{rl} & 4x - 5y + 7z = 1 \\ \text{minus} & 4x + \tfrac{8}{3}y + 4z = 4 \\ \hline & -\tfrac{23}{3}y + 3z = -3 \end{array} \qquad \text{[E]}$$

Then we eliminate x from equation [C] by multiplying equation [D] by the coefficient of x in equation [C] and subtracting it from [C]:

$$\begin{array}{rl} & 2x + 3y - 2z = 6 \\ \text{minus} & 2x + \tfrac{4}{3}y + 2z = 2 \\ \hline & \tfrac{5}{3}y - 4z = 4 \end{array} \qquad \text{[F]}$$

We can repeat the procedure in order to eliminate y from equations [E] and [F]. Thus, dividing equation [E] by the coefficient of y gives:

$$y - \tfrac{9}{23}z = \tfrac{9}{23} \qquad \text{[G]}$$

Then we eliminate y from [F] and [G] by multiplying equation [G] by the coefficient of y in equation [F] and subtracting it from [F]:

$$\begin{array}{rl} & \tfrac{5}{3}y - \;\; 4z = 4 \\ \text{minus} & \tfrac{5}{3}y - \tfrac{45}{69}z = \tfrac{45}{69} \\ \hline & -\tfrac{231}{69}z = \tfrac{231}{69} \end{array} \qquad \text{[H]}$$

Equation [H] thus gives $z = -1$.

If we substitute this value of z in equation [G] we obtain $y = 0$. If we substitute these values of y and z in equation [A] we obtain $x = 2$.

The above may seem a rather cumbersome way of arriving at the values of the variables. It is, however, the basis of a general procedure which can be used with n equations in n unknowns. As the number of equations and variables increases, the number of steps involved increases. To cope with this a routine process is required that can be systematically applied. The procedure is termed *Gaussian elimination*. The procedure can be formally stated for the three simultaneous equations:

$$a_{11}x_1 + a_{12}x_2 + a_{13}x_3 = b_1 \qquad \text{[A]}$$

$$a_{21}x_1 + a_{22}x_2 + a_{23}x_3 = b_2 \qquad \text{[B]}$$

$$a_{31}x_1 + a_{32}x_2 + a_{33}x_3 = b_3 \qquad \text{[C]}$$

as:

1 Divide equation [A] by the coefficient of x_1 in equation [A], i.e. a_{11}, to give:

$$x_1 + \frac{a_{12}}{a_{11}}x_2 + \frac{a_{13}}{a_{11}}x_3 = \frac{b_1}{a_{11}} \qquad \text{[D]}$$

If the coefficient is zero then choose another of the equations and make it equation [A].

2 Eliminate x_1 from equation [B] by subtracting from it equation [D] when multiplied by a_{21}:

$$0 + \left(a_{22} - a_{21}\frac{a_{12}}{a_{11}}\right)x_2 + \left(a_{23} - a_{21}\frac{a_{13}}{a_{11}}\right)x_3 = b_2 - a_{21}\frac{b_1}{a_{11}} \qquad \text{[E]}$$

3 Eliminate x_1 from equation [C] by subtracting from it equation [D] when multiplied by a_{31}.

$$0 + \left(a_{32} - a_{31}\frac{a_{12}}{a_{11}}\right)x_2 + \left(a_{33} - a_{31}\frac{a_{13}}{a_{11}}\right)x_3 = b_3 - a_{31}\frac{b_1}{a_{11}} \qquad \text{[F]}$$

4 x_3 can then be obtained from the pair of simultaneous equations [E] and [F] by dividing equation [E] by the coefficient of x_2 in that equation to give:

$$x_2 + \frac{a_{23} - a_{21}\frac{a_{13}}{a_{11}}}{a_{22} - a_{21}\frac{a_{12}}{a_{11}}}x_3 = \frac{b_2 - a_{21}\frac{b_1}{a_{11}}}{a_{22} - a_{21}\frac{a_{12}}{a_{11}}} \qquad \text{[G]}$$

This equation is then multiplied by the coefficient of x_2 in equation [F] and subtracted from equation [F]. The result is the value for x_3.

$$0 + \left[a_{33} - a_{31}\frac{a_{13}}{a_{11}} - \left(a_{32} - a_{31}\frac{a_{12}}{a_{11}}\right)\left(\frac{a_{23} - a_{21}\frac{a_{13}}{a_{11}}}{a_{22} - a_{21}\frac{a_{12}}{a_{11}}}\right)\right]x_3$$

$$= b_3 - a_{31}\frac{b_1}{a_{11}} - \left(a_{32} - a_{31}\frac{a_{12}}{a_{11}}\right)\left(\frac{b_2 - a_{21}\frac{b_1}{a_{11}}}{a_{22} - a_{21}\frac{a_{12}}{a_{11}}}\right) \qquad \text{[H]}$$

5 Substitution of the value of x_3 in either equation [E] or [G] or [F] will give x_2. Substitution of the values of x_2 and x_3 in equation [A] or [D] or [B] or [C] will give x_1.

Revision

2 Obtain the values of the variables in the following sets of equations by means of Gaussian elimination:

(a) $2x - 4y + 2z = 0$, $3x + y - z = 2$, $4x + 2y - 2z = 2$,

(b) $2x + 3y + z = 5$, $5x - y + 2z = 12$, $3x + 3y - z = 5$,

(c) $3x + 2y + 3z = 7$, $2x + 3y + 2z = 3$, $4x - y + z = 7$

16.3 Matrices

Since in Gaussian elimination all that we are interested in are the coefficients and constants, there is no need to write anything other than them and so we can tabulate the coefficients and constants in a rectangular array of numbers. Thus the equations:

$$a_{11}x_1 + a_{12}x_2 + a_{13}x_3 = b_1 \qquad [A]$$

$$a_{21}x_1 + a_{22}x_2 + a_{23}x_3 = b_2 \qquad [B]$$

$$a_{31}x_1 + a_{32}x_2 + a_{33}x_3 = b_3 \qquad [C]$$

can be represented by:

$$\begin{bmatrix} a_{11} & a_{12} & a_{13} & b_1 \\ a_{21} & a_{22} & a_{23} & b_2 \\ a_{31} & a_{32} & a_{33} & b_3 \end{bmatrix} \qquad [2]$$

The term *matrix*, plural matrices, is used for a rectangular array of numbers and the above is termed the *augmented matrix* for the system of equations. The term augmented is used because the matrix includes both the coefficients and the constants. The *coefficient matrix* is one with just the coefficients.

When we now apply Gaussian elimination to the augmented matrix the sequence of operations becomes:

1 Divide the first row of the matrix by the term in the first column of that row, i.e. a_{11}.

2 Subtract a multiple of the new first row from the second row to give a 0 in the first column and so a new second row.

3 Subtract a multiple of the new first row from the third row to give a 0 in the first column and so a new third row.

4 Divide the new second row by the term in its second column and give another new second row.

5 Subtract a multiple of this new second row from the new third row to give a 0 in the second column.

6 Divide this new third row by the number in its third column to give a 1 in that column and hence represent an equation giving the x_3 variable in terms of a constant. The result of all these operations should be a matrix with a diagonal of 1s down from the top left corner with a triangle of 0s in the lower left corner.

To illustrate the above, consider the three equations used to illustrate Gaussian elimination in the previous section, namely:

$$3x + 2y + 3z = 3 \qquad [A]$$

$$4x - 5y + 7z = 1 \qquad [B]$$

$$2x + 3y - 2z = 6 \qquad [C]$$

The augmented matrix is:

$$\begin{bmatrix} 3 & 2 & 3 & 3 \\ 4 & -5 & 7 & 1 \\ 2 & 3 & -2 & 6 \end{bmatrix}$$

Dividing the first row by the number in the first column of that row gives:

$$\begin{bmatrix} 1 & \frac{2}{3} & 1 & 1 \\ 4 & -5 & 7 & 1 \\ 2 & 3 & -2 & 6 \end{bmatrix}$$

Subtracting a multiple of the first row from the second row to give a zero in the first column of the second row:

$$\begin{bmatrix} 1 & \frac{2}{3} & 1 & 1 \\ 0 & -\frac{23}{3} & 3 & -3 \\ 2 & 3 & -2 & 6 \end{bmatrix}$$

Subtracting a multiple of the first row from the third row to give a zero in the first column of the third row:

$$\begin{bmatrix} 1 & \frac{2}{3} & 1 & 1 \\ 0 & -\frac{23}{3} & 3 & -3 \\ 0 & \frac{5}{3} & -4 & 4 \end{bmatrix}$$

Dividing the second row by the number in the second column of that row:

$$\begin{bmatrix} 1 & \frac{2}{3} & 1 & 1 \\ 0 & 1 & -\frac{9}{23} & \frac{9}{23} \\ 0 & \frac{5}{3} & -4 & 4 \end{bmatrix}$$

Subtracting a multiple of the second row from the third row to give a 0 in the second column:

$$\begin{bmatrix} 1 & \frac{2}{3} & 1 & 1 \\ 0 & 1 & -\frac{9}{23} & \frac{9}{23} \\ 0 & 0 & \frac{-231}{69} & \frac{231}{69} \end{bmatrix}$$

Dividing this third row by $-\frac{69}{231}$ gives:

$$\begin{bmatrix} 1 & \frac{2}{3} & 1 & 1 \\ 0 & 1 & -\frac{9}{23} & \frac{9}{23} \\ 0 & 0 & 1 & -1 \end{bmatrix}$$

Thus we have obtained a row of the form 0 0 1 ? and so $z = -1$. We can obtain the values of the other variables by substituting this value for z in the original equations. However, we can obtain it more easily by using the matrix. To obtain y we need to obtain a row of the form 0 1 0 ?. We can do this by subtracting $-\frac{9}{23}$ times the third row from the second row:

$$\begin{bmatrix} 1 & \frac{2}{3} & 1 & 1 \\ 0 & 1 & 0 & 0 \\ 0 & 0 & 1 & -1 \end{bmatrix}$$

Thus $y = 0$. To obtain x we need a first row of the form 1 0 0 ? and obtain this by subtracting multiples of the second and third rows from the first row. Thus subtracting $\frac{2}{3}$ times the second row and 1 times the third row gives:

$$\begin{bmatrix} 1 & 0 & 0 & 2 \\ 0 & 1 & 0 & 0 \\ 0 & 0 & 1 & -1 \end{bmatrix}$$

Thus $x = 2$.

The aim of the manipulations of the rows of the augmented matrix is to get it into the above type of form. In this form the equations represented by each row are:

$$1x = 2, \quad 1y = 0, \quad 1z = -1$$

and so are the solutions.

In carrying out Gaussian elimination, problems can arise if the coefficient of a variable that is selected to be divided at a particular stage is zero. When carrying out this operation by hand we can rearrange the sequence of the equations, i.e. the rows, so that this does not occur. When the process is handled by a computer it needs a rule to determine what should be done. The rule used is that the interchange occurs with the row having the largest numerical coefficient (minus signs are ignored). This is known as *partial pivoting*. The following example illustrates this.

Example

Solve the following equations by Gaussian elimination:

$$x_1 + 2x_2 + x_3 = 5,\ \ 2x_1 + 4x_2 + x_3 = 11,\ \ x_1 - x_2 + 2x_3 = 1$$

The augmented matrix is:

$$\begin{bmatrix} 1 & 2 & 1 & 5 \\ 2 & 4 & 1 & 11 \\ 1 & -1 & 2 & 1 \end{bmatrix}$$

Subtracting 2 times the first row from the second row and 1 times the first row from the third row gives:

$$\begin{bmatrix} 1 & 2 & 1 & 5 \\ 0 & 0 & -1 & 1 \\ 0 & -3 & 1 & -4 \end{bmatrix}$$

The second row has a 0 in the second column. Thus, interchanging the second and third rows:

$$\begin{bmatrix} 1 & 2 & 1 & 5 \\ 0 & -3 & 1 & -4 \\ 0 & 0 & -1 & 1 \end{bmatrix}$$

Multiplying the third row by -1 gives:

$$\begin{bmatrix} 1 & 2 & 1 & 5 \\ 0 & -3 & 1 & -4 \\ 0 & 0 & 1 & -1 \end{bmatrix}$$

Thus $x_3 = -1$. If we subtract the third row from the second row:

$$\begin{bmatrix} 1 & 2 & 1 & 5 \\ 0 & -3 & 0 & -3 \\ 0 & 0 & 1 & -1 \end{bmatrix}$$

Dividing the second row by –3 gives:

$$\begin{bmatrix} 1 & 2 & 1 & 5 \\ 0 & 1 & 0 & 1 \\ 0 & 0 & 1 & -1 \end{bmatrix}$$

Thus $x_2 = 1$. If we subtract 2 times the second row and 1 times the third row from the first row:

$$\begin{bmatrix} 1 & 0 & 0 & 4 \\ 0 & 1 & 0 & 1 \\ 0 & 0 & 1 & -1 \end{bmatrix}$$

Thus $x_1 = 4$.

Example

Solve the following set of four equations:

$$x_1 + 3x_2 + x_3 = 6,\ 2x_1 + 3x_2 + x_3 - x_4 = 4,$$

$$3x_1 - 2x_2 + 4x_4 = 11,\ -x_1 + x_2 - 3x_3 - x_4 = 1$$

The augmented matrix is:

$$\begin{bmatrix} 1 & 3 & 1 & 0 & 6 \\ 2 & 3 & 1 & -1 & 4 \\ 3 & -2 & 0 & 4 & 11 \\ -1 & 1 & -3 & -1 & 1 \end{bmatrix}$$

Subtracting 2 times the first row from the second row, 3 times the first row from the third row and –1 times the first row from the fourth row gives:

$$\begin{bmatrix} 1 & 3 & 1 & 0 & 6 \\ 0 & -3 & -1 & -1 & -8 \\ 0 & -11 & -3 & 4 & -7 \\ 0 & 4 & -2 & -1 & 7 \end{bmatrix}$$

Dividing the second row by –3 gives:

$$\begin{bmatrix} 1 & 3 & 1 & 0 & 6 \\ 0 & 1 & \frac{1}{3} & \frac{1}{3} & \frac{8}{3} \\ 0 & -11 & -3 & 4 & -7 \\ 0 & 4 & -2 & -1 & 7 \end{bmatrix}$$

Subtracting −11 times the second row from the third row and 4 times the second row from the fourth row:

$$\begin{bmatrix} 1 & 3 & 1 & 0 & 6 \\ 0 & 1 & \frac{1}{3} & \frac{1}{3} & \frac{8}{3} \\ 0 & 0 & \frac{2}{3} & \frac{23}{3} & \frac{67}{3} \\ 0 & 0 & -\frac{10}{3} & -\frac{7}{3} & -\frac{11}{3} \end{bmatrix}$$

Dividing the third row by $\frac{2}{3}$ gives:

$$\begin{bmatrix} 1 & 3 & 1 & 0 & 6 \\ 0 & 1 & \frac{1}{3} & \frac{1}{3} & \frac{8}{3} \\ 0 & 0 & 1 & \frac{23}{2} & \frac{67}{2} \\ 0 & 0 & -\frac{10}{3} & -\frac{7}{3} & -\frac{11}{3} \end{bmatrix}$$

Subtracting $-\frac{10}{3}$ times the third row from the fourth row gives:

$$\begin{bmatrix} 1 & 3 & 1 & 0 & 6 \\ 0 & 1 & \frac{1}{3} & \frac{1}{3} & \frac{8}{3} \\ 0 & 0 & 1 & \frac{23}{2} & \frac{67}{2} \\ 0 & 0 & 0 & \frac{216}{6} & \frac{648}{6} \end{bmatrix}$$

Dividing the fourth row by $\frac{216}{6}$ gives:

$$\begin{bmatrix} 1 & 3 & 1 & 0 & 6 \\ 0 & 1 & \frac{1}{3} & \frac{1}{3} & \frac{8}{3} \\ 0 & 0 & 1 & \frac{23}{2} & \frac{67}{2} \\ 0 & 0 & 0 & 1 & 3 \end{bmatrix}$$

Subtracting $\frac{23}{2}$ times the fourth row from the third row gives:

$$\begin{bmatrix} 1 & 3 & 1 & 0 & 6 \\ 0 & 1 & \frac{1}{3} & \frac{1}{3} & \frac{8}{3} \\ 0 & 0 & 1 & 0 & -1 \\ 0 & 0 & 0 & 1 & 3 \end{bmatrix}$$

Subtracting $\frac{1}{3}$ times the third row and $\frac{1}{3}$ times the fourth row from the second row gives:

$$\begin{bmatrix} 1 & 3 & 1 & 0 & 6 \\ 0 & 1 & 0 & 0 & 2 \\ 0 & 0 & 1 & 0 & -1 \\ 0 & 0 & 0 & 1 & 3 \end{bmatrix}$$

Subtracting 3 times the second row and 1 times the third row from the first row gives:

$$\begin{bmatrix} 1 & 0 & 0 & 0 & 1 \\ 0 & 1 & 0 & 0 & 2 \\ 0 & 0 & 1 & 0 & -1 \\ 0 & 0 & 0 & 1 & 3 \end{bmatrix}$$

Thus the solutions are $x_1 = 1$, $x_2 = 2$, $x_3 = -1$ and $x_4 = 3$.

Revision

3 Write the set of equations corresponding to the following augmented matrix if the variables are x_1, x_2 and x_3:

$$\begin{bmatrix} 2 & 1 & -1 & 2 \\ 1 & 2 & 1 & 4 \\ 3 & -1 & 3 & 1 \end{bmatrix}$$

4 Write the augmented matrix corresponding to the following set of equations:

$$x_1 + 4x_2 - x_3 = 4,\ \ 2x_1 + 3x_2 - x_3 = 1,\ \ 3x_1 + x_2 + 2x_3 = 5$$

5 What are the values of the variables x_1, x_2 and x_3 described by the following matrix:

$$\begin{bmatrix} 1 & 0 & 0 & 3 \\ 0 & 1 & 0 & -2 \\ 0 & 0 & 1 & 4 \end{bmatrix}$$

6 Write the augmented matrices for the following sets of equations and then solve them by Gaussian elimination:

(a) $2x + y + z = 5$, $3x - 2y + 2z = 5$, $x + 2y + 2z = 7$,

(b) $x + y + z = 1$, (b) $2x - 2y - z = 6$, (c) $2x + 2y + 3z = 2$,

(c) $x + 4y + 2z = 7$, (b) $2x + 2y - z = 7$, $3x - y - 2z = 3$,

(d) $2x + 3y + 2z = 11$, $4x - 2y + 3z = -4$, $3x + 3y + 5z = 16$,

(e) $x_1 + x_2 - x_3 + x_4 = 2$, $x_1 + 2x_2 - x_3 + 2x_4 = 4$, $2x_1 + 2x_2 + x_4 = 5$,

$x_1 + 2x_2 - 2x_3 = 7$

16.3.1 Existence of solutions

There are three possible outcomes when solving a set of simultaneous equations:

1 There is a unique solution.
2 There is no solution.
3 There are an infinite number of solutions.

If there is no solution the set of simultaneous equations is said to be *inconsistent* and if there is at least one solution, *consistent*. For consistent equations, if the values for the variables are substituted into one of the equations, the equation will balance and this will be true for all the equations in the set. With Gaussian elimination and augmented matrices, a consistent equation will end up with a row of the form:

$$0 \quad 0 \quad 1 \quad c \text{ or } 0 \quad 1 \quad 0 \quad c \text{ or } 1 \quad 0 \quad 0 \quad 0 \quad c$$

where c is some constant. Then the variable equals c. However, if the row is of the form:

$$0 \quad 0 \quad 0 \quad c$$

then there is no solution since $c \neq 0$. If there is a row of the form:

$$0 \quad 0 \quad 0 \quad 0$$

then this row is of no help in determining a solution. This occurs when a row is just a multiple of the row subtracted from it and so one equation is just a multiple of the other, e.g. $x + y + z = 1$ and $2x + 2y + 2z = 2$. There is then an infinite number of solutions.

Example

Determine whether the set of equations giving the following augmented matrix has a unique solution:

$$\begin{bmatrix} 1 & -2 & 3 & 4 \\ 2 & -3 & -1 & 2 \\ 4 & -7 & 5 & 10 \end{bmatrix}$$

If we take twice the first row from the second row and four times the first row from the third row we obtain:

$$\begin{bmatrix} 1 & -2 & 3 & 4 \\ 0 & 1 & -7 & -6 \\ 0 & 1 & -7 & -6 \end{bmatrix}$$

Subtracting the second row from the third row gives:

$$\begin{bmatrix} 1 & -2 & 3 & 4 \\ 0 & 1 & -7 & -6 \\ 0 & 0 & 0 & 0 \end{bmatrix}$$

Because of the all zero row, we have an infinite number of solutions.

Revision

7 Determine whether the sets of equations giving the following augmented matrices have unique solutions:

(a) $\begin{bmatrix} 2 & -4 & 8 & -4 \\ 0 & 3 & 1 & 4 \\ 0 & 0 & 1 & -3 \end{bmatrix}$, (b) $\begin{bmatrix} 1 & 3 & 2 & 10 \\ 0 & 4 & 1 & 6 \\ 0 & 4 & 1 & 6 \end{bmatrix}$, (c) $\begin{bmatrix} 3 & 6 & 9 & 12 \\ 2 & 5 & 7 & 9 \\ 1 & 2 & 3 & 4 \end{bmatrix}$,

(d) $\begin{bmatrix} 1 & 3 & 5 & 7 \\ 2 & -3 & 6 & 4 \\ 2 & 6 & 10 & 20 \end{bmatrix}$

16.4 Translating applications into equations

The following are a few examples of applications which can be expressed as systems of two or three simultaneous equations. Chapter 21 includes examples relating particularly to circuit analysis.

Example

The sum of two angles x and y is 90°. If one of the angles is to be 10° more than seven times the other angle, determine the angles.

The above gives the equations:

$$x + y = 90 \text{ and } x - 7y = 10$$

Subtracting the second equation from the first one gives:

$$8y = 80$$

Hence $y = 10°$. Substituting this value into either of the initial equations gives $x = 80°$.

We could have written this pair of equations as an augmented matrix and solved it by Gaussian elimination:

$$\begin{bmatrix} 1 & 1 & 90 \\ 1 & -7 & 10 \end{bmatrix} \begin{bmatrix} 1 & 1 & 90 \\ 0 & -8 & -80 \end{bmatrix} \begin{bmatrix} 1 & 0 & 80 \\ 0 & 1 & 10 \end{bmatrix}$$

Example

A box contains a total of 95 castings, there being two types of castings with one casting weighing 1 kg and the other casting 3 kg. If the total weight of the castings is 135 kg, how many of each casting are in the box?

If x is the number of one casting and y the number of the other in the box, for the quantities of castings we have the equation:

$$x + y = 95$$

and for the mass:

$$x + 3y = 135$$

Subtracting the first equation from the second equation gives:

$$2y = 40$$

Hence $y = 20$. Substituting this value into one of the initial equations gives $x = 75$.

We could have solved the above equations by writing them as an augmented matrix and using Gaussian elimination:

$$\begin{bmatrix} 1 & 1 & 95 \\ 1 & 3 & 135 \end{bmatrix} \begin{bmatrix} 1 & 1 & 95 \\ 0 & 2 & 40 \end{bmatrix} \begin{bmatrix} 1 & 0 & 75 \\ 0 & 1 & 20 \end{bmatrix}$$

Example

A solution containing 15% of a particular salt is to be mixed with another solution containing 40% of that salt to give a 25% concentration solution. How much of each of the solutions should be used if 5 litres of the 25% solution are to be produced?

If x is the number of litres of the 15% solution and y the number of litres of the 40% solution then obtain the required volume we have:

$$x + y = 5$$

and to obtain the required concentration:

$$0.15x + 0.40y = 0.25 \times 5 = 1.25$$

Subtracting 0.15 times the first equation from the second equation gives:

$$0.25y = 0.50$$

Hence $y = 2$. Substituting this value in one of the initial equations gives $x = 3$.

We could have solved the equations by writing them as an augmented matrix and using Gaussian elimination:

$$\begin{bmatrix} 1 & 1 & 5 \\ 0.15 & 0.40 & 1.25 \end{bmatrix} \begin{bmatrix} 1 & 1 & 5 \\ 0 & 0.25 & 0.50 \end{bmatrix} \begin{bmatrix} 1 & 0 & 3 \\ 0 & 1 & 2 \end{bmatrix}$$

Revision

Solve the following by writing sets of linear equations.

8 If the sum of two numbers is 50 and their difference is 16, determine the numbers.

9 20 litres of a solution with a concentration of 12% is to be produced by mixing two other solutions, one with a concentration of 5% and the other with a concentration of 15%. Determine how much of each solution should be used.

10 The sum of the angles in a triangle is 180°. Angle A is to be one-third angle B and the other angle C is to be twice the sum of A and B. Determine the angles.

11 Solution *A* has a concentration of 25%, solution B a concentration of 40% and solution C a concentration of 50%. 50 litres of a 32% concentration are to be produced using twice as much of the 25% solution as the 40% solution. How much of A, B and C should be used?

Problems

1 Solve the following matrices for the variables x_1, x_2 and x_3:

(a) $\begin{bmatrix} 1 & 0 & 0 & 2 \\ 0 & 1 & 0 & -5 \\ 0 & 0 & 1 & 4 \end{bmatrix}$, (b) $\begin{bmatrix} 1 & 0 & 0 & -2 \\ 0 & 1 & 0 & 6 \\ 0 & 0 & 1 & 2 \end{bmatrix}$, (c) $\begin{bmatrix} 1 & 0 & 0 & 7 \\ 0 & 1 & 0 & -6 \\ 0 & 0 & 1 & -2 \end{bmatrix}$

2 Write the augmented matrices for the following sets of equations and then solve them by Gaussian elimination:

(a) $2x + y - 2z = 2$, $x + 3y + z = 15$, $3x - y + 4z = 7$,

(b) $2x + 3y + z = -2$, $4x + 2y + 3z = 1$, $5x - 2y + 3z = 6$,

(c) $3x - 5y + z = 5$, $2x + 4y - 3z = -4$, $5x - 6y - z = 0$,

(d) $x + 2y - 3z = 5$, $3x + y - z = 13$, $-2x + 3y + z = -1$,

(e) $x_1 + 2x_2 - x_3 + x_4 = 4$, $x_1 + 3x_2 + x_3 - 2x_4 = 0$, $2x_1 + x_2 - 3x_3 + 2x_4 = 3$,

$2x_1 - 2x_2 - x_3 + 4x_4 = 10$,

(f) $x_1 + 3x_2 - 2x_3 + x_4 = -1$, $2x_1 + 2x_2 + x_3 + 4x_4 = 11$, $x_1 + 4x_3 - x_4 = 4$,

$2x_2 - x_3 + 3x_4 = 10$

3 The following are augmented matrices for sets of linear equations. Determine whether the systems have a unique solution, infinitely many solutions, or no solution:

$$\text{(a)}\begin{bmatrix} 1 & 5 & -2 & 4 \\ 0 & 2 & 5 & -3 \\ 0 & 2 & 5 & 6 \end{bmatrix}, \text{(b)}\begin{bmatrix} 1 & -1 & 0 & 4 \\ 0 & 1 & 2 & 4 \\ 0 & 2 & 4 & 8 \end{bmatrix}, \text{(c)}\begin{bmatrix} 1 & 2 & 1 & 5 \\ 1 & -1 & 2 & 1 \\ 2 & 4 & 1 & 11 \end{bmatrix}$$

4 The sum of two angles is 180°. If one of the angles is 30° less than four times the other angle, determine the angles.

5 In a triangle, angle A is to be 70° more than angle B and angle C 10° more than three times angle B. Determine the angles.

6 A manufacturer produces three products A, B and C. They require assembly, testing and packaging. Product A requires 45 minutes for assembly, 30 minutes for testing and 10 minutes for packaging. Product B requires 60 minutes for assembly, 45 minutes for testing and 15 minutes for packaging. Product C requires 90 minutes for assembly, 45 minutes for testing and 15 minutes for packaging. The assembly line operates for 1065 minutes per day, the test facility for 750 minutes per day and the packaging team for these products for 225 minutes per day. How many of each product can be produced per day?

7 A company produces three grades A, B and C of a material. Customer 1 purchases 2 kg of A, 5 kg of B and 7 kg of C and pays £65. Customer 2 purchases 3 kg of A, 10 kg of B and 1 kg of C and pays £72. Customer 3 purchases 1 kg of A, 8 kg of B and 6 kg of C and pays £70. Determine the costs of each grade.

8 Determine the value of the constant a for which the following equations will have a unique solution:

$$x - 2y + 3z = 2,\ 2x - y + 2z = 3,\ x + y + az = 4$$

17 Matrices

17.1 Introduction

A *matrix* is a rectangular array of numbers. In Chapter 16 such arrays were used to represent sets of simultaneous equations. There are, however, many other instances where data is arranged as a regular array of numbers. For example, a spreadsheet indicating the output in number of items of a product produced by three machines per shift is such an array:

$$\begin{array}{lc} & \text{Machines} \\ & \begin{array}{ccc} 1 & 2 & 3 \end{array} \\ \text{Shift 1} & \left[\begin{array}{ccc} 300 & 245 & 302 \\ 180 & 210 & 270 \end{array}\right] \\ \text{Shift 2} & \end{array}$$

Matrices can be considered a type of mathematical object, just like real numbers and complex numbers are forms of mathematical objects. We thus need to consider how they are to be added, subtracted, multiplied, etc. This chapter is concerned with developing the properties of such arrays, whether they represent sets of simultaneous equations or not.

This chapter is written assuming Chapter 16, though it could equally have been written to precede it. As with Chapter 16, algebraic dexterity is assumed.

17.2 Notation and terminology

A matrix is defined as being a rectangular array of numbers, the numbers in the array being termed entries.

There is no arithmetical connection between the elements and the matrix as a whole has no numerical value. The arrays are written between [ ] brackets. The term *row* is used for a horizontal line of numbers and *column* for a vertical line of numbers. The *size* of a matrix is specified in terms of the number or rows and columns it contains, an array with p rows and q columns being termed a $p \times q$ matrix. Figure 17.1 shows some examples of matrices. A matrix having the same number of rows as columns, as in (a), is termed a *square matrix*. A matrix having just a single column but more than one row, as in (b), is termed a *column matrix* or *column vector*. A matrix having just a single row but more than one column, as in (c), is termed a *row matrix* or *row vector*.

$$\underset{(a)}{\begin{bmatrix} 1 & 2 \\ -3 & 4 \end{bmatrix}} \quad \underset{(b)}{\begin{bmatrix} 1 \\ 4 \end{bmatrix}} \quad \underset{(c)}{\begin{bmatrix} 2 & 5 \end{bmatrix}}$$

Figure 17.1 *(a) 2 × 2, (b) 2 × 1, (c) 1 × 2 matrices*

Generally bold capital letters are used in print to denote matrices though lower case bold letters are often used for column and row vectors, e.g.

$$\mathbf{A} = \begin{bmatrix} 1 & 2 & 3 \\ -1 & 2 & 4 \\ 3 & -2 & 5 \end{bmatrix} \quad \mathbf{a} = \begin{bmatrix} 1 \\ 2 \\ 3 \end{bmatrix}$$

Lower case letters, not in bold print, tend to be used for entries in matrices with suffix notation being used to refer to particular elements in a matrix, the suffixes indicating the row followed by the column, e.g.

$$\begin{bmatrix} a_{11} & a_{12} & a_{13} & a_{14} \\ a_{21} & a_{22} & a_{23} & a_{24} \\ a_{31} & a_{32} & a_{33} & a_{34} \\ a_{41} & a_{42} & a_{43} & a_{44} \end{bmatrix}$$

The term *unit matrix* or *identity matrix*, symbol **I**, is used for a matrix with all the entries along the main diagonal from left to right 1 and all the other entries 0, e.g.

$$\begin{bmatrix} 1 & 0 & 0 \\ 0 & 1 & 0 \\ 0 & 0 & 1 \end{bmatrix}$$

A *null matrix*, symbol **0**, is a matrix in which all the entries are 0, e.g.

$$\begin{bmatrix} 0 & 0 & 0 \\ 0 & 0 & 0 \\ 0 & 0 & 0 \end{bmatrix}$$

17.3 Matrix arithmetic

Two matrices are said to be *equal* if they are the same size, i.e. have the same number of rows and same number of columns, and the corresponding entries in the two matrices are the same. Thus for the matrices:

$$\mathbf{A} = \begin{bmatrix} 1 & 2 \\ 3 & 4 \end{bmatrix} \quad \mathbf{B} = \begin{bmatrix} 1 & 2 \\ 3 & 5 \end{bmatrix} \quad \mathbf{C} = \begin{bmatrix} 1 & 2 & 0 \\ 3 & 4 & 0 \end{bmatrix}$$

A does not equal **B** because, though the sizes are the same, the entries are different. **A** does not equal **C** because the sizes are different. In order for the matrices **D** and **E** to be equal, where:

$$\mathbf{D} = \begin{bmatrix} 2 & 3 \\ -4 & 5 \end{bmatrix} \text{ and } \mathbf{E} = \begin{bmatrix} 2 & a \\ b & 5 \end{bmatrix}$$

then we must have $a = 3$ and $b = -4$.

17.3.1 Addition of matrices

Addition of matrices is defined in the following way:

> *If* **A** *and* **B** *are matrices of the same size, then the sum* **A** + **B** *is the matrix obtained by adding together the corresponding entries in the two matrices.*

Matrices of different sizes *cannot* be added. It is easy to check that:

$$\mathbf{A} + \mathbf{B} = \mathbf{B} + \mathbf{A} \qquad [1]$$

and:

$$\mathbf{A} + (\mathbf{B} + \mathbf{C}) = (\mathbf{A} + \mathbf{B}) + \mathbf{C} \qquad [2]$$

It is also easy to check that if **A** is a matrix and **0** a zero matrix of the same size, then:

$$\mathbf{A} + \mathbf{0} = \mathbf{0} + \mathbf{A} = \mathbf{A} \qquad [3]$$

Example

Add the matrices **A** and **B**, where:

$$\mathbf{A} = \begin{bmatrix} 2 & 5 & -1 \\ 0 & 4 & 2 \\ 3 & 1 & -6 \end{bmatrix} \text{ and } \mathbf{B} = \begin{bmatrix} -1 & 4 & 0 \\ 2 & 4 & 3 \\ -3 & 3 & 1 \end{bmatrix}$$

The matrices are the same size so they can be added. Thus adding corresponding entries gives:

$$\mathbf{A} + \mathbf{B} = \begin{bmatrix} 2-1 & 5+4 & -1+0 \\ 0+2 & 4+4 & 2+3 \\ 3-3 & 1+3 & -6+1 \end{bmatrix} = \begin{bmatrix} 1 & 9 & -1 \\ 2 & 8 & 5 \\ 0 & 4 & -5 \end{bmatrix}$$

Revision

1 Determine, where possible, (a) **A** + **B**, (b) **B** + **C**, (c) **C** + **D**, (d) **A** + **D**.

$$\mathbf{A} = \begin{bmatrix} 1 & 3 \\ 2 & 5 \\ 1 & 2 \end{bmatrix} \qquad \mathbf{B} = \begin{bmatrix} -1 & 0 & 4 \\ 3 & 1 & 2 \\ 2 & 4 & -1 \end{bmatrix}$$

$$\mathbf{C} = \begin{bmatrix} 2 & 3 & 4 \\ -2 & -1 & 0 \\ 0 & 3 & 2 \end{bmatrix} \qquad \mathbf{D} = \begin{bmatrix} 2 & 5 & 1 \\ 1 & 2 & -3 \end{bmatrix}$$

17.3.2 Multiplication by a constant

If we add two identical matrices **A** then the result is a matrix with all the entries having twice the value. We can consider this to be represented by 2**A**. Thus:

$$\mathbf{A} + \mathbf{A} = 2\mathbf{A}$$

If we had **A** + **A** + **A** then we would have 3**A**.

> *If* **A** *is any matrix and c is a constant, then the product c***A** *is the matrix obtained by multiplying each entry of* **A** *by c.*

The term *scalar* is often used for the constant c. It is easy to show that, for constants c and d:

$$(c + d)\mathbf{A} = c\mathbf{A} + d\mathbf{A} \qquad [4]$$

and:

$$c(\mathbf{A} + \mathbf{B}) = c\mathbf{A} + c\mathbf{B} \qquad [5]$$

If we multiply some matrix **A**, e.g.

$$\mathbf{A} = \begin{bmatrix} 2 & -1 & -3 \\ 1 & 2 & 5 \\ -2 & 1 & 3 \end{bmatrix}$$

by −1, then:

$$(-1)\mathbf{A} = \begin{bmatrix} -2 & 1 & 3 \\ -1 & -2 & -5 \\ 2 & -1 & -3 \end{bmatrix}$$

The matrix −**A** is used to denote (−1)**A**.

> *If* **A** *is any matrix then* −**A** *is the matrix whose entries are the negatives of the corresponding entries in* **A**.

If **A** is any matrix and **0** the zero matrix of the same size, then it is evident that:

$$\mathbf{A} + (-\mathbf{A}) = 0 \qquad [6]$$

If **A** and **B** are matrices of the same size, then the *difference* between **A** and **B**, i.e. **A** − **B**, is defined by:

$$\mathbf{A} - \mathbf{B} = \mathbf{A} + (-\mathbf{B}) \qquad [7]$$

Example

Determine for the following matrices (a) 2**B**, (b) –**A**, (c) **A** – **B**.

$$\mathbf{A} = \begin{bmatrix} -2 & 1 & 3 \\ 3 & -1 & -2 \end{bmatrix} \quad \mathbf{B} = \begin{bmatrix} 1 & 4 & -2 \\ 3 & -1 & 1 \end{bmatrix}$$

(a) Each of the entries is doubled to give:

$$2\mathbf{B} = \begin{bmatrix} 2 & 4 & -4 \\ 6 & -2 & 2 \end{bmatrix}$$

(b) Each of the entries is multiplied by –1 to give:

$$\mathbf{A} = \begin{bmatrix} 2 & -1 & -3 \\ -3 & 1 & 2 \end{bmatrix}$$

(c) Subtracting matrices involves subtracting corresponding entries. Thus:

$$\mathbf{A} - \mathbf{B} = \begin{bmatrix} -3 & -3 & 5 \\ 0 & 0 & -3 \end{bmatrix}$$

Example

Simplify the following matrix by removing a factor from it.

$$\mathbf{A} = \begin{bmatrix} 2 & 6 & -8 \\ 4 & -2 & 6 \end{bmatrix}$$

Each of the entries in the matrix is divisible by 2, thus we can write:

$$\mathbf{A} = 2\begin{bmatrix} 1 & 3 & -4 \\ 2 & -1 & 3 \end{bmatrix}$$

Revision

2 Determine for the matrices given (a) 2**A**, (b) **A** – **B**, (c) 2(**A** + **B**).

$$\mathbf{A} = \begin{bmatrix} 1 & -2 & 4 \\ 2 & -4 & 2 \end{bmatrix} \quad \mathbf{B} = \begin{bmatrix} 2 & -2 & 3 \\ -1 & -2 & 5 \end{bmatrix}$$

3 Simplify the following matrices by removing factors from them:

$$\text{(a)}\begin{bmatrix} 3 & 6 \\ -9 & 3 \end{bmatrix}, \text{(b)}\begin{bmatrix} -4 & 10 & -6 \\ -8 & 2 & 6 \end{bmatrix}, \text{(c)}\begin{bmatrix} a & ab & a^2 \\ 4a & 2a^2 & -ab \end{bmatrix}$$

17.3.3 Multiplication of matrices

It might seem that since the addition of two matrices is accomplished by adding corresponding entries, multiplication of two matrices should be just the multiplication of corresponding entries. This is *not* the case.

Consider two simultaneous equations:

$$a_{11}x_1 + a_{12}x_2 = c_1 \text{ and } a_{21}x_1 + a_{22}x_2 = c_2$$

where x_1 and x_2 are the variables and the a and c terms constants. We write matrices; one each for the coefficients, the variables and the constants:

$$\mathbf{A} = \begin{bmatrix} a_{11} & a_{12} \\ a_{21} & a_{22} \end{bmatrix}, \quad \mathbf{x} = \begin{bmatrix} x_1 \\ x_2 \end{bmatrix}, \quad \mathbf{c} = \begin{bmatrix} c_1 \\ c_2 \end{bmatrix}$$

Multiplication of matrices is defined so that:

$$\mathbf{Ax} = \mathbf{c} \qquad [8]$$

This means:

$$\begin{bmatrix} a_{11} & a_{12} \\ a_{21} & a_{22} \end{bmatrix}\begin{bmatrix} x_1 \\ x_2 \end{bmatrix} = \begin{bmatrix} c_1 \\ c_2 \end{bmatrix} \qquad [9]$$

We must have a multiplication procedure which generates the two simultaneous equations. We do this by using the concept of a *row–column product*. If we have the row $[\ a \quad b\ ]$ in one matrix and the column:

$$\begin{bmatrix} A \\ C \end{bmatrix}$$

in the other matrix, then the row–column product is $aA + bC$. For two 2×2 matrices we have a product matrix which has the first entry in the first row given by the row–column product for the first row in the first matrix and the first column in the second matrix. The second entry in the first row of the product is the row–column product for the first row in the first matrix and the second column in the second matrix. The second row of the product matrix is generated in a similar way but using the second row of the first matrix. Thus:

$$\begin{bmatrix} a & b \\ c & d \end{bmatrix}\begin{bmatrix} A & B \\ C & D \end{bmatrix} = \begin{bmatrix} aA+bC & aB+bD \\ cA+dC & cB+dD \end{bmatrix} \qquad [10]$$

Applying this to the matrices [9] for the simultaneous equations gives:

$$\begin{bmatrix} a_{11}x_1 + a_{12}x_2 \\ a_{21}x_1 + a_{22}x_2 \end{bmatrix} = \begin{bmatrix} c_1 \\ c_2 \end{bmatrix}$$

For these two matrices to be equal, corresponding elements must be equal. Thus we end up with the two simultaneous equations $a_{11}x_1 + a_{12}x_2 = c_1$ and $a_{21}x_1 + a_{22}x_2 = c_2$.

In general, the multiplication procedure is stated as:

If **A** *is an* $m \times r$ *matrix and* **B** *is an* $r \times n$ *matrix, the product* **AB** *is the* $m \times n$ *matrix obtained by multiplying the entries in the i row of* **A** *with the corresponding entries in the j column of* **B** *and adding them to give the entry in row i and column j of* **AB**.

Thus, for example, for the following matrices:

$$\begin{bmatrix} \cdot & \cdot & \cdot \\ a_{21} & a_{22} & a_{23} \\ \cdot & \cdot & \cdot \end{bmatrix} \begin{bmatrix} \cdot & \cdot & b_{13} \\ \cdot & \cdot & b_{23} \\ \cdot & \cdot & b_{33} \end{bmatrix} = \begin{bmatrix} \cdot & \cdot & \cdot \\ \cdot & \cdot & c_{23} \\ \cdot & \cdot & \cdot \end{bmatrix}$$

we have $c_{23} = a_{21}b_{13} + a_{22}b_{23} + a_{23}b_{33}$.

For multiplication of two matrices to be possible, we must have the number of columns in the first matrix equal to the number of rows in the second matrix. If this is not the case, the product of the two is undefined.

To illustrate matrix multiplication, consider the product **AB** where:

$$\mathbf{A} = \begin{bmatrix} 1 & 2 \\ 3 & 4 \end{bmatrix} \qquad \mathbf{B} = \begin{bmatrix} 5 & 6 \\ 7 & 8 \end{bmatrix}$$

Then:

$$\mathbf{AB} = \begin{bmatrix} 1\times5+2\times7 & 1\times6+2\times8 \\ 3\times5+4\times7 & 3\times6+4\times8 \end{bmatrix} = \begin{bmatrix} 19 & 22 \\ 43 & 50 \end{bmatrix}$$

With matrix multiplication, unlike in real number algebra where $ab = ba$, we have:

$$\mathbf{AB} \neq \mathbf{BA} \qquad [11]$$

Thus, with the above example:

$$\mathbf{BA} = \begin{bmatrix} 5\times1+6\times3 & 5\times2+6\times4 \\ 7\times1+8\times3 & 7\times2+8\times4 \end{bmatrix} = \begin{bmatrix} 23 & 34 \\ 31 & 46 \end{bmatrix}$$

and so confirming equation [11]. Also with multiplication:

$$\mathbf{A(BC)} = \mathbf{(AB)C} \qquad [12]$$

$$\mathbf{A(B + C)} = \mathbf{AB} + \mathbf{BC} \qquad [13]$$

$$\mathbf{(B + C)A} = \mathbf{BA} + \mathbf{CA} \qquad [14]$$

With real number algebra, if $ab = 0$ then either a or b must be 0. This is not true with matrices. For example, if we have:

$$\mathbf{A} = \begin{bmatrix} 2 & -4 \\ -1 & 2 \end{bmatrix} \text{ and } \mathbf{B} = \begin{bmatrix} 2 & 2 \\ 1 & 1 \end{bmatrix}$$

then:

$$\mathbf{AB} = \begin{bmatrix} 2\times2+(-4)\times1 & 2\times2+(-4)\times1 \\ (-1)\times2+2\times1 & (-1)\times2+2\times1 \end{bmatrix} = \begin{bmatrix} 0 & 0 \\ 0 & 0 \end{bmatrix}$$

The product is **0** without **A** or **B** being zero.

Example

Determine the product matrix **AB** where:

$$\mathbf{A} = \begin{bmatrix} 3 & 4 \end{bmatrix} \quad \mathbf{B} = \begin{bmatrix} 2 \\ -5 \end{bmatrix}$$

We have one row in the first matrix and one column in the second matrix, the product is thus a 1 × 1 matrix. Using the rule for multiplication given above:

$$\mathbf{AB} = \begin{bmatrix} 3\times2+4\times(-5) \end{bmatrix} = [-14]$$

Example

Determine the products **AB** and **(AB)C** where:

$$\mathbf{A} = \begin{bmatrix} 1 & 2 \\ 3 & 4 \\ 0 & 1 \end{bmatrix} \quad \mathbf{B} = \begin{bmatrix} 4 & 3 \\ 2 & 1 \end{bmatrix} \quad \mathbf{C} = \begin{bmatrix} 1 & 0 \\ 2 & 3 \end{bmatrix}$$

For the product matrix **AB** we have two columns in **A** and two rows in **B**. Thus we can obtain a product. The product will be a 3 × 2 matrix since **A** has three rows and **B** two columns. Using the rule for multiplication given above:

$$\mathbf{AB} = \begin{bmatrix} 1\times4+2\times2 & 1\times3+2\times1 \\ 3\times4+4\times2 & 3\times3+4\times1 \\ 0\times4+1\times2 & 0\times3+1\times1 \end{bmatrix} = \begin{bmatrix} 8 & 5 \\ 20 & 13 \\ 2 & 1 \end{bmatrix}$$

and:

$$(\mathbf{AB})\mathbf{C} = \begin{bmatrix} 8 & 5 \\ 20 & 13 \\ 2 & 1 \end{bmatrix} \begin{bmatrix} 1 & 0 \\ 2 & 3 \end{bmatrix}$$

$$= \begin{bmatrix} 8\times1+5\times2 & 8\times0+5\times3 \\ 20\times1+13\times2 & 20\times0+13\times3 \\ 2\times1+1\times2 & 2\times0+1\times3 \end{bmatrix} = \begin{bmatrix} 18 & 15 \\ 46 & 39 \\ 4 & 3 \end{bmatrix}$$

Revision

4 Determine, if possible, the product **AB** for the following matrices:

(a) $\mathbf{A} = \begin{bmatrix} 3 & 4 \\ 5 & -1 \end{bmatrix}$, $\mathbf{B} = \begin{bmatrix} 1 & 2 \\ 1 & -1 \end{bmatrix}$,

(b) $\mathbf{A} = \begin{bmatrix} 4 & -1 & 0 \\ 2 & 5 & -3 \end{bmatrix}$, $\mathbf{B} = \begin{bmatrix} -3 \\ 3 \\ 1 \end{bmatrix}$,

(c) $\mathbf{A} = \begin{bmatrix} 1 & -1 & 2 \\ 3 & 4 & -4 \\ 2 & 1 & 3 \end{bmatrix}$, $\mathbf{B} = \begin{bmatrix} 0 & 2 & -3 \\ 2 & 5 & 1 \\ 1 & 2 & 3 \end{bmatrix}$,

(d) $\mathbf{A} = \begin{bmatrix} 1 & 2 & -3 & 4 \\ 3 & -1 & 5 & 2 \end{bmatrix}$, $\mathbf{B} = \begin{bmatrix} 1 & 2 \\ 3 & 4 \end{bmatrix}$,

(e) $\mathbf{A} = \begin{bmatrix} 1 & 1 & 1 \\ 3 & 3 & 3 \end{bmatrix}$, $\mathbf{B} = \begin{bmatrix} 4 & 1 \\ -4 & -3 \\ 3 & 1 \end{bmatrix}$,

(f) $\mathbf{A} = \begin{bmatrix} 1 & 2 & 4 \\ 2 & 4 & 8 \end{bmatrix}$, $\mathbf{B} = \begin{bmatrix} 2 \\ 3 \\ -2 \end{bmatrix}$

5 For the following matrices, determine $(\mathbf{A} + \mathbf{B})(\mathbf{A} - \mathbf{B})$:

$$\mathbf{A} = \begin{bmatrix} 1 & 2 \\ -2 & 3 \end{bmatrix} \quad \mathbf{B} = \begin{bmatrix} 3 & -4 \\ 2 & -1 \end{bmatrix}$$

17.3.4 Powers of matrices

The matrix $\mathbf{A}^2$ indicates the product **AA**. Similarly $\mathbf{A}^3$ indicates **AAA**. We can only obtain powers of square matrices, i.e. where the number of rows equals the number of columns, since for other matrices multiplication is not possible.

Example

Determine $\mathbf{A}^2$ if:

$$\mathbf{A} = \begin{bmatrix} 2 & 1 \\ 4 & -1 \end{bmatrix}$$

Using the rules of multiplication given above:

$$\mathbf{A}^2 = \mathbf{AA} = \begin{bmatrix} 2 & 1 \\ 4 & -1 \end{bmatrix}\begin{bmatrix} 2 & 1 \\ 4 & -1 \end{bmatrix}$$

$$= \begin{bmatrix} 2\times2+1\times4 & 2\times1+1\times(-1) \\ 4\times2+(-1)\times4 & 4\times1+(-1)\times(-1) \end{bmatrix} = \begin{bmatrix} 8 & 1 \\ 4 & 5 \end{bmatrix}$$

Revision

6 For the following matrix determine $\mathbf{A}^2$:

$$\mathbf{A} = \begin{bmatrix} 2 & -1 \\ 0 & 3 \end{bmatrix}$$

17.3.5 Multiplication with identity matrices

Multiplication of a square matrix **A** by the same size identity matrix **I** plays the same role in matrix algebra as multiplying by 1 in conventional arithmetic.

$$\mathbf{AI} = \mathbf{IA} = \mathbf{A} \qquad [15]$$

For example, if we have:

$$\mathbf{A} = \begin{bmatrix} 1 & 2 & 3 \\ 4 & 5 & 6 \\ 7 & 8 & 9 \end{bmatrix}$$

then:

$$\mathbf{AI} = \begin{bmatrix} 1 & 2 & 3 \\ 4 & 5 & 6 \\ 7 & 8 & 9 \end{bmatrix}\begin{bmatrix} 1 & 0 & 0 \\ 0 & 1 & 0 \\ 0 & 0 & 1 \end{bmatrix} = \begin{bmatrix} 1 & 2 & 3 \\ 4 & 5 & 6 \\ 7 & 8 & 9 \end{bmatrix} = \mathbf{A}$$

and:

$$\mathbf{IA} = \begin{bmatrix} 1 & 0 & 0 \\ 0 & 1 & 0 \\ 0 & 0 & 1 \end{bmatrix}\begin{bmatrix} 1 & 2 & 3 \\ 4 & 5 & 6 \\ 7 & 8 & 9 \end{bmatrix} = \begin{bmatrix} 1 & 2 & 3 \\ 4 & 5 & 6 \\ 7 & 8 & 9 \end{bmatrix} = \mathbf{A}$$

Revision

7 Determine $\mathbf{A}^2 - 7\mathbf{I}$ if:

$$\mathbf{A} = \begin{bmatrix} -1 & 3 \\ 2 & 1 \end{bmatrix}$$

8 Determine $\mathbf{A}^3 + 7\mathbf{I}$ if:

$$\mathbf{A} = \begin{bmatrix} 2 & 1 \\ 4 & -1 \end{bmatrix}$$

17.4 Inverse matrix

For the two simultaneous equations:

$$a_{11}x_1 + a_{12}x_2 = c_1 \text{ and } a_{21}x_1 + a_{22}x_2 = c_2$$

where x_1 and x_2 are the variables and the a and c terms constants, we can write three matrices; one for the coefficients, one for the variables and one for the constants:

$$\mathbf{A} = \begin{bmatrix} a_{11} & a_{12} \\ a_{21} & a_{22} \end{bmatrix}, \quad \mathbf{x} = \begin{bmatrix} x_1 \\ x_2 \end{bmatrix}, \quad \mathbf{c} = \begin{bmatrix} c_1 \\ c_2 \end{bmatrix}$$

Multiplication of matrices has been defined so that:

$$\mathbf{Ax} = \mathbf{c} \qquad [16]$$

and so:

$$\begin{bmatrix} a_{11} & a_{12} \\ a_{21} & a_{22} \end{bmatrix}\begin{bmatrix} x_1 \\ x_2 \end{bmatrix} = \begin{bmatrix} c_1 \\ c_2 \end{bmatrix}$$

How can we obtain **x** from equation [16]?

With ordinary number algebra, if $ax = c$ then we can solve for x by obtaining $x = c/a = (1/a)c = a^{-1}c$, this is provided a is not zero since division by zero is not permitted. What we effectively have done is multiply both sides of the equation by a^{-1}, i.e. $a^{-1} \times ax = a^{-1} \times c$, and since $a^{-1} \times a = 1$ we have $x = a^{-1}c$. To obtain a^{-1} we inverted a and so it would be feasible to talk of a^{-1} as the inverse of a and defined by $a^{-1} \times a = 1$.

With matrices we can adopt a similar procedure. Multiplying both sides of equation [16] by the inverse of matrix **A**, denoted by $\mathbf{A}^{-1}$, gives:

$$\mathbf{A}^{-1}\mathbf{A}\mathbf{x} = \mathbf{A}^{-1}\mathbf{c}$$

We define the inverse of a matrix in a similar way to ordinary number algebra, i.e.

$$\mathbf{A}\mathbf{A}^{-1} = \mathbf{I} \qquad [17]$$

where **I** is the identity matrix. Thus, since $\mathbf{xI} = \mathbf{x}$ (equation [15]), then:

$$\mathbf{x} = \mathbf{A}^{-1}\mathbf{c} \qquad [18]$$

To obtain the **x** matrix, we have to multiply the **c** matrix by the inverse of the **A** matrix.

The following are definitions of what constitutes an *invertible matrix* and the *inverse*:

> *If* **A** *is a square matrix and a matrix* **B** *can be found such that* **AB** = **I**, *then* **B** *is called the inverse of* **A** *and is written as* $\mathbf{A}^{-1}$.

> *If* **A** *is a square matrix, then* **A** *is said to be invertible or non-singular if there is a matrix* $\mathbf{A}^{-1}$ *such that* $\mathbf{A}\mathbf{A}^{-1} = \mathbf{I}$.

To illustrate the above, consider the matrices:

$$\mathbf{A} = \begin{bmatrix} 1 & 2 \\ 2 & 3 \end{bmatrix} \text{ and } \mathbf{B} = \begin{bmatrix} -3 & 2 \\ 2 & 1 \end{bmatrix}$$

B is the inverse of **A** since **AB** = **I**:

$$\mathbf{AB} = \begin{bmatrix} 1 & 2 \\ 2 & 3 \end{bmatrix}\begin{bmatrix} -3 & 2 \\ 2 & 1 \end{bmatrix} = \begin{bmatrix} 1 & 0 \\ 0 & 1 \end{bmatrix}$$

Thus matrix **A** is invertible.

Now consider the matrix:

$$\mathbf{A} = \begin{bmatrix} 1 & 3 \\ -2 & -6 \end{bmatrix}$$

If this is to be invertible then we must have a matrix **B** such that **AB** = **I**:

$$\begin{bmatrix} 1 & 3 \\ -2 & -6 \end{bmatrix}\begin{bmatrix} b_{11} & b_{12} \\ b_{21} & b_{21} \end{bmatrix} = \begin{bmatrix} 1 & 0 \\ 0 & 1 \end{bmatrix}$$

Multiplying the matrices gives:

$$\begin{bmatrix} 1b_{11}+3b_{21} & 1b_{12}+3b_{22} \\ (-2)\times b_{11}+(-6)\times b_{21} & (-2)\times b_{12}+(-6)\times b_{22} \end{bmatrix} = \begin{bmatrix} 1 & 0 \\ 0 & 1 \end{bmatrix}$$

The two matrices will only be equal if:

$$b_{11}+3b_{21}=1,\ \ b_{12}+3b_{22}=0,\ \ -2b_{11}-6b_{21}=0,\ \ -2b_{12}-6b_{22}=1$$

But if we look at the first and third of the above equations, and the second and the forth, the system of equations is inconsistent. Thus there can be no matrix **B** which is the inverse of **A**. The term *singular* is often used.

Revision

9 For the following pairs of matrices verify that **B** is the inverse of **A**:

(a) $\mathbf{A}=\begin{bmatrix} 1 & 2 \\ 2 & 3 \end{bmatrix}$, $\mathbf{B}=\begin{bmatrix} -3 & 2 \\ 2 & -1 \end{bmatrix}$, (b) $\mathbf{A}=\begin{bmatrix} 13 & 5 \\ 5 & 2 \end{bmatrix}$, $\mathbf{B}=\begin{bmatrix} 2 & -5 \\ -5 & 13 \end{bmatrix}$

17.4.1 Inversion of a 2 × 2 matrix

Consider the matrix:

$$\mathbf{A}=\begin{bmatrix} a & b \\ c & d \end{bmatrix} \qquad [19]$$

and suppose its inverse is:

$$\mathbf{A}^{-1}=\begin{bmatrix} p & q \\ r & s \end{bmatrix}$$

Then we must have:

$$\begin{bmatrix} a & b \\ c & d \end{bmatrix}\begin{bmatrix} p & q \\ r & s \end{bmatrix}=\begin{bmatrix} 1 & 0 \\ 0 & 1 \end{bmatrix}$$

and so:

$$\begin{bmatrix} ap+br & aq+bs \\ cp+dr & cq+ds \end{bmatrix}=\begin{bmatrix} 1 & 0 \\ 0 & 1 \end{bmatrix}$$

For these matrices to be equal we must have:

$$ap+br=1,\ \ aq+bs=0,\ \ cp+dr=0,\ \ cq+ds=1$$

The third equation gives $r = -cp/d$ and substituting this in the first equations gives:

$$p = \frac{d}{ad - bc}$$

and:

$$r = -\frac{cp}{d} = -\frac{c}{ad - bc}$$

The second equation gives $s = -aq/b$ and substituting this in the fourth equation gives:

$$q = -\frac{b}{ad - bc}$$

and:

$$s = -\frac{aq}{b} = \frac{a}{ad - bc}$$

Thus the inverse is:

$$\mathbf{A}^{-1} = \begin{bmatrix} p & q \\ r & s \end{bmatrix} = \begin{bmatrix} \frac{d}{ad - bc} & -\frac{b}{ad - bc} \\ -\frac{c}{ad - bc} & \frac{a}{ad - bc} \end{bmatrix}$$

We can take the factor $1/(ad - bc)$ out of the matrix to give the inverse of the matrix in equation [19] as:

$$\mathbf{A}^{-1} = \frac{1}{ad - bc} \begin{bmatrix} d & -b \\ -c & a \end{bmatrix} \qquad [20]$$

The expression $ad - bc$ is known as the *determinant* of the matrix in equation [19]. See Chapter 18 for a discussion of determinants. The inverse is thus a matrix with a and d swapped over, the signs of b and c changed, and divided by the factor $(ad - bc)$.

A matrix is not always invertible. Consider the matrix discussed in the previous section:

$$\mathbf{A} = \begin{bmatrix} 1 & 3 \\ -2 & -6 \end{bmatrix}$$

The determinant is $1(-6) - 3(-2) = 0$. Dividing by 0 is not possible and so the matrix is not invertible. Thus:

A matrix is singular, i.e. not invertible, if its determinant is zero.

Example

Determine the inverse of the matrix:

$$\mathbf{A} = \begin{bmatrix} 1 & -2 \\ 4 & 3 \end{bmatrix}$$

Using equation [20]:

$$\mathbf{A}^{-1} = \frac{1}{1 \times 3 - 4(-2)} \begin{bmatrix} 3 & 2 \\ -4 & 1 \end{bmatrix} = \frac{1}{11} \begin{bmatrix} 3 & 2 \\ -4 & 1 \end{bmatrix}$$

Revision

10 Determine the inverses of the following matrices:

$$\text{(a)} \begin{bmatrix} 1 & 5 \\ 1 & 6 \end{bmatrix}, \text{(b)} \begin{bmatrix} 2 & -3 \\ 1 & 4 \end{bmatrix}, \text{(c)} \begin{bmatrix} 4 & -3 \\ -1 & 1 \end{bmatrix}, \text{(d)} \begin{bmatrix} 3 & 2 \\ 4 & 1 \end{bmatrix}$$

11 State whether each of the following matrices has an inverse:

$$\text{(a)} \begin{bmatrix} 4 & 8 \\ 2 & 4 \end{bmatrix}, \text{(b)} \begin{bmatrix} -2 & 3 \\ 1 & 2 \end{bmatrix}, \text{(c)} \begin{bmatrix} 0 & 1 \\ 1 & 1 \end{bmatrix}, \text{(d)} \begin{bmatrix} 3 & 9 \\ 2 & 6 \end{bmatrix}$$

17.4.2 The inverse by row operations

Consider the equation $\mathbf{Ax} = \mathbf{c}$, which we could equally well write in the form $\mathbf{Ax} = \mathbf{Ic}$ with $\mathbf{I}$ being the identity matrix. If we multiply both sides of the equation by $\mathbf{A}^{-1}$:

$$\mathbf{A}^{-1}\mathbf{Ax} = \mathbf{A}^{-1}\mathbf{c}$$

and hence:

$$\mathbf{Ix} = \mathbf{A}^{-1}\mathbf{c}$$

We can think of this operation has having transformed the $\mathbf{A}$ matrix on the left-hand side of the equation to an $\mathbf{I}$ matrix and an $\mathbf{I}$ matrix on the right-hand side of the equation to the $\mathbf{A}^{-1}$ matrix.

When we have linear equations we can multiply both sides of them by the same number without affecting the equality. Likewise, with systems of simultaneous linear equations, we can add or subtract equations and still maintain the equality. This is the basis of the elimination process for solving sets of linear equations and, in particular, the Gaussian method (see Chapter 16). We can do similar operations on rows in matrices, maintaining equality by doing the same operations to comparable rows in matrices on both sides of the equals sign.

Thus to obtain an inverse $\mathbf{A}^{-1}$ we transform the **A** matrix to an **I** matrix by row manipulation and carry out the same manipulations on an **I** matrix. The result is that it becomes transformed into the $\mathbf{A}^{-1}$ matrix. The following example illustrates this.

Matrix **A** Matrix **I**

$$\begin{bmatrix} 1 & 2 \\ 2 & 3 \end{bmatrix} \quad \begin{bmatrix} 1 & 0 \\ 0 & 1 \end{bmatrix}$$

Subtract twice the first row from the second row:

$$\begin{bmatrix} 1 & 2 \\ 0 & -1 \end{bmatrix} \quad \begin{bmatrix} 1 & 0 \\ -2 & 1 \end{bmatrix}$$

Multiply the second row by −1:

$$\begin{bmatrix} 1 & 2 \\ 0 & 1 \end{bmatrix} \quad \begin{bmatrix} 1 & 0 \\ 2 & -1 \end{bmatrix}$$

Subtract twice the second row from the first row:

$$\begin{bmatrix} 1 & 0 \\ 0 & 1 \end{bmatrix} \quad \begin{bmatrix} -3 & 2 \\ 2 & -1 \end{bmatrix}$$

The result is that the **A** matrix has been transformed into the **I** matrix and the **I** matrix into the $\mathbf{A}^{-1}$ matrix.

Generally the above pairs of matrices are written in the form of an augmented matrix, the division between the two being generally indicated by a dashed line:

$$\left[\begin{array}{cc|cc} 1 & 2 & 1 & 0 \\ 2 & 3 & 0 & 1 \end{array}\right]$$

If during row manipulation we end up with the matrix under transformation to the **I** matrix having a row of all zeros, then the matrix is not invertible.

Example

Determine the inverse of the matrix:

$$\begin{bmatrix} 1 & 1 & 1 \\ 0 & 1 & 2 \\ 0 & 1 & 1 \end{bmatrix}$$

Writing this, with the same size identity matrix, as an augmented matrix and then carrying out row manipulations to transform the above matrix into the identity matrix we have:

$$\left[\begin{array}{ccc|ccc} 1 & 1 & 1 & 1 & 0 & 0 \\ 0 & 1 & 2 & 0 & 1 & 0 \\ 0 & 1 & 1 & 0 & 0 & 1 \end{array}\right]$$

Subtracting the second row from the third row:

$$\left[\begin{array}{ccc|ccc} 1 & 1 & 1 & 1 & 0 & 0 \\ 0 & 1 & 2 & 0 & 1 & 0 \\ 0 & 0 & -1 & 0 & -1 & 1 \end{array}\right]$$

Multiplying the third row by −1:

$$\left[\begin{array}{ccc|ccc} 1 & 1 & 1 & 1 & 0 & 0 \\ 0 & 1 & 2 & 0 & 1 & 0 \\ 0 & 0 & 1 & 0 & 1 & -1 \end{array}\right]$$

Subtracting twice the third row from the second row:

$$\left[\begin{array}{ccc|ccc} 1 & 1 & 1 & 1 & 0 & 0 \\ 0 & 1 & 0 & 0 & -1 & 2 \\ 0 & 0 & 1 & 0 & 1 & -1 \end{array}\right]$$

Subtracting the second row from the first row:

$$\left[\begin{array}{ccc|ccc} 1 & 0 & 1 & 1 & 1 & -2 \\ 0 & 1 & 0 & 0 & -1 & 2 \\ 0 & 0 & 1 & 0 & 1 & -1 \end{array}\right]$$

Subtracting the third row from the first row:

$$\left[\begin{array}{ccc|ccc} 1 & 0 & 0 & 1 & 0 & -1 \\ 0 & 1 & 0 & 0 & -1 & 2 \\ 0 & 0 & 1 & 0 & 1 & -1 \end{array}\right]$$

The inverse is thus:

$$\begin{bmatrix} 1 & 0 & -1 \\ 0 & -1 & 2 \\ 0 & 1 & -1 \end{bmatrix}$$

Revision

12 Determine the inverses of the following matrices by means of row operations:

(a) $\begin{bmatrix} 6 & 7 \\ 1 & 2 \end{bmatrix}$, (b) $\begin{bmatrix} 5 & 3 \\ 1 & 2 \end{bmatrix}$, (c) $\begin{bmatrix} 1 & 1 & 1 \\ 1 & -1 & 2 \\ 3 & 2 & 0 \end{bmatrix}$, (d) $\begin{bmatrix} 2 & -1 & 0 \\ -1 & 1 & 0 \\ 1 & -1 & 1 \end{bmatrix}$,

(e) $\begin{bmatrix} 1 & 0 & 2 \\ 0 & 1 & 2 \\ 1 & 2 & 0 \end{bmatrix}$, (f) $\begin{bmatrix} 2 & 1 & 1 \\ 1 & 0 & -1 \\ 1 & 3 & 2 \end{bmatrix}$

17.5 Using inverses to solve systems of linear equations

The inverse provides a method of solving sets of linear equations. For example, if we have the simultaneous equations:

$$2x + y = 3 \text{ and } x + 4y = 1$$

we can write them in matrix form as:

$$\begin{bmatrix} 2 & 1 \\ 1 & 4 \end{bmatrix}\begin{bmatrix} x \\ y \end{bmatrix} = \begin{bmatrix} 3 \\ 1 \end{bmatrix}$$

This is an equation of the form $\mathbf{Ax} = \mathbf{c}$ and for which we can write $\mathbf{x} = \mathbf{A}^{-1}\mathbf{c}$. The inverse can be determined by one of the methods given earlier as:

$$\text{inverse of} \begin{bmatrix} 2 & 1 \\ 1 & 4 \end{bmatrix} = \frac{1}{7}\begin{bmatrix} 4 & -1 \\ -1 & 2 \end{bmatrix}$$

Hence:

$$\begin{bmatrix} x \\ y \end{bmatrix} = \frac{1}{7}\begin{bmatrix} 4 & -1 \\ -1 & 2 \end{bmatrix}\begin{bmatrix} 3 \\ 1 \end{bmatrix}$$

We now have the multiplication of two matrices. Thus:

$$\begin{bmatrix} x \\ y \end{bmatrix} = \frac{1}{7}\begin{bmatrix} 4 \times 3 - 1 \times 1 \\ -1 \times 3 + 2 \times 1 \end{bmatrix} = \frac{1}{7}\begin{bmatrix} 11 \\ -1 \end{bmatrix}$$

Thus $x = 11/7$ and $y = -1/7$.

Example

Solve the simultaneous linear equations $x_1 + x_3 = 5$, $2x_1 + x_2 = -2$ and $x_2 - x_3 = 3$.

Writing the equations as matrices gives:

$$\begin{bmatrix} 1 & 0 & 1 \\ 2 & 1 & 0 \\ 0 & 1 & -1 \end{bmatrix} \begin{bmatrix} x_1 \\ x_2 \\ x_3 \end{bmatrix} = \begin{bmatrix} 5 \\ -2 \\ 3 \end{bmatrix}$$

This is an equation of the form $\mathbf{Ax} = \mathbf{c}$ and for which we can write $\mathbf{x} = \mathbf{A}^{-1}\mathbf{c}$. The inverse can be determined by one of the methods given earlier as:

$$\text{inverse of} \begin{bmatrix} 1 & 0 & 1 \\ 2 & 1 & 0 \\ 0 & 1 & -1 \end{bmatrix} = \begin{bmatrix} -1 & 1 & -1 \\ 2 & -1 & 2 \\ 2 & -1 & 1 \end{bmatrix}$$

Thus:

$$\begin{bmatrix} x_1 \\ x_2 \\ x_3 \end{bmatrix} = \begin{bmatrix} -1 & 1 & -1 \\ 2 & -1 & 2 \\ 2 & -1 & 1 \end{bmatrix} \begin{bmatrix} 5 \\ -2 \\ 3 \end{bmatrix} = \begin{bmatrix} -10 \\ 18 \\ 15 \end{bmatrix}$$

and so $x_1 = -10$, $x_2 = 18$ and $x_3 = 15$.

Revision

13 Solve, by using inverses, the following sets of simultaneous equations:

(a) $x_1 + x_2 = 3$, $2x_1 - 3x_2 = -4$,

(b) $x_1 + x_2 + x_3 = 2$, $x_1 - x_2 + 2x_3 = -1$, $3x_1 + 2x_2 = 5$,

(c) $2x_1 + 4x_2 + x_3 = -11$, $2x_1 - 3x_2 + 5x_3 = 21$, $-x_1 + 3x_2 - 2x_3 = -16$,

(d) $x_1 + x_2 + 5x_3 = 2$, $x_1 + 2x_2 + 10x_3 = 0$, $4x_1 + x_2 + x_3 = 10$

Problems

1 For the following matrices determine, where possible:

(a) $\mathbf{A} + \mathbf{B}$, (b) $\mathbf{A} + \mathbf{C}$, (c) $\mathbf{C} + \mathbf{B}$, (d) $\mathbf{C} + \mathbf{E}$, (e) $\mathbf{E} + \mathbf{C}$, (f) $\mathbf{B} + \mathbf{C} + \mathbf{E}$,

(g) $2\mathbf{A}$, (h) $3\mathbf{C}$, (i) $-\mathbf{D}$, (j) $\mathbf{D} - \mathbf{A}$, (k) $\mathbf{B} - \mathbf{C}$, (l) $2(\mathbf{B} + \mathbf{E})$, (m) $\mathbf{A} - \mathbf{B}$

$$\mathbf{A} = \begin{bmatrix} 2 & 3 \\ 4 & -2 \end{bmatrix} \quad \mathbf{B} = \begin{bmatrix} 1 & -2 & 5 \\ 0 & 2 & -2 \end{bmatrix} \quad \mathbf{C} = \begin{bmatrix} 1 & 4 & 2 \\ 2 & -1 & 3 \end{bmatrix}$$

$$\mathbf{D} = \begin{bmatrix} 4 & 2 \\ 1 & 3 \end{bmatrix} \quad \mathbf{E} = \begin{bmatrix} 2 & 5 & 1 \\ 2 & 3 & 3 \end{bmatrix}$$

2 Simplify the following matrices by removing factors from them:

$$\text{(a)}\begin{bmatrix} ab & b^2 & 3b^2 \\ -2b & 4b^2 & -ab \\ a^2b & 2b & 3ab \end{bmatrix}, \text{(b)}\begin{bmatrix} -12 & 4 & -8 \\ 4 & 16 & -20 \end{bmatrix}$$

3 Determine, where possible, the product **AB** for the following matrices:

$$\text{(a) } \mathbf{A} = \begin{bmatrix} 1 & 1 \\ -1 & 1 \end{bmatrix}, \ \mathbf{B} = \begin{bmatrix} 1 & 0 & 1 \\ 0 & 1 & 0 \end{bmatrix},$$

$$\text{(b) } \mathbf{A} = \begin{bmatrix} 2 & -2 \\ 2 & 2 \end{bmatrix}, \ \mathbf{B} = \begin{bmatrix} 1 & 2 & -3 \\ -2 & 1 & 0 \\ 3 & 3 & 2 \end{bmatrix},$$

$$\text{(c) } \mathbf{A} = \begin{bmatrix} 1 & 2 \\ 3 & 4 \end{bmatrix}, \ \mathbf{B} = \begin{bmatrix} 1 & 0 & -1 \\ 0 & 1 & 1 \end{bmatrix},$$

$$\text{(d) } \mathbf{A} = \begin{bmatrix} 2 & 3 & 4 \\ 4 & 3 & 2 \end{bmatrix}, \ \mathbf{B} = \begin{bmatrix} 1 & 2 \\ 1 & 0 \\ 4 & 4 \end{bmatrix},$$

$$\text{(e) } \mathbf{A} = \begin{bmatrix} 1 & 0 & 1 \\ 0 & 1 & 1 \\ 1 & 1 & 0 \end{bmatrix}, \ \mathbf{B} = \begin{bmatrix} 0 & 0 & 1 \\ 0 & 1 & 0 \\ 1 & 0 & 0 \end{bmatrix},$$

$$\text{(f) } \mathbf{A} = \begin{bmatrix} 0 & 2 & -3 \\ 1 & 2 & 3 \\ -1 & -2 & 4 \end{bmatrix}, \ \mathbf{B} = \begin{bmatrix} 1 & -1 & 2 \\ 3 & 4 & -4 \\ 2 & 1 & 3 \end{bmatrix},$$

$$\text{(g) } \mathbf{A} = \begin{bmatrix} 3 & 1 & 1 \\ 2 & 4 & 2 \\ -3 & 2 & -1 \end{bmatrix}, \ \mathbf{B} = \begin{bmatrix} 1 \\ 2 \\ 3 \end{bmatrix},$$

$$\text{(h) } \mathbf{A} = \begin{bmatrix} 1 \\ 2 \\ 3 \end{bmatrix}, \ \mathbf{B} = \begin{bmatrix} 1 & -2 & 0 \\ 3 & 1 & 2 \\ -2 & 3 & 1 \end{bmatrix}$$

4 Determine **AB** and **BA** for:

$$\mathbf{A} = \begin{bmatrix} 2 & 1 \\ 3 & 4 \end{bmatrix} \quad \mathbf{B} = \begin{bmatrix} 0 & 0 \\ 0 & 0 \end{bmatrix}$$

5 Determine $-\mathbf{A}$, $\mathbf{A}+\mathbf{B}$, $\mathbf{A}-2\mathbf{B}$, $\mathbf{AB}$ and $\mathbf{A}^2$ for:

$$\mathbf{A} = \begin{bmatrix} 4 & -1 & 3 \\ 2 & 5 & 3 \\ 6 & 2 & 1 \end{bmatrix} \quad \mathbf{B} = \begin{bmatrix} 1 & -3 & 2 \\ 5 & 0 & 3 \\ -5 & 2 & 1 \end{bmatrix}$$

6 Determine $\mathbf{A}+\mathbf{B}$, $2\mathbf{A}$, $\mathbf{A}^2$, $\mathbf{AB}$ and $\mathbf{B}^2$ for:

$$\mathbf{A} = \begin{bmatrix} 3 & -1 \\ 2 & -5 \end{bmatrix} \quad \mathbf{B} = \begin{bmatrix} 3 & 2 \\ 7 & 5 \end{bmatrix}$$

7 Show that $\mathbf{A}^2 - 3\mathbf{A} - 4\mathbf{I} = \mathbf{0}$ when:

$$\mathbf{A} = \begin{bmatrix} 2 & -3 \\ -2 & 1 \end{bmatrix}$$

8 Determine, using determinants, the inverses of the following matrices:

$$\text{(a)} \begin{bmatrix} 2 & -3 \\ 5 & -1 \end{bmatrix}, \text{(b)} \begin{bmatrix} 1 & 2 \\ 4 & 9 \end{bmatrix}, \text{(c)} \begin{bmatrix} 1 & -2 \\ -3 & -4 \end{bmatrix}, \text{(d)} \begin{bmatrix} -1 & 5 \\ 2 & -3 \end{bmatrix}$$

9 State, by considering determinants, whether the following matrices have inverses:

$$\text{(a)} \begin{bmatrix} 2 & 6 \\ 1 & 3 \end{bmatrix}, \text{(b)} \begin{bmatrix} -2 & 3 \\ 4 & 6 \end{bmatrix}, \text{(c)} \begin{bmatrix} 5 & -2 \\ 15 & -6 \end{bmatrix}, \text{(d)} \begin{bmatrix} -10 & -5 \\ 4 & -2 \end{bmatrix}$$

10 Determine the inverses of the following matrices by means of row operations:

$$\text{(a)} \begin{bmatrix} 1 & 4 \\ 2 & 3 \end{bmatrix}, \text{(b)} \begin{bmatrix} 1 & 0 \\ 2 & 1 \end{bmatrix}, \text{(c)} \begin{bmatrix} 1 & -2 & 1 \\ 3 & 0 & 2 \\ 2 & 4 & 1 \end{bmatrix}, \text{(d)} \begin{bmatrix} 1 & 2 & 1 \\ 0 & 1 & 1 \\ 1 & 0 & 1 \end{bmatrix},$$

$$\text{(e)} \begin{bmatrix} 2 & -3 & -1 \\ 0 & 1 & 3 \\ 1 & 0 & 2 \end{bmatrix}, \text{(f)} \begin{bmatrix} 1 & -1 & 2 \\ 1 & 2 & 1 \\ -4 & -1 & 2 \end{bmatrix}$$

11 Solve, by using inverses, the following sets of simultaneous equations:

(a) $x_1 + 3x_2 = 10$, $3x_1 - 2x_2 = 5$,

(b) $2x_1 + 4x_2 = -8$, $3x_1 - x_2 = 9$,

(c) $3x_1 + 4x_2 - 2x_3 = 5$, $6x_1 + 7x_2 - 5x_3 = 2$, $7x_1 + 3x_2 - 2x_3 = 6$,

(d) $x_1 + 4x_2 + 3x_3 = 2$, $2x_1 + 6x_2 + 5x_3 = 3$, $3x_1 + 8x_2 + x_3 = 7$,

(e) $x_1 - 4x_2 - 7x_3 = 16$, $2x_1 + x_2 + 2x_3 = 3$, $3x_1 - x_2 - x_3 = 13$

12 Prove that:

$$\begin{bmatrix} \cos\theta & \sin\theta \\ -\sin\theta & \cos\theta \end{bmatrix}\begin{bmatrix} \cos\phi & \sin\phi \\ -\sin\phi & \cos\phi \end{bmatrix} = \begin{bmatrix} \cos(\theta+\phi) & \sin(\theta+\phi) \\ -\sin(\theta+\phi) & \cos(\theta+\phi) \end{bmatrix}$$

18 Determinants

18.1 Introduction

The term *determinant* was introduced in chapter 17, in the discussion of the inversion of a 2 × 2 matrix (Section 17.4.1), as the denominator in the factor multiplying a matrix. The principle of a determinant is thus related to that of a square matrix and is a crucial element in the solution of systems of linear equations. This chapter is about determinants, their properties and use in the solution of systems of linear equations. The Cramer rule is developed as a useful tool for the solution of linear equations.

The approach to determinants adopted in this chapter follows from the discussion of matrices in Chapter 17. As with that chapter, dexterity with algebra is assumed. Examples of the use of determinants in the solution of simultaneous equations in electrical circuit analysis are given in Chapter 21.

18.2 Evaluating determinants

If we consider two simultaneous equations:

$$a_{11}x_1 + a_{12}x_2 = c_1 \text{ and } a_{21}x_1 + a_{22}x_2 = c_2 \qquad [1]$$

then we can represent them by the matrix equation $\mathbf{Ax} = \mathbf{c}$ with the solution $\mathbf{x} = \mathbf{A}^{-1}\mathbf{c}$. The inverse of the coefficients matrix $\mathbf{A}$, a 2 × 2 matrix, is:

$$\text{inverse of}\begin{bmatrix} a_{11} & a_{12} \\ a_{21} & a_{22} \end{bmatrix} = \frac{1}{a_{11}a_{22} - a_{12}a_{21}}\begin{bmatrix} a_{22} & -a_{12} \\ -a_{21} & a_{11} \end{bmatrix} \qquad [2]$$

The denominator in the multiplying factor, i.e. $a_{11}a_{22} - a_{12}a_{21}$, is termed the *determinant* of the matrix $\mathbf{A}$ and is written as:

$$\det \mathbf{A} = \begin{vmatrix} a_{11} & a_{12} \\ a_{21} & a_{22} \end{vmatrix} = a_{11}a_{22} - a_{12}a_{21} \qquad [3]$$

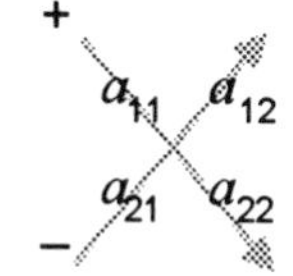

Figure 18.1 *Value of a determinant*

The determinant is written with the entries as in the matrix but between | | instead of [ ], the above determinant being termed a *second-order determinant* because it has two rows and two columns. *A determinant has a value, unlike a matrix which is just an array of numbers.* The number can be determined by the use of equation [3], though Figure 18.1 gives a useful way of realising that equation. The value is the product of the entries on the + arrow, downward from left to right, minus the product of the entries on the – arrow, upward from left to right. For example:

$$\begin{vmatrix} 4 & -2 \\ 1 & 3 \end{vmatrix} = +4(4 \times 3) - (1 \times (-2)) = 14$$

The solution of the two simultaneous equations [1] can thus be written as:

$$\begin{bmatrix} x_1 \\ x_2 \end{bmatrix} = \frac{1}{\det \mathbf{A}} \begin{bmatrix} a_{22} & -a_{12} \\ -a_{21} & a_{11} \end{bmatrix} \begin{bmatrix} c_1 \\ c_2 \end{bmatrix} \qquad [4]$$

$$= \frac{1}{\det \mathbf{A}} \begin{bmatrix} a_{22}c_1 - a_{12}c_2 \\ -a_{21}c_1 + a_{11}c_2 \end{bmatrix}$$

and thus:

$$x_1 = \frac{a_{22}c_1 - a_{12}c_2}{\det \mathbf{A}} \qquad [5]$$

$$x_2 = \frac{a_{11}c_2 - a_{21}c_1}{\det \mathbf{A}} \qquad [6]$$

The numerators of equations [5] and [6] can also be expressed as determinants and thus x_1 and x_2 become the ratios of two determinants:

$$x_1 = \frac{\begin{vmatrix} c_1 & a_{12} \\ c_2 & a_{22} \end{vmatrix}}{\begin{vmatrix} a_{11} & a_{12} \\ a_{21} & a_{22} \end{vmatrix}} \qquad [7]$$

$$x_2 = \frac{\begin{vmatrix} a_{11} & c_1 \\ a_{21} & c_2 \end{vmatrix}}{\begin{vmatrix} a_{11} & a_{12} \\ a_{21} & a_{22} \end{vmatrix}} \qquad [8]$$

If det **A**, i.e. $a_{11}a_{22} - a_{12}a_{21}$, is not zero then the system of equations has a unique solution. If det **A** = 0 and the numerators in equations [7] and [8] are also zero, then the system has infinitely many solutions. If det **A** = 0 and the numerators are not zero, then the system has no solution.

As an aid to remembering the above equations, note that the denominator is the determinant of the coefficient matrix and the numerator is the same except that the coefficients of the unknown are replaced by the constants from the right-hand side of the equation. The equations are known as *Cramer's rule*. Here it is only applied to two simultaneous equations, but can be applied to any number of simultaneous equations where we have n variables and n linear equations (see Section 18.2.1 for its application to three simultaneous equations). In general the rule can be stated as:

If $\mathbf{Ax} = \mathbf{c}$ *describes a system of n linear equations with n variables such that det* $\mathbf{A} \neq 0$*, then the system has a unique solution of:*

$$x_1 = \frac{\det A_1}{\det A}, \quad x_2 = \frac{\det A_2}{\det A}, \quad \dots \quad x_n = \frac{\det A_n}{\det A} \qquad [9]$$

where $\mathbf{A}_j$ *is the matrix obtained by replacing the entries in the jth column of A by the entries in the* **c** *matrix.*

Equations [7] and [8] are often combined to give the rule in another form:

$$\frac{x_1}{\begin{vmatrix} c_1 & a_{12} \\ c_2 & a_{22} \end{vmatrix}} = \frac{x_2}{\begin{vmatrix} a_{11} & c_1 \\ a_{21} & c_2 \end{vmatrix}} = \frac{1}{\begin{vmatrix} a_{11} & a_{12} \\ a_{21} & a_{22} \end{vmatrix}} \qquad [10]$$

Example

Determine the value of the determinant:

$$\begin{vmatrix} 2 & 5 \\ -1 & 3 \end{vmatrix}$$

Using the relationship given in equation [3], i.e. the form expressed in Figure 18.1, then:

value of the determinant $= +(2 \times 3) - (-1 \times 5) = 11$

Example

Solve the simultaneous equations $2x_1 + 3x_2 = 8$, $5x_1 - x_2 = 3$.

Using Cramer's rule:

$$x_1 = \frac{\begin{vmatrix} 8 & 3 \\ 3 & -1 \end{vmatrix}}{\begin{vmatrix} 2 & 3 \\ 5 & -1 \end{vmatrix}} = \frac{-8-9}{-2-15} = 1$$

$$x_2 = \frac{\begin{vmatrix} 2 & 8 \\ 5 & 3 \end{vmatrix}}{\begin{vmatrix} 2 & 3 \\ 5 & -1 \end{vmatrix}} = \frac{6-40}{-2-15} = 2$$

Revision

1 Determine the values of the following determinants:

(a) $\begin{vmatrix} 1 & 2 \\ 3 & 4 \end{vmatrix}$, (b) $\begin{vmatrix} 3 & 5 \\ -2 & 4 \end{vmatrix}$, (c) $\begin{vmatrix} -3 & 4 \\ -2 & -5 \end{vmatrix}$, (d) $\begin{vmatrix} 4 & a+2 \\ -2 & a-1 \end{vmatrix}$

2 By using Cramer's rule, solve the following sets of simultaneous equations:

(a) $x_1 - 2x_2 = 4$, $2x_1 + x_2 = 3$, (b) $3x_1 + x_2 = 6$, $x_1 + 2x_2 = 7$,

(c) $x_1 - 2x_2 = 2$, $2x_1 + x_2 = 9$, (d) $2x_1 + 3x_2 = 4$, $5x_1 - x_2 = -7$

18.2.1 Determinants of 3 × 3 matrices

If we have three simultaneous equations:

$$a_{11}x_1 + a_{12}x_2 + a_{13}x_3 = c_1, \quad a_{21}x_1 + a_{22}x_2 + a_{23}x_3 = c_2$$

$$\text{and } a_{31}x_1 + a_{32}x_2 + a_{33}x_3 = c_2 \qquad [11]$$

then the coefficients matrix **A** is:

$$\begin{bmatrix} a_{11} & a_{12} & a_{13} \\ a_{21} & a_{22} & a_{23} \\ a_{31} & a_{32} & a_{33} \end{bmatrix}$$

and the determinant of the matrix is:

$$\det \mathbf{A} = a_{11}a_{22}a_{33} + a_{12}a_{23}a_{31} + a_{13}a_{21}a_{32} - a_{31}a_{22}a_{13} - a_{32}a_{23}a_{13} - a_{33}a_{21}a_{12} \qquad [12]$$

As with the 2 × 2 matrix, a useful way of remembering equation [12] is by drawing diagonal lines through the matrix. So that the diagonal lines each pass through three entries, the first two columns are repeated after the third column to give the array shown in Figure 18.2. The lines passing diagonally downwards from left to right give positive products and the lines passing diagonally upwards from left to right give negative products.

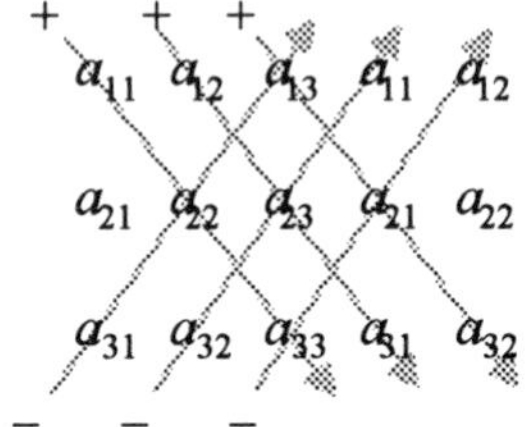

Figure 18.2 *Value of a determinant*

This method of determining the value of determinants becomes cumbersome with more than a 3×3 matrix and another method is used (see Section 18.3).

As with the 2×2 matrix, we can write $\mathbf{x} = \mathbf{A}^{-1}\mathbf{c}$ and by evaluating the inverse of **A** show that Cramer's rule applies.

Example

Determine the determinant of the matrix:

$$\mathbf{A} = \begin{bmatrix} 5 & 2 & 3 \\ 1 & 2 & 0 \\ 0 & 1 & 1 \end{bmatrix}$$

Using the rule given in Figure 18.2 we write the array as:

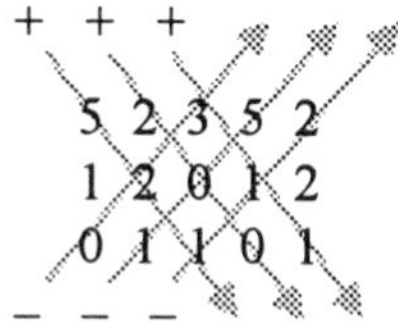

and so:

$$\det \mathbf{A} = (5)(2)(1) + (2)(0)(0) + (3)(1)(1) - (0)(2)(3) - (1)(0)(5) - (1)(1)(2)$$

$$= 10 + 3 - 2 = 11$$

Example

Using Cramer's rule, solve the following simultaneous equations:

$$2x_1 + x_2 + x_3 = 7,\ 2x_2 - x_3 = 1,\ x_1 + 3x_2 - 2x_3 = 1$$

The coefficients matrix is:

$$\begin{bmatrix} 2 & 1 & 1 \\ 0 & 2 & -1 \\ 1 & 3 & -2 \end{bmatrix}$$

and its determinant is:

$$\det \mathbf{A} = (2)(2)(-2) + (1)(-1)(1) + (1)(0)(3) - (1)(2)(1) - (3)(-1)(2) - (-2)(0)(1)$$

$$= -8 - 1 - 2 + 6 = -5$$

Using Cramer's rule:

$$x_1 = \frac{\begin{vmatrix} 7 & 1 & 1 \\ 1 & 2 & -1 \\ 1 & 3 & -2 \end{vmatrix}}{-5} = \frac{-28-1+3-2+21+2}{-5} = 1$$

$$x_2 = \frac{\begin{vmatrix} 2 & 7 & 1 \\ 0 & 1 & -1 \\ 1 & 1 & -2 \end{vmatrix}}{-5} = \frac{-4-7+0-1+2+0}{-5} = 2$$

$$x_3 = \frac{\begin{vmatrix} 2 & 1 & 7 \\ 0 & 2 & 1 \\ 1 & 3 & 1 \end{vmatrix}}{-5} = \frac{4+1+0-14-6-0}{-5} = 3$$

Revision

3 Determine the values of the following determinants:

$$\text{(a)} \begin{vmatrix} 2 & 3 & 5 \\ 2 & 1 & -3 \\ 1 & 3 & 4 \end{vmatrix}, \text{(b)} \begin{vmatrix} 2 & 1 & 4 \\ 1 & 9 & 3 \\ 4 & -1 & 6 \end{vmatrix}, \text{(c)} \begin{vmatrix} 6 & 2 & 5 \\ 10 & 2 & 6 \\ 5 & 1 & 4 \end{vmatrix}, \text{(d)} \begin{vmatrix} 4 & 3 & 2 \\ 3 & 2 & 1 \\ 4 & -4 & 0 \end{vmatrix}$$

4 Solve the following sets of simultaneous equations:

(a) $x_1 + x_2 + 2x_3 = 1$, $2x_1 - x_2 - 3x_3 = 3$, $4x_1 + x_2 + x_3 = 5$,

(b) $2x_1 + 3x_2 - 2x_3 = 1$, $-x_1 + 3x_2 + 2x_3 = 7$, $3x_1 - x_2 + x_3 = 8$,

(c) $2x_1 + x_2 + 4x_3 = 5$, $-2x_1 - 2x_2 + 5x_3 = 1$, $x_1 + 3x_2 - 2x_3 = 6$

18.3 Minors and cofactors

Consider a 3 × 3 matrix:

$$\begin{bmatrix} a_{11} & a_{12} & a_{13} \\ a_{21} & a_{22} & a_{23} \\ a_{31} & a_{32} & a_{33} \end{bmatrix}$$

The determinant of the matrix is (see equation [12]):

$$\det \mathbf{A} = a_{11}a_{22}a_{33} + a_{12}a_{23}a_{31} + a_{13}a_{21}a_{32} - a_{31}a_{22}a_{13} - a_{32}a_{23}a_{13} - a_{33}a_{21}a_{12}$$

We can rearrange these terms in a number of different ways. For example, if we extract the first row terms we can write:

$$\det \mathbf{A} = a_{11}(a_{22}a_{33} - a_{23}a_{32}) - a_{12}(a_{21}a_{33} - a_{23}a_{31}) + a_{13}(a_{21}a_{32} - a_{22}a_{31})$$

The quantities in the brackets are 2 × 2 determinants. Thus we can write the equation as:

$$\det \mathbf{A} = a_{11}\begin{vmatrix} a_{22} & a_{23} \\ a_{32} & a_{33} \end{vmatrix} - a_{12}\begin{vmatrix} a_{21} & a_{23} \\ a_{31} & a_{33} \end{vmatrix} + a_{13}\begin{vmatrix} a_{21} & a_{22} \\ a_{31} & a_{32} \end{vmatrix} \qquad [13]$$

Equation [13] involves multiplying each of the entries in the first row by 2 × 2 determinants, these being the determinants obtained from the determinant for **A** by deleting the first row and the column containing the particular entry:

$$\det \mathbf{A} = a_{11}\begin{bmatrix} & a_{22} & a_{23} \\ & a_{32} & a_{33} \end{bmatrix} - a_{12}\begin{bmatrix} a_{21} & & a_{23} \\ a_{31} & & a_{33} \end{bmatrix}$$

$$+ a_{13}\begin{bmatrix} a_{21} & a_{22} & \\ a_{31} & a_{32} & \end{bmatrix}$$

Each of the 2 × 2 determinants is called a *minor* in the determinant of **A** of the entry by which it is multiplied. Similar expressions for the 3 × 3 determinant can be obtained with the minors taken of the entries in other rows or columns. Thus if we take the initial equation:

$$\det \mathbf{A} = a_{11}a_{22}a_{33} + a_{12}a_{23}a_{31} + a_{13}a_{21}a_{32} - a_{31}a_{22}a_{13} - a_{32}a_{23}a_{13} - a_{33}a_{21}a_{12}$$

and rearrange it so that we extract each of the entries in the first column:

$$\det \mathbf{A} = a_{11}(a_{22}a_{33} - a_{32}a_{23}) - a_{21}(a_{12}a_{33} - a_{32}a_{13}) + a_{31}(a_{12}a_{23} - a_{22}a_{13})$$

$$= a_{11}\begin{bmatrix} & & \\ & a_{22} & a_{23} \\ & a_{32} & a_{33} \end{bmatrix} - a_{21}\begin{bmatrix} & a_{12} & a_{13} \\ & & \\ & a_{32} & a_{33} \end{bmatrix}$$

$$+ a_{31}\begin{bmatrix} & a_{12} & a_{13} \\ & a_{22} & a_{23} \\ & & \end{bmatrix}$$

In a similar way we can write the determinant in terms of the minors associated with any row or column.

Each minor of an entry has a + or − sign associated with it, the sign depending on the position of the entry. For a 3 × 3 determinant the signs are chosen according to the following chart:

$$\begin{vmatrix} + & - & + \\ - & + & - \\ + & - & + \end{vmatrix}$$

The minor with its attached sign is termed a *cofactor*. Thus:

A determinant is equal to the sum of the products of the entries of a row, or column, with its cofactors.

This process of expressing a determinant as such a sum is termed the *cofactor expansion of a row or column.*

We can apply the cofactor expansion to 3 × 3 determinants and higher. Thus with a 4 × 4 determinant:

$$\det \mathbf{A} = \begin{vmatrix} a_{11} & a_{12} & a_{13} & a_{14} \\ a_{21} & a_{22} & a_{23} & a_{24} \\ a_{31} & a_{32} & a_{33} & a_{34} \\ a_{41} & a_{42} & a_{43} & a_{44} \end{vmatrix}$$

when we expand it by the first row we obtain:

$$\det \mathbf{A} = a_{11}\begin{vmatrix} a_{22} & a_{23} & a_{24} \\ a_{32} & a_{33} & a_{34} \\ a_{42} & a_{43} & a_{44} \end{vmatrix} - a_{12}\begin{vmatrix} a_{21} & a_{23} & a_{24} \\ a_{31} & a_{33} & a_{34} \\ a_{41} & a_{43} & a_{44} \end{vmatrix}$$

$$+ a_{13}\begin{vmatrix} a_{21} & a_{22} & a_{24} \\ a_{31} & a_{32} & a_{34} \\ a_{41} & a_{42} & a_{44} \end{vmatrix} - a_{14}\begin{vmatrix} a_{21} & a_{22} & a_{23} \\ a_{31} & a_{32} & a_{33} \\ a_{41} & a_{42} & a_{43} \end{vmatrix} \qquad [14]$$

The signs attached to the minors are according to the following chart:

$$\begin{vmatrix} + & - & + & - & + & \cdots \\ - & + & - & + & - & \cdots \\ + & - & + & - & + & \cdots \\ - & + & - & + & - & \cdots \\ + & - & + & - & + & \cdots \\ \vdots & \vdots & \vdots & \vdots & \vdots & \end{vmatrix}$$

Example

Evaluate the following determinant using cofactor expansion by (a) the first row, (b) the first column:

$$\det \mathbf{A} = \begin{vmatrix} 2 & 3 & 4 \\ 1 & 2 & -3 \\ 1 & 1 & 5 \end{vmatrix}$$

(a) Expanding by the first row gives:

$$\det \mathbf{A} = 2\begin{vmatrix} 2 & -3 \\ 1 & 5 \end{vmatrix} - 3\begin{vmatrix} 1 & -3 \\ 1 & 5 \end{vmatrix} + 4\begin{vmatrix} 1 & 2 \\ 1 & 1 \end{vmatrix}$$

$$= 2(10 + 3) - 3(5 + 3) + 4(1 - 2) = -2$$

(b) Expanding by the first column gives:

$$\det \mathbf{A} = 2\begin{vmatrix} 2 & -3 \\ 1 & 5 \end{vmatrix} - 1\begin{vmatrix} 3 & 4 \\ 1 & 5 \end{vmatrix} + 1\begin{vmatrix} 3 & 4 \\ 2 & -3 \end{vmatrix}$$

$$= 2(10 + 3) - 1(15 - 4) + 1(-9 - 8) = -2$$

Again the result is –2.

Example

Evaluate the following 4×4 determinant:

$$\det \mathbf{A} = \begin{vmatrix} 4 & 3 & 2 & -1 \\ 2 & 1 & 0 & 3 \\ 1 & -1 & 0 & 2 \\ 2 & 2 & 0 & 1 \end{vmatrix}$$

It is simplest if we expand by the third column since three of its four entries are zero and so there is less multiplying to do. Thus:

$$\det \mathbf{A} = 2\begin{vmatrix} 2 & 1 & 3 \\ 1 & -1 & 2 \\ 2 & 2 & 1 \end{vmatrix} + 0 + 0 + 0$$

We can evaluate this by either using the rule given earlier for 3×3 determinants or further cofactoring:

$$\det \mathbf{A} = 2(-2 + 4 + 6 + 6 - 8 - 1) = 10$$

Revision

5 Evaluate the following determinants:

(a) $\begin{vmatrix} 5 & 0 & 4 \\ 1 & 2 & 0 \\ 0 & 1 & 1 \end{vmatrix}$, (b) $\begin{vmatrix} 3 & -1 & 2 \\ 1 & 4 & 0 \\ 3 & 5 & 0 \end{vmatrix}$, (c) $\begin{vmatrix} 1 & 2 & 3 \\ 0 & 2 & 3 \\ 0 & 1 & 1 \end{vmatrix}$, (d) $\begin{vmatrix} 2 & -2 & -2 \\ 3 & 4 & 2 \\ 0 & -1 & -1 \end{vmatrix}$,

(e) $\begin{vmatrix} 3 & 7 & -1 & 3 \\ 2 & 5 & 0 & 2 \\ 0 & 2 & 6 & 3 \\ 1 & 2 & 0 & 3 \end{vmatrix}$, (f) $\begin{vmatrix} 3 & 0 & 0 & 0 \\ -1 & 1 & 4 & 2 \\ -2 & 0 & 3 & 2 \\ -4 & 0 & 1 & 5 \end{vmatrix}$, (g) $\begin{vmatrix} 1 & 4 & 6 & -1 \\ 0 & 1 & 2 & 3 \\ 2 & -3 & 1 & 4 \\ 0 & 1 & 2 & 3 \end{vmatrix}$

18.4 Properties of determinants

The following are properties of determinants which can be of use in simplifying calculations involved in evaluating them.

1 *Interchanging rows*
Interchanging two rows changes the sign of the determinant, e.g.

$$\begin{vmatrix} a & b \\ c & d \end{vmatrix} = -\begin{vmatrix} c & d \\ a & b \end{vmatrix}$$

$$ad - bc = -(bc - ad)$$

2 *Interchanging columns*
Interchanging two columns changes the sign of the determinant, e.g.

$$\begin{vmatrix} a & b \\ c & d \end{vmatrix} = -\begin{vmatrix} b & a \\ d & c \end{vmatrix}$$

$$ad - bc = -(bc - ad)$$

3 *Interchanging the rows with the columns*
Interchanging the rows and columns does not affect the value of the determinant, e.g.

$$\begin{vmatrix} a & b \\ c & d \end{vmatrix} = \begin{vmatrix} a & c \\ b & d \end{vmatrix}$$

$$ad - bc = ad - bc$$

4 *Multiplication of a row or column by a constant*
Multiplying a row by a constant k multiplies the value of the determinant by k, e.g.

$$\begin{vmatrix} ka & kb \\ c & d \end{vmatrix} = k\begin{vmatrix} a & b \\ c & d \end{vmatrix}$$

$$kad - kbc = k(bc - ad)$$

5 *Addition or subtraction of rows or columns*
Adding or subtracting one or a multiple of one row from another, or one column from another, does not affect the value of the determinant. Consider:

$$\det \mathbf{A} = \begin{vmatrix} a_1 & a_2 & a_3 \\ b_1 & b_2 & b_3 \\ c_1 & c_2 & c_3 \end{vmatrix}$$

If we add k times the third row to the second row we obtain:

$$\det \mathbf{B} = \begin{vmatrix} a_1 & a_2 & a_3 \\ b_1 + kc_1 & b_2 + kc_2 & b_3 + kc_3 \\ c_1 & c_2 & c_3 \end{vmatrix}$$

Evaluation of this gives:

$$a_1b_2c_3 + ka_1c_2c_3 + a_2b_3c_1 + ka_2c_3c_1 + a_3b_1c_2 + ka_3c_1c_2$$
$$- c_1b_2a_3 - kc_1c_2a_3 - c_2b_3a_1 - kc_2c_3a_1 - c_3b_1a_2 - kc_3c_1a_2$$

The sum of the terms without k give the value of determinant **A**. The terms including k are:

$$ka_1c_2c_3 + ka_2c_3c_1 + ka_3c_1c_2 - kc_1c_2a_3 - kc_2c_3a_1 - kc_3c_1a_2$$

But these have the value 0. Hence det **A** = det **B**.

6 *Equal rows or columns*
A determinant with two equal rows, or two equal columns, has a zero value. For example, if we have:

$$\det \mathbf{A} = \begin{vmatrix} a_1 & b_1 & c_1 \\ a_1 & b_1 & c_1 \\ a_2 & b_2 & c_2 \end{vmatrix}$$

and subtract row 1 from row 2 we have:

$$\det \mathbf{A} = \begin{vmatrix} a_1 & b_1 & c_1 \\ 0 & 0 & 0 \\ a_2 & b_2 & c_2 \end{vmatrix}$$

and so the determinant has zero value.

7 *A row or column is a multiple of another*
When a row or column is a multiple of another then the determinant has a value of zero. This is because if we extract the multiplication factor from the determinant we are left with two rows or columns the same and hence, as in 6 above, the determinant will have zero value.

Example

After simplification, evaluate the following determinants:

$$\text{(a)}\begin{vmatrix} 6 & 9 \\ 12 & 16 \end{vmatrix},\ \text{(b)}\begin{vmatrix} 1 & 2 & 3 \\ 2 & 4 & 6 \\ 12 & 13 & 14 \end{vmatrix},\ \text{(c)}\begin{vmatrix} 0 & 3 & 6 \\ 2 & 3 & 4 \\ 1 & 2 & 2 \end{vmatrix}$$

(a) We can extract a factor of 3 from the first row to give:

$$3\begin{vmatrix} 2 & 3 \\ 12 & 16 \end{vmatrix}$$

We can then extract a factor of 4 from the second row:

$$3\times 4\begin{vmatrix} 2 & 3 \\ 3 & 4 \end{vmatrix}$$

Hence the value is $12(8 - 9) = -12$.
(b) The second row is twice the first row. Thus the determinant has a value of zero.
(c) If we multiply the third row by 2 and subtract it from the third row we obtain:

$$\mathbf{A} = \begin{vmatrix} 0 & 3 & 6 \\ 2 & 3 & 4 \\ 0 & 1 & 0 \end{vmatrix}$$

The presence of so many zeros simplifies the calculation. Hence $\mathbf{A} = 6$.

Revision

6 After simplification, evaluate the following determinants:

$$\text{(a)}\begin{vmatrix} 12 & 12 & 12 \\ 50 & 100 & 50 \\ -35 & -35 & 70 \end{vmatrix},\ \text{(b)}\begin{vmatrix} 2 & -1 & 5 \\ 0 & 4 & 2 \\ 2 & -1 & 4 \end{vmatrix},\ \text{(c)}\begin{vmatrix} 4 & 12 & -6 \\ 0 & 0 & 0 \\ 5 & -3 & -7 \end{vmatrix},$$

(d) $\begin{vmatrix} 1 & 1 & 1 \\ 1 & 2 & 3 \\ 1 & 3 & 6 \end{vmatrix}$, (e) $\begin{vmatrix} -2 & 4 & 6 \\ 9 & 3 & 5 \\ 10 & -20 & 10 \end{vmatrix}$, (f) $\begin{vmatrix} 29 & 1 & -2 \\ 28 & 1 & -2 \\ -30 & -4 & 5 \end{vmatrix}$

7 By inspection determines the values of the following determinants if:

$$\begin{vmatrix} 1 & 0 & 1 \\ 2 & 1 & 2 \\ 3 & 4 & 1 \end{vmatrix} = -2$$

(a) $\begin{vmatrix} 4 & 0 & 4 \\ 2 & 1 & 1 \\ 3 & 4 & 1 \end{vmatrix}$, (b) $\begin{vmatrix} 2 & 1 & 2 \\ 1 & 0 & 1 \\ 3 & 4 & 1 \end{vmatrix}$, (c) $\begin{vmatrix} 3 & 1 & 3 \\ 2 & 1 & 2 \\ 3 & 4 & 1 \end{vmatrix}$

18.5 Cofactor and adjoint matrices

Consider three simultaneous equations with three variables:

$$a_{11}x_1 + a_{12}x_2 + a_{13}x_3 = c_1, \quad a_{21}x_1 + a_{22}x_2 + a_{23}x_3 = c_2$$

$$\text{and } a_{31}x_1 + a_{32}x_2 + a_{33}x_3 = c_2 \qquad [15]$$

The coefficient matrix is:

$$\mathbf{A} = \begin{bmatrix} a_{11} & a_{12} & a_{13} \\ a_{21} & a_{22} & a_{23} \\ a_{31} & a_{32} & a_{33} \end{bmatrix} \qquad [16]$$

Expressing the determinant of the matrix as the cofactor expansion of the first row (see equation [13]) gives:

$$\det \mathbf{A} = a_{11}\begin{vmatrix} a_{22} & a_{23} \\ a_{32} & a_{33} \end{vmatrix} - a_{12}\begin{vmatrix} a_{21} & a_{23} \\ a_{31} & a_{33} \end{vmatrix} + a_{13}\begin{vmatrix} a_{21} & a_{22} \\ a_{31} & a_{32} \end{vmatrix} \qquad [17]$$

If we represent the cofactor of a_{11} by A_{11}, the cofactor of a_{12} by A_{12} and the cofactor of a_{13} by A_{13} then we can write equation [17] as:

$$\det \mathbf{A} = a_{11}A_{11} + a_{12}A_{12} + a_{13}A_{13} \qquad [18]$$

We could have expressed the determinant as the cofactor expansion of the second row as:

$$\det \mathbf{A} = a_{21}A_{21} + a_{22}A_{22} + a_{23}A_{23} \qquad [19]$$

Likewise the cofactor expansion of the third row is:

$$\det \mathbf{A} = a_{31}A_{31} + a_{32}A_{32} + a_{33}A_{33} \qquad [20]$$

We can express equations [18], [19] and [20] as a *cofactor matrix*:

$$\text{cofactor } \mathbf{A} = \begin{bmatrix} A_{11} & A_{12} & A_{13} \\ A_{21} & A_{22} & A_{23} \\ A_{31} & A_{32} & A_{33} \end{bmatrix} \qquad [21]$$

If we interchange the rows and the columns we have a matrix called the *adjoint* or *adjugate matrix* (this action of interchanging rows and columns is called *transposing*). Thus:

$$\text{adj } \mathbf{A} = \begin{bmatrix} A_{11} & A_{21} & A_{31} \\ A_{12} & A_{22} & A_{32} \\ A_{13} & A_{23} & A_{33} \end{bmatrix} \qquad [22]$$

Consider the product of the matrix **A** and its adjoint:

$$\mathbf{A}(\text{adj } \mathbf{A}) = \begin{bmatrix} a_{11} & a_{12} & a_{13} \\ a_{21} & a_{22} & a_{23} \\ a_{31} & a_{32} & a_{33} \end{bmatrix} \begin{bmatrix} A_{11} & A_{21} & A_{31} \\ A_{12} & A_{22} & A_{32} \\ A_{13} & A_{23} & A_{33} \end{bmatrix}$$

If we take the product to be some matrix **B** with:

$$\mathbf{B} = \begin{bmatrix} b_{11} & b_{12} & b_{13} \\ b_{21} & b_{22} & b_{23} \\ b_{31} & b_{32} & b_{33} \end{bmatrix}$$

then we must have:

$$b_{11} = a_{11}A_{11} + a_{12}A_{12} + a_{13}A_{13}$$

$$b_{12} = a_{11}A_{21} + a_{12}A_{22} + a_{13}A_{23}$$

$$b_{13} = a_{11}A_{31} + a_{12}A_{32} + a_{13}A_{33}$$

$$b_{21} = a_{21}A_{11} + a_{22}A_{12} + a_{23}A_{13}$$

$$b_{22} = a_{21}A_{21} + a_{22}A_{22} + a_{23}A_{23}$$

$$b_{23} = a_{21}A_{31} + a_{22}A_{32} + a_{23}A_{33}$$

$$b_{31} = a_{31}A_{11} + a_{32}A_{12} + a_{33}A_{13}$$

$$b_{32} = a_{31}A_{21} + a_{32}A_{22} + a_{33}A_{23}$$

$$b_{33} = a_{31}A_{31} + a_{32}A_{32} + a_{33}A_{33}$$

b_{11}, b_{22} and b_{33} are the entries of **A** multiplied by the cofactors of the same entries and so, being equations of the form given in equation [18], are each equal to det **A**. With the other b terms, the coefficients and the cofactors come from different rows of **A**. Because of this they all have zero value. To illustrate this, consider b_{21}:

$$b_{21} = a_{21}A_{11} + a_{22}A_{12} + a_{23}A_{13}$$

$$= a_{21}(a_{22}a_{33} - a_{32}a_{23}) - a_{22}(a_{21}a_{33} - a_{31}a_{23}) + a_{23}(a_{21}a_{32} - a_{31}a_{23}) = 0$$

Thus we can write for the product of a matrix and its adjoint:

$$\mathbf{A}(\text{adj }\mathbf{A}) = \begin{bmatrix} \det\mathbf{A} & 0 & 0 \\ 0 & \det\mathbf{A} & 0 \\ 0 & 0 & \det\mathbf{A} \end{bmatrix} = \det\mathbf{A}\begin{bmatrix} 1 & 0 & 0 \\ 0 & 1 & 0 \\ 0 & 0 & 1 \end{bmatrix}$$

Hence:

$$\mathbf{A}(\text{adj }\mathbf{A}) = (\det\mathbf{A})\mathbf{I}$$

But $\mathbf{AA}^{-1} = \mathbf{I}$ (Chapter 17, equation [17]), thus the inverse matrix $\mathbf{A}^{-1}$ is given by:

$$\mathbf{A}^{-1} = \frac{\text{adj }\mathbf{A}}{\det\mathbf{A}} \qquad [23]$$

Note that if det **A** = 0 that there can be no inverse.

Example

Determine (a) the cofactor matrix, (b) the adjoint matrix and (c) the inverse of the matrix:

$$\mathbf{A} = \begin{bmatrix} -1 & 3 & 1 \\ 2 & 1 & -2 \\ 3 & 5 & 0 \end{bmatrix}$$

(a) The cofactor matrix is:

$$\begin{bmatrix} \begin{vmatrix} 1 & -2 \\ 5 & 0 \end{vmatrix} & -\begin{vmatrix} 2 & -2 \\ 3 & 0 \end{vmatrix} & \begin{vmatrix} 2 & 1 \\ 3 & 5 \end{vmatrix} \\ -\begin{vmatrix} 3 & 1 \\ 5 & 0 \end{vmatrix} & \begin{vmatrix} -1 & 1 \\ 3 & 0 \end{vmatrix} & -\begin{vmatrix} -1 & 3 \\ 3 & 5 \end{vmatrix} \\ \begin{vmatrix} 3 & 1 \\ 1 & -2 \end{vmatrix} & -\begin{vmatrix} -1 & 1 \\ 2 & -2 \end{vmatrix} & \begin{vmatrix} -1 & 3 \\ 2 & 1 \end{vmatrix} \end{bmatrix} = \begin{bmatrix} 10 & 6 & 7 \\ 5 & -3 & 14 \\ -7 & 0 & -7 \end{bmatrix}$$

(b) The adjoint matrix is obtained by transposing:

$$\text{adj } \mathbf{A} = \begin{bmatrix} 10 & 5 & -7 \\ 6 & -3 & 0 \\ 7 & 14 & -7 \end{bmatrix}$$

(c) The determinant of **A** is given by det **A** = –18 + 10 – 3 – 10 = –21. Hence, using equation [23]:

$$\mathbf{A}^{-1} = \frac{\text{adj } \mathbf{A}}{\det \mathbf{A}} = -\frac{1}{21}\begin{bmatrix} 10 & 5 & -7 \\ 6 & -3 & 0 \\ 7 & 14 & -7 \end{bmatrix}$$

Revision

8 Determine the cofactor matrix, the adjoint matrix and the inverse matrix for each of the following:

$$\text{(a)} \begin{bmatrix} 3 & 1 & 2 \\ 5 & -4 & 1 \\ -1 & 2 & 1 \end{bmatrix}, \text{(b)} \begin{bmatrix} 1 & 2 & 4 \\ 0 & -1 & -2 \\ 1 & 3 & 2 \end{bmatrix}, \text{(c)} \begin{bmatrix} 2 & -3 & 1 \\ 5 & 4 & 1 \\ 2 & -2 & -1 \end{bmatrix}$$

Problems

1 Determine the values of the following determinants:

$$\text{(a)} \begin{vmatrix} 2 & -1 \\ -2 & 3 \end{vmatrix}, \text{(b)} \begin{vmatrix} -1 & -2 \\ 3 & -5 \end{vmatrix}, \text{(c)} \begin{vmatrix} \sqrt{2} & \sqrt{3} \\ 2 & \sqrt{6} \end{vmatrix}, \text{(d)} \begin{vmatrix} a & a+1 \\ 2 & 3 \end{vmatrix},$$

$$\text{(e)} \begin{vmatrix} 1 & 4 & -1 \\ 6 & 13 & -3 \\ 3 & 1 & 0 \end{vmatrix}, \text{(f)} \begin{vmatrix} 4 & 3 & 2 \\ 6 & 2 & 1 \\ 1 & -1 & 0 \end{vmatrix}, \text{(g)} \begin{vmatrix} 6 & 1 & 3 \\ 1 & 3 & 1 \\ 5 & 0 & 4 \end{vmatrix}, \text{(h)} \begin{vmatrix} 3 & 2 & 4 \\ 12 & 1 & 11 \\ 8 & 3 & 5 \end{vmatrix}$$

$$\text{(i)} \begin{vmatrix} 1 & 2 & 3 & 4 \\ 2 & 4 & 6 & 8 \\ -2 & 3 & -4 & 5 \\ -1 & -2 & 0 & 0 \end{vmatrix}, \text{(j)} \begin{vmatrix} 7 & -1 & 0 & 0 \\ 2 & 1 & 0 & 0 \\ 3 & 14 & 3 & -3 \\ 6 & -6 & 2 & 1 \end{vmatrix}, \text{(k)} \begin{vmatrix} 1 & 1 & 0 & 7 \\ 2 & 2 & 0 & -2 \\ 4 & 1 & -3 & 0 \\ 2 & 9 & 6 & 2 \end{vmatrix},$$

$$\text{(l)} \begin{vmatrix} 1 & 2 & -3 & 1 & 2 \\ 4 & 0 & 2 & 3 & 1 \\ 5 & 0 & 0 & -2 & 4 \\ 0 & 0 & 0 & 4 & 10 \\ 0 & 0 & 0 & 2 & -7 \end{vmatrix}$$

20 Iterative methods

20.1 Introduction

The Gaussian elimination method (see Chapter 16) for the solution of sets of linear equations is an exact method and will give unique solutions if they exist. However, a problem with the method is, if there are many equations and variables, that the number of computations can be very high. An alternative method which can be used and reduces the number of computations required is an *iterative method*. This starts with an initial approximation to a solution and then generates a succession of better and better approximations. In this chapter Jacobi and Gauss-Seidel iteration methods are discussed.

Algebraic dexterity is assumed and, apart from the discussion of convergence, could follow on from Part 1. Only in the discussion of convergence is any knowledge of matrices assumed.

20.2 Jacobi iteration

To illustrate this method of solving linear equations, consider the following two simultaneous equations with two unknowns:

$$2x + y = 4 \quad \text{and } x + 4y = 9$$

We rewrite these equations so that the first equation expresses x in terms of the other variable and the second equation expresses y in terms of the other variable:

$$x = -\tfrac{1}{2}y + 2 \qquad [1]$$

$$y = -\tfrac{1}{4}x + 2.25 \qquad [2]$$

We then make a guess as to the possible values of x and y. Suppose we take $x = 0$ and $y = 1$. We now substitute these values into the right-hand sides of equations [1] and [2] to give newer estimates of x and y:

$$x = -\tfrac{1}{2} \times 1 + 2 = 1.5, \quad y = -\tfrac{1}{4} \times 0 + 2.25 = 2.25$$

We then substitute these new values into the right-hand sides of equations [1] and [2] to give newer estimates of x and y:

$$x = -\tfrac{1}{2} \times 2.25 + 2 = 0.875, \quad y = -\tfrac{1}{4} \times 1.5 + 2.25 = 1.5$$

We then again substitute these values into the right-hand sides of equations [1] and [2] to give newer estimates of x and y:

$$x = -\tfrac{1}{2} \times 1.5 + 2 = 1.25, \quad y = -\tfrac{1}{4} \times 0.875 + 2.25 = 2.03$$

$$\mathbf{A} = \begin{bmatrix} 8 & 10 \\ -5 & -7 \end{bmatrix}$$

2 Determine the characteristic equations and eigenvalues of the following matrices:

(a) $\begin{bmatrix} 5 & 4 \\ 4 & 5 \end{bmatrix}$, (b) $\begin{bmatrix} 2 & -3 \\ -4 & 6 \end{bmatrix}$, (c) $\begin{bmatrix} 2 & 3 \\ 4 & 6 \end{bmatrix}$, (d) $\begin{bmatrix} 1 & -1 & 0 \\ 1 & 2 & 1 \\ -2 & 1 & -1 \end{bmatrix}$,

(e) $\begin{bmatrix} 1 & 2 & 1 \\ 2 & 1 & 1 \\ 1 & 1 & 2 \end{bmatrix}$, (f) $\begin{bmatrix} 8 & -5 \\ 5 & 8 \end{bmatrix}$, (g) $\begin{bmatrix} 1 & -5 \\ 2 & 2 \end{bmatrix}$, (h) $\begin{bmatrix} -5 & 2 \\ 2 & -4 \end{bmatrix}$

3 Determine the eigenvalues and associated eigenvectors for the following matrices:

(a) $\begin{bmatrix} -2 & 2 \\ -1 & -5 \end{bmatrix}$, (b) $\begin{bmatrix} -5 & 0 \\ 1 & 2 \end{bmatrix}$, (c) $\begin{bmatrix} 0 & 2 \\ -3 & 5 \end{bmatrix}$, (d) $\begin{bmatrix} 0 & -1 & 0 \\ 0 & 0 & -1 \\ 1 & 0 & 0 \end{bmatrix}$,

(e) $\begin{bmatrix} 0 & -1 & 1 \\ -1 & 0 & 1 \\ 1 & 1 & 0 \end{bmatrix}$

$$\frac{d^2y}{dx^2} - 2\frac{dy}{dx} + y = 0$$

is $y = A\ e^x + Bx\ e^x$, determine the solution if $y = 1$ at $x = 0$ and $dy/dx = -1$ at $x = 0$.

From the initial condition $y = 1$ at $x = 0$ we have, when substituting these values in the general solution, $1 = A + 0$. Thus $A = 1$. If we differentiate the general solution to give $dy/dx = A\ e^x + Bx\ e^x + B\ e^x$ and substitute the initial condition $dy/dx = 0$ at $x = 0$, then $-1 = A + 0 + B$ and so $B = -2$. Thus the solution is $y = e^x - 2x\ e^x$.

Revision

1 Determine the unique solutions for the following differential equations given the general solutions and initial conditions:

(a) $\frac{d^2y}{dx^2} = 3$, $y = \frac{3}{2}x^2 + Ax + B$, $y = 2$ and $\frac{dy}{dx} = 4$ at $x = 0$,

(b) $\frac{d^2y}{dx^2} + \frac{dy}{dx} - 6y = 0$, $y = A\ e^{2x} + B\ e^{-3x}$, $y = 1$ and $\frac{dy}{dx} = 0$ at $x = 0$,

(c) $\frac{d^2y}{dx^2} + 4\frac{dy}{dx} + 4y = 0$, $y = A\ e^{-2x} + Bx\ e^{-2x}$, $y = 1$ and $\frac{dy}{dx} = 0$ at $x = 0$

30.3 Second-order homogeneous linear differential equations

Consider a homogeneous linear ordinary second-order differential equation, the basic form being:

$$a_2\frac{d^2y}{dx^2} + a_1\frac{dy}{dx} + a_0y = 0 \qquad [6]$$

where a_2, a_1 and a_0 are constants. In the case of a homogeneous linear first-order differential equation with constant coefficients:

$$a_1\frac{dy}{dx} + a_0y = 0$$

we have $dy/dx = -(a_0/a_1)y$ and thus, by separation of the variables, the solution is $\ln y = -(a_0/a_1)x + A$ or $y = C\ e^{kx}$, where $k = -(a_0/a_1)$. To solve the constant coefficient second-order differential equation it seems reasonable to consider that it might have a solution of the form $y = A\ e^{sx}$, where A and s are constants. Thus, trying this as a solution, the second-order differential equation [6] becomes:

$$a_2As^2\ e^{sx} + a_1As\ e^{sx} + a_0A\ e^{sx} = 0$$

Since the exponential function is never zero we must have, if $y = A\ e^{sx}$ is to be a solution:

$$a_2s^2 + a_1s + a_0 = 0 \qquad [7]$$

Equation [7] is called the *auxiliary equation* or *characteristic equation* associated with the differential equation [6]. This quadratic equation has the roots:

$$s = \frac{-a_1 \pm \sqrt{a_1^2 - 4a_2a_0}}{2a_2} \qquad [8]$$

The roots of the auxiliary equation, as given by equation [8], can be:

1 *Two distinct real roots if* $a_1^2 > 4a_2a_0$
The general solution to the differential equation is then:

$$y = A\ e^{s_1x} + B\ e^{s_2x} \qquad [9]$$

Example

Determine the general solution of the differential equation:

$$\frac{d^2y}{dx^2} - \frac{dy}{dx} - 6y = 0$$

Trying $y = A\ e^{sx}$ as a solution gives the auxiliary equation:

$$s^2 - s + 6 = 0$$

which factors as $(s - 3)(s + 2) = 0$ and so $s_1 = 3$ and $s_2 = -2$. Thus the general solution is:

$$y = A\ e^{3x} + B\ e^{-2x}$$

2 *Two equal real roots if* $a_1^2 = 4a_2a_0$
This gives $s_1 = s_2 = -a_1/2a_2$. In order to have a solution with two arbitrary constants we *cannot* have a general solution of:

$$y = A\ e^{s_1x} + B\ e^{s_2x} = (A + B)\ e^{sx} = C\ e^{sx}$$

since this can be reorganised to imply only one constant. Thus we try a solution of the form $y = Bx\ e^{sx}$. Then, since $dy/dx = B\ e^{sx} + Bsx\ e^{sx}$ and $d^2y/dx^2 = 2Bs\ e^{sx} + Bs^2x\ e^{sx}$, substituting into equation [6] gives:

$$a_2(2s + s^2x) + a_1(1 + sx) + a_0x = 0$$

$$(a_2s^2 + a_1s + a_0)x + (2a_2s + a_1) = 0$$

But $a_2s^2 + a_1s + a_0 = 0$ is the auxiliary equation and so the first term is zero. Also $s = -a_1/2a_2$ and so the second term is zero. Thus $y = Bx\,e^{sx}$ is a solution. The general solution is thus:

$$y = A\,e^{s_1x} + Bx\,e^{s_2x} \qquad [10]$$

Example

Determine the general solution of the differential equation:

$$\frac{d^2y}{dx^2} + 8\frac{dy}{dx} + 16y = 0$$

Trying $y = A\,e^{sx}$ as a solution gives the auxiliary equation:

$$s^2 + 8s + 16 = 0$$

This factors as $(s + 4)(s + 4) = 0$ and so we have two roots of $s = -4$. The solution is thus of the form given in equation [10]:

$$y = A\,e^{-4x} + Bx\,e^{-4x}$$

3 *Two distinct complex roots if* $a_1^2 < 4a_2a_0$
With this condition, equation [8] can be written as:

$$s = \frac{-a_1 \pm j\sqrt{4a_2a_0 - a_1^2}}{2a_2} = \alpha \pm j\beta$$

where $\alpha = -(a_1/2a_2)$ and $\beta = \sqrt{(a_0/a_2) - (a_1/a_2)^2/4}$. Thus the general solution is:

$$y = A\,e^{(\alpha + j\beta)x} + B\,e^{(\alpha - j\beta)x}$$

This can be written as:

$$y = A\,e^{\alpha x}\,e^{j\beta x} + B\,e^{\alpha x}\,e^{-j\beta x} = e^{\alpha x}(A\,e^{j\beta x} + B\,e^{-j\beta x})$$

Euler's formula (equation [2], Chapter 8) enables the above equation to be written as:

$$y = e^{\alpha x}\left[A(\cos\beta x + j\sin\beta x) + B(\cos\beta x - j\sin\beta x)\right]$$

$$= e^{\alpha x}\left[(A + B)\cos\beta x + j(A - B)\sin\beta x\right]$$

$$= e^{\alpha x}\left[C\cos\beta x + D\sin\beta x\right] \qquad [11]$$

Example

Determine the general solution of the differential equation:

$$\frac{d^2y}{dx^2} - 2\frac{dy}{dx} + 5y = 0$$

and the particular solution if $y = 1$ and $dy/dx = 2$ at $x = 0$.

Trying $y = A\,e^{sx}$ as a solution gives the auxiliary equation:

$$s^2 - 2s + 5 = 0$$

This has roots:

$$s = \frac{2 \pm \sqrt{4-20}}{2} = 1 \pm j2$$

The general solution will thus be of the form given for equation [11]:

$$y = e^x(C \cos 2x + D \sin 2x)$$

With $y = 1$ when $x = 0$ we have $1 = 1(C + 0)$ and thus $C = 1$. Differentiating the solution gives:

$$\frac{dy}{dx} = e^x(2C \sin 2x - 2D \cos 2x) + e^x(C \cos 2x + D \sin 2x)$$

With $dy/dx = 2$ at $x = 0$ we have $2 = -2D + C$ and so $D = -½$. Thus the particular solution is:

$$y = e^x(\cos 2x + ½ \sin 2x)$$

Revision

2 Determine the general solutions of the following differential equations:

(a) $2\frac{d^2y}{dx^2} + \frac{dy}{dx} - y = 0$, (b) $\frac{d^2y}{dx^2} - 6\frac{dy}{dx} + 9y = 0$,

(c) $\frac{d^2y}{dx^2} - 10\frac{dy}{dx} + 25y = 0$, (d) $\frac{d^2y}{dx^2} - 4\frac{dy}{dx} + 5y = 0$,

(e) $\frac{d^2y}{dx^2} - 2\frac{dy}{dx} + 5y = 0$, (f) $\frac{d^2y}{dx^2} + 3\frac{dy}{dx} - 10y = 0$

3 Determine the particular solutions of the following differential equations:

(a) $\dfrac{d^2y}{dx^2}+6\dfrac{dy}{dx}+5y=0$, $y=0$ and $\dfrac{dy}{dx}=3$ at $x=0$,

(b) $\dfrac{d^2y}{dx^2}-6\dfrac{dy}{dx}+25y=0$, $y=3$ and $\dfrac{dy}{dx}=1$ at $x=0$,

(c) $4\dfrac{d^2y}{dx^2}+12\dfrac{dy}{dx}+9y=0$, $y=2$ and $\dfrac{dy}{dx}=1$ at $x=0$

30.3.1 A common form of homogeneous differential equation

Second-order differential equations of the following type occur quite often in engineering and science:

$$\frac{d^2y}{dx^2}\pm\omega^2y=0 \qquad [12]$$

If we try $y = A\,e^{sx}$ as a solution we obtain the auxiliary equation $s^2 \pm \omega^2 = 0$.

When we have the positive sign $s^2 = -\omega^2$ and $s = \pm j\omega$. This gives the general solution:

$$y=A\cos\omega t+B\sin\omega t \qquad [13]$$

When we have the negative sign $s^2 = \omega^2$ and so $s = \pm\omega$. This gives the general solution $y = C\,e^{\omega x} + D\,e^{\omega x}$. This result can be written in another form. Since $\cosh\omega x = ½(e^{\omega x} + e^{-\omega x})$ and $\sinh\omega x = ½(e^{\omega x} + e^{-\omega x})$ then the general solution is:

$$y=C(\cosh\omega x+\sinh\omega x)+D(\cosh\omega x-\sinh\omega x)$$

$$=(C+D)\cosh\omega x+(C-D)\sinh\omega x$$

$$=E\cosh\omega x+F\sinh\omega x \qquad [14]$$

30.4 Second-order non-homogeneous linear differential equations

Consider a non-homogeneous linear second-order differential equation with constant coefficients a_2, a_1 and a_0 with $f(x)$ being some function of x, often being referred to as the *forcing function*, applied to the system:

$$a_2\frac{d^2y}{dx^2}+a_1\frac{dy}{dx}+a_0y=f(x) \qquad [15]$$

In considering non-homogeneous first-order equations in Chapter 29 (Section 29.4), the technique used was to consider the general solution for the non-homogeneous equation to be the sum of the complementary function and the particular integral, the complementary function being the solution of the homogeneous form of the differential equation and the particular integral a particular solution of the non-homogeneous equation.

The same technique can be used with higher-order linear differential equations. The following examples illustrate its use with second-order differential equations.

If we have a linear non-homogeneous differential equation of the form $a_2\,\mathrm{d}^2y/\mathrm{d}x^2 + a_1\,\mathrm{d}y/\mathrm{d}x + a_0y = f(x)$ *then it has a general solution which is equal to the sum of the complementary function and the particular integral. The complementary function is obtained by solving the equivalent homogeneous differential equation, i.e.* $f(x) = 0$, *and the particular integral by considering the form of the* $f(x)$ *function and trying a particular solution of a similar form but which contains undetermined coefficients.*

Right-hand side of non-homogeneous equation	*Trial function, with A, B, C, etc. being undetermined coefficients*
Constant	A
Polynomial	$A + Bx + Cx^2 + \ldots$
Exponential	$A\,\mathrm{e}^{kx}$
Sine or cosine	$A \sin kx + B \cos kx$

Note: if the right-hand side is a sum of more than one term then the trial solution is the sum of the trial functions for these terms.

Example

Determine the general solution of the differential equation:

$$\frac{\mathrm{d}^2y}{\mathrm{d}x^2} - 5\frac{\mathrm{d}y}{\mathrm{d}x} + 6y = x^2$$

To obtain the complementary function we consider the equivalent homogeneous differential equation, i.e.

$$\frac{\mathrm{d}^2y}{\mathrm{d}x^2} - 5\frac{\mathrm{d}y}{\mathrm{d}x} + 6y = 0$$

Trying $y = A\,\mathrm{e}^{sx}$ as a solution gives the auxiliary equation:

$$s^2 - 5s + 6 = 0$$

This can be factored as $(s - 3)(s - 2) = 0$ and so $s_1 = 3$ and $s_2 = 2$. The complementary function is thus:

$$y_c = A\,\mathrm{e}^{3x} + B\,\mathrm{e}^{2x}$$

To find the particular integral with x^2 we try a solution of the form $y = C + Dx + Ex^2$. Substituting this into the non-homogeneous differential equation gives:

$$2E - 5(2Ex + D) + 6(C + Dx + Ex^2) = x^2$$

Equating coefficients of x^2 gives $6E = 1$ and so $E = 1/6$. Equating coefficients of x gives $-10E + 6D = 0$ and so $D = 10/36 = 5/18$. Equating constants gives $2E - 5D + 6C = 0$ and so $C = 19/108$. Thus the particular integral is:

$$y_p = \tfrac{19}{108} + \tfrac{5}{18}x + \tfrac{1}{6}x^2$$

Thus the general solution is:

$$y = y_c + y_p = A\,e^{3x} + B\,e^{2x} + \tfrac{19}{108} + \tfrac{5}{18}x + \tfrac{1}{6}x^2$$

Example

Determine the general solution of the differential equation:

$$\frac{d^2y}{dx^2} + \frac{dy}{dx} - 2x = 3\,e^{2x}$$

The corresponding homogeneous differential equation is:

$$\frac{d^2y}{dx^2} + \frac{dy}{dx} - 2y = 0$$

Trying $y = A\,e^{sx}$ as a solution gives the auxiliary equation:

$$s^2 + s - 2 = 0$$

This can be factored as $(s + 2)(s - 1) = 0$ and so the roots are $s_1 = -2$ and $s_2 = 1$. Thus the complementary function is:

$$y_c = A\,e^{-2x} + B\,e^{x}$$

For the particular integral with an exponential forcing function we try a solution of the form $y = C\,e^{kx}$. Substituting this into the non-homogeneous differential equation gives:

$$k^2C\,e^{kx} + kC\,e^{kx} - 2C\,e^{kx} = 3\,e^{2x}$$

Thus we must have $k = 2$ for equality of the exponentials and for the coefficients $(k^2 + k - 2)C = 3$ and hence $C = 3/4$. Hence the particular integral is:

$$y_p = \tfrac{3}{4}\,e^{2x}$$

and the general solution is:

$$y = y_c + y_p = A\,e^{-2x} + B\,e^{x} + \tfrac{3}{4}\,e^{2x}$$

Example

Determine the general solution of the differential equation:

$$3\frac{d^2y}{dx^2}+\frac{dy}{dx}-2y=2\cos x$$

The corresponding homogeneous differential equation is:

$$3\frac{d^2y}{dx^2}+\frac{dy}{dx}-2y=0$$

Trying $y = A\,e^{sx}$ as a solution gives the auxiliary equation:

$$3s^2+s-2=0$$

This can be factored as $(3s-2)(s+1)=0$ and so the roots are $s_1 = 2/3$ and $s_2 = -1$. Thus the complementary function is:

$$y_c = A\,e^{2x/3} + B\,e^{-x}$$

For the particular integral we try a solution of the form $y = C\cos kx + D\sin kx$. Substituting this into the non-homogeneous differential equation gives:

$$3(-C\cos kx - D\sin kx) + (-C\sin kx + D\cos kx)$$
$$-2(C\cos kx + D\sin kx) = 2\cos x$$

For equality of the cosines we must have $k = 1$ and $-3C + D - 2C = 2$. Equating coefficients of the sines gives $-3D - C - 2D = 0$. Thus we have $C = -5/13$ and $D = 1/13$. The particular integral is thus:

$$y_p = -\tfrac{5}{13}\cos x + \tfrac{1}{13}\sin x$$

The general solution is thus:

$$y = y_c + y_p = A\,e^{2x/3} + B\,e^{-x} - \tfrac{5}{13}\cos x + \tfrac{1}{13}\sin x$$

Revision

4 Determine the general solutions of the following differential equations:

(a) $\dfrac{d^2y}{dx^2}-4y=2\,e^{3x}$, (b) $\dfrac{d^2y}{dx^2}+3\dfrac{dy}{dx}+4y=3x+2$,

(c) $\dfrac{d^2y}{dx^2}-6\dfrac{dy}{dx}+8y=3\cos x$, (d) $\dfrac{d^2y}{dx^2}-5\dfrac{dy}{dx}+6y=e^{x}$,

(e) $\frac{d^2y}{dx^2} - 5\frac{dy}{dx} + 6y = 2\ e^x - 3\ e^{-x}$, (f) $\frac{d^2y}{dx^2} - \frac{dy}{dx} - 2y = 5 \sin 2x$

30.4.1 Exceptional cases of particular integrals

There are situations when the obvious form of function to be tried to obtain the particular integral yields no result because when it is substituted in the differential equation we obtain 0 = 0. This occurs when the right-hand side of the non-homogeneous differential equation consists of a function that is also a term in the complementary function. To illustrate this, consider the differential equation:

$$\frac{d^2y}{dx^2} + \frac{dy}{dx} - 2y = e^{-2x}$$

The complementary function is $y = A\ e^{-2x} + B\ e^x$. For the particular integral, if we try the solution $y = A\ e^{kx}$ we obtain:

$$4A\ e^{kx} - 2A\ e^{kx} - 2A\ e^{kx} = e^{-2x}$$

and so no solution for A. In such cases we have to try something different.

The basic rule is to multiply the trial solution by x.

Thus we try $y = Ax\ e^{kx}$. This gives, for the above differential equation:

$$(-2A\ e^{kx} + 4Ax\ e^{kx} - 2A\ e^{kx}) + (A\ e^{kx} - 2Ax\ e^{kx}) - 2Ax\ e^{kx} = e^{-2x}$$

Thus $k = -2$ and $-2A + 4Ax - 2A + A - 2Ax - 2Ax = 1$. Equating constants gives $3A = 1$, equating the x coefficients gives $0 = 0$, and so the particular integral is $y = \frac{1}{3}\ e^{-2x}$.

Example

Determine the general solution of the differential equation:

$$\frac{d^2y}{dx^2} - 3\frac{dy}{dx} - 10y = 4 - e^{-2x}$$

The corresponding homogeneous differential equation is:

$$\frac{d^2y}{dx^2} - 3\frac{dy}{dx} - 10y = 0$$

Trying $y = A\ e^{sx}$ as a solution gives the auxiliary equation:

$$s^2 - 3s - 10 = 0$$

This can be factored as $(s-5)(s+2)=0$ and so the complementary function is $y_c = A\ e^{5x} + B\ e^{-2x}$. The right-hand side of the non-homogeneous differential equation is the sum of two terms for which the trial functions would be C and $Dx\ e^{kx}$. We thus try the sum of these. Thus:

$$Dk^2x\ e^{kx} + Dk\ e^{kx} + Dk\ e^{kx} - 3Dkx\ e^{kx} - 3D\ e^{kx} - 10(C + Dx\ e^{kx})$$
$$= 4 - e^{-2x}$$

Equating exponential terms gives $k = -2$, $4Dx - 2D - 2D + 6Dx - 3D - 10Dx = -1$ and so $D = 1/7$. Equating constants gives $-10C = 4$ and so $C = -4/10$. Thus the particular integral is $-(4/10) + (1/7)\ e^{2x}$. The general solution is therefore:

$$y = A\ e^{5x} + B\ e^{-2x} - \tfrac{4}{10} + \tfrac{1}{7}x\ e^{-2x}$$

Revision

5 Determine the general solutions of the following differential equations:

(a) $\dfrac{d^2y}{dx^2} - 2\dfrac{dy}{dx} - 8y = 3\ e^{-2x}$, (b) $\dfrac{d^2y}{dx^2} + 9y = 5 \sin 3x$

Problems

1 Determine the unique solutions for the following differential equations given the general solutions and initial conditions:

(a) $\dfrac{d^2y}{dx^2} - y = 0$, $y = A\ e^{x} + B\ e^{-x}$, $y = 0$ and $\dfrac{dy}{dx} = 5$ at $x = 0$,

(b) $\dfrac{d^2y}{dx^2} - 3\dfrac{dy}{dx} + 2y = 0$, $y = A\ e^{x} + B\ e^{2x}$, $y = 1$ and $\dfrac{dy}{dx} = 0$ at $x = 0$,

(c) $\dfrac{d^2y}{dx^2} + 2\dfrac{dy}{dx} + y = 0$, $y = A\ e^{-x} + Bx\ e^{-x}$, $y = 2$ and $\dfrac{dy}{dx} = -1$ at $x = 0$

2 Determine the general solutions of the following differential equations:

(a) $\dfrac{d^2y}{dx^2} - 4\dfrac{dy}{dx} + 5y = 0$, (b) $\dfrac{d^2y}{dx^2} + 2\dfrac{dy}{dx} - 8y = 0$,

(c) $\dfrac{d^2y}{dx^2} + 3\dfrac{dy}{dx} - 4y = 0$, (d) $\dfrac{d^2y}{dx^2} + 2\dfrac{dy}{dx} + 4y = 0$,

(e) $\dfrac{d^2y}{dx^2} + 6\dfrac{dy}{dx} + 9y = 0$, (f) $9\dfrac{d^2y}{dx^2} + 6\dfrac{dy}{dx} + y = 0$

3 Determine the particular solutions of the following differential equations:

(a) $\frac{d^2y}{dx^2} - 7\frac{dy}{dx} + 12y = 0$, $y = 3$ and $\frac{dy}{dx} = 2$ at $x = 0$,

(b) $\frac{d^2y}{dx^2} + 2\frac{dy}{dx} + y = 0$, $y = 1$ and $\frac{dy}{dx} = 1$ at $x = 0$,

(c) $\frac{d^2y}{dx^2} - 6\frac{dy}{dx} + 25y = 0$, $y = 1$ and $\frac{dy}{dx} = 7$ at $x = 0$

4 Determine the general solutions of the following differential equations:

(a) $\frac{d^2y}{dx^2} - 3\frac{dy}{dx} + 2y = 3 - 2x^2$, (b) $\frac{d^2y}{dx^2} - 3\frac{dy}{dx} + 2y = 2\,e^{-2x}$,

(c) $\frac{d^2y}{dx^2} - 3\frac{dy}{dx} + 2y = 5\sin 2x$, (d) $\frac{d^2y}{dx^2} - 5\frac{dy}{dx} + 6y = x^2$,

(e) $\frac{d^2y}{dx^2} - 2\frac{dy}{dx} + y = e^x$, (f) $\frac{d^2y}{dx^2} - \frac{dy}{dx} - 6y = e^{3x}$

(g) $\frac{d^2y}{dx^2} + 2\frac{dy}{dx} - 3y = \sin 2x + 5\,e^{-3x}$, (h) $\frac{d^2y}{dx^2} + 4\frac{dy}{dx} + 3y = 2\,e^{-x}$,

(i) $\frac{d^2y}{dx^2} + y = 2\cos x$, (j) $\frac{d^2y}{dx^2} + 2\frac{dy}{dx} = x$

5 Determine the particular solution of the following differential equation given that $y = -2$ and $dy/dx = -3$ when $x = 0$:

$$\frac{d^2y}{dx^2} - 4y = 5\,e^{3x}$$

6 Determine the particular solution of the following differential equation given that $y = 1$ and $dy/dx = 0$ when $x = 0$:

$$\frac{d^2y}{dx^2} + 9y = \sin 2x$$

31 Numerical methods

31.1 Introduction

Consider the differential equation:

$$\frac{dy}{dx} = e^{-x^2}$$

We can separate the variables to give:

$$\int dy = \int e^{-x^2}\, dx$$

Thus to obtain a solution we need the antiderivative of e^{-x^2}. But there is no elementary function which differentiates to give that function. We can, however, evaluate the integral by using a numerical method such as the trapezium rule or Simpson's rule (see Chapter 26). There are many other differential equations which can arise and are like this. In such situations we have to adopt a numerical method.

This chapter is a basic consideration of numerical methods for the solution of ordinary differential equations, the methods considered being Euler's method and the Runge-Kutta method. The solution of first-order differential equations is considered and then the extension of the methods to higher-order differential equations. Chapter 28 and a knowledge of the Taylor series from Chapter 5 are assumed.

31.2 Euler's method

dy/dx gives the slope of the tangent at a point on the graph produced when y is plotted against x. Suppose we have a first-order differential equation $dy/dx = f(x, y)$ and we have the initial value of y_0 at $x = 0$. If we put these initial values into the differential equation we can obtain the slope of the tangent at $x = 0$, say $(dy/dx)_{x=0}$. Then if we want to find the value of y at $x = 1h$, i.e. y_1, we can make the approximating assumption that the graph line described by the tangent at $x = 0$ represents the differential equation for this interval (Figure 31.1). Thus:

$y_1 = y_0$ + the amount by which function is estimated as increasing by in going from $x = 0$ to $x = h$

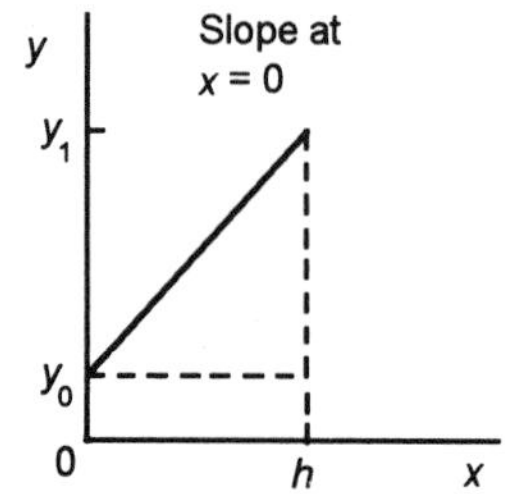

Figure 31.1 *Determining a point from initial conditions*

The amount by which function increases by in going from $x = 0$ to $x = h$ is the amount by which the tangent rises in that interval. Thus:

$$y_1 = y_0 + h\left(\frac{dy}{dx}\right)_{x=0}$$

We can now put this value of y_1 at $x = 1h$ in the differential equation and obtain a value of the slope of the tangent at y_1, say $(dy/dx)_{x=1h}$. If we assume

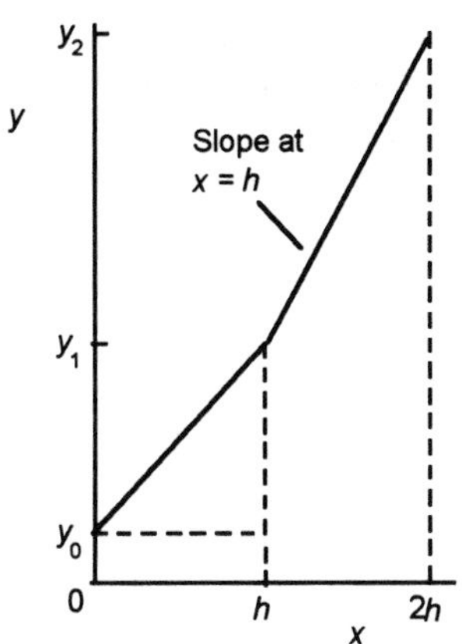

Figure 31.2 *Determining a point from conditions at* $x = h$

that this tangent describes the graph for the next interval h, i.e. from $x = 1h$ to $x = 2h$ (Figure 31.2), then the estimated value of y_2 at $x = 2h$ is:

$y_2 = y_1$ + the amount by which function is estimated as increasing by in going from $x = 1h$ to $x = 2h$

and so:

$$y_2 = y_1 + h\left(\frac{dy}{dx}\right)_{x=1h}$$

We can, in this way, systematically build up the solution graph by taking it step by step and obtaining values of y at x = 0, h, $2h$, $3h$, etc. This step-by-step method of solving a first-order differential equation is known as *Euler's method.*

Euler's method of solving a first-order differential equation given initial values is to compute successive approximations y_1, y_2, y_3, etc. at successive x values which increase in steps of h by the use of:

$$y_{n+1} = y_n + h\left(\frac{dy}{dx}\right)_{x=nh} \qquad [1]$$

Example

Use Euler's method with a step size in x of $h = 1$ to solve the first-order differential equation $dy/dx = x$ given that $y = 1$ at $x = 0$.

Initially we have $y_0 = 1$ at $x = 0$. This gives an initial slope of 0. Using equation [1] gives an estimate of y_1 at $x = 1h = 1$ of:

$$y_1 = 1 + 1(0) = 1$$

The slope is 1 at $x = 1$. Thus, using equation [1], the estimate of y_2 at $x = 2h = 2$ is:

$$y_2 = 1 + 1(1) = 2$$

The slope is 2 at $x = 2$. Thus, using equation [1], the estimate of y_3 at $x = 3h = 3$ is:

$$y_3 = 2 + 1(2) = 4$$

The slope is 3 at $x = 3h$. Thus, using equation [1], the estimate of y_4 at $x = 4h = 4$ is:

$$y_4 = 4 + 1(3) = 4$$

We can continue for further steps.

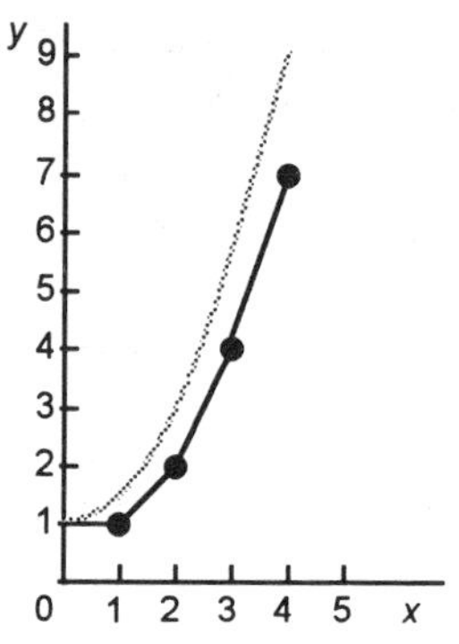

Figure 31.3 *Example*

Figure 31.3 shows the solution graph. The differential equation could have been solved by separation of the variables, the solution then being, with the given initial conditions, $y = \frac{1}{2}x^2 + 1$. This line is also shown on the graph. Notice that as the number of steps increases the error increases.

Revision

1 Use Euler's method to solve the following differential equations, using the step size indicated and tabulating the results for the first four steps:

(a) $\dfrac{dy}{dx} = x + y,\ y = 1$ at $x = 1,\ h = 0.1$,

(b) $\dfrac{dy}{dx} = y + 2x,\ y = 1$ at $x = 0,\ h = 0.2$,

(c) $\dfrac{dy}{dx} = 3yx^2,\ y = 1$ at $x = 1,\ h = 0.1$

31.2.1 Errors with Euler's method

The error is the difference between the value given by Euler's method and the actual result obtained by analytical means. Consider the solution by Euler's method of the differential equation $dy/dx = x + y$ with $y = 1$ at $x = 0$. Table 31.1 shows the results obtained with step sizes of 0.1, 0.02 and 0.01 and how the values obtained at specific x values compare with the value given analytically, the analytical solution being $y = 2\,e^x - x - 1$.

Table 31.1 *Solving* $dy/dx = x + y$

x	y with $h = 0.1$	y with $h = 0.02$	y with $h = 0.01$	Analytical value
0	1.000 0	1.000 0	1.000 0	1.000 0
0.1	1.100 0	1.108 2	1.109 2	1.110 3
0.2	1.220 0	1.238 0	1.240 4	1.242 8
0.3	1.362 0	1.391 7	1.395 7	1.399 7
0.4	1.528 2	1.571 9	1.577 7	1.583 6
0.5	1.721 0	1.781 2	1.789 3	1.797 4

The following points emerge from a consideration of the errors given for the different step sizes in Table 31.1:

1 As the number of steps increases the error increases.

2 As the step size is reduced the error is reduced, the error being roughly proportional to h.

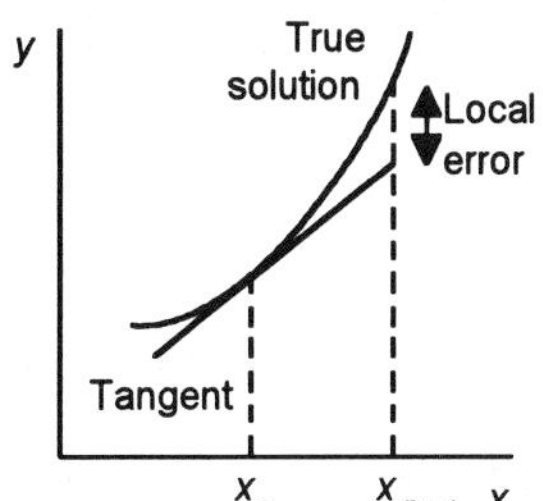

Figure 31.4 *Local error*

The error in Euler's method occurs because the tangent line at x_n is used to predict the value of y at x_{n+1} and unless the function is a straight line it will depart from the true value and give a local error (Figure 31.4). This error not only affects the value of y at x_{n+1} but the following values predicted for y since the errors are accumulative. The local error is reduced at each step by the use of smaller steps. However, the number of computations required when very small steps are involved can be very large. With a step size of 0.1 one step is required to get from $x = 0$ to $x = 0.1$, with step size 0.2 five steps are needed, with step size 0.01 ten steps. With a step size of 0.001 the error would be significantly reduced but 100 steps would be required.

31.2.2 Euler's method in terms of the Taylor series

We can use a polynomial (Chapter 5, equation [18]) to give an approximation to a function. The Taylor series (see Section 5.4) gives the value of y_{nh} at $x = nh$ as:

$$y_{n+1} = y_n + h\left(\frac{dy}{dx}\right)_{x=nh} + \frac{h^2}{2!}\left(\frac{d^2y}{dx^2}\right)_{x=nh} + \frac{h^3}{3!}\left(\frac{d^3y}{dx^3}\right)_{x=nh} + \dots \qquad [2]$$

But the first two terms give the Euler equation [1]. Thus the other terms can be considered to be the error that occurs as a result of using the Euler equation. Considering a step size of h, then if we consider the error with Euler's equation as arising from just the third term, then at each step a local error proportional to h^2 is produced. The total error after a number of steps will be the sum of the errors incurred at each step. Thus some value of y at a particular x value will have required x/h steps from the initial value and so the error will be proportional $(x/h)h^2$ and thus h. Table 31.1 bears this out.

The accuracy can thus be improved by using more terms. However, such a method of solution is not simple since it requires the calculation of second- and higher-order derivatives. A more simple method, called the *Runge-Kutta method*, involves a modification of the Taylor series to make the calculations simpler.

31.3 The Runge-Kutta method

The *Runge-Kutta method* gives a result which is equivalent to using the Taylor polynomial without the need to take derivatives. With the Euler method, the value of y after one step is determined using the slope of the tangent at the beginning of the step. The Runge-Kutta method determines the slope of tangents at a number of points in the interval and then uses an estimate of the average value of the slope over the interval for determining the value of y after a step. The following outlines the method for what is termed *order four*, it effectively using the Taylor series to the fourth order, though the theory of this is not discussed here but a simple graphical approach adopted.

Consider a differential equation which is a function of x and y, i.e. $dy/dx = f(x, y)$. If we start with initial values x_0 and y_0 and use the Euler method to

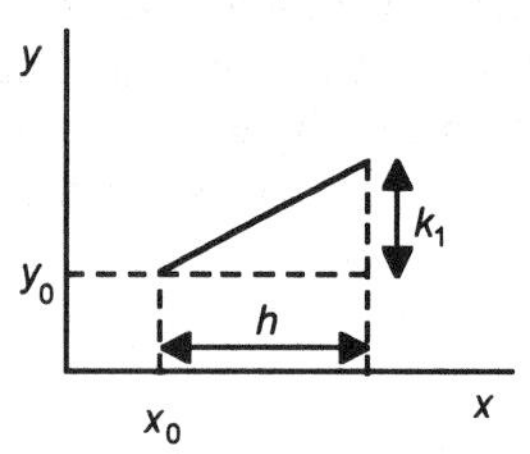

Figure 31.5 k_1

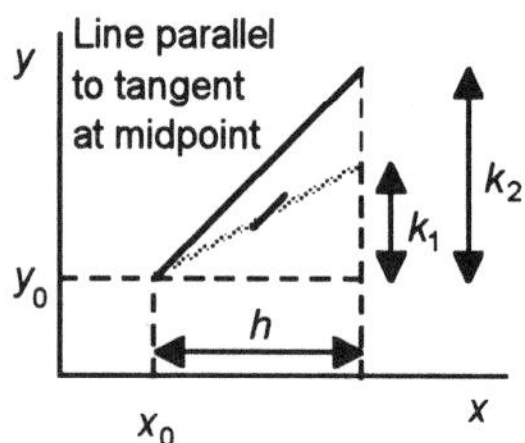

Figure 31.6 k_2

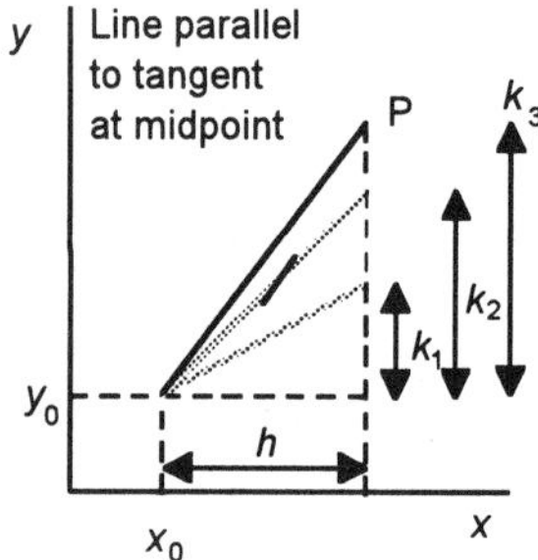

Figure 31.7 k_3

determine the value of y after a step size h and obtain $y_0 + k_1$ we have a tangent of slope k_1/h (Figure 31.5). Thus:

$$\frac{k_1}{h} = f(x_0, y_0) \quad [3]$$

The x value at the midpoint of the step will be $x_0 + h/2$ and the estimated y value $y_0 + k_1/2$. If we substitute these values into the differential equation we can arrive at a value for the slope of the tangent at this midpoint of $f(x_0 + h/2, y_0 + k_1/2)$. If we use this slope to determine the value of y (Figure 31.6) after the step as $y_0 + k_2$ then k_2/h becomes the estimate of the derivative:

$$\frac{k_2}{h} = f(x_0 + h/2, y_0 + k_1/2) \quad [4]$$

The x value at the midpoint of the step will be $x_0 + h/2$ and the new estimated y value $y_0 + k_2/2$. If we substitute these values into the differential equation we can arrive at a value for the slope of the tangent at this midpoint of $f(x_0 + h/2, y_0 + k_2/2)$. If we use this slope to determine the value of y (Figure 31.7) after the step as $y_0 + k_3$ then k_3/h becomes the estimate of the derivative:

$$\frac{k_3}{h} = f(x_0 + h/2, y_0 + k_2/2) \quad [5]$$

The co-ordinates of point P are then $(x_0 + h, y_0 + k_3)$. An estimate of the slope of the tangent at this point is $f(x_0 + h, y_0 + k_3)$. Then:

$$\frac{k_4}{h} = f(x_0 + h,\ y_0 + k_3) \quad [6]$$

To summarise: four estimates have been made for the rise of approximating line over the step, k_1 based on the estimate of the slope at the left-hand end of the step, k_2 and k_3 based on estimates of the slope at the midpoint of the step and k_4 based on the estimate of the slope at the right-hand end of the step. A weighted average is then taken of these estimates, the estimates for mid-step being weighted twice as much as those for the start and end of the step. Thus:

$$\text{rise over step } h = \frac{k_1 + 2k_2 + 2k_3 + k_4}{6}$$

Hence, in general:

$$y_n = y_{n-1} + \frac{k_1 + 2k_2 + 2k_3 + k_4}{6} \quad [7]$$

This method is considerably more accurate than the Euler method and is very widely used. The accuracy is such that quite large step sizes can be taken and still yield acceptable accuracy.

$$\int \sec^2 y \frac{dy}{dx}\, dx = \int 1\, dx$$

or:

$$\int \sec^2 y\, dy = \int 1\, dx$$

Hence $\tan y = x + A$. Since $y = \pi/4$ when $x = 0$ then $\tan \pi/4 = A$ and so $A = 1$. Thus $\tan y = x + 1$ or $y = \tan^{-1}(x + 1)$.

Revision

1 Solve, by separation of the variables, the following differential equations:

$$\text{(a) } \frac{dy}{dx} = \frac{1}{x}, \text{ (b) } \frac{dy}{dx} = \cos \tfrac{1}{2}x, \text{ (c) } \frac{dy}{dx} = y^2, \text{ (d) } \frac{dy}{dx} = -2y,$$

$$\text{(e) } \frac{dy}{dx} = 2x(y^2 + 1), \text{ (f) } \frac{dy}{dx} = 3x^2\, e^{-y}$$

2 Determine the solution of $dy/dx = 2xy^2$ if $y = ½$ when $x = 0$.

3 A capacitor of capacitance C which has been charged to a voltage V_0 is discharged through a resistance R. Determine how the voltage v_C across the capacitor changes with time t if $dv_C/dt = -V/RC$.

4 The rate at which radioactivity decays with time t is given by the differential equation $dN/dt = -kN$, where N is the number of radioactive atoms present at time t. If at time $t = 0$ the number of radioactive atoms is N_0, solve the differential equation and show how the number of radioactive atoms varies with time.

29.3 Integrating factor

Consider the differential equation $dy/dx = 2xy$. If we multiply both sides of the equation by $1/y$ we obtain:

$$\frac{1}{y}\frac{dy}{dx} = 2x \quad \text{or} \quad \frac{1}{y}\, dy = 2x\, dx$$

which we can integrate to give $\ln y = x^2 + A$. $1/y$ is recognisable as the derivative of $\ln y$, $2x$ as the derivative of x^2. What we have is:

$$\frac{d}{dx}(\ln y) = \frac{d}{dx}(x^2)$$

It is because each side of the equation is recognisable as a derivative with respect to x that we can carry out the integration. To get the equation into this form we had to multiply each side of it by the factor $1/y$. The $1/y$ is termed an *integrating factor*.

The integrating factor is the factor used to multiply each side of a differential equation so as to put it into a form that enables each side to be recognisable as a derivative.

Consider the linear first-order differential equation:

$$\frac{dy}{dx} + Py = Q \qquad [3]$$

where P and Q are functions of x. If we multiply equation [3] by an integrating factor I then:

$$I\frac{dy}{dx} + IPy = IQ \qquad [4]$$

We want to make the left-hand side of equation [4] into a form recognisable as a derivative. Since:

$$\frac{d}{dx}(Iy) = I\frac{dy}{dx} + y\frac{dI}{dx} \qquad [5]$$

then to make the left-hand side of equation [4] equal to the left-hand side of equation [5] we need to have:

$$y\frac{dI}{dx} = IPy$$

This means, if we cancel the y from both sides of the equation and solve the differential equation by separation of the variables:

$$\int \frac{1}{I}\frac{dI}{dx}\,dx = \int P\,dx$$

$$\int \frac{1}{I}\,dI = \int P\,dx$$

$$\ln I = \int P\,dx + A$$

$$I = e^{\int P\,dx + A} = e^{A}\,e^{\int P\,dx} = C\,e^{\int P\,dx}$$

where C is a constant. In order to clearly indicate that P is a function of x, the integrating factor is generally written as

$$I = C\,e^{\int P(x)\,dx} \qquad [6]$$

The procedure for use of an integrating factor to solve a differential equation is thus:

1 If the equation is not in the standard form of $dy/dx + Py = Q$, write it in that form. For example:

$$x\frac{dy}{dx} + y = 2$$

would be written as:

$$\frac{dy}{dx} + \frac{y}{x} = \frac{2}{x}$$

2 Identify P and then determine the integrating factor I by the use of equation [6]. Thus for the above equation $P = 1/x$.

3 Multiply both sides of the differential equation by the integrating factor. The left-hand side of the equation is then the derivative of Iy.

4 Integrate and so obtain the general solution as:

$$Iy = \int IQ \, dx + A$$

Example

Solve the differential equation $dy/dx + y = x$ using an integrating factor.

If we compare this with equation [3] then $P = 1$. The integrating factor is then given by equation [6] as:

$$I = C\,e^{\int 1\,dx} = C\,e^x$$

Multiplying both sides of the differential equation by the integrating factor then gives:

$$C\,e^x\frac{dy}{dx} + C\,e^x y = C\,e^x x$$

The left-hand side of the equation is known to be the derivative of the product Iy (equation [5]). Thus:

$$\frac{d}{dx}(C\,e^x y) = C\,e^x x$$

The constant C can be cancelled. Thus integrating, using integration by parts (Section 25.4.2) with respect to x, gives:

$$e^x y = x\,e^x - e^x + A$$

$$y = x - 1 + A\,e^{-x}$$

Example

Solve the differential equation $dy/dx - y = e^{2x}$ using an integrating factor.

If we compare this with equation [3] then $P = -1$. The integrating factor is then given by equation [6] as:

$$I = C\,\mathrm{e}^{\int -1\,\mathrm{d}x} = C\,\mathrm{e}^{-x}$$

Multiplying both sides of the differential equation by the integrating factor then gives:

$$C\,\mathrm{e}^{-x}\frac{\mathrm{d}y}{\mathrm{d}x} - C\,\mathrm{e}^{-x}y = C\,\mathrm{e}^{-x}\,\mathrm{e}^{2x} = C\,\mathrm{e}^{x}$$

The left-hand side of the equation is known to be the derivative of the product Iy (equation [5]). Thus:

$$\frac{\mathrm{d}}{\mathrm{d}x}(C\,\mathrm{e}^{-x}y) = C\,\mathrm{e}^{x}$$

Cancelling C from both sides of the equation and integrating gives:

$$\mathrm{e}^{-x}y = \mathrm{e}^{x} + A$$

$$y = \mathrm{e}^{2x} + A\,\mathrm{e}^{x}$$

Revision

5 Solve the following differential equations by the use of integrating factors:

(a) $\dfrac{\mathrm{d}y}{\mathrm{d}x} - 3y = \mathrm{e}^{2x}$, (b) $\dfrac{\mathrm{d}y}{\mathrm{d}x} - \dfrac{y}{x} = 1 + \dfrac{1}{x}$, (c) $x\dfrac{\mathrm{d}y}{\mathrm{d}x} + y = \mathrm{e}^{x}$,

(d) $\dfrac{\mathrm{d}y}{\mathrm{d}x} + y\tan x = \cos^2 x$, over the interval $-\pi/2 < x < \pi/2$

29.4 Complementary function and particular integral

Suppose we have the first-order differential equation $\mathrm{d}y/\mathrm{d}x + y = 5$. Such an equation is termed non-homogeneous (see Section 28.4). The differential equation has the solution $y = C\,\mathrm{e}^{-x} + 5$ (obtained by using an integrating factor). Now suppose we have the corresponding homogeneous differential equation, i.e. $\mathrm{d}y/\mathrm{d}x + y = 0$. This has the solution $y = C\,\mathrm{e}^{-x}$ (obtained by separation of the variables). Thus the solution of the non-homogeneous equation is equal to the sum of the general solution of the corresponding homogeneous equation plus another term. The general solution of the homogeneous differential equation is called the *complementary function* and the term added to it for the non-homogeneous solution is called the *particular integral*. The particular integral is a particular solution of the non-homogeneous equation. Thus for the above differential equation this is $y = 5$. If this is substituted into the non-homogeneous equation we obtain $0 + 5 = 5$, confirming that it is indeed a particular solution.

If we have a linear non-homogeneous differential equation of the form $\mathrm{d}y/\mathrm{d}x + Py = Q$ *then it has a general solution which is equal to the sum of the complementary function and the particular integral.*

This is a very useful result because it means that we only need find the general solution of the homogeneous equation and can get away with finding any particular solution of the full non-homogeneous equation. The result applies to any linear non-homogeneous equation. What we are doing is letting $y = u + v$. Substituting this into the non-homogeneous differential equation gives:

$$\frac{\mathrm{d}}{\mathrm{d}x}(u+v) + P(u+v) = Q$$

$$\left(\frac{\mathrm{d}u}{\mathrm{d}x} + Pu\right) + \left(\frac{\mathrm{d}v}{\mathrm{d}x} + Pv\right) = Q$$

Because the equation is linear we can have:

$$\frac{\mathrm{d}u}{\mathrm{d}x} + Pu = 0$$

with a solution which is the complementary function and:

$$\frac{\mathrm{d}v}{\mathrm{d}x} + Pv = Q$$

with a solution which is the particular integral.

For a particular integral:

Assume a form of particular integral which contains undetermined coefficients and determine the value of the coefficients by substitution in the differential equation.

For the particular integral, functions of the same form as the right-hand side of the non-homogeneous differential equation are tried. If the right-hand side of the non-homogeneous differential equation is a constant then $y = A$ is tried, the form $a + bx + cx^2 + \ldots$ then $y = A + Bx + Cx^2 + \ldots$ is tried, an exponential then $y = A\,\mathrm{e}^{kx}$ is tried, a sine or cosine then $y = A \sin \omega x + B \cos \omega x$ is tried, A, B, C and ω being constants.

Example

Solve the linear differential equation $\mathrm{d}y/\mathrm{d}x + y = 2x$.

The homogeneous form of the differential equation is $\mathrm{d}y/\mathrm{d}x + y = 0$. We can determine its solution by separation of the variables:

$$\frac{1}{y}\frac{\mathrm{d}y}{\mathrm{d}x} = -1$$

$$\int \frac{1}{y}\frac{\mathrm{d}y}{\mathrm{d}x}\,\mathrm{d}x = -\int 1\,\mathrm{d}x$$

$$\int \frac{1}{y}\,\mathrm{d}y = -\int 1\,\mathrm{d}x$$

$$\ln y = -x + A$$

$$y = \mathrm{e}^{-x+A} = \mathrm{e}^{A}\,\mathrm{e}^{-x} = C\,\mathrm{e}^{-x}$$

The above is thus the complementary function. Because the right-hand side of the non-homogeneous differential equation is $2x$, for the particular integral we try $y = A + Bx$. With this solution the non-homogeneous differential equation becomes:

$$B + A + Bx = 2x$$

Equating coefficients of x gives $B = 2$. Equating constant terms gives $A + B = 0$ and so $A = -2$. Thus the particular integral is $y = -2 + 2x$. The solution of the non-homogeneous equation is thus:

$$y = C\,\mathrm{e}^{-x} - 2 + 2x$$

Example

Solve the linear differential equation $\mathrm{d}y/\mathrm{d}x + 2y = x^2$.

The homogeneous form of the differential equation is $\mathrm{d}y/\mathrm{d}x + 2y = 0$. We can determine its solution by separation of the variables. However, such an equation is the form which gives an exponential decay. Thus we might guess that the solution is of the form $y = C\,\mathrm{e}^{-kx}$. If we try this solution then:

$$-kC\,\mathrm{e}^{-kx} + 2C\,\mathrm{e}^{-kx} = 0$$

Hence $k = 2$ and so the solution is $y = C\,\mathrm{e}^{-2x}$. Since the right-hand side of the non-homogeneous differential equation is x^2, for the particular integral we try $y = D + Ex + Fx^2$. With this as the solution the differential equation becomes:

$$E + 2Fx + 2(D + Ex + Fx^2) = x^2$$

Equating the coefficients of x^2 gives $2F = 1$ and so $F = \frac{1}{2}$. Equating the coefficients of x gives $2F + 2E = 0$ and so $E = -\frac{1}{2}$. Equating the constants gives $E + 2D = 0$ and so $D = \frac{1}{4}$. Thus the particular integral is $y = \frac{1}{4} - \frac{1}{2}x + \frac{1}{2}x^2$. The solution of the non-homogeneous equation is thus:

$$y = C\,\mathrm{e}^{-2x} + \tfrac{1}{4} - \tfrac{1}{2}x + \tfrac{1}{2}x^2$$

Revision

6 Determine particular integrals for the following differential equations:

(a) $\frac{dy}{dx} + y = 5$, (b) $\frac{dy}{dx} + y = 2x + 5$, (c) $\frac{dy}{dx} + y = x^2 + 2x + 5$,

(d) $\frac{dy}{dx} + y = e^{5x}$, (e) $\frac{dy}{dx} + y = \sin 5x$

7 Solve the following differential equations by determining the complementary functions and particular integrals:

(a) $\frac{dy}{dx} + 2y = 2x$, (b) $\frac{dy}{dx} + 2y = e^x$, (c) $\frac{dy}{dx} + 2y = 2x - 7$

Problems

1 Solve, by separation of the variables, the following differential equations:

(a) $\frac{dy}{dx} = 4 + 3x^2$, (b) $\frac{dy}{dx} = 2y^2$, (c) $\frac{dy}{dx} = 3x^2 e^{-y}$, (d) $\frac{dy}{dx} = \frac{1}{2y}$,

(e) $x^2 \frac{dy}{dx} - y + 1 = 0$, (f) $\frac{dy}{dx} = x^2 y$

2 When a steady voltage V is applied to a circuit consisting of a resistance R in series with inductance L, determine how the current i changes with time t if $L\,di/dt + Ri = V$ and $i = 0$ when $t = 0$.

3 Determine the solution of $dy/dx = 2 - y$ if $y = 1$ when $x = 0$.

4 A stone freely falls from rest and is subject to air resistance which is proportional to its velocity. Derive the differential equation describing its motion and hence determine how its velocity v varies with time t if $v = 0$ at $t = 0$. Take the acceleration due to gravity as 10 m/s^2.

5 For a belt drive, the difference in tension T between the slack and tight sides of the belt over a pulley is related to the angle of lap θ on the pulley by $dT/d\theta = \mu T$, where μ is the coefficient of friction. Solve the differential equation if $T = T_0$ when $\theta = 0°$.

6 A rectangular tank is initially full of water. The water, however, leaks out through a small hole in the base at a rate proportional to the square root of the depth of the water. If the tank is half empty after one hour, how long must elapse before it is completely empty?

7 A hot object cools at a rate proportional to the difference between its temperature and that of its surroundings. If it initially is at 75°C and

cooling at a rate of 2° per minute, what will be its temperature after 15 minutes if the surroundings are at a temperature of 15°C?

8 A sphere of ice melts so that its volume V changes at the rate given by $dV/dt = -4\pi k r^2$, where k is a constant and r is the radius at time t after it began to melt. Show that, if R is the initial radius, $r = R\,e^{-kt}$.

9 Solve the following differential equations by the use of integrating factors:

$$\text{(a) } x\frac{dy}{dx} - 4y = 4x^2, \text{ (b) } (x^2+1)\frac{dy}{dx} + 3xy = 6x,$$

$$\text{(c) } (x^2-1)\frac{dy}{dx} + 2y = x+1, \text{ (d) } \frac{dy}{dx} + y\cot x = x\cot x,$$

$$\text{(e) } x\frac{dy}{dx} + y = x, \text{ (f) } \cos x\frac{dy}{dx} + 2y\sin x = \cos^2 x$$

10 A tank initially contains 10 m³ of water in which 50 kg of salt has been dissolved. Brine containing 3 kg of salt per cubic metre is run in at a rate of 4 m³ per minute. The mixture is continually stirred to maintain a uniform brine concentration and 2 m³ per minute are taken from the tank. Show that the amount of salt x in the brine in the tank is described by:

$$\frac{dx}{dt} + \frac{2x}{10+2t} = 12$$

and hence determine the brine concentration after 30 minutes.

11 Solve the following differential equations by determining the complementary functions and particular integrals:

$$\text{(a) } \frac{dy}{dx} + 3y = 1, \text{ (b) } \frac{dy}{dx} - y = 4\,e^{3x}, \text{ (c) } \frac{dy}{dx} + y = x$$

30 Second-order differential equations

30.1 Introduction

The order of a differential equation is the order of the highest derivative it contains. Chapter 29 was concerned with first-order ordinary differential equations, this chapter extends the coverage to second-order ordinary differential equations. Second-order ordinary differential equations occur frequently in engineering and science. This chapter assumes Chapters 28 and 29, and hence Part 6.

As a simple illustration of second-order ordinary differential equations, consider the displacement y of a freely falling object in a vacuum as a function of time t. Such an object falls with the acceleration due to gravity g and is described by the second-order differential equation (see Section 28.2.1):

$$\text{acceleration} = \frac{d^2y}{dt^2} = g \qquad [1]$$

Another example is the displacement y of an object when freely oscillating with simple harmonic motion when there is damping, this being described by the second-order differential equation:

$$m\frac{d^2y}{dt^2} + c\frac{dy}{dt} + ky = 0 \qquad [2]$$

If the oscillating object is not left freely to oscillate when some external force is applied, say $F \sin \omega t$, then we have:

$$m\frac{d^2y}{dt^2} + c\frac{dy}{dt} + ky = F\sin\omega t \qquad [3]$$

With a series electrical circuit containing resistance R, capacitance C and inductance L, the potential difference v_C across the capacitor when it is allowed to discharge is described by the second-order differential equation (see Section 28.2.2):

$$LC\frac{d^2v_C}{dt^2} + RC\frac{dv_C}{dt^2} + v_C = 0 \qquad [4]$$

If such a circuit has a voltage V applied to it we have:

$$LC\frac{d^2v_C}{dt^2} + RC\frac{dv_C}{dt^2} + v_C = V \qquad [5]$$

In general, a linear second-order differential equation has the form:

$$a_2\frac{d^2y}{dx^2}+a_1\frac{dy}{dx}+a_0y=b$$

where a_2, a_1, a_0 and b are functions of x, b often being termed the forcing function.

30.2 Arbitrary constants

Consider an object falling freely with the acceleration due to gravity g. If we take g to be 10 m/s^2 then the differential equation (equation [1]) becomes:

$$\frac{d^2y}{dt^2}=10$$

If we integrate both sides of the equation with respect to t we have:

$$\int\frac{d^2y}{dt^2}\,dt=\int 10\,dt$$

$$\frac{dy}{dt}=10t+A$$

where A is the constant of integration. If we now integrate this equation with respect to t:

$$\int\frac{dy}{dt}\,dt=\int(10t+A)\,dt$$

$$y=5t^2+At+B$$

where B is the constant arising from this integration. Thus the above general solution for the second-order differential equation has two arbitrary constants. With all second-order differential equations there will be two arbitrary constants because two integrations are needed to obtain the solution.

With an n-order differential equation there will be n arbitrary constants because n integrations are required to obtain the solution.

Because there are two arbitrary constants with a second-order differential equation, two sets of values are needed to determine them. This is generally done by specifying two initial conditions: the value of the solution and the value of the derivative at a single point. Thus we might have the initial conditions that $y = 20$ at $t = 0$ and $dy/dt = 0$ at $x = 0$.

Example

If the general solution to the differential equation:

If we again substitute these values into the right-hand sides of equations [1] and [2]:

$$x = -\tfrac{1}{2} \times 2.03 + 2 = 0.99, \quad y = -\tfrac{1}{4} \times 1.25 + 2.25 = 1.94$$

If we again substitute these values into the right-hand sides of equations [1] and [2]:

$$x = -\tfrac{1}{2} \times 1.94 + 2.25 = 1.03, \quad y = -\tfrac{1}{4} \times 0.99 + 2.25 = 2.00$$

If we consider the estimates obtained we have:

for x: 0, 1.5, 0.875, 1.25, 0.99, 1.03
for y: 1, 2.25, 1.5, 2.03, 1.94, 2.00

with x apparently converging after six substitutions (termed *iterations*) to the value 1.0 and y to the value 2.0. Figure 20.1 illustrates this convergence.

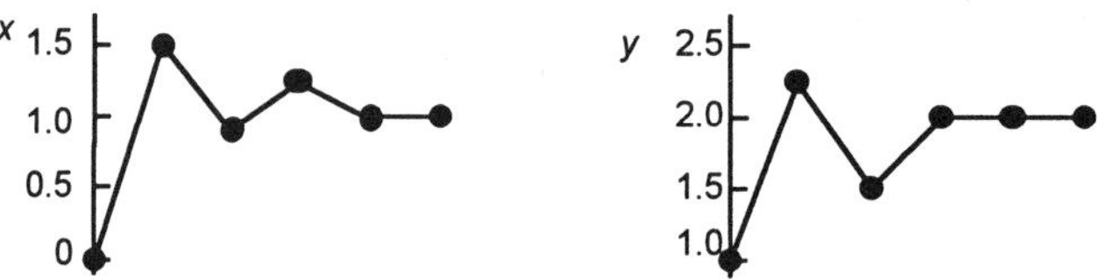

Figure 20.1 *x and y values converging to 1 and 2*

This method of iteration is known as *Jacobi iteration*. In general the steps to be followed are:

1 Rewrite the first equation so that it expresses x_1 in terms of the other variables, the second equation so that it expresses x_2 in terms of the other variables, the third equation so that it expresses x_3 in terms of the other variables, and so on for the n equations with n variables. Convergence is helped if, wherever possible, the equations with the largest coefficient of the variable are used.

2 If approximate values of the variables are known or can be guessed, substitute them into the right-hand sides of the equations and obtain revised estimates of the variables. If no initial approximate values of the variables are available then generally the variables are tried with each having the value 0.

3 Take the revised estimates obtained in 2 and substitute them into the right-hand side of the equations and obtain further revised estimates.

4 Repeat 3 until the differences in successive estimates are small enough to indicate that the required accuracy of result has been obtained.

The results do not always converge and this is discussed in Section 19.4.

Example

Use the Jacobi iteration method to solve the following set of simultaneous equations:

$$4x + 2y + 2z = 8,\ x + 2y - z = 5,\ x + y + 2z = 1$$

Rewriting the equations:

$$x = -0.5y - 0.5z + 2 \qquad [3]$$

$$y = -0.5x + 0.5z + 2.5 \qquad [4]$$

$$z = -0.5x - 0.5y + 0.5 \qquad [5]$$

As we have no information about approximate values we will take the initial estimates of x, y and z to be each 0. Putting these estimates in equations [3], [4] and [5] gives:

$$x = 2,\ y = 2.5,\ z = 0.5$$

Putting these estimates in equations [3], [4], [5] gives:

$$x = -1.25 - 0.25 + 2 = 0.5$$

$$y = -1 + 0.25 + 2.5 = 1.75$$

$$z = -1 - 1.25 + 0.5 = -1.75$$

Putting these estimates in equations [3], [4], [5] gives:

$$x = -0.875 + 0.875 + 2 = 2$$

$$y = -0.25 - 0.875 + 2.5 = 1.375$$

$$z = -0.25 - 0.875 + 0.5 = -0.625$$

Putting these estimates into equations [3], [4], [5] gives:

$$x = -0.687\,5 + 0.312\,5 + 2 = 1.625$$

$$y = -1 - 0.312\,5 + 2.5 = 1.187\,5$$

$$z = -1 - 0.687\,5 + 0.5 = -1.187\,5$$

Putting these estimates into equations [3], [4], [5] gives:

$$x = -0.593\ 75 + 0.593\ 75 + 2 = 2$$

$$y = -0.812\ 5 - 0.593\ 75 + 2.5 = 1.093\ 75$$

$$z = -0.812\ 5 - 0.593\ 75 + 0.5 = -0.906\ 25$$

The results appear to be converging to $x = 2$, $y = 1$ and $z = -1$.

Revision

1 Use the Jacobi iteration method to solve the following sets of simultaneous equations:

(a) $2x + y = 7$, $x - 2y = 1$, (b) $2x + y = 1$, $x - 3y = 5$,

(c) $4x + y = 8$, $x + 3y = 7$, (d) $2x + y = 5$, $x + 4y = -1$,

(e) $4x + y + z = 9$, $x + 3y - z = 4$, $x - 2y + 4z = 9$,

(f) $3x + 2y - z = -1$, $x + 4y + z = 5$, $x + y + 2z = -2$

20.3 Gauss-Seidel iteration

With Jacobi iteration a set of values of the variables is calculated at each iteration and then used to obtain the next set. However, convergence can be improved if after the first equation has been solved the new value is used immediately in the calculation of the second variable instead of completing the calculation of the variables using the values from the previous set. This method of iteration is called *Gauss-Seidel iteration* and the procedure is:

1 Rewrite the first equation so that it expresses x_1 in terms of the other variables, the second equation so that it expresses x_2 in terms of the other variables, the third equation so that it expresses x_3 in terms of the other variables, and so on for the n equations with n variables. Convergence is helped if, wherever possible, the equations with the largest coefficient of the variable are used.

2 If approximate values of the variables are known or can be guessed, substitute them into the right-hand sides of the first equation and obtain a revised estimate of the first variable. If no initial approximate values of the variables are available then generally the variables are tried with each having the value 0.

3 Take the revised estimate obtained in 2 and substitute it, with the estimates of the other variables, into the right-hand side of the second of the equations and obtain a revised estimate of the second variable.

4 Using this newer estimated value of the second variable and the older estimated values of the other variables, use the third equation to obtain a revised estimate of the third variable.

5 Keep on using the latest values of the variables to obtain new values for each of the variables.

6 Repeat 5 until the differences in successive estimates are small enough to indicate that the required accuracy of result has been obtained.

The following calculation illustrates this method with the pair of equations [1] and [2] used to illustrate the previous section:

$$2x + y = 4 \text{ and } x + 4y = 9$$

We rewrite these equations so that the first equation expresses x in terms of the other variable and the second equation expresses y in terms of the other variable:

$$x = -\tfrac{1}{2}y + 2 \qquad [6]$$

$$y = -\tfrac{1}{4}x + 2.25 \qquad [7]$$

We then make a guess as to the possible value of y. Suppose we take $y = 1$. We now substitute these values into the right-hand side of equation [6] to give newer estimates of x:

$$x = -\tfrac{1}{2} \times 1 + 2 = 1.5$$

We use this estimate in equation [7] to give a newer estimate of y:

$$y = -\tfrac{1}{4} \times 1.5 + 2.25 = 1.875$$

We use this estimate in equation [6] to give a newer estimate of x:

$$x = -\tfrac{1}{2} \times 1.875 + 2 = 1.0625$$

We use this estimate in equation [7] to give a newer estimate of y:

$$y = -\tfrac{1}{4} \times 1.0625 + 2.25 = 1.984375$$

We use this estimate in equation [6] to give a newer estimate of x:

$$x = -\tfrac{1}{2} \times 1.984375 + 2 = 1.0078125$$

We use this estimate in equation [6] to give a newer estimate of y:

$$y = -\tfrac{1}{4} \times 1.0078125 + 2.25 = 1.9980469$$

The results have converged to $x = 1$ and $y = 2$ with less steps required than with the Jacobi iteration.

Example

Use the Jacobi iteration method to solve the following set of simultaneous equations (these being the ones used [3], [4], [5] in the example for the Jacobi iteration):

$$4x + 2y + 2z = 8, \quad x + 2y - z = 5, \quad x + y + 2z = 1$$

Rewriting the equations:

$$x = -0.5y - 0.5z + 2 \qquad [8]$$

$$y = -0.5x + 0.5z + 2.5 \qquad [9]$$

$$z = -0.5x - 0.5y + 0.5 \qquad [10]$$

Taking the initial estimates of y and z to be 0 and putting these into equation [8]:

$$x = 2$$

Putting this estimate of x with the initial estimate of z in equation [9]:

$$y = -1 + 0 + 2.5 = 1.5$$

Putting these latest values of x and y in equation [10]:

$$z = -1 - 0.75 + 0.5 = -1.25$$

Putting these latest values of y and z in equation [8]:

$$x = -0.75 + 0.625 + 2 = 1.875$$

Putting these latest values of x and z in equation [9]:

$$y = -0.937\,5 - 0.625 + 2.5 = 0.937\,5$$

Putting these latest values of x and y in equation [10]:

$$z = -0.937\,5 - 0.468\,75 + 0.5 = -0.906\,25$$

Putting these latest values of y and z in equation [8]:

$$x = -0.468\,75 + 0.453\,125 + 2 = 1.984\,375$$

Putting these latest values of x and z in equation [9]:

$$y = -0.992\ 187\ 5 - 0.453\ 125 + 2.5 = 1.054\ 687\ 5$$

Putting these latest values of x and y in equation [10]:

$$z = -0.992\ 187\ 5 - 0.527\ 343\ 7 + 0.5 = -1.019\ 531\ 2$$

The results appear to be converging to $x = 2$, $y = 1$ and $z = -1$.

Revision

2 Repeat revision problem 1 using Gauss-Seidel iteration.

19.4 Convergence

In some cases, the Jacobi and Gauss-Seidel methods do not give results which converge, regardless of the number of iterations carried out. What are the conditions under which convergence shall occur?

As part of the sequence of instructions given with the iteration methods the statement is included in step 1, where variables are expressed each in terms of the other variables, that the equation with the largest coefficient of a variable should be used. In the coefficient matrix for the set of equations, this makes the diagonal entry larger in magnitude than the magnitudes of the other coefficients either side. Such a system is said to be *diagonally dominant*. For example, with:

$$4x + 2y + 2z = 8,\quad x + 2y - z = 5,\quad x + y + 2z = 1$$

if we use the first equation for x, the second for y and the third for z we have the coefficients matrix:

$$\begin{bmatrix} 4 & 2 & 2 \\ 1 & 2 & -1 \\ 1 & 1 & 2 \end{bmatrix}$$

The diagonal 4, 2, 2 is dominant. If, however, we had used the equations in a different order we would not have had diagonal dominance, e.g.

$$x + 2y - z = 5,\quad x + y + 2z = 1,\ 4x + 2y + 2z = 8$$

$$\begin{bmatrix} 1 & 2 & -1 \\ 1 & 1 & 2 \\ 4 & 2 & 2 \end{bmatrix}$$

Diagonal dominance guarantees that convergence will occur. If the system is not diagonally dominant then convergence may or may not occur.

Problems

1 Use (i) Jacobi iteration, (ii) Gauss-Seidel iteration to solve the following sets of simultaneous equations, rearranging if necessary the sequence of equations in a set to ensure convergence:

(a) $2x + y = 5$, $x + 3y = 5$, (b) $x + 4y = 11$, $2x + y = 1$,

(c) $3x + y = 11$, $x + 2y = 7$, (d) $x - 2y = 13$, $2x + y = 6$,

(e) $3x - y + 2z = 9$, $x + 3y + z = 10$, $x + 2y - 3z = 4$,

(f) $x + 3y + 2z = 7$, $3x + y + 2z = 1$, $x + y + 3z = 4$,

(g) $5x + y - 2z = 10$, $2x - 3y + z = -6$, $x + y + 2z = 10$

(h) $x + y + 4z = 2$, $5x + y + z = 3$, $x + 3y + 2z = -7$

21 Application: Circuit analysis

21.1 Introduction

This chapter illustrates the application of linear equation techniques, matrices and determinants in the solution of electrical circuit problems involving Kirchhoff's laws. Chapters 16, 17 and 18 are assumed.

Kirchhoff's laws can be stated as:

1 *Kirchhoff's current law*
At any instant of time, the algebraic sum of the currents at any junction in a circuit is zero, i.e. the current entering a junction equals the current leaving it.

2 *Kirchhoff's voltage law*
At any instant of time, around any closed path in a circuit, the algebraic sum of the voltage drops is zero.

In this chapter two systematic approaches to electrical network analysis based on Kirchhoff's laws are considered, node voltage analysis and mesh current analysis.

21.2 Node voltage analysis

Node voltage analysis is the term used to describe a technique of systematically applying Kirchhoff's current law to electrical networks. The term *node* is used for the points at which two or more circuit elements are connected. Thus for the circuit shown in Figure 21.1 we have nodes as the points labelled 0, 1 and 2. Note that 0 refers to the junction of the two vertical branches at their bases, for convenience the diagram shows them joined by a conductor with no circuit element present. Each branch of the circuit is connected between two nodes. If we take node 0 as the reference node then we express all the voltages at the other nodes with respect to this reference node. The voltage thus specified for any non-reference node is thus the node to reference node voltage. The voltage across any circuit branch is the difference in voltages of its two end nodes.

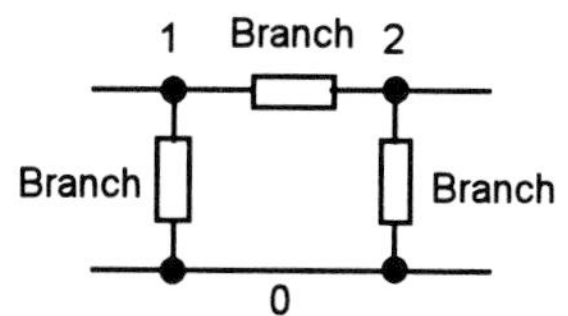

Figure 21.1 *Nodes*

The node voltage analysis procedure can be stated as:

1 Label the nodes on a circuit diagram, selecting one as the reference node.

2 Write Kirchhoff's current law equations for all non-reference nodes in the circuit. If there are n nodes then there will be $n - 1$ equations.

3 Eliminate the currents by substitution using the branch current-voltage relations.

4 Solve the resulting $(n - 1)$ simultaneous equations.

To illustrate the application of the above, consider the circuit shown in Figure 12.2. The nodes have been labelled 0, 1 and 3 with node 0 being taken as the reference node. For node 1 we must have the algebraic sum of the currents equal to zero and thus:

i – current leaving node through R_1
– current leaving node through $R_2 = 0$ [1]

The current leaving node 1 through R_2 is the current entering node 2. Thus for node 2 we can write:

current through R_2 – current leaving node through R_3
– current leaving node through $R_4 = 0$ [2]

The current leaving node 2 through R_4 is the current entering node 3. Thus for node 3 we can write:

current through R_4 – current leaving node through R_5
– current leaving node through $R_6 = 0$ [3]

We will label the voltage at node 1 with reference to the reference node as v_1, that at node 2 with reference to the reference node as v_2 and that at node 3 with reference to the reference node as v_3. Then the current passing from node 1 to node 0 is v_1/R_1, node 1 to node 2 is $(v_1 - v_2)/R_2$, from node 2 to node 0 is v_3/R_3, from node 2 to node 3 is $(v_2 - v_3)/R_4$, from node 3 to node 0 is v_3/R_5 and v_3/R_6. Substituting these values in equations [1], [2] and [3] gives:

$$i - \frac{v_1}{R_1} - \frac{v_1 - v_2}{R_2} = 0 \qquad [4]$$

$$\frac{v_1 - v_2}{R_2} - \frac{v_2}{R_3} - \frac{v_2 - v_3}{R_4} = 0 \qquad [5]$$

$$\frac{v_2 - v_3}{R_4} - \frac{v_3}{R_5} - \frac{v_3}{R_6} = 0 \qquad [6]$$

We now have to solve the three simultaneous equations [4], [5] and [6].

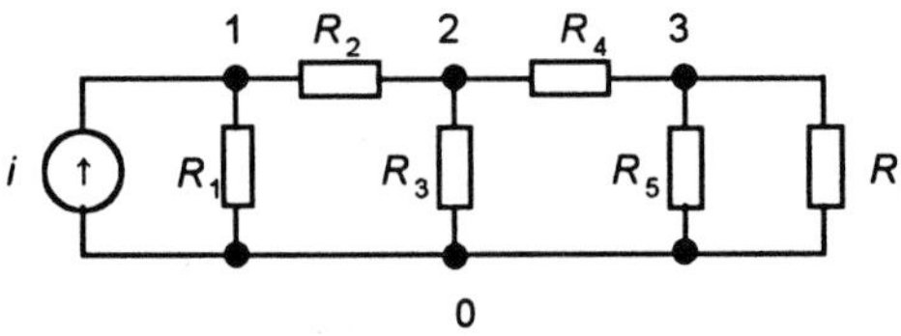

Figure 21.2 *Circuit for node analysis*

It is often more convenient to use conductances G rather than resistances; G being $1/R$. The equations then become:

$$i - G_1v_1 - G_2(v_1 - v_2) = 0 \qquad [7]$$

$$G_2(v_1 - v_2) - G_3v_2 - G_4(v_2 - v_3) = 0 \qquad [8]$$

$$G_4(v_2 - v_3) - G_5v_3 - G_6v_3 = 0 \qquad [9]$$

Rearranging the terms in the above equations gives:

$$(G_1 + G_2)v_1 - G_2v_2 = i \qquad [10]$$

$$G_2v_1 - (G_2 + G_3 + G_4)v_2 + G_4v_3 = 0 \qquad [11]$$

$$G_4v_2 - (G_4 + G_5 + G_6)v_3 = 0 \qquad [12]$$

In matrix form we can write these equations as:

$$\begin{bmatrix} G_1+G_2 & -G_2 & 0 \\ G_2 & -(G_1+G_2+G_3) & G_4 \\ 0 & G_4 & -(G_4+G_5+G_6) \end{bmatrix} \begin{bmatrix} v_1 \\ v_2 \\ v_3 \end{bmatrix} = \begin{bmatrix} i \\ 0 \\ 0 \end{bmatrix}$$

[13]

Such equations can be solved in a number of ways, e.g. by Gaussian elimination, by Cramer's rule or by using the inverse matrix.

In the above analysis of Figure 21.2 there was just a current source. Consider how the equation for a node is altered if a branch contains a voltage source, Figure 21.3 showing the relevant part of such a circuit. With node 0 as the reference node and v_1 the voltage at node 1:

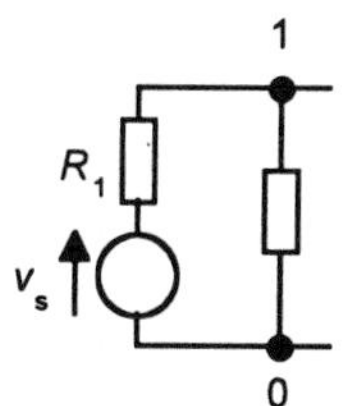

Figure 21.3 *Voltage sourc*

$$v_1 = R_1i_1 + v_s$$

where i_1 is the current through resistor R_1 and so the current entering node 1 from that branch of the circuit. We can thus consider there to be a current source of:

$$i_1 = \frac{v_1 - v_s}{R_1} \qquad [13]$$

and proceed as before. Nodes that are connected together by just a voltage source have the voltage difference between the nodes fixed by this voltage.

Example

Determine the node voltages at each of the nodes of the circuit given in Figure 21.4.

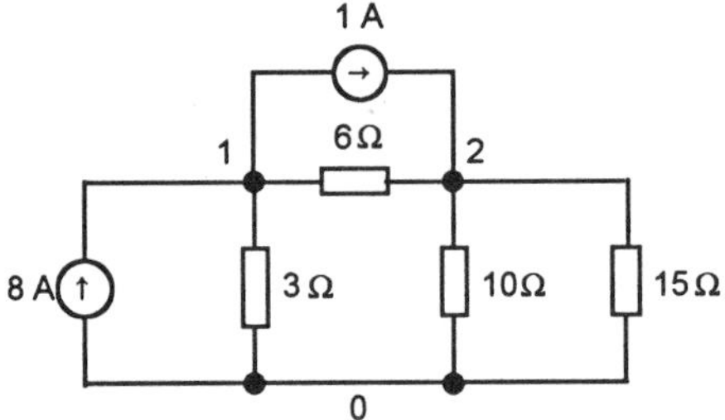

Figure 21.4 *Example*

With node 0 as the reference node, the voltage at node 1 as v_1 and that at node 2 as v_2, we have:

$$8 - 1 - \tfrac{1}{3}v_1 - \tfrac{1}{6}(v_1 - v_2) = 0$$

$$1 + \tfrac{1}{6}(v_1 - v_2) - \tfrac{1}{10}v_2 - \tfrac{1}{15}v_2 = 0$$

These can be rearranged to give:

$$\left(\tfrac{1}{3} + \tfrac{1}{6}\right)v_1 - \tfrac{1}{6}v_2 = 7$$

$$\tfrac{1}{6}v_1 - \left(\tfrac{1}{6} + \tfrac{1}{10} + \tfrac{1}{15}\right)v_2 = -1$$

The matrix equation for these two simultaneous equations is thus:

$$\begin{bmatrix} \frac{1}{2} & -\frac{1}{6} \\ \frac{1}{6} & -\frac{1}{3} \end{bmatrix}\begin{bmatrix} v_1 \\ v_2 \end{bmatrix} = \begin{bmatrix} 7 \\ -1 \end{bmatrix}$$

This equation can be solved in a number of ways; consider here the use of Cramer's rule (see Section 18.2 and equation [10]):

$$\frac{v_1}{\begin{vmatrix} 7 & -\frac{1}{6} \\ -1 & -\frac{1}{3} \end{vmatrix}} = \frac{v_2}{\begin{vmatrix} \frac{1}{2} & 7 \\ \frac{1}{6} & -1 \end{vmatrix}} = \frac{1}{\begin{vmatrix} \frac{1}{2} & -\frac{1}{6} \\ \frac{1}{6} & -\frac{1}{3} \end{vmatrix}}$$

$$\frac{v_1}{-\frac{15}{6}} = \frac{v_2}{-\frac{10}{6}} = \frac{1}{-\frac{5}{36}}$$

Hence v_1 = 18 V and v_2 = 12 V.

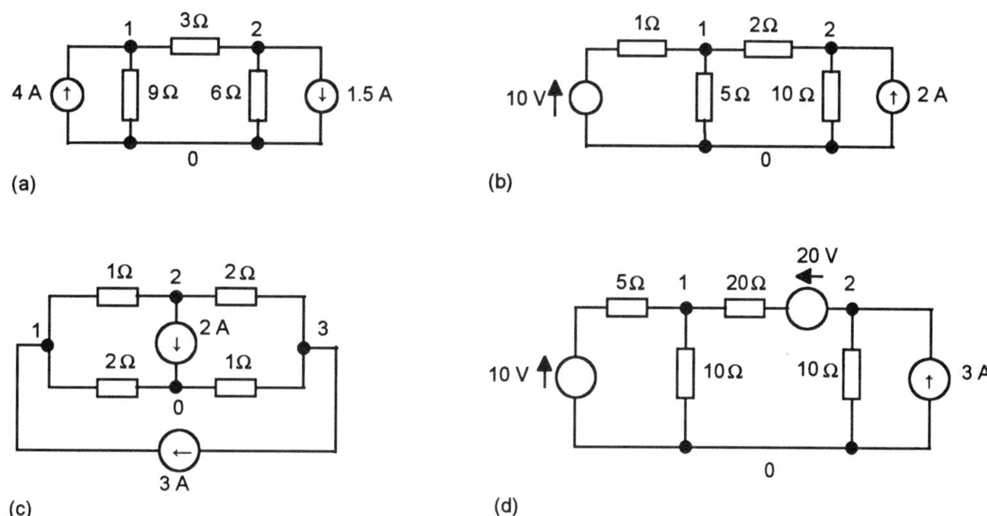

Figure 21.5 *Revision problem 1*

Revision

1 Determine the voltages at each of the nodes in the circuits given in Figure 21.5, in each case node 0 being taken as the reference node.

21.3 Mesh current analysis

Mesh current analysis is a systematic way of applying Kirchhoff's voltage law to planar networks. Planar networks are those that can be drawn so the entire network is on a plane surface without any two of its branches crossing each other. A *mesh* is a set of circuit branches that form a loop but is only used for those loops that do not enclose any other branch of the circuit. Circulating round each mesh we identify a current. The same direction of circulation must be chosen for each mesh current, a usual convention being to have all the currents circulating in a clockwise direction. As a consequence, every branch current is then either equal to the mesh current or the difference between two mesh currents. Figure 21.6 shows a simple circuit having two meshes with mesh currents i_1 and i_2. The current through R_1 is then the mesh current i_1 while the current through R_2 is the difference between the mesh currents i_1 and i_2.

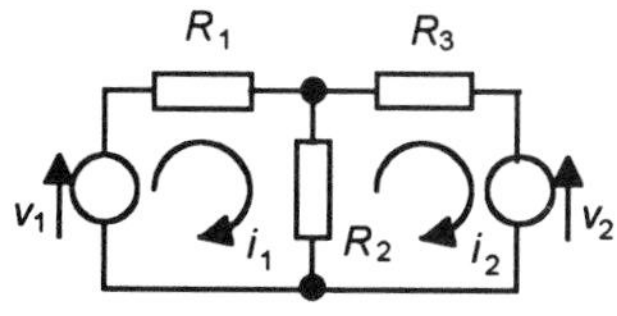

Figure 21.6 *Meshes*

If we apply Kirchhoff's voltage law to each mesh in Figure 21.6 then we have (note that the direction of the potential difference across an element is in the opposite direction to the current through it):

$$-v_1 + R_1 i_1 + R_2(i_1 - i_2) = 0 \qquad [14]$$

$$v_2 + R_3 i_2 + R_2(i_2 - i_1) = 0 \qquad [15]$$

If we group the current terms in the equation then equations [14] and [15] become:

$$(R_1 + R_2)i_1 - R_2 i_2 = v_1 \qquad [16]$$

$$-R_2 i_1 + (R_2 + R_3)i_2 = -v_2 \qquad [17]$$

This pair of simultaneous equations can be written in matrix form as:

$$\begin{bmatrix} R_1 + R_2 & -R_2 \\ -R_2 & R_2 + R_3 \end{bmatrix} \begin{bmatrix} i_1 \\ i_2 \end{bmatrix} = \begin{bmatrix} v_1 \\ -v_2 \end{bmatrix} \qquad [18]$$

The general procedure for carrying out mesh analysis of a network involves the following steps:

1 Identify the meshes on the circuit diagram and assign a mesh current circulating in a clockwise direction to each.

2 Apply Kirchhoff's voltage law to each mesh, writing the voltages across circuit elements in terms of the branch currents by the use of Ohm's law. If there is a current source it can either be just considered as a mesh current or converted to an equivalent voltage source (see the example after the next).

3 Solve the resulting set of simultaneous equations.

Example

Determine by mesh analysis the currents in each of the branches of the circuit shown in Figure 21.7.

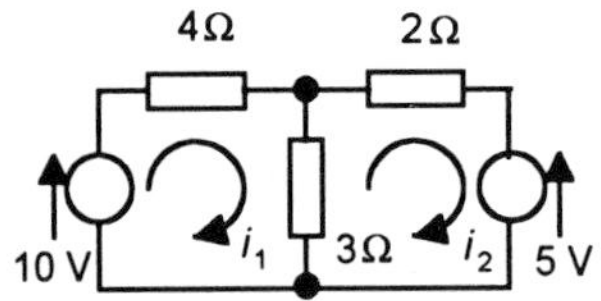

Figure 21.7 *Example*

The mesh currents have been indicated on the figure. Hence applying Kirchhoff's voltage law to each mesh gives:

$$-10 + 4i_1 + 3(i_1 - i_2) = 0$$

$$5 + 2i_2 + 3(i_2 - i_1) = 0$$

These equations can be rewritten as:

$$7i_1 - 3i_2 = 10$$

$$-3i_1 + 5i_2 = -5$$

Writing these equations in matrix form gives:

$$\begin{bmatrix} 7 & -3 \\ -3 & 5 \end{bmatrix} \begin{bmatrix} i_1 \\ i_2 \end{bmatrix} = \begin{bmatrix} 10 \\ -5 \end{bmatrix}$$

Using Cramer's rule (see Section 18.2 and equation [10]):

$$\frac{i_1}{\begin{vmatrix} 10 & -3 \\ -5 & 5 \end{vmatrix}} = \frac{i_2}{\begin{vmatrix} 7 & 10 \\ -3 & -5 \end{vmatrix}} = \frac{1}{\begin{vmatrix} 7 & -3 \\ -3 & 5 \end{vmatrix}}$$

$$\frac{i_1}{35} = \frac{i_2}{-5} = \frac{1}{26}$$

Hence $i_1 = 35/26 = 1.35$ A and $i_2 = -5/26 = -0.19$ A. The current through the 4 Ω resistor is i_1 and so 1.35 A, through the 2 Ω resistor is i_2 and so −0.19 A and through the 3 Ω resistor is $(i_1 - i_2) = 1.35 + 0.19 = 1.54$ A in the direction of the i_1 current.

Alternatively we might have obtained the currents from the equations by using Gaussian elimination (see Section 16.3), the augmented matrix being:

$$\begin{bmatrix} 7 & -3 & 10 \\ -3 & 5 & -5 \end{bmatrix}$$

Adding 3/7th of the first row to the second row gives:

$$\begin{bmatrix} 7 & -3 & 10 \\ 0 & \frac{26}{7} & -\frac{5}{7} \end{bmatrix}$$

Adding 21/26th of the second row to first row gives:

$$\begin{bmatrix} 7 & 0 & \frac{245}{26} \\ 0 & \frac{26}{7} & -\frac{5}{7} \end{bmatrix}$$

Hence, as before, $i_1 = 35/26$ A and $i_2 = -5/26$ A.

Example

Using mesh analysis, determine the currents through each of the resistors in the circuit shown in Figure 21.8.

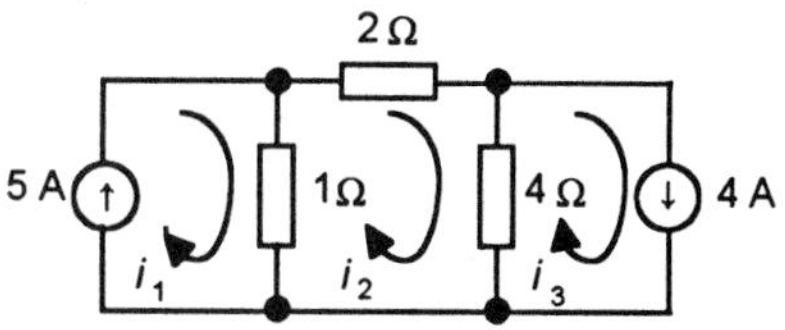

Figure 21.8 *Example*

The mesh current i_1 must be 5 A and the mesh current i_3 4 A. Thus the only mesh we have to consider is the one with current i_2. For this mesh:

$$2i_2 + 4(i_2 - 4) + 1(i_2 - 5) = 0$$

Hence $i_2 = 3$ A.

Alternatively we could have solved this by replacing the current sources by equivalent voltage sources. The 5 A current has 1 Ω in parallel with it and so produces a potential difference of 5 V across it. We can thus replace the current source by a voltage source of 5 V in series with the 1 Ω resistor. Likewise we can replace the 4 A source by a voltage source of 16 V in series with the 4 Ω resistor. The circuit now looks like Figure 21.9. Hence:

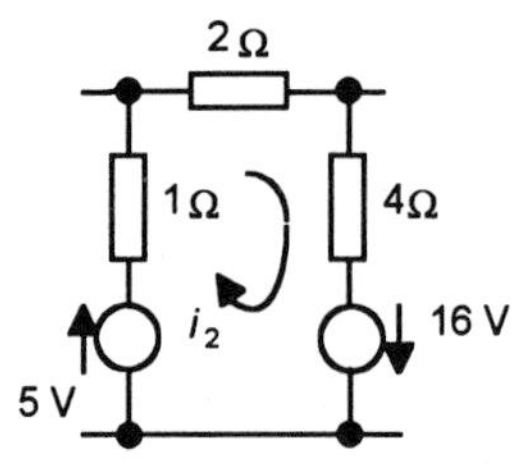

Figure 21.9 *Example*

$$-5 + 1i_2 + 2i_2 + 4i_2 - 16 = 0$$

and, as before, $i_2 = 3$ A.

Revision

2 Using mesh analysis, determine the currents in the branches of the circuits shown in Figure 21.10.

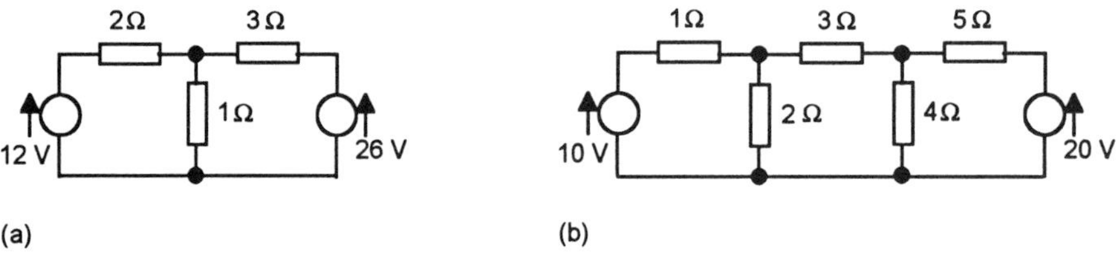

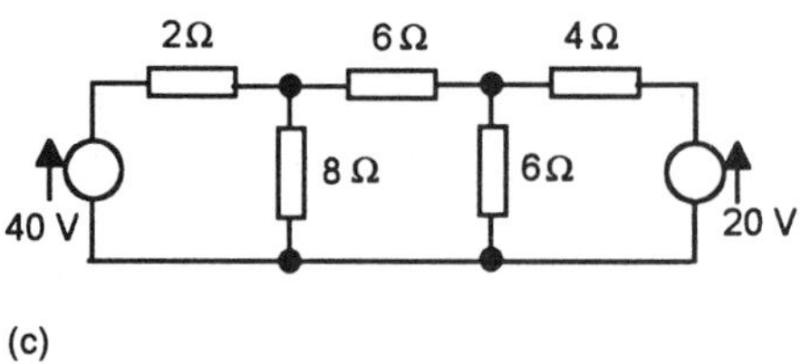

Figure 21.10 *Revision problem 2*

Problems

1 Using the node voltage technique, determine the voltages at each of the nodes in the circuits given in Figure 21.11, in each case node 0 being taken as the reference node.

2 Using mesh analysis, determine the currents in each of the branches of the circuits shown in Figure 21.12.

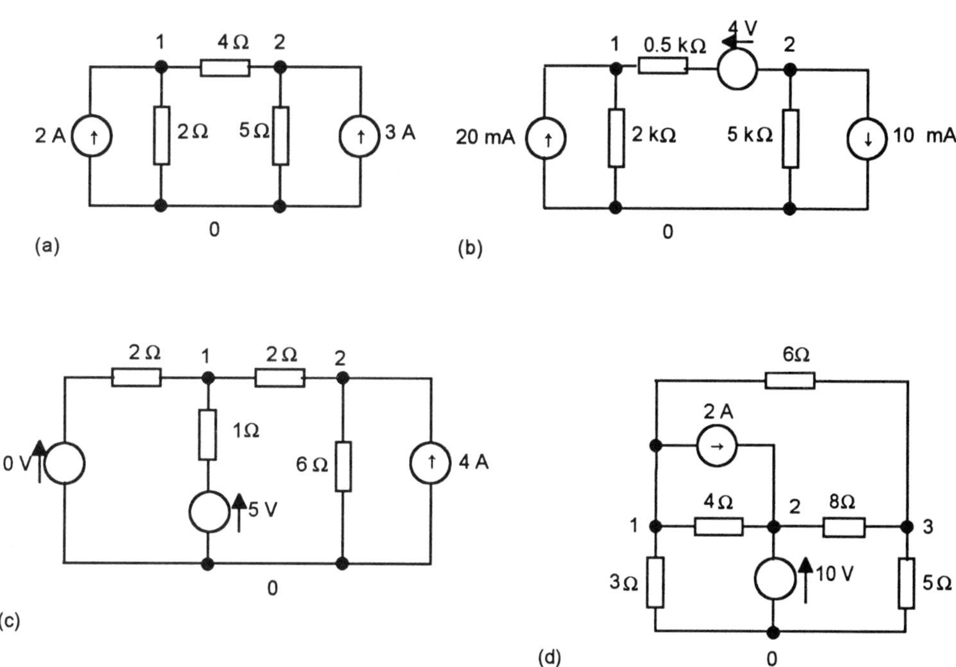

Figure 21.11 *Problem 1*

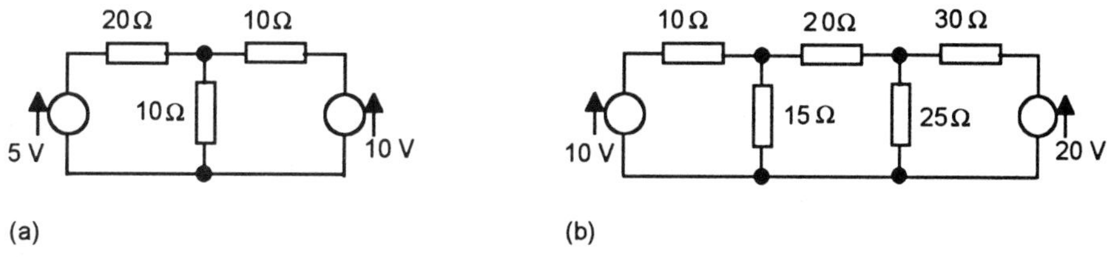

Figure 21.12 *Problem 2*

Part 6
Differentiation and integration

The aims of this part are to enable the reader to:

- Determine the derivatives of powers, exponentials, logarithmic, trigonometric and hyperbolic functions.
- Use sums of functions, products of functions, quotients of functions, the chain rule, implicit differentiation, inverse functions, parametric differentiation and logarithmic functions techniques in the determination of derivatives.
- Apply differentiation to the solution of maxima and minima problems.
- Use numerical differentiation techniques.
- Develop partial derivatives and solve problems involving them.
- Determine the integrals of powers, exponentials, trigonometric and hyperbolic functions.
- Use sums of functions, substitution, integration by parts and partial fractions in the determination of integrals.
- Use numerical integration techniques.
- Apply calculus methods to the solutions of problems involving areas, volumes, moments, means and r.m.s. values.

This part assumes Part 1 Functions. While differentiation and integration are introduced from first principles, it is assumed that this is in the nature of revision and the greater emphasis is on developing the techniques of differentiation and integration. Chapter 22 is the basic chapter on differentiation on which Chapters 23, 24 and 25 are based. Chapters 26 and 27, the application chapter, assume Chapter 25.

22 Differentiation

22.1 Introduction

Differentiation is a mathematical technique for determining the rate at which functions change. It might be used to determine the value of the tangent to a curve, the rate at which the current in a circuit is changing at some instant of time, or perhaps the rate at which the displacement of a particle is changing, i.e. the velocity, at some instant of time, or maximum and minimum values of functions.

This chapter reviews the basic principle of differentiation and introduces the techniques of differentiating functions. It assumes a knowledge of functions (Chapters 1, 2 and 3), the concept of a limit (Chapter 5), for Section 22.10 vectors (Part 3) and algebraic dexterity.

22.2 Graphical approach to differentiation

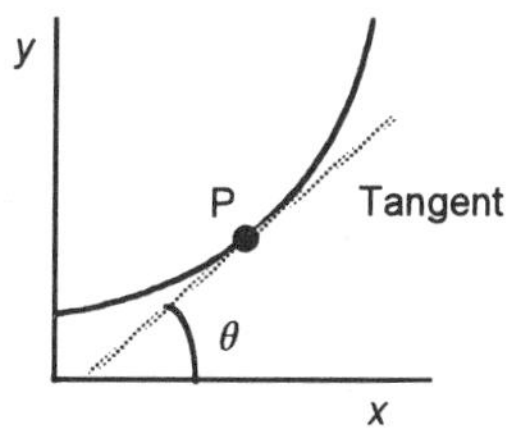

Figure 22.1 *Tangent*

Consider the problem of finding the slope of a graph of a function at some point P on its graph, for example that shown in Figure 22.1. The *slope* or *gradient* of a curve at some point is defined as being the quantity tan θ, and the line giving this angle is termed the *tangent*. The tangent line is the rate at which the function is changing at the point. We can obtain an approximation to the tangent by drawing a line through the point of tangency P and a second point Q on the curve, as in Figure 22.2. If Q is some distance from P then the slope of PQ will not be close to that of the tangent at P. However, if we take a succession of points Q closer and closer to P (Figure 22.2) then the slope of PQ will become more and more equal to that of the tangent as Q approaches P. We can thus define the slope of the tangent line as being the *limiting value* of the slope of PQ as Q approaches P. Algebraically we can define this as: if P has the co-ordinates (x, y) and Q $(x + \delta x, y + \delta y)$, where δx and δy are the changes in x and y of Q with respect to P (the symbol δ or Δ is used for 'an increment in'), then since the slope of PQ is $\delta y/\delta x$ the slope of the tangent is the limiting value of $\delta y/\delta x$ as δx tends to zero, i.e.

$$\text{slope of tangent} = \lim_{\delta x \to 0} \frac{\delta y}{\delta x} \qquad [1]$$

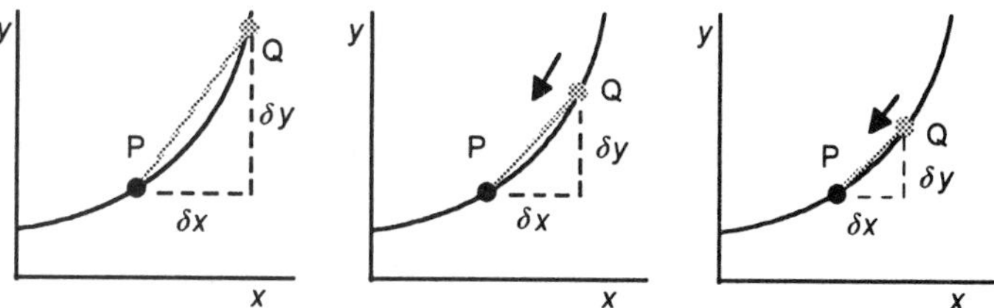

Figure 22.2 *As Q tends to P then the slope tends to the tangent value*

If the graph is of $y = f(x)$ and P is the point (x, y), i.e. $(x, f(x))$, with Q the point $(x + dx, y + \delta y)$, i.e. $(x + \delta x, f(x + \delta x))$, then equation [1] becomes:

$$\text{slope of tangent} = \lim_{\delta x \to 0} \frac{f(x+\delta x) - f(x)}{\delta x} \qquad [2]$$

Example

Determine the slope of the tangent to the curve $y = f(x) = x^2$ at the point on the curve (1, 1).

Using equation [2]:

$$\text{slope of tangent} = \lim_{\delta x \to 0} \frac{(x+\delta x)^2 - x^2}{\delta x}$$

$$= \lim_{\delta x \to 0} \frac{x^2 + 2x(\delta x) + (\delta x)^2 - x^2}{\delta x}$$

$$= \lim_{\delta x \to 0} (2x + \delta x)$$

As δx tends to zero so the slope of the tangent tends to the value $2x$. With $x = 1$ then the slope of the tangent is 2.

Revision

1 Determine the slopes of the tangents at the point (1, 1) for graphs of the functions: (a) $f(x) = 2x + 4$, (b) $f(x) = 3x^2 + 2x$.

22.2.1 Derivatives

A special notation is used to indicate the result obtained for $\delta y/\delta x$ as $\delta x \to 0$. The result is termed the *derivative of y with respect to x*, or the *derivative of f(x)* and denoted by:

$$\frac{dy}{dx}, \text{ or } \frac{df(x)}{dx}, \text{ or } \frac{d}{dx}f(x), \text{ or } f'(x) \text{ or } \dot{f}(x) \text{ or } D(x)$$

Note that d is used for the limiting value and not δ. The d symbols cannot be cancelled from the numerator and denominator, you have to consider that d/dx, f' or $\dot{f}$ or D is used to indicate that what follows should be differentiated. It is termed an *operator* in that it indicates that one function is to be operated on to produce another one. Thus we can write:

$$\frac{d}{dx}f(x) = \lim_{\delta x \to 0} \frac{f(x+\delta x) - f(x)}{\delta x} \qquad [3]$$

The process of determining the derivative is called *differentiation*.

Example

Determine the derivative of y with respect to x of $y = x^3 + 2x$.

For the above function we have $f(x) = x^3 + 2x$ and thus:

$$f(x + \delta x) = (x + \delta x)^3 + 2(x + \delta x)$$

$$= x^3 + 3x(\delta x)^2 + 3x^2(\delta x) + (\delta x)^3 + 2x + 2(\delta x)$$

Hence:

$$f(x + \delta x) - f(x) = 3x(\delta x)^2 + 3x^2(\delta x) + (\delta x)^3 + 2(\delta x)$$

and so, using equation [3]:

$$\frac{\mathrm{d}}{\mathrm{d}x}f(x) = \lim_{\delta x \to 0} \frac{f(x+\delta x) - f(x)}{\delta x} = \frac{3x(\delta x)^2 + 3x^2(\delta x) + (\delta x)^3 + 2(\delta x)}{\delta x}$$

$$= \lim_{\delta x \to 0} \left[3x(\delta x) + 3x^2 + (\delta x)^2 + 2\right] = 3x^2 + 2$$

Revision

2 Determine the derivative of y with respect to x of the following functions: (a) $y = x^2 + 3x$, (b) $y = 2x^2 + 5$.

22.2.2 Existence of derivatives

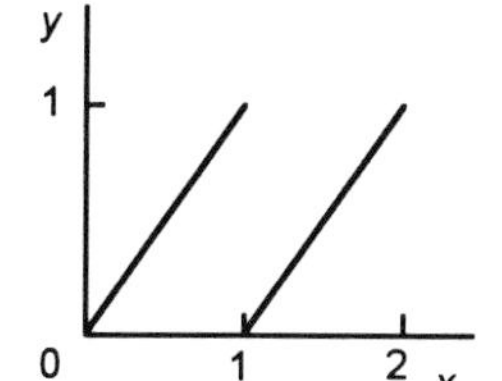

Figure 22.3 *Discontinuous function*

Since we can interpret the derivative as representing the slope of the tangent to a graph of a function at a particular point, this means with a continuous function, i.e. a function which has values of y which smoothly and continuously change as x changes for all values of x, that we have derivatives for all values of x. However, with a discontinuous graph there will be some values of x for which we can have no derivative. For example, with the graph shown in Figure 22.3, there is no derivative for $x = 1$.

22.3 Common derivatives

The following illustrates how some commonly used functions can be differentiated, a table of commonly used derivatives being given the end of the section.

22.3.1 Derivative of a constant

Consider $f(x) = c$, where c is a constant. Then $f(x + \delta x) = c$ and so, since $f(x + \delta x) - f(x) = c - c = 0$, equation [3] gives:

$$\frac{\mathrm{d}}{\mathrm{d}x}c = \lim_{\delta x \to 0} \frac{0}{\delta x} = 0$$

The derivative of a constant is zero.

22.3.2 Derivative of x^n

If we have a function $f(x) = x^n$ then $f(x + \delta x) = (x + \delta x)^n$. Expanding this by the use of the binomial theorem gives:

$$(x+\delta x)^n = x^n + nx^{n-1}(\delta x) + \frac{n(n-1)}{2!}x^{n-2}(\delta x)^2 + \ldots + (\delta x)^n$$

Hence, using equation [3]:

$$\frac{\mathrm{d}}{\mathrm{d}x}f(x) = \lim_{\delta x \to 0}\left[\frac{nx^{n-1}(\delta x) + \frac{n(n-1)}{2!}x^{n-2}(\delta x)^2 + \ldots + (\delta x)^n}{\delta x}\right]$$

and so:

$$\frac{\mathrm{d}}{\mathrm{d}x}(x^n) = nx^{n-1} \qquad [4]$$

This relationship applies for positive, negative and fractional values of n.

Example

Determine the derivative of the functions (a) $y = x^{3/2}$, (b) $y = x^{-4}$.

(a) Using equation [4]:

$$\frac{\mathrm{d}y}{\mathrm{d}x} = \tfrac{3}{2}x^{\frac{3}{2}-1} = \tfrac{3}{2}x^{\frac{1}{2}}$$

(b) Using equation [4]:

$$\frac{\mathrm{d}y}{\mathrm{d}x} = -4x^{-4-1} = -4x^{-5}$$

Revision

3 Determine the derivatives of: (a) $y = x^4$, (b) $y = x^{5/3}$, (c) $y = x^{-2}$, (d) $y = \sqrt{x}$.

22.3.3 Derivatives of trigonometric functions

Consider the function $f(x) = \sin x$. Then $f(x + \delta x) - f(x) = \sin(x + \delta x) - \sin x$.

Since (see Section 3.4.1) sin P – sin Q = 2 cos ½(P + Q) sin ½(P – Q), then the derivative can be written as:

$$\frac{\mathrm{d}}{\mathrm{d}x}(\sin x) = \lim_{\delta x \to 1} \frac{2\cos(x + \frac{1}{2}\delta x)\sin\frac{1}{2}\delta x}{\delta x}$$

For small angles sin ½ $\delta x \approx$ ½ δx and so (2 sin ½ δx)/δx tends to 1 as $\delta x \to 0$. The cosine term will tend to cos x as $\delta x \to 0$. Thus:

$$\frac{\mathrm{d}}{\mathrm{d}x}(\sin x) = \cos x$$

If we had considered the function sin ax then we would have obtained:

$$\frac{\mathrm{d}}{\mathrm{d}x}(\sin ax) = a\cos ax \qquad [5]$$

In a similar manner we can derive:

$$\frac{\mathrm{d}}{\mathrm{d}x}(\cos ax) = -a\sin ax \qquad [6]$$

The derivatives of tan x, cosec ax, sec ax and cot ax can be derived using the quotient rule (see Section 22.4.3).

$$\frac{\mathrm{d}}{\mathrm{d}x}(\tan ax) = a\sec^2 ax \qquad [7]$$

$$\frac{\mathrm{d}}{\mathrm{d}x}(\operatorname{cosec} ax) = -a\operatorname{cosec} ax\cot ax \qquad [8]$$

$$\frac{\mathrm{d}}{\mathrm{d}x}(\sec ax) = a\sec ax\tan ax \qquad [9]$$

$$\frac{\mathrm{d}}{\mathrm{d}x}(\cot ax) = -a\operatorname{cosec}^2 ax \qquad [10]$$

Example

Determine the derivatives of (a) sin $2x$, (b) cos $3x$.

(a) Using equation [5]:

$$\frac{\mathrm{d}}{\mathrm{d}x}(\sin 2x) = 2\cos 2x$$

(b) Using equation [6]:

$$\frac{\mathrm{d}}{\mathrm{d}x}(\cos 3x) = -3\sin 3x$$

Revision

4 Determine the derivatives of (a) sin $5x$, (b) cos $2x$, (c) tan $3x$.

5 The variation of the current i in an electrical circuit with time t is given by the equation $i = \sin 314t$. Derive an equation for the rate of change of current with time.

6 The variation of the displacement x of an oscillating object with time t is given by $y = \cos 50t$. Derive an equation indicating how the velocity, i.e. the rate of change of displacement with time, varies with time.

22.3.4 Derivatives of exponential functions

Consider the function $y = e^x$. Since $f(x + \delta x) = e^{x+\delta x} = e^x\, e^{\delta x}$ then:

$$f(x + dx) - f(x) = e^x\, e^{\delta x} - e^x = e^x(e^{\delta x} - 1)$$

Exponentials can be expressed as a series (see Table 5.1) and thus:

$$e^{\delta x} = 1 + \delta x + \frac{(\delta x)^2}{2!} + \frac{(\delta x)^3}{3!} + \dots$$

$$\frac{d}{dx}e^x = \lim_{\delta x \to 0} \frac{e^x\left(\delta x + \frac{(\delta x)^2}{2!} + \frac{(\delta x)^3}{3!} + \dots\right)}{\delta x}$$

$$= \lim_{\delta x \to 0} e^x\left(1 + \frac{\delta x}{2!} + \frac{(\delta x)^2}{3!} + \dots\right)$$

Hence:

$$\frac{d}{dx}e^x = e^x \qquad [12]$$

If we had the function $y = e^{ax}$ then:

$$\frac{d}{dx}e^{ax} = a\, e^{ax} \qquad [13]$$

Example

Determine the derivative of $y = e^{2x}$.

Using equation [13]:

$$\frac{d}{dx}e^{2x} = 2\, e^{2x}$$

Table 22.1 *Derivatives*

	Function	Derivative		Function	Derivative
1	constant	0	8	$\sec(at + b)$	$a \sec(at + b) \tan(at + b)$
2	t^n	nt^{n-1}	9	$\operatorname{cosec}(at + b)$	$-a \operatorname{cosec}(at + b) \cot(at + b)$
3	e^{at}	$a e^{at}$	10	$\cot(at + b)$	$-a \operatorname{cosec}^2(at + b)$
4	$\ln t$	$1/t$	11	$\sinh(at + b)$	$a \cosh(at + b)$
5	$\sin(at + b)$	$a \cos(at + b)$	12	$\cosh(at + b)$	$a \sinh(at + b)$
6	$\cos(at + b)$	$-a \sin(at + b)$	13	$\tanh(at + b)$	$a \operatorname{sech}^2(at + b)$
7	$\tan(at + b)$	$a \sec^2(at + b)$			

Revision

7 Determine the derivatives of (a) $y = e^{4x}$, (b) $y = e^{-3x}$.

22.3.5 Common derivatives

Table 22.1 lists derivatives of commonly encountered functions.

22.4 Differentiation rules

In this section the basic rules are developed for the differentiation of constant multiples, sums, products and quotients of functions and the chain rule for functions of functions.

22.4.1 Constant multiples

Consider a multiple of some function, e.g. $cf(x)$ where c is a constant. Using equation [2]:

$$\frac{d}{dx}cf(x) = \lim_{\delta x \to 0} \frac{cf(x+\delta x) - cf(x)}{\delta x} = c \lim_{\delta x \to 0} \frac{f(x+\delta x) - f(x)}{\delta x} = c\frac{d}{dx}f(x) \qquad [14]$$

The derivative of some function multiplied by a constant is the same as the constant multiplying the derivative of the function.

Example

Determine the derivatives of (a) $4x^2$, (b) $2 \sin 3x$, (c) $y = \dfrac{1}{3\sqrt{x}}$.

(a) Using equations [4] and [14]:

$$\frac{d}{dx}4x^2 = 4 \times 2x = 8x$$

(b) Using equations [5] and [14]:

$$\frac{\mathrm{d}}{\mathrm{d}x}(2\sin 3x) = 2 \times 3 \sin 3x = 6\sin 3x$$

(c) Using equations [4] and [14]:

$$\frac{\mathrm{d}}{\mathrm{d}x}\left(\frac{1}{3\sqrt{x}}\right) = \frac{\mathrm{d}}{\mathrm{d}x}\left(\tfrac{1}{3}x^{-1/2}\right) = \tfrac{1}{3} \times \left(-\tfrac{1}{2}\right)x^{-3/2} = -\tfrac{1}{6}x^{-3/2}$$

Revision

8 Determine the derivatives of (a) $y = 2x^{3/2}$, (b) $y = 4\cos 2x$, (c) $y = 5\,\mathrm{e}^{2x}$.

22.4.2 Sums of functions

Consider a function which can be considered to be a sum of a number of other functions, e.g. $y = f(x) + g(x)$. Using equation [2]:

$$\frac{\mathrm{d}}{\mathrm{d}x}[f(x) + g(x)] = \lim_{\delta x \to 0} \frac{[f(x+\delta x) + g(x+\delta x)] - [f(x) + g(x)]}{\delta x}$$

$$= \lim_{\delta x \to 0}\left[\frac{f(x+\delta x) - f(x)}{\delta x} + \frac{g(x+\delta x) - g(x)}{\delta x}\right]$$

$$= \lim_{\delta x \to 0}\left[\frac{f(x+\delta x) - f(x)}{\delta x}\right] + \lim_{\delta x \to 0}\left[\frac{g(x+\delta x) - g(x)}{\delta x}\right]$$

$$= \frac{\mathrm{d}}{\mathrm{d}x}f(x) + \frac{\mathrm{d}}{\mathrm{d}x}g(x) \qquad [15]$$

The derivative of the sum of two differentiable functions is the sum of their derivatives.

Consider the differentiation of the hyperbolic function $y = \sinh x$. This function (see Section 3.5) can be written as $\frac{1}{2}(\mathrm{e}^{x} - \mathrm{e}^{-x})$. Thus:

$$\frac{\mathrm{d}}{\mathrm{d}x}(\sinh x) = \frac{\mathrm{d}}{\mathrm{d}x}\tfrac{1}{2}(\mathrm{e}^{x} - \mathrm{e}^{-x}) = \tfrac{1}{2}(\mathrm{e}^{x} + \mathrm{e}^{x}) = \cosh x$$

In a similar way we can differentiate sinh *ax* and cosh *ax*, obtaining:

$$\frac{\mathrm{d}}{\mathrm{d}x}(\sinh ax) = a\cosh ax \qquad [16]$$

$$\frac{\mathrm{d}}{\mathrm{d}x}(\cosh ax) = a\sinh ax \qquad [17]$$

The hyperbolic function tanh *ax* can be differentiated using the quotient rule (see Section 22.4.3).

Example

Determine the derivatives of:

(a) $y = 2x^3 + x^2$, (b) $y = \sin x + \cos 2x$, (c) $y = e^{4x} + x$

(a) Using the rule given in equation [15]:

$$\frac{dy}{dx} = 6x^2 + 2x$$

(b) Using the rule given in equation [15]:

$$\frac{dy}{dx} = \cos x - 2\sin 2x$$

(c) Using the rule given in equation [15]:

$$\frac{dy}{dx} = 4\,e^{4x} + 1$$

Revision

9 Determine the derivatives of:

(a) $y = x^2 + 2x + 1$, (b) $y = x^2 + \sin 3x$, (c) $y = \pi x^2 + 2\pi x$,

(d) $y = 2\sin x + 3\sin 3x$, (e) $y = 2\,e^{-3x} + 3\,e^{-5x}$, (f) $y = 2(e^x + 1)$

22.4.3 The product rule

Consider a function $y = f(x)g(x)$ which is the product of two other differentiable functions, e.g. $y = x \sin x$. Using equation [2]:

$$\frac{d}{dx}[f(x)g(x)] = \lim_{\delta x \to 0} \frac{f(x+\delta x)g(x+\delta x) - f(x)g(x)}{\delta x}$$

We can simplify this by adding and subtracting the same quantity to the numerator, namely $f(x + \delta x)g(x)$:

$$\frac{d}{dx}[f(x)g(x)]$$

$$= \lim_{\delta x \to 0} \frac{f(x+\delta x)g(x+\delta x) + f(x+\delta x)g(x) - f(x+\delta x)g(x) - f(x)g(x)}{\delta x}$$

$$= \lim_{\delta x \to 0} \left[f(x+\delta x)\frac{g(x+\delta x) - g(x)}{\delta x} + g(x)\frac{f(x+\delta x) - f(x)}{\delta x} \right]$$

$$= f(x)\frac{d}{dx}g(x) + g(x)\frac{d}{dx}f(x) \qquad [18]$$

This is often written in terms of u and v, where $u = f(x)$ and $v = g(x)$:

$$\frac{\mathrm{d}}{\mathrm{d}x}uv = u\frac{\mathrm{d}v}{\mathrm{d}x} + v\frac{\mathrm{d}u}{\mathrm{d}x}$$

The derivative of the product of two differentiable functions is the sum of the first function multiplied by the derivative of the second function and the second function multiplied by the derivative of the first function.

Example

Determine the derivatives of the following functions:

(a) $y = x \sin x$, (b) $y = x^2 e^{3x}$, (c) $y = (2 + x)^2$, (d) $y = x\, e^x \sin x$

(a) Using the rule given in equation [18]:

$$\frac{\mathrm{d}y}{\mathrm{d}x} = x\frac{\mathrm{d}}{\mathrm{d}x}(\sin x) + \sin x\frac{\mathrm{d}}{\mathrm{d}x}x = x\cos x + \sin x$$

(b) Using the rule given in equation [18]:

$$\frac{\mathrm{d}y}{\mathrm{d}x} = x^2\frac{\mathrm{d}}{\mathrm{d}x}e^{3x} + e^{3x}\frac{\mathrm{d}}{\mathrm{d}x}x^2 = 3x^2\, e^{3x} + 2x\, e^{3x}$$

(c) This can be written as $(2 + x)(2 + x)$ and so, using equation [18]:

$$\frac{\mathrm{d}y}{\mathrm{d}x} = (2+x)\frac{\mathrm{d}}{\mathrm{d}x}(2+x) + (2+x)\frac{\mathrm{d}}{\mathrm{d}x}(2+x) = 2(2+x)$$

(d) This product has three terms and to use equation [18] we have to carry out the differentiation in two stages. Thus if we first consider $x\, e^x$ as one term and the $\sin x$ as the other term:

$$\frac{\mathrm{d}y}{\mathrm{d}x} = x\, e^x\frac{\mathrm{d}}{\mathrm{d}x}\sin x + \sin x\frac{\mathrm{d}}{\mathrm{d}x}(x\, e^x) = x\, e^x\cos x + \sin x\frac{\mathrm{d}}{\mathrm{d}x}(x\, e^x)$$

We can then use equation [18] to evaluate the derivative of $x\, e^x$.

$$\frac{\mathrm{d}}{\mathrm{d}x}(x\, e^x) = x\frac{\mathrm{d}}{\mathrm{d}x}e^x + e^x\frac{\mathrm{d}}{\mathrm{d}x}x = x\, e^x + e^x$$

Hence:

$$\frac{\mathrm{d}y}{\mathrm{d}x} = x\, e^x\cos x + x\, e^x + e^x$$

Revision

10 Determine the derivatives of:

(a) $y = 2x(1 + x^2)$, (b) $y = (x + 1)(x + 3)$, (c) $y = x\,\mathrm{e}^{4x}$,

(d) $y = \sin x \cos 2x$, (e) $y = \mathrm{e}^{-2x}(x + 5)$, (f) $y = \mathrm{e}^x \sin x \cos x$

22.4.4 The quotient rule

Consider obtaining the derivative of a function which is the quotient of two other functions, e.g. $f(x)/g(x)$. Using equation [2]:

$$\frac{\mathrm{d}}{\mathrm{d}x}\left(\frac{f(x)}{g(x)}\right) = \lim_{\delta x \to 0} \frac{\dfrac{f(x+\delta x)}{g(x+\delta x)} - \dfrac{f(x)}{g(x)}}{\delta x}$$

$$= \lim_{\delta x \to 0} \frac{g(x)f(x+\delta x) - f(x)g(x+\delta x)}{\delta x g(x)g(x+\delta x)}$$

Adding and subtracting $f(x)g(x)$ to the numerator enables the above equation to be simplified:

$$\frac{\mathrm{d}}{\mathrm{d}x}\left(\frac{f(x)}{g(x)}\right) = \lim_{\delta x \to 0} \frac{g(x)f(x+\delta x) + f(x)g(x) - f(x)g(x) - f(x)g(x+\delta x)}{\delta x g(x)g(x+\delta x)}$$

$$= \frac{\lim\limits_{\delta x \to 0} \dfrac{g(x)\left[f(x+\delta x) - f(x)\right]}{\delta x} - \lim\limits_{\delta x \to 0} \dfrac{f(x)\left[g(x+\delta x) - g(x)\right]}{\delta x}}{\lim\limits_{\delta x \to 0}\left[g(x)g(x+\delta x)\right]}$$

$$= \frac{g(x)\dfrac{\mathrm{d}}{\mathrm{d}x}f(x) - f(x)\dfrac{\mathrm{d}}{\mathrm{d}x}g(x)}{\left[g(x)\right]^2} \qquad [19]$$

This is often written in terms of u and v, where $u = f(x)$ and $v = g(x)$:

$$\frac{\mathrm{d}}{\mathrm{d}x}\left(\frac{u}{v}\right) = \frac{v\dfrac{\mathrm{d}u}{\mathrm{d}x} - u\dfrac{\mathrm{d}v}{\mathrm{d}x}}{v^2}$$

Note that if we have just the reciprocal of some function, i.e. $1/g(x)$, then we have $f(x) = 1$ and so equation [19] gives:

$$\frac{\mathrm{d}}{\mathrm{d}x}\left[\frac{1}{g(x)}\right] = -\frac{1}{\left[g(x)\right]^2}\frac{\mathrm{d}}{\mathrm{d}x}g(x) \qquad [20]$$

Equation [19] can be used to determine the derivative of $\tan x$, since $\tan x = \sin x/\cos x$. Thus $f(x) = \sin x$ and $g(x) = \cos x$. Hence:

$$\frac{d}{dx}(\tan x) = \frac{\cos x \frac{d}{dx}\sin x - \sin x \frac{d}{dx}\cos x}{\cos^2 x} = \frac{\cos^2 x + \sin^2 x}{\cos^2 x} = \frac{1}{\cos^2 x} \qquad [21]$$

Likewise, equation [19] can be used to determine the derivative of tanh x.

$$\frac{d}{dx}(\tanh x) = \frac{d}{dx}\left(\frac{\sinh x}{\cosh x}\right) = \frac{\cosh x \cosh x - \sinh x \sinh x}{\cosh^2 x} = \frac{1}{\cosh^2 x} \qquad [22]$$

Example

Determine the derivative of $y = (2x^2 + 5x)/(x + 3)$.

Using equation [19] with $f(x) = 2x^2 + 5x$ and $g(x) = x + 3$:

$$\frac{dy}{dx} = \frac{(x+3)(4x+5) - (2x^2 + 5x)(1)}{(x+3)^2} = \frac{2x^2 + 12x + 15}{(x+3)^2}$$

Example

Determine the derivative of $y = x\ e^x/\cos x$.

This example requires the use of both the quotient and product rules for differentiation. Using equation [19] with $f(x) = x\ e^x$ and $g(x) = \cos x$:

$$\frac{dy}{dx} = \frac{\cos x \frac{d}{dx}(x\ e^x) - x\ e^x(-\sin x)}{\cos^2 x}$$

Now using equation [18] to obtain the derivative for the product $x\ e^x$:

$$\frac{dy}{dx} = \frac{\cos x(x\ e^x + e^x) - x\ e^x(-\sin x)}{\cos^2 x} = \frac{e^x(x\cos x + \cos x + x\sin x)}{\cos^2 x}$$

Revision

11 Determine the derivatives of the following functions:

(a) $y = \frac{2x}{x+3}$, (b) $y = \frac{x^2 - 4}{x+2}$, (c) $y = \frac{1-x}{(1+x)^2}$, (d) $y = \frac{1-\cos x}{\sin x}$

22.4.5 The chain rule

Suppose we have $y = \cos x^4$ and, in order to differentiate it, write it in the form $y = \cos u$ and $u = x^4$. We can then obtain dy/du and du/dx, but how from them do we obtain dy/dx?

Consider the function $y = f(u)$ where $u = g(x)$ and the obtaining of the derivative of $y = f(g(x))$. For $u = g(x)$ a small increase of δx in the value of x

causes a corresponding small increase of δu in the value of u. But $y = f(u)$ and so the small increase δu causes a correspondingly small increase of δy in the value of y. We can write, since the δu terms cancel:

$$\frac{\delta y}{\delta x} = \frac{\delta y}{\delta u} \times \frac{\delta u}{\delta x}$$

Thus:

$$\frac{dy}{dx} = \lim_{\delta x \to 0} \left(\frac{\delta y}{\delta u} \times \frac{\delta u}{\delta x} \right) = \lim_{\delta x \to 0} \left(\frac{\delta y}{\delta u} \right) \lim_{\delta x \to 0} \left(\frac{\delta u}{\delta x} \right)$$

and so:

$$\frac{dy}{dx} = \frac{dy}{du} \times \frac{du}{dx} \qquad [23]$$

This is known as the *function of a function rule* or the *chain rule*.

The chain rule can be used to determine the derivative a function such as $y = \sin x^n$, for n being positive or negative or fractional. Let $u = x^n$ and so consequently $y = \sin u$. Then $du/dx = nx^{n-1}$ and $dy/du = \cos u$. Hence, using the chain rule (equation [23]) we have $dy/dx = \cos u \times nx^{n-1} = nx^{n-1} \cos x^n$.

Another application of the chain rule is to determine the derivatives of functions of the form $y = (ax + b)^n$, $y = e^{ax+b}$, $y = \sin(ax + b)$, etc. With such functions we let $u = ax + b$ and so then we have, for the three examples, $y = u^n$, $y = e^u$, $y = \sin u$. Then we have $du/dx = a$ and $dy/du = nx^{n-1}$, $dy/du = e^u$, $dy/du = \cos u$. Using the chain rule we then obtain $dy/dx = anu^{n-1}$, $dy/dx = a\,e^{ax+b}$ and $dy/dx = a \cos u$. Thus, for the three examples, we have:

$$\frac{dy}{dx} = an(ax+b)^{n-1}, \quad \frac{dy}{dx} = a\,e^{ax+b}, \quad \frac{dy}{dx} = a\cos(ax+b), \quad \text{etc.}$$

Example

Determine the derivative of $y = (2x - 5)^4$.

Let $u = 2x - 5$ and so $y = u^4$. Then $du/dx = 2$ and $dy/du = 4u^3$ and so, using equation [23]:

$$\frac{dy}{dx} = \frac{dy}{du} \times \frac{du}{dx} = 4u^3 \times 2 = 8u^3 = 8(2x-5)^3$$

Example

Determine the derivative of $y = \sin x^3$.

Let $u = x^3$ and so $y = \sin u$. Then $dy/du = 3x^2$ and $dy/du = \cos u$ and so, using equation [23]:

$$\frac{dy}{dx} = \frac{dy}{du} \times \frac{du}{dx} = 3x^2 \times \cos u = 3x^2 \cos x^3$$

Example

Determine the derivative of:

$$y = \sqrt{\frac{x^2}{x^2+1}}$$

Let $u = x^2/(x^2 + 1)$ and so $y = u^{1/2}$. Then, using the quotient rule:

$$\frac{du}{dx} = \frac{(x^2+1)2x - x^2(2x)}{(x^2+1)^2}$$

Using the chain rule:

$$\frac{dy}{dx} = \tfrac{1}{2}u^{-1/2} \times \frac{(x^2+1)2x - x^2(2x)}{(x^2+1)^2}$$

$$= \tfrac{1}{2}\left(\frac{x^2}{x^2+1}\right)^{-1/2} \frac{(x^2+1)2x - x^2(2x)}{(x^2+1)^2}$$

$$= \frac{x^{1/2}}{(x^2+1)^{3/2}} = \sqrt{\frac{x}{(x^2+1)^3}}$$

Revision

12 Determine the derivatives of the following functions:

(a) $y = (3x + 2)^5$, (b) $y = \sin(5x + 2)$, (c) $y = (x^2 + 1)^3$,

(d) $y = (x^2 - 6x + 1)^5$, (e) $y = \sqrt{6 - x^3}$, (f) $y = \dfrac{1}{(6x^3 - x)^4}$

(g) $y = \left(\dfrac{x-3}{x+3}\right)^4$, (h) $y = x\sin x^2$, (i) $y = (x + \sin x)^2$

22.5 Inverse functions

If we have a function y which is a continuous function of x then the derivative, i.e. the slope of the tangent to a graph of y plotted against x, is dy/dx. However, if we have x as a continuous function of y then the derivative, i.e. the slope of the tangent to a graph of x plotted against y, is dx/dy. How are these derivatives related? We might, for example, have $y = x^2$ and so $dy/dx = 2x$. For the inverse function $x = \sqrt{y}$ and $dx/dy = \tfrac{1}{2}y^{-1/2}$.

If we have a function $y = f(x)$ then we can write for the inverse $x = g(y)$. Thus $x = g\{f(x)\}$. Differentiating both sides of this equation with respect to x, using the chain rule for the right-hand side, gives:

$$1 = \frac{dx}{dy} \times \frac{dy}{dx}$$

Hence, with $y = f(x)$, the derivatives of the inverse function can be derived by using:

$$\frac{dy}{dx} = \frac{1}{\frac{dx}{dy}} \qquad [24]$$

For example, for $y = x^2$ we have $dy/dx = 2x$. For the inverse function $x = \sqrt{y}$ and $dx/dy = \frac{1}{2}y^{-1/2}$. Then $dy/dx = 1/(\frac{1}{2}y^{-1/2}) = 2\sqrt{y} = 2x$.

Example

Determine dy/dx for the function described by the equation $x = y^2 + 2y$.

It is easier to obtain dx/dy from the equation and thus the problem is tackled by doing that operation first. Thus $dx/dy = 2y + 2$. Then, using equation [24]:

$$\frac{dy}{dx} = \frac{1}{2y+2}$$

Revision

13 Determine dy/dx for the functions described by the following equations:

(a) $y^2 + 6y + 3 = x$, (b) $y^3 + 2y = x$

22.5.1 Logarithmic functions

Consider the function $y = \ln x$. We can write this as $x = e^y$. Differentiating x with respect to y (equation [12]) gives:

$$\frac{dx}{dy} = e^y = x$$

Hence, using equation [24]:

$$\frac{d}{dx}(\ln x) = \frac{1}{x} \qquad [25]$$

Note that since x must be positive for $\ln x$ to have any meaning (see Section 3.3), equation [25] only applies for positive values of x.

Example

Determine the derivative of $y = e^{-x} \ln x$.

Using the product rule then:

$$\frac{dy}{dx} = e^{-x}\left(\frac{1}{x}\right) - \ln x\,(e^{-x}) = e^{-x}\left(\frac{1}{x} - \ln x\right)$$

Revision

14 Determine the derivatives of (a) $y = \ln 2x$, (b) $y = \ln(1 + e^{2x})$.

22.5.2 Inverse trigonometric functions

Consider the inverse function $y = \sin^{-1} x$. This function is restricted to the range $-\pi/2$ to $+\pi/2$ (see Section 3.4.5). We can rewrite the function as $x = \sin y$ and so $dx/dy = \cos y$. Hence the derivative of the inverse sine is:

$$\frac{d}{dx}(\sin^{-1}x) = \frac{1}{\cos y}$$

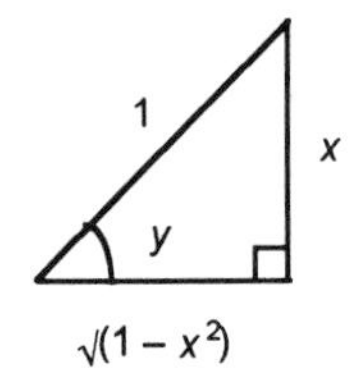

Figure 22.4 $y = sin^{-1} x$

To express $\cos y$ in terms of x we draw a right-angled triangle (Figure 22.4) with sides such that it has y as the angle with a sine of x. Using the Pythagoras theorem gives the base of such a triangle as $\sqrt{(1 - x^2)}$. Thus $\cos y = \sqrt{(1 - x^2)}/1$ and so:

$$\frac{d}{dx}(\sin^{-1}x) = \frac{1}{\sqrt{1 - x^2}} \qquad [26]$$

For the function $y = \cos^{-1} x$, with y restricted to values between 0 and π, then $x = \cos y$ and $dx/dy = -\sin y$. Thus $dy/dx = -1/\sin y$. To express $\sin y$ in terms of x we draw a right-angled triangle (Figure 22.5) with sides such that it has y as the angle with a cosine of x. Using the Pythagoras theorem gives the perpendicular of such a triangle as $\sqrt{(1 - x^2)}$ and thus we have $\sin y = \sqrt{(1 - x^2)}/1$. Hence:

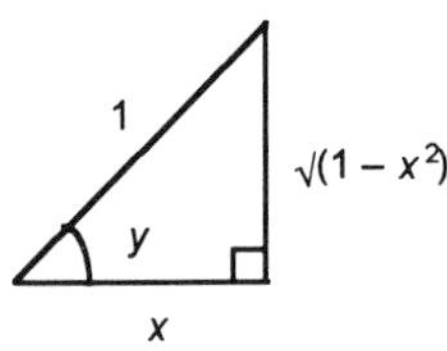

Figure 22.5 $y = cos^{-1} x$

$$\frac{d}{dx}(\cos^{-1}x) = -\frac{1}{\sqrt{1 - x^2}} \qquad [27]$$

For the function $y = \tan^{-1} x$, with y restricted to values between $-\pi/2$ and $+\pi/2$, then $x = \tan y$ and $dx/dy = \sec^2 y$. To express $\sec y$ in terms of x we draw a right-angled triangle (Figure 22.6) with sides such that it has y as the angle with a tangent of x. Using the Pythagoras theorem gives the hypotenuse of such a triangle as $\sqrt{(1 + x^2)}$ and thus we have $\sec y = 1/\cos y = \sqrt{(1 + x^2)}/1$. Hence:

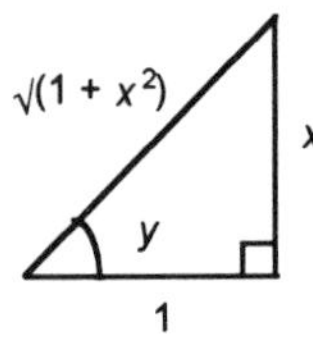

Figure 22.6 $y = tan^{-1} x$

$$\frac{d}{dx}(\tan^{-1}x) = \frac{1}{1 + x^2} \qquad [28]$$

In a similar manner, provided we place restrictions on the inverses, we can obtain:

$$\frac{d}{dx}(\mathrm{cosec}^{-1}x) = -\frac{1}{x\sqrt{x^2 - 1}} \qquad [29]$$

$$\frac{d}{dx}(\sec^{-1}x) = \frac{1}{x\sqrt{x^2 - 1}} \qquad [30]$$

$$\frac{d}{dx}(\cot^{-1}x) = -\frac{1}{1+x^2} \qquad [31]$$

Example

Determine dy/dx for (a) $y = \sin^{-1} 2x$, (b) $y = \sin^{-1} x^2$, (c) $y = x \sin^{-1} x$.

(a) Let $u = 2x$ and so $y = \sin^{-1} u$. Then $du/dx = 2$ and, using equation [26], $dy/du = 1/\sqrt{(1 - u^2)}$. Then, using the chain rule:

$$\frac{dy}{dx} = \frac{dy}{du} \times \frac{du}{dy} = \frac{2}{\sqrt{1-u^2}} = \frac{2}{\sqrt{1-4x^2}}$$

(b) Let $u = x^2$ and so $y = \sin^{-1} u$. Then $du/dx = 2x$ and, using equation [26], $dy/du = 1/\sqrt{(1 - u^2)}$. Then, using the chain rule:

$$\frac{dy}{dx} = \frac{dy}{du} \times \frac{du}{dy} = \frac{2x}{\sqrt{1-u^2}} = \frac{2x}{\sqrt{1-x^4}}$$

(c) This is the derivative of a product. Thus using the product rule:

$$\frac{dy}{dx} = x\frac{d}{dx}(\sin^{-1}x) + \sin^{-1}x = \frac{x}{\sqrt{1-x^2}} + \sin^{-1}x$$

Revision

15 Determine dy/dx for the following functions:

(a) $y = \sin^{-1}\frac{3}{4}x$, (b) $y = \tan^{-1}\sqrt{x}$, (c) $y = \sin 3x \sin^{-1} 3x$,

(d) $y = (\tan^{-1} 2x)^3$, (e) $y = \cot^{-1} x^2$, (f) $y = x^2 \cos^{-1} (x - 1)$

22.5.3 Inverse hyperbolic functions

Consider the inverse hyperbolic function $y = \sinh^{-1} x$. We can write this as $x = \sinh y$ and hence $dx/dy = \cosh y$. Thus $dy/dx = 1/\cosh y$. But $\cosh^2 y - \sinh^2 y = 1$ (Chapter 3, equation [69]). Thus:

$$\frac{d}{dx}(\sinh^{-1}x) = \frac{1}{\cosh y} = \frac{1}{\sqrt{1+\sinh^2 y}} = \frac{1}{\sqrt{1+x^2}} \qquad [32]$$

For the inverse function $y = \cosh^{-1} x$ we can write $x = \cosh y$ and hence $dx/dy = \sinh y$. Thus:

$$\frac{d}{dx}(\cosh^{-1}x) = \frac{1}{\sinh y} = \frac{1}{\sqrt{\cosh^2 x - 1}} = \frac{1}{\sqrt{x^2 - 1}} \quad [33]$$

For the inverse function $y = \tanh^{-1} x$ we can write $x = \tanh y$ and hence $dx/dy = \text{sech}^2 x$. Using $1 - \tanh^2 x = \text{sech}^2 x$ (Chapter 3, equation [70]):

$$\frac{d}{dx}(\tanh^{-1}x) = \frac{1}{\text{sech}^2 x} = \frac{1}{1 - \tanh^2 x} = \frac{1}{1 - x^2} \quad [34]$$

Example

Determine dy/dx for $y = \sinh^{-1} 3x$.

Let $u = 3x$ and so $y = \sinh^{-1} u$. Then $du/dx = 3$ and $dy/du = 1/\sqrt{(1 + u^2)}$. Then, using the chain rule:

$$\frac{dy}{dx} = \frac{dy}{du} \times \frac{du}{dx} = \frac{3}{\sqrt{1 + u^2}} = \frac{3}{\sqrt{1 + 9x^2}}$$

Revision

16 Determine dy/dx for the following functions:

(a) $y = \sinh^{-1} x/3$, (b) $y = \cosh^{-1} 3x$, (c) $y = \tanh^{-1} 3x$

22.6 Implicit differentiation

An equation in the form $y = 2x^2 + 5$ is said to be an *explicit equation* in that y is directly stated as a function of x. However, relationships are sometimes given indirectly, the term used is *implicitly*. For example, we might have $x^2 + y^2 = 9$. With such a relationship, y is not stated uniquely in terms of x (see Section 1.2.4 for a discussion of functions). It is possible to differentiate such an implicit relationship without solving the equation to give y explicitly. For implicit relationships we determine dy/dx by finding the derivatives of both sides of the equation with respect to x, the procedure being termed *implicit differentiation*. Thus, for $x^2 + y^2 = 9$:

$$\frac{d}{dx}(x^2) + \frac{d}{dx}(y^2) = \frac{d}{dx}(9)$$

To evaluate the derivative of y^2 we use the chain rule, letting $u = y^2$:

$$\frac{d}{dx}(y^2) = \frac{du}{dx} = \frac{du}{dy} \times \frac{dy}{dx} = 2y\frac{dy}{dx}$$

Thus we have:

$$2x + 2y\frac{dy}{dx} = 0$$

and hence $dy/dx = -x/y$.

Example

Use implicit differentiation to find $\mathrm{d}y/\mathrm{d}x$ when $x^3 + 2y^2 + 2x - y = 10$.

Differentiating with respect to x:

$$\frac{\mathrm{d}}{\mathrm{d}x}(x^3) + \frac{\mathrm{d}}{\mathrm{d}x}(2y^2) + \frac{\mathrm{d}}{\mathrm{d}x}(2x) - \frac{\mathrm{d}}{\mathrm{d}x}(y) = \frac{\mathrm{d}}{\mathrm{d}x}(10)$$

To determine the derivative of $2y^2$, let $u = 2y^2$. Then:

$$\frac{\mathrm{d}}{\mathrm{d}x}(2y^2) = \frac{\mathrm{d}u}{\mathrm{d}y} \times \frac{\mathrm{d}y}{\mathrm{d}x} = 4y\frac{\mathrm{d}y}{\mathrm{d}x}$$

and so:

$$3x^2 + 4y\frac{\mathrm{d}y}{\mathrm{d}x} + 2 - \frac{\mathrm{d}y}{\mathrm{d}x} = 0$$

Hence:

$$\frac{\mathrm{d}y}{\mathrm{d}x} = -\frac{3x^2 + 2}{4y + 2}$$

Revision

17 Use implicit differentiation to determine $\mathrm{d}y/\mathrm{d}x$ when:

(a) $9x^2 + 4y^2 = 25$, (b) $2x^6 + y^4 - 9xy = 10$, (c) $\sin(x + y) = y + 2$

22.7 Parametric differentiation

In some situations we have both y and x dependent on some other variable. This third variable is termed a *parameter*. For example, we might have $x = 5 \cos \theta$ and $y = 5 \sin \theta$ with θ as the parameter, or perhaps $x = 10t$ and $y = 5t^2$ with t as the parameter. While we could take these pairs of equations, eliminate the parameter, obtain y in terms of x and so obtain $\mathrm{d}y/\mathrm{d}x$, it is not always possible and often may be simpler to tackle the problem in a different way. The method, termed *parametric differentiation*, involves the use of the chain rule.

For $x = 10t$ and $y = 5t^2$ we can directly obtain $\mathrm{d}x/\mathrm{d}t = 10$ and $\mathrm{d}y/\mathrm{d}t = 10t$. The chain rule gives:

$$\frac{\mathrm{d}y}{\mathrm{d}x} = \frac{\mathrm{d}y}{\mathrm{d}t} \times \frac{\mathrm{d}t}{\mathrm{d}x}$$

and so:

$$\frac{\mathrm{d}y}{\mathrm{d}x} = (5t^2) \times (\tfrac{1}{10}) = 0.5t^2$$

Example

Use parametric differentiation to obtain dy/dx when $x = 5 \cos \theta$ and $y = 5 \sin \theta$.

We have $dx/d\theta = -5 \sin \theta$ and $dy/d\theta = 5 \cos \theta$. Thus, using the chain rule:

$$\frac{dy}{dx} = \frac{dy}{d\theta} \times \frac{d\theta}{dx} = (5 \cos \theta) \times \left(-\frac{1}{5 \sin \theta}\right) = -\frac{\cos \theta}{\sin \theta}$$

Revision

18 Use parametric differentiation to obtain dy/dx when:

(a) $x = 4t, y = t^2$, (b) $x = 3 \cos \theta, y = 4 \sin \theta$, (c) $x = 2t, y = t^2 + 2t + 1$,

(d) $x = a(\theta - \sin \theta), y = a(1 - \cos \theta)$

22.8 Logarithmic differentiation

To differentiate a product consisting of two terms the product rule can be used. For example, $y = x \sin x$ can be differentiated using the product rule:

$$\frac{dy}{dx} = x\frac{d}{dx}(\sin x) + \sin x\frac{d}{dx}(x) = x \cos x + \sin x$$

If we need to differentiate a product consisting of, say, three terms then we can use the product rule twice. An alternative to using the product rule, and which is often simpler and can be used with other types of functions, is to use *logarithmic differentiation*. This involves differentiating the natural logarithm of the function. The situations in which such a manipulation is useful is for functions involving products and/or quotients, exponents and those which have a variable base expression raised to a variable power, e.g. x^x.

To illustrate the use of logarithmic differentiation, consider the example discussed earlier, i.e. $y = x \sin x$. Taking logarithms:

$$\ln y = \ln x + \ln(\sin x)$$

Differentiating with respect to x:

$$\frac{d}{dx}(\ln y) = \frac{d}{dx}(\ln x) + \frac{d}{dx}\left[\ln(\sin x)\right]$$

For the first term, if we let $u = \ln y$ and use the chain rule:

$$\frac{d}{dx}(\ln y) = \frac{du}{dx} = \frac{du}{dy} \times \frac{dy}{dx} = \frac{1}{y}\frac{dy}{dx}$$

For the last term, if we let $u = \sin x$ then the chain rule gives:

$$\frac{d}{dx}\left[\ln(\sin x)\right] = \frac{d}{dx}(\ln u) = \frac{d}{du}(\ln u) \times \frac{du}{dx} = \frac{1}{u}\cos x = \frac{\cos x}{\sin x}$$

Hence:

$$\frac{1}{y}\frac{dy}{dx} = \frac{1}{x} + \frac{\cos x}{\sin x}$$

and so, as before:

$$\frac{dy}{dx} = y\left(\frac{1}{x} + \frac{\cos x}{\sin x}\right) = x\sin x\left(\frac{1}{x} + \frac{\cos x}{\sin x}\right) = \sin x + x\cos x$$

Example

Determine dy/dx for $y = e^{x^2-3}$.

Taking logarithms:

$$\ln y = x^2 - 3$$

Differentiating with respect to x:

$$\frac{d}{dx}(\ln y) = \frac{d}{dx}(x^2 - 3)$$

For the first term, if we let $u = \ln y$ and use the chain rule:

$$\frac{d}{dx}(\ln y) = \frac{du}{dx} = \frac{du}{dy} \times \frac{dy}{dx} = \frac{1}{y}\frac{dy}{dx}$$

and thus:

$$\frac{1}{y}\frac{dy}{dx} = 2x$$

$$\frac{dy}{dx} = 2xy = 2x\,e^{x^2-3}$$

Revision

19 Using logarithmic differentiation, determine dy/dx for:

(a) $y = x^x$, (b) $y = (x + 3)^x$, (c) $y = x^3\,e^{4x}(1 + x)^5$, (d) $y = \sqrt{1 + x^2}\,\sin^2 x$

22.9 Higher-order derivatives

The derivative dy/dx of the function $y = f(x)$ is itself a function and may also be differentiated. The derivative of a derivative is called the *second derivative*. You can think of dy/dx giving the function which describes how the slope of the tangent to a graph of the function varies with x and can be plotted as a graph of slope against x. Thus the second derivative describes

how the slope of the tangents to this second graph vary with x. For example, if we have the displacement of an object described as a function of time, then the first derivative gives the rate of change of displacement with time, i.e. the velocity. The second derivative gives the rate of change of the velocity with time, i.e. the acceleration.

The second derivative can be written in a number of forms:

$$\frac{d}{dx}\left(\frac{dy}{dx}\right) \text{ or } \frac{d^2y}{dx^2} \text{ or } \frac{d^2}{dx^2}f(x) \text{ or } f''(x) \text{ or } \ddot{f}(x) \text{ or } D^2x$$

If the second derivative is then differentiated we obtain the *third derivative*, it being written as:

$$\frac{d}{dx}\left(\frac{d^2y}{dx^2}\right) \text{ or } \frac{d^3y}{dx^3} \text{ or } \frac{d^3}{dx^3}f(x) \text{ or } f'''(x) \text{ or } \dddot{f}(x) \text{ or } D^3x$$

This in turn may be differentiated to give a *fourth derivative*, then a *fifth derivative*, and so on.

Example

Determine the second derivative of the function $y = x^3$.

The first derivative is:

$$\frac{dy}{dx} = 3x^2$$

The second derivative is:

$$\frac{d^2y}{dx^2} = 6x$$

Revision

20 Determine the second derivatives of the following functions:

(a) $y = \frac{1}{x}$, (b) $y = \sin 5x$, (c) $y = \tan 2x$, (d) $y = \sqrt{1 + 2x}$

21 Determine the first derivative, second derivative, third derivative and the fourth derivative of (a) $y = x^4 + 2x^3$, (b) $y = \sin 2x$.

22.10 Applications of differentiation

The derivative dy/dx of a function is a measure of the slope of the tangent to a graph of the function. It can also be considered as being a measure of the rate at which y increases with respect to x. Thus with regard to the displacement x of an object, with x being a function of time t, then the rate of change of displacement with time is the velocity v and the rate of change of velocity with time is the acceleration a. We thus have:

$$v = \frac{dx}{dt} \text{ and } a = \frac{dv}{dx} = \frac{d}{dx}\left(\frac{dx}{dt}\right) = \frac{d^2x}{dt^2} \qquad [35]$$

As a further illustration of a rate, electrical current i is the rate at which charge q passes a point:

$$i = \frac{dq}{dt} \qquad [36]$$

For a capacitor the charge q on a plate is given by $q = Cv$, where C is the capacitance and v the potential difference across it. Thus:

$$i = \frac{d}{dt}(Cv) = C\frac{dv}{dt} \qquad [37]$$

With an inductor in an electrical circuit, the potential difference v across it is related to the current i by:

$$v = L\frac{di}{dt} \qquad [38]$$

where L is the inductance.

Example

The displacement x of a particle from some fixed point is described by $x = A \cos \omega t$. Derive equations for the velocity v and acceleration a.

The motion is simple harmonic motion. Using equations [35], the velocity is given by:

$$v = \frac{dx}{dt} = -A\omega \sin \omega t$$

and the acceleration by:

$$a = \frac{dv}{dt} = -A\omega^2 \cos \omega t = -\omega^2 x$$

Example

The voltage v in volts across a capacitor is continuously adjusted so that it varies with time t in seconds according to the equation $v = 3t$. How does the current vary with time for a capacitance of 4 mF?

Using equation [37]:

$$i = C\frac{dv}{dt} = C\frac{d}{dt}(3t) = 3C = 3 \times 0.004 = 0.012 \text{ A}$$

The current is thus constant and maintained at this value.

Revision

22 The distance x in metres of a car travelling along a straight road from its start after a time t in seconds is given by $x = 4t^2 + 2t$. Determine (a) the velocity and (b) the acceleration after a time of 5 s.

23 The distance x travelled by a body in a straight line from a fixed point varies as the square root of the time t elapses. Derive relationships for the velocity and the acceleration.

24 The voltage v, in volts, across a capacitor varies with time t, in seconds, according to the equation $v = 5 \sin 2t$. Determine how the current will vary with time for a capacitance of 2 mF.

25 The current i, in amps, through an inductor of inductance 0.5 H varies with time t, in seconds, according to the equation $i = 4t$. Determine the voltage across the inductor.

26 Determine the rate of change of the area of a square with respect to the length of a side when the length is 50 mm.

27 The kinetic energy E of a rotating flywheel depends on the angular velocity ω of the wheel and is given by $E = \frac{1}{2}I\omega^2$. determine the rate of change of kinetic energy with respect to angular velocity.

22.10.1 Maxima and minima

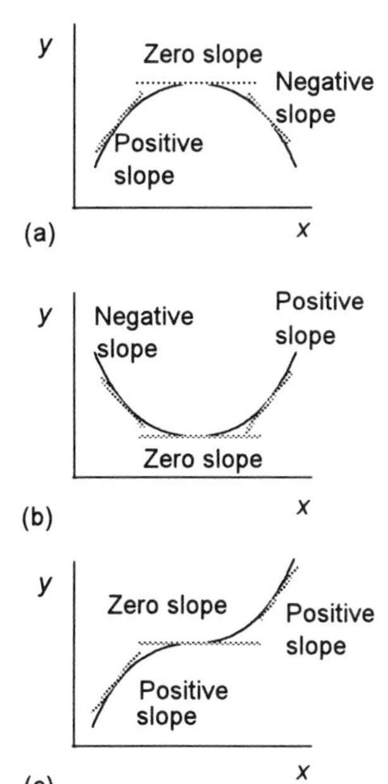

Figure 22.7 *Stationary points*

If tangents are drawn to a graph of the function $y = f(x)$ then the slopes of the tangents are given by dy/dx (Figure 22.7). If dy/dx is greater than 0 then the tangents have positive value slopes and so y is increasing as x increases. If dy/dx is less than 0 then the tangents have negative value slopes and so y is decreasing as x increases. If dy/dx is 0 then the tangent is parallel to the x-axis and so y is neither increasing nor decreasing. Such points are termed *stationary points* or *critical points* or *turning points*.

When, at some point, $dy/dx = 0$ and the derivative changes from being positive prior to the point to negative afterwards (Figure 22.7(a)) then we have a *maximum*. When the derivative changes from being negative prior to the point to positive after it (Figure 22.7(b)) we have a *minimum*. When the derivative prior to the point is positive and after it is positive (Figure 22.7(c)), or prior to it is negative and after it is negative, we have a *point of inflexion*.

Since d^2y/dx^2 is the rate of change of dy/dx with x then when the derivative dy/dx changes from positive to negative it is decreasing and so d^2y/dx^2 must be negative. When the derivative dy/dx changes from negative to positive it is increasing and so d^2y/dx^2 must be positive. Thus at a maximum d^2y/dx^2 is negative and at a minimum positive.

Example

Determine and identify the form of the turning points on the graph of the function $y = 2x^3 - 3x^2$.

The slopes of the tangents are given by:

$$\frac{dy}{dx} = 6x^2 - 6x$$

The slope is zero when $6x^2 - 6x = 0$, i.e. $x = 0$ or $x = 1$. Substituting these values into the original equation gives the y values of the turning points and so the turning points are (0, 0) and (1, –1). We can determine the form of these turning points by obtaining the second derivative:

$$\frac{d^2x}{dx^2} = 12x - 6$$

When $x = 0$ the second derivative is negative and so the turning point (0, 0) is a maximum. When $x = 1$ the second derivative is positive and so the turning point (1, –1) is a minimum.

Example

Determine and identify the form of the turning points on the graph of the function $y = x^4 - 2x^3$.

Differentiating the function gives:

$$\frac{dy}{dx} = 4x^3 - 6x^2 = x^2(4x - 6)$$

The derivative is zero when $x = 0$ or $x = 1.5$. The y values for these points can be obtained by substituting these values of x in the original equation, so giving the turning points as (0, 0) and (1.5, –1.7). The second derivative is:

$$\frac{d^2y}{dx^2} = 12x^2 - 12x$$

When $x = 0$, the second derivative is 0 and so tells us nothing about the turning point. If we consider the value of the first derivative prior to and after the point, say at $x = -0.5$ and $x = +0.5$, then the derivative has negative values prior to and after the point. The point (0, 0) is thus a point of inflexion. When $x = 1.5$ then the second derivative is +9 and so, because it is positive, the point (1.5, –1.7) is a minimum.

Example

If the sum of two numbers is 40, determine the values of the numbers which will give the minimum value for the sum of their squares.

Let the two numbers be x and y. Then $x + y = 40$. We have to find the minimum value of the sum S of the squares:

$$S = x^2 + y^2 = x^2 + (40 - x)^2 = 2x^2 - 80x + 1600$$

Differentiating gives:

$$\frac{dS}{dx} = 4x - 80$$

The derivative is zero when $x = 20$. We can check that this gives the minimum by taking the second derivative:

$$\frac{d^2S}{dx^2} = 4$$

Since this is positive for all values of x, the point is a minimum. Thus the two numbers are 20 and 20.

Revision

29 Determine and identify the form of the turning points on graphs of the following functions:

(a) $y = x^2 - 5x + 6$, (b) $y = x^3 + x^2 - 2$, (c) $y = x^4 - x^2$, (d) $y = x^3$

30 A 100 cm length of wire is to be bent to form two squares, one with side x and the other with side y. Determine the values of x and y which give the minimum area enclosed by the squares.

31 The rate r at which a chemical reaction proceeds depends on the quantity x of a chemical and is given by $r = k(a - x)(b + x)$. Determine the maximum rate.

32 A cylinder has a radius r and height h with the sum of the radius and height being 2 m. Determine the radius giving the maximum volume.

33 A rectangle is to have an area of 36 cm^2. Determine the lengths of the sides which will give a minimum value for the perimeter.

34 An open tank is to be constructed with a square base and vertical sides and to be able to hold, when full to the brim, 32 m^3 of water. Determine the dimensions of the tank if the area of sheet metal used is to be a minimum.

22.10.2 Small changes

Consider a function $y = f(x)$ when x changes to $x + \delta x$. The change δy in the value of y is:

$$\delta y = f(x + \delta x) - f(x)$$

Hence we can write:

$$\delta y = \frac{f(x + \delta x) - f(x)}{\delta x}\delta x$$

In the limit as $\delta x \to 0$, the fraction term becomes the derivative $f'(x)$ and we can write dy for δy and dx for δx:

$$\mathrm{d}y = f'(x)\,\mathrm{d}x \qquad [39]$$

dx is termed the *differential*. As an approximation when we have small changes in x and y:

$$\delta y \approx f'(x)\,\delta x \qquad [40]$$

This equation is widely used to determine the change in y that results from a small change in x.

Example

Determine an approximate value for $\sqrt{9.02}$.

Let $y = f(x) = \sqrt{x}$. We can consider the effect on y of the small change in x from 9 to 9.02, i.e. we let $\delta x = 0.02$. The derivative of $\sqrt{x}$ is $1/2\sqrt{x}$. Thus, using equation [40]:

$$\delta y \approx \frac{1}{2\sqrt{x}}\delta x \approx \frac{1}{2\sqrt{9}} \times 0.02 = 0.003\,3$$

Thus the approximate value is $\sqrt{9} + 0.003\,3 = 3.003\,3$.

Example

A cube has a side length which has been measured as 6 ± 0.05 cm. Determine the maximum possible error in the volume when calculated from this measurement.

The volume V of a cube of side x is given by $V = x^3$. Thus dV/d$x = 3x^2$. Hence, using equation [40], the change in volume δV produced by a change in side of δx is:

$$\delta V \approx 3x^2\,\delta x \approx 3 \times 6^2 \times 0.05 \approx 5.4$$

The maximum possible error in the volume is thus ±5.4 cm³.

Revision

35 Determine approximate values for (a) 2.02^2, (b) $\sqrt{15.92}$.

36 The radius of a sphere is measured as 5 ± 0.04 cm. Determine the maximum possible error in the volume calculated.

37 If the radius of a sphere is increased from 5 cm to 5.01 cm, estimate the increase in the surface area.

38 Determine an approximate value of:

$$\frac{2}{\sqrt{0.99} + 0.99^2}$$

22.11 Differentiating vectors

Consider a point P having a position vector **r** and moving along a curved path (Figure 22.8). Since **r** is a function of time we can write it as $\mathbf{r}(t)$. After a time δt the point is at Q and has a position vector $\mathbf{r}(t + \delta t)$. $\overrightarrow{PQ}$ then represents the displacement vector of the object during the time δt and is the difference between the position vectors $\mathbf{r}(t)$ and $\mathbf{r}(t + \delta t)$. The average velocity during this time interval is thus:

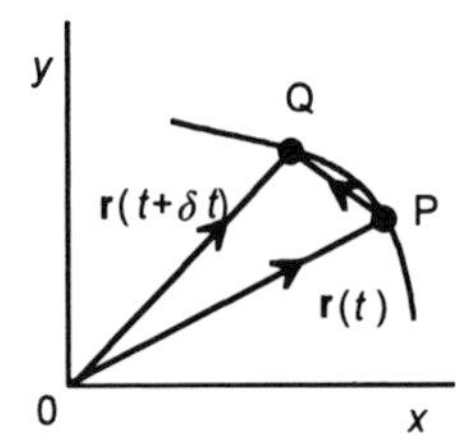

Figure 22.8 *Position vector of moving point*

$$\text{average velocity} = \frac{\overrightarrow{PQ}}{\delta t} = \frac{\mathbf{r}(t+\delta t) - \mathbf{r}(t)}{\delta t}$$

The instantaneous velocity **v** is then:

$$\mathbf{v} = \lim_{\delta t \to 0} \frac{\mathbf{r}(t+\delta t) - \mathbf{r}(t)}{\delta t} = \frac{\mathrm{d}\mathbf{r}}{\mathrm{d}t} \qquad [41]$$

The x and y co-ordinates of the point P are given by $\mathbf{r} = x\mathbf{i} + y\mathbf{j}$ (see Section 10.4). Since they depend on time we can write them as $\mathbf{r}(t) = x(t)\mathbf{i} + y(t)\mathbf{j}$. For the point Q we can write $\mathbf{r}(t + \delta t) = x(t + \delta t)\mathbf{i} + y(t + \delta t)\mathbf{j}$. Thus:

$$\mathbf{v} = \lim_{\delta t \to 0} \frac{x(t+\delta t)\mathbf{i} + y(t+\delta t)\mathbf{j} - x(t)\mathbf{i} - y(t)\mathbf{j}}{\delta t}$$

and so:

$$\mathbf{v} = \frac{\mathrm{d}\mathbf{r}}{\mathrm{d}t} = \frac{\mathrm{d}x}{\mathrm{d}t}\mathbf{i} + \frac{\mathrm{d}y}{\mathrm{d}t}\mathbf{j} \qquad [42]$$

With three dimensions we have:

$$\mathbf{v} = \frac{\mathrm{d}\mathbf{r}}{\mathrm{d}t} = \frac{\mathrm{d}x}{\mathrm{d}t}\mathbf{i} + \frac{\mathrm{d}y}{\mathrm{d}t}\mathbf{j} + \frac{\mathrm{d}z}{\mathrm{d}t}\mathbf{k} \qquad [43]$$

For the acceleration **a**, i.e. the rate of change of the velocity and so the second derivative of the displacement vector, we have:

$$\mathbf{a} = \frac{d^2\mathbf{r}}{dt^2} = \frac{d^2x}{dt^2}\mathbf{i} + \frac{d^2y}{dt^2}\mathbf{j} + \frac{d^2z}{dt^2}\mathbf{k} \qquad [44]$$

Example

If the position vector **r** equals $2t^2\mathbf{i} + 4t\mathbf{j}$, determine $d\mathbf{r}/dt$.

Using equation [42]:

$$\frac{d\mathbf{r}}{dt} = 4t\mathbf{i} + 4\mathbf{j}$$

Revision

39 Determine $d\mathbf{r}/dt$ and $d^2\mathbf{r}/dt^2$ when:

(a) $\mathbf{r} = 4t^2\mathbf{i} + (\cos 2t)\mathbf{j}$, (b) $\mathbf{r} = (3t^2 + 2)\mathbf{i} + 1\mathbf{j}$

22.11.1 Rules for the differentiation of vectors

If **a** and **b** are vectors and c a scalar constant, we can show, in a similar manner to that used for establishing the rules for differentiation earlier in this chapter:

1 *Product of a constant scalar and a vector*

$$\frac{d}{dt}(c\mathbf{a}) = c\frac{d\mathbf{a}}{dt} \qquad [45]$$

2 *Sum of two vectors*

$$\frac{d}{dt}(\mathbf{a} + \mathbf{b}) = \frac{d\mathbf{a}}{dt} + \frac{d\mathbf{b}}{dt} \qquad [46]$$

3 *Product of a variable scalar and a vector*

$$\frac{d}{dt}(c\mathbf{a}) = c\frac{d\mathbf{a}}{dt} + \mathbf{a}\frac{dc}{dt} \qquad [47]$$

4 *Scalar product*

$$\frac{d}{dt}(\mathbf{a} \cdot \mathbf{b}) = \mathbf{a} \cdot \frac{d\mathbf{b}}{dt} + \mathbf{b} \cdot \frac{d\mathbf{a}}{dt} \qquad [48]$$

5 *Vector product*

$$\frac{\mathrm{d}}{\mathrm{d}t}(\mathbf{a} \times \mathbf{b}) = \mathbf{a} \times \frac{\mathrm{d}\mathbf{b}}{\mathrm{d}t} + \frac{\mathrm{d}\mathbf{a}}{\mathrm{d}t} \times \mathbf{b} \qquad [49]$$

The derivative of the vector and scalar products is thus formed in the same way as the derivative of the product of two scalar functions. However, with the vector product the order of **a** and **b** must remain unaltered.

Example

If $\mathbf{a} = 6t^2\mathbf{i} + 4t\mathbf{j} - 3\mathbf{k}$ and $\mathbf{b} = 2t^2 - t\mathbf{j} + 1\mathbf{k}$, determine the derivative of the vector product $\mathbf{a} \times \mathbf{b}$.

Using equation [49]:

$$\frac{\mathrm{d}}{\mathrm{d}t}(\mathrm{a} \times \mathrm{b}) = (6t^2\mathbf{i} + 4t\mathbf{j} - 3\mathbf{k}) \times (4t\mathbf{i} - 1\mathbf{j}) + (12t\mathbf{i} + 4\mathbf{j}) \times (2t^2\mathbf{i} - t\mathbf{j} + 1\mathbf{k})$$

$$= -3\mathbf{i} - 12\mathbf{i} - 22t^2\mathbf{k} + 4\mathbf{i} - 12t\mathbf{j} - 20t^2\mathbf{k}$$

$$= 1\mathbf{i} - 24t\mathbf{j} - 42t^2\mathbf{k}$$

Alternatively we could have first evaluated the vector product and then differentiated the result:

$$\mathbf{a} \times \mathbf{b} = (6t^2\mathbf{i} + 4t\mathbf{j} - 3\mathbf{k}) \times (2t^2\mathbf{i} - t\mathbf{j} + 1\mathbf{k}) = t\mathbf{i} - 12t^2\mathbf{j} - 14t^3\mathbf{k}$$

This can be differentiated using equation [43] to give the required derivative as $1\mathbf{i} - 24\mathbf{j} - 42t^2\mathbf{k}$.

Revision

40 If $\mathbf{a} = 3t\mathbf{i} - t^2\mathbf{j}$ and $\mathbf{b} = 2t^2\mathbf{i} + 3\mathbf{j}$ determine the derivatives of (a) $(\mathbf{a} \cdot \mathbf{b})$ and (b) $(\mathbf{a} \times \mathbf{b})$.

41 If $\mathbf{a} = 2\mathbf{i} - 5t^2\mathbf{j}$ and $\mathbf{b} = 3t\mathbf{i} + t^3\mathbf{j}$, determine the derivatives of (a) $(\mathbf{a} \cdot \mathbf{b})$ and (b) $(\mathbf{a} \times \mathbf{b})$.

42 For a particle moving with constant speed in a circular path, the position vector is $\mathbf{r} = r(\mathbf{i} \cos \omega t + \mathbf{j} \sin \omega t)$. Determine (a) the velocity and (b) the acceleration.

Problems

1 Determine the derivatives of the following functions:

(a) $y = x^5$, (b) $y = 2x^{-4}$, (c) $y = -3x^2$, (d) $y = \frac{1}{2}x$, (e) $y = 2\pi x^2$,

(f) $y = \tan 3x$, (g) $y = 5 \cos 2x$, (h) $y = 4\, e^{x/2}$, (i) $y = 2\, e^{-2x}$, (j) $y = 3\, e^{3x}$,

(k) $y = \dfrac{5}{3\sqrt{x}}$, (l) $y = \dfrac{6}{5x^2}$, (m) $y = \dfrac{7}{\sqrt{3x}}$, (n) $y = \dfrac{5}{(2x)^3}$, (o) $y = \sqrt{3x}$,

(p) $y = (3x - 2x^2)(5 + 4x)$, (q) $y = 5x \sin x$, (r) $x\, e^{x/2}$, (s) $(x^2 + 1) \sin x$,

(t) $y = \dfrac{2x+1}{x-6}$, (u) $y = \dfrac{x+1}{\sqrt{x}}$, (v) $y = \dfrac{\sin x}{x}$, (w) $y = \dfrac{e^{2x}}{x^2+1}$,

(x) $y = \dfrac{\sinh 2x}{\cosh 3x}$, (y) $y = \dfrac{1}{2-7x}$, (z) $y = \dfrac{1}{\sqrt{3-x}}$, (aa) $y = (x+5)^3$,

(ab) $y = \sqrt{\dfrac{x^2-1}{x^2+1}}$, (ac) $y = \dfrac{x}{\sqrt[3]{x^2+4}}$, (ad) $y = x \sin^2 x$, (ae) $y = e^{\ln 3x}$,

(af) $y = \sin^{-1} 5x$, (ag) $y = \tan^{-1} x^{1/2}$, (ah) $y = \sec^{-1} e^{2x}$, (ai) $y = (\tan^{-1} x)^2$,

(aj) $y = \sin^{-1} x + x\sqrt{1-x^2}$, (ak) $y = \sin^{-1} x + \cos^{-1} x$, (al) $y = \coth^{-1} x$,

(am) $y = \cosh^{-1} 4x$, (an) $y = x \sinh^{-1} x$

2 Use implicit differentiation to determine d*y*/d*x* when:

(a) $y^2 + x^2 + xy = 5$, (b) $2x^2 + y^2 + 3xy - 4y = 10$, (c) $x^3 - y^3 + y = 3$,

(d) $x^2 - y + \ln y = 0$

3 Use parametric differentiation to obtain d*y*/d*x* when:

(a) $x = t^3 + t$, $y = t^4 + t + 1$, (b) $x = \sin 2t$, $y = \cos 4t$,

(c) $x = t^3 + \sin 2\pi t$, $y = t + e^t$

4 Use logarithmic differentiation to obtain d*y*/d*x* when:

(a) $y = (2x^2 + 3)^x$, (b) $y = x^x \sqrt{x}$, (c) $y = x^{\sqrt{x}}$, (d) $y = x^{\sqrt{x+1}}$

5 Determine the second derivatives of the following functions:

(a) $y = x^2 + 2x$, (b) $y = \sin 2x$, (c) $y = \dfrac{1}{x^2}$, (d) $y = 3x^4 - x^2 - \dfrac{1}{x}$,

(e) $y = x^4 + 2x^3 - 8x + 5$, (f) $y = \dfrac{x+1}{\sqrt{x}}$

6 Determine the velocity and acceleration after a time of 2 s for an object which has a displacement x which is a function of time t and given by $x = 12 + 15t - 2t^2$, with t being in seconds.

7 Determine the velocity and acceleration at a time t for an object which has a displacement x in metres given by $x = 3 \sin 2t + 3 \cos 3t$, t being in seconds.

8 The voltage v, in volts, across a capacitor of capacitance 2 mF varies with time t, in seconds, according to the equation $v = 3 \sin 5t$. Determine how the current varies with time.

9 The current i, in amps, through an inductor of inductance 0.05 H varies with time t, in seconds, according to the equation $i = 10(1 - e^{-100t})$. Determine how the potential difference across the inductor varies with time.

10 The volume of a cone is one-third the product of the base area and the height. For a cone with a height equal to the base radius, determine the rate of change of cone volume with respect to the base radius.

11 The volume of a sphere of radius r is $\frac{4}{3}\pi r^3$. Determine the rate of change of the volume with respect to the radius.

12 With the Doppler effect, the frequency f_o heard by an observer when a sound source of frequency f_s is moving away from the observer with a velocity v is given by $f_o = f_s/(1 + v/c)$, where c is the velocity of sound. Determine the rate of change of the observed frequency with respect to the velocity.

13 A parabola is defined by the parametric equations $x = 5t^2$ and $y = 10t$. Determine the slope of the tangent to the parabola at $t = 2$.

14 An object follows an elliptical path which is defined by the parametric equations $x = 2 \cos \theta$ and $y = 3 \sin \theta$. Determine the slope of the tangent to the ellipse at $\theta = \pi/3$.

15 The length L of a metal rod is a function of temperature T and is given by the equation $L = L_0(1 + aT + bT^2)$. Determine an equation for the rate of change of length with temperature.

16 Determine and identify the form of the turning points on graphs of the following functions:

(a) $y = x^2 - 4x + 3$, (b) $y = x^3 - 6x^2 + 9x + 3$, (c) $y = x^5 - 5x$,

(d) $y = \sin x$ for x values between 0 and 2π, (e) $y = 2x^3 + 3x^2 - 12x + 3$

17 A cylindrical container, open at one end, has a height of h m and a base radius of r m. The total surface area of the container is to be 3π m^2. Determine the values of h and r which will make the volume a maximum.

18 A cylindrical metal container, open at one end, has a height of h cm and a base radius of r cm. It is to have an internal volume of 64π cm^3. Determine the dimensions of the container which will require the minimum area of metal sheet in its construction.

19 The bending moment M of a uniform beam of length L at a distance x from one end is given by $M = \frac{1}{2}wLx - \frac{1}{2}wx^2$, where w is the weight per unit length of beam. Determine the value of x at which the bending moment is a maximum.

20 The deflection y of a beam of length L at a distance x from one end is found to be given by $y = 2x^4 - 5Lx^3 + 2L^2x^2$. Determine the values of x at which the deflection is a maximum.

21 An electrical circuit has a non-ideal voltage source of voltage V in series with internal resistance r and is connected to a load of resistance R. The power transferred to the load by the voltage source is $V^2R/(R + r)$. Determine the value of R which results in maximum power transfer.

22 Determine the maximum value of the alternating voltage described by the equation $v = 40 \cos 1000t + 15 \sin 1000t$ V.

23 The intensity of illumination from a point light source of intensity I at a distance d from it is I/d^2. Determine the point along the line between two sources 10 m apart at which the intensity of illumination is a minimum if one of the sources has eight times the intensity of the other.

24 Determine the maximum rate of change with time of the alternating current $i = 10 \sin 1000t$ mA, the time t being in seconds.

25 The deflection y of a propped cantilever of length L at a distance x from the fixed end is given by:

$$y = \frac{1}{EI}\left(\frac{5wLx^3}{48} - \frac{wL^2x^2}{16} - \frac{wx^4}{24}\right)$$

where w is the weight per unit length and E and I are constants. Determine the value of x at which the deflection is a maximum.

26 The e.m.f. E produced by a thermocouple depends on the temperature T and is given by $E = aT + bT^2$. Determine the temperature at which the e.m.f. is a maximum.

27 The horizontal range R of a projectile projected with a velocity v at an angle θ to the horizontal is given by $R = (v^2/g) \sin 2\theta$. Determine the angle at which the range is a maximum for a particular velocity.

28 Determine an approximate value for $y = 2x^3 - 5x^2 + 2x + 5$ when $x =$ 1.003.

29 Determine an approximate value for $y = x^4 - 5x^3 + 3x - 4$ when x has the value 0.997.

30 The sides of a cube have lengths of 3.2 ± 0.1 cm. Estimate the maximum possible error in the calculated volume.

31 Determine $d\mathbf{r}/dt$ and $d^2\mathbf{r}/dt^2$ when:

(a) $\mathbf{r} = 2t^2\mathbf{i} + (\sin 2t)\mathbf{j}$, (b) $\mathbf{r} = 3t^2\mathbf{i} + (t + 1)\mathbf{j}$

32 If $\mathbf{a} = 2t\mathbf{i} + t^2\mathbf{j} + 2\mathbf{k}$ and $\mathbf{b} = t^2\mathbf{i} - 2\mathbf{j} + t\mathbf{k}$, determine the derivative with respect to t of (a) $\mathbf{a} \cdot \mathbf{b}$, (b) $\mathbf{a} \times \mathbf{b}$.

33 The position vector of a particle is given by $\mathbf{r} = 2\,e^{-t/5}\mathbf{i} + t\mathbf{k} + 2\mathbf{j}$. Determine (a) the velocity and (b) the acceleration.

23 Numerical differentiation

23.1 Introduction

The methods given in Chapter 22 for the determination of derivatives assume that we have a differentiable algebraic expression. In some situations we do not have this. Perhaps as the outcome of an experiment we have only a set of data points. For example, an experiment might be used to determine the times t taken for a freely falling object to cover certain measured distances x. The outcome is thus a series of distance and time points and we perhaps require to determine how the downward velocity dx/dt varies with time. In Chapter 5 a method was developed to give polynomials which gave an approximate fit to a set of data points. This chapter extends that to develop derivatives which give an approximate fit to a set of data points.

Chapter 5 and the basic techniques of differentiation from Chapter 22 are assumed.

23.2 Two-point equation

The derivative is defined in terms of the slope of a line joining two points on a graph of the function (see Section 22.2), the derivative being the value of the slope as the distance between the points tends to zero (Chapter 22, equation [2]). For some function of x, if we have just two data points then we take the slope of the line joining the two points as the approximation to the derivative (Figure 23.1). Thus for $x = a$ and $x = a + h$ the derivative at point P is approximated by:

$$\frac{dy}{dx} \approx \frac{f(a+h) - f(a)}{h} \qquad [1]$$

This is termed a *two-point equation.*

In general we can describe a function by means of the *Taylor polynomial* $p(x)$ (see equation [26], Chapter 5):

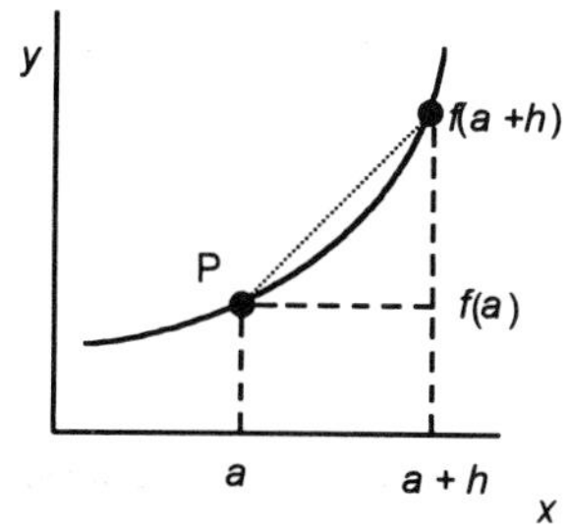

Figure 23.1 *Derivative approximation*

$$p(x) = f(a) + \frac{(x-a)}{1!}f'(a) + \frac{(x-a)^2}{2!}f''(a) + \frac{(x-a)^3}{3!}f'''(a) + \ldots + \frac{(x-a)^n}{n!}f^{(n)}(a) \qquad [2]$$

At $x = a + h$ the Taylor equation gives:

$$p(a+h) = f(a) + \frac{h}{1!}f'(a) + \frac{h^2}{2!}f''(a) + \frac{h^3}{3!}f'''(a) + \ldots + \frac{h^n}{n!}f^{(n)}(a) \qquad [3]$$

If we accept the polynomial to adequately approximate to the function, i.e. $p(a + h) \approx f(a + h)$, then rearranging equation [3] gives:

$$f'(a) = \frac{f(a+h) - f(a)}{h} - \frac{h}{2}f''(a) + \ldots \qquad [4]$$

The first term on the right-hand side of the equation is the two-point equation. Thus we can regard the two-point equation as giving the derivative with an error of $-\frac{1}{2}hf''(a)$ plus terms due to higher derivatives. The smaller the value of h the smaller will be the error term.

Example

The following values were obtained from measurements of the distance x fallen by a freely falling body as a function of time t: $x = 50$ cm for $t = 0.1$ s, $x = 200$ cm for $t = 0.2$ s. Using the two-point equation, estimate the value of the derivative at $t = 0.1$ s.

Using equation [1]:

$$\frac{dy}{dx} \approx \frac{f(a+h) - f(a)}{h} = \frac{200 - 50}{0.2 - 0.1} = 1500 \text{ m/s}$$

Example

Determine, using the two-point equation, the derivative of the function $y = \ln x$ for $x = 1.5$ with an increment of 0.1.

Using equation [1]:

$$\frac{dy}{dx} \approx \frac{f(a+h) - f(a)}{h} = \frac{\ln 1.6 - \ln 1.5}{0.1} = 0.645$$

This compares with the derivative obtained using the standard formula of $1/x = 0.667$ (see Section 22.5.1).

Revision

1 Use the two-point equation to estimate the derivative at the first of each of the set of points given:

(a) $y = 4.30$ at $x = 2.0$, $y = 4.51$ at $x = 3.0$,

(b) $y = 4.92$ at $x = 1.0$, $y = 5.62$ at $x = 1.2$

2 Determine, using the two-point equation, the derivative of the function $y = x^2$ at $x = 1.3$ for an increment of 0.1.

23.3 Three-point equation

We can improve the accuracy of the estimate of the derivative at point P on a graph of a function by considering three points: point P at $x = a$ and a point either side of it at $a - h$ and $a + h$ (Figure 23.2). The slope of the line joining these two outer points is taken to be the approximation to the slope of the tangent at point P and so the required derivative. Thus:

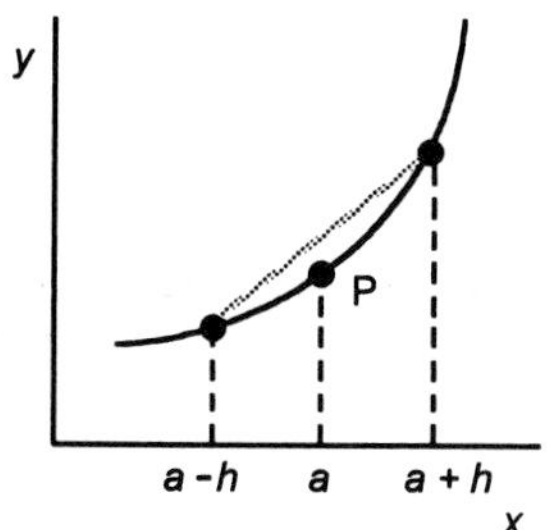

Figure 22.2 *Derivative approximation*

$$\frac{dy}{dx} \approx \frac{f(a+h) - f(a-h)}{2h} \qquad [5]$$

This is termed the *three-point equation*.

Taylor's equation [2] for $x = a + h$ is:

$$p(a+h) = f(a) + \frac{h}{1!}f'(a) + \frac{h^2}{2!}f''(a) + \frac{h^3}{3!}f'''(a) + \ldots + \frac{h^n}{n!}f^{(n)}(a) \qquad [6]$$

and for $x = a - h$ is:

$$p(a-h) = f(a) - \frac{h}{1!}f'(a) + \frac{h^2}{2!}f''(a) - \frac{h^3}{3!}f'''(a) + \ldots + \frac{h^n}{n!}f^{(n)}(a) \qquad [7]$$

If we take the polynomial to be a reasonable approximation to the function, then equation [7] subtracted from equation [6] gives:

$$f(a+h) - f(a-h) = 2hf'(a) + \frac{2h^3}{6}f'''(a) + \ldots$$

and so:

$$\frac{dy}{dx} = f'(a) = \frac{f(a+h) - f(a-h)}{2h} - \tfrac{1}{6}h^2 f'''(a) + \ldots \qquad [8]$$

The first term on the right-hand side of the equation is the three-point equation and the remaining terms represent the error in using that equation. The error can thus be considered to be proportional to h^2, this comparing with an error proportional to h for the two-point equation. If h is less than 1 then the h^2 error is less than the h error and so the three-point equation gives a more accurate result than the two-point equation.

Example

Using the three-point equation, estimate the derivative at the point $x = 2$ for the following data: $y = 4.12$ at $x = 1.8$, $y = 5.64$ at $x = 2.2$.

Using equation [5]:

$$\frac{dy}{dx} \approx \frac{f(a+h) - f(a-h)}{2h} = \frac{5.64 - 4.12}{2 \times 0.2} = 3.8$$

Example

Determine, using the three-point equation, the derivative of the function $y = \ln x$ for $x = 1.5$ with an increment of 0.1.

Using equation [5]:

$$\frac{dy}{dx} \approx \frac{f(a+h)-f(a-h)}{2h} = \frac{\ln 1.6 - \ln 1.4}{2 \times 0.1} = 0.668$$

This compares with the derivative obtained using the standard formula of $1/x = 0.667$ (see Section 22.5.1) and 0.645 obtained in an earlier example using the two-point equation.

Revision

3 Use the three-point equation to estimate the derivative at the midpoint of each of the set of points given:

(a) $y = 2.30$ at $x = 1.0$, $y = 2.51$ at $x = 2.0$,

(b) $y = 4.12$ at $x = 1.0$, $y = 4.62$ at $x = 1.2$

2 Determine, using the three-point equation, the derivative of the function $y = x^2$ at $x = 1.3$ for an increment of 0.1.

23.4 Five-point equation

Consider the derivative for a point P at $x = a$ when we use data values for $x = a - h$ and $x = a + h$ and also $x = a - 2h$ and $x = a + 2h$. Taylor's equation gives for the $x = a + h$ and $x = a - h$ points:

$$p(a+h) = f(a) + \frac{h}{1!}f'(a) + \frac{h^2}{2!}f''(a) + \frac{h^3}{3!}f'''(a) + \frac{h^4}{4!}f^{iv}(a) + \frac{h^5}{5!}f^{v}(a) + \dots$$

$$p(a-h) = f(a) - \frac{h}{1!}f'(a) + \frac{h^2}{2!}f''(a) - \frac{h^3}{3!}f'''(a) + \frac{h^4}{4!}f^{iv}(a) - \frac{h^5}{5!}f^{v}(a) + \dots$$

Hence, assuming that the polynomial gives a reasonable approximation to the function:

$$f(a+h) - f(a-h) = 2hf'(a) + \frac{2h^3}{6}f'''(a) + \frac{2h^5}{5!}f^{v}(a) + \dots$$

and so we can write:

$$f'(a) = \frac{f(a+h)-f(a-h)}{2h} - \frac{1}{6}h^2 f'''(a) - \frac{1}{120}h^4 f^{v}(a) + \dots \qquad [9]$$

For the $x = a + 2h$ and $x = a - 2h$ terms, the Taylor equation gives:

$$p(a+2h) = f(a) + \frac{2h}{1!}f'(a) + \frac{4h^2}{2!}f''(a) + \frac{8h^3}{3!}f'''(a) + \frac{16h^4}{4!}f^{iv}(a) + \frac{32h^5}{5!}f^{v}(a) + \ldots$$

$$p(a-2h) = f(a) - \frac{2h}{1!}f'(a) + \frac{4h^2}{2!}f''(a) - \frac{8h^3}{3!}f'''(a) + \frac{16h^4}{4!}f^{iv}(a) - \frac{32h^5}{5!}f^{v}(a) + \ldots$$

Hence, assuming that the polynomial gives a reasonable approximation to the function:

$$f(a+2h) - f(a-2h) = 4hf'(a) + \frac{16h^3}{6}f'''(a) + \frac{64h^5}{120}f^{v}(a) + \ldots$$

and so:

$$f'(a) = \frac{f(a+2h) - f(a-2h)}{4h} - \frac{4}{6}h^2f'''(a) - \frac{16}{120}h^4f^{v}(a) + \ldots \qquad [10]$$

We can eliminate the h^2 term by subtracting equation [10] from four times equation [9] to give:

$$4f'(a) - 3f'(a) = 4\left[\frac{f(a+h) - f(a-h)}{2h}\right] - \left[\frac{f(a+2h) - f(a-2h)}{4h}\right] + \tfrac{1}{10}h^4f^{v}(a) + \ldots \qquad [11]$$

Hence:

$$f'(a) = \frac{f(a-2h) - 8f(a-h) + 8f(a+h) - f(a+2h)}{12h} + \tfrac{1}{30}h^4f^{v}(a) + \ldots \qquad [12]$$

Alternatively we can write equation [11] as:

$$f'(a) = \tfrac{1}{3}\left[4(\text{slope}_{2h}) - (\text{slope}_{h})\right] + \tfrac{1}{30}h^4f^{v}(a) + \ldots \qquad [13]$$

Equations [12] and [13] are termed the *five-point equations*.

Example

Use the five-point equation to determine the derivative at $x = 2.0$ for the function described by the following data:

y	10.889	12.703	14.778	17.149	19.855
x	1.8	1.9	2.0	2.1	2.2

Using equation [12]:

$$f'(x) = \frac{10.889 - 8 \times 12.703 + 8 \times 17.149 - 19.885}{12 \times 0.1} = 22.143$$

Note that the above data was for the function $y = x\ e^x$. This has the derivative $(x + 1)\ e^x$ and so the value at $x = 2.0$ of 22.167.

Revision

5 Using the five-point equation, determine the derivative of the function $y = \sin x$ at $x = 0.90$ with an increment of 0.1.

6 Using the five-point equation, determine the derivative at $x = 0.5$ of the function giving the following data:

y	0.18	0.32	0.50	0.72	0.98
x	0.3	0.4	0.5	0.6	0.7

Problems

1 Using the two-point equation, determine the derivative at $x = 0$ of the function giving the following data:

y	0.904 8	1.000 0	1.105 2
x	−0.1	0	0.1

2 Using the two-point equation, determine the derivative at $x = 3$ of the function $y = \ln x$ with an interval of 0.1 being used.

3 Using the three-point equation, determine the derivative at $x = 1$ of the function $y = 2 \sin x$ with an interval of 0.1 being used.

4 Using the three-point equation, determine the derivative at $x = 0.6$ of the function $y = \sin(e^x - 2)$ with an interval of 0.1 being used.

5 Using the five-point equation, determine the derivative at $x = 0.5$ of the function giving the following data:

y	0.740 8	0.670 3	0.606 5	0.548 8	0.496 6
x	0.3	0.4	0.5	0.6	0.7

6 Using the five-point equation, determine the derivative at $x = 2.3$ of the function $y = \tan x$ with an interval of 0.1 being used.

24 Partial differentiation

24.1 Introduction

For a function of a single variable $f(x)$ the derivative measures the rate at which the values of the function change as x changes. When we have functions of more than one variable $y = f(x_1, x_2, \ldots x_n)$ we can also be concerned with the rate at which the values of the function change but the issue now is with respect to which of the variables $x_1, x_2, \ldots x_n$. The term *partial derivative* is used for the rate at which the value of a function changes as one of the variables changes and all the other variables are held constant. Thus we obtain one partial derivative for each of the variables.

This chapter considers the notation, derivation and geometrical interpretation of partial derivatives. Chapter 22 on differentiation is assumed.

24.2 Partial differentiation

Situations often occur where we have a function of more than one variable. For example, the volume V of a cylinder of height h and radius r is given by $V = \pi r^2 h$. Thus the volume is a function of two independent variables r and h. We can write this as $V = f(r, h)$. As an another example, the power P developed when a current i passes through a resistor or resistance R is given by $P = i^2 R$. Thus the power is a function of two independent variables i and R, $P = f(i, R)$.

If we have $z = f(x, y)$ then a *partial derivative* is the rate at which the value of a function changes as one of the variables changes and all the other variables are held constant. In effect it is an ordinary derivative with respect to just one of the variables. In order to indicate that it is a partial derivative and there are other variables, a special sign ∂ is used instead of d in the derivative. Thus we have for $z = f(x, y)$ the partial derivatives:

$$\frac{\partial z}{\partial x} \text{ and } \frac{\partial z}{\partial y}$$

Since the partial derivatives are the ordinary derivatives with respect to just one variable:

$$\frac{\partial z}{\partial x} = \lim_{\delta x \to 0} \frac{f(x+\delta x, y) - f(x, y}{\delta x}$$

$$\frac{\partial z}{\partial y} = \lim_{\delta x \to 0} \frac{f(x, y+\delta x) - f(x, y}{\delta y} \qquad [1]$$

The rules and techniques used to obtain ordinary derivatives can thus be applied to the obtaining of partial derivatives. For example, we can use the quotient and product rules. The following examples illustrate the applications of such rules and techniques.

Example

If $z = x^3y + 2x^2 + y$, determine the partial derivatives $\partial z/\partial x$ and $\partial z/\partial y$ at the point $x = 1, y = 2$.

For $\partial z/\partial x$ we consider y to be a constant. Then:

$$\frac{\partial z}{\partial x} = 3x^2y + 4x$$

At the point (1, 2) the partial derivative has the value $3 \times 1 \times 1 + 4 \times 1 = 7$. For $\partial z/\partial y$ we consider x to be a constant. Then:

$$\frac{\partial z}{\partial y} = x^3 + 1$$

At the point (1, 2) the partial derivative has the value $1 + 1 = 2$.

Example

For the function $z = xy\,\mathrm{e}^x$, determine the partial derivatives $\partial z/\partial x$ and $\partial z/\partial y$.

For $\partial z/\partial x$ we consider y to be a constant. Then, using the product rule:

$$\frac{\partial z}{\partial x} = y(\mathrm{e}^x \times 1 + x\,\mathrm{e}^x) = y\,\mathrm{e}^x(1 + x)$$

For $\partial z/\partial y$ we consider x to be a constant. Then:

$$\frac{\partial z}{\partial y} = x\,\mathrm{e}^x$$

Example

For the function $z = \sin(x^2 + 2y)$, determine the partial derivatives $\partial z/\partial x$ and $\partial z/\partial y$.

For $\partial z/\partial x$ we consider y to be a constant. Letting $u = x^2 + 2y$, then $\partial u/\partial x = 2x$ and $\partial z/\partial u = \cos u$. Thus:

$$\frac{\partial z}{\partial x} = 2x\cos(x^2 + 2y)$$

For $\partial z/\partial y$ we consider x to be a constant. Letting $u = x^2 + 2y$, then $\partial u/\partial y = 2$ and $\partial z/\partial u = \cos u$. Thus:

$$\frac{\partial z}{\partial y} = 2\cos(x^2 + 2y)$$

Revision

1 Determine the partial derivatives $\partial z/\partial x$ and $\partial z/\partial y$ for the following functions:

(a) $z = x^3 + 4x^2y^3 + y^2 + 9$, (b) $z = x^2 + xy + y^2$, (c) $z = \sqrt{x+y}$,

(d) $z = \tan(2x + 3y)$, (e) $z = e^{x+2y}$

24.2.1 Higher order derivatives

Just as we have ordinary derivatives of derivatives, so we can have partial derivatives of derivatives. Thus if $z = f(x, y)$ we can determine $\partial z/\partial x$ and $\partial z/\partial y$ and then the second derivatives:

$$\frac{\partial}{\partial x}\left(\frac{\partial z}{\partial x}\right) = \frac{\partial^2 z}{\partial x^2},\ \frac{\partial}{\partial y}\left(\frac{\partial z}{\partial x}\right) = \frac{\partial^2 z}{\partial y\,\partial x},$$

$$\frac{\partial}{\partial y}\left(\frac{\partial z}{\partial y}\right) = \frac{\partial^2 z}{\partial y^2},\ \frac{\partial}{\partial x}\left(\frac{\partial z}{\partial y}\right) = \frac{\partial^2 z}{\partial x\,\partial y} \qquad [2]$$

There are four second derivatives. As the following example shows:

$$\frac{\partial}{\partial y}\left(\frac{\partial z}{\partial x}\right) = \frac{\partial^2 z}{\partial y\,\partial x} = \frac{\partial}{\partial x}\left(\frac{\partial z}{\partial y}\right) = \frac{\partial^2 z}{\partial x\,\partial y} \qquad [3]$$

Example

Determine the second derivatives of the function $z = 4x^2y + 2xy^3$.

The first derivative with respect to x, with y constant, is:

$$\frac{\partial z}{\partial x} = 8xy + 2y^3$$

The second derivative of $\partial z/\partial x$ with respect to x, with y constant is:

$$\frac{\partial^2 z}{\partial x^2} = 8y$$

The second derivative of $\partial z/\partial x$ with respect to y, with x constant is:

$$\frac{\partial^2 z}{\partial y\,\partial x} = 8x + 6y^2$$

The first derivative with respect to y, with x constant is:

$$\frac{\partial z}{\partial y} = 4x^2 + 6xy^2$$

The second derivative of $\partial z/\partial y$ with respect to y, with x constant, is:

$$\frac{\partial^2 z}{\partial y^2} = 12xy$$

The second derivative of $\partial z/\partial y$ with respect to x, with y constant, is:

$$\frac{\partial^2 z}{\partial x\,\partial y} = 8x + 6y^2$$

This is the same as the second derivative of $\partial z/\partial x$ with respect to y.

Example

If $V^2 = x^2 + y^2 + z^2$, show that:

$$\frac{\partial^2 V}{\partial x^2} + \frac{\partial^2 V}{\partial y^2} + \frac{\partial^2 V}{\partial z^2} = \frac{2}{V}$$

With $V = (x^2 + y^2 + z^2)^{1/2}$, if we let $u = x^2 + y^2 + z^2$ then $\partial u/\partial x = 2x$ and $\partial V/\partial u = ½u^{-1/2}$ and so:

$$\frac{\partial V}{\partial x} = \frac{x}{(x^2 + y^2 + z^2)^{1/2}}$$

Using the quotient rule:

$$\frac{\partial^2 V}{\partial x^2} = \frac{(x^2 + y^2 + z^2)^{1/2} \times 1 - x\dfrac{x}{(x^2 + y^2 + z^2)^{1/2}}}{(x^2 + y^2 + z^2)}$$

$$= \frac{y^2 + z^2}{(x^2 + y^2 + z^2)^{3/2}}$$

In a similar way:

$$\frac{\partial^2 V}{\partial y^2} = \frac{x^2 + z^2}{(x^2 + y^2 + z^2)^{3/2}} \quad \text{and} \quad \frac{\partial^2 V}{\partial z^2} = \frac{x^2 + y^2}{(x^2 + y^2 + z^2)^{3/2}}$$

Thus:

$$\frac{\partial^2 V}{\partial x^2} + \frac{\partial^2 V}{\partial y^2} + \frac{\partial^2 V}{\partial z^2} = \frac{(y^2 + z^2) + (x^2 + z^2) + (x^2 + y^2)}{(x^2 + y^2 + z^2)^{3/2}}$$

$$= \frac{2}{(x^2 + y^2 + z^2)^{1/2}} = \frac{2}{V}$$

Revision

2 Determine $\partial^2 z/\partial x^2$, $\partial^2 z/\partial y^2$ and $\partial^2 z/\partial x\,\partial y$ for the following functions:

(a) $z = 2x^2 - 5y^3 + 3$, (b) $z = 4x^2y^3 + 3x^3 + 6y^2$, (c) $z = e^{-x} \sin y$,

(d) $z = \frac{x-y}{x+y}$, (e) $z = 3x^2 + y^2 + e^{xy}$, (f) $z = x^2y - \cos 3xy^2$

3 If $V = \ln(x^2 + y^2)$, show that:

$$\frac{\partial^2 V}{\partial x^2} + \frac{\partial^2 V}{\partial y^2} = 0$$

24.3 The chain rule

The chain rule can be extended to cases where z is a function of x and y with x and y themselves being functions of a single independent variable t. For example, we might have $z = x^2y + y^2$ with $x = \sin t$ and $y = e^t$. In such a case, a change in t will give rise to changes in both x and y and so:

$$\frac{\partial z}{\partial t} = \lim_{\delta t \to 0} \frac{f(x+\delta x, y+\delta y) - f(x,y)}{\delta t} \qquad [4]$$

If we subtract and add the quantity $f(x, y + \delta y)$ to the numerator:

$$\frac{\partial z}{\partial t} = \lim_{\delta t \to 0} \frac{f(x+\delta x, y+\delta y) - f(x,y+\delta y) + f(x,y+\delta y) - f(x,y)}{\delta t}$$

$$= \lim_{\delta t \to 0} \frac{f(x+\delta x, y+\delta y) - f(x,y+\delta y)}{\delta x}\frac{\delta x}{\delta t}$$

$$+ \lim_{\delta t \to 0} \frac{f(x+\delta x, y+\delta y) - f(x+\delta x,y)}{\delta y}\frac{\delta y}{\delta t}$$

$$= \frac{\partial z}{\partial x}\frac{dx}{dt} + \frac{\partial z}{\partial y}\frac{dy}{dt} \qquad [5]$$

If we have the situation where z is a function of x and y with x and y both being functions of s and t then similar analysis yields:

$$\frac{\partial z}{\partial s} = \frac{\partial z}{\partial x}\frac{\partial x}{\partial s} + \frac{\partial z}{\partial y}\frac{\partial y}{\partial s} \qquad [6]$$

$$\frac{\partial z}{\partial t} = \frac{\partial z}{\partial x}\frac{\partial x}{\partial t} + \frac{\partial z}{\partial y}\frac{\partial y}{\partial t} \qquad [7]$$

Example

Determine $\partial z/\partial t$ when $z = x^2y + y^2$ and $x = \sin t$ and $y = e^t$.

Using equation [5]:

$$\frac{\partial z}{\partial t} = \frac{\partial z}{\partial x}\frac{dx}{dt} + \frac{\partial z}{\partial y}\frac{dy}{dt} = 2xy\cos t + (x^2 + 2y)\,e^t$$

Example

Determine $\partial z/\partial r$ and $\partial z/\partial h$ when $z = xy$ with $x = s^2 + t^2$ and $y = s/t$.

With t held constant, equation [6] gives:

$$\frac{\partial z}{\partial s} = \frac{\partial z}{\partial x}\frac{\partial x}{\partial s} + \frac{\partial z}{\partial y}\frac{\partial y}{\partial s} = y(2s) + x\left(\frac{1}{t}\right)$$

With s held constant, equation [7] gives:

$$\frac{\partial z}{\partial t} = \frac{\partial z}{\partial x}\frac{\partial x}{\partial t} + \frac{\partial z}{\partial y}\frac{\partial y}{\partial t} = y(2t) + x\left(\frac{-s}{t^2}\right)$$

Example

The height and radius of a cylinder are increasing at the rate of 2 cm/s and 1 cm/s respectively. At what rate is the volume of the cylinder increasing when the height is 10 cm and the radius 4 cm?

The volume V of the cylinder is related to the height h and the radius r by $V = \pi r^2 h$ with r and h being functions of time t. Using equation [5]:

$$\frac{\partial V}{\partial t} = \frac{\partial V}{\partial r}\frac{\mathrm{d}r}{\mathrm{d}t} + \frac{\partial V}{\partial h}\frac{\mathrm{d}h}{\mathrm{d}t} = 2\pi rh\frac{\mathrm{d}r}{\mathrm{d}t} + \pi r^2\frac{\mathrm{d}h}{\mathrm{d}t}$$

$$= 2\pi \times 4 \times 10 \times 1 + \pi \times 4^2 \times 2 = 112\pi \text{ cm}^3/\text{s}$$

Revision

4 Determine $\partial z/\partial t$ for $z = \sin(3x - y)$ when $x = 2t^2$ and $y = t^2 - 2t$.

5 Determine $\partial z/\partial t$ for $z = x^2 + y^2$ when $x = \mathrm{e}^t$ and $y = \mathrm{e}^{-t}$.

6 Determine $\partial z/\partial t$ and $\partial z/\partial s$ for $z = \mathrm{e}^{xy}$ when $x = 3t + 2s$ and $y = 4t - 2s$.

7 Determine $\partial z/\partial t$ and $\partial z/\partial s$ for $z = \mathrm{e}^x \cos y$ when $x = s^2 - t^2$ and $y = 2st$.

8 A right-angled triangle has a perpendicular of height y, a base of z and an hypotenuse of x. Determine the rate at which z is changing when $x = 5$ cm and $y = 3$ cm if x increases at 4 cm/s and y at 2 cm/s.

9 A box with sides of length x, y and z has the x length increasing at the rate of 2 cm/s, the y length increasing at the rate of 3 cm/s and the z length decreasing at the rate of 4 cm/s. Determine the rate of change of the volume when $x = y = 20$ cm and $z = 40$ cm.

10 If $z = x^3 \sin 2y$, determine the rate of change of z when x is 2 units and increasing at 0.4 units/s and y is $\pi/6$ and increasing as 0.1 rad/s.

24.4 The total differential

Consider a function $z = f(x, y)$ with two variables x and y. If x increases by an increment δx and y increases by an increment δx, then the corresponding incremental change in z of δz is given by:

$$\delta z = f(x + \delta x, y + \delta y) - f(x, y)$$

If we subtract $f(x, y + \delta y)$ from the first term and add it to the second term we have:

$$\delta z = [f(x + \delta x, y + \delta y) - f(x, y + \delta y)] + [f(x, y + \delta y) - f(x, y)]$$

Hence we can write:

$$\delta z = \frac{f(x+\delta x, y+\delta y) - f(x, y+\delta y)}{\delta x}\delta x + \frac{f(x, y+\delta y) - f(x, y)}{\delta y}\delta y$$

In the limit we could replace the fractional terms in the above equation by the partial derivatives, δx, δy and δz by dx, dy and dz, and so:

$$\mathrm{d}z = \frac{\partial f(x,y)}{\partial x}\mathrm{d}x + \frac{\partial f(x,y)}{\partial y}\mathrm{d}y \qquad [9]$$

dz is called the *differential*, the term *total differential* often being used when the differential refers to the variation of the function with respect to all the independent variables. If we had just one variable, i.e. $z = f(x)$ then the above equation becomes (see Section 22.9.2):

$$\mathrm{d}z = f'(x)\,\mathrm{d}x \qquad [10]$$

where $f'(x)$ is the derivative dz/dx of the function. If we assume δx and δy are small we can approximate equations [9] and [10] to:

$$\delta z \approx \frac{\partial f(x,y)}{\partial x}\delta x + \frac{\partial f(x,y)}{\partial y}\delta y \qquad [11]$$

and:

$$\delta z \approx f'(x)\,\delta x \qquad [12]$$

Such equations are widely used in considering the effects of small changes. For example, we can use them to determine the greatest possible overall error that is produced in a quantity as a result of errors in its variables. Thus if the side of a cube is measured as 2.4 ± 0.1 cm, we can use equation [12] to compute the resulting greatest possible error in the volume. If x is the length of a side then the volume $V = x^3$ and dV/d$x = 3x^2$. Thus:

$$\delta V \approx (3x^2)\,\delta x \approx 3 \times 2.4^2 \times 0.1 \approx 1.9 \text{ cm}^3$$

If we had a rectangle with sides measured as 4.2 ± 0.1 cm and 8.1 ± 0.1 cm, then we can use equation [11] to determine the greatest possible error in the area derived from these measurements. Area $A = xy$, where x and y are the lengths of the sides of the rectangle. Since $\partial A/\partial x = y$ and $\partial A/\partial y = x$:

$$\delta A \approx y\,\delta x + x\,\delta y \approx 8.1 \times 0.1 + 4.2 \times 0.1 \approx 1.23 \text{ cm}^2$$

Example

Determine the total differential dz of the function $z = f(x, y) = xy^2$.

The partial derivative of the function with respect to x is y^2 and the partial derivative with respect to y is $2xy$. Thus, using equation [9]:

$$\mathrm{d}z = y^2\,\mathrm{d}x + 2xy\,\mathrm{d}y$$

Example

Determine the greatest possible error in the volume of a cylinder calculated from measurements of its radius as 6 ± 0.01 cm and its height as 20 ± 0.1 cm.

Since the volume V is given by $V = \pi r^2 h$, where r is the radius and h the height, then equation [11] gives:

$$\delta V \approx 2\pi rh\,\delta r + \pi r^2\,\delta h$$

$$\approx 2\pi \times 6 \times 20 \times 0.01 + \pi \times 6^2 \times 0.1 \approx 18.8 \text{ cm}^3$$

Revision

11 Determine the total differential dz of the following functions:

(a) $z = x^2y^3$, (b) $z = 3x/y$, (c) $z = x^y$, (d) $z = \sin(x + y)$

12 Determine the percentage increase in the volume of a cylinder if the radius increases by 2% and the height decreases by 1%.

13 A box has a height with a possible error of 3% of the height and a square base with a side with a possible error of 2% of the length. Determine the maximum possible percentage error in the volume.

14 The area A of a triangle is given by $A = \frac{1}{2}ac \sin B$, where B is the angle between the sides a and c. If measurements give $a = 8 \pm 0.2$ cm, $b = 12 \pm 0.2$ cm and $B = 45° \pm 3°$, estimate the maximum possible error in the area calculated from these measurements.

15 The periodic time T of a simple pendulum is given by $T = 2\pi\sqrt{(L/g)}$. Show that $\delta T/T = \frac{1}{2}(\delta L/L - \delta g/g)$.

24.5 Geometric interpretation

In two-dimensional geometry the sets of points satisfying the equation $x = a$ are in a line parallel to the x-axis, the points satisfying $y = b$ being in a line parallel to the y-axis (Figure 24.1). With three dimensions the set of points satisfying the equation $x = a$ form a vertical plane parallel to the y-z plane, the points satisfying $y = b$ forming a vertical plane parallel to the x-z plane (Figure 24.2).

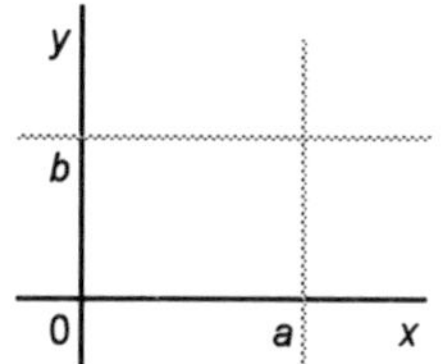

Figure 24.1 *Two-dimensions*

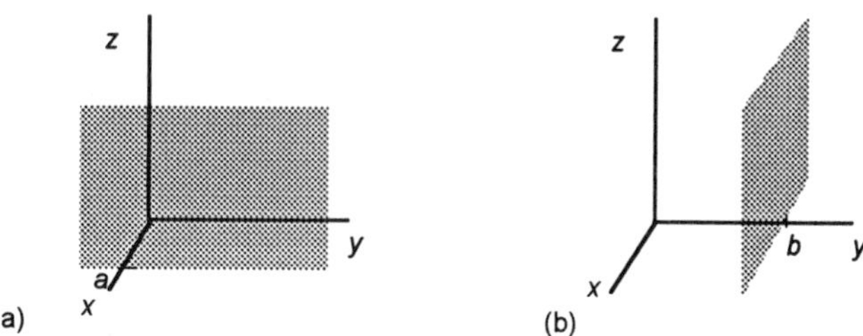

Figure 24.2 *Three dimensions: parts of the planes satisfying (a) $x = a$ and (b) $y = b$*

Now consider the geometric interpretation of the partial derivatives of a function of two variables $z = f(x, y)$ at some point where $x = a$ and $y = b$. If x is held constant as $x = a$ then the point (a, y, z) will lie on the curve that is the intersection of the surface specified by the function $z = f(x, y)$ and the plane $x = a$ (Figure 24.3(a)). Then $\partial z/\partial y$ is the slope of the tangent in the y direction at the point of intersection. Likewise, $\partial z/\partial x$ is the slope of the tangent in the x direction at the point of intersection of the plane $y = b$ and the surface (Figure 24.3(b)). The values of $\partial z/\partial y$ and $\partial z/\partial x$ at a point thus represent the slope of the surface in the x and y directions at that point.

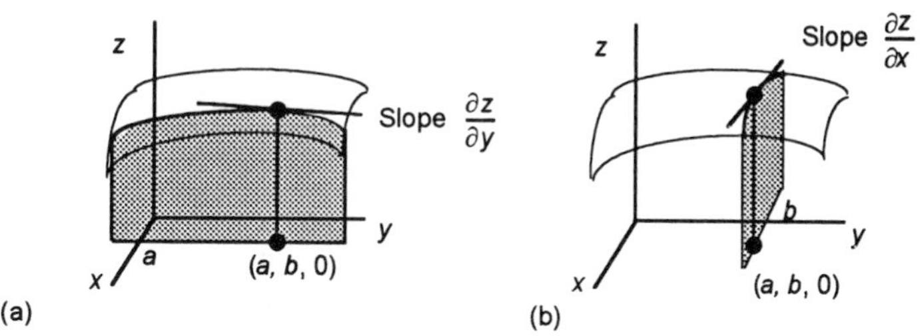

Figure 24.3 *The partial differentials as the slopes of tangents*

Example

Determine the slope of the surface given by $z = f(x, y) = x^2 + y^2 + 2$ at the point (1, 2, 7) in the x and y directions.

In the x direction:

$$\text{slope} = \frac{\partial z}{\partial x} = 2x = 2$$

In the y direction:

$$\text{slope} = \frac{\partial z}{\partial y} = 2y = 4$$

Revision

16 Determine the slopes of the surfaces in the x and y directions given by the following functions at the specified points:

(a) $z = xy^2 + 2xy + y^3$ at $(-1, 2, 0)$, (b) $z = xy^2$ at $(1, 2, 4)$,

(c) $z = x^2 + xy + xy^2$ at $(1, 1, 3)$

24.5.1 Maxima and minima

Figure 24.4(a) shows a function $f(x, y)$ with a *maximum*, the slope of the tangent lines in both the x and y directions being zero at that point. Thus, if this point is (a, b):

$$\frac{\partial}{\partial x}f(x,y) = 0 \text{ and } \frac{\partial}{\partial y}f(x,y) = 0 \text{ at } (a, b) \qquad [13]$$

and $f(a, b)$ is greater than $f(x, y)$ at all points in its immediate locality. Figure 24.4(b) shows a function $f(x, y)$ with a *minimum*, the slope of the tangent lines on both the x and y directions being zero at that point. Thus if the point is (a, b):

$$\frac{\partial}{\partial x}f(x,y) = 0 \text{ and } \frac{\partial}{\partial y}f(x,y) = 0 \text{ at } (a, b) \qquad [14]$$

and $f(a, b)$ is less than $f(x, y)$ at all points in its immediate locality.

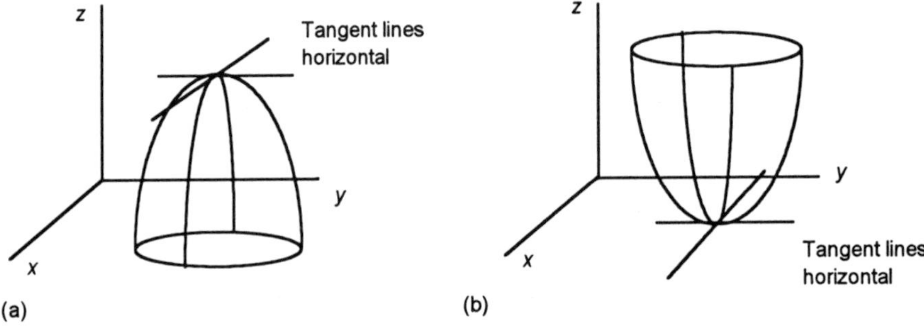

Figure 24.4 *(a) A maximum, (b) a minimum*

For a maximum or a minimum a necessary condition is that there is zero slope in both the x and y directions. Such points are termed *stationary points*. For the point to be maximum $f(a, b)$ must be greater than $f(x, y)$ at all points in its immediate locality; for a minimum $f(a, b)$ must be less than $f(x, y)$ at all points in its immediate locality. The condition that there is zero slope in both the x and y directions, but not the condition that $f(a, b)$ is less or greater than $f(x, y)$ at all points in its immediate locality, is also met by *saddle points*. For the saddle point shown in Figure 24.5, for the y direction we have a minimum but for the x direction a maximum,

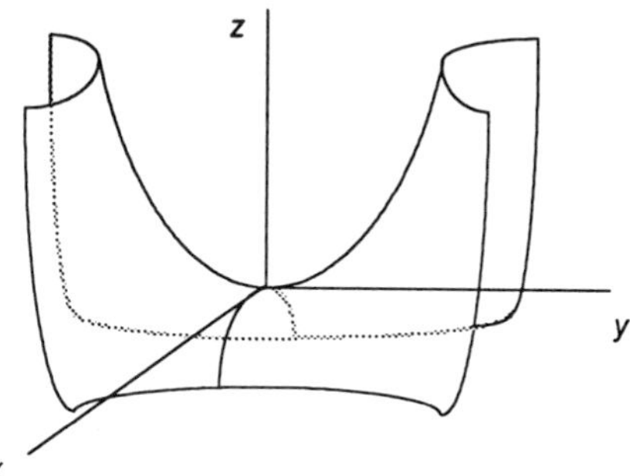

Figure 24.5 *A saddle point*

As with functions of a single variable, we can test whether a point with zero slope in the x and y directions is a maximum or minimum by using the second derivatives (no proof is given here):

1 *Maximum*
   (a, b) is a maximum if:

$$\frac{\partial^2}{\partial x^2}f(a,b)\frac{\partial^2}{\partial y^2}f(a,b)-\left[\frac{\partial^2}{\partial x\,\partial y}f(a,b)\right]^2>0$$

   and:

$$\frac{\partial^2}{\partial x^2}f(a,b)<0 \text{ or } \frac{\partial^2}{\partial y^2}f(a,b)<0$$

2 *Minimum*
   (a, b) is a minimum if:

$$\frac{\partial^2}{\partial x^2}f(a,b)\frac{\partial^2}{\partial y^2}f(a,b)-\left[\frac{\partial^2}{\partial x\,\partial y}f(a,b)\right]^2>0$$

   and:

$$\frac{\partial^2}{\partial x^2}f(a, b) > 0 \text{ or } \frac{\partial^2}{\partial y^2}f(a, b) > 0$$

3 *Saddle*
(a, b) is a saddle point if:

$$\frac{\partial^2}{\partial x^2}f(a, b)\frac{\partial^2}{\partial y^2}f(a, b) - \left[\frac{\partial^2}{\partial x \partial y}f(a, b)\right]^2 < 0$$

Example

Determine the stationary points of $z = 2x^2 - 2x + xy + 3y + y^2$ and whether they are minimum, maximum or saddle points.

Since:

$$\frac{\partial z}{\partial x} = 4x - 2 + y \text{ and } \frac{\partial z}{\partial y} = x + 3 + 2y$$

there will be stationary points when $4x - 2 + y = 0$ and $x + 3 + 2y = 0$. If we subtract four times the second equation from the first we obtain $-14 - 7y = 0$ and so $y = -2$. Substituting this in either of the equations gives $x = 1$. These values give $z = -4$ and so the only stationary point is $(1, -2, -4)$. To apply the second partial derivatives test we need:

$$\frac{\partial^2 z}{\partial x^2} = 4, \ \frac{\partial^2 z}{\partial y} = 2, \ \frac{\partial^2 z}{\partial x \partial y} = 1$$

and so:

$$\frac{\partial^2}{\partial x^2}f(a, b)\frac{\partial^2}{\partial y^2}f(a, b) - \left[\frac{\partial^2}{\partial x \partial y}f(a, b)\right]^2 = 4 \times 2 - 1 = 7$$

It is greater than 0 and so there is a maximum or a minimum. Since both $\partial^2 z/\partial x^2$ and $\partial^2 z/\partial y^2$ are greater than 0 the point is a minimum.

Example

A rectangular box with an open top is to be made from thin sheet metal with an internal volume of 1 m^3. What must be the dimensions of the sides for the minimum amount of sheet metal to be used?

If the box has sides of lengths y and z and a height x, then the volume $V = xyz$ and the area A of sheet metal used is $A = 2xz + 2zy + xy$. We can write the area in terms of the volume by substituting V/xy for z. Thus:

$$A = \frac{2V}{y} + \frac{2V}{x} + xy$$

and so:

$$\frac{\partial A}{\partial x} = -\frac{2V}{x^2} + y \text{ and } \frac{\partial A}{\partial y} = -\frac{2V}{y^2} + x$$

Equating these to zero gives $-2V + x^2y = 0$ and $-2V + y^2x = 0$. These simultaneous equations give $x = y = (2V)^{1/3} = 1.26$ m. This gives $Z = V/xy = 0.63$ m.

We can check that these values give a minimum by using the second partial derivative test.

$$\frac{\partial^2 A}{\partial x^2} = \frac{4V}{x^3}, \frac{\partial^2 A}{\partial y^2} = \frac{4V}{y^3}, \frac{\partial^2 A}{\partial x \partial y} = 1$$

and so:

$$\frac{\partial^2}{\partial x^2}f(a,b)\frac{\partial^2}{\partial y^2}f(a,b) - \left[\frac{\partial^2}{\partial x \partial y}f(a,b)\right]^2 = \frac{4V}{x^3}\frac{4V}{y^3} - 1$$

With $x = y = (2V)^{1/3}$, this gives $4 - 1 = 3$ and so as it is greater than 0 we have either a maximum or a minimum. Since both $\partial^2 z/\partial x^2$ and $\partial^2 z/\partial y^2$ are greater than 0 the point is a minimum.

Revision

17 Determine the stationary points of the following functions and whether they are maximum, minimum or saddle points:

(a) $z = 2x^2 + 8x - 6y + y^2 + 20$, (b) $z = x^2 + 3xy + y^2$,

(c) $z = 2x^2 + 2xy + 2x + y^2 - 3$, (d) $z = 2x^3 - 6x^2 - 8y^2 + 2$,

(e) $z = x^2 + \frac{2}{x} + \frac{2}{y} + y^2$

18 A company finds that the profit P from selling x items of product A and y items of product B per month is $P = -2x^2 + xy + 49y - y^2 + 2000$. What values of x and y will maximise the profit?

19 Determine the three positive numbers which have a sum of 30 and a maximum product.

24.5.2 Contour maps

One way we can visualise a function of two variables $z = f(x, y)$ is by a *contour map* in which the values of z are assigned to the (x, y) points. An example of this is a geographical map where the height above sea level is a function of position (x, y). The contour lines on such a map are plots of constant height lines. A contour map of the surface $z = f(x, y)$ depicts the

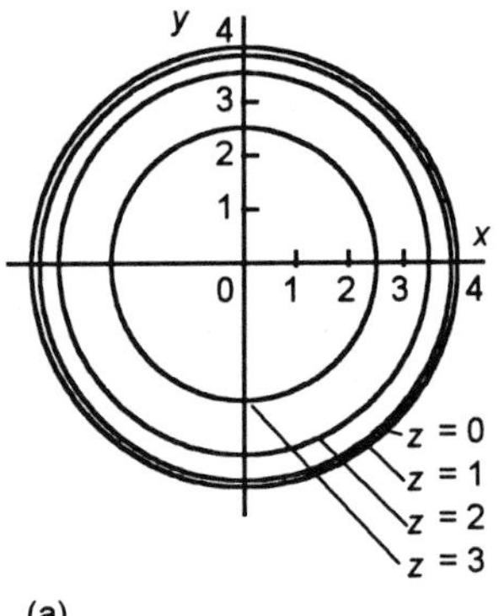

(a)

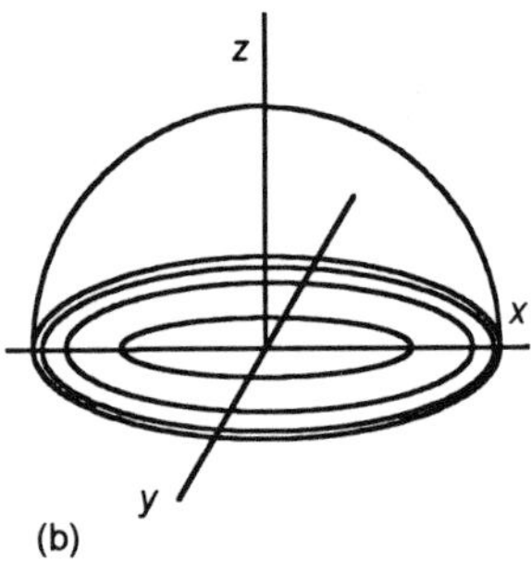

(b)

Figure 24.6 *Contour map*

variation of z with respect to x and y by the spacing between lines of constant height.

Consider the function $z = f(x, y) = \sqrt{(16 - x^2 - y^2)}$. For each contour line we let $z = f(x, y)$ = a constant and sketch the results in the x-y plane. With $z = 0$: when $x = 0$ then $y = 4$, when $y = 0$ then $x = 4$. In fact, the contour line is a circle of radius 4. With $z = 1$: when $x = 0$ then $y = \sqrt{15}$, when $y = 0$ then $x = \sqrt{15}$. In fact, the contour line is a circle of radius $\sqrt{15}$. With $z = 2$: when $x = 0$ then $y = \sqrt{12}$, when $y = 0$ then $x = \sqrt{12}$. In fact, the contour line is a circle of radius $\sqrt{12}$. With $z = 3$: when $x = 0$ then $y = \sqrt{7}$, when $y = 0$ then $x = \sqrt{7}$. In fact, the contour line is a circle of radius $\sqrt{7}$. Figure 24.6(a) shows the resulting contour lines.

If we examine the function for stationary points then:

$$\frac{\partial z}{\partial x} = \frac{x}{\sqrt{16 - x^2 - y^2}} \text{ and } \frac{\partial z}{\partial y} = \frac{y}{\sqrt{16 - x^2 - y^2}}$$

Equating these to zero gives a stationary point at (0, 0). The second derivative test can be carried out to show that the point is a maximum. Figure 24.6(b) shows the form of the surface with its contour map superimposed.

As another example, consider the function $z = f(x, y) = y^2 - x^2$. With $z = 0$: when $x = 0$ then $y = 0$, with $x = 1$ then $y = \pm 1$, with $x = 2$ then $y = \pm 2$, with $x = -1$ then $y = \pm 1$, with $x = -2$ then $y = \pm 2$. The contour lines are thus straight lines through the origin. With $z = 4$ we have $4 = y^2 - x^2$ which is the equation of a hyperbola with intercept on the x-axis of 2. With all the z values, other than 0, we obtain hyperbolas. Figure 24.7(a) shows the contour map.

If we examine the surface for stationary points:

$$\frac{\partial z}{\partial x} = -2x \text{ and } \frac{\partial z}{\partial y} = 2y$$

A stationary point thus ocurs at (0, 0). Carrying out the second derivative test:

$$\frac{\partial^2 z}{\partial x^2} = -2, \ \frac{\partial^2 z}{\partial y^2} = 2, \ \frac{\partial^2 z}{\partial x \, \partial y} = 0$$

Thus:

$$\frac{\partial^2}{\partial x^2} f(a, b) \frac{\partial^2}{\partial y^2} f(a, b) - \left[\frac{\partial^2}{\partial x \, \partial y} f(a, b) \right]^2$$

is less than 0. Thus we have a saddle point. Figure 24.7(b) shows the shape of the surface.

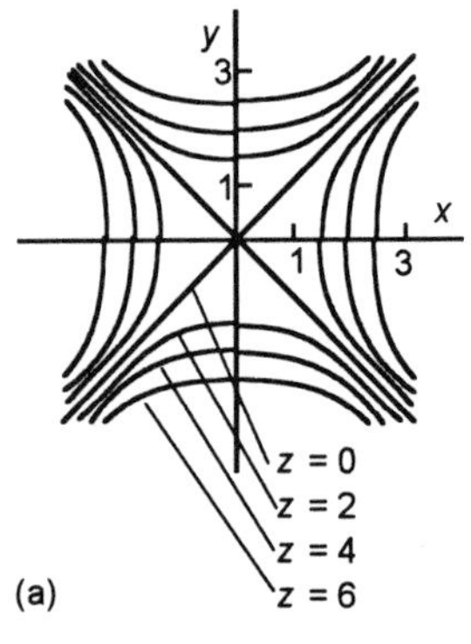

(a)

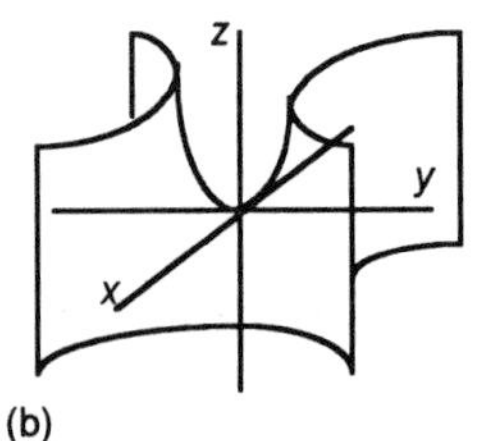

(b)

Figure 24.7 *Contour lines*

Revision

20 Sketch the contour lines, indicating any stationary points for:

(a) $z = xy$, (b) $z = (x-1)^2 + (y-2)^2$

Problems

1 Determine the partial derivatives $\partial z/\partial x$ and $\partial z/\partial y$ for the following functions:

(a) $z = xy^2 + 3x^2y - x + 4$, (b) $z = 3x + 5y + 2$, (c) $z = (2x + 5y)^3$,

(d) $z = \dfrac{x-y}{x+y}$, (e) $z = \tan^{-1}\left(\dfrac{x}{y}\right)$, (f) $z = 3\,\mathrm{e}^x + 2\,\mathrm{e}^y + x^2y^3 + 3$,

(g) $z = \sin(x^2 - 3y)$, (h) $z = \mathrm{e}^{xy}\cos x$, (i) $y = \ln xy$, (j) $z = \mathrm{e}^{-5t}\sin(x - 2t)$

2 If $z = \ln(\mathrm{e}^x + \mathrm{e}^y)$, show that:

$$\frac{\partial z}{\partial x} + \frac{\partial z}{\partial y} = 1$$

3 Determine $\partial^2 z/\partial x^2$, $\partial^2 z/\partial y^2$ and $\partial^2 z/\partial x\,\partial y$ for the following functions:

(a) $z = 2x^2 + xy - y^2$, (b) $z = xy + \sin x$, (c) $z = 5x + 7y$,

(d) $z = x\sin y + y\sin x$, (e) $z = \sin xy$

4 If $z = (x^2 + y^2)/(x + y)$, show that:

$$x^2\frac{\partial^2 z}{\partial x^2} + 2xy\frac{\partial^2 z}{\partial x\,\partial y} + y^2\frac{\partial^2 z}{\partial y^2} = 0$$

5 If $r^2 = (x-a)^2 + (y-b)^2 + (z-c)^2$, show that:

$$\frac{\partial^2 r}{\partial x^2} + \frac{\partial^2 r}{\partial y^2} + \frac{\partial^2 r}{\partial z^2} = \frac{2}{r}$$

6 If $z = \mathrm{e}^x(\sin y + \cos y)$, show that:

$$\frac{\partial^2 z}{\partial x^2} + \frac{\partial^2 z}{\partial y^2} = 0$$

7 If $V = (Ar^n + B/r^n)\cos(n\theta - a)$, show that:

$$\frac{\partial^2 V}{\partial r^2} + \frac{1}{r}\frac{\partial V}{\partial r} + \frac{1}{r^2}\frac{\partial^2 V}{\partial \theta^2} = 0$$

8 Determine $\partial z/\partial t$ for $z = x^2y^3$ when $x = 2t^2$ and $y = 2t$.

9 Determine $\partial z/\partial t$ for $z = x^4y^4$ when $x = \cos 2t$ and $y = \sin 3t$.

10 Determine $\partial z/\partial t$ and $\partial z/\partial s$ for $z = x^2 + y^2$ with $x = s + t$ and $y = s - t$.

11 Determine $\partial z/\partial t$ and $\partial z/\partial s$ for $z = x^2 - y^2$ with $x = s \cos t$ and $y = s \sin t$.

12 The area A of a triangle with angle B between two sides of lengths a and c is given by $A = \frac{1}{2}ac \sin B$. If a is increasing at the rate of 2 mm/s, c is decreasing at 4 mm/s and B is increasing as 0.01 rad/s, determine the rate of change of the area when $a = 30$ mm, $c = 40$ mm and B is $\pi/6$ radians.

13 Determine the total differentials dz of the following functions:

(a) $z = x^2y$, (b) $z = x^3 + y^2$, (c) $z = 2x + \cos y$, (d) $z = x \ln y$

14 The pressure p, volume V and temperature T of an ideal gas are related by the equation $pV = RT$, where R is a constant. Determine the total differential dp in terms of p, V and T.

15 A cylinder has a radius which is measured as 2 ± 0.1 cm and a height as 20 ± 0.2 cm. Estimate the maximum possible error in the volume calculated from these measurements.

16 The area A of a triangle is given by $A = \frac{1}{2}ac \sin B$, where B is the angle between the sides a and c. Sides b and c have an error of ±1% and the angle B is 45° with an error of ±1%. Determine the maximum possible percentage error in the computed area.

17 The deflection at the centre of a particular loaded beam is proportional to WL^3/d^4. What is the percentage increase in the deflection if W is increased by 2%, L increased by 3% and d decreased by 2%.

18 The height of a mast is calculated from the measured elevation h of the top of the mast at a measured horizontal distance x from its base. Determine the error in the calculated height due to errors δh and δx in the measurements.

19 Determine the slopes of the surfaces in the x and y directions given by the following functions at the specified points:

(a) $z = x/y$ at (1, 1, 1), (b) $z = x\,e^y + y^2\,e^x$ at (0, 1, 1),

(c) $z = \dfrac{xy}{x - y}$ at (2, −2, −1)

20 Determine the stationary points of the following functions and whether they are maximum, minimum or saddle points:

(a) $z = 2x^2 - 4x - 12y + 3y^2 + 13$, (b) $z = -2x^2 + 4x + 2y - 2xy - y^2$,

(c) $z = x^3 - 3xy + y^3$, (d) $z = -x^2 - xy - y^2 + 2$, (e) $z = x^4 + 2x^2y^2 + y^4$

21 A company finds that the profit P from selling a x items of product A and y items of product B per month is given by:

$$P = 8x + 10y - 0.01(x^2 + xy + y^2) - 5000$$

What values of x and y will maximise the profit?

22 Determine the three positive numbers which have a sum of 30 and the sum of the squares is a minimum.

23 A rectangular box with an open top is to be made from thin sheet metal with an internal volume of 32 m^3. What must be the dimensions of the sides for the minimum amount of sheet metal to be used?

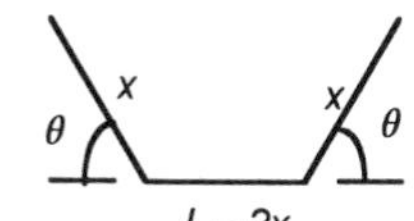

Figure 24.8 *Problem 24*

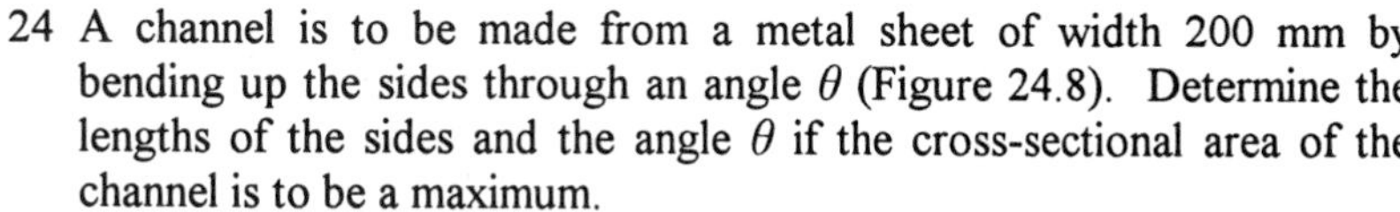

24 A channel is to be made from a metal sheet of width 200 mm by bending up the sides through an angle θ (Figure 24.8). Determine the lengths of the sides and the angle θ if the cross-sectional area of the channel is to be a maximum.

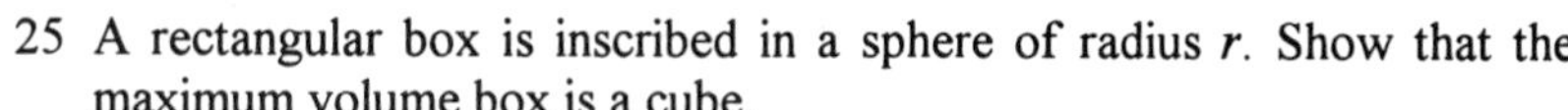

25 A rectangular box is inscribed in a sphere of radius r. Show that the maximum volume box is a cube.

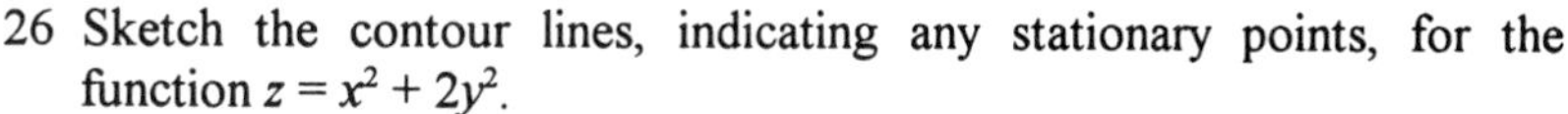

26 Sketch the contour lines, indicating any stationary points, for the function $z = x^2 + 2y^2$.

25 Integration

25.1 Introduction

This chapter introduces integration with the definite integral being considered as giving information about the area under a curve and the indefinite integral as the inverse of differentiation. The integrals of common functions and the basic techniques of integration, i.e. those of substitution, integration by parts and the use of partial fractions, are developed. Chapter 26 considers numerical integration and Chapter 27 gives some applications of integration.

Chapter 23 on differentiation, familiarity with functions and the principles of partial fractions (Chapter 2) are assumed.

25.2 Antiderivative

If we are given some function we can differentiate it to give the derivative dy/dx, it being the rate at which y changes with respect to x. Now consider the reverse situation where we are given the rate at which y changes with respect to x as some function $f(x)$ and want to find the function $F(x)$ that could have produced it. We talk of finding the *antiderivative* of the rate.

A function $F(x)$ is called an antiderivative of $f(x)$ if $\frac{d}{dx}F(x) = f(x)$.

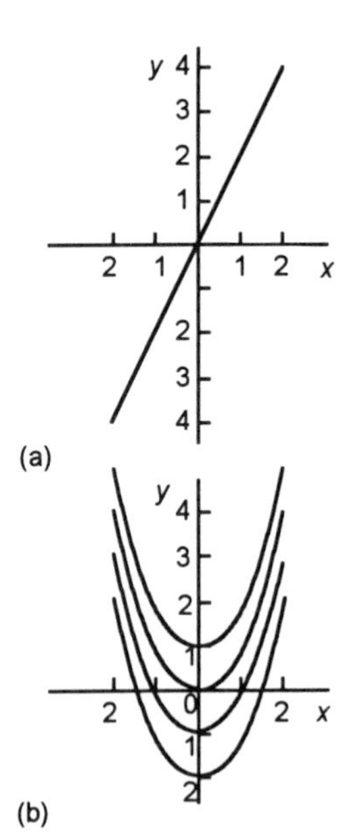

Figure 25.1 *Each of the graphs in (b) has the same slope at the same value of x and hence the same slope function (a)*

For example, we might be given the derivative as $3x^2$. From our knowledge of how powers are differentiated we might suggest that the antiderivative is x^3. This is, however, only one possible solution. If we differentiate x^3, or $x^3 + 1$, or $x^3 + 2$, or indeed x^3 plus any constant, we obtain $3x^2$. Thus, in general, the antiderivative of $3x^2$ is $3x^2 + C$, where C is some constant. This applies whenever we obtain antiderivatives, whatever the function.

If $F(x)$ is a particular antiderivative of $f(x)$ then all the antiderivatives are given by $F(x) + C$, where C is a constant.

An antiderivative is more usually called an *indefinite integral* and the process of deriving it *integration*.

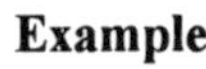

Example

If $f(x) = 2x$, plot graphs of the antiderivatives of the function.

$F(x) = 2x$ (Figure 25.1(a)). If we differentiate $f(x) = x^2 + C$ we can obtain $F(x)$ and so $x^2 + C$ is the equation of the antiderivatives. Figure 25.1(b) shows part of the family of curves for the antiderivative $f(x)$, the constant C resulting in a family of functions, all of which have a derivative of $2x$.

Example

Determine the antiderivatives of sin $2x$.

We have to find y such that $dy/dx = \sin 2x$. If we differentiate cos $2x$ we obtain $-2 \sin 2x$. To obtain an answer of sin $2x$ after differentiating we need to differentiate $-\frac{1}{2} \cos 2x$. Thus an antiderivative of sin $2x$ is $-\frac{1}{2} \cos 2x$. We could also obtain sin $2x$ as a result of differentiating $-\frac{1}{2} \cos 2x + C$, where C is a constant. Thus the antiderivatives of sin $2x$ are $-\frac{1}{2} \cos 2x + C$.

Revision

1 Determine the antiderivatives of (a) e^{2x}, (b) $1/x$, (c) $2x$.

25.2.1 Table of antiderivatives

Since antidifferentiation is the inverse of differentiation, a table of derivatives can be read backwards in order to give antiderivatives. Table 25.1 illustrates this for some commonly encountered functions.

Table 25.1 *Antiderivatives*

Function $f(x)$	Antiderivative $F(x)$
a (a constant)	$ax + C$
x^n, except $n = 1$	$\frac{1}{n+1}x^{n+1} + C$
$\frac{1}{x}$	$\ln x + C$ if $x > 0$, $\ln(-x) + C$ if $x < 0$
e^{ax}	$\frac{1}{a}e^{ax} + C$
$\sin ax$	$-\frac{1}{a}\cos ax + C$
$\cos ax$	$\frac{1}{a}\sin ax + C$

25.2.2 Properties of antiderivatives

If we differentiate $x^2 + x$ we differentiate each term independently to obtain $2x + 1$. An antiderivative of $2x + 1$ is thus $x^2 + x$, with the antiderivative being taken of each term independently.

The antiderivative of a sum of two functions equals the sum of the individual antiderivatives.

If we differentiate $5x^2$ we differentiate the x^2 and multiply it by the constant of 5 to give $5(2x)$. An antiderivative of $5 \times 2x$ is thus $5x^2$.

The antiderivative of af(x) equals the antiderivative of f(x) multiplied by the constant a.

Example

Determine the antiderivatives of $e^{-4x} + 2x + 1$.

If we differentiate e^{ax} we obtain $a\,e^{ax}$. Thus an antiderivative of the e^{-4x} term is $-\frac{1}{4}\,e^{-4x}$. If we differentiate x^2 we obtain $2x$. Thus an antiderivative of $2x$ is x^2. If we differentiate x we obtain 1. Thus an antiderivative of 1 is x. Thus a particular antiderivative of $e^{-4x} + 2x + 1$ is $-\frac{1}{4}\,e^{-4x} + x^2 + x$. All the antiderivatives are $-\frac{1}{4}\,e^{-4x} + x^2 + x + C$, where C is a constant.

Revision

2 Determine the antiderivatives of:

(a) $5x^4 + 4x^3 + 3x^2$, (b) $2 \cos 2x + 4 \sin 2x$, (c) $2x^{3/2} - 4x^{1/2}$

25.3 Definite integral

Consider the problem of finding the area under a curve $y = f(x)$ between the vertical lines $x = a$ and $x = b$ and the x-axis (Figure 25.2). We can determine the area by dividing the area into a number of equal width vertical strips (Figure 25.3), each of which we can effectively consider to be rectangular. Then the area under the curve is the sum of the areas of each of these rectangular strips. A strip has an area of approximately $y\,\delta x$, where y is the effective value of the function for the strip and δx its width. The total area A under the curve between $x = a$ and $x = b$ is thus:

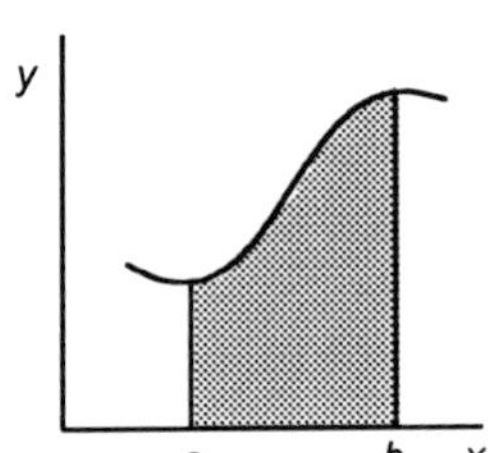

Figure 25.2 *Area under a curve*

$$A \approx \sum_{x=a}^{x=b} y\,\delta x \text{ or } \sum_{x=a}^{x=b} f(x)\,\delta x$$

The smaller we make the width of the strips the better the sum of the strip areas will approximate to the actual area under the curve. In the limit, when $\delta x \to 0$, we can write:

$$A = \lim_{\delta x \to 0} \sum_{x=a}^{x=b} f(x)\,\delta x \qquad [1]$$

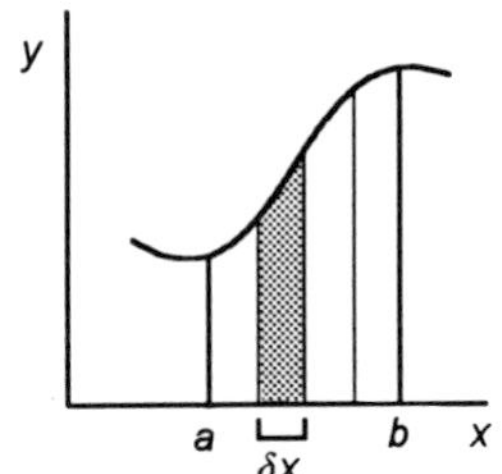

Figure 25.3 *Area of an element*

A special sign is used to indicate the summation with $\delta x \to 0$, the summation being written as:

$$A = \int_a^b f(x)\,dx \qquad [2]$$

The sign is really just an extended letter S to indicate summation. The expression is called a *definite integral* and equation [2] is read as 'the integral of $f(x)$ dx from a to b'. The operation of determining the area is termed integration and the function to be integrated $f(x)$ is termed the

integrand. x is the name of the variable of integration with a being the *lower limit* and b the *upper limit* between which the integration is carried out.

The area between $x = a$ and $x = b$ can be considered to be the total area $A(b)$ under the curve from $x = 0$ to $x = b$ minus the area $A(a)$ from $x = 0$ to $x = a$. Thus:

$$A = \int_a^b f(x)\, dx = A(b) - A(a)$$

The expression $A(b) - A(a)$ is generally written using a special square-bracket notation as $\left[A(x)\right]_a^b$ which is evaluated as $[A(x)]^b - [A(x)]^a$.

Note that since:

$$\int_a^b f(x)\, dx = A(b) - A(a) \quad \text{and} \quad \int_b^a f(x)\, dx = A(a) - A(b)$$

when we interchange the limits, the sign of the integral changes.

$$\int_a^b f(x)\, dx = -\int_b^a f(x)\, dx \qquad [3]$$

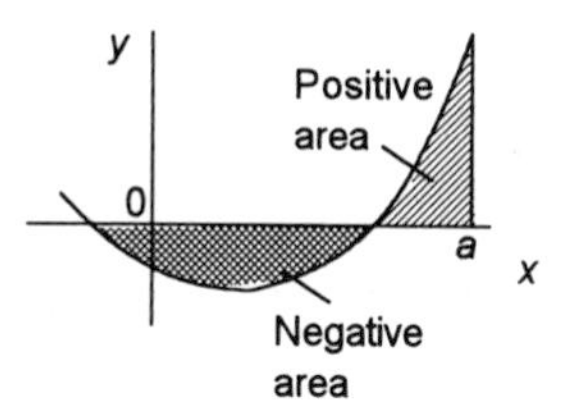

Figure 25.4 *Signs of areas*

The area under the graph in Figure 25.3 is all above the x-axis. If for a certain range of x, the graph is below the x-axis (Figure 25.4), then the values of y in $\Sigma y\, \delta x$ will be negative and so its contribution to the summation will be negative. Thus, for Figure 25.4, the integration between the limits 0 and a will include a negative term for the negative area element and a positive term for the positive area element. The result will be the difference between the two areas.

25.3.1 Definite and indefinite integrals

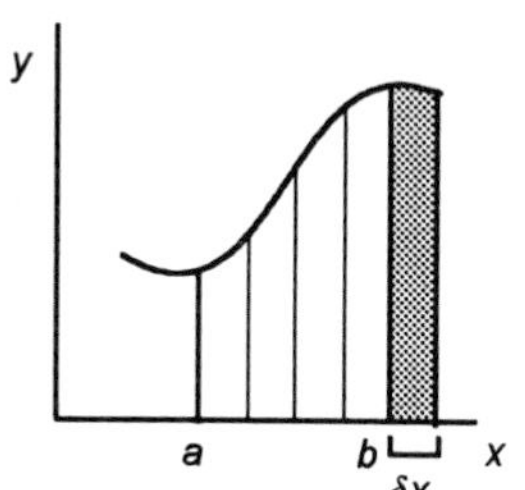

Figure 25.5 *Area increased by an element*

Suppose we increase the area under the curve by one strip (Figure 25.5). The area δA of a strip under the curve is approximated by $f(x)\, \delta x$. Thus we can write for the rate of change of area with x:

$$\frac{\delta A}{\delta x} \approx f(x)$$

In the limit when $\delta x \to 0$, then:

$$\frac{dA}{dx} = f(x)$$

Thus A is the antiderivative of $f(x)$. But the integration of the above equation gives the area under a curve. We can thus write, when we have no limits specified for the area:

$$A = \int f(x)\, dx$$

The notation $\int f(x)\,\mathrm{d}x$ *with no limits specified stands for any anti-derivative of f(x).*

$\int_a^b f(x)\,\mathrm{d}x$ is termed a definite integral because it takes a definite value, representing an area under a curve. When we evaluate such an integral there is no constant in the answer. However, the indefinite integral $\int f(x)\,\mathrm{d}x$ does not represent a number but an antiderivative which is a function. When we evaluate such an integral there will be a constant in the answer.

$\int_a^b f(x)\,\mathrm{d}x$ *gives the area under the curve of the function f(x) between* $x = a$ *and* $x = b$ *and:*

$$\int_a^b f(x)\,\mathrm{d}x = \left[F(x)\right]_a^b = F(b) - F(a) \qquad [4]$$

where F(x) is any antiderivative of f(x) and F(b) and F(a) are particular values of the antiderivative.

Example

Evaluate (a) $\int_2^3 x^3\,\mathrm{d}x$, (b) $\int_{-1}^2 5\,\mathrm{d}x$, (c) $\int_{-1}^3 2x\,\mathrm{d}x$

(a) Using equation [4] and the antiderivative for a power given in Table 25.1:

$$\int_2^3 x^3\,\mathrm{d}x = \left[\frac{x^4}{4}\right]_2^3 = \frac{3^4}{4} - \frac{2^4}{4} = \frac{65}{4}$$

(b) Using equation [4] and the antiderivative for a constant given in Table 25.1:

$$\int_{-1}^2 5\,\mathrm{d}x = [5x]_{-1}^2 = 5 \times 2 - 5 \times (-1) = 15$$

(c) Using equation [4] and the antiderivative for a power given in Table 25.1:

$$\int_{-1}^3 2x\,\mathrm{d}x = \left[2\frac{x^2}{2}\right]_{-1}^3 = 9 - (-1)^2 = 8$$

Example

Find the areas under the curve $y = x^3$ between (a) $x = 0$ and $x = 1$, (b) $x = -1$ and $x = 1$.

(a) Figure 25.6 shows the graph. Using equation [4] and the antiderivative for a power given in Table 25.1, the area is:

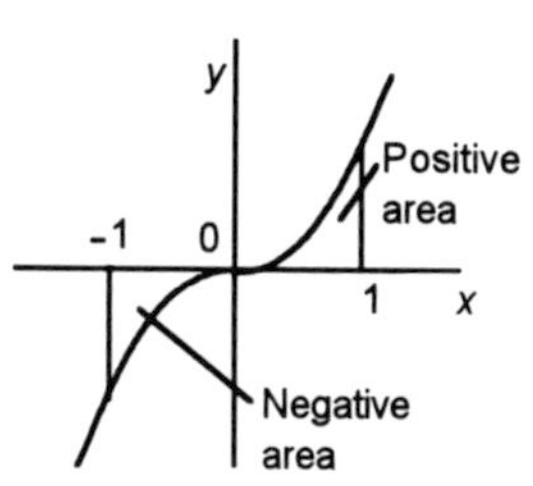

Figure 25.6 *Example*

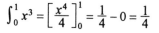

$$\int_0^1 x^3 = \left[\frac{x^4}{4}\right]_0^1 = \frac{1}{4} - 0 = \frac{1}{4}$$

(b) The 'mathematical area' or 'signed area', i.e. the area taking into account the sign of y, is:

$$\int_{-1}^1 x^3 = \left[\frac{x^4}{4}\right]_{-1}^1 = \frac{1}{4} - \frac{1}{4} = 0$$

The area is zero because the area between $x = -1$ and $x = 0$ is negative. What we have is the sum of two areas:

$$\int_{-1}^1 x^3 = \int_{-1}^0 x^3 \, dx + \int_0^1 x^3 \, dx$$

$$\int_{-1}^0 x^3 \, dx = \left[\frac{x^4}{4}\right]_{-1}^0 = 0 - \frac{1}{4} = -\frac{1}{4}$$

$$\int_0^1 x^3 \, dx = \left[\frac{x^4}{4}\right]_0^1 = \frac{1}{4} - 0 = \frac{1}{4}$$

The sum of the two areas is thus zero. If we want the total area, regardless of sign, between the axis and the curve for any function then we have to determine the positive and negative elements separately and then, ignoring the sign, add them. For this curve this gives ½.

Revision

3 Determine the areas under the following curves between the specified limits and the x-axis:

(a) $y = x^2$ between $x = 0$ and $x = 2$,

(b) $y = e^x$ between $x = 0$ and $x = 2$,

(c) $y = 3x$ between $x = -1$ and $x = 2$,

(d) $y = 3$ between $x = -1$ and $x = 2$,

(e) $y = 2/x$ between $x = 1$ and $x = 2$

25.3.2 Basic properties of definite integrals

The following are some basic properties of definite integrals.

1 *Multiplication by a constant*

$$\int_a^b kf(x) \, dx = k \int_a^b f(x) \, dx \qquad [5]$$

All this is saying is that if you multiply a function by a constant you multiply the area under the function curve by the same factor.

2 *Addition of functions*

$$\int_a^b [f(x)+g(x)]\,\mathrm{d}x = \int_a^b f(x)\,\mathrm{d}x + \int_a^b g(x)\,\mathrm{d}x \qquad [6]$$

If you have a curve which is the sum of two functions then the area under the curve is the sum of the areas under the curve due to each function.

3 *Integration between limits of a and a*

$$\int_a^a f(x)\,\mathrm{d}x = 0 \qquad [7]$$

The definite integral over an interval of length zero is zero. This corresponds to the fact that a rectangle of width zero has zero area.

4 *Integration between limits of a and b with an intermediate limit c*

$$\int_a^b f(x)\,\mathrm{d}x = \int_a^c f(x)\,\mathrm{d}x + \int_c^b f(x)\,\mathrm{d}x \qquad [8]$$

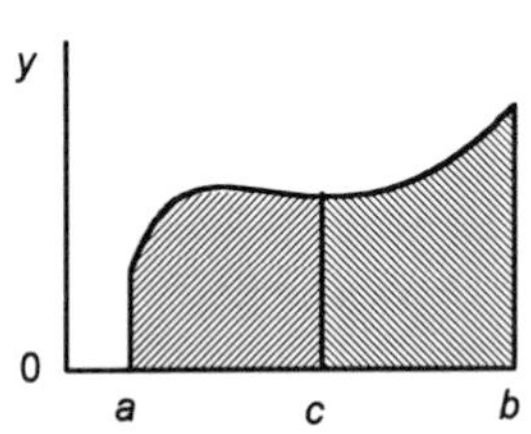

Figure 25.7 *The sum of two areas*

Figure 25.7 illustrates this. The area between limits a and b is equal to the sum of the areas between c and a and b and c. This property is particularly useful in integrating a continuous function which is defined in two pieces as the example that follows illustrates.

Example

Determine the area specified by the integral $\int_{-1}^{2} |x|\,\mathrm{d}x$.

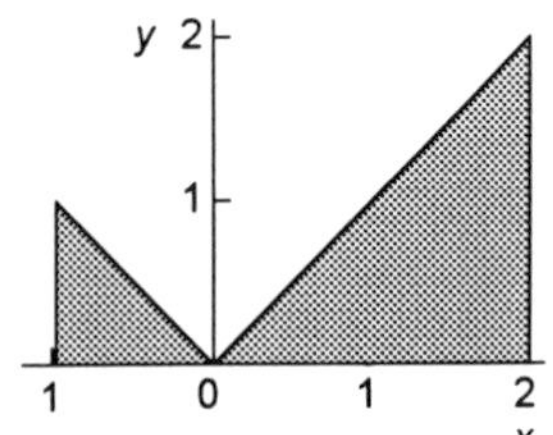

Figure 25.8 *Example*

$y = |x|$ means that regardless of the sign of x, only positive values of y result. Figure 25.8 shows the graph between the specified limits. Using equation [8] we can write:

$$\int_{-1}^{2} |x|\,\mathrm{d}x = \int_{-1}^{0} |x|\,\mathrm{d}x + \int_{0}^{2} |x|\,\mathrm{d}x = \int_{-1}^{0} (-x)\,\mathrm{d}x + \int_{0}^{2} x\,\mathrm{d}x$$

$$= \left[-\frac{x^2}{2}\right]_{-1}^{0} + \left[\frac{x^2}{2}\right]_{0}^{2} = 0 - \left(-\frac{1}{2}\right) + \frac{4}{2} + 0 = \frac{5}{2}$$

Note: there is no antiderivative formula for $|x|$.

Example

Determine the area specified by the integral $\int_{-1}^{3} (x^2 - 2x - 3)\,\mathrm{d}x$.

Using equation [6]:

$$\int_{-1}^{3}(x^2-2x-3)\,dx=\int_{-1}^{3}x^2\,dx-\int_{-1}^{3}2x\,dx-\int_{-1}^{3}3\,dx$$

$$=\left[\frac{x^3}{3}\right]_{-1}^{3}-\left[2\frac{x^2}{2}\right]_{-1}^{3}-[3x]_{-1}^{3}$$

$$=9-9-9-(-\tfrac{1}{3}-1+3)=-10\tfrac{2}{3}$$

Example

Determine the area of the region bounded above by the graph of $y = e^x$ and below by $y = 1/x$ for the interval between $x = 1$ and $x = 2$.

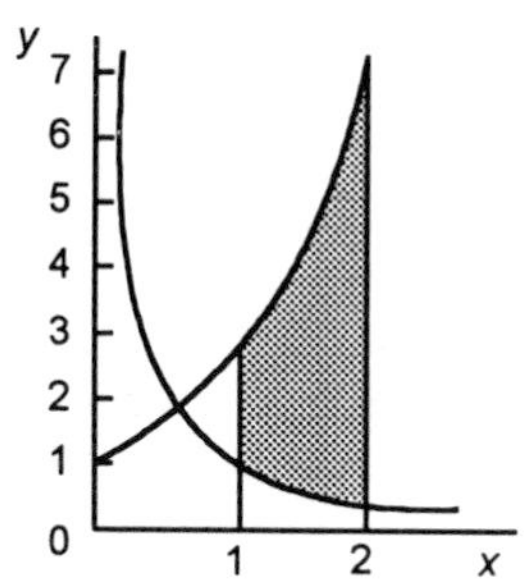

Figure 25.9 *Example*

Figure 25.9 shows the area. We need to determine the area under the $y = e^x$ graph between the specified limits and then subtract from it the area under the $y = 1/x$ graph with the same limits. The area under the e^x graph is:

$$\int_{1}^{2}e^x\,dx$$

and that under the $1/x$ graph:

$$\int_{1}^{2}\frac{1}{x}\,dx$$

Thus the required area is:

$$\int_{1}^{2}e^x\,dx-\int_{1}^{2}\frac{1}{x}\,dx=[e^x-\ln x]_1^2=(e^2-\ln 2)-(e^1-\ln 1)=3.98$$

Revision

4 Evaluate the following definite integrals:

(a) $\int_{0}^{2}(x+4)\,dx$, (b) $\int_{1}^{5}(x^2-6)\,dx$, (c) $\int_{-2}^{2}|x|\,dx$, (d) $\int_{1}^{4}|x-3|\,dx$

5 Determine the areas bounded by graphs of the following functions and between the specified ordinates:

(a) $y = x + 1$, $y = 1 - 2x$, $x = 0$ and $x = 2$,

(b) $y = x^2$, $y = x - 5$, $x = -1$ and $x = 2$

6 The graph of the function $y = 6 - x - x^2$ cuts the x-axis in two points. Determine the area bounded by the curve and the x-axis and between these points of intersection with the x-axis.

7 Determine the area bounded by graphs of $y = x^3$ and $y = x$. Hint: find the points of intersection and determine the difference in areas between these points.

8 Determine the area bounded by the curve $y^2 = x$ and the y-axis and between the limits $y = 0$ and $y = 3$. Hint: treat this as x being a function of y and hence the area as being the definite integral of $f(y)$.

25.3.3 Improper integrals

If a definite integral is integrated between limits for which one or both are infinite or the integrand becomes infinite at some point between the limits, then the integral is said to be *improper*. Consider the integral of $1/x^2$ between 1 and infinity:

$$\int_1^\infty \frac{1}{x^2}\,\mathrm{d}x = \left[-\frac{1}{x}\right]_1^\infty = 0 - (-1) = 1$$

Because the area is finite (Figure 25.10), the area can readily be evaluated. We can think of the area under such a curve in terms of the integral between 1 and some finite limit b and consider what happens as b tends to an infinite value.

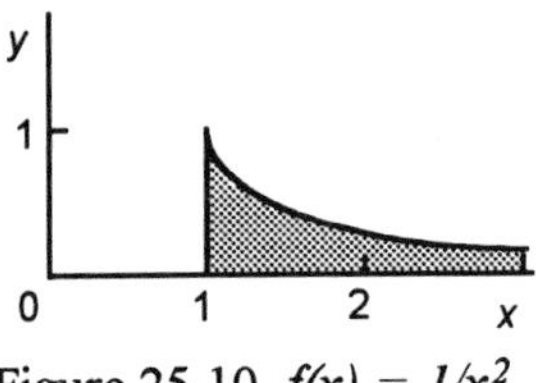

Figure 25.10 $f(x) = 1/x^2$

$$\int_1^\infty \frac{1}{x^2}\,\mathrm{d}x = \lim_{b\to\infty}\left(\int_1^b \frac{1}{x^2}\,\mathrm{d}x\right) = \lim_{b\to\infty}\left[-\frac{1}{x}\right]_1^b = \lim_{b\to\infty}\left(-\frac{1}{b}+1\right)$$

As b gets bigger and bigger, the contribution of the $1/b$ term to the area becomes less and less. Thus in the limit the area converges to 1.

Now consider the integral of $1/x$ between 1 and infinity:

$$\int_1^\infty \frac{1}{x}\,\mathrm{d}x = [\ln x]_1^\infty = \ln\infty - \ln 1$$

Considered in terms of the integral between 1 and some limit b as b tends to an infinite value:

$$\int_1^\infty \frac{1}{x}\,\mathrm{d}x = \lim_{b\to\infty}\left(\int_1^b \frac{1}{x}\,dx\right) = \lim_{b\to\infty}[\ln x]_1^b = \lim_{b\to\infty}(\ln b - \ln 1)$$

The logarithm becomes infinite as b becomes infinite and so the area is infinite. In the limit the area does not converge to a finite value.

In the case of functions which go off to infinity at some point between their limits, we can consider the area under the curve in two parts, from the lower limit up to the ordinate at which it goes off to infinity and then from that ordinate to the upper limit.

Sometimes we need to evaluate an integral from $-\infty$ to $+\infty$. If the function is continuous for all values between these limits then we evaluate such an integral in two parts, from $-\infty$ to 0 and then from 0 to $+\infty$.

$$\int_{-\infty}^\infty f(x)\,\mathrm{d}x = \int_{-\infty}^0 f(x)\,\mathrm{d}x + \int_0^\infty f(x)\,\mathrm{d}x \qquad [5]$$

Consider the integral of e^x between $-\infty$ and $+\infty$:

$$\int_{-\infty}^{\infty} e^x \, dx = \int_{-\infty}^{0} e^x \, dx + \int_{0}^{\infty} e^x \, dx$$

The first integral gives:

$$\int_{-\infty}^{0} e^x \, dx = [e^x]_{-\infty}^{0} = e^0 - e^{-\infty} = 1 - 0$$

The second integral gives:

$$\int_{0}^{\infty} e^x \, dx = [e^x]_{0}^{\infty} = e^{\infty} - e^0 = \infty - 1$$

Thus the first integral gives an area of 1 and the second integral goes off to infinity. Thus the integral of e^x between $-\infty$ and $+\infty$ does not converge.

Revision

9 Determine the values, if they exist, of the following definite integrals:

$$\text{(a)} \int_{0}^{\infty} 2x \, dx, \text{ (b)} \int_{1}^{\infty} \frac{1}{\sqrt{x}} \, dx, \text{ (c)} \int_{-\infty}^{0} e^x \, dx, \text{ (d)} \int_{-\infty}^{\infty} e^{2x} \, dx$$

25.4 Techniques of integration

Techniques for integrating functions discussed in this section are *integration by substitution*, *integration by parts* and *integration using partial fractions*.

25.4.1 Integration by substitution

This involves simplifying integrals by making a substitution. The term *integration by change of variable* is often used since the variable has to be changed as a result of the substitution. The aim of making a substitution is to put the integral into a simpler form for integration.

The method requires an integral of $f(x)$ to be put in terms of $f(u)$, where u is some function of x, say $u = g(x)$. We also have to change the variable dx to du and this is done by differentiating u with respect to x. Thus du/dx = $g'(x)$ and so we can substitute $(1/g'(x))$ du for dx. Thus:

$$\int f(x) \, dx = \int f(u) \frac{1}{g'(x)} \, du \qquad [6]$$

There are no general rules for finding suitable substitutions. The following show some of the more commonly used substitutions.

1 Integral of the form $\int f(ax + b) \, dx$
Let $u = ax + b$. Then, since du/dx = a we have dx = $(1/a)$ du and so substituting for $ax + b$ and dx we have:

$$\frac{1}{a}\int f(u)\,\mathrm{d}u$$

Example

Determine the indefinite integral $\int (4x + 1)^3\,\mathrm{d}x$.

If we let $u = 4x + 1$ then $\mathrm{d}u/\mathrm{d}x = 4$ and $\mathrm{d}x = \frac{1}{4}\,\mathrm{d}u$:

$$\int (4x+1)^3\,\mathrm{d}x = \int u^3 \frac{1}{4}\,\mathrm{d}u = \frac{1}{4}\frac{u^4}{4} + C = \frac{1}{16}(4x+1)^4 + C$$

2 Integral of the form $\int xf(ax^2 + b)\,\mathrm{d}x$
Let $u = ax^2 + b$. Then $\mathrm{d}u/\mathrm{d}x = 2ax$ and so $x\,\mathrm{d}x = (1/2a)\,\mathrm{d}u$. Making substitutions, the integral can be written as:

$$\frac{1}{2a}\int f(u)\,\mathrm{d}u$$

The above is an example of where the function to be integrated is a product of x^{n-1} and some function of x^n or $(ax^n + b)$, the substitution $u = x^n$ or $u = ax^n + b$ being used.

Example

Determine the indefinite integral $\int \frac{x}{3x^2+4}\,\mathrm{d}x$.

If we let $u = 3x^2 + 4$, then $\mathrm{d}u/\mathrm{d}x = 6x$ and so $x\,\mathrm{d}x = (1/6)\,\mathrm{d}u$. Hence:

$$\int \frac{x}{3x^2+4}\,\mathrm{d}x = \int \frac{1}{u}\frac{1}{6}\,\mathrm{d}u = \frac{1}{6}\int \frac{1}{u}\,\mathrm{d}u$$

$$= \frac{1}{6}\ln|u| + C = \frac{1}{6}\ln(3x^2+4) + C$$

The modulus sign is used with the integration of $1/u$ because no assumption is made at that stage as to whether u is positive or negative. The sign is dropped when the substitution is made because $3x^2 + 4$ is always positive.

3 Integrals of the form $\int \cos^m ax \sin^n ax\,\mathrm{d}x$, when n is odd.
Let $u = \cos ax$. Then $\mathrm{d}u/\mathrm{d}x = -a \sin ax$ and $\sin ax\,\mathrm{d}x = -(1/a)\,\mathrm{d}u$. Since n is odd, $\sin^{n-1}$ is even and we can replace it using $\sin^2 ax = 1 - \cos^2 ax$.

Example

Determine the indefinite integral $\int \cos^2 x \sin^3\,\mathrm{d}x$.

If we let $u = \cos x$, then $\mathrm{d}u/\mathrm{d}x = -\sin x$ and so $\sin x\,\mathrm{d}x = -\mathrm{d}u$. The integral then can be written as:

$$\int \cos^2 x \sin^3 x \, dx = \int \cos^2 x \sin^2 x \sin x \, dx$$

$$= \int \cos^2 x (1 - \cos^2 x) \sin x \, dx$$

$$= \int u^2 (1 - u^2) \, du = \int (u^2 - u^4) \, du$$

$$= \frac{u^3}{3} - \frac{u^5}{5} + C = \tfrac{1}{3} \cos^3 x - \tfrac{1}{5} \cos^5 x + C$$

4 Integral of the form $\int \cos^m ax \sin^n ax \, dx$, when m is odd.
Let $u = \sin ax$. Then $du/dx = a \cos ax$ and $\cos ax \, dx = (1/a) \, du$. Since m is odd $\cos^{m-1}$ is even and we can replace this using $\cos^2 ax = 1 - \sin^2 ax$.

Example

Determine the indefinite integral $\int \cos x \sin^2 x \, dx$.

Let $u = \sin x$. Then $du/dx = \sin x$ and so $\sin x \, dx = du$. The integral can then be written as:

$$\int \cos x \sin^2 x \, dx = \int \sin^2 x \cos x \, dx$$

$$= \int u^2 \, du = \frac{u^3}{3} + C = \tfrac{1}{3} \sin^3 x + C$$

5 Integral of the form $\int \cos^m ax \sin^n ax \, dx$, when m and n are both even or both odd.
The integral should be rewritten using relationships such as:

$$\sin^2 x = \tfrac{1}{2}(1 - \cos 2x), \ \cos^2 x = \tfrac{1}{2}(1 + \cos 2x),$$

$$\sin x \cos x = \tfrac{1}{2} \sin 2x$$

before integrating using a suitable substitution.

Example

Determine the indefinite integral $\int \cos^4 x \sin^2 x \, dx$.

The integral can be rewritten as:

$$\int \cos^4 x \sin^2 x \, dx = \int \tfrac{1}{4}(1 + \cos 2x)^2 \tfrac{1}{2}(1 - \cos 2x) \, dx$$

$$= \int \tfrac{1}{8}(1 + \cos 2x - \cos^2 2x - \cos^3 2x) \, dx$$

This requires further rewriting to give:

$$\int \cos^4 x \sin^2 x \, dx = \int \tfrac{1}{8} \, dx + \tfrac{1}{8} \int \cos 2x \, dx - \tfrac{1}{8} \int \tfrac{1}{2}(1 + \cos 4x) \, dx$$
$$- \tfrac{1}{8} \int \cos^2 2x \cos 2x \, dx$$

$$= \int \tfrac{1}{8} \, dx + \tfrac{1}{8} \int \cos 2x \, dx - \int \tfrac{1}{16} \, dx$$
$$- \tfrac{1}{16} \int \cos 4x \, dx - \tfrac{1}{8} \int (1 - \sin^2 2x) \cos 2x \, dx$$

$$= \int \tfrac{1}{8} \, dx + \tfrac{1}{8} \int \cos 2x \, dx - \int \tfrac{1}{16} \, dx$$
$$- \tfrac{1}{16} \int \cos 4x \, dx - \tfrac{1}{8} \int \cos 2x \, dx$$
$$+ \tfrac{1}{8} \int \sin^2 2x \cos 2x \, dx$$

This last integral can be determined using the substitution $u = \sin 2x$.

$$\tfrac{1}{8} \int \sin^2 2x \cos 2x \, dx = \tfrac{1}{8} \int \tfrac{1}{2} u^2 \, du = \tfrac{1}{48} u^3 + C = \tfrac{1}{48} \sin^3 2x + C$$

Hence:

$$\int \cos^4 x \sin^2 x \, dx = \tfrac{1}{8} x + \tfrac{1}{16} \sin 2x - \tfrac{1}{16} x - \tfrac{1}{32} \sin 4x - \tfrac{1}{16} \sin 2x$$
$$+ \tfrac{1}{48} \sin^3 2x + C$$

6 Integrals involving $\sqrt{(a^2 - x^2)}$ terms
Let $x = a \sin \theta$. Then $\sqrt{(a^2 - x^2)} = \sqrt{(a^2 - a^2 \sin^2 \theta)} = a \cos \theta$, since we have $1 - \sin^2 \theta = \cos^2 \theta$. Since $dx/d\theta = a \cos \theta$ then $dx = a \cos \theta \, d\theta$. Alternatively we could let $x = a \tanh \theta$.

Example

Determine the indefinite integral $\int \sqrt{(1 - x^2)} \, dx$.

Let $x = \sin h$. Then $\sqrt{(1 - x^2)} = \sqrt{(1 - \sin^2 \theta)} = \cos \theta$. Since $dx/d\theta = \cos \theta$ then $dx = \cos \theta \, d\theta$. Thus the integral becomes:

$$\int \sqrt{(1 - x^2)} \, dx = \int \cos \theta \cos \theta \, d\theta = \int \cos^2 \theta \, d\theta$$

Since $\cos 2\theta = 2 \cos^2 \theta - 1$ (see Section 3.4.1), we have:

$$\int \tfrac{1}{2}(1 + \cos 2\theta) \, d\theta = \tfrac{1}{2} \theta + \tfrac{1}{4} \sin 2\theta + C$$

Back substitution using $\theta = \sin^{-1} x$ gives:

$$\tfrac{1}{2} \sin^{-1} x + \tfrac{1}{4} \sin(2 \sin^{-1} x) + C$$

However, a simpler expression is obtained if we first replace the $\sin 2\theta$ using $\sin 2\theta = 2 \sin \theta \cos \theta = 2 \sin\theta \sqrt{(1 - \sin^2 \theta)}$.

$$\tfrac{1}{2} \theta + \tfrac{1}{4} \sin 2\theta + C = \tfrac{1}{2} \sin^{-1} x + \tfrac{1}{2} x \sqrt{1 - x^2} + C$$

7 Integrals involving $\sqrt{(a^2 + x^2)}$ terms
Let $x = a \tan \theta$. Then $\sqrt{(a^2 + x^2)} = \sqrt{(a^2 + a^2 \tan^2 \theta)} = a \sec \theta$, since we have $\tan^2 \theta + 1 = \sec^2 \theta$ (see Section 3.4.1). $dx/d\theta = \sec^2 \theta$ and so $dx = \sec^2 \theta \, d\theta$. Alternatively we could let $x = a \sinh \theta$.

Example

Determine the indefinite integral $\int \frac{1}{x^2+4} \, dx$.

Let $x = 2 \tan \theta$. Then $dx/d\theta = 2 \sec^2 \theta$ and so:

$$\int \frac{1}{x^2+4} \, dx = \int \frac{1}{4\tan^2\theta+4} 2\sec^2\theta \, d\theta = \int \frac{1}{4\sec^2\theta} 2\sec^2\theta \, d\theta$$

$$= \int \tfrac{1}{2} \, d\theta = \tfrac{1}{2}\theta + C = \tfrac{1}{2}\tan^{-1}\frac{x}{2} + C$$

8 Integrals involving $\sqrt{(x^2 - a^2)}$ terms
Let $x = a \sec \theta$. Then $\sqrt{(x^2 - a^2)} = \sqrt{(a^2 \sec^2 \theta - a^2)} = a \tan \theta$, since we have $\tan^2 \theta = \sec^2 \theta - 1$ (see Section 3.4.1). $dx/d\theta = a \sec \theta \tan \theta$ and so $dx = a \sec \theta \tan \theta \, d\theta$. Alternatively we could let $x = a \cosh \theta$, the following example illustrating this substitution.

Example

Determine the indefinite integral $\int \sqrt{x^2-9} \, dx$.

Let $x = 3 \cosh \theta$. Then $\sqrt{(x^2 - 3^2)} = \sqrt{(3^2 \cosh^2 \theta - 3^2)} = 3 \sinh \theta$, since we have $\sinh^2 \theta = \cosh^2 \theta - 1$ (see Section 3.5.4). $dx/d\theta = 3 \sinh \theta$ and so $dx = 3 \sinh \theta \, d\theta$. Thus:

$$\int \sqrt{x^2-9} \, dx = \int 3\sinh\theta \times 3\sinh\theta \, d\theta = \int 9\sinh^2\theta \, d\theta$$

$$= 9\int \tfrac{1}{2}(\cosh 2\theta - 1) \, d\theta = \tfrac{9}{2}(\tfrac{1}{2}\sinh 2\theta - \theta) + C$$

$$= \tfrac{9}{2}(\sinh\theta\cosh\theta - \theta) + C$$

$$= \tfrac{9}{2}\left(\tfrac{1}{3}\sqrt{x^2-3}\right)\left(\frac{x}{3}\right) - \tfrac{9}{2}\cosh^{-1}\frac{x}{3} + C$$

$$= \tfrac{1}{2}x\sqrt{x^2-3} - \tfrac{9}{2}\cosh^{-1}\frac{x}{3} + C$$

9 Integrals containing $\sin x$, $\cos x$, $\tan x$
Let $u = \tan \tfrac{1}{2}x$. Then $du/dx = \tfrac{1}{2} \sec^2 \tfrac{1}{2}x$. But $\sec^2 x = 1 + \tan^2 x$, thus $du/dx = \tfrac{1}{2}(1 + \tan^2 x) = \tfrac{1}{2}(1 + u^2)$. Thus $dx = 2 \, du/(1 + u^2)$. The trigonometric functions can all be expressed in terms of u. Thus:

$$\sin x = 2\sin\frac{x}{2}\cos\frac{x}{2} = 2\sin\frac{x}{2}\cos\frac{x}{2}\frac{\cos\frac{x}{2}}{\cos\frac{x}{2}} = \frac{2\tan\frac{x}{2}}{\sec^2\frac{x}{2}}$$

$$= \frac{2\tan\frac{x}{2}}{1+\tan^2\frac{x}{2}} = \frac{2u}{1+u^2}$$

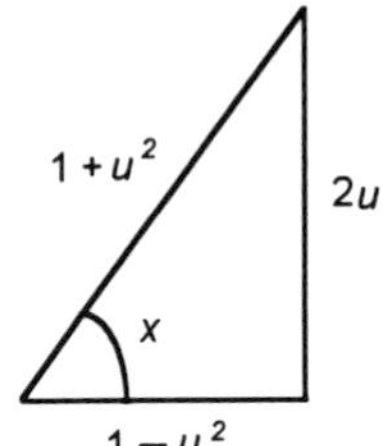

Figure 25.11 *Angle x*

Figure 25.11 shows the right-angled triangle with such an angle. Hence:

$$\cos x = \frac{1-u^2}{1+u^2}$$

$$\tan x = \frac{2u}{1-u^2}$$

Note that integration of the squares of trigonometric functions can be obtained by using trigonometric identities to put the functions in non-squared form. Thus:

$$\int \sin^2 x \,\mathrm{d}x = \int \tfrac{1}{2}(1-\cos 2x)\,\mathrm{d}x = \tfrac{1}{2}(x - \tfrac{1}{2}\sin 2x) + C$$

$$\int \cos^2 x \,\mathrm{d}x = \int \tfrac{1}{2}(1+\cos 2x)\,\mathrm{d}x = \tfrac{1}{2}(x + \tfrac{1}{2}\sin 2x) + C$$

$$\int \tan^2 x \,\mathrm{d}x = \int(\sec^2 x - 1)\,\mathrm{d}x = \tan x - x + C$$

Example

Determine the indefinite integral $\int \frac{1}{\sin x}\,\mathrm{d}x$.

Let $u = \tan \tfrac{1}{2}x$, then $\mathrm{d}u/\mathrm{d}x = \tfrac{1}{2}\sec^2 \tfrac{1}{2}x = \tfrac{1}{2}(1 + \tan^2 \tfrac{1}{2}x) = \tfrac{1}{2}(1 + u^2)$ and replacing $\sin x$ by $2u/(1 + u^2)$:

$$\int \frac{1}{\sin x}\,\mathrm{d}x = \int \frac{1+u^2}{2u}\,\frac{2}{1+u^2}\,\mathrm{d}u = \int \frac{1}{u}\,\mathrm{d}u$$

$$= \ln|u| + C = \ln\left|\tan\frac{x}{2}\right| + C$$

The above has discussed the substitution procedure with indefinite integrals where the variable was changed from x to u. When we have definite integrals we can do the same procedure and take account of the limits of integration at the end *after* reversing the substitution. The limits are in terms of values of x. However, it is often simpler to express the limits in terms of u and take account of the limits *before* reversing the substitution. To illustrate this, consider the integration of $\cos^3 x$ between the limits 0 and $\tfrac{1}{2}\pi$. If we let $u = \sin x$ then $\mathrm{d}u/\mathrm{d}x = \cos x$ and so $\cos x\,\mathrm{d}x = \mathrm{d}u$. When $x = 0$ then $u = 0$ and when $x = \tfrac{1}{2}\pi$ then $u = 1$. Thus the integral can be written as:

$$\int_0^{\pi/2} \cos^3 x \, dx = \int_0^{\pi/2} \cos^2 x \cos x \, dx = \int_0^{\pi/2} (1 - \sin^2 x) \cos x \, dx$$

$$= \int_0^1 (1 - u^2) \, du = \left[u - \frac{u^3}{3} \right]_0^1 = \frac{2}{3}$$

Revision

10 Determine the following indefinite integrals using the substitutions indicated:

(a) $\int 3(3x+1)^5 \, dx$ using $u = 3x+1$, (b) $\int \sin 2x \, dx$ using $u = 2x$,

(c) $\int \tan x \, dx$ using $u = \cos x$, (d) $\int \frac{1}{\sqrt{4-x^2}} \, dx$ using $x = 2 \sin \theta$,

(e) $\int \frac{e^x}{1+e^x} \, dx$ using $u = 1 + e^x$, (f) $\int e^{x^3} x^2 \, dx$ using $u = x^3$,

(g) $\int \frac{1}{1+\sin x} \, dx$ using $u = \tan \frac{x}{2}$, (h) $\int \sin^2 x \cos^5 x \, dx$ using $u = \sin x$,

(i) $\int \sin^2 x \cos^3 x \, dx$ using $u = \sin x$

11 Evaluate the following definite integrals using the substitutions indicated:

(a) $\int_2^3 \frac{x}{(1+x^2)^2} \, dx$ using $u = 1 + x^2$,

(b) $\int_0^{\pi/6} \sin^3 x \cos x \, dx$ using $u = \sin x$,

(c) $\int_0^{\pi/4} \sin x \cos^5 x \, dx$ using $u = \cos x$,

(d) $\int_0^3 x\sqrt{x^2+16} \, dx$ using $u = x^2 + 16$,

(e) $\int_0^1 \frac{x}{\sqrt{1+x^2}} \, dx$ using $u = 1 + x^2$

25.4.2 Integration by parts

The product rule for differentiation gives (see Section 22.4.2 and equation [18]):

$$\frac{d}{dx}[f(x)g(x)] = f(x)\frac{d}{dx}g(x) + g(x)\frac{d}{dx}f(x)$$

Integrating both sides of this equation with respect to x gives:

$$\int \frac{\mathrm{d}}{\mathrm{d}x}[f(x)g(x)]\,\mathrm{d}x = \int f(x)\frac{\mathrm{d}}{\mathrm{d}x}g(x)\,\mathrm{d}x + \int g(x)\frac{\mathrm{d}}{\mathrm{d}x}f(x)\,\mathrm{d}x$$

Hence:

$$\int f(x)\frac{\mathrm{d}}{\mathrm{d}x}g(x)\,\mathrm{d}x = \int \frac{\mathrm{d}}{\mathrm{d}x}[f(x)g(x)]\,\mathrm{d}x - \int g(x)\frac{\mathrm{d}}{\mathrm{d}x}f(x)\,\mathrm{d}x$$

$$= f(x)g(x) - \int g(x)\frac{\mathrm{d}}{\mathrm{d}x}f(x)\,\mathrm{d}x \qquad [7]$$

This is the formula for *integration by parts*. This is often written in terms of $u = f(x)$ and $v = g(x)$ as:

$$\int u\frac{\mathrm{d}v}{\mathrm{d}x}\,\mathrm{d}x = uv - \int v\frac{\mathrm{d}u}{\mathrm{d}x}\,\mathrm{d}x \qquad [8]$$

With a definite integral the equation becomes:

$$\int_a^b u\frac{\mathrm{d}v}{\mathrm{d}x}\,\mathrm{d}x = [uv]_a^b - \int_a^b v\frac{\mathrm{d}u}{\mathrm{d}x}\,\mathrm{d}x \qquad [9]$$

Example

Determine the indefinite integral $\int x\,\mathrm{e}^x\,\mathrm{d}x$.

The integrand consists of the product of two factors. If we let $u = x$ and $\mathrm{d}v/\mathrm{d}x = \mathrm{e}^x$, then $v = \int \mathrm{e}^x\,\mathrm{d}x$ and equation [8] gives:

$$\int x\,\mathrm{e}^x\,\mathrm{d}x = x\,\mathrm{e}^x - \int(\mathrm{e}^x)(1)\,\mathrm{d}x = x\,\mathrm{e}^x - \mathrm{e}^x + C$$

Example

Determine the indefinite integral $\int \mathrm{e}^x \sin x\,\mathrm{d}x$.

Let $u = \mathrm{e}^x$ and $\mathrm{d}v/\mathrm{d}x = \sin x$. Then $v = \int \sin x\,\mathrm{d}x = -\cos x$. Hence, using equation [8] gives:

$$\int \mathrm{e}^x \sin x\,\mathrm{d}x = \mathrm{e}^x(-\cos x) - \int(-\cos x)(\mathrm{e}^x)\,\mathrm{d}x$$

$$= -\mathrm{e}^x \cos x + \int \mathrm{e}^x \cos x\,\mathrm{d}x + C$$

Applying integration by parts again, with $u = \mathrm{e}^x$ and $\mathrm{d}v/\mathrm{d}x = \cos x$. Then $v = \int \cos x\,\mathrm{d}x = \sin x$. Hence, using equation [8] gives:

$$\int \mathrm{e}^x \sin x\,\mathrm{d}x = -\mathrm{e}^x \cos x + \mathrm{e}^x \sin x - \int \mathrm{e}^x \sin x\,\mathrm{d}x + C$$

Thus:

$$\int e^x \sin x \, dx = \tfrac{1}{2}(-e^x \cos x + e^x \sin x + C)$$

Example

Determine the definite integral $\int_0^1 x^2 e^x \, dx$.

Let $u = x^2$ and $dv/dx = e^x$. Then $v = \int e^x \, dx = e^x$. Thus, using equation [9]:

$$\int_0^1 x^2 e^x \, dx = [x^2 e^x]_0^1 - \int_0^1 e^x(2x) \, dx$$

Applying integration by parts again, with $u = x$ and $dv/dx = e^x$. Then $v = \int e^x \, dx = e^x$. Thus, using equation [9]:

$$\int_0^1 x^2 e^x \, dx = [x^2 e^x]_0^1 - 2[x e^x]_0^1 + 2\int_0^1 e^x \, dx$$

$$= [x^2 e^x]_0^1 - 2[x e^x]_0^1 + 2[e^x]_0^1$$

$$= e^1 - 2e^1 + 2e^1 - 2e^0 = e - 2$$

Revision

12 Using the method of integration by parts, determine the following indefinite integrals:

(a) $\int x \ln x \, dx$, (b) $\int x^2 e^{3x} \, dx$, (c) $\int x \cos x \, dx$, (d) $\int x^2 \sin 2x \, dx$

13 Using the method of integration by parts, evaluate the following definite integrals:

(a) $\int_0^{\pi/2} 2x \sin x \, dx$, (b) $\int_0^{\pi/4} x \cos 2x \, dx$, (c) $\int_1^2 (x-1)^2 \ln x \, dx$

25.4.3 Integration by partial fractions

The technique of expressing a rational function as the sum of its partial fractions has been dealt with in Section 2.6. Expressing integrands in terms of partial fractions can often simplify integration.

Example

Determine the indefinite integral $\int \frac{1}{x^2 - 1} \, dx$.

The fraction $1/(x^2 - 1)$ can be written as:

$$\frac{1}{(x-1)(x+1)} = \frac{A}{x-1} + \frac{B}{x+1} = \frac{A(x+1)+B(x-1)}{(x-1)(x+1)}$$

Hence, equating coefficients of x gives $A + B = 0$ and equating integers gives $A - B = 1$. Thus $A = ½$ and $B = -½$. Hence the integral can be expressed as:

$$\int \frac{1}{x^2-1}\,\mathrm{d}x = \tfrac{1}{2}\int \frac{1}{x-1}\,\mathrm{d}x - \tfrac{1}{2}\int \frac{1}{x+1}\,\mathrm{d}x$$

We can determine these integrals by substitution. Thus if we let $u = x - 1$ then $\mathrm{d}u/\mathrm{d}x = 1$ and so:

$$\int \frac{1}{x-1}\,\mathrm{d}x = \int \frac{1}{u}\,\mathrm{d}u = \ln|u| = \ln|x-1| + A$$

Likewise the integral of $1/(x + 1)$ is $\ln|x + 1| + B$. Hence:

$$\int \frac{1}{x^2-1}\,\mathrm{d}x = \tfrac{1}{2}\ln|x-1| + \tfrac{1}{2}\ln|x+1| + C$$

Example

Determine the indefinite integral $\int \frac{x^3}{x-2}\,\mathrm{d}x$.

This fraction has a numerator of higher degree than the denominator. If the degree of the numerator is equal to or higher than the degree of the denominator, the numerator must be divided by the denominator until the remainder is of lower degree than the denominator. Thus:

$$\begin{array}{r|l} & x^2 + 2x + 4 \\ \hline x-2 & x^3 \\ & \underline{x^3 - 2x^2} \\ & \quad 2x^2 \\ & \quad \underline{2x^2 - 4x} \\ & \qquad 4x \\ & \qquad \underline{4x - 8} \\ & \qquad\quad 8 \end{array}$$

Hence the integral becomes:

$$\int \frac{x^3}{x-2}\,\mathrm{d}x = \int\left(x^2 + 2x + 4 + \frac{8}{x-2}\right)\mathrm{d}x$$

$$= \frac{x^3}{3} + x^2 + 4x + 8\ln|x-2| + C$$

Example

Determine the indefinite integral $\int \frac{1}{x(x^2+1)}\,\mathrm{d}x$.

Expressed as partial fractions the integrand is:

$$\frac{1}{x(x^2+1)} = \frac{A}{x} + \frac{Bx+C}{x^2+1} = \frac{A(x^2+1)+(Bx+C)x}{x(x^2+1)}$$

Equating the constant terms gives $A = 1$. Equating the coefficients of x gives $C = 0$. Equating the coefficients of x^2 gives $A + B = 0$, and so $B = -1$. Thus the integral becomes:

$$\int \frac{1}{x(x^2+1)}\,\mathrm{d}x = \int \left(\frac{1}{x} - \frac{x}{x^2+1}\right)\,\mathrm{d}x$$

The integration of $1/(x^2 + 1)$ can be carried out by using a substitution. Let $u = x^2 + 1$ and so $\mathrm{d}u/\mathrm{d}x = 2x$. Thus:

$$\int \frac{x}{x^2+1}\,\mathrm{d}x = \int \frac{1}{2u}\,\mathrm{d}u = \tfrac{1}{2}\ln|u| + C = \tfrac{1}{2}\ln|x^2+1| + C$$

and so:

$$\int \frac{1}{x(x^2+1)}\,\mathrm{d}x = \ln|x| - \tfrac{1}{2}\ln|x^2+1| + C$$

Example

Determine the indefinite integral $\int \frac{x^2}{(x^2+1)^2}\,\mathrm{d}x$.

Expressed as a partial fraction the integrand is:

$$\frac{x^2}{(x^2+1)^2} = \frac{Ax+B}{x^2+1} + \frac{Cx+D}{(x^2+1)^2} = \frac{(Ax+B)(x^2+1)+(Cx+D)}{(x^2+1)^2}$$

Equating the constant terms gives $B + D = 0$, equating the coefficients of x gives $A + C = 0$, equating the coefficients of x^2 gives $B = 1$, and equating the coefficients of x^3 gives $A = 0$. Hence $A = 0$, $B = 1$, $C = 0$ and $D = -1$. Thus:

$$\int \frac{x^2}{(x^2+1)^2}\,\mathrm{d}x = \int \left(\frac{1}{x^2+1} - \frac{1}{(x^2+1)^2}\right)\,\mathrm{d}x$$

The first integral gives $\tan^{-1} x + A$. The second integral can be evaluated using the substitution $x = \tan\theta$. This gives $\mathrm{d}x/\mathrm{d}\theta = \sec^2\theta$ and $x^2 + 1 = \tan^2\theta + 1 = \sec^2\theta$. Thus:

$$\int \frac{1}{(x^2+1)^2}\,\mathrm{d}x = \int \frac{1}{\sec^4\theta}\sec^2\theta\,\mathrm{d}\theta = \int \cos^2\theta\,\mathrm{d}\theta$$

$$= \tfrac{1}{2}\int (1+\cos 2\theta)\,\mathrm{d}\theta$$

$$= \frac{\theta}{2} + \frac{\sin 2\theta}{4} + B = \frac{\theta}{2} + \tfrac{1}{2}\sin\theta\cos\theta + B$$

$$= \tfrac{1}{2}\tan^{-1}x + \tfrac{1}{2}\left(\frac{x}{\sqrt{x^2+1}}\right)\left(\frac{1}{\sqrt{x^2+1}}\right) + B$$

Hence:

$$\int \frac{x^2}{(x^2+1)^2}\,dx = \tan^{-1}x - \left[\tfrac{1}{2}\tan^{-1}x + \tfrac{1}{2}\left(\frac{x}{x^2+1}\right)\right] + C$$

$$= \tfrac{1}{2}\tan^{-1}x - \tfrac{1}{2}\left(\frac{x}{x^2+1}\right) + C$$

Revision

14 Determine the following indefinite integrals:

(a) $\int \frac{x^2}{x+3}\,dx$, (b) $\int \frac{x}{x^2-1}\,dx$, (c) $\int \frac{5x-4}{(x-6)(x-2)}\,dx$,

(d) $\int \frac{4x+3}{(x-3)^2}\,dx$, (e) $\int \frac{x^2+1}{x(x-2)(x+2)}\,dx$,

(f) $\int \frac{1}{(x+1)(x^2+2x+2)}\,dx$, (g) $\int \frac{1}{(x-1)^2(x^2+1)}\,dx$,

(h) $\int \frac{x^2+x+2}{(x^2+2)^2}\,dx$

25.5 Integration of vectors

Since integration is the inverse of differentiation, if we have $d\mathbf{a}/dt = \mathbf{b}$ then:

$$\int \mathbf{b}\,dt = \mathbf{a} + \mathbf{c} \qquad [10]$$

where **c** is an arbitrary constant vector. If a vector is written as $f(t)\mathbf{i} + g(t)\mathbf{j} + h(t)\mathbf{k}$, then, since **i**, **j** and **k** are constant vectors:

$$\int [f(t)\mathbf{i} + g(t)\mathbf{j} + h(t)\mathbf{k}]\,dt = \left(\int f(t)\,dt\right)\mathbf{i} + \left(\int g(t)\,dt\right)\mathbf{j} + \left(\int h(t)\,dt\right)\mathbf{k} \qquad [11]$$

Example

If the velocity **v** of a particle at a time t is given by $6 \sin 2t\,\mathbf{i} - 4\mathbf{j} + 8t^2\mathbf{k}$, determine how the displacement **r** varies with time is $\mathbf{v} = d\mathbf{r}/dt$ and $\mathbf{r} = 0$ when $t = 0$.

r is the integral of **r** dt and so is given by:

$$\mathbf{r} = \left(\int 6\sin 2t\,\mathrm{d}t\right)\mathbf{i} - \left(\int 4\,\mathrm{d}t\right)\mathbf{j} + \left(\int 8t^2\,\mathrm{d}t\right)\mathbf{k}$$

$$= -3\cos 2t\mathbf{i} - 4t\mathbf{j} + \tfrac{8}{3}t^3\mathbf{k} + \mathbf{c}$$

Since $\mathbf{r} = 0$ when $t = 0$, then $0 = -3\mathbf{i} + \mathbf{c}$ and so $\mathbf{c} = 3\mathbf{i}$. Thus:

$$\mathbf{r} = (3 - 3\cos 2t)\mathbf{i} - 4t\mathbf{j} + \tfrac{8}{3}t^3\mathbf{k}$$

Revision

15 Determine the following integrals:

(a) $\int (3t^2\mathbf{i} + 4t\mathbf{j})\,\mathrm{d}t$, (b) $\int [t^2\mathbf{i} + (1 + 2t)\mathbf{j}]\,\mathrm{d}t$

25.6 Table of integrals

Table 25.2 is a table of commonly encountered standard integrals.

Table 25.2 *Standard integrals*

Function $f(x)$	$\int f(x)\,\mathrm{d}x$
a (a constant)	$ax + C$
x^n, except $n = 1$	$\frac{1}{n+1}x^{n+1} + C$
$\frac{1}{x}$	$\ln x + C$ if $x > 0$, $\ln(-x) + C$ if $x < 0$ i.e. $\ln\lvert x\rvert + C$ if $x \neq 0$
e^{ax}	$\frac{1}{a}\mathrm{e}^{ax} + C$
$\sin ax$	$-\frac{1}{a}\cos ax + C$
$\cos ax$	$\frac{1}{a}\sin ax + C$
$\tan ax$	$\frac{1}{a}\ln\lvert\sec ax\rvert + C = -\frac{1}{a}\ln\lvert\cos ax\rvert + C$
$\operatorname{cosec} ax$	$\frac{1}{a}\ln\lvert\operatorname{cosec} ax - \cot ax\rvert + C$
$\sec ax$	$\frac{1}{a}\ln\lvert\sec ax + \tan ax\rvert + C$
$\cot ax$	$\frac{1}{a}\ln\lvert\sin ax\rvert + C$
$\sin^{-1} ax$	$x\sin^{-1}ax + \frac{1}{a}\sqrt{1 - a^2x^2}$
$\cos^{-1} ax$	$x\cos^{-1}ax - \frac{1}{a}\sqrt{1 - a^2x^2}$
$\tan^{-1} ax$	$x\tan^{-1}ax - \frac{1}{2a}\ln\sqrt{1 + a^2x^2}$
$\cot^{-1} ax$	$x\cot^{-1}ax + \frac{1}{2a}\ln(1 + a^2x^2) + C$

Function $f(x)$	$\int f(x)\,dx$
$\sin^2 ax$	$\frac{x}{2} - \frac{\sin 2ax}{4a} + C$
$\cos^2 ax$	$\frac{x}{2} + \frac{\sin 2ax}{4a} + C$
$\sinh x$	$\cosh x + C$
$\cosh x$	$\sinh x + C$
$\tanh x$	$\ln \lvert\cosh x\rvert + C$
$\coth x$	$\ln \lvert\sinh x\rvert + C$
$\sinh^{-1} x$	$x\sinh^{-1}x - \sqrt{x^2+1} + C$
$\cosh^{-1} x$	$x\cosh^{-1}x - \sqrt{x^2-1} + C$
$\tanh^{-1} x$	$x\tanh^{-1}x + \ln\sqrt{x^2-1} + C$
$\coth^{-1} x$	$x\coth^{-1}x + \ln\sqrt{x^2-1} + C$
$\frac{1}{x^2+a^2}$	$\frac{1}{a}\tan^{-1}\frac{x}{a} + C$
$\frac{1}{\sqrt{a^2-x^2}}$	$\sin^{-1}\frac{x}{a} + C$
$\frac{1}{x^2-a^2}$	$\frac{1}{2a}\ln\left\lvert\frac{x-a}{x+a}\right\rvert + C$

Problems

1 Determine the antiderivatives of the following:

(a) 4, (b) $2x^3$, (c) $2x^3 + 5x$, (d) $x^{2/3} - 3x^{1/2}$, (e) $4 + \cos 5x$, (f) $2\,e^{-3x}$,

(g) $4\,e^{x/2} + x^2 + 2$, (h) $4/x$

2 Determine the areas under the following curves between the specified limits and the x-axis:

(a) $y = 4x^3$ between $x = 1$ and $x = 2$,

(b) $y = x$ between $x = 0$ and $x = 4$,

(c) $y = 1/x$ between $x = 1$ and $x = 3$,

(d) $y = x^3 - 3x^2 - 2x + 2$ between $x = -1$ and $x = 2$,

(e) $y = x^2 - x - 2$ between $x = -1$ and $x = 2$,

(f) $y = x^2 - 1$ between $x = -1$ and $x = 2$,

(g) the area between $x = 0$ and $x = 2$ for the curve defined by $y = x^2$ between $x = 0$ and $x = 1$ and by $y = 2 - x$ between $x = 1$ and $x = 2$.

3 Determine the areas bounded by graphs of the following functions and between the specified ordinates:

(a) $y = 9 - x^2$, $y = -2$, $x = -2$ and $x = 2$,

(b) $y = 4$, $y = x^2$, $x = 0$ and $x = 1$

4 Determine the geometrical area enclosed between the graph of the function $y = x(x-1)(x-2)$ and the x-axis.

5 Determine the area bounded by graphs of $y = x^3$ and $y = x^2$.

6 Determine the area bounded by the graph of $y = \sin x$, the x-axis and the line $x = \pi/2$.

7 Determine the area bounded by graphs of $y = x^2 - 2x + 2$ and $y = 4 - x$.

8 Determine the values, if they exist, of the following definite integrals:

(a) $\int_{-\infty}^{1} x \, dx$, (b) $\int_{1}^{\infty} \frac{1}{x^3} \, dx$, (c) $\int_{0}^{\infty} e^{-3x} \, dx$, (d) $\int_{-\infty}^{-1} x^4 \, dx$

9 Determine the following indefinite integral by using the given substitutions:

(a) $\int (x^2+1)x^3 \, dx$ using $u = x^2 + 1$, (b) $\int 2\, e^{4x-1} \, dx$ using $u = 4x - 1$,

(c) $\int \frac{x+1}{\sqrt{2x+1}} \, dx$ using $u = \sqrt{2x+1}$, (d) $\int x \sin x^2 \, dx$ using $u = x^2$,

(e) $\int \sqrt{x^2+4} \, dx$ using $x = 2 \sinh u$, (f) $\int x\sqrt{x-1} \, dx$ using $u = \sqrt{x-1}$,

(g) $\int \sec x \, dx$ using $u = \tan \frac{1}{2}x$, (h) $\int \frac{1}{\sqrt{9-x^2}} \, dx$ using $x = 3 \sin \theta$,

(i) $\int \sin^2 2x \cos^3 2x \, dx$ using $u = \sin x$

10 Determine the following indefinite integrals by making appropriate substitutions:

(a) $\int x\sqrt{x+2} \, dx$, (b) $\int \frac{1}{(x^2+1)^{3/2}} \, dx$, (c) $\int \sin^3 x \, dx$,

(d) $\int \frac{1}{\sqrt{1-4x^2}} \, dx$, (e) $\int x\sqrt{x^2+2} \, dx$, (f) $\int \frac{1}{4+25x^2} \, dx$,

(g) $\int \cos^3 x \sin^4 x \, \mathrm{d}x$, (h) $\int \tan^3 x \sec^2 x \, \mathrm{d}x$, (i) $\int \frac{1}{5+4\cos x} \, \mathrm{d}x$

11 By making appropriate substitutions, evaluate the following definite integrals:

(a) $\int_0^1 \frac{1}{2-x} \, \mathrm{d}x$, (b) $\int_0^1 \frac{3x^2}{(x^3+9)^2} \, \mathrm{d}x$, (c) $\int_0^1 \frac{x^2}{\sqrt{1-x^2}} \, \mathrm{d}x$,

(d) $\int_{-1}^1 x^2 \sqrt{2-x^2} \, \mathrm{d}x$, (e) $\int_0^2 \frac{1}{4+x^2} \, \mathrm{d}x$

12 Using the method of integration by parts, determine the following indefinite integrals:

(a) $\int x^2 \ln x \, \mathrm{d}x$, (b) $\int x \, \mathrm{e}^{2x} \, \mathrm{d}x$, (c) $\int x^3 \cos x \, \mathrm{d}x$, (d) $\int x \sin 5x \, \mathrm{d}x$,

(e) $\int x \ln 3x \, \mathrm{d}x$, (f) $\int \sin^2 x \, \mathrm{d}x$

13 Using the method of integration by parts, evaluate the following definite integrals:

(a) $\int_0^{\pi/2} x \cos x \, \mathrm{d}x$, (b) $\int_0^{\pi/2} x \cos^2 x \, \mathrm{d}x$, (c) $\int_0^{\pi} (\pi - x) \cos x \, \mathrm{d}x$

14 Determine the following indefinite integrals:

(a) $\int \frac{x^2}{2x-3} \, \mathrm{d}x$, (b) $\int \frac{x}{1-2x} \, \mathrm{d}x$, (c) $\int \frac{x^2}{2x^2+x-3} \, \mathrm{d}x$,

(d) $\int \frac{x^2}{(x^2-1)(2x+1)} \, \mathrm{d}x$, (e) $\int \frac{x+1}{x(x-2)(x+2)} \, \mathrm{d}x$,

(f) $\int \frac{3x-1}{(2x+1)(x-1)} \, \mathrm{d}x$, (g) $\int \frac{2x^3+3x^2-3}{2x^2-x-1} \, \mathrm{d}x$,

(h) $\int \frac{1}{(x-2)(x-3)} \, \mathrm{d}x$, (i) $\int \frac{5x^2+20x+6}{x(x+1)^2} \, \mathrm{d}x$,

(j) $\int \frac{2x^3-4x-8}{x(x-1)(x^2+4)} \, \mathrm{d}x$, (k) $\int \frac{1}{x^2(x^2+1)} \, \mathrm{d}x$

15 If $\mathbf{r} = (t - t^2)\mathbf{i} + 2t^2\mathbf{j} + 4\mathbf{k}$, determine $\int \mathbf{r} \, \mathrm{d}t$.

26 Numerical integration

26.1 Introduction

Situations can occur in engineering and science where functions have to be integrated and they are only specified by a set of values, perhaps the outcome of some experiment. In other situations, there can be the need to evaluate definite integrals for functions that have no antiderivative or whose antiderivative is not easily obtained. In these types of situations, numerical integration can be used.

The definite integral $\int_a^b f(x)\,dx$ is equal to the signed area, i.e. the area taking into account that areas above the axis are positive and areas under it negative, between $y = f(x)$ and the x-axis and between the vertical lines $x = a$ and $x = b$ (see Section 25.3). Thus methods based on the determination of areas can be used to evaluate a definite integral.

The simplest way of determining a definite integral from data values is to plot the data on a graph and estimate the area under the graph. This might just involve 'counting the squares' under the graph. More sophisticated techniques are the mid-ordinate rule, the trapezium rule and Simpson's rule. This chapter discusses these methods.

This chapter assumes the concept of the definite integral (see Chapter 25) and the principle of approximating a function by means of a polynomial (see Section 5.4).

26.2 Mid-ordinate rule

This method involves assuming that we represent a function by a horizontal line, i.e. $y = A$, with A being a constant. Thus if we have just two points, as in Figure 26.1, a straight line is drawn through the value of the function at the mid-point between the two values, i.e. at $\frac{1}{2}(b - a)$. The estimated area is then taken as being the value of the function y_1 at the mid-ordinate multiplied by $(b - a)$. Thus:

$$\int_a^b f(x)\,dx \approx (b-a)y_1 \qquad [1]$$

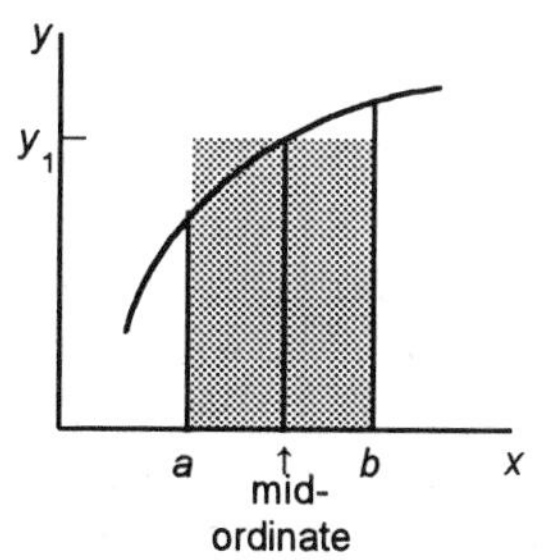

Figure 26.1 *Mid-ordinate rule*

Just replacing the area under a function by a rectangle in this way gives a simple approximation. We can improve on this if we subdivide the area under the function between the ordinates concerned into more than one rectangular strip and then apply the mid-ordinate rule to each strip separately. The estimate of the definite integral is then the sum of the areas of the rectangular strips. Figure 26.2 illustrates this with three strips. With n strips between ordinates $x = a$ and $x = b$, each has a width of $(b - a)/n$ and we have:

$$\int_a^b f(x)\,dx \approx \frac{(b-a)}{n}\left(y_1 + y_2 + y_3 + \ldots + y_n\right) \qquad [2]$$

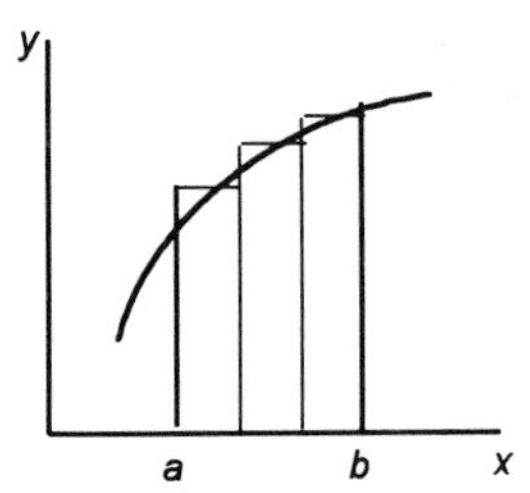

Figure 26.2 *Mid-ordinate rule*

Thus the estimated area is the product of the width of the strips multiplied by the sum of the mid-ordinate values.

Example

Evaluate $\int_0^2 (x^2 + 1)\,dx$ using the mid-ordinate rule when the interval is divided into (a) just one strip, (b) two strips, (c) four strips.

(a) With just one strip the strip width is 2 – 0 = 2. The mid-ordinate occurs at $x = 1$ and the value of the function at this value is $1^2 + 1 = 2$. Hence, using equation [1]:

$$\int_0^2 (x^2 + 1)\,dx \approx 2 \times 2 \approx 4$$

(b) With two strips the strip width is (2 – 0)/2 = 1. The mid-ordinates occur at $x = 0.5$ and $x = 1.5$ and the values of the function at these values are $0.5^2 + 1 = 1.25$ and $1.5^2 + 1 = 3.25$. Hence, using equation [2]:

$$\int_0^2 (x^2 + 1)\,dx \approx 1(1.25 + 3.25) \approx 4.5$$

(c) With four strips the strip width is (2 – 0)/4 = 0.5. The mid-ordinates occur at $x = 0.25$, $x = 0.75$, $x = 1.25$ and $x = 1.75$. The values of the functions at these values are $0.25^2 + 1 = 1.062\,5$, $0.75^2 + 1 = 1.562\,5$, $1.25^2 + 1 = 2.562\,5$ and $1.75^2 + 1 = 4.062\,5$. Hence, using equation [2]:

$$\int_0^2 (x^2 + 1)\,dx \approx 0.5(1.062\,5 + 1.562\,5 + 2.562\,5 + 4.062\,5) \approx 4.625$$

Note that analytically the indefinite integral has the value 4.667. Thus as the number of strips is increased, the accuracy improves.

Revision

1 Use the mid-ordinate rule with (i) two strips, (ii) four strips to evaluate the following integrals:

$$\text{(a) } \int_1^3 \frac{1}{\sqrt{x}}\,dx, \text{ (b) } \int_0^1 \frac{1}{x^2 + 1}\,dx, \text{ (c) } \int_{-1}^1 e^x\,dx$$

26.3 Trapezium rule

This method involves assuming that we represent a function by an inclined line, i.e. $y = A + Bx$ with A and B being constants, joining the tops of two ordinates (Figure 26.3). The area of the resulting trapezium is half the sum of the parallel sides multiplied by its width and so for the trapezium in Figure 26.3 the area is $\frac{1}{2}(y_a + y_b)(b - a)$. Thus we have for the estimate of the definite integral:

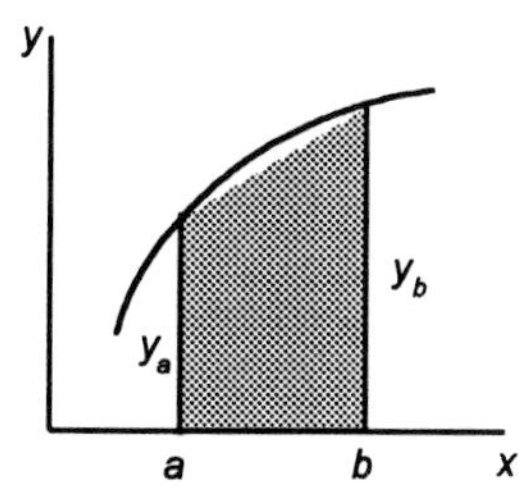

Figure 26.3 *Trapezium rule*

$$\int_a^b f(x)\,dx \approx (b-a)\tfrac{1}{2}(y_a + y_b) \qquad [3]$$

A better value for the integral can be obtained if the area beneath the curve is divided into a number of equal width strips, rather than just the one strip (Figure 26.4). With n strips, each will have a width of $(b-a)/n$. If we then determine the area of each trapezium-shaped strip we obtain:

$$\int_a^b f(x)\,dx \approx \frac{b-a}{n}\left[\tfrac{1}{2}(y_a + y_1) + \tfrac{1}{2}(y_1 + y_2) + \ldots + \tfrac{1}{2}(y_n + y_b)\right]$$

$$\approx \frac{b-a}{n}\left[\tfrac{1}{2}(y_a + y_b) + y_1 + y_2 + \ldots + y_n\right] \qquad [4]$$

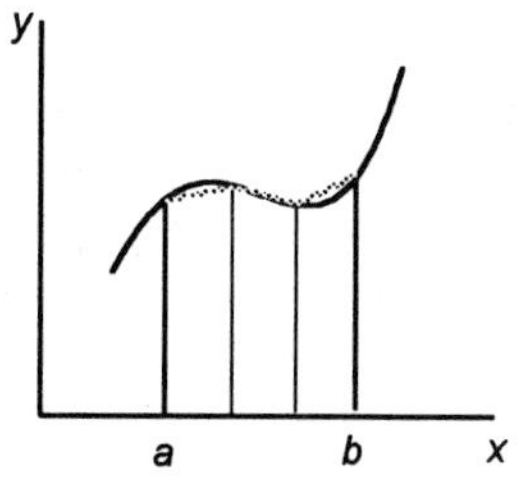

Figure 26.4 *Trapezium rule*

i.e. the width of a strip multiplied by half the sum of the outer ordinates and the sum of the other ordinates.

Example

Using the trapezium rule with four strips, estimate the value of the definite integral $\int_0^\pi \sin x\,dx$.

The width of each strip will be $(\pi - 0)/4 = \pi/4$. There will thus be ordinates at 0, $\pi/4$, $\pi/2$, $3\pi/4$ and π. The values of the function at these ordinates are $\sin 0 = 0$, $\sin \pi/4 = 0.707$, $\sin \pi/2 = 1$, $\sin 3\pi/4 = 0.707$ and $\sin \pi = 0$. Thus, using equation [4]:

$$\int_0^\pi \sin x\,dx \approx \frac{\pi}{4}\left[\tfrac{1}{2}(0+0) + 0.707 + 1 + 0.707\right] \approx 1.896$$

Revision

2 Using the trapezium rule, evaluate the following definite integral:

(a) $\int_0^1 \frac{1}{1+x^2}\,dx$ with $n = 4$, (b) $\int_0^1 x^2\,dx$ with $n = 3$,

(c) $\int_1^4 \frac{1}{x}\,dx$ with $n = 6$, (d) $\int_0^4 \cos\left(\frac{\pi x}{4}\right)\,dx$ with $n = 4$

26.4 Simpson's rule

With the mid-ordinate rule the horizontal line $y = A$ through the mid-ordinate was used to delineate the upper extent of the area of a strip. With the trapezium rule it was the sloping line $y = A + Bx$. *Simpson's rule* involves using the curved line $y = A + Bx + Cx^2$.

The area under the function $y = A + Bx + Cx^2$ between $x = x_1$ and $x = x_2$ (Figure 26.5) is given by the definite integral:

$$\int_{x_1}^{x_2} (A + Bx + Cx^2)\,dx = \left[Ax + \tfrac{1}{2}Bx^2 + \tfrac{1}{3}Cx^3\right]_{x_1}^{x_2}$$

$$= A(x_2 - x_1) + \tfrac{1}{2}B(x_2^2 - x_1^2) + \tfrac{1}{3}C(x_2^3 - x_1^3)$$

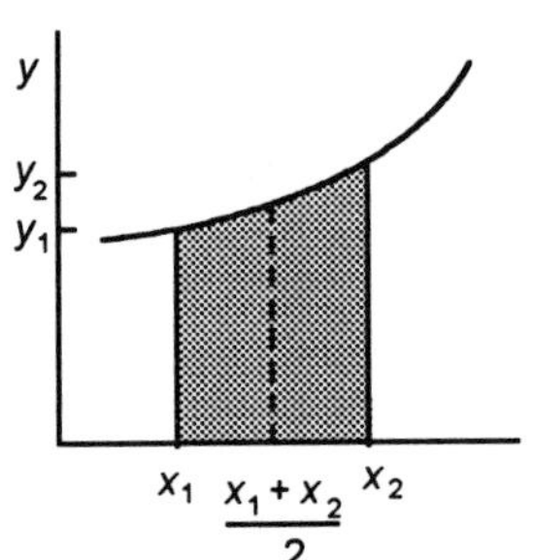

Figure 26.5 *Area of strip*

$$= \tfrac{1}{6}(x_2 - x_1)[6A + 3B(x_1 + x_2) + 2C(x_1^2 + x_1x_2 + x_2^2)]$$

$$= \tfrac{1}{6}(x_2 - x_1)\left[(A + Bx_1 + Cx_1^2) + 4\left\{A + B\left(\frac{x_1 + x_2}{2}\right) + C\left(\frac{x_1 + x_2}{2}\right)^2\right\} + (A + Bx_2 + Cx_2^2)\right]$$

Since $y_1 = A + Bx_1 + Cx_1^2$, $y_2 = A + Bx_2 + Cx_2^2$ and y_m at the midpoint between x_1 and x_2, i.e. $(x_1 + x_2)/2$, is $A + B(x_1 + x_2)/2 + C\{(x_1 + x_2)/2\}^2$, then:

$$\int_{x_1}^{x_2} (A + Bx + Cx^2)\,\mathrm{d}x = \tfrac{1}{6}(x_2 - x_1)\left[y_1 + 4y_m + y_2\right] \qquad [5]$$

If we now consider the area under some function divided into n strips (Figure 26.6), with n being an even number, then the area of each double strip is approximated by the use of equation [5]. The reason for the double strip is to enable the central ordinate of each strip to give the y_m value in equation [5]. The width of each double strip is $2(b - a)/n$. Thus for the integration between the limits a and b, when equation [5] is applied to each pair of strips:

$$\int_a^b f(x)\,\mathrm{d}x \approx \tfrac{1}{6}\frac{2(b-a)}{n}\left(y_a + 4y_1 + y_2\right) + \tfrac{1}{6}\frac{2(b-a)}{n}\left(y_2 + 4y_3 + y_4\right) + \tfrac{1}{6}\frac{2(b-a)}{n}\left(y_4 + 4y_5 + y_b\right)$$

$$\approx \tfrac{1}{3}\frac{b-a}{n}\left[y_a + 4y_1 + 2y_2 + 4y_3 + 2y_4 + 4y_5 + y_b\right] \qquad [6]$$

The estimate is thus one third the width of a strip multiplied by the sum of the first ordinate, four times the odd ordinates, two times the even ordinates and the last ordinate. This is known as *Simpson's rule*.

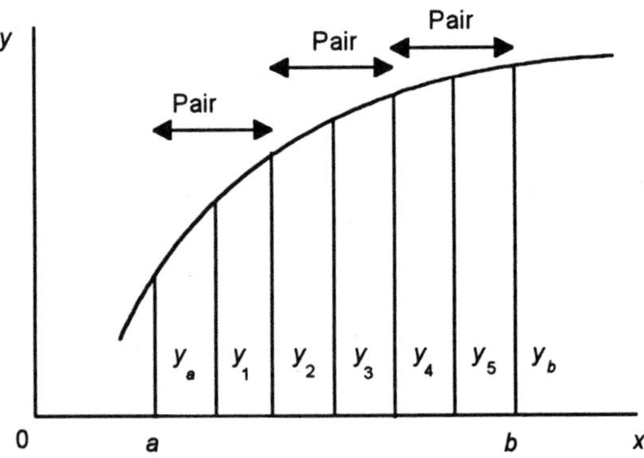

Figure 26.6 *Simpson's rule*

Example

Use Simpson's rule with four strips to evaluate the integral $\int_1^3 \frac{1}{\sqrt{x}}\,dx$.

Each strip will have a width of (3 – 1)/4 = 0.5. Thus we have ordinates at x = 1.0, 1.5, 2.0, 2.5, 3.0. The values of the function at these ordinates are 1.0, 0.816, 0.707, 0.632 and 0.577. The 0.806 and 0.632 are values for odd ordinates and the 0.707 for an even ordinate. Thus equation [6] gives:

$$\int_1^3 \frac{1}{\sqrt{x}}\,dx \approx \tfrac{1}{3} \times 0.5(1.0 + 4 \times 0.816 + 2 \times 0.707$$
$$+ 4 \times 0.632 + 0.577)$$
$$\approx 1.464$$

Example

Use Simpson's rule with four strips to evaluate $\int_0^\pi \sin x\,dx$.

With four strips the width of each strip is $(b - a)/n = \pi/4$. The ordinates thus occur at $x = 0$, $\pi/4$, $\pi/2$, $3\pi/4$ and π. The values of the function at these values are 0, 0.707, 1, 0.707 and 0. Thus equation [6] gives:

$$\int_0^\pi \sin x\,dx \approx \tfrac{1}{3}\tfrac{\pi}{4}(0 + 4 \times 0.707 + 2 \times 1 + 4 \times 0.707 + 0) \approx 2.004$$

Revision

3 Use Simpson's rule to evaluate the following integrals:

(a) $\int_0^1 \frac{1}{1+x^2}\,dx$ with $n = 4$, (b) $\int_0^1 e^{2x}\,dx$ with $n = 4$,

(c) $\int_1^4 \frac{1}{x}\,dx$ with $n = 6$, (d) $\int_2^4 \frac{1}{1-x^2}\,dx$ with $n = 8$

26.5 Comparison of methods

The error of a value obtained by numerical integration is the true value of that integral minus the calculated value. As an indication of the effect of the method used and the number of strips considered, consider Table 26.1 which shows the results of the evaluation of the following integral using different numbers of strips and different methods:

$$\int_1^3 \frac{1}{\sqrt{x}}\,dx$$

The true value of the integral is 1.464 102.

Table 26.1 *Numerical integration*

Number of strips	Strip width	Mid-ordinate rule		Trapezium rule		Simpson's rule	
		Result	Error	Result	Error	Result	Error
2	1	1.448	+0.016	1.496	–0.032	1.468 592	–0.007 576
4	0.5	1.460	+0.004	1.472	–0.008	1.464 562	–0.000 460
8	0.25	1.463	+0.001	1.466	–0.002	1.464 137	–0.000 036

With the mid-ordinate rule, doubling the number of strips considered, i.e. halving the width of strips, reduces the error by a factor of 4. Thus the error can be considered to be proportional to the reciprocal of the square of the number of strips. With the trapezium rule, doubling the number of strips considered reduces the error by a factor of 4. Thus the error can be considered to be proportional to the reciprocal of the square of the number of strips. With Simpson's rule, the convergence to the true value is much greater than with the other rules, fewer strips being required to obtain an accurate result. Doubling the number of strips considered reduces the error by a factor of about 16. Thus the error can be considered to be proportional to the square of the fourth power of the number of strips. These error factors are not specific to the integral here considered but apply in general.

Problems

1 Use the mid-ordinate rule with eight strips to evaluate the following integrals:

(a) $\int_1^5 (x^2 - 2x + 2)\,dx$, (b) $\int_1^2 \frac{1}{x^2+1}\,dx$

2 Using the trapezium rule, evaluate the following definite integral:

(a) $\int_1^3 \frac{1}{x+2}\,dx$ with $n = 4$, (b) $\int_0^{\pi/2} \frac{1}{\sin x + 1}\,dx$ with $n = 6$,

(c) $\int_1^2 \frac{1}{x^2}\,dx$ with $n = 4$, (d) $\int_0^2 \sqrt{x - x^2}\,dx$ with $n = 4$

3 Use Simpson's rule to evaluate the following integrals:

(a) $\int_1^3 \frac{1}{x+2}\,dx$ with $n = 4$, (b) $\int_0^2 x^3\,dx$ with $n = 4$,

(c) $\int_1^2 \frac{1}{x^2}\,dx$ with $n = 4$, (d) $\int_0^{\pi/2} \sqrt{\sin x}\,dx$ with $n = 6$,

(e) $\int_1^3 \sqrt{x^4 + 1}\,dx$ with $n = 6$, (f) $\int_1^4 \frac{1}{x^3}\,dx$ with $n = 6$

27 Application: Areas, volumes, moments and means

27.1 Introduction

This chapter illustrates the use of the definite integral in the determination of areas (plane areas or the surface area of solids), volumes, moments and means. It is concerned with the evaluation of definite integrals and thus assumes Chapter 25.

The product of a force and its perpendicular distance from some axis is termed its *moment* about that force. Such moments are important in the consideration of the equilibrium of bodies. There are, however, other forms of moment which are used in engineering and science. For example, in connection with the rotation of a body we have the *moment of inertia*, this involving the sum of the products of a large number of elements of mass and the square of their distances from some axis. As another example of a moment, the analysis of the stresses developed in the bending of beams requires taking into account the shape and area of cross-section of the beam. The analysis involves considering the cross-section as being made up of a large number of small elements of area, each element being at a different distance from an axis, with the requirement to sum the products of the areas and their distances from the axis. Such a sum is referred to as the *first moment of area*. *Moments* involve the consideration of products of small strips of area or mass and their distances, or distances squared, from some axis.

The *mean* value of a number of values is the sum of the values divided by the number considered. Likewise we can consider some continuous function effectively as a lot of values for which we can determine a mean. In connection with some physical quantities, e.g. the power developed by an electrical current, we are not concerned with the mean value but the square root of the mean values of the squares of the values, this being referred to as the *root-mean-square value*.

27.2 Areas and volumes

If we require the area generated between the curve of a function $y = f(x)$ and the x-axis and between ordinates $x = a$ and $x = b$ (Figure 27.1), then we can obtain it from the definite integral since:

$$\text{area} = \int_a^b f(x)\,dx \qquad [1]$$

The area is a *signed area* in that areas with positive values of y are positive and negative values of y are negative. Thus if we require the total area between a function and the axis when there are both positive and negative

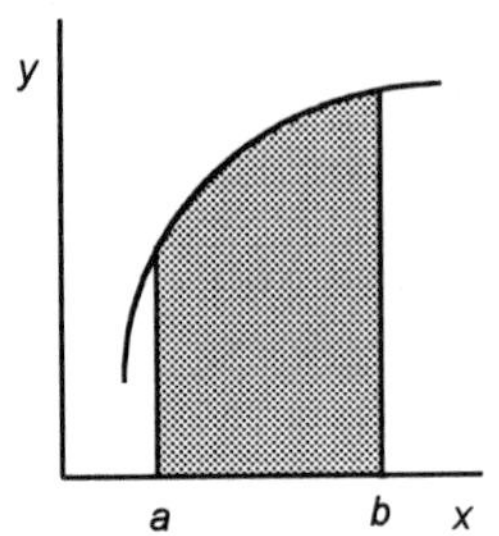

Figure 27.1 *Area under curve*

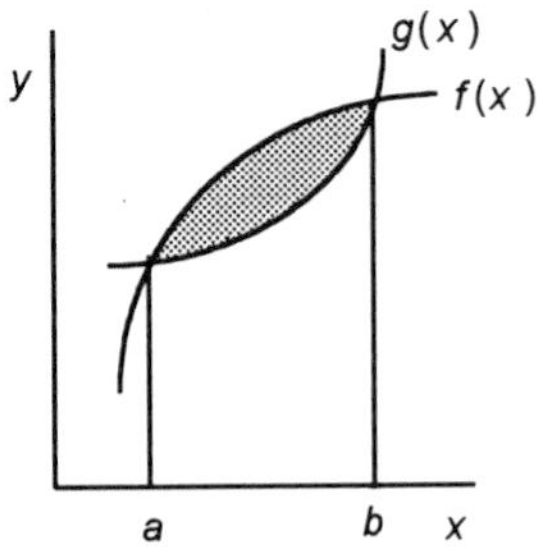

Figure 27.2 *Area between curves*

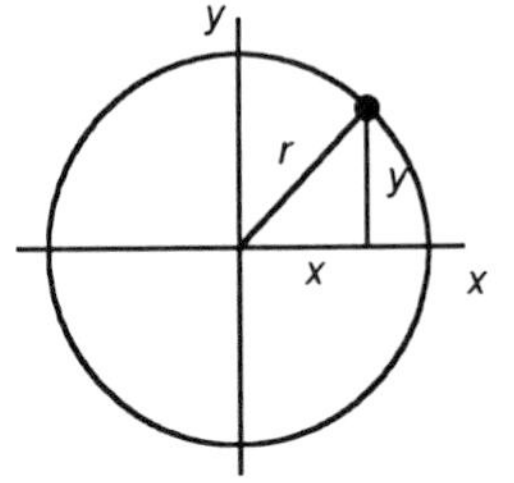

Figure 27.3 *Example*

areas, we have to determine each area separately so that we can disregard the signs in taking the sum.

When we require the area between two curves (Figure 27.2) between particular ordinates we can consider the problem as being the determination of the area between one curve and the axis between the ordinates minus the area between the other curve and the same axis, i.e.

$$\text{area between curves} = \int_a^b f(x)\,dx - \int_a^b g(x)\,dx = \int_a^b [f(x) - g(x)]\,dx \quad [2]$$

Example

Determine the equation for the area of a circle.

Figure 27.3 shows a circle with centre at the origin. A point on the circumference of the circle will have co-ordinates x and y related, by applying the Pythagoras theorem, to the radius r by:

$$x^2 + y^2 = r^2$$

Consider the area of the positive quadrant of the circle. This is:

$$\text{area} = \int_0^r y\,dx = \int_0^r \sqrt{r^2 - x^2}\,dx$$

We can solve this integral by using the substitution $x = r \sin \theta$. Then $dx/d\theta = r \cos \theta$. When $x = 0$ then $\theta = 0$ and when $x = r$ then $\theta = \pi/2$. Thus:

$$\text{area} = \int_0^{\pi/2} \sqrt{r^2 - r^2 \sin^2\theta} \times r\cos\theta\,d\theta = r^2 \int_0^{\pi/2} \cos^2\theta\,d\theta$$

$$= r^2 \int_0^{\pi/2} \tfrac{1}{2}(1 + \cos 2\theta)\,d\theta = \frac{r^2}{2}\left[\theta - \tfrac{1}{2}\sin 2\theta\right]_0^{\pi/2} = \frac{\pi r^2}{4}$$

Each of the quadrants will have the same area and thus the total area of the circle is πr^2.

Revision

For more revision problems see Section 25.3, problems 3, 5, 6, 7, 8.

1 For a circle of radius 3, determine the area cut off from the circle by a chord whose distance from the centre is 2.

2 Determine the area bounded by the following curves:

(a) $y = x$ and $y = 2 - x^2$, (b) $y = x^2$ and $y = x + 2$,

(c) $y = \sqrt{x}$ and $y = x - 2$

27.2.1 Volumes

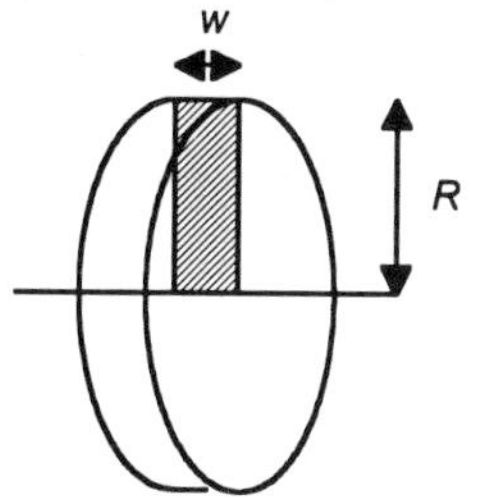

Figure 27.4 *Generating a disc*

Many solid shapes can be considered to be *solids of revolution*. Such solids have circular cross-sections and can be considered to be generated by the rotation of an area about an axis. You can think of these solids as perhaps being manufactured on a lathe, the stock being shaped by rotating against a fixed tool; or perhaps a potter working a ball of clay into a suitable shape on a potter's wheel, the wheel rotating the clay and allowing the potter to press a tool, or fingers, against it as it rotates.

Figure 27.4 shows how we can regard a disc as a solid produced by rotating a rectangular area about an axis. The rotated rectangle has an area of Rw and the resulting disc a volume of $\pi R^2 w$. We can consider any solid of revolution to be made up of discs. Figure 27.5 illustrates this, showing just three of the discs in the solid formed by rotation of an area. The volume of the solid is then the sum of the volumes of all the discs that are produced by rotating all the rectangles of width δx and radius y which occur between the limits $x = a$ and $x = b$. The volume of a disc is $\pi y^2\,\delta x$ and so the sum is:

$$\text{volume of solid} = \sum \pi y^2\,\delta x \text{ of all the discs between the limits}$$

In the limit as $\delta x \to 0$:

$$\text{volume of solid} = \int_a^b \pi y^2\,\mathrm{d}x \qquad [3]$$

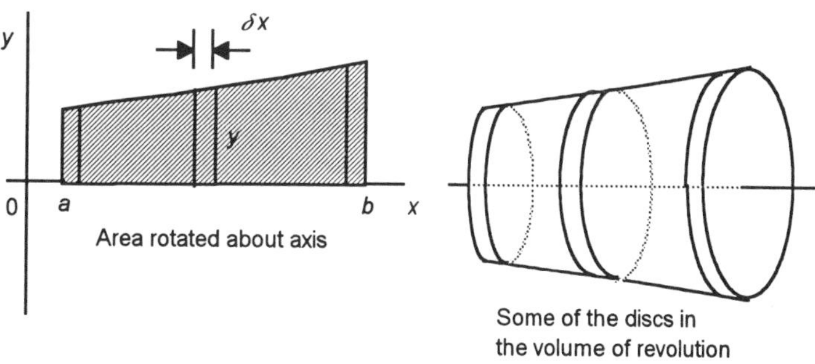

Figure 27.5 *Generating a solid of revolution*

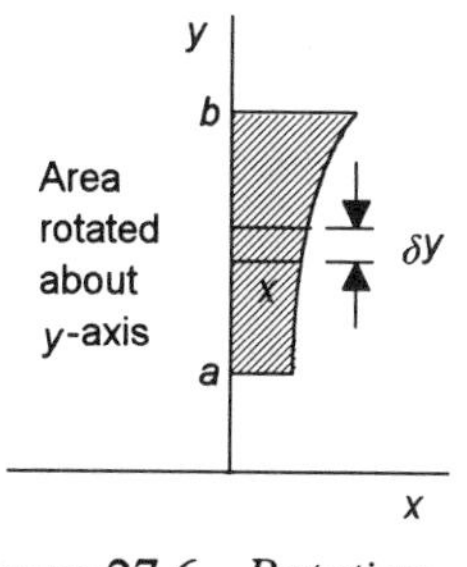

Figure 27.6 *Rotation about the y-axis*

If the solid of revolution had been formed by a revolution around the y-axis (Figure 27.6) then each disc has a radius x and width δy and so a volume of $\pi x^2\,\delta y$. Hence the volume of the solid is the sum of all the discs between the limits and as $\delta x \to 0$ becomes:

$$\text{volume of solid} = \int_a^b \pi x^2\,\mathrm{d}y \qquad [4]$$

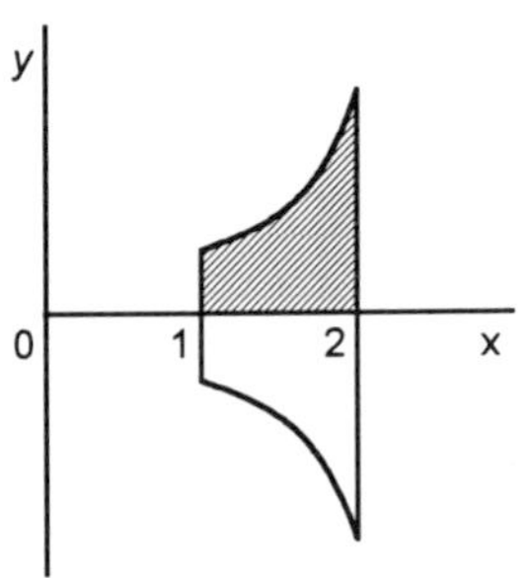

Figure 27.7 *Example*

Example

Determine the volume generated by rotating about the x-axis the area between the function $y = e^x$, the x-axis and the ordinates $x = 1$ and $x = 2$.

Figure 27.7 shows the cross-section of the generated solid and the area rotated. Using equation [3]:

$$\text{volume of solid} = \int_a^b \pi y^2 \, dx = \int_1^2 \pi (e^x)^2 \, dx = \left[\pi \frac{e^{2x}}{2}\right]_1^2$$

$$= \frac{\pi}{2}(e^4 - e^2) = 74.2 \text{ cubic units}$$

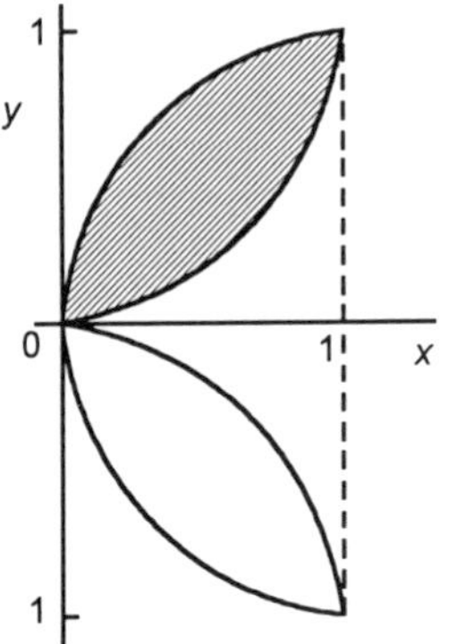

Figure 27.8 *Example*

Example

Determine the volume of the solid generated by rotating the region bounded by the functions $y = \sqrt{x}$ and $y = x^2$ about the x-axis.

Figure 27.8 shows the cross-section of the generated solid and the area rotated. The graphs of the two functions cross at $x = 0$ and $x = 1$ and so these give the limits of the area rotated. We can consider the volume generated as being that of $y = \sqrt{x}$ between these limits rotated about the x-axis minus that of $y = x^2$ between the same limits rotated about the x-axis. Thus, using equation [3]:

$$\text{volume} = \int_0^1 \pi(\sqrt{x})^2 \, dx - \int_0^1 \pi(x^2)^2 \, dx = \left[\pi \frac{x^2}{2}\right]_0^1 - \left[\pi \frac{x^5}{5}\right]_0^1$$

The volume is thus $3\pi/10$ cubic units.

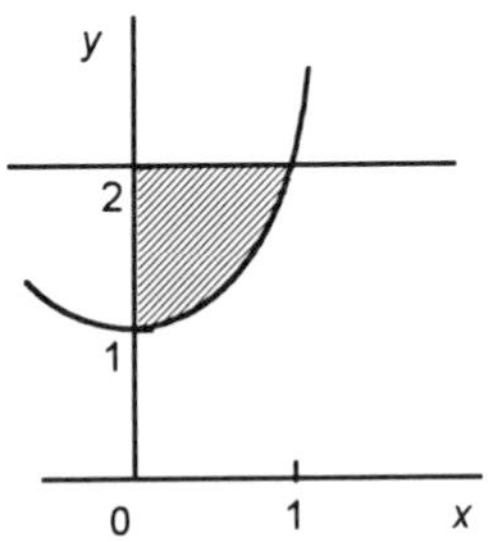

Figure 27.9 *Example*

Example

Determine the volume generated by rotating about the y-axis the area defined by $y \geq x^2 + 1$, with x greater than or equal to 0 and y less than or equal to 2.

Figure 27.9 shows the area. For rotation about the y-axis we use equation [4] to give:

$$\text{volume} = \int_a^b \pi x^2 \, dy = \int_1^2 \pi(y - 1) \, dy = \left[\pi \frac{y^2}{2} - \pi y\right]_1^2 = \frac{\pi}{2} \text{ units}^3$$

Example

Determine the volume generated by rotating about the line $y = 3$ the area bounded by $y = 4x - x^2$ and $y = 3$.

Figure 27.10 shows the area. The graphs of $y = 4x - x^2$ and $y = 3$ intersect at the points (1, 3) and (3, 3). If we consider the rectangular

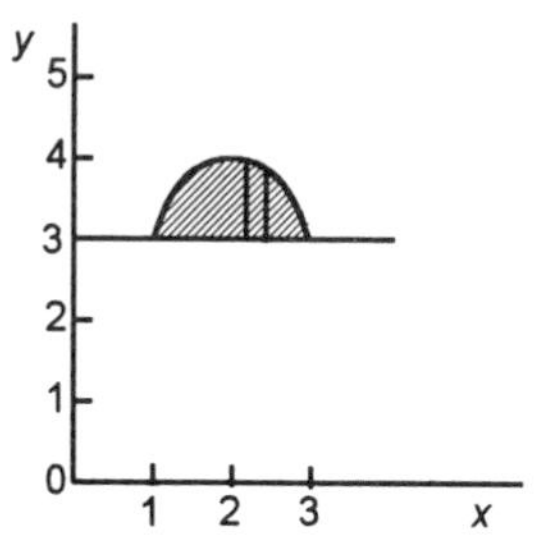

Figure 27.10 *Example*

element of area indicated, then when rotated it gives a disc of area $\pi(y-3)^2\,\delta x$. Thus the total volume is:

$$\text{volume} = \int_1^3 \pi(y-3)^2\,dx = \int_1^3 \pi(4x-x^2-3)^2\,dx$$

$$= \pi\int_1^3 (x^4 - 8x^3 + 22x^2 - 24x + 9)\,dx$$

$$= \pi\left[\frac{x^5}{5} - 2x^4 + \frac{22x^3}{3} - 12x^2 + 9x\right]_1^3 = \tfrac{16}{15}\pi \text{ cubic units}$$

Revision

3 Determine the volumes generated by rotating about the x-axis the areas between the following functions and the x-axis:

(a) $y = 2x$ between $x = 0$ and $x = 2$, (b) $y = e^{-x}$ between $x = 1$ and $x = 2$,

(c) $y = \dfrac{10}{x^2}$ between $x = 2$ and $x = 5$

4 Determine the volumes generated by rotating about the x-axis the area bounded by the following functions:

(a) $y = \frac{1}{x}$, $y = 0$ between $x = 1$ and $x = 4$, (b) $y = \sqrt{x}$, $y = x$

5 Determine the volume generated by rotating about the line $x = 1$ the area bounded by $y = x^3$, $y = 0$ and $x = 1$.

6 Determine the volume generated by rotating about the y-axis the area bounded by $y = \sqrt{x}$, $y = 0$ and $x = 4$.

27.2.2 Lengths of curves

Consider the problem of determining the length of the arc of some curve, such as that between points A and B in Figure 27.11. We can take the length of the arc between A and B as being the sum of a large number of small arcs lengths δs.

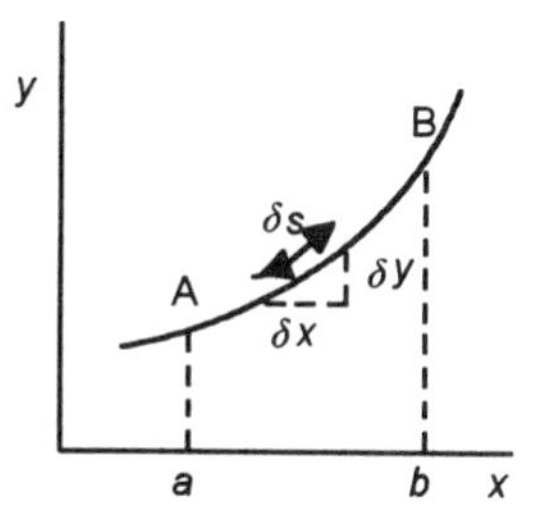

Figure 27.11 *Length of a curve*

arc length between A and B = $\sum \delta s$ elements between A and B

For a small enough arc length we can write, using the Pythagoras theorem:

$$(\delta s)^2 = (\delta x)^2 + (\delta y)^2$$

Dividing by $(\delta x)^2$ gives:

$$\left(\frac{\delta s}{\delta x}\right)^2 = 1 + \left(\frac{\delta y}{\delta x}\right)^2$$

and hence:

$$\frac{\delta s}{\delta x} = \sqrt{1 + \left(\frac{\delta y}{\delta x}\right)^2} \qquad [5]$$

Thus, in the limit as the segments of arc considered tend to zero, the arc length between A and B is:

$$\text{arc length} = \int_{\mathrm{A}}^{\mathrm{B}} \mathrm{d}s = \int_a^b \frac{\mathrm{d}s}{\mathrm{d}x}\,\mathrm{d}x = \int_a^b \sqrt{1 + \left(\frac{\mathrm{d}y}{\mathrm{d}x}\right)^2}\,\mathrm{d}x \qquad [6]$$

We can obtain a similar equation for the arc length in terms of y:

$$\text{arc length} = \int_{\mathrm{A}}^{\mathrm{B}} \mathrm{d}s = \int_c^d \frac{\mathrm{d}s}{\mathrm{d}y}\,\mathrm{d}y = \int_c^d \sqrt{1 + \left(\frac{\mathrm{d}x}{\mathrm{d}y}\right)^2}\,\mathrm{d}y \qquad [7]$$

When the curve is defined in terms of parametric equations, say $x = f(t)$ and $y = g(t)$, then the equation $(\delta s)^2 = (\delta x)^2 + (\delta y)^2$ can be divided by $(\delta t)^2$ to give:

$$\left(\frac{\delta s}{\delta t}\right)^2 = \left(\frac{\delta x}{\delta t}\right)^2 + \left(\frac{\delta y}{\delta t}\right)^2 \qquad [8]$$

and hence, in the limit:

$$\text{arc length} = \int_{\mathrm{A}}^{\mathrm{B}} \mathrm{d}s = \int_{t_1}^{t_2} \frac{\mathrm{d}s}{\mathrm{d}t}\,\mathrm{d}t = \int_{t_1}^{t_2} \sqrt{\left(\frac{\mathrm{d}x}{\mathrm{d}t}\right)^2 + \left(\frac{\mathrm{d}y}{\mathrm{d}t}\right)^2}\,\mathrm{d}t \qquad [9]$$

Example

Determine the arc length of the curve $y = x^{3/2}$ between $x = 0$ and $x = 1$.

Since $y = x^{3/2}$ then $\mathrm{d}y/\mathrm{d}x = \frac{3}{2}x^{1/2}$. Thus, using equation [6]:

$$\text{arc length} = \int_a^b \sqrt{1 + \left(\frac{\mathrm{d}y}{\mathrm{d}x}\right)^2}\,\mathrm{d}x = \int_0^1 \sqrt{1 + \left(\tfrac{3}{2}x^{1/2}\right)^2}\,\mathrm{d}x$$

$$= \frac{1}{2}\int_0^1 \sqrt{4 + 9x}\,\mathrm{d}x$$

We can solve this integral by substitution. Let $u = 4 + 9x$ and thus as a consequence $\mathrm{d}u/\mathrm{d}x = 9$. When $x = 1$ then $u = 13$ and when $x = 0$ we have $u = 4$. Thus:

$$\text{arc length} = \tfrac{1}{2}\int_4^{13} u^{1/2}\frac{1}{9}\,du = \frac{1}{18}\left[\frac{2u^{3/2}}{3}\right]_4^{13} = 1.44 \text{ units}$$

Revision

7 Determine the lengths of the curves of graphs of the following functions:

(a) $y = x^{4/3}$ from $x = 0$ to $x = 1$, (b) $y = \ln(1 - x^2)$ from $x = 0$ to $x = ¼$,

(c) $y = e^x$ from $x = 0$ to $x = 2\pi$

8 Determine the length of the curve of the graph described by the parametric equations $x = 2\cos^3\theta$ and $y = 2\sin^3\theta$ between $\theta = 0$ and $\theta = \pi/2$.

27.2.3 Surface areas of solids

Consider the determination of the area of the surface of a curved surface of a solid of revolution. We can divide the curved surface into discs (Figure 27.12), each of which contributes a surface area element of the disc circumference multiplied by the disc width, i.e. $2\pi y\,\delta s$. Thus the total surface area of all the discs between A and B is:

$$\text{surface area} = \sum 2\pi y\,\delta s \text{ for all strips between A and B}$$

As $\delta s \to 0$, then:

$$\text{surface area} = \int_A^B 2\pi y\,ds$$

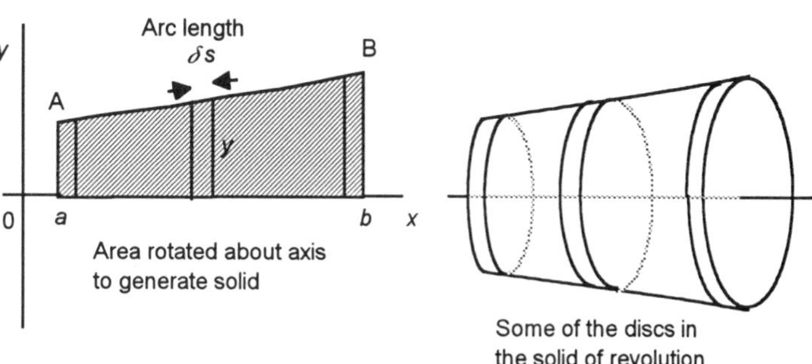

Figure 27.12 *Surface area of a solid of revolution*

We can write this as:

$$\text{surface area} = \int_a^b 2\pi y \frac{ds}{dx}\, dx$$

Using equation [5] we thus have:

$$\text{surface area} = \int_a^b 2\pi y \sqrt{1 + \left(\frac{dy}{dx}\right)^2}\, dx \qquad [10]$$

When the rotated curve is defined by parametric equations $x = f(t)$ and $y = g(t)$ then using equation [8]:

$$\text{surface area} = \int_{t_1}^{t_2} 2\pi y \frac{ds}{dt}\, dt = \int_{t_1}^{t_2} 2\pi y \sqrt{\left(\frac{dx}{dt}\right)^2 + \left(\frac{dy}{dt}\right)^2}\, dt \qquad [11]$$

Example

Determine the area of the surface formed by rotating about the x-axis the graph of $y = x^3$ between $x = 0$ and $x = 1$.

For $y = x^3$ we have $dy/dx = 3x^2$. Thus, using equation [10]:

$$\text{surface area} = \int_a^b 2\pi y \sqrt{1 + \left(\frac{dy}{dx}\right)^2}\, dx$$

$$= 2\pi \int_0^1 x^3 \sqrt{1 + (3x^2)^2}\, dx = 2\pi \int_0^1 x^3 \sqrt{1 + 9x^4}\, dx$$

We can evaluate this integral by substitution. Let $u = 1 + 9x^4$. Then $du/dx = 36x^3$ and so:

$$\text{surface area} = 2\pi \int_1^{10} x^3 u^{1/2} \frac{1}{36x^3}\, dx = \frac{\pi}{18} \int_1^{10} u^{1/2}\, du$$

$$= \frac{\pi}{18}\left[\tfrac{2}{3} u^{3/2}\right]_1^{10} = 3.56 \text{ square units}$$

Revision

9 Determine the areas of the curved surfaces of the solids formed by rotation about the x-axis of the following functions:

(a) $y = x^3/3$ between $x = 0$ and $x = 3$, (b) $y = x^2$ between $x = 1$ and $x = 2$,

(c) $y = \sin x$ between $x = 0$ and $x = \pi$

10 Determine the surface area of a mirror with a parabolic surface generated by the rotation about the x-axis of the parabola $y^2 = 4x$ between $x = 0$ and $x = 2$.

11 Determine the surface area of the surface formed by rotation about the x-axis of one arch of the cycloid $x = \theta - \sin\theta$, $y = 1 - \cos\theta$.

27.3 Moments

The *moment* of a force about an axis is defined as the product of the force and the perpendicular distance of its line of action from the axis. When an object is in equilibrium under the action of a number of parallel forces, then the sum of the moments of the forces about an axis must be zero, otherwise it would rotate.

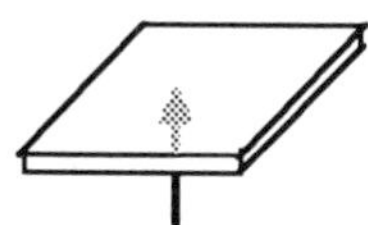

Figure 27.13 *Centre of gravity*

Consider a thin flat sheet of some material. The *centre of gravity* of the sheet is the point on which it can be rested and balance (Figure 27.13). We can think of the entire weight of the object as being concentrated at that point. If we consider the sheet to be made of a large number of small strip elements of mass at different distances from an axis (Figure 27.14) then the weight of each element will give rise to a moment about that axis. Thus the total moment due to all the weight elements is $\delta w_1 x_1 + \delta w_2 x_2 + \delta w_3 x_3 + \ldots$. If a single weight W at a distance $\bar{x}$ is to give the same moment, then:

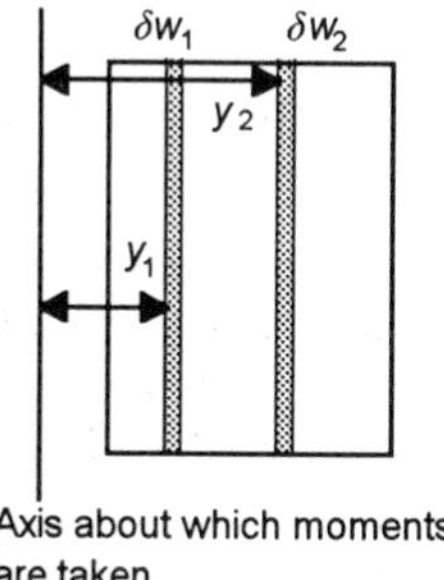

Figure 27.14 *Moments of elements*

$$W\bar{x} = \sum \delta w\, x \text{ for all the strips in the sheet}$$

Thus the distance of the centre of gravity from the chosen axis is:

$$\bar{x} = \frac{\sum_{i=1}^{n} \delta w_i x_i}{W}$$

The total weight is the sum of all the mass elements and so we can write:

$$\bar{x} = \frac{\sum_{i=1}^{n} \delta w_i x_i}{\sum_{i=1}^{n} \delta w_i} \qquad [12]$$

Assuming the acceleration due to gravity is constant, the weights are proportional to the masses and so we can talk of the *centre of mass* of the body. Its distance from the chosen axis is thus:

$$\bar{x} = \frac{\sum_{i=1}^{n} \delta m_i x_i}{\sum_{i=1}^{n} \delta m_i} \qquad [13]$$

where δm represents the mass of an elemental strip.

For a thin flat plate of uniform density, the mass of an element is proportional to its area. We then refer to the centre of mass as being at the *centroid*. The distance of the centroid from the chosen axis is thus:

$$\bar{x} = \frac{\sum_{i=1}^{n} \delta a_i x_i}{\sum_{i=1}^{n} \delta a_i} \qquad [14]$$

where δa represents the area of an elemental strip. The product of an area and its distance from an axis is known as the *first moment of area* of that area about the axis. Thus the centroid distance from an axis is the sum of the first moments of all the area elements divided by the sum of all the areas of the elements.

Consider a plate with an area bounded by the function $f(x)$, the x-axis, $x = a$ and $x = b$, as in Figure 27.15. An elemental area strip has an area $y_i\,\delta x$. Thus if we make the elemental area strips infinitely thin:

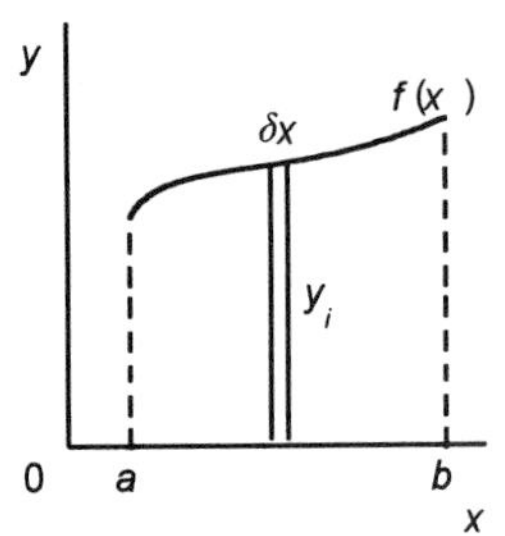

Figure 27.15 *Area element*

$$\bar{x} = \frac{\int_a^b yx\,\mathrm{d}x}{\int_a^b y\,\mathrm{d}x} = \frac{\int_a^b f(x)x\,\mathrm{d}x}{\int_a^b f(x)\,\mathrm{d}x} \qquad [15]$$

The term $\int_a^b yx\,\mathrm{d}x = \int_a^b f(x)x\,\mathrm{d}x$ is the *first moment of area* for the area about the y-axis. Thus the distance of the centroid from the y-axis is the first moment of area for the area about that axis divided by the area.

Equation [15] only locates the line parallel to the y-axis along which the centroid must lie. We also need to find its position from the x-axis to define the centroid point. The elemental area strip shown in Figure 27.15 has a rectangular shape. The point of balance of such a shape is at its midpoint, i.e. $\frac{1}{2}y_i$. Thus, since we can think of all the area as effectively being at this midpoint:

$$\bar{y} = \frac{\sum_{i=1}^{n} \delta a_i \frac{1}{2} y_i}{\sum_{i=1}^{n} \delta a_i} \qquad [16]$$

and thus, when we make the strips infinitely thin:

$$\bar{y} = \frac{\frac{1}{2}\int_a^b y^2\,\mathrm{d}x}{\int_a^b y\,\mathrm{d}x} = \frac{\frac{1}{2}\int_a^b [f(x)]^2\,\mathrm{d}x}{\int_a^b f(x)\,\mathrm{d}x} \qquad [17]$$

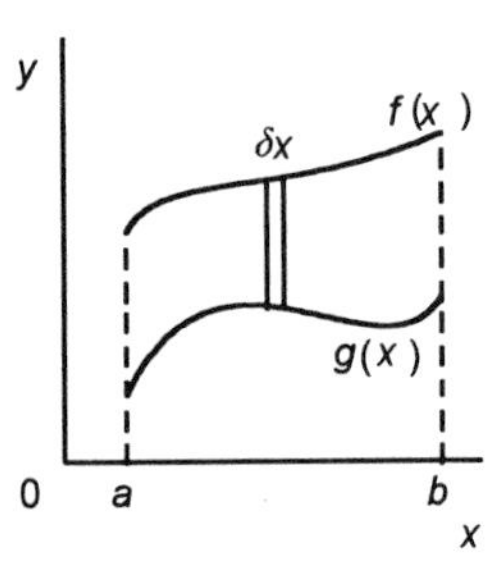

Figure 27.16 *Area element*

Thus the distance of the centroid from the x-axis is the first moment of area about that axis for the area divided by the total area.

Consider a plate with an area bounded by the functions $f(x)$ and $g(x)$ and by $x = a$ and $x = b$, as in Figure 27.16. An elemental area strip has an area $\delta a = [f(x) - g(x)]\,\delta x$. As $\delta x \to 0$:

$$\bar{x} = \frac{\int_a^b [f(x) - g(x)]x \, dx}{\int_a^b [f(x) - g(x)] \, dx} \qquad [17]$$

The term $\int_a^b [f(x) - g(x)]x\,d$ is the first moment of area for the area about the y-axis.

Equation [17] locates the line parallel to the y-axis along which the centroid must lie. For its distance from the x-axis, since the elemental area strip shown in Figure 27.16 has a rectangular shape, the point of balance of such a shape is at its midpoint, i.e. $\frac{1}{2}y_i$. Since $\delta a = [f(x) - g(x)]\,\delta x$ and $\frac{1}{2}y = \frac{1}{2}[f(x) + g(x)]$, as $\delta x \to 0$:

$$\bar{y} = \frac{\frac{1}{2}\int_a^b [f(x) - g(x)][f(x) + g(x)] \, dx}{\int_a^b [f(x) - g(x)] \, dx} = \frac{\frac{1}{2}\int_a^b [f(x)^2 - g(x)^2] \, dx}{\int_a^b [f(x) - g(x)] \, dx} \qquad [18]$$

The term $\frac{1}{2}\int_a^b [f(x) - g(x)][f(x) + g(x)]$ is the first moment of area about the x-axis for the area.

In a similar manner we can derive the equations for the position of the centroid for an area specified between a curve $x = f(y)$ and the y-axis and between $y = a$ and $y = b$ (Figure 27.17). The first moment of area about the x-axis for an area element is $\delta a\, y$ and since $\delta a = x\,\delta y$ the first moment is $xy\,\delta y$. Thus, as $\delta x \to 0$:

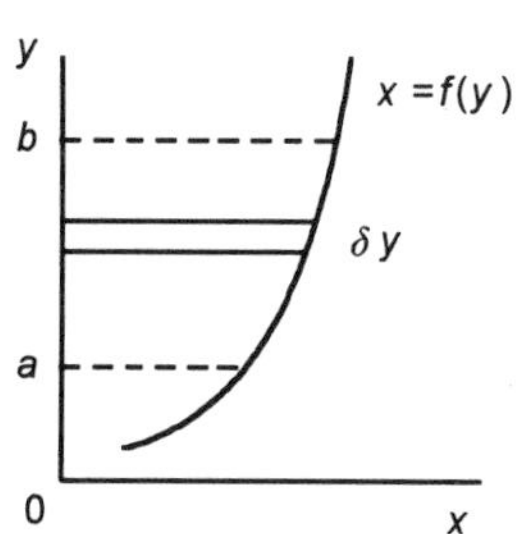

Figure 27.17 *Area element*

$$\bar{y} = \frac{\int_a^b xy \, dy}{\int_a^b x \, dy} \qquad [19]$$

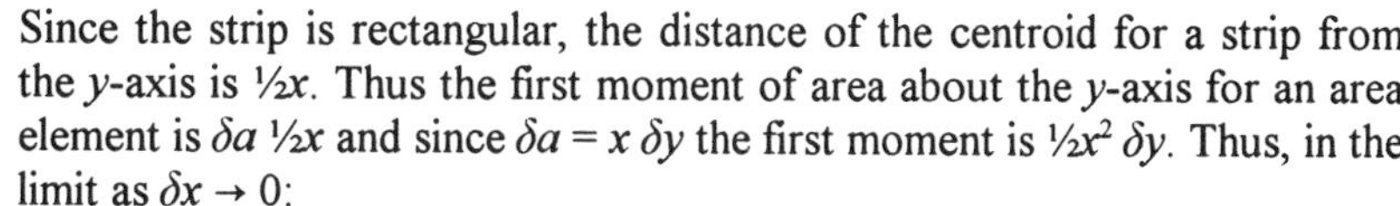

Since the strip is rectangular, the distance of the centroid for a strip from the y-axis is $\frac{1}{2}x$. Thus the first moment of area about the y-axis for an area element is $\delta a\, \frac{1}{2}x$ and since $\delta a = x\,\delta y$ the first moment is $\frac{1}{2}x^2\,\delta y$. Thus, in the limit as $\delta x \to 0$:

$$\bar{x} = \frac{\frac{1}{2}\int_a^b x^2 \, dy}{\int_a^b x \, dy} \qquad [19]$$

Example

Determine the position of the centroid of the area between the function $y = 4 - x^2$ and the x-axis.

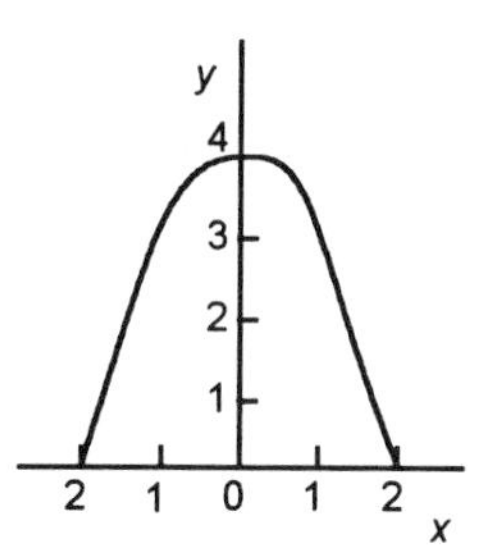

Figure 27.18 *Example*

Figure 27.18 shows the area, the graph crossing the x-axis when $4 - x^2 = 0$, i.e. $x = \pm 2$. Using equation [15] for the distance from the y-axis:

$$\bar{x} = \frac{\int_a^b yx \, dx}{\int_a^b y \, dx} = \frac{\int_{-2}^2 (4 - x^2)x \, dx}{\int_{-2}^2 (4 - x^2) \, dx} = \frac{\left[2x - \frac{x^4}{4}\right]_{-2}^2}{\left[4x - \frac{x^3}{3}\right]_{-2}^2} = 0$$

This should have been expected since the area is symmetrically disposed about the y-axis. For the distance of the centroid from the x-axis, equation [17] gives:

$$\bar{y} = \frac{\frac{1}{2}\int_a^b y^2\,dx}{\int_a^b y\,dx} = \frac{\frac{1}{2}\int_{-2}^{2}(4-x^2)^2\,dx}{\int_{-2}^{2}(4-x^2)\,dx} = \frac{\frac{1}{2}\int_{-2}^{2}(16-8x^2+x^4)\,dx}{\int_{-2}^{2}(4-x^2)\,dx}$$

$$= \frac{\frac{1}{2}\left[16x - \frac{8x^3}{3} + \frac{x^5}{5}\right]_{-2}^{2}}{\left[4x - \frac{x^3}{3}\right]_{-2}^{2}} = \frac{8}{5}$$

The centroid thus lies at the point (0, 8/5).

Example

Determine the position of the centroid for the area bounded by the functions $y = x^2$ and $y = x + 2$.

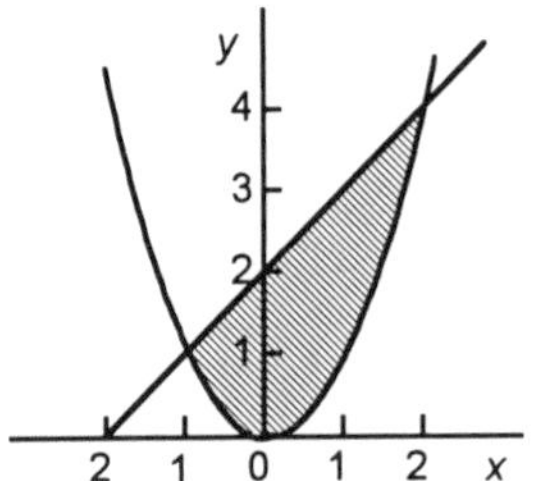

Figure 27.19 *Example*

Figure 27.19 shows the area, the graphs crossing when $x^2 = x + 2$, i.e. $x^2 - x - 2 = 0 = (x - 2)(x + 1)$ and so at $x = 2$ and $x = -1$. Using equation [17], the position of the centroid from the y-axis is:

$$\bar{x} = \frac{\int_a^b [f(x) - g(x)]x\,dx}{\int_a^b [f(x) - g(x)]\,dx} = \frac{\int_{-1}^{2}(x+2-x^2)x\,dx}{\int_{-1}^{2}(x+2-x^2)\,dx}$$

$$= \frac{\left[\frac{x^3}{3} + x^2 - \frac{x^4}{4}\right]_{-1}^{2}}{\left[\frac{x^2}{2} + 2x - \frac{x^3}{3}\right]_{-1}^{2}} = \frac{1}{2}$$

Using equation [18], the position of the centroid from the x-axis is:

$$\bar{y} = \frac{\frac{1}{2}\int_a^b [f(x)^2 - g(x)^2]\,dx}{\int_a^b [f(x) - g(x)]\,dx} = \frac{\frac{1}{2}\int_{-1}^{2}[(x+2)^2 - x^4]\,dx}{\int_{-1}^{2}(x+2-x^2)\,dx}$$

$$= \frac{\frac{1}{2}\int_{-1}^{2}(x^2+4x+4-x^4)\,dx}{\int_{-1}^{2}(x+2-x^2)\,dx} = \frac{\frac{1}{2}\left[\frac{x^3}{3} + 2x^2 + 4x - \frac{x^5}{5}\right]_{-1}^{2}}{\left[\frac{x^2}{2} + 2x - \frac{x^3}{3}\right]_{-1}^{2}} = \frac{8}{5}$$

Thus the centroid is at $(\frac{1}{2}, \frac{8}{5})$.

Revision

12 Determine the positions of the centroids of the areas bounded by:

(a) $y = x^2$ between $x = 1$ and $x = 3$, (b) $y = 4x$ between $x = 0$ and $x = 3$,

(c) $y = x^2$ and $y = 4$, (d) $y = \sqrt{x}$, $y = x$ and $x \geq 0$

27.3.1 Theorems of Pappus

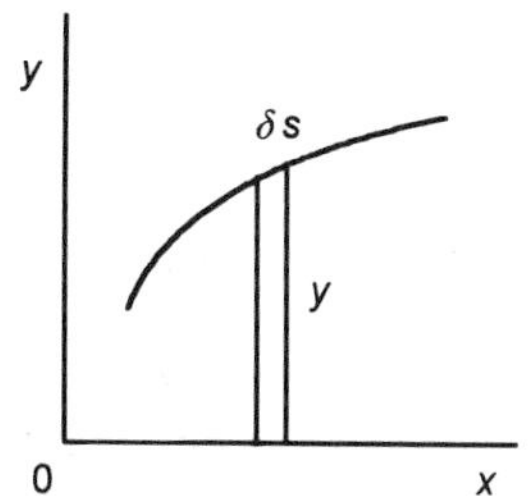

Figure 27.20 *Rotation of a curve element*

Consider an element δs of some curve (Figure 27.20). The area swept out by revolving that element about the x-axis is the surface area of a disc of radius y and is thus $2\pi y\ \delta s$. The total surface area generated by rotating a length of curve s is, when $\delta s \to 0$:

$$\text{area} = \int_0^s 2\pi y \, \mathrm{d}s = 2\pi \int_0^s y \, \mathrm{d}s$$

A line can be considered to be an area which has a constant width w. Thus for a line element of length δs the area is $w\ \delta s$. The first moment of this area is $yw\ \delta s$ and so, if we consider the sum of all these line elements in a line of length s, as $\delta s \to 0$ the distance of the centroid from the x-axis is:

$$\bar{y} = \frac{\int_0^s yw \, \mathrm{d}s}{\int_0^s w \, \mathrm{d}s} = \frac{w \int_0^s y \, \mathrm{d}s}{w \int_0^s \mathrm{d}s} = \frac{\int_0^s y \, \mathrm{d}s}{s}$$

Thus the surface area generated by rotating the line is:

$$\text{area} = s \times 2\pi\bar{y} \qquad [20]$$

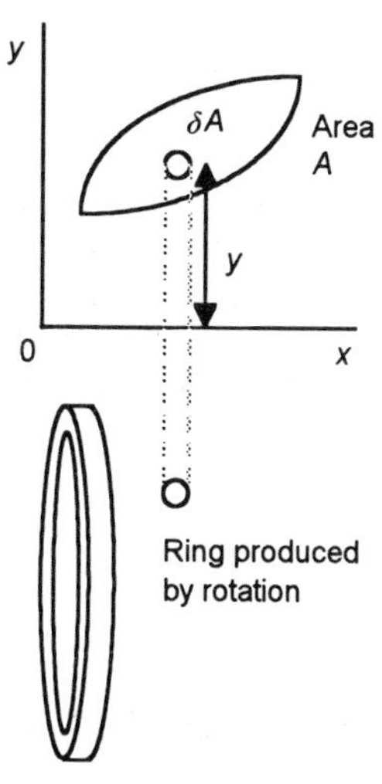

Figure 27.21 *Rotation of an area element*

This is known as the *first theorem of Pappus*:

If a curve is rotated about an axis, the area of the surface generated is equal to the length of the curve multiplied by the distance travelled by the centroid of the curve.

Consider an element δA of some area A being rotated about the x-axis (Figure 27.21). The volume generated by the rotation is the volume of an anchor ring with a cross-sectional area of δA and radius y. It is thus $2\pi y\ \delta A$. The total volume swept out by the rotation of the entire area is thus, when $\delta A \to 0$:

$$\text{volume} = \int_0^A 2\pi y \, \mathrm{d}A = 2\pi \int_0^A y \, \mathrm{d}A$$

The first moment of the element of area about the x-axis is $y\ \delta A$. Thus distance of the centroid from the x-axis for the entire area is, as $\delta A \to 0$:

$$\bar{y} = \frac{\int_0^A y\,\mathrm{d}A}{A}$$

Thus the volume generated by the rotation of the area is:

$$\text{volume} = A \times 2\pi\bar{y} \qquad [21]$$

This is known as the *second theorem of Pappus*:

If an area is rotated about an axis which does not cut the area, then the volume generated is equal to the area rotated multiplied by the distance travelled by the centroid.

Example

Determine the position of the centroid of a wire bent into a semicircle.

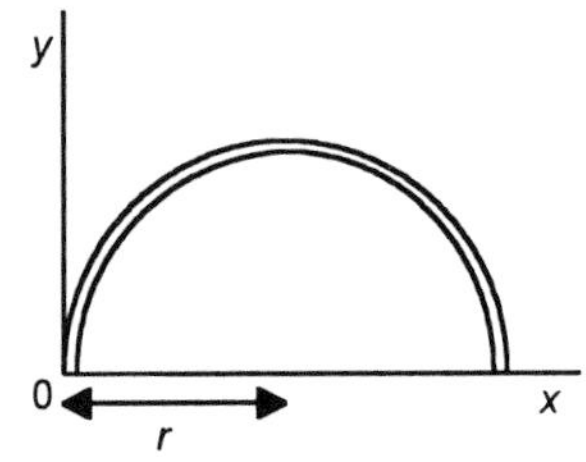

Figure 27.22 *Example*

Figure 27.22 shows the situation. Because it is symmetrical, the centroid will lie a distance r from the y-axis. We can determine the position of the centroid from the x-axis by using the first theorem of Pappus. The length of curve being rotated is $2\pi r$ and the rotation will generate a sphere with a surface area of $4\pi r^2$. The centroid will be rotated through a distance $2\pi\bar{y}$. Thus:

$$4\pi r^2 = 2\pi r \times 2\pi\bar{y}$$

Hence $\bar{y} = 2r/\pi$.

Example

Determine the volume of the torus formed by rotating the circular region specified by the equation $(x - 3)^2 + y^2 = 1$ about the y-axis.

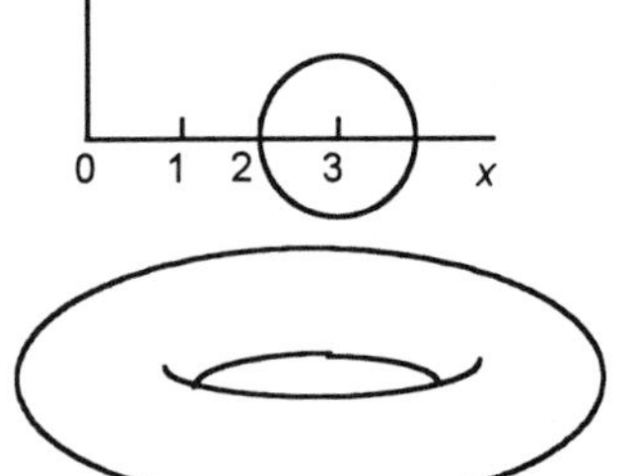

Figure 27.23 *Example*

Figure 27.23 shows the circular region and the form of the shape generated. The centroid of the circular region is on the x-axis and at a distance of 3 from the origin and is rotated through $2\pi 3$. The circular region has a radius of 1 and thus an area of $\pi 1^2$. Using the second theorem of Pappus, the volume generated by rotating the circular area about the y-axis is:

$$\text{volume generated} = \pi 1^2 \times 2\pi 3 = 6\pi^2 \text{ cubic units}$$

Revision

13 Determine the position of the centroid of a semicircular area of radius r.

14 An equilateral triangle with sides of length L is rotated about its base. Determine the surface area and volume of the generated solid. Note: the centroid of a triangle is one-third of its height above the base.

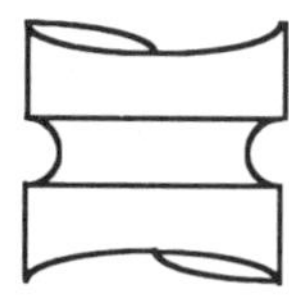

Figure 27.24 *Problem 15*

15 A cylinder of diameter 200 mm has a semicircular groove of diameter 30 mm cut round its circumference (Figure 27.24). Determine the volume of material removed. Hint: you will need the result of problem 13.

16 The area between the curve $y = 2x^2$ and the x-axis and bounded by $x = 0$ and $x = 3$ is rotated about the x-axis. Determine the position of the centroid.

27.3.2 Moments of inertia

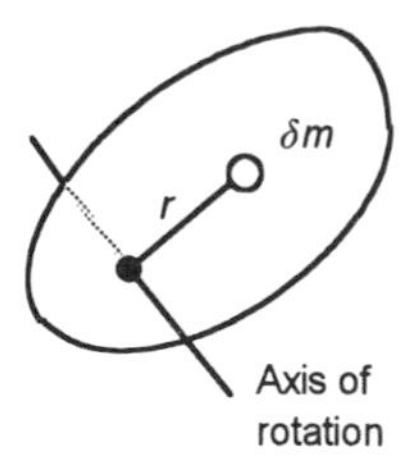

Figure 27.25 *Rotation of a rigid body*

Consider a rigid body rotating with a constant angular acceleration α about some axis (Figure 27.25). We can consider the body to be made up of small elements of mass δm. For such an element a distance r from the axis of rotation we have a linear acceleration of $a = r\alpha$. Thus the force acting on the element is $\delta m \times r\alpha$. The moment of this force is thus $Fr = r^2\alpha\ \delta m$. The total moment, i.e. torque T, due to all the elements of mass in the body is thus:

$$T = \sum r^2 \alpha\ \delta m \text{ for all the elements}$$

Thus if we have elements of mass at radial distance from 0 to R, in the limit as $\delta m \to 0$:

$$T = \int_0^R r^2 \alpha\ \mathrm{d}m$$

Since α is a constant we can write the above equation as:

$$T = \left(\int_0^R r^2\ \mathrm{d}m\right)\alpha = I\alpha \qquad [22]$$

where *I is the moment of inertia*. The moment of inertia is sometimes referred to as the *second moment of mass*, the word second occurring because we have r^2 instead of just the r that occurs in first moments. The examples that follow show how equation [22] can be used for the calculation of moments of inertia.

It is often useful to consider the moment of inertia of a body in terms of an equivalent, imaginary, body for which we consider all the mass M to be concentrated at a point a distance k from the axis. Then the moment of inertia is Mk^2, with k being termed the *radius of gyration*. Thus for a slender rod of length L, the moment of inertia about an axis through its centre and at right angles to its length is $\frac{1}{12}ML^2$ and so it has a radius of gyration given by $Mk^2 = \frac{1}{12}ML^2$ and thus $k = \sqrt{\frac{1}{12}}\,L$.

Situations often occur where the moment of inertia is known for a body about some particular axis and is required about some other axis. There are two theorems:

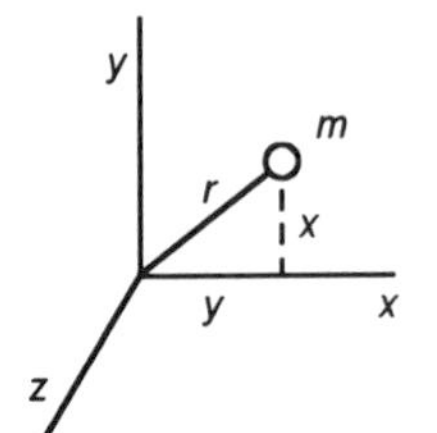

Figure 27.26 *Perpendicular axes*

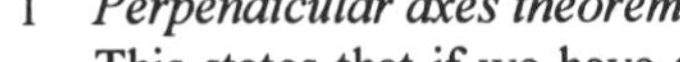

1 *Perpendicular axes theorem*
This states that if we have an object with a plane area in the x-y plane then the moment of inertia I_z about an axis at right angles to the plane is $I_z = I_x + I_y$, where I_x is the moment of inertia about the x-axis and I_y the moment of inertia about the y-axis.

The moment of inertia about the z-axis of a small point mass m in the x-y plane at a distance r from the origin (Figure 27.26) is mr^2. But $r^2 = x^2 + y^2$ and so $I_z = m(x^2 + y^2) = mx^2 + my^2 = I_x + I_y$.

2 *Parallel axes theorem*
This states that the moment of inertia with respect to any axis is equal to the moment of inertia with respect to a parallel axis through the centroid plus the product of the mass and the square of the distance between the two axes.

The moment of inertia I_c of an element of mass δm of a body about an axis through its centroid is $x^2\ \delta m$, where x is its distance from the centroid. If we now consider its moment of inertia I_d about a parallel axis a distance d away, then its moment of inertia is $(x + d)^2\ \delta m = x^2\ \delta m + 2dx\ \delta m + d^2\ \delta m$. Thus the moment of inertia of all the elements in the solid is:

$$I_d = \int x^2\ \mathrm{d}m + 2d \int x\ \mathrm{d}m + d^2 \int \mathrm{d}m$$

The first of these terms is the moment of inertia through the centroid. The second term is the sum of all the moments of the elements of mass about the centroid and so is zero. The sum of all the elements of mass is the total mass M. Thus $I_d = I_c + Md^2$.

Example

Determine the moment of inertia of a uniform disc about an axis through its centre and at right angles to its plane.

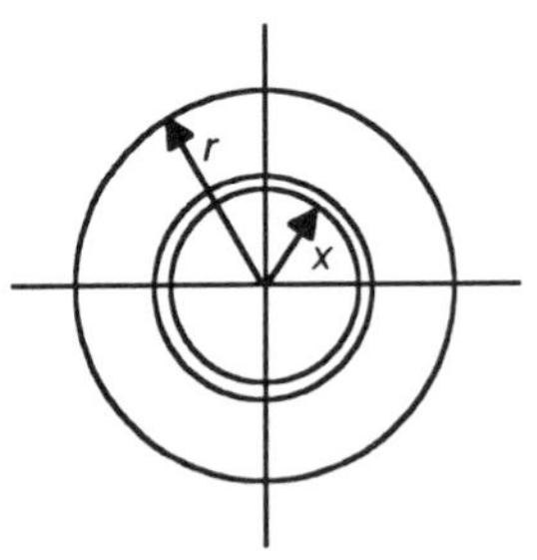

Figure 27.27 *Example*

Figure 27.27 shows the disc with an element of mass being chosen as a disc with a radius x and width δx. The element is a strip of length $2\pi x$ and so an area of $2\pi x\ \delta x$. If the mass of the disc is m per unit area, then the mass of the element is $\delta m = 2\pi mx\ \delta x$. The moment of inertia of the element is $x^2\ \delta m = 2\pi mx^3\ \delta x$. Thus the moment of inertia of the disc is:

$$I = \int_0^r 2\pi mx^3\ \mathrm{d}x = 2\pi m\left[\frac{x^4}{4}\right]_0^r = \tfrac{1}{2}\pi mr^4$$

Example

Determine the moment of inertia about a diameter for a sphere.

Consider a sphere of radius r and mass per unit volume m. If we take a thin slice of thickness δx of the sphere perpendicular to the diameter about which the moment of inertia is to be determined and a distance x

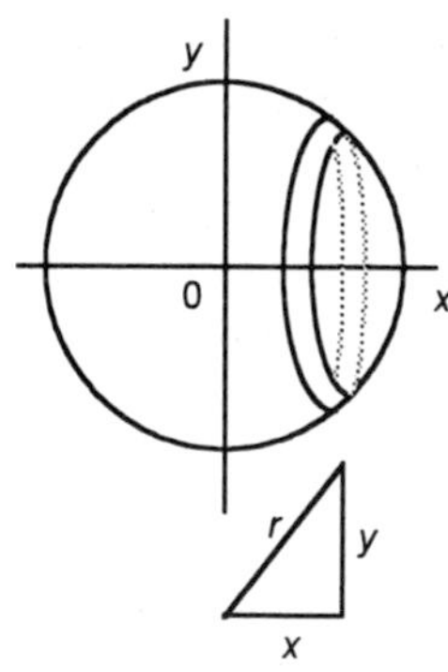

Figure 27.28 *Example*

from the sphere centre (Figure 27.28), then with the slice radius y we have an element of volume $\pi y^2\,\delta x$ and hence mass $\pi m y^2\,\delta x$. The moment of inertia of a disc is ½mass × radius² (see the previous example) and thus the moment of inertia of the slice is $\frac{1}{2}(\pi m y^2\,\delta x)y^2$ and the moment of inertia of the sphere as the sum of all the slices as $\delta x \to 0$ is:

$$I = \int_{-r}^{r} \tfrac{1}{2}\pi m y^4 \, dx$$

Since $r^2 = y^2 + x^2$:

$$I = \tfrac{1}{2}\pi m \int_{-r}^{r} (r^2 - x^2)^2 \, dx = \tfrac{1}{2}\pi m \int_{-r}^{r} (r^4 - 2r^2x^2 + x^4) \, dx$$

$$= \tfrac{1}{2}\pi m \left[r^4 x - \tfrac{2}{3}r^2x^3 + \tfrac{1}{5}x^5 \right]_{-r}^{r} = \tfrac{8}{15}\pi m r^5$$

Since the total mass M of the sphere is $\frac{4}{3}\pi m r^3$ then $I = \frac{2}{5}Mr^2$.

Example

The moment of inertia of a uniform disc of mass M and radius r about a diameter is $Mr^2/4$. Determine the moment of inertia about a tangent to the disc.

This requires the use of the parallel axes theorem. The required axis is parallel to the axis through the centroid and a distance r from it. Thus the moment of inertia about a tangent is $Mr^2/4 + Mr^2 = 5Mr^2/4$.

Revision

17 Determine the moment of inertia of a uniform rectangular sheet of mass M, length $2L$ and breadth $2b$, about an axis through its centre and parallel to its breadth.

18 Determine the moment of inertia of a uniform rod of length $2L$ and mass M about an axis through its centre and perpendicular to its length.

19 Determine the moment of inertia about the x-axis of a solid of mass M which is formed by rotating the area under the curve $y = x \tan\theta$ between $x = 0$ and $x = h$ about the x-axis.

20 Determine the moment of inertia of a right circular cone of mass M, radius r and height h about its axis. Hint: treat the cone as a solid of revolution produced by rotating the line $y = (r/h)x$ about the x-axis between $x = 0$ and $x = h$.

21 The moment of inertia of a uniform rectangular sheet of mass M, length L and base b, about an axis parallel to the base and through its centroid, is $ML^2/3$. Determine the moment of inertia of the rectangle about the base.

27.3.3 Second moments of area

Integrals of the form $\int y^2 \, dA$ are encountered in a number of situations in engineering and science, e.g. the bending of beams, and are referred to as *second moments of area*. Because they are similar to the integrals encountered with moments of inertia, they are sometimes referred to as the *area moment of inertia*. The second moments of area can be derived in a similar manner to the moments of inertia and the perpendicular and parallel axes theorems can be applied.

Example

Determine the second moment of area of a rectangular area about an axis (a) through the centroid and parallel to one side, (b) through the centroid and perpendicular to the plane of the area, (c) in the plane through the base.

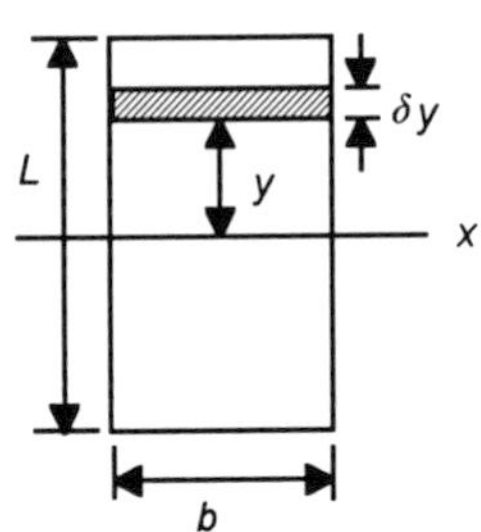

Figure 27.29 *Example*

(a) Figure 27.29 shows the area and the x-axis about which the second moment of area is to be taken. Consider a strip of width δy a distance y from the x-axis. The strip has an area $\delta A = b\,\delta y$. The second moment of area of the strip about the x-axis is $y^2\,\delta A = y^2 b\,\delta y$. Thus, as $\delta y \to 0$, the second moment of area for the rectangle is:

$$I_x = \int_{-L/2}^{L/2} y^2 b \, dy = b\left[\frac{y^3}{3}\right]_{-L/2}^{L/2} = \tfrac{1}{12} bL^3$$

(b) The second moment of area about the x-axis is as derived above. The second moment of area about the y-axis, an axis though the centre and parallel to the side of length L, is similar and $Lb^3/12$. The second moment of area about the z-axis which is perpendicular to the plane and through the centre is given by the perpendicular axes theorem as:

$$I_z = I_x + I_y = \tfrac{1}{12} bL^3 + \tfrac{1}{12} Lb^3 = \tfrac{1}{12} bL(L^2 + b^2)$$

Since the area is bL and the square of the length of the diagonal $L^2 + b^2$, the moment of inertia is one twelfth of the product of the area and the square of the diagonal.

(c) The second moment of area about the base of the rectangle is given by the parallel axes theorem as that moment through the centroid plus Ad^2, where d is the distance of the base from the centroid. Thus:

$$I = \tfrac{1}{12} bL^3 + bL\left(\tfrac{1}{2} L\right)^2 = \tfrac{1}{3} bL^3$$

Revision

22 Determine the second moment of area of a circle of radius r about a diameter.

23 Determine the second moment of area of the area bounded by the curve $y = 2\sqrt{x}$, the x-axis and $x = 1$ about (a) the y-axis, (b) the x-axis.

27.4 Means

The *mean* of a set of numbers is their sum divided by the number of numbers summed. The *mean value of a function* between $x = a$ and $x = b$ is the mean value of all the ordinates between these limits. Suppose we divide the area into n equal width strips (Figure 27.30), then if the values of the mid-ordinates of the strips are $y_1, y_2, \dots y_n$ the mean value is:

$$\text{mean value of } y = \frac{y_1 + y_2 + \dots + y_n}{n}$$

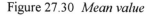

If δx is the width of the strips, then $n\,\delta x = b - a$. Thus:

$$\text{mean value of } y = \frac{(y_1 + y_2 + \dots + y_n)\delta x}{b - a}$$

Figure 27.30 *Mean value*

Hence, as $\delta x \to 0$:

$$\text{mean value of } y = \frac{1}{b-a}\int_a^b y\,\mathrm{d}x \qquad [23]$$

Since the sum of all the $y\,\delta x$ terms is the area under the graph between $x = a$ and $x = b$:

$$\text{mean value of } y = \frac{\text{area under graph}}{b - a}$$

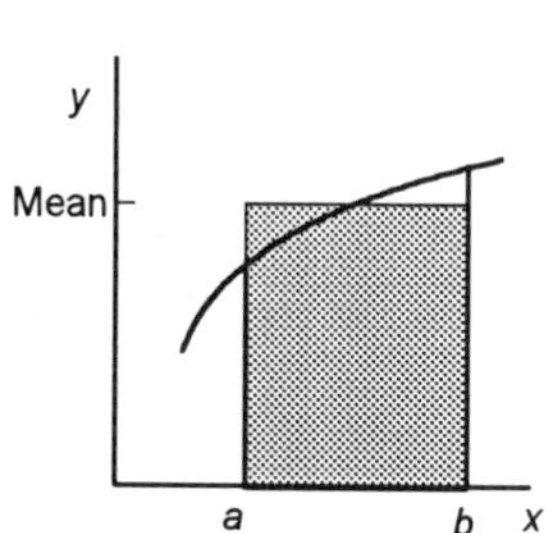

Figure 27.31 *Mean value rectangle*

But the product of the mean value and $(b - a)$ is the area of a rectangle of height equal to the mean value and width $(b - a)$. Figure 27.31 shows this mean value rectangle.

Example

Determine the mean value of the function $y = \sin x$ between $x = 0$ and $x = \pi$.

Using equation [23]:

$$\text{mean value of function} = \frac{1}{b-a}\int_a^b y\,\mathrm{d}x = \frac{1}{\pi - 0}\int_0^\pi \sin x\,\mathrm{d}x$$

$$= \frac{1}{\pi}[-\cos x]_0^\pi = \frac{2}{\pi} = 0.637$$

Revision

24 Determine the mean values of the following functions between the specified limits:

(a) $y = 2x$ between $x = 0$ and $x = 1$,

(b) $y = x^2$ between $x = 1$ and $x = 4$,

(c) $y = 3x^2 - 2x$ between $x = 1$ and $x = 4$,

(d) $y = \cos^2 x$ between $x = 0$ and $x = 2\pi$,

(e) $y = 3 \sin x + 2$ from $x = 0$ to $x = \pi$,

(f) $y = e^x$ between $x = 1$ and $x = 4$

25 With simple harmonic motion, the displacement x of an object is related to the time t by $x = A \cos \omega t$. Determine the mean value of the displacement during one-quarter of an oscillation, i.e. between when $\omega t = 0$ and $\omega t = \pi$.

26 The number N of radioactive atoms in a sample is a function of time t, being given by $N = N_0 e^{-\lambda t}$. Determine the mean number of radioactive atoms in the sample between $t = 0$ and $t = 1/\lambda$.

27.4.1 Root-mean-square values

The power dissipated by an alternating current i when passing through a resistance R is i^2R. The mean power dissipated over a time interval from $t = 0$ to $t = T$ will thus be:

$$\text{mean power} = \frac{1}{T-0}\int_0^T i^2 R \, dt = \frac{R}{T}\int_0^T i^2 \, dt$$

If we had a direct current I generating the same power then we would have:

$$I^2R = \frac{R}{T}\int_0^T i^2 \, dt$$

and:

$$I = \sqrt{\frac{1}{T}\int_0^T i^2 \, dt} \qquad [24]$$

This current I is known as the *root-mean-square* current. There are other situations in engineering and science where we are concerned with determining root-mean-square quantities. The procedure is thus to determine the mean value of the squared function over the required interval and then take the square root.

Example

Determine the root-mean-square current value of the alternating current $i = I \sin \omega t$ over the time interval $t = 0$ to $t = 2\pi/\omega$.

Using equation [24]:

$$I = \sqrt{\frac{1}{T}\int_0^T i^2\,\mathrm{d}t} = \sqrt{\frac{\omega}{2\pi}\int_0^{2\pi/\omega} I^2 \sin^2 \omega t\,\mathrm{d}t}$$

$$= \sqrt{\frac{I^2\omega}{2\pi}\int_0^{2\pi/\omega} \tfrac{1}{2}(1 - \cos 2\omega t)\,\mathrm{d}t} = \sqrt{\frac{I^2\omega}{4\pi}\left[t - \frac{1}{2\omega}\sin 2\omega t\right]_0^{2\pi/\omega}}$$

$$= \frac{I}{\sqrt{2}}$$

Example

Determine the root-mean-square value of the waveform shown in Figure 27.32 over a period of 0 to 2 s.

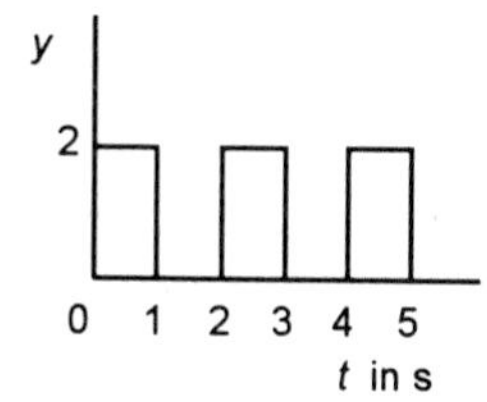

Figure 27.32 *Example*

From $t = 0$ to $t = 1$ s the waveform is described by $y = 2$. From $t = 1$ s to $t = 2$ s the waveform is described by $y = 0$. Thus the root-mean-square value is given by:

$$y_{\text{rms}} = \sqrt{\tfrac{1}{2}\left(\int_0^1 4\,\mathrm{d}t + \int_1^2 0\,\mathrm{d}t\right)} = \sqrt{\tfrac{1}{2}[4t]_0^1} = \sqrt{2}$$

Revision

27 Determine the root-mean-square values of the following functions between the specified limits:

(a) $y = x^2$ from $x = 1$ to $x = 3$, (b) $y = x$ from $x = 0$ to $x = 2$,

(c) $y = \sin x + 1$ from $x = 0$ to $x = 2\pi$, (d) $y = \sin 2x$ from $x = 0$ to $x = \pi$,

(e) $y = \mathrm{e}^x$ from $x = -1$ to $x = +1$

28 Determine the root-mean-square value of a half-wave rectified sinusoidal voltage. Between the times $t = 0$ and $t = \pi/\omega$ the equation is $v = V \sin \omega t$ and between $t = \pi/\omega$ and $t = 2\pi/\omega$ we have $v = 0$.

Problems

1 For a circle of radius 4, determine the area cut off from the circle by a chord whose distance from the centre is 3.

2 Determine the area of the region bounded by the graphs of $y = x - 2$ and $y = 4 - x^2$.

3 Determine the volumes generated by rotating the areas between the following functions and the x-axis about the x-axis:

(a) $y = x^3$ between $x = 0$ and $x = 3$, (b) $y = 5x$ between $x = 1$ and $x = 4$,

(c) $y = 2 \sin x + 3$ between $x = 0$ and $x = \pi$

4 Determine the volume generated by rotating about the line $x = 4$ the area bounded by $y = \sqrt{x}$, $y = 0$ and $x = 4$.

5 Determine the volume generated by rotating about the line $y = 6$ the area bounded by the graphs of $y = x^2$ and $y = 4x - x^2$.

6 An ellipsoid is formed by rotation about the x-axis of the area bounded by:

$$y = b\sqrt{1 - \frac{x^2}{a^2}}$$

$x = -a$ and $x = +a$. Determine the volume.

7 Show that the volume of a right-circular cone of radius r and height h is given by $\frac{1}{3}\pi r^2 h$. Hint: such a cone can be generated by rotating about the x-axis the area under the line $y = (r/h)x$ from $x = 0$ to $x = h$.

8 Show that the volume of a sphere of radius r is $\frac{4}{3}\pi r^3$. Hint: rotate about the x-axis that part of the circle $x^2 + y^2 = r^2$ which is in the first quadrant and then deduce from this the volume of the entire sphere.

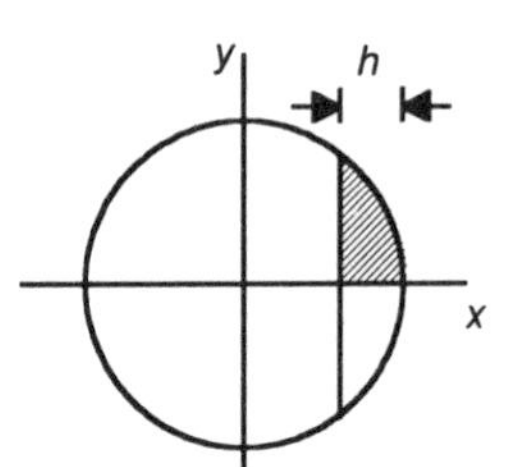

Figure 27.33 *Problem 9*

9 A spherical storage tank of radius r contains a liquid which at the deepest has a depth of h. Determine the volume of the liquid. Hint: consider the rotation of the shaded area shown in Figure 27.33 about the x-axis.

10 Determine the volume obtained by rotating the area under an arch of the cycloid described by $x = a(\theta - \sin \theta)$, $y = a(1 - \cos \theta)$ about the x-axis.

11 Determine the lengths of the curves of graphs of the following functions:

(a) $y = 40 \cosh(x/40)$ from $x = 0$ to 20, (b) $y^2 = x^3$ from $x = 0$ to $x = 8$,

(c) $y = \ln(\cos x)$ from $x = 0$ to $x = \pi/4$, (d) $y = e^x$ from $x = \frac{3}{4}$ to $x = \frac{4}{3}$,

(e) $x = t^2$, $y = \frac{1}{2}t^2$ from $x = 0$ to $x = \sqrt{2}$

12 An electric cable is suspended between two pylons 200 m apart. The cable hangs in the shape of a catenary with an equation given by

$y = 150 \cosh(x/150)$. Determine the length of the cable between the pylons.

13 Determine the length of one arch of the cycloid given by the equations $x = a(\theta - \sin\theta)$, $y = a(1 - \cos\theta)$.

14 Determine the areas of the curved surfaces of the solids formed by rotation about the x-axis of the following functions:

(a) $y = \cosh x$ from $x = 0$ to $x = 1$, (b) $y = 4\sqrt{x}$ from $x = 5$ to $x = 12$,

(c) $y = 2\sqrt{x}$ from $x = 0$ to $x = 15$, (d) $y = x^3$ from $x = 0$ to $x = 1$,

15 A right-circular cone of base radius r and height h is generated by rotation about the x-axis of the function $y = rx/h$ between the limits $x = 0$ and $x = h$. Determine the surface area of the cone.

16 Determine the positions of the centroids of the areas bounded by:

(a) $y = \sin 2x$, the x-axis and between $x = 0$ and $x = \pi/2$,

(b) $y = x$, $y = 2x$ and $x = 2$, (c) $y = x^2$, the x-axis and $x = 2$,

(d) $y = x^2$, $y^2 = 8x$, (e) $y = x^3$, the x-axis and $x = 1$, (f) $y = \frac{1}{2}x^2$ and $y = 8$

17 Determine (a) the position of the centroid of one half loop the curve $a^2y^2 = x^2(a^2 - x^2)$ and (b) the volume of the solid generated by rotating one loop of it. Use the theorems of Pappus.

18 A square of side L is rotated about an axis in its plane through a corner and which is at right angles to the diagonal through the corner. Using the theorems of Pappus, determine (a) the surface area and (b) volume of the generated solid.

19 Determine the volume of the torus formed by rotating the circular region specified by the equation $(x - 5)^2 + y^2 = 1$ about the y-axis.

20 Determine the volume of the solid formed by rotating the region bounded by $y = x$, $y = 4$ and $x = 0$ about the x-axis.

21 Determine the moment of inertia for a uniform triangular sheet of mass M, base b and height h about (a) an axis through the centroid and parallel to the base and (b) about the base. The centroid is at one-third the height.

22 Determine the moment of inertia of a flat circular ring with an inner radius r, outer radius $2r$ and mass M about an axis through its centre and at right angles to its plane.

23 Determine the radius of gyration about the x-axis of a plane shape bounded by $y^2 = 4ax$, the x-axis and $x = b$.

24 Determine the moment of inertia of a uniform square sheet of mass M and side L about (a) an axis through its centre and in its plane, (b) an axis in its plane a distance d from its centre..

25 Determine the second moment of area of a triangular area of base b and height h about an axis (a) through the centroid and parallel to the base, (b) through the base. The centroid is at one-third the height.

26 Determine the mean values of the following functions between the specified limits:

(a) $y = 3 \sin 2x$ from $x = 0$ to $x = \frac{1}{2}\pi$,

(b) $y = x - 2$ from $x = -1$ to $x = 0$,

(c) $y = (2x - 1)^2$ from $x = 0$ to $x = 1$,

(d) $y = \sin^2 x$ between $x = 0$ and $x = \pi$,

(e) $y = 10 \sin x$ between $x = 0$ and $x = \pi$

27 Determine the mean value of a voltage v between times $t = 0$ and $t = \pi/100$ if $v = 10 \sin 250t$.

28 Determine the root-mean-square values of the following functions between the specified limits:

(a) $y = 3x + 2$ from $x = 0$ to $x = 2$,

(b) $y = 2x - 2x^2$ from $x = 0$ to $x = 2$,

(c) $y = 5 \sin 2x$ from $x = 0$ to $x = \pi$,

(d) $y = 2 \sin x$ from $x = 0$ to $x = 2\pi$,

(e) $y = e^x + 1$ from -1 to 1

29 Two voltages are described by the equations $v_1 = V_1 \sin(\omega t + a_1)$ and $v_2 = V_2 \sin(2\omega t + a_2)$. Determine the root-mean-square values of each voltage alone and sum of their voltages.

30 Determine the root-mean-square value of the periodic triangular voltage shown in Figure 27.34 over one complete period T.

31 Determine the root-mean-square value of the periodic rectangular voltage shown in Figure 27.35 over one complete period of 2 s.

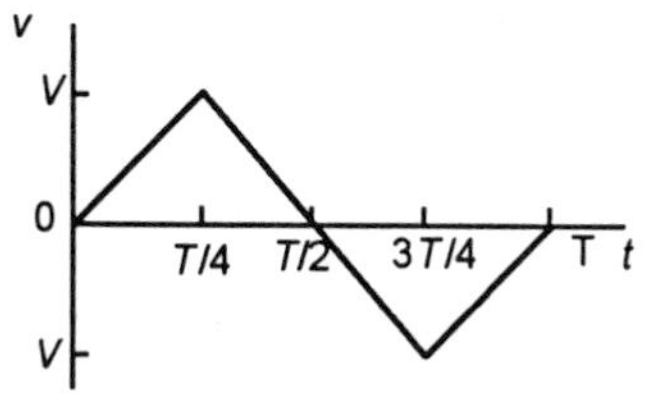

Figure 27.34 *Problem 30*

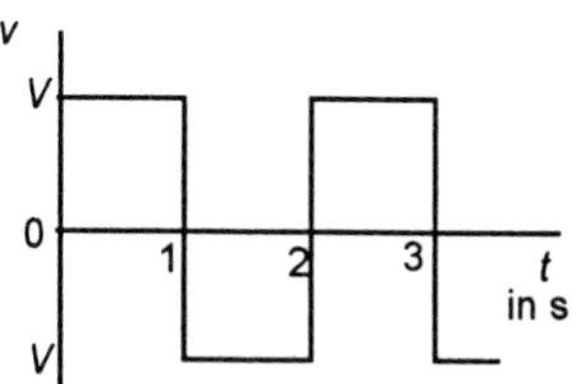

Figure 27.35 *Problem 31*

Part 7 Ordinary differential equations

The aims of this part are to enable the reader to:

- Develop differential equations to model systems.
- Solve first-order differential equations by separation of variables, integrating factors and the use of complementary functions and particular integrals.
- Solve second-order differential equations.
- Solve differential equations by numerical methods.
- Solve differential equations relating to dynamic responses of systems such as measurement systems.

This part assumes Part 6 Differentiation and Integration and consequently Part 1 Functions. In considering methods used to solve differential equations, the Laplace transform has not been included in the part but is dealt with separately in Part 9. Chapter 31 on numerical methods of solving differential equations makes some use of the Taylor series (Chapter 5).

28 Modelling with ordinary differential equations

28.1 Introduction

This chapter introduces ordinary differential equations and shows how they can be used to model the behaviour of systems in engineering and science. Chapter 32 extends this modelling to the behaviour of circuit and system dynamic responses.

A *differential equation* is an equation involving derivatives of a function. Thus examples of differential equations are:

$$\frac{dy}{dx} + 2y = 5 \text{ and } \frac{d^2y}{dx^2} + 3\frac{dy}{dx} + 2y = 5$$

The term *ordinary differential equation* is used when there is only one independent variable, the above examples having only y as a function of x and so being ordinary differential equations. If two or more independent variables occur, the equation is termed a *partial differential equation*, e.g.

$$\frac{\partial z}{\partial x} + \frac{\partial z}{\partial y} + 2z = 0 \text{ and } \frac{\partial^2 z}{\partial x^2} + \frac{\partial^2 z}{\partial y^2} = 0$$

Differential equations arise from such situations as the motion of projectiles, the cooling of a solid or liquid, transient currents and voltages in electrical circuits, the rate of decay of radioactive materials and oscillations with mechanical or electrical systems.

This chapter introduces the idea of developing ordinary differential equations to describe systems, with Chapters 29, 30 and 31 dealing with the solutions of such equations. Chapter 22 and the concept of linear equations from Chapter 16 are assumed.

28.2 Modelling with differential equations

In developing mathematical models we:

1 Consider the real world situation and formulate it in mathematical terms. This can involve making assumptions about which factors are involved and the relationships between variables.

2 Carry out mathematical analysis on the model, e.g. considering what it forecasts as happening with particular inputs.

3 Use the results of the mathematical analysis to lead to an interpretation in the context of the real world situation.

The following illustrate how ordinary differential equations can be evolved as mathematical models of some simple systems, later chapters showing how such models react to different inputs.

28.2.1 Mechanical systems

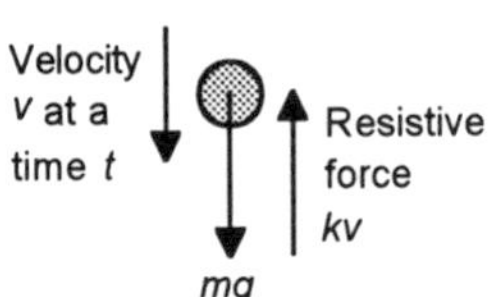

Figure 28.1 *Body falling in air*

Consider a freely falling body of mass m in air (Figure 28.1). The gravitational force acting on the body is mg, where g is the acceleration due to gravity. Opposing the movement of the body through the air is air resistance. Assuming that the air resistance force is proportional to the velocity v, the net force F acting on the body is $mg - kv$, where k is a constant. But Newton's second law gives the net force F acting on a body as the product of its mass m and acceleration a, i.e. $F = ma$. But acceleration is the rate of change of velocity v with time t. Thus we can write:

$$F = m\frac{\mathrm{d}v}{\mathrm{d}t} = mg - kv$$

and so the differential equation describing this system is:

$$m\frac{\mathrm{d}v}{\mathrm{d}t} + kv = mg \qquad [1]$$

This is a *first-order differential equation* because the highest derivative is just $\mathrm{d}v/\mathrm{d}t$. It describes how the velocity varies with time.

The order of a differential equation is equal to the order of the highest derivative that appears in the equation.

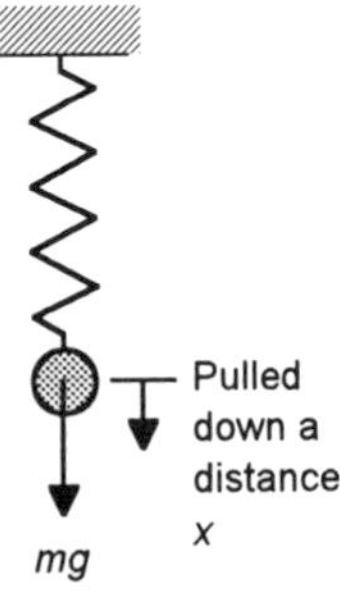

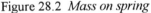

Figure 28.2 *Mass on spring*

Consider another mechanical system, an object of mass m suspended from a support by a spring (Figure 28.2). When the mass is placed on the spring it stretches by d. Assuming Hooke's law, and so the displacement proportional to the force exerted by the spring, at equilibrium we have $mg = kd$. Now if we pull the body down a distance x from this equilibrium position, the net restoring force acting on the body is $mg - kd - kx = -kx$. The body when released is thus acted on by this force and, since Newton's second law gives $F = ma$ and acceleration is the rate of change of velocity with time, with velocity being the rate of change of displacement with time:

$$F = m\frac{\mathrm{d}^2x}{\mathrm{d}t^2} = -kx$$

and so:

$$m\frac{\mathrm{d}^2x}{\mathrm{d}t^2} + kx = 0 \qquad [2]$$

This is a *second-order differential equation* because the highest derivative is d^2x/dt^2. It described the resulting oscillations of the body after it has been released.

Revision

1 Write differential equations relating:

(a) The velocity v and time t for an object of mass m thrown vertically upwards against air resistance proportional to the square of its velocity.

(b) The displacement x of a mass m on a spring when the mass is pulled down from its equilibrium position and released with there being a damping force proportional to the velocity.

28.2.2 Electrical systems

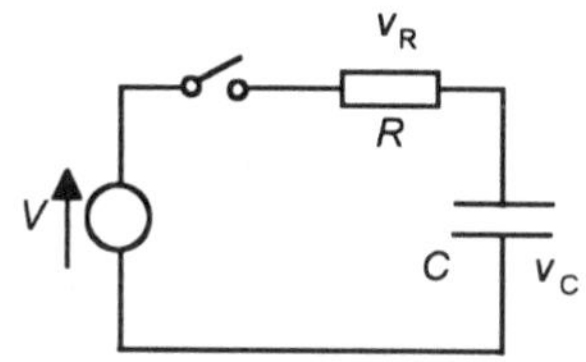

Figure 28.3 *Series RC circuit*

When a pure capacitor has a potential difference v applied across it, the charge q on the plates is given by $q = Cv$, where C is the capacitance. Current i is the rate of movement of charge and so:

$$i = \frac{dq}{dt} = C\frac{dv}{dt} \qquad [3]$$

Thus if we have an electrical circuit containing a resistor in series with a capacitor (Figure 28.3) we have the supply voltage V equal to the sum of the voltages across the resistor and capacitor:

$$V = v_R + v_C = Ri + v_C$$

and thus using equation [3]:

$$RC\frac{dv_C}{dt} + v_C = V \qquad [4]$$

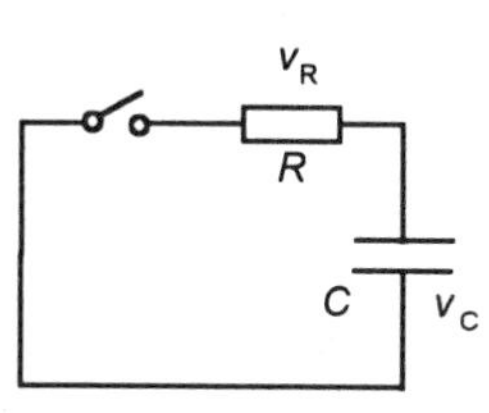

Figure 28.4 *RC discharge circuit*

This differential equation describes how the voltage across the capacitor changes with time from when the switch is closed.

When a charged capacitor discharges through a resistance (Figure 28.4) then $v_R + v_C = 0$ and so:

$$RC\frac{dv_C}{dt} + v_C = 0 \qquad [5]$$

This differential equation describes how the voltage across the capacitor changes with time from when the switch is closed.

When a pure inductor has a current i flowing through it, then the induced e.m.f. produced in the component is proportional to the rate of change of current, the induced e.m.f. being $-L\ di/dt$ where L is the inductance. If the component has only inductance and no resistance, then there can be no

potential drop across the component due to the current through the resistance and thus to maintain the current through the inductor the voltage source must supply a potential difference v which just cancels out the induced e.m.f. Thus the potential difference across an inductor is:

$$v = L\frac{\mathrm{d}i}{\mathrm{d}t} \qquad [6]$$

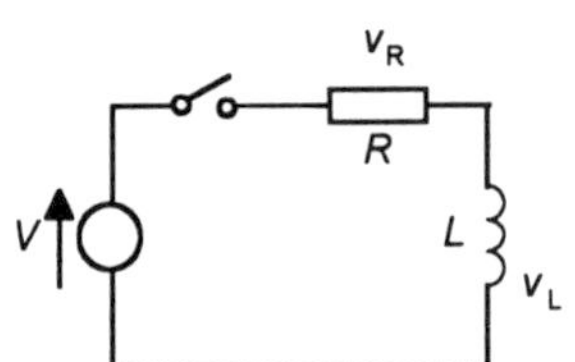

Figure 28.5 *Series RL circuit*

If we have an electrical circuit containing an inductor in series with a resistor (Figure 28.5) then, when the supply voltage V is applied, we have $V = v_L + v_R$. Thus, using equation [6]:

$$L\frac{\mathrm{d}i}{\mathrm{d}t} + Ri = V$$

The steady state current I will be attained when the current ceases to change with time. We then have $RI = V$ and so the equation can be written as:

$$\frac{L}{R}\frac{\mathrm{d}i}{\mathrm{d}t} + i = I \qquad [7]$$

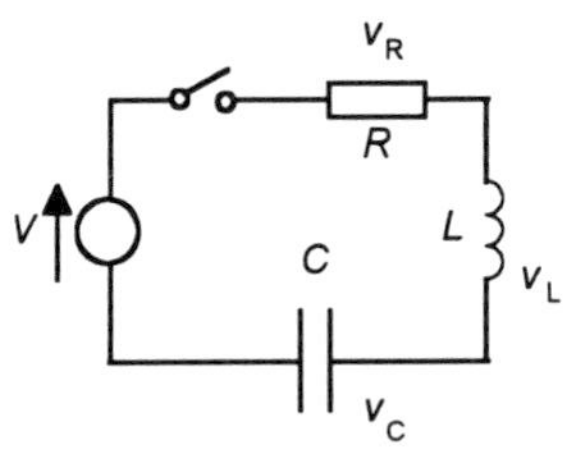

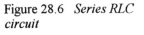

Figure 28.6 *Series RLC circuit*

Consider now a circuit including a resistor, a capacitor and an inductor in series (Figure 28.6). When the switch is closed the supply voltage v is applied across the three components and $V = v_R + v_L + v_C$. Thus, using equation [6]:

$$Ri + L\frac{\mathrm{d}i}{\mathrm{d}t} + v_C = V$$

Since $i = C\,\mathrm{d}v_C/\mathrm{d}t$ (equation [3]), then:

$$LC\frac{\mathrm{d}^2 v_C}{\mathrm{d}t^2} + RC\frac{\mathrm{d}v_C}{\mathrm{d}t} + v_C = V \qquad [8]$$

This second order differential equation describes how the voltage across the capacitor varies with time.

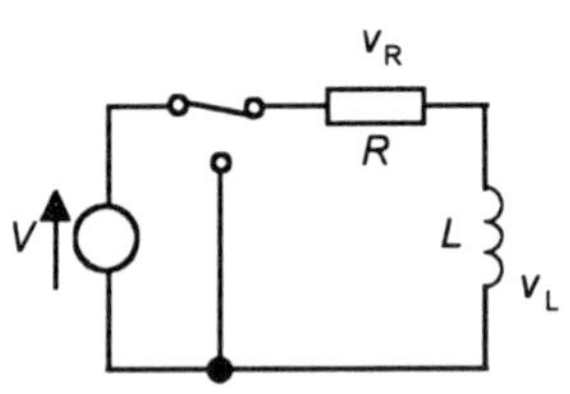

Figure 28.7 *Revision problem 2*

Revision

2 Derive the differential equation for the circuit shown in Figure 28.7 showing how the current through the resistor varies with time when, following the application of the supply voltage to the circuit to give a current, the switch is closed and the circuit consists of just the resistor and inductor.

28.2.3 Hydraulic systems

Consider an open tank into which liquid can enter at the top through one pipe and leave at the base through another (Figure 28.8). If the liquid enters

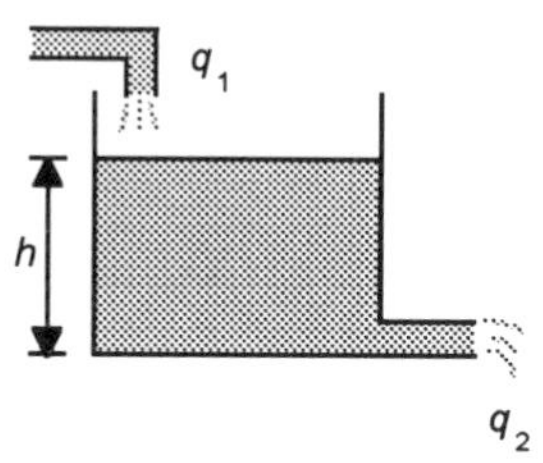

Figure 28.8 *Liquid level in a tank*

at the rate of a volume of q_1 per second and leaves at the rate of q_2 per second, then the rate at which the volume V of liquid in the tank changes with time is:

$$\frac{\mathrm{d}V}{\mathrm{d}t} = q_1 - q_2$$

But $V = Ah$, where A is the cross-sectional area of the tank and h the height of the liquid in the tank. Thus:

$$\frac{\mathrm{d}(Ah)}{\mathrm{d}t} = A\frac{\mathrm{d}h}{\mathrm{d}t} = q_1 - q_2$$

The rate at which liquid leaves the tank, when flowing from the base of the tank into the atmosphere, is given by Torricelli's theorem as $q_2 = \sqrt{(2gh)}$. Thus the differential equation can be written as:

$$A\frac{\mathrm{d}h}{\mathrm{d}t} + \sqrt{2gh} = q_1 \qquad [9]$$

Revision

3 Derive a differential equation showing how the height of liquid in a tank open to the atmosphere varies with time when liquid leaks from its base.

28.3 Solving differential equations

The differential equation $\mathrm{d}y/\mathrm{d}x = 2$ describes a straight line with a constant gradient of 2 (Figure 28.9). There are, however, many possible graphs which fit this specification, the family of such lines having equations of the form $y = 2x + A$, where A is a constant. These are all solutions for the differential equation.

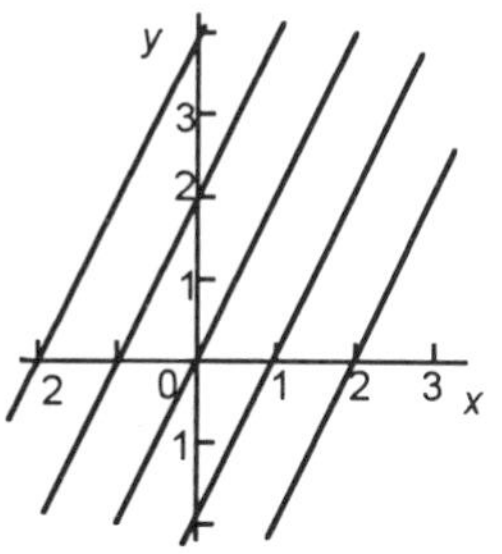

Figure 28.9 *General solution*

The term solution is used with a differential equation for the relationship between the dependent and independent variables such that the differential equation is satisfied for all values of the independent variable.

Thus the differential equation $\mathrm{d}y/\mathrm{d}x = 2$ has many solutions given by $y = 2x + A$, this being termed the *general solution*. Only if constraints are specified which enable constants like A to be evaluated will there be just one solution, this being then termed a *particular solution*. The term *initial conditions* are used for the constraints if specified at $y = 0$ and *boundary conditions* if specified at some other value of y. Thus if, for a general solution $y = 2x + A$, we have the initial condition that $y = 0$ when $x = 0$ then A is 0 and so the particular solution is $y = 2x$.

Example

Verify that $y = e^x$ is a particular solution of the differential equation $\mathrm{d}y/\mathrm{d}x = y$.

If $y = e^x$ then $dy/dx = e^x$. Thus for all values of y we have $dy/dx = y$ and so $y = e^x$ is a solution.

Example

$y = A\,e^x + B\,e^{2x}$ is a general solution of the differential equation:

$$\frac{d^2y}{dx^2} - 3\frac{dy}{dx} + 2y = 0$$

Determine the particular solution for the boundary conditions $y = 3$ when $x = 0$ and $dy/dx = 5$ when $x = 0$.

For $y = A\,e^x + B\,e^{2x}$ with $y = 3$ when $x = 0$ we have $3 = A + B$. With $y = A\,e^x + B\,e^{2x}$ we have $dy/dx = A\,e^x + 2B\,e^{2x}$ and thus with $dy/dx = 5$ when $x = 0$ we have $5 = A + 2B$. This pair of simultaneous equations gives $A = 1$ and $B = 2$. Thus the particular solution is:

$$y = e^x + 2\,e^{2x}$$

We can check that this is a valid solution by substituting it in the differential equation:

$$e^x + 8\,e^{2x} - 3(e^x + 4\,e^{2x}) + 2(e^x + 2\,e^{2x}) = 0$$

Revision

4 Verify that that the following are solutions of the given differential equations:

(a) $y = 2\sin 3x$ for $\frac{d^2y}{dx^2} + 9y = 0$, (b) $y = e^{2x}$ for $\frac{dy}{dx} - 2y = 0$,

(c) $y = \frac{1}{A - x}$ for $\frac{dy}{dx} - y^2 = 0$

5 For the following general solutions of differential equations, verify that they are solutions and determine the particular solution for the given initial/boundary conditions:

(a) $y = A\,e^x$ for $\frac{dy}{dx} - y = 0$, $y = 2$ at $x = 0$,

(b) $y = A\,e^x - 1$ for $\frac{dy}{dx} - y = 1$, $y = 3$ at $x = 0$,

(c) $y = A\cos x + B\sin x$ for $\frac{d^2y}{dx^2} + y = 0$, $y = 0$ at $x = 0$ and $y = 3$ at $x = \frac{1}{2}\pi$

6 For an electrical circuit with a capacitor in series with a resistor the differential equation relating the voltage across the capacitor with time when a constant voltage V is applied to the circuit is:

$$RC\frac{\mathrm{d}v_C}{\mathrm{d}t} + v_C = V$$

Verify that $v_C = A\,\mathrm{e}^{-t/RC} + V$ is the general solution and determine the particular solution if $v_C = 0$ when $t = 0$.

28.4 Linear differential equations

In Section 16.1.1 a *linear equation* was defined as one not involving any products or roots of variables, all variables only occurring to the first power, and no variables occurring in trigonometric, logarithmic or exponential functions. A differential equation is said to be linear if the dependent variable and all its derivatives only occur to the first power, there are no products of terms involving the dependent variable, e.g. no terms like $y\ \mathrm{d}y/\mathrm{d}x$, and there are no functions of the dependent variable or its derivatives which are in trigonometric, logarithmic or exponential form. For example:

$$\frac{\mathrm{d}y}{\mathrm{d}x} + y = x^2$$

is linear. It does not matter that the independent variable x is raised to the power 2, the dependent variable and its derivative are not. Examples of non-linear differential equations are:

$$\frac{\mathrm{d}y}{\mathrm{d}x} + y^2 = 0 \text{ and } \left(\frac{\mathrm{d}y}{\mathrm{d}x}\right)^2 + y = 0$$

Linear equations can be classified as *homogeneous* or *non-homogeneous*. If all the terms containing the dependent variable are moved to the left-hand side of the equals sign, then an equation is said to be homogeneous if there is then just a zero on the right-hand side. For example:

$$\frac{\mathrm{d}y}{\mathrm{d}x} + 4y = 0$$

is homogeneous but:

$$\frac{\mathrm{d}y}{\mathrm{d}x} + 4y = 5$$

is not.

An important property of linear equations is the *additive property* of solutions. Suppose we have the linear equation $y = 2x$ and we let $y = u + v$. Then $u + v = 2x$. Because the equation is linear we can have $u = 2x$ and $v = 0$. We can treat linear differential equations in a similar manner. Thus if y and z are some functions of x:

$$\frac{d}{dx}(y+z) = \frac{dy}{dx} + \frac{dz}{dx}$$

The differential equation of the combined function is the sum of the differential equations of the two separate functions. We can use this to obtain a general solution to a linear differential equation as the sum of two parts called the complementary function and the particular integral. This is discussed in more detail in the next chapter.

Problems

1 Derive differential equations to represent the following situations:

(a) The velocity v of a boat of mass m on still water in terms of time t after the engines are switched off if the drag forces acting on the boat are proportional to the velocity.

(b) The velocity v of an object falling from rest in air if the drag forces are proportional to the square of the velocity.

(c) The intensity I of a beam of light emerging from a block of glass in terms of the thickness x of the glass if the intensity decreases at a rate proportional to the block thickness.

(d) The rate at which the pressure p at the base of a tank changes with time if liquid of density ρ enters the tank at the volume rate of q_1 and leaves at the rate of q_2.

2 Verify that the following are solutions of the given differential equations:

(a) $y = \cos 2x$ for $\frac{d^2y}{dx^2} + 4y = 0$,

(b) $y = 2\sqrt{x} - \sqrt{x}\ \ln x$ for $4x^2\frac{d^2y}{dx^2} + y = 0$,

(c) $y = e^x \cos x$ for $\frac{d^2y}{dx^2} - 2\frac{dy}{dx} + 2y = 0$,

(d) $y = 2\ e^x + 3x\ e^x$ for $\frac{d^2y}{dx^2} + 3\frac{dy}{dx} + 2y = 0$

3 For the following general solutions of differential equations, verify that they are solutions and determine the particular solution for the given boundary conditions:

(a) $y = A\ e^x + Bx\ e^x$ for $\frac{d^2y}{dx^2} - 2\frac{dy}{dx} + y = 0$, $y = 0$ and $\frac{dy}{dx} = 1$ at $x = 0$,

(b) $y = A\sin\omega t + B\cos\omega t$ for $\dfrac{d^2y}{dt^2} + \omega^2 y = 0$, $y = 2$ and $\dfrac{dy}{dt} = 1$ at $t = 0$,

(c) $y = (A + x^2)\,e^{-x}$ for $\dfrac{dy}{dx} + y = 2x\,e^{-x}$, $y = 2$ at $x = 0$

4 The differential equation relating the deflection y with distance x from the fixed end of a cantilever with a uniformly distributed load is:

$$\frac{d^2y}{dx^2} = -\frac{w}{2EI}(L^2 - 2Lx + x^2)$$

The general solution is given as:

$$y = -\frac{w}{2EI}\left(\tfrac{1}{2}L^2x^2 - \tfrac{1}{3}Lx^3 + \tfrac{1}{12}x^4\right) + Ax + B$$

Verify that this is the general solution and determine the particular solution for $y = 0$ and $dy/dx = 0$ at $x = 0$.

29 First-order differential equations

29.1 Introduction

First-order differential equations are often used to model the behaviour of engineering systems. For example, the exponential growth system where the rate of change $\mathrm{d}N/\mathrm{d}t$ of some quantity is proportional to the quantity N present can be represented by:

$$\frac{\mathrm{d}N}{\mathrm{d}t} = kN$$

or exponential decay, e.g. radioactivity, where the rate at which a quantity decreases is proportional to the quantity present:

$$\frac{\mathrm{d}N}{\mathrm{d}t} = -kN$$

Such differential equations are of the form:

$$\frac{\mathrm{d}y}{\mathrm{d}x} = f(y) \qquad [1]$$

Another form of differential equation is illustrated by the growth of the voltage across a capacitor in an electrical circuit having a capacitor in series with a resistor (see Chapter 28, equation [4]):

$$RC\frac{\mathrm{d}v_C}{\mathrm{d}t} + v_C = V$$

Such equations are of the form:

$$\frac{\mathrm{d}y}{\mathrm{d}x} + Py = Q \qquad [2]$$

where P and Q are constants or functions of x.

This chapter concerns methods that can be used for the solution of such differential equations, the methods being separation of variables, integrating factors and complementary functions and particular integrals. Chapters 22, 25 and 28 are assumed.

29.2 Separation of variables

A first-order equation is said to be *separable* if the variables x and y can be separated. To solve such equations we simply separate the variables and then integrate both sides of the equation with respect to x. The following shows solutions of the various forms taken by separable equations:

1 Equations of the form $\frac{dy}{dx} = f(x)$
If we integrate both sides of the equation with respect to x:

$$\int \frac{dy}{dx}\,dx = \int f(x)\,dx$$

This is equivalent to separating the variables and writing:

$$\int dy = \int f(x)\,dx$$

Example

Solve the differential equation $dy/dx = 2x$.

Integrating both sides of the equation with respect to x:

$$\int \frac{dy}{dx}\,dx = \int 2x\,dx$$

which is equivalent to:

$$\int dy = \int 2x\,dx$$

and thus $y = x^2 + A$.

2 Equations of the form $\frac{dy}{dx} = f(y)$
This can be rearranged to give:

$$\frac{1}{f(y)}\frac{dy}{dx} = 1$$

Integrating both sides with respect to x:

$$\int \frac{1}{f(y)}\frac{dy}{dx}\,dx = \int 1\,dx$$

This is equivalent to separating the variables:

$$\int \frac{1}{f(y)}\,dy = \int 1\,dx$$

Example

Solve the differential equation $dy/dx = 2y$.

Rearranging the equation and integrating with respect to x:

$$\int \frac{1}{y}\frac{dy}{dx}\,dx = \int 2\,dx$$

which is equivalent to:

$$\int \frac{1}{y}\, dx = \int 2\, dx$$

Thus $\ln y = 2x + A$. Taking exponentials of both sides of the equation enables us to write it as $y = e^{2x+A} = e^{2x} e^{A} = B\, e^{2x}$, where B is a constant.

3 Equations of the form $g(y)\frac{dy}{dx} = f(x)$
Integrating both sides of the equation with respect to x gives:

$$\int g(y)\frac{dy}{dx}\, dx = \int f(x)\, dx$$

This is equivalent to:

$$\int g(y)\, dy = \int f(x)\, dx$$

Example

Solve the differential equation $dy/dx = 2x/y$.

This can be rearranged and integrated with respect to x:

$$\int y\frac{dy}{dx}\, dx = \int 2x\, dx$$

This is equivalent to:

$$\int y\, dy = \int 2x\, dx$$

Thus $\frac{1}{2}y^2 = x^2 + A$.

4 Equations of the form $\frac{dy}{dx} = f(x)g(y)$
This can be rearranged and integrated with respect to x to give:

$$\int \frac{1}{g(y)}\frac{dy}{dx}\, dx = \int f(x)\, dx$$

This is equivalent to:

$$\int \frac{1}{g(y)}\, dy = \int f(x)\, dx$$

Example

Solve the differential equation $dy/dx = 2yx$.

This can be rearranged and integrated with respect to x:

$$\int \frac{1}{y}\frac{dy}{dx}\,dx = \int 2x\,dx$$

This is equivalent to:

$$\int \frac{1}{y}\,dy = \int 2x\,dx$$

Thus $\ln y = x^2 + A$.

Equations which are not of any of the above forms may often be put into one of the forms by a *change of variable*. As an illustration, consider the differential equation $dy/dx = y/(y+x)$. This can be written as:

$$\frac{dy}{dx} = \frac{\frac{y}{x}}{\frac{y}{x}+1}$$

If we let $v = y/x$ then $y = vx$ and $dy/dx = v + x\,dv/dx$. Thus the above equation can be written as:

$$v + x\frac{dv}{dx} = \frac{v}{v+1}$$

$$x\frac{dv}{dx} = \frac{v}{v+1} - v = -\frac{v^2}{v+1}$$

$$\frac{v+1}{v^2}\frac{dv}{dx} = -\frac{1}{x}$$

Integrating with respect to x:

$$\int\left(\frac{1}{v} + \frac{1}{v^2}\right)\frac{dv}{dx}\,dx = -\int \frac{1}{x}\,dx$$

This is equivalent to:

$$\int\left(\frac{1}{v} + \frac{1}{v^2}\right)dv = -\int \frac{1}{x}\,dx$$

Hence $\ln v - (1/v) = -\ln x + A$ and so $\ln(y/x) - (x/y) = -\ln x + A$.

Example

Solve the differential equation $dy/dx = \cos^2 y$ if $y = \pi/4$ when $x = 0$.

We can write the equation as:

$$\sec^2 y\,\frac{dy}{dx} = 1$$

Hence, integrating both sides with respect to x:

2 By using Cramer's rule, solve the following sets of equations:

(a) $2x_1 + x_2 = 3$, $x_1 - 4x_2 = -3$, (b) $3x_1 + x_2 = 5$, $4x_1 - 3x_2 = 11$,

(c) $3x_1 + 2x_3 = 3$, $-x_1 + x_2 = 4$, (d) $x_1 + 2x_2 = 8$, $3x_1 - 4x_2 = 4$,

(e) $2x_1 + 3x_2 - x_3 = 3$, $-x_1 + 2x_3 = 3$, $x_1 + 2x_2 + x_3 = 5$,

(f) $4x_1 + 2x_2 + x_3 = 4$, $-2x_1 + x_2 + 3x_3 = 11$, $x_1 + 4x_2 + 2x_3 = 15$,

(g) $x_1 + 5x_2 - x_3 = 10$, $2x_1 - 3x_2 + x_3 = 1$, $x_1 + 2x_2 - 2x_3 = 9$,

(h) $x_1 + 2x_2 + x_3 + 2x_4 = 8$, $x_2 + 3x_3 + x_4 = 1$, $x_2 - 3x_3 + 3x_4 = 11$,

$-x_1 + x_2 + 2x_3 = 5$,

(i) $2x_1 + x_2 + 2x_4 = 2$, $3x_1 - 3x_2 - x_4 = -21$, $x_2 + 3x_3 + x_4 = 2$,

$x_1 + 2x_2 - 3x_3 = 3$

3 Determine the values of the following determinants given that:

$$\begin{vmatrix} a & b & c \\ d & e & f \\ g & h & i \end{vmatrix} = 4$$

(a) $\begin{vmatrix} a & d & g \\ b & e & h \\ c & f & i \end{vmatrix}$, (b) $\begin{vmatrix} a & b & c \\ 2d & 2e & 2f \\ 3g & 3h & 3i \end{vmatrix}$, (c) $\begin{vmatrix} a+2 & b+2 & c+2 \\ d & e & f \\ g & h & i \end{vmatrix}$,

(d) $\begin{vmatrix} a-d & b-e & c-f \\ d & e & f \\ g & h & i \end{vmatrix}$, (e) $\begin{vmatrix} a+3 & b & c \\ d+3 & e & f \\ g+3 & h & i \end{vmatrix}$, (f) $\begin{vmatrix} b & a & c \\ e & d & f \\ h & c & i \end{vmatrix}$

4 By considering the rows in the following determinant, show that:

$$\begin{vmatrix} 1 & 1 & 1 \\ a & b & c \\ b+c & a+c & a+b \end{vmatrix} = 0$$

5 Determine the cofactor matrix, the adjoint matrix and the inverse matrix for each of the following:

(a) $\begin{bmatrix} 3 & -4 & 2 \\ 2 & -3 & -1 \\ 1 & 2 & -4 \end{bmatrix}$, (b) $\begin{bmatrix} 1 & 0 & 0 \\ 0 & 0 & 1 \\ 0 & 1 & 0 \end{bmatrix}$, (c) $\begin{bmatrix} 2 & 1 & 1 \\ -1 & 1 & 0 \\ 1 & -1 & 1 \end{bmatrix}$

19 Eigenvalues and eigenvectors

19.1 Introduction

This chapter is an extension of the earlier consideration of matrices with a brief consideration of eigenvalues and eigenvectors and assumes that the earlier chapters in this part have been covered. *Eigenvalues* are numbers produced from matrix equations of the form $\mathbf{Ax} = \lambda\mathbf{x}$ and *eigenvectors* are the associated values of $\mathbf{x}$. Such types of equations arise in many aspects of engineering and science. In particular they occur in the vibration of mechanical systems, as the following illustrates.

19.1.1 Vibration of a mechanical system

Consider a system consisting of a mass m tethered between two supports by identical springs of stiffness k (Figure 19.1(a)). When the mass is displaced a distance x from its equilibrium position then one of the springs is compressed by this amount and the other extended by the same amount. The resulting force on the mass is thus $2kx$ in the opposite direction to the displacement. Applying Newton's second law then:

$$ma = -2kx$$

where a is the acceleration of the mass at that displacement. Suppose we now have the system shown in Figure 19.1(b) where we have two masses tethered by identical springs. For mass m when displaced by x_1 from its equilibrium position we have the spring to its left changed in length by x_1 and the spring to its right changed in length by $x_1 - x_2$, where x_2 is the displacement of the $2m$ mass. Thus, if a_1 is the acceleration of mass m we have:

$$ma_1 = -(2kx_1 - kx_2)$$

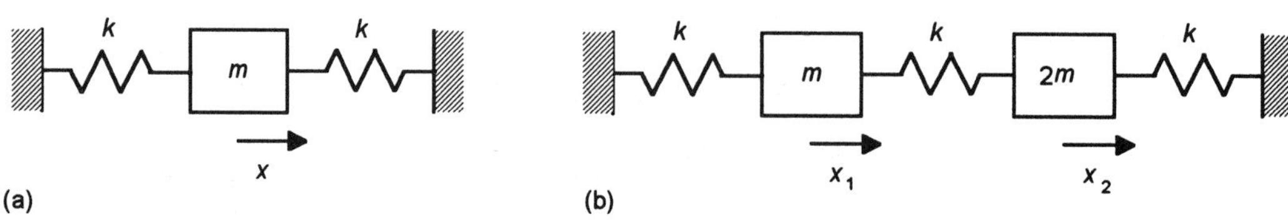

Figure 19.1 *Mechanical systems*

For the $2m$ mass we have:

$$2ma_2 = -(kx_1 - 2kx_2)$$

with a_2 being the acceleration of the $2m$ mass. Thus for the two mass system we have two simultaneous equations. We can write these in matrix notation as:

$$\begin{bmatrix} m & 0 \\ 0 & 2m \end{bmatrix}\begin{bmatrix} a_1 \\ a_2 \end{bmatrix} = \begin{bmatrix} -2k & k \\ k & -2k \end{bmatrix}\begin{bmatrix} x_1 \\ x_2 \end{bmatrix}$$

But if the motion of each mass is a simple harmonic then the acceleration is proportional to the displacement ($a = -\omega^2 x$). Thus the above equation can be written as:

$$-\omega^2\begin{bmatrix} m & 0 \\ 0 & 2m \end{bmatrix}\begin{bmatrix} x_1 \\ x_2 \end{bmatrix} = \begin{bmatrix} -2k & k \\ k & -2k \end{bmatrix}\begin{bmatrix} x_1 \\ x_2 \end{bmatrix}$$

Thus if we let $\lambda = \omega^2$ then we have the equation:

$$\begin{bmatrix} 2k & -k \\ -k & 2k \end{bmatrix}\begin{bmatrix} x_1 \\ x_2 \end{bmatrix} = \lambda\begin{bmatrix} m & 0 \\ 0 & 2m \end{bmatrix}\begin{bmatrix} x_1 \\ x_2 \end{bmatrix} \qquad [1]$$

and so an equation of the form:

$$\mathbf{Ax} = \lambda\mathbf{x} \qquad [2]$$

19.2 Eigenvalue and eigenvector

The following are the basic definitions of *eigenvalue* and *eigenvector*:

For a relationship of the form:

$$\mathbf{Ax} = \lambda\mathbf{x} \qquad [3]$$

if **A** *is an* $n \times m$ *matrix, the number* λ *is called an eigenvalue of* **A** *and the resulting non-zero* $n \times 1$ *matrix* **x** *which satisfies this equation for some value of* λ *is called the eigenvector of* **A** *associated with the eigenvalue* λ.

Example

For the following matrix **A**, verify that 3 is an eigenvalue with associated eigenvector $\begin{bmatrix} 1 \\ 1 \end{bmatrix}$:

$$\mathbf{A} = \begin{bmatrix} -2 & 5 \\ 3 & 0 \end{bmatrix}$$

For equation [3] to be valid we must have:

$$\begin{bmatrix} -2 & 5 \\ 3 & 0 \end{bmatrix}\begin{bmatrix} 1 \\ 1 \end{bmatrix} = 3\begin{bmatrix} 1 \\ 1 \end{bmatrix}$$

$$\begin{bmatrix} -2+5 \\ 3 \end{bmatrix} = 3\begin{bmatrix} 1 \\ 1 \end{bmatrix}$$

and so the values quoted are verified.

Revision

1 For the following matrix **A** verify that 3 is an eigenvalue with associated eigenvector $\begin{bmatrix} -1 \\ 3 \end{bmatrix}$:

$$\mathbf{A} = \begin{bmatrix} 0 & -1 \\ 6 & 5 \end{bmatrix}$$

2 For the following matrix **A** verify that 1 is an eigenvalue with associated eigenvector $\begin{bmatrix} 1 \\ 0 \\ 0 \end{bmatrix}$:

$$\mathbf{A} = \begin{bmatrix} 1 & -1 & 0 \\ 0 & 1 & 1 \\ 0 & 0 & -1 \end{bmatrix}$$

19.2.1 Finding eigenvalues

Consider a matrix:

$$\mathbf{A} = \begin{bmatrix} a_{11} & a_{12} \\ a_{21} & a_{22} \end{bmatrix} \qquad [4]$$

and suppose that $\mathbf{x} = \begin{bmatrix} x \\ y \end{bmatrix}$ is an eigenvector with associated eigenvalue λ.

Then we must have:

$$\mathbf{Ax} = \lambda\mathbf{x}$$

and so:

$$\begin{bmatrix} a_{11} & a_{12} \\ a_{21} & a_{22} \end{bmatrix}\begin{bmatrix} x \\ y \end{bmatrix} = \lambda\begin{bmatrix} x \\ y \end{bmatrix}$$

This describes the pair of simultaneous equations:

$$a_{11}x + a_{12}y = \lambda x \text{ and } a_{21}x + a_{22}y = \lambda y$$

These equations can be written as:

$$(a_{11} - \lambda)x + a_{12}y = 0 \text{ and } a_{21}x + (a_{22} - \lambda)y = 0$$

Thus we can rearrange the matrix equation to give:

$$\begin{bmatrix} a_{11} - \lambda & a_{12} \\ a_{21} & a_{22} - \lambda \end{bmatrix}\begin{bmatrix} x \\ y \end{bmatrix} = \begin{bmatrix} 0 \\ 0 \end{bmatrix} \qquad [5]$$

Effectively what we have done is subtract $\lambda\mathbf{I}$ from $\mathbf{A}$ (note that we can only subtract matrices of the same size from each other), i.e.

$$\begin{bmatrix} a_{11} & a_{12} \\ a_{21} & a_{22} \end{bmatrix}\begin{bmatrix} x \\ y \end{bmatrix} - \lambda\begin{bmatrix} 1 & 0 \\ 0 & 1 \end{bmatrix}\begin{bmatrix} x \\ y \end{bmatrix} = 0 \qquad [6]$$

Thus equation [5] can be written as:

$$\mathbf{Ax} - \lambda\mathbf{Ix} = (\mathbf{A} - k\mathbf{I})\mathbf{x} = 0 \qquad [7]$$

If $\mathbf{x}$ is not to be zero then we must have (see Section 18.2) the determinant of the matrix $(\mathbf{A} - \lambda\mathbf{I})$ equal to zero:

$$|\mathbf{A} - \lambda\mathbf{I}| = 0 \qquad [8]$$

and so:

$$\begin{vmatrix} a_{11} - \lambda & a_{12} \\ a_{21} & a_{22} - \lambda \end{vmatrix} = 0 \qquad [9]$$

$$(a_{11} - \lambda)(a_{22} - \lambda) - a_{21}a_{12} = 0 \qquad [10]$$

This equation [10], whose roots are the eigenvalues, is called the *characteristic equation*.

Example

Determine the characteristic equation and eigenvalues of:

$$\mathbf{A} = \begin{bmatrix} 0 & 1 \\ 1 & 0 \end{bmatrix}$$

Since:

$$\mathbf{A} - \lambda\mathbf{I} = \begin{bmatrix} 0 & 1 \\ 1 & 0 \end{bmatrix} - \lambda\begin{bmatrix} 1 & 0 \\ 0 & 1 \end{bmatrix} = \begin{bmatrix} 0-\lambda & 1 \\ 1 & 0-\lambda \end{bmatrix}$$

the characteristic equation is given by equation [8] as:

$$|\mathbf{A} - \lambda\mathbf{I}| = \begin{vmatrix} 0-\lambda & 1 \\ 1 & 0-\lambda \end{vmatrix} = \lambda^2 - 1 = 0$$

The characteristic equation is $\lambda^2 - 1 = 0$ and so the eigenvalues $\lambda = +1$ and $\lambda = -1$.

Example

Determine the characteristic equation and eigenvalues of:

$$\mathbf{A} = \begin{bmatrix} 1 & -1 & 0 \\ 0 & 1 & 1 \\ 0 & 0 & -1 \end{bmatrix}$$

Since:

$$\mathbf{A} - \lambda\mathbf{I} = \begin{bmatrix} 1 & -1 & 0 \\ 0 & 1 & 1 \\ 0 & 0 & -1 \end{bmatrix} - \lambda\begin{bmatrix} 1 & 0 & 0 \\ 0 & 1 & 0 \\ 0 & 0 & 1 \end{bmatrix}$$

the characteristic equation is given by equation [8] as:

$$|\mathbf{A} - \lambda\mathbf{I}| = \begin{vmatrix} 1-\lambda & -1 & 0 \\ 0 & 1-\lambda & 1 \\ 0 & 0 & -1-\lambda \end{vmatrix}$$

This gives the characteristic equation $-(1 - \lambda)(1 - \lambda)(1 + \lambda) = 0$ and so eigenvalues of +1, +1 and −1.

Example

Determine the characteristic equation and the eigenvalues of:

$$\mathbf{A} = \begin{bmatrix} 0 & 1 \\ -1 & 0 \end{bmatrix}$$

Since:

$$\mathbf{A} - \lambda\mathbf{I} = \begin{bmatrix} 0 & 1 \\ -1 & 0 \end{bmatrix} - \lambda \begin{bmatrix} 1 & 0 \\ 0 & 1 \end{bmatrix} = \begin{bmatrix} -\lambda & 1 \\ -1 & -\lambda \end{bmatrix}$$

the characteristic equation is given by equation [8] as:

$$|\mathbf{A} - \lambda\mathbf{I}| = \begin{vmatrix} -\lambda & 1 \\ -1 & -\lambda \end{vmatrix} = \lambda^2 + 1 = 0$$

This has the solution $\lambda = \pm\sqrt{-1} = \pm \mathrm{j}$. The eigenvalues are complex.

Revision

3 Determine the characteristic equations and eigenvalues of the following matrices:

$$\text{(a)} \begin{bmatrix} 3 & 2 \\ 1 & 2 \end{bmatrix}, \text{(b)} \begin{bmatrix} 3 & 2 \\ -1 & 0 \end{bmatrix}, \text{(c)} \begin{bmatrix} 1 & 3 \\ 2 & 2 \end{bmatrix}, \text{(d)} \begin{bmatrix} 1 & 0 & 0 \\ 0 & 0 & 1 \\ 0 & 1 & 0 \end{bmatrix},$$

$$\text{(e)} \begin{bmatrix} 4 & 3 \\ -3 & 4 \end{bmatrix}, \text{(f)} \begin{bmatrix} 1 & 3 \\ 2 & 1 \end{bmatrix}$$

19.2.2 Finding eigenvectors

For any matrix, say:

$$\mathbf{A} = \begin{bmatrix} a_{11} & a_{12} \\ a_{21} & a_{22} \end{bmatrix}$$

the characteristic equation $|\mathbf{A} - \lambda\mathbf{I}| = 0$ can be used to find the eigenvalues. The eigenvector associated with eigenvalue λ can then be found by substituting the eigenvalues in the equation:

$$\begin{bmatrix} a_{11} & a_{12} \\ a_{21} & a_{22} \end{bmatrix} \begin{bmatrix} x \\ y \end{bmatrix} = \lambda \begin{bmatrix} x \\ y \end{bmatrix} \qquad [11]$$

or

$$\begin{bmatrix} a_{11} & a_{12} \\ a_{21} & a_{22} \end{bmatrix} \begin{bmatrix} x \\ y \end{bmatrix} - \lambda \begin{bmatrix} 1 & 0 \\ 0 & 1 \end{bmatrix} \begin{bmatrix} x \\ y \end{bmatrix} = 0 \qquad [12]$$

$$\begin{bmatrix} a_{11}-\lambda & a_{12} \\ a_{21} & a_{22}-\lambda \end{bmatrix}\begin{bmatrix} x \\ y \end{bmatrix}=\begin{bmatrix} 0 \\ 0 \end{bmatrix} \qquad [13]$$

To find the eigenvectors we substitute each value of λ into the matrix equation. The result is a system of linear equations which have to be solved. We generally only need one eigenvector corresponding to each eigenvalue. However, once we have obtained one eigenvector associated with a particular eigenvalue, any non-zero multiple of this eigenvector is also an eigenvector. The following example illustrates these points.

If **x** *is an eigenvector of* **A** *corresponding to a particular eigenvalue, then there are also eigenvalues* c**x** *corresponding to the same eigenvalue for all values of c that are not zero.*

Example

Determine the eigenvalues and associated eigenvectors of the matrix:

$$\mathbf{A}=\begin{bmatrix} 3 & 5 \\ 1 & -1 \end{bmatrix}$$

The characteristic equation is given by equation [8] as:

$$|\mathbf{A}-\lambda\mathbf{I}|=\begin{vmatrix} 3-\lambda & 5 \\ 1 & -1-\lambda \end{vmatrix}=(3-\lambda)(-1-\lambda)-5=0$$

Hence $\lambda^2-2k-8=(\lambda-4)(\lambda+2)=0$ and $\lambda=4$ or -2. Equation [12] gives, for $\lambda=4$:

$$\begin{bmatrix} 3-4 & 5 \\ 1 & -1-4 \end{bmatrix}\begin{bmatrix} x \\ y \end{bmatrix}=\begin{bmatrix} 0 \\ 0 \end{bmatrix}$$

and hence:

$$\begin{bmatrix} -x+5y \\ x-5y \end{bmatrix}\begin{bmatrix} x \\ y \end{bmatrix}=\begin{bmatrix} 0 \\ 0 \end{bmatrix}$$

If x and y are not to be zero we require values of x and y that make $-x+5y$ and $x-5y=0$. Possible eigenvectors are thus:

$$\begin{bmatrix} 5 \\ 1 \end{bmatrix}, \text{ or } \begin{bmatrix} 10 \\ 2 \end{bmatrix}, \text{ or } \begin{bmatrix} -5 \\ -1 \end{bmatrix}, \text{ etc.}$$

For the other eigenvalue $\lambda=-2$ we have:

$$\begin{bmatrix} 3+2 & 5 \\ 1 & -1+2 \end{bmatrix} \begin{bmatrix} x \\ y \end{bmatrix} = \begin{bmatrix} 0 \\ 0 \end{bmatrix}$$

$$\begin{bmatrix} 5x+5y \\ x+y \end{bmatrix} \begin{bmatrix} x \\ y \end{bmatrix} = \begin{matrix} 0 \\ 0 \end{matrix}$$

Possible eigenvectors are:

$$\begin{bmatrix} -1 \\ 1 \end{bmatrix}, \text{ or} \begin{bmatrix} -2 \\ 2 \end{bmatrix}, \text{ or} \begin{bmatrix} 1 \\ -1 \end{bmatrix}, \text{ etc.}$$

Example

Determine the eigenvalues and associated eigenvectors for:

$$\begin{bmatrix} 2 & 0 & 0 \\ 1 & 0 & 2 \\ 0 & 0 & 3 \end{bmatrix}$$

The characteristic equation is given by equation [8] as:

$$|\mathbf{A} - \lambda \mathbf{I}| = \begin{vmatrix} 2-\lambda & 0 & 0 \\ 1 & -\lambda & 2 \\ 0 & 0 & 3-\lambda \end{vmatrix} = 0$$

Hence $(2 - \lambda)\lambda(3 - \lambda) = 0$ and so $\lambda = 2$, 0 and 3. For $\lambda = 0$:

$$\begin{bmatrix} 2-0 & 0 & 0 \\ 1 & -0 & 2 \\ 0 & 0 & 3-0 \end{bmatrix} \begin{bmatrix} x \\ y \\ z \end{bmatrix} = \begin{bmatrix} 0 \\ 0 \\ 0 \end{bmatrix}$$

Thus we have:

$$2x = 0,\ x + 2z = 0,\ 3z = 0$$

These give $x = z = 0$ with no indication of any restriction on the value of y. Thus a possible eigenvector is:

$$\begin{bmatrix} 0 \\ 1 \\ 0 \end{bmatrix}$$

For $\lambda = 2$:

$$\begin{bmatrix} 2-2 & 0 & 0 \\ 1 & -2 & 2 \\ 0 & 0 & 3-2 \end{bmatrix}\begin{bmatrix} x \\ y \\ z \end{bmatrix} = \begin{bmatrix} 0 \\ 0 \\ 0 \end{bmatrix}$$

Thus we have:

$$0 = 0,\ x - 2y + 2z = 0,\ z = 0$$

Thus we must have $x = 2y$ and so a possible eigenvector is:

$$\begin{bmatrix} 2 \\ 1 \\ 0 \end{bmatrix}$$

For $\lambda = 3$:

$$\begin{bmatrix} 2-3 & 0 & 0 \\ 1 & -3 & 2 \\ 0 & 0 & 3-3 \end{bmatrix}\begin{bmatrix} x \\ y \\ z \end{bmatrix} = \begin{bmatrix} 0 \\ 0 \\ 0 \end{bmatrix}$$

Thus we have:

$$-x = 0,\ x - 3y + 2z = 0,\ 0 = 0$$

Hence we must have $3y = 2z$ and so a possible eigenvector is:

$$\begin{bmatrix} 0 \\ 2 \\ 3 \end{bmatrix}$$

Revision

4 Determine the eigenvalues and associated eigenvectors for the following matrices:

(a) $\begin{bmatrix} 3 & 2 \\ 1 & 2 \end{bmatrix}$, (b) $\begin{bmatrix} 3 & 0 \\ 8 & -1 \end{bmatrix}$, (c) $\begin{bmatrix} 0 & 0 \\ 0 & 0 \end{bmatrix}$, (d) $\begin{bmatrix} 1 & 2 & 1 \\ 2 & 1 & 1 \\ 1 & 1 & 2 \end{bmatrix}$

Problems

1 For the following matrix **A** verify that 3 is an eigenvalue with associated eigenvector $\begin{bmatrix} 2 \\ -1 \end{bmatrix}$:

The Runge-Kutta methods involves determining for the first step:

1 k_1 by means of equation [3].

2 k_2 by means of equation [4].

3 k_3 by means of equation [5].

4 k_4 by means of equation [6].

5 The value of y at the end of the step by means of equation [7].

Then repeating the above for the next and following steps.

Example

Using the fourth-order Runge-Kutta method with a step size of 0.1, solve the differential equation $dy/dx = x + y$ if $y = 1$ when $x = 0$.

Using equation [3]:

$$k_1 = hf(x_0, y_0) = 0.1(0 + 1) = 0.1$$

Using equation [4]:

$$k_2 = hf(x_0 + h/2, y_0 + k_1/2) = 0.1(0.05 + 1.05) = 0.11$$

Using equation [5]:

$$k_3 = hf(x_0 + h/2, y_0 + k_2/2) = 0.1(0.05 + 1.055) = 0.110\,5$$

Using equation [6]:

$$k_4 = hf(x_0 + h, y_0 + k_3) = 0.1(0.10 + 1.110\,5) = 0.1210\,5$$

Thus, using equation [7]:

$$y_1 = y_0 + \frac{k_1 + 2k_2 + 2k_3 + k_4}{6}$$

$$= 1 + \frac{0.1 + 2 \times 0.11 + 2 \times 0.1105 + 0.12105}{6} = 1.1103\,4$$

This procedure can be repeated for each step. Table 31.2 gives the results and also shows the true values computed analytically. Compare the results with those given in Table 31.1 for the same differential equation solved by Euler's method.

Table 31.2 *Runge-Kutta solution for* $\mathrm{d}y/\mathrm{d}x = x + y$

x	Runga-Kutta value	Analytical value
0	1.000 00	1.000 00
0.1	1.110 34	1.110 34
0.2	1.242 81	1.242 81
0.3	1.399 72	1.399 72
0.4	1.583 65	1.538 65
0.5	1.797 44	1.797 44

Revision

2 Use the fourth-order Runga-Kutta method to determine the value of y at the required x value for the following differential equations, using the step size indicated:

(a) $\dfrac{\mathrm{d}y}{\mathrm{d}x} = x^2 + y^2$ at $x = 0.5$, with $y = 1$ when $x = 0$, $h = 0.1$,

(b) $\dfrac{\mathrm{d}y}{\mathrm{d}x} = y - 2$ at $x = 0.4$, with $y = 1$ when $x = 0$, $h = 0.2$,

(c) $\dfrac{\mathrm{d}y}{\mathrm{d}x} = 1 + y - x^2$ at $x = 0.4$, with $y = 0.5$ when $x = 0$, $h = 0.2$

31.4 Second-order differential equations

Consider the simple second-order differential equation for the displacement x with time t of a freely falling object when the motion is one of constant acceleration of about 10 m/s^2:

$$\frac{\mathrm{d}^2x}{\mathrm{d}t^2} = 10$$

We can transform this second-order differential equation into two first-order differential equations if we introduce another variable $v = \mathrm{d}x/\mathrm{d}t$. Then $\mathrm{d}^2x/\mathrm{d}t^2 = \mathrm{d}v/\mathrm{d}t$ and so we now have a pair of first-order differential equations:

$$\frac{\mathrm{d}v}{\mathrm{d}t} = 10 \text{ and } v = \frac{\mathrm{d}x}{\mathrm{d}t}$$

We can solve the first of these equations by separation of the variables:

$$v = 10t + A \qquad [8]$$

To solve the second equation we need to write it so that we only have two variables:

$$v = \frac{\mathrm{d}x}{\mathrm{d}t} = \frac{\mathrm{d}x}{\mathrm{d}v}\frac{\mathrm{d}v}{\mathrm{d}t} = 10\frac{\mathrm{d}x}{\mathrm{d}v}$$

We can then solve this by separation of the variables to give:

$$\tfrac{1}{2}v^2 = 10x + B$$

Hence using equation [8] to eliminate v gives:

$$\tfrac{1}{2}(10t + A)^2 = 10x + B \text{ and so } x = 5t^2 + Ct + D$$

Any second-order differential equation can be transformed into a pair of simultaneous first-order equations. What we have is $d^2y/dx^2 = f(x, y, dy/dx)$ with initial conditions $y = y_0$ at $x = x_0$ and $dy/dx = v_0$ at $x = x_0$. We then introduce the new variable $v = dy/dx$ and obtain $dv/dx = d^2y/dx^2$.

Likewise a third-order differential equation can be transformed into three first-order differential equations, a fourth-order differential equation into four first-order differential equations, etc. This form of transformation enables higher-order differential equations to be solved by the numerical methods discussed earlier in this chapter for first-order differential equations.

Example

Transform the second-order differential equation:

$$\frac{d^2y}{dx^2} + 4\frac{dy}{dx} + y = 0$$

into a pair of first-order differential equations.

Let $v = dy/dx$, then:

$$\frac{d^2y}{dx^2} = \frac{dv}{dx} = \frac{dv}{dy}\frac{dy}{dx} = v\frac{dv}{dy}$$

and thus the first-order equations are:

$$v = \frac{dy}{dx} \text{ and } v\frac{dv}{dy} + 4v + y = 0$$

Revision

3 Transform the following second-order differential equations into pairs of first-order differential equations:

(a) $\dfrac{d^2y}{dx^2} + y = 0$, (b) $\dfrac{d^2y}{dx^2} + 3\dfrac{dy}{dx} = 2$

31.4.1 Euler's method with second-order differential equations

First we convert the second-order differential equation into a pair of first-order differential equations by introducing another variable v equal to the first derivative. The procedure is then as outlined earlier for the Euler method but two first-order equations have to be solved step by step. Thus equation [1] becomes:

$$y_{n+1} = y_n + h\left(\frac{\mathrm{d}y}{\mathrm{d}x}\right)_{x=nh} \qquad [9]$$

and:

$$v_{n+1} = v_n + h\left(\frac{\mathrm{d}v}{\mathrm{d}x}\right)_{x=nh} \qquad [10]$$

Example

Using Euler's method with a step size of 0.1 solve the second-order differential equation:

$$\frac{\mathrm{d}^2y}{\mathrm{d}x^2} + \frac{\mathrm{d}y}{\mathrm{d}x} - 2y = 0$$

if $y = 3/2$ at $x = 0$ and $\mathrm{d}y/\mathrm{d}x = 0$ at $x = 0$.

If we let $v = \mathrm{d}y/\mathrm{d}x$ then we obtain the pair of simultaneous equations:

$$\frac{\mathrm{d}y}{\mathrm{d}x} = v \text{ and } \frac{\mathrm{d}v}{\mathrm{d}x} = 2y - v$$

Starting with the initial values of $y = 3/2$ at $x = 0$ and $v = 0$ at $x = 0$ and using equations [9] and [10]:

$$y_1 = y_0 + hv_0 = 1.5 + 0.1(0) = 1.5$$

$$v_1 = v_0 + h(2y_0 - v_0) = 0 + 0.1(3 - 0) = 0.3$$

We then repeat the process for the next step:

$$y_2 = y_1 + hv_1 = 1.5 + 0.1(0.3) = 1.53$$

$$v_2 = v_1 + h(2y_1 - v_1) = 0.3 + 0.1(3 - 0.3) = 0.57$$

The process is then repeated for the next step:

$$y_3 = y_2 + hv_2 = 1.53 + 0.1(0.57) = 1.587$$

$$v_3 = v_2 + h(2y_2 - v_2) = 0.57 + 0.1(3.06 - 0.57) = 0.819$$

The process can then be repeated for further steps.

Revision

4 Using Euler's method with a step size of 0.1 determine the value of y at $x = 0.3$ for the second-order differential equation:

$$\frac{d^2y}{dx^2} - 2\frac{dy}{dx} + 2y = 0$$

if $y = 1$ at $x = 0$ and $dy/dx = 1$ at $x = 0$.

31.4.2 Runge-Kutta method with second-order differential equations

First we convert the second-order differential equation into a pair of first-order differential equations by introducing another variable v equal to the first derivative. The procedure is then as outlined earlier for the Runga-Kutta method but two first-order equations have to be solved step by step. Thus the fourth-order Runge-Kutta equation [7] becomes:

$$y_n = y_{n-1} + \frac{k_1 + 2k_2 + 2k_3 + k_4}{6} \qquad [11]$$

$$v_n = v_{n-1} + \frac{l_1 + 2l_2 + 2l_3 + l_4}{6} \qquad [12]$$

Example

Using the fourth-order Runge-Kutta method with a step size of 0.1 solve the second-order differential equation:

$$\frac{d^2y}{dx^2} + \frac{dy}{dx} - 2y = 0$$

if $y = 3/2$ at $x = 0$ and $dy/dx = 0$ at $x = 0$.

If we let $v = dy/dx$ then we obtain the pair of simultaneous equations:

$$\frac{dy}{dx} = v \text{ and } \frac{dv}{dx} = 2y - v$$

Starting with the initial values of $y_0 = 3/2$ at $x = 0$ and $v_0 = 0$ at $x = 0$ and estimating the values of k and l using equations [3], [4], [5] and [6] we have:

$$k_1 = hf(x_0, v_0) = 0.1(0) = 0$$

$$l_1 = hg(x_0, y_0, v_0) = 0.1(2 \times 1.5 - 0) = 0.3$$

$k_2 = hf(x_0 + h/2, v_0 + l_1/2) = 0.1(0.15) = 0.015$

$l_2 = hg(x_0 + h/2, y_0 + k_1/2, v_0 + l_1/2) = 0.1(2 \times 1.5 - 0.15) = 0.285$

$k_3 = hf(x_0 + h/2, v_0 + l_2/2) = 0.1(0.285/2) = 0.014\ 25$

$l_3 = hg(x_0 + h/2, y_0 + k_2/2, v_0 + l_2/2) = 0.1(2 \times 1.507\ 5 - 0.142\ 5)$
$= 0.287\ 25$

$k_4 = hf(x_0 + h, v_0 + l_3) = 0.1(0.287\ 25) = 0.028\ 725$

$l_4 = hg(x_0 + h, y_0 + k_3, v_0 + l_3) = 0.1(2 \times 1.514\ 25 - 0.287\ 25)$
$= 0.274\ 125$

Thus, using equation [7]:

$$y_1 = y_0 + \frac{k_1 + 2k_2 + 2k_3 + k_4}{6}$$

$$= 1.5 + \frac{0 + 2 \times 0.015 + 2 \times 0.014\ 25 + 0.028\ 725}{6}$$

$$= 1.514\ 54$$

$$v_1 = y_0 + \frac{l_1 + 2l_2 + 2l_3 + l_4}{6}$$

$$= 0 + \frac{0.3 + 2 \times 0.2875 + 2 \times 0.28725 + 0.274\ 125}{6}$$

$$= 0.287\ 27$$

This procedure can then be repeated for further steps.

Revision

5 Using the fourth-order Runge-Kutta method with a step size of 0.1 to determine the value of y at x = 0.1 for the second-order differential equation:

$$\frac{d^2y}{dx^2} - \frac{dy}{dx} + 2y = 0$$

if $y = 1$ at $x = 0$ and $dy/dx = 1$ at $x = 0$.

Problems

1 Use Euler's method to solve the following differential equations, using the step size indicated and tabulating the results for the first four steps:

(a) $\frac{dy}{dx} = x^2 + y^2$, $y = 1$ at $x = 0$, $h = 0.1$,

(b) $\frac{dy}{dx} = y + 1$, $y = 1$ at $x = 0$, $h = 0.1$,

(c) $\frac{dy}{dx} = xy^2$, $y = 1$ at $x = 0$, $h = 0.1$

2 The rate at which the potential difference v_C across a capacitor decreases with time t when it discharges through a resistor is described by the differential equation $dv_C/dt = -v_C$. If $v_C = 2$ V at $t = 0$, use Euler's method to solve the differential equation for the first four steps with a step size of 0.1 s.

3 The rate at which the current i in an inductive circuit changes with time t after a constant voltage is applied at $t = 0$ is described by the differential equation $di/dt = 5 - \frac{1}{2}i$. Use Euler's method to solve the differential equation for the first four steps with a step size of 0.2 s, the current being 0 at $t = 0$.

4 Use the fourth-order Runga-Kutta method to determine the value of y at the required x value for the following differential equations, using the step size indicated:

(a) $\frac{dy}{dx} = y - x$ at $x = 0.2$, with $y = 2$ at $x = 0$, $h = 0.1$,

(b) $\frac{dy}{dx} = x^2 + x - y$ at $x = 0.4$, with $y = 0$ at $x = 0$, $h = 0.2$,

(c) $\frac{dy}{dx} = y + e^{-x}$ at $x = 1.2$, with $y = 0$ at $x = 1$, $h = 0.1$

5 Using Euler's method with a step size of 0.1 determine the value of y at $x = 0.3$ for the second-order differential equation:

$$\frac{d^2y}{dx^2} - y = 0$$

if $y = 2$ at $x = 0$ and $dy/dx = 0$ at $x = 0$.

6 Repeat problem 5 using the fourth-order Runge-Kutta method.

7 Using Euler's method with a step size of 0.1 determine the value of y at $x = 0.3$ for the second-order differential equation:

$$\frac{d^2y}{dx^2} - 3\frac{dy}{dx} = e^x$$

if $y = 1$ at $x = 0$ and $dy/dx = 0$ at $x = 0$.

8 Repeat problem 7 using the fourth-order Runge-Kutta method.

32 Application: Dynamic response of systems

32.1 Introduction

This chapter illustrates how differential equations are involved in modelling the dynamic responses of systems. For example, if the input signal to a measurement system suddenly changes, the output will not instantaneously change to the new value but some time will elapse before it reaches a steady-state value. If the voltage applied to an electrical circuit suddenly changes to a new value, the current in the circuit will not change instantly to the new value but some time will elapse before it reaches the steady new value. If a continually changing signal is applied to a system, the response of the system may lag behind the input. The way in which a system reacts to input changes is termed its *dynamic characteristic*.

This chapter assumes Chapters 28, 29 and 30, taking the basic models developed for systems in Chapter 28 and considering what happens when there are inputs to such systems.

32.2 Zero-order systems

Consider a measurement system element such as a potentiometer (Figure 32.1). The input to the element is a displacement of the sliding contact and the output is a voltage which is related to the position of the sliding contact. If y is the input and x the output, then, if the resistance is distributed uniformly along the length of the slide wire, the relationship between the output and input can be described by the equation:

$$x = ky \qquad [1]$$

where k is a constant. The output is directly proportional to the input.

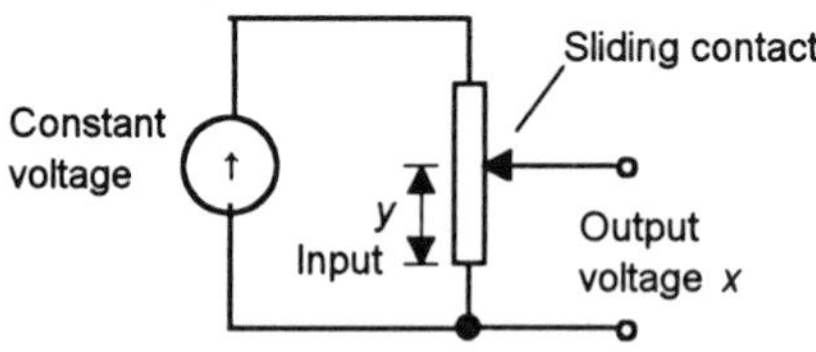

Figure 32.1 *Input and output for a potentiometer*

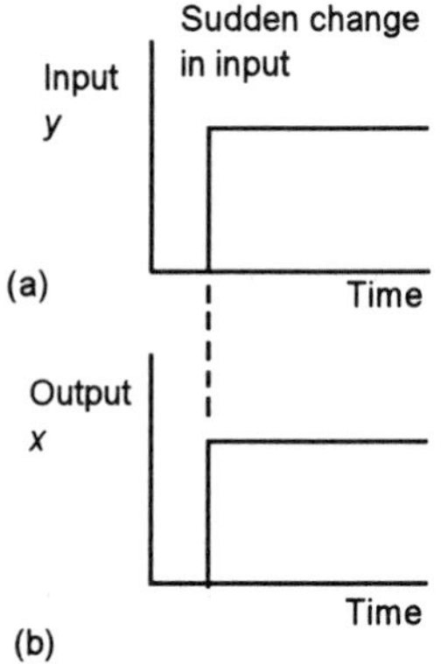

Figure 32.2 *Step input to a zero-order system*

With the above mathematical model for the potentiometer there is no time term involved in the equation. Thus we should expect that the output reacts instantaneously to any change in input. Thus for a *step input*, i.e. a sudden abrupt change in the input (Figure 32.2(a)), the output is likewise a step (Figure 32.2(b)), since at every instant of time the output has to be

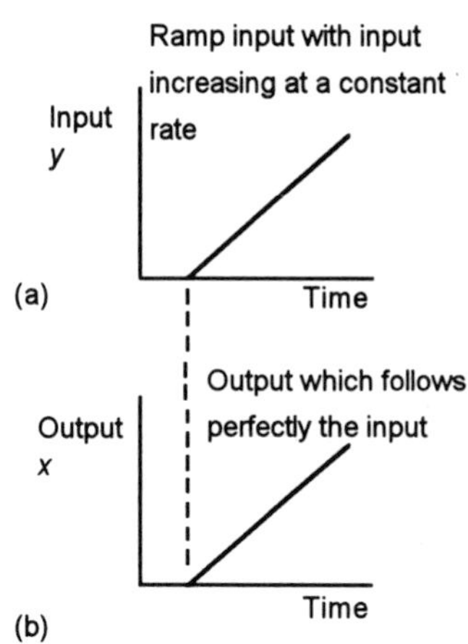

Figure 32.3 *Ramp input*

directly proportional to the input. Whatever the form of the input, the output is expected to follow it perfectly with no distortion or time lag. An element which has this characteristic is termed a *zero-order element*.

With a zero-order element, if we have a *ramp* change as the input, i.e. an input which increases at a constant rate with time, then the output follows the changing input perfectly. Figure 32.3 shows how the input and output changes might appear. Zero-order elements are ideal or perfect elements, no lag occurring between the input and the output. If we have a sinusoidal input then we obtain a sinusoidal output which is perfectly in phase with the input, no lag occurring.

With an electrical circuit having purely resistance, we have the relationships between the input voltage v and the circuit current i of $v = Ri$, where R is a constant. This equation is a zero-order equation and thus if there is a step voltage input to the circuit the current rises immediately to its new value. If there is a ramp voltage input the current follows immediately the change with no time lag. If we have a sinusoidal input then we obtain a sinusoidal output which is perfectly in phase with the input, no lag occurring.

32.3 First-order systems

Systems are said to be *first order* when the relationship between the input and the output involves the rate at which the output changes, i.e. just the first derivative and no higher derivative. To illustrate this, consider a thermometer at temperature T_0 inserted into a liquid at a temperature T_1. We can thus think of the thermometer being subject to a step input, i.e. the input abruptly changes from T_0 to T_1. The thermometer will then, over a period of time, change its temperature until it becomes T_1. Thus we have a measurement system, the thermometer, which has a step input and an output which changes from T_0 to T_1 over some time. How does the output, i.e. the reading of the thermometer T, vary with time.

The rate at which energy enters the thermometer from the liquid is proportional to the difference in temperature between the liquid and the thermometer. Thus, at some instant of time when the temperature of the thermometer is T, we can write:

$$\frac{\mathrm{d}Q}{\mathrm{d}t} = h(T_1 - T)$$

where h is a constant called the *heat transfer coefficient*. For a thermometer with a specific heat capacity c and a mass m, the relationship between heat input Q and the consequential temperature change is:

$$Q = mc \text{ (temperature change)}$$

When the rate at which heat enters the thermometer is $\mathrm{d}Q/\mathrm{d}t$, we can write for the rate at which the temperature changes:

$$\frac{\mathrm{d}Q}{\mathrm{d}t} = mc\frac{\mathrm{d}T}{\mathrm{d}t}$$

Thus:

$$mc\frac{dT}{dt} = h(T_1 - T)$$

We can rewrite this with all the output terms on one side of the equals sign and the input on the other, thus:

$$mc\frac{dT}{dt} + hT = hT_1 \qquad [2]$$

We no longer have a simple relationship between the input and output but a relationship which involves time. The form of this equation is typical of first-order systems.

We can solve this equation by separation of the variables:

$$\int \frac{1}{T_1 - T}\,dT = \int \frac{h}{mc}\,dt$$

$$-\ln(T_1 - T) = (h/mc)t + A$$

where A is a constant. This can be rewritten as:

$$T_1 - T = e^{A}\,e^{t/\tau} = C\,e^{t/\tau}$$

where $\tau = mc/h$ and is termed the *time constant*. The time constant can be defined as the value of the time which makes the exponential term become e^{-1}. $T = T_0$ at $t = 0$ and so $C = T_1 - T_0$. Thus:

$$T = T_1 + (T_0 - T_1)\,e^{-t/\tau} \qquad [3]$$

The first term is the *steady-state value*, i.e. the value that will occur after sufficient time has elapsed for all transients to die away, and the second term a transient one which changes with time, eventually becoming zero. Figure 32.4 shows graphically how the temperature T indicated by the thermometer changes with time.

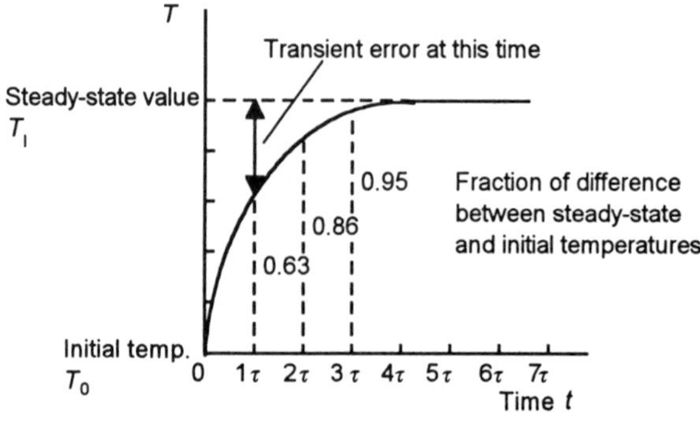

Figure 32.4 *Response of a first-order system to a step input*

After a time equal to one time constant the output has reached about 63% of the way to the steady-state temperature, after a time equal to two time constants the output has reached about 86% of the way, after three time constants about 95% and after about four time constants it is virtually equal to the steady-state value. The error at any instant is the difference between what the thermometer is indicating and what the temperature actually is. Thus:

$$\text{error} = T - T_1 = (T_0 - T_1)\,e^{-t/\tau} \qquad [4]$$

This error changes with time and eventually will become zero. Thus it is a transient error.

If a thermometer is required to be fast reacting and quickly attain the temperature being measured, it needs to have a small time constant. Since $\tau = mc/h$, this means a thermometer with a small mass, a small thermal capacity and a large heat transfer coefficient. If we compare a mercury-in-glass thermometer with a thermocouple, then the smaller mass and specific heat capacity of the thermocouple will give it a smaller time constant and hence a faster response to temperature changes.

Example

A thermometer indicates a temperature of 20°C when it is suddenly immersed in a liquid at a temperature of 60°C. If the thermometer has a time constant of 5 s what will its readings be after (a) 5 s, (b) 10 s, (c) 15 s.

The temperature T of the thermometer varies with time according to equation [3]:

$$T = T_1 + (T_0 - T_1)\,e^{-t/\tau} = 60 - 40\,e^{-t/5}$$

After 5 s the thermometer reading will have reached about 63% of the way to the steady-state value, after 10 s about 86%, after 15 s about 95% and after 20 s it is virtually at the steady-state value. Thus after 5 s the reading is 45.3°C, after 10 s it is 54.6°C, after 15 s it is 58.0°C.

Example

A thermometer which behaves as a first-order element has a time constant of 15 s. Initially it reads 20°C. What will be the time taken for the temperature to rise to 90% of the steady-state value when it is immersed in a liquid of temperature 100°C, i.e. a temperature of 92°C?

Equation [3], $T = T_1 + (T_0 - T_1)\,e^{-t/\tau}$, can be rearranged to give:

$$\frac{T - T_1}{T_1 - T_0} = e^{-t/\tau}$$

With $T - T_0$ as 90% of $T_1 - T_0$, then we have $T - T_1$ as 10% of $T_1 - T_0$ and thus:

$$0.10 = e^{-t/15}$$

Taking logarithms gives $-2.30 = -t/15$ and so $t = 34.5$ s.

Revision

1 For a circuit containing resistance R in series with capacitance C, the the potential difference v_C across the capacitor varies with time, being given by $v_C = V - V\,e^{-t/RC}$. What is the time constant for the circuit?

32.3.1 First-order systems and step inputs

In general, a first-order system subject to a step input has a differential equation which can be written in the form:

$$a_1 \frac{dx}{dt} + a_0 x = y \qquad [5]$$

where x is the output, t the time and y the step input. Alternatively it can be written as:

$$\tau \frac{dx}{dt} + x = \frac{1}{a_0} y \qquad [6]$$

where τ is the time constant and is given by (a_1/a_0). The solution of the differential equation for a step input from some initial value to final value at time $t = 0$ is of the form:

$$x = \text{steady-state value} + (\text{initial value} - \text{steady-state value})\ e^{-t/\tau} \qquad [7]$$

Table 32.1 shows the percentage of the response, i.e. (x – initial value)/ (steady – initial values) × 100%, that will have been achieved after various multiples of the time constant. The percentage dynamic error is (steady-state value – x)/(steady – initial value) × 100%.

Table 32.1 *First-order system response*

Time	% response	% dynamic error
0	0.0	100.0
1τ	63.2	36.8
2τ	86.5	13.5
3τ	95.0	5.0
4τ	98.2	1.8
5τ	99.3	0.7
∞	100.0	0.0

Example

An electrical circuit consisting of resistance R in series with an initially uncharged capacitor of capacitance C has an input of a step voltage V at time $t = 0$. Determine (a) how the potential difference across the capacitor will change with time and (b) with $R = 1\ \text{M}\Omega$, $C = 4\ \mu\text{F}$ and a step voltage of 12 V, the potential difference across the capacitor after 2 s.

(a) The differential equation for the system (equation [4], Chapter 28) is:

$$RC\frac{\mathrm{d}v_C}{\mathrm{d}t} + v_C = V$$

This equation can be solved by separation of variables. However, since it is the same form as equation [5] we can recognise that the solution must be of the form given by equation [7] with the time constant being RC:

$$x = \text{steady-state value} + (\text{initial value} - \text{steady-state value})\ \mathrm{e}^{-t/\tau}$$

$$v_C = V + (0 - V)\ \mathrm{e}^{-t/\tau} = V(1 - \mathrm{e}^{-t/\tau})$$

(b) The time constant is $RC = 1 \times 10^6 \times 4 \times 10^{-6} = 4$ s. Thus after 2 s:

$$v_C = 12(1 - \mathrm{e}^{-2/4}) = 4.72\ \text{V}$$

Revision

2 A 1000 μF capacitor has been charged to a potential difference of 12 V. At time $t = 0$ it is discharged through a 20 kΩ resistor. What will be the potential difference across the capacitor after 2 s?

3 Determine how the circuit current varies with time when there is a step voltage V input to a circuit having an inductance L in series with resistance R.

32.3.2 Ramp inputs to first-order systems

Consider now the response of a thermometer to a ramp input, i.e. a steadily increasing temperature input. If the thermometer is initially at a temperature of T_0 and the temperature T_1 of the liquid in which the thermometer is immersed is given by $T_1 - T_0 = at$, where a is a constant and t the time, i.e. the temperature is rising at a steady rate from the initial reading, then the differential equation becomes:

$$mc\frac{\mathrm{d}T}{\mathrm{d}t} + hT = hT_1 = h(at + T_0) \qquad [8]$$

or:

$$\tau\frac{dT}{dt} + T = at + T_0 \qquad [9]$$

with $\tau = mc/h$. We can solve this non-homogeneous differential equation by obtaining the complementary function and the particular integral. For the homogeneous form of the equation:

$$\tau\frac{dT}{dt} + T = 0$$

separation of the variables leads to the complementary solution $T = A\ e^{-t/\tau}$. For the particular integral we try a solution $T = B + Ct$. Substituting this in the non-homogeneous equation gives $\tau C + B + Ct = aT + T_0$. Hence $C = a$ and $\tau C + B = T_0$ to give $B = T_0 - \tau a$. Thus the solution of the differential equation is:

$$T = A\ e^{-t/\tau} + T_0 - \tau a + at$$

Since $T = T_0$ at $t = 0$, $T_0 = A + T_0 - \tau a + 0$ and so $A = \tau a$ and:

$$T = a(\tau\ e^{-t/\tau} + t - \tau) + T_0 \qquad [10]$$

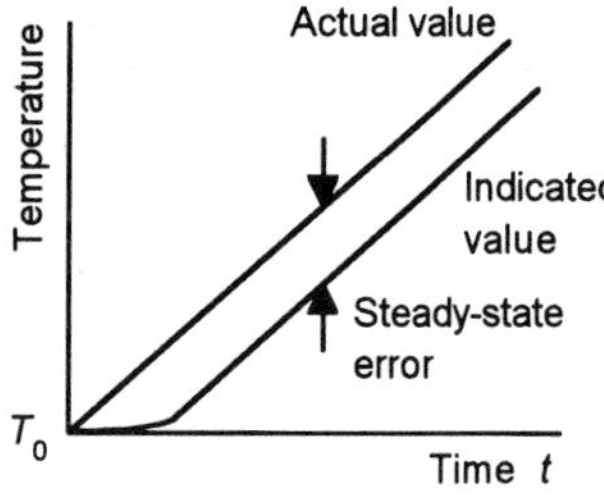

Figure 32.5 *Steady-state error*

The measurement error at any instant is:

$$\text{error} = T - T_1 = T - at - T_0 = a\tau\ e^{-t/\tau} - a\tau \qquad [11]$$

The first term varies with the time t and eventually will die away. It is thus the *transient error*. The second term is a constant and so is an error that is always going to be present. It is termed the *steady-state error*. The steady-state error means that even when the thermometer reaches the steady-state value it will always indicate a temperature which lags behind the actual temperature. Figure 32.5 illustrates this.

Example

A thermocouple has a time constant of 10 s and is subject to a temperature which is changing at the rate of 5°C/s. What will be (a) the error in the temperature indicated by the thermocouple after 20 s and (b) the steady-state error?

(a) Equation [11] gives:

$$\text{error} = a\tau\ e^{-t/\tau} - a\tau = 5 \times 10\ e^{-20/10} - 5 \times 1 = 1.8°\text{C}$$

(b) The steady state error is $a\tau$ = 5°C.

32.4 Second-order systems

There are many measurement systems which behave in a similar manner to a damped mass on a spring when subject to inputs. An obvious example of such a system is a spring balance for the measurement of force. Another example is a diaphragm pressure gauge. Figure 32.6 illustrates the basic features of such systems.

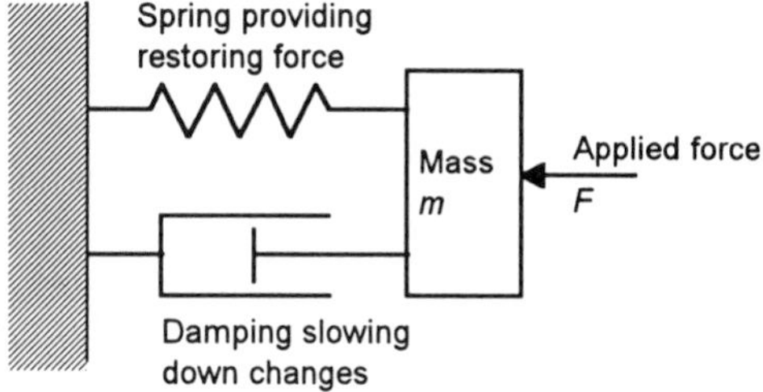

Figure 32.6 *Mass, spring, damper system*

Consider the system when a force F is applied to the mass. The net force applied to the mass is the applied force F minus the force resulting from the compressing, or stretching, of the spring and the force from the damper. The force resulting from compressing the spring is proportional to the change in length x of the spring, i.e. kx with k being a constant termed the spring stiffness. The damper can be thought of as a piston moving in a cylinder, though it can take many other forms. The force arising from the damping is proportional to the rate at which the displacement of the piston is changing, i.e. $c\,\mathrm{d}x/\mathrm{d}t$ with c being a constant. Thus:

$$\text{net force applied to mass} = F - kx - c\frac{\mathrm{d}x}{\mathrm{d}t}$$

This net force will cause the mass to accelerate. Thus:

$$m\frac{\mathrm{d}^2x}{\mathrm{d}t^2} = F - kx - c\frac{\mathrm{d}x}{\mathrm{d}t}$$

We can write this as:

$$m\frac{\mathrm{d}^2x}{\mathrm{d}t^2} + c\frac{\mathrm{d}x}{\mathrm{d}t} + kx = F \qquad [12]$$

In the absence of damping and a force F, we have $m\,\mathrm{d}^2x/\mathrm{d}t^2 + kx = 0$. This has the solution (see equations [12] and [13] in Chapter 30):

$$y = A\cos\sqrt{\frac{k}{m}}\,t + B\sin\sqrt{\frac{k}{m}}\,t$$

The term *natural angular frequency* ω_n is given to:

$$\omega_n = \sqrt{\frac{k}{m}} \qquad [13]$$

If we define a constant ζ, termed the *damping ratio*, by:

$$\zeta = \frac{c}{2\sqrt{mk}} \qquad [14]$$

then we can write the second-order differential equation [12] as:

$$\frac{1}{\omega_n^2}\frac{d^2x}{dt^2} + \frac{2\zeta}{\omega_n}\frac{dx}{dt} + x = \frac{F}{k} \qquad [15]$$

Consider a step input such that the applied force jumps from zero to F at time $t = 0$. We can solve the differential equation by determining the complementary function and the particular integral. For the homogeneous form of the differential equation we try a solution of the form $x = A\,e^{st}$. This produces the auxiliary equation:

$$\frac{1}{\omega_n^2}s^2 + \frac{2\zeta}{\omega_n}s + 1 = 0$$

$$s^2 + 2\omega_n\zeta s + \omega_n^2 = 0$$

This has the roots:

$$s = \frac{-2\omega_n\zeta \pm \sqrt{4\omega_n^2\zeta^2 - 4\omega_n^2}}{2} = -\omega_n\zeta \pm \omega_n\sqrt{\zeta^2 - 1} \qquad [16]$$

and leads to the following complementary function solutions:

1 *Damping ratio with a value between 0 and 1*
This gives two complex roots.

$$s = -\omega_n\zeta \pm j\sqrt{1-\zeta^2}$$

If we let $\omega = \omega_n\sqrt{1-\zeta^2}$ then $s = -\omega_n \pm j\omega$ and so (see Section 30.3 and equation [11]) we obtain:

$$x = e^{-\zeta\omega_n t}(P\cos\omega t + Q\sin\omega t) \qquad [17]$$

This can be expressed in an alternative form, since for the sine of a sum we can write $\sin(\omega t + \phi) = \sin\omega t\cos\phi + \cos\omega t\sin\phi$. If we let P and Q represent the opposite sides of a right-angled triangle of angle ϕ (Figure 32.7), then $\sin\phi = P/\sqrt{(P^2 + Q^2)}$ and $\cos\phi = P/\sqrt{(P^2 + Q^2)}$ and so:

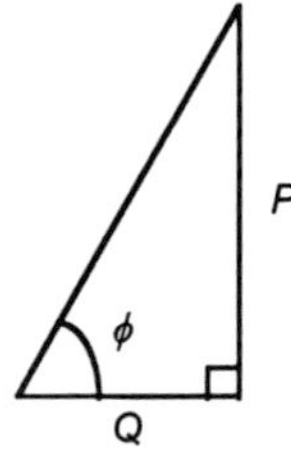

Figure 32.7 *Angle ϕ*

$$x = \sqrt{P^2 + Q^2}\; e^{-\zeta\omega_n t}\sin(\omega t + \phi)$$

$$x = C\,\mathrm{e}^{-\zeta\omega_n t}\sin\left(\omega t + \phi\right) \qquad [18]$$

where C is a constant and ϕ a phase difference. This describes a sinusoidal oscillation which is damped, the exponential term being the damping factor which gradually reduces the amplitude of the oscillation. Such a motion is said to be *under-damped.*

2 *Damping ratio with the value 1*
This gives two equal roots $s_1 = s_2 = -\omega_n$ and thus (see Section 30.3):

$$x = (At + B)\,\mathrm{e}^{-\omega_n t} \qquad [19]$$

where A and B are constants. This describes a situation where no oscillations occur but x exponentially changes with time. Such a motion is said to be *critically damped.*

3 *Damping ratio greater than 1*
This gives two real roots (see Section 30.3) $s_1 = -\omega_n\zeta + \omega_n\sqrt{(\zeta^2 - 1)}$ and $s_2 = -\omega_n\zeta - \omega_n\sqrt{(\zeta^2 - 1)}$, thus:

$$x = A\,\mathrm{e}^{s_1 t} + B\,\mathrm{e}^{s_2 t} \qquad [20]$$

where A and B are constants. This describes a situation where no oscillations occur but x exponentially changes with time, taking longer to reach the steady-state value than the critically damped motion. Such a motion is said to be *over-damped.*

When we have a step input then we can try for the particular integral $x = A$. Substituting this in equation [15] gives $0 + 0 + A = F/k$. Thus the particular integral is $x = F/k$. Thus the solutions corresponding to the different degrees of damping are:

1 *Under-damped*

$$x = C\,\mathrm{e}^{-\zeta\omega_n t}\sin\left(\omega t + \phi\right) + \frac{F}{k} \qquad [21]$$

2 *Critically damped*

$$x = (At + B)\,\mathrm{e}^{-\omega_n t} + \frac{F}{k} \qquad [22]$$

3 *Over-damped*

$$x = A\,\mathrm{e}^{s_1 t} + B\,\mathrm{e}^{s_2 t} + \frac{F}{k} \qquad [23]$$

As t tends to an infinite value, in all cases the response tends to a steady-state value of F/k.

Figure 32.8 shows the form the solution of the second order differential equation takes for different values of the damping ratio. The output is plotted as a multiple of the steady-state value F/k. Instead of just giving the output variation with time t, the axis used is $\omega_n t$. This is because t and ω_n always appear as the product $\omega_n t$ and using this product makes the graph applicable for any value of ω_n.

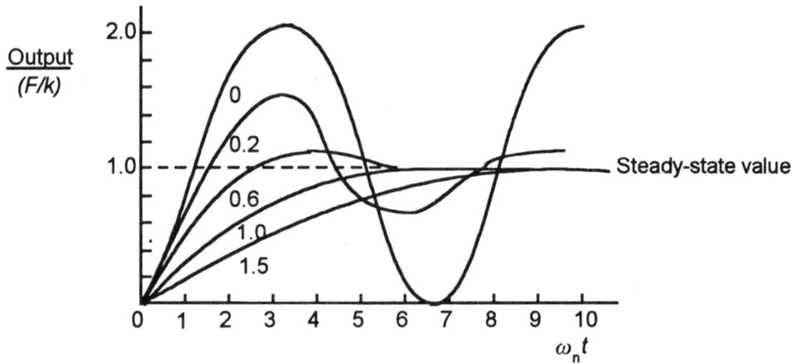

Figure 32.8 *Response of second-order system to step input for different damping factors*

Revision

4 An object of mass 1 kg is suspended from a rigid support by a vertical spring of stiffness 4 N/m. Determine how the displacement of the object varies with time when the object is pulled down from its initial position and released to freely move if the object is subject to a damping force of five times its velocity?

5 An object of mass 1 kg is suspended from a rigid support by a vertical spring of stiffness 9 N/m. The object is pulled down for an initial displacement of 0.2 m and then released with zero initial velocity. Determine how the displacement of the object varies with time when (a) there is no damping, (b) the damping is twice the velocity of the object.

Problems

1 A sensor behaves as a capacitance of 2 μF in series with a 1 MΩ resistance. As such the relationship between its input y and output x is given by:

$$2\frac{dx}{dt} + x = y$$

How will the output vary with time when the input is (a) a unit step input at time $t = 0$, (b) a ramp input described by $y = 4t$?

2 A system is specified as being first order with a time constant of 10 s and a steady-state value of 5. How will the output of the system vary with time when subject to a step input?

3 A sensor is first order with a time constant of 1 s. What will be the percentage dynamic error after (a) 1 s, (b) 2 s, from a unit step input signal to the sensor?

4 How long must elapse for the dynamic error of a sensor subject to a step input to drop below 5% if the sensor is first order with a time constant of 4 s?

5 A thermometer originally indicates a temperature of 20°C and is then suddenly inserted into a liquid at 45°C. The thermometer has a time constant of 2 s. (a) Derive a differential equation showing how the thermometer reading is related to the temperature input and (b) give its solution showing how the thermometer reading varies with time.

6 An element has an output x for an input y related by the differential equation:

$$\frac{dx}{dt} + 2x = y$$

Determine how the output varies with time when there is a ramp input of $y = 3t$.

7 A second-order system has a natural angular frequency of 2.0 rad/s and a damped angular frequency of 1.8 rad/s. What is the damping factor?

8 Determine the natural angular frequency and damping factor for a second-order system with input y and output x described by the following differential equation:

$$0.02\frac{d^2x}{dt^2} + 0.20\frac{dx}{dt} + 0.50x = y$$

9 A sensor can be considered to be a mass-damper-spring system with a mass of 10 g and a spring of stiffness 1.0 N/mm. Determine the natural angular frequency and the damping constant required for the damping element if the system is to be critically damped.

10 Determine whether the second-order system described by the following differential equation is under-damped, critically damped or over-damped when subject to a step input y:

$$\frac{d^2x}{dt^2} + 5\frac{dx}{dt} + 6x = y$$

11 An object of mass 1 kg is suspended from a rigid support by a vertical spring of stiffness 9 N/m. What is the damping force per unit velocity which would be needed to give critical damping?

12 Determine the natural angular frequency and damping force per unit velocity for a system having its displacement x with time t described by the following second-order differential equation:

$$\frac{d^2x}{dt^2} + 4\frac{dx}{dt} + 7x = 0$$

13 An object of mass 1 kg is suspended from a rigid support by a vertical spring of stiffness 9 N/m. If there is a damping force of $1v$ opposing the motion of the object, determine how the displacement varies with time when the object is given an initial displacement of 0.2 m and an initial velocity of –0.3 m/s.

14 The angular displacement θ of a door controlled by a hydraulic damping mechanism is described by the differential equation:

$$\frac{d^2\theta}{dt^2} + 5\frac{d\theta}{dt} + 4\theta = 0$$

Determine how the angular displacement varies with time t when there is an initial displacement of $\pi/3$ and zero initial angular velocity.

15 An electrical circuit having resistance R, inductance L and capacitance C in series with a step voltage source V at t = 0 has the potential difference across the capacitor v_C described by the differential equation:

$$LC\frac{d^2v_C}{dt^2} + RC\frac{dv_C}{dt} + v_C = V$$

Show that the three possible solutions are:

$$v = A\,e^{s_1 t} + B\,e^{s_2 t} + V,\ s = -\frac{R}{2L} \pm \sqrt{\left(\frac{R}{2L}\right)^2 - \frac{1}{LC}}$$

$$v = (A + Bt)\,e^{-Rt/2L} + V$$

$$v = e^{-Rt/2L}(A\cos\omega t + B\sin\omega t) + V,\ \omega = \sqrt{\frac{1}{LC} - \left(\frac{R}{2L}\right)^2}$$

Part 8
Fourier series

The aims of this part are to enable the reader to:

- Represent periodic and non-periodic waveforms as Fourier series.
- Examine waveforms for symmetry and hence reduce the labour of determining Fourier series.
- Represent Fourier series by exponentials.
- Use the Fourier transform to determine the response of systems to inputs.
- Analyse waveforms to determine their harmonic components.
- Determine circuit responses to waveforms represented by Fourier series.

The Fourier series is a way of representing a non-sinusoidal waveform as the sum of sinusoidal waveforms. The part assumes Chapters 2, 5, 6, differentiation and integration methods from Part 6 and, for the application, Chapter 29. Chapter 36 is an introduction to the Fourier transform, this chapter assuming Chapters 34 and 35.

33 Fourier series

33.1 Introduction

Series are often used to represent a function. For example, we can represent the exponential function as:

$$e^x = 1 + x + \frac{x^2}{2!} + \frac{x^3}{3!} + \dots$$

This chapter introduces the Fourier series and its use to represent periodic waveforms. Chapter 34 extends this to a consideration of non-periodic waveforms.

This chapter assumes the ability to integrate (Chapter 25), manipulate trigonometric functions (Chapter 2) and series (Chapter 5) and a knowledge of how functions can be used to represent waveforms.

33.1.1 The periodic sinusoidal waveform

A function is said to be *periodic* if its function value repeats at regular intervals of the independent variable. The interval between repetitions is termed the *period.* An example of a periodic signal is the sinusoidal signal $y = A \sin \omega t$, where y is the value of the function at a time t, the value at time $t = 0$ being zero. A is the amplitude. Such a function is periodic and repeats itself every cycle, i.e. 360° or 2π radians. ω is the angular frequency and is equal to $2\pi f$ with f being the frequency. If the frequency of the waveform is doubled then the equation becomes $y = A \sin 2\omega t$. If the frequency is trebled then the equation becomes $y = A \sin 3\omega t$.

A periodic function is sometimes expressed in terms of a series of different sinusoidal signals. The component with the angular frequency ω is termed the *first harmonic* or *fundamental*, the one with 2ω the *second harmonic*, 3ω the *third harmonic*, etc. Thus $y = A \sin \omega t$ is the first harmonic, $y = A \sin 2\omega t$ the second harmonic, $y = A \sin 3\omega t$ the third harmonic, etc.

If the above sinusoidal signals had not started off with zero values at time $t = 0$ but with a phase angle ϕ then for the single, double and treble frequencies with different amplitudes and phase angles we would have:

$$y = A_1 \sin\left(\omega t + \phi_1\right),\ y = A_2 \sin\left(2\omega t + \phi_2\right),\ y = A_3 \sin\left(3\omega t + \phi_3\right)$$

33.1.2 Useful angle relationships

The following are some angle relationships which are useful in this chapter (see Section 3.4.3):

$$\sin \omega t = -\cos(\omega t + \pi/2) = \cos(\omega t - \pi/2) = -\sin(\omega t + \pi) = -\sin(\omega t - \pi) \quad [1]$$

$$\cos \omega t = -\cos(\omega t + \pi) = -\cos(\omega t - \pi) = \sin(\omega t + \pi/2) = -\sin(\omega t - \pi/2) \quad [2]$$

Useful simplifications (see Section 3.4.1, equations [57], [58], [59] and [60]) are:

$$A\cos(\omega t - \phi) = a\cos\omega t + b\sin\omega t \quad [3]$$

$$A\cos(\omega t + \phi) = a\cos\omega t - b\cos\omega t \quad [4]$$

$$A\sin(\omega t + \phi) = a\sin\omega t + b\cos\omega t \quad [5]$$

$$A\sin(\omega t - \phi) = a\sin\omega t - b\cos\omega t \quad [6]$$

with $\phi = \tan^{-1} b/a$ and $A = \sqrt{a^2 + b^2}$.

Revision

1 Express (a) $y = 10 \sin 3t$ as a cosine, (b) $y = 2 \sin 4t + 3 \cos 4t$ as a single cosine function.

33.1.3 Useful integrals

In considering Fourier series we are often concerned with integrals of sine and cosine functions over a period $T = 2\pi/\omega$. The following shows the evaluation of such integrals (see Section 3.4.1 for the trigonometric identities used):

$$\int_0^T \sin n\omega t \, dt = -\left[\frac{\cos n\omega t}{n}\right]_0^T = -\frac{1}{n}(\cos 2n\pi - \cos 0) = 0 \quad [7]$$

The above equation is valid for all values of n.

$$\int_0^T \cos n\omega t \, dt = \left[\frac{\sin n\omega t}{n}\right]_0^T = \frac{1}{n}(\sin 2n\pi - \sin 0) = 0 \quad [8]$$

The above equation is for $n \neq 0$. When $n = 0$ we have the integral of cos 0 and so 1. Thus the integral then has the value T.

$$\int_0^T \sin n\omega t \cos m\omega t \, dt = \int_0^T \tfrac{1}{2}\{\sin(n+m)\omega t + \sin(n-m)\omega t\} \, dt$$

$$= \tfrac{1}{2}\left[-\frac{1}{n+m}\cos(n+m)\omega t + \frac{1}{n-m}\sin(n-m)\omega t\right]_0^T = 0 \quad [9]$$

The above is valid for all values of n and m.

$$\int_0^T \sin n\omega t \sin m\omega t \, dt = \int_0^T \tfrac{1}{2}\{\cos(n-m)\omega t - \cos(n+m)\omega t\} \, dt$$

$$= \tfrac{1}{2}\left[\frac{1}{n-m}\sin(n-m)\omega t - \frac{1}{n+m}\sin(n+m)\omega t\right]_0^T = 0 \qquad [10]$$

The above is the result for $n \neq m$. For $n = m \neq 0$ we have:

$$\int_0^T \sin^2 n\omega t \, dt = \tfrac{1}{2}\int_0^T (1 - \cos 2n\omega t) \, dt = \tfrac{1}{2}\left[t - \frac{\sin 2n\omega t}{2n\omega}\right]_0^T = \frac{T}{2} \qquad [11]$$

Also:

$$\int_0^T \cos n\omega t \cos m\omega t \, dt = \int_0^T \tfrac{1}{2}\{\cos(n+m)\omega t + \cos(n-m)\omega t\} \, dt$$

$$= \tfrac{1}{2}\left[\frac{1}{n+m}\sin(n+m)\omega t + \frac{1}{n-m}\sin(n-m)\omega t\right]_0^T = 0 \qquad [12]$$

The above is the result for $n \neq m$. For $n = m \neq 0$ we have:

$$\int_0^T \cos^2 n\omega t \, dt = \tfrac{1}{2}\int_0^T (1 + \cos 2n\omega t) \, dt = \tfrac{1}{2}\left[t + \frac{\sin 2n\omega t}{2n\omega}\right]_0^T = \frac{T}{2} \qquad [13]$$

The above are said to be the *orthogonal relationships* for the sine and cosine functions.

Note that sometimes the integrals are taken for a unit angular frequency. With $\omega = 1$ we have $T = 2\pi$. Thus the integrals are taken between 0 and 2π. The results are basically the same.

33.2 The Fourier series

In 1822, Jean Baptiste Fourier showed that a periodic function, whatever its waveform, could be built up from a series of sinusoidal waves of multiples of a basic frequency. Thus any periodic signal can be represented by a constant signal of size A_0, i.e. a d.c. term, plus sines of multiples of a basic frequency, each of the sine terms having possibly a different amplitude and phase angle:

$$y = A_0 + A_1 \sin(\omega t + \phi_1) + A_2 \sin(2\omega t + \phi_2) + A_3 \sin(3\omega t + \phi_3) + \ldots + A_n \sin(n\omega t + \phi_n) \qquad [14]$$

This is an infinite series. For the series to converge and so represent the function, certain conditions have to be met. These are termed the *Dirichlet conditions*. These are that for each value of t there is only one value of y and that y must be continuous or piecewise continuous with a finite number of finite discontinuities within the periodic interval. Fortunately most functions of interest to engineers meet these conditions.

The Fourier series is concisely expressed as:

$$y = A_0 + \sum_{n=1}^{\infty} A_n \sin(n\omega t + \phi_n) \quad [15]$$

Since sin $(A + B)$ = sin A cos B + cos A sin B, we can write:

$$A_n \sin(n\omega t + \phi_n) = A_n \sin\phi_n \cos n\omega t + A_n \cos\phi_n \sin n\omega t$$

If we represent the non-time varying terms A_n sin ϕ_n and A_n cos ϕ_n by constants a_n and b_n we can then write:

$$A_n \sin(n\omega t + \phi_n) = a_n \cos n\omega t + b_n \sin n\omega t$$

and the Fourier series equation can be written as:

$$y = \tfrac{1}{2}a_0 + a_1 \cos\omega t + a_2 \cos 2\omega t + \ldots + a_n \cos n\omega t$$
$$+ b_1 \sin\omega t + b_2 \sin 2\omega t + \ldots + b_n \sin n\omega t \quad [16]$$

with, for convenience, a_0 being taken as $2A_0$ (see later for the reason). Hence we can write the Fourier series equation as:

$$y = \tfrac{1}{2}a_0 + \sum_{n=1}^{\infty} a_n \cos n\omega t + \sum_{n=1}^{\infty} b_n \sin n\omega t \quad [17]$$

The a and b terms are called the *Fourier coefficients.*

Since we have $a_n = A_n$ sin ϕ_n and $b_n = A_n$ cos ϕ_n then:

$$\phi_n = \tan^{-1}\left(\frac{a_n}{b_n}\right) \quad [18]$$

and, since $a_n^2 + b_n^2 = A_n^2 \sin^2\phi_n + A_n^2 \cos^2\phi_n = A_n^2$, we have:

$$A_n = \sqrt{a_n^2 + b_n^2} \quad [19]$$

33.2.1 Establishing the Fourier coefficients

Now consider how we can establish the Fourier coefficients for a waveform. Suppose we have the Fourier series in the form of equation [16]:

$$y = \tfrac{1}{2}a_0 + a_1 \cos\omega t + a_2 \cos 2\omega t + \ldots + a_n \cos n\omega t$$
$$+ b_1 \sin\omega t + b_2 \sin 2\omega t + \ldots + b_n \sin n\omega t$$

If we integrate both sides of the equation over one period T of the fundamental, the integral for each cosine and sine term will be the area under the graph of that expression for one cycle and thus zero. A consequence of this is that the only term which is not zero when we integrate the equation is the integral of the a_0 term. Thus, integrating over one period T gives:

$$\int_0^T y\,dt = \int_0^T \tfrac{1}{2}a_0\,dt = \tfrac{1}{2}a_0 T$$

$$a_0 = \frac{2}{T}\int_0^T y\,dt \qquad [20]$$

We can obtain the a_1 term by multiplying the equation by cos ωt and then integrating over one period. Thus the equation becomes:

$$y\cos\omega t = \tfrac{1}{2}a_0\cos\omega t + a_1\cos\omega t\cos\omega t + a_2\cos\omega t\cos 2\omega t + \dots + b_1\cos\omega t\sin\omega t + b_2\cos\omega t\sin 2\omega t + \dots$$

$$= \tfrac{1}{2}a_0\cos\omega t + a_1\cos^2\omega t + a_2\cos\omega t\cos 2\omega t + \dots + b_1\cos\omega t\sin\omega t + b_2\cos\omega t\sin 2\omega t + \dots$$

The integration over a period T of all the terms involving sin ωt and cos ωt will be zero. Thus we are only left with the $\cos^2 \omega t$ term and so, using equation [13]:

$$\int_0^T y\cos\omega t\,dt = \int_0^T a_1\cos^2\omega t\,dt = \tfrac{1}{2}a_1 T$$

$$a_1 = \frac{2}{T}\int_0^T y\cos\omega t\,dt \qquad [21]$$

In general, multiplying the equation by cos $n\omega t$ gives:

$$a_n = \frac{2}{T}\int_0^T y\cos n\omega t\,dt \qquad [22]$$

This equation gives for $n = 0$ the equation given earlier for a_0. This would not have been the case if the first term in the Fourier series had been written as a_0 instead of $a_0/2$. This simplification is thus the reason for the ½ being used with a_0 in the defining equation for the series.

In a similar way, multiplying the equation by sin ωt and integrating over a period enables us to obtain the b coefficients. Thus:

$$y\sin\omega t = \tfrac{1}{2}a_0\sin\omega t + a_1\sin\omega t\cos\omega t + a_2\sin\omega t\cos 2\omega t + \dots + b_1\sin\omega t\sin\omega t + b_2\sin\omega t\sin 2\omega t + \dots$$

$$= \tfrac{1}{2}a_0\sin\omega t + a_1\sin\omega t\cos\omega t + a_2\sin\omega t\cos 2\omega t + \dots + b_1\sin^2\omega t + b_2\sin\omega t\sin 2\omega t + \dots$$

The integration over a period T of all the terms involving sin ωt and cos ωt will be zero and so, using equation [11]:

$$\int_0^T y\sin\omega t\,dt = \int_0^T b_1\sin^2\omega t\,dt = \tfrac{1}{2}b_1 T$$

$$b_1 = \frac{2}{T}\int_0^T y\sin\omega t\,dt \qquad [23]$$

In general, multiplying the equation by sin $n\omega t$ and integrating gives:

$$b_n = \frac{2}{T}\int_0^T y \sin n\omega t \, dt \qquad [24]$$

33.3 Fourier series for common waveforms

The following illustrates how the Fourier series can be established for a number of common waveforms.

33.3.1 Rectangular waveform

Consider the *rectangular waveform* shown in Figure 33.1. It can be described as:

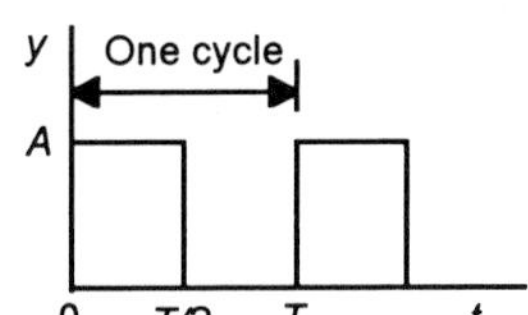

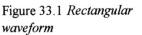

Figure 33.1 *Rectangular waveform*

$y = A$ for $0 \le t < T/2$
$y = 0$ for $T/2 \le t < T$, period T

Now consider the determination of the coefficients. Equation [20] for a_0:

$$a_0 = \frac{2}{T}\int_0^T y \, dt$$

has an integral which is the area under the graph of y against t for the period T. Since this area is $AT/2$, we have $a_0 = A$. To obtain a_n we use equation [22]:

$$a_n = \frac{2}{T}\int_0^T y \cos n\omega t \, dt$$

Since y has the value A up to $T/2$ and is zero thereafter, we can write the above equation in two parts as:

$$a_n = \frac{2}{T}\int_0^{T/2} A \cos n\omega t \, dt + \frac{2}{T}\int_{T/2}^T 0 \cos n\omega t \, dt$$

The value of the second integral is 0 and so:

$$a_n = \frac{2}{T}\left[\frac{A}{n\omega}\sin n\omega t\right]_0^{T/2}$$

Since $\omega = 2\pi/T$ then the sine term is sin $2n\pi t/T$. Thus with $t = T/2$ we have sin $n\pi$ which is zero and since sin 0 = 0, we have $a_n = 0$.

For the b_n terms we use equation [24]:

$$b_n = \frac{2}{T}\int_0^T y \sin n\omega t \, dt$$

Since we have $y = A$ from 0 to $T/2$ and then $y = 0$ for the remainder of the period, this equation can be written in two parts as:

$$b_n = \frac{2}{T}\int_0^{T/2} A \sin n\omega t \, dt + \frac{2}{T}\int_{T/2}^T 0 \sin n\omega t \, dt$$

The value of the second integral is 0 and so:

$$b_n = \frac{2}{T}\left[-\frac{A}{n\omega}\cos n\omega t\right]_0^{T/2} = \frac{A}{\pi n}(1 - \cos n\pi)$$

Hence:

$$b_1 = \frac{A}{\pi}(1 - \cos\pi) = \frac{2A}{\pi}, \quad b_2 = \frac{A}{2\pi}(1 - \cos 2\pi) = 0$$

$$b_3 = \frac{A}{3\pi}(1 - \cos 3\pi) = \frac{2A}{3\pi}, \quad \text{etc.}$$

Thus the Fourier series for the rectangular waveform can be written as:

$$y = A\left(\frac{1}{2} + \frac{2}{\pi}\sin\omega t + \frac{2}{3\pi}\sin 3\omega t + \ldots\right) \qquad [25]$$

Note that only odd harmonics are present..

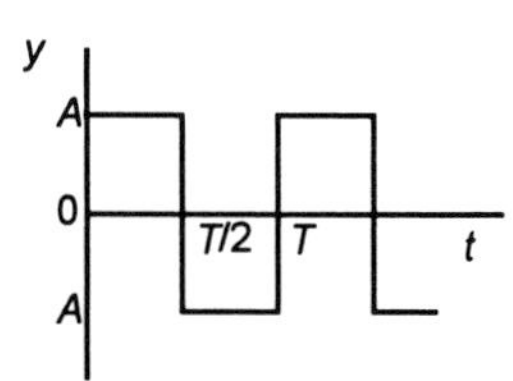

Figure 33.2 *Revision problem 2*

Revision

2 Determine the Fourier series for the waveform shown in Figure 33.2.

33.3.2 Sawtooth waveform

Consider the *sawtooth waveform* shown in Figure 33.3. It can be described by:

$$y = At/T \text{ for } 0 \le t < T, \text{ period } T$$

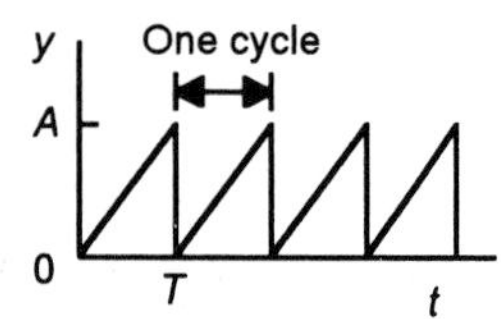

Figure 33.3 *Sawtooth waveform*

To determine a_0 we use equation [20]:

$$a_0 = \frac{2}{T}\int_0^T y\,dt$$

The integral is the area under the graph of y against t between 0 and time T. This is $AT/2$ and so $a_0 = A$. To obtain a_n we use equation [22]:

$$a_n = \frac{2}{T}\int_0^T y\cos n\omega t\,dt$$

Since $\omega = 2\pi/T$ and $y = At/T$ then:

$$a_n = \frac{2}{T}\int_0^T \frac{At}{T}\cos\frac{2\pi nt}{T}\,dt$$

Using integration by parts gives:

$$a_n = \frac{2}{T}\left[\frac{At}{2\pi n}\sin\frac{2\pi nt}{T} + \frac{At}{4\pi^2 n^2}\cos\frac{2\pi nt}{T}\right]_0^T$$

$$= \frac{2A}{T}\left[\frac{T}{4\pi^2 n^2} - \frac{T}{4\pi^2 n^2}\right] = 0$$

The values of a_n are zero for all values other than a_0. The values of b_n can be found by using equation [24]:

$$b_n = \frac{2}{T}\int_0^T y \sin n\omega t \, dt = \frac{2}{T}\int_0^T \frac{At}{T} \sin\frac{2\pi nt}{T} \, dt$$

Integration by parts gives:

$$b_n = \frac{2}{T}\left[-\frac{At}{2\pi n}\cos\frac{2\pi nt}{T} + \frac{At}{4\pi^2 n^2}\sin\frac{2\pi nt}{T}\right]_0^T$$

$$= \frac{2A}{T}\left[-\frac{T}{2\pi n}\right]_0^T = -\frac{A}{\pi n}$$

The Fourier series for the sawtooth waveform is thus:

$$y = \frac{A}{2} - \frac{A}{\pi}\sin\omega t - \frac{A}{2\pi}\sin 2\omega t - \frac{A}{3\pi}\sin 3\omega t - \dots \qquad [26]$$

We can write this, using equation [1], as:

$$y = \frac{A}{2} + \frac{A}{\pi}\cos(\omega t + \frac{\pi}{2}) + \frac{A}{2\pi}\cos(2\omega t + \frac{\pi}{2}) + \frac{A}{3\pi}\cos(3\omega t + \frac{\pi}{2}) + \dots$$

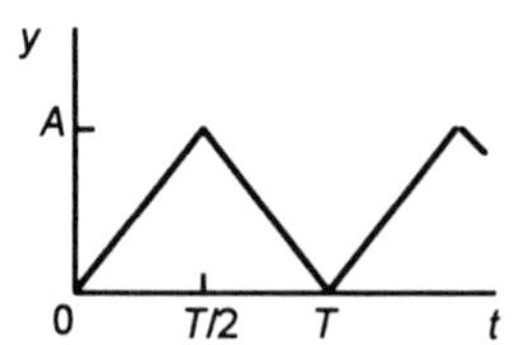

Figure 33.4 *Revision problem 3*

Revision

3 Determine the Fourier series for the waveform in Figure 33.4.

33.3.3 Half-wave rectified sinusoid

Consider a half-rectified sinusoidal waveform of period T (Figure 33.5). This can be described by:

$y = A \sin \omega t = A \sin 2\pi t/T$ for $0 \le t \le T/2$
$y = 0$ for $T/2 \le t < T$

We can determine a_0 by using equation [20]:

$$a_0 = \frac{2}{T}\int_0^T y \, dt = \frac{2}{T}\left(\int_0^{T/2} A \sin\omega t \, dt + \int_{T/2}^T 0 \, dt\right)$$

$$= -\frac{2A}{T\omega}[\cos\omega t]_0^{T/2} = \frac{2A}{\pi}$$

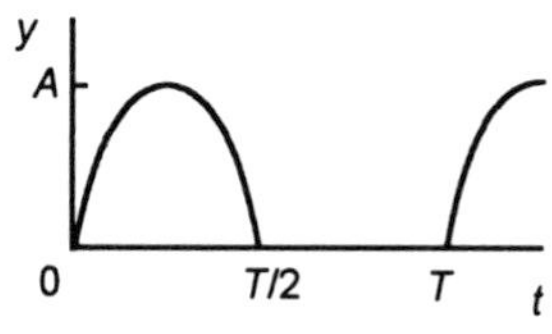

Figure 33.5 *Half-wave rectified sinusoid*

Equation [22] can be used to determine a_n:

$$a_n = \frac{2}{T}\int_0^T y\cos n\omega t\,\mathrm{d}t = \frac{2}{T}\left(\int_0^{T/2} A\sin\omega t\cos n\omega t\,\mathrm{d}t + \int_{T/2}^T 0\,\mathrm{d}t\right)$$

Since 2 sin A cos A = sin(A + B) + sin(A – B):

$$a_n = \frac{A}{T}\int_0^{T/2}\left[\sin(1+n)\omega t + \sin(1-n)\omega t\right]\mathrm{d}t$$

For $n = 1$ we have:

$$a_1 = \frac{A}{T}\int_0^{T/2}\sin(1+1)\omega t\,\mathrm{d}t = -\frac{A}{2T\omega}[\cos\omega t]_0^{T/2} = 0$$

For $n > 1$ we have:

$$a_n = \frac{A}{T}\left[-\frac{1}{(1+n)\omega}\cos(1+n)\omega t - \frac{1}{(1-n)\omega}\cos(1-n)\omega t\right]_0^{T/2}$$

For even values of n we have cos(1 + n)π = –1 and cos(1 – n)π = –1 and so:

$$a_{n\text{ even}} = \frac{A}{T}\left(\frac{1}{(1+n)\omega} + \frac{1}{(1+n)\omega} + \frac{1}{(1-n)\omega} + \frac{1}{(1-n)\omega}\right)$$

$$= \frac{A}{\pi}\left(\frac{2}{1+n} + \frac{2}{1-n}\right) = \frac{2A}{\pi(1-n^2)}$$

For odd values, other than 1, of n we have cos(1 + n)π = 1 and cos(1 – n)π = 1. This gives:

$$a_{n\text{ odd}} = \frac{A}{T}\left(-\frac{1}{(1+n)\omega} + \frac{1}{(1+n)\omega} - \frac{1}{(1-n)\omega} + \frac{1}{(1-n)\omega}\right) = 0$$

The values of b_n can be found using equation [24]:

$$b_n = \frac{2}{T}\int_0^T y\sin n\omega t\,\mathrm{d}t = \frac{2}{T}\int_0^{T/2} A\sin\omega t\sin n\omega t\,\mathrm{d}t$$

Since 2 sin A sin B = cos(A – B) – cos(A + B):

$$b_n = \frac{A}{T}\int_0^{T/2}\left[\cos(1-n)\omega t - \cos(1+n)\omega t\right]\mathrm{d}t$$

For $n = 1$ we have:

$$b_n = \frac{A}{T}\int_0^{T/2}\left[1 - \cos(1+n)\omega t\right]\mathrm{d}t = \frac{A}{t}\left[t - \frac{1}{2\omega}\sin 2\omega t\right]_0^{T/2} = \frac{A}{2}$$

For $n > 1$ we have:

$$b_n = \frac{A}{T}\left[\frac{1}{(1-n)\omega}\sin(1-n)\omega t - \frac{1}{(1+n)\omega}\sin(1+n)\omega t\right]_0^{T/2}$$

For other values of n, since $\sin(1 - n)\pi = 0$ and $\sin(1 + n)\pi = 0$, we have $b_n = 0$.

The Fourier series for the half-wave rectified sinusoid is thus:

$$y = \frac{A}{\pi} - \frac{2A}{3\pi}\cos 2\omega t - \frac{2A}{15\pi}\cos 4\omega t + \ldots + \frac{A}{2}\sin\omega t \qquad [27]$$

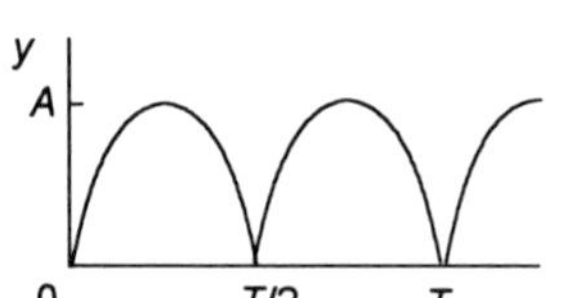

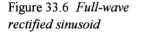

Figure 33.6 *Full-wave rectified sinusoid*

Revision

4 Determine the Fourier series for the full-wave rectified sinusoid (Figure 33.6).

33.4 Shift of origin

The Fourier series for the rectangular waveform shown in Figure 33.7(a) is:

$$y = \frac{4A}{\pi}\left[\sin\omega t + \tfrac{1}{3}\sin 3\omega t + \tfrac{1}{5}\sin 5\omega t + \ldots\right] \qquad [28]$$

Now consider the waveform in Figure 33.7(b). This is the waveform in (a) with the time origin shifted to the right by $\pi/2$. If we work out the Fourier series for this waveform we find that it is equation [28] with t replaced by $(t + \pi/2)$.

$$y = \frac{4A}{\pi}\left[\sin\left(\omega t + \frac{\pi}{2}\right) + \tfrac{1}{3}\sin 3\omega\left(t + \frac{\pi}{2}\right) + \tfrac{1}{5}\sin 5\omega\left(t + \frac{\pi}{2}\right) + \ldots\right]$$

and so:

$$y = \frac{4A}{\pi}\left[\cos\omega t - \tfrac{1}{3}\cos 3\omega t + \tfrac{1}{5}\cos 5\omega t - \ldots\right] \qquad [29]$$

Thus we have the rule:

Shifting the time origin of a waveform to the right by h means replacing t by (t + θ) in the Fourier series. Shifting the time origin to the left by h means replacing t by (t – θ).

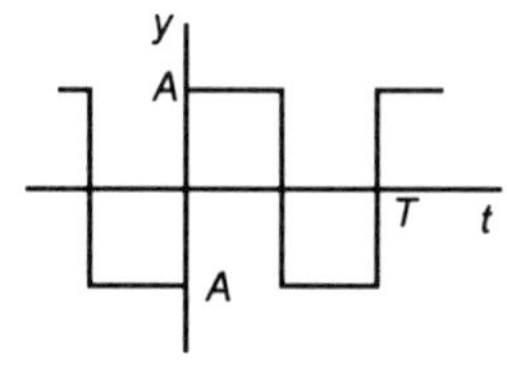

(a)

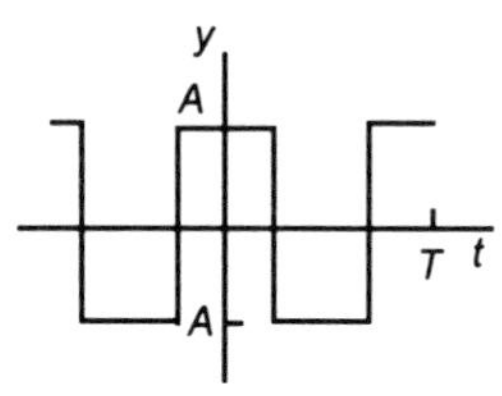

(b)

Now consider the waveform in Figure 33.7(c). This is that in (a) shifted vertically by A, i.e. the waveform in (a) plus A. The Fourier series is then that of (a) plus A:

$$y = A + \frac{4A}{\pi}\left[\sin\omega t + \tfrac{1}{3}\sin 3\omega t + \tfrac{1}{5}\sin 5\omega t + \ldots\right] \qquad [30]$$

This gives the rule:

Shifting the time axis vertically downwards adds to the Fourier series the amount of the shift, shifting upwards subtracts.

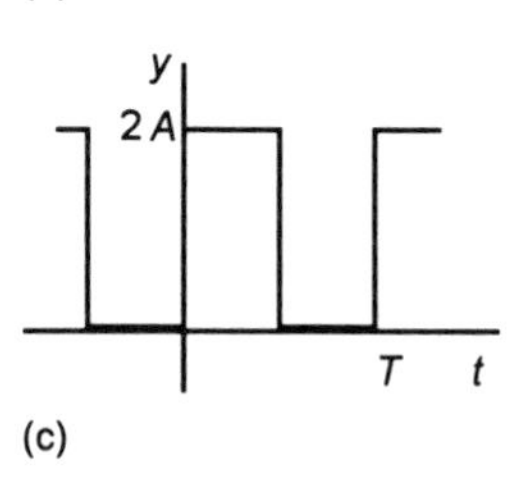

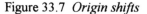

(c)

Figure 33.7 *Origin shifts*

Revision

5 Using equation [28], determine the Fourier series for the waveform shown in Figure 33.8.

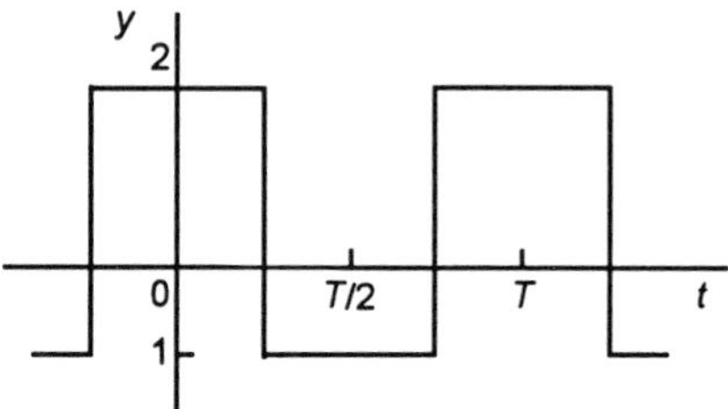

Figure 33.8 *Revision problem 5*

33.5 Waveform symmetry

As will be apparent from the above examples, certain terms are not always present in a Fourier series. Consideration of whether functions have odd or even symmetry (see Section 1.3.2) about the origin enables us to determine the presence or otherwise of terms.

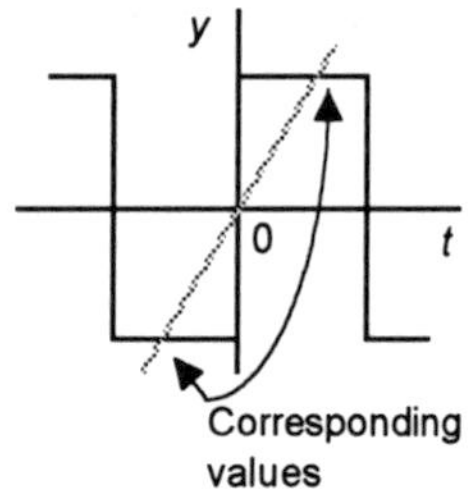

Figure 33.9 *Odd symmetry*

1 *Odd symmetry*
A function with odd symmetry is defined as having $f(-t) = -f(t)$. This means that the function value for a particular positive value of time is equal in magnitude but of opposite sign to that for the corresponding negative value of that time. Thus $y = f(x) = x^3$ is an odd function since $f(-2) = -8 = -f(2)$. For every point on the waveform for positive times there is a corresponding point on the waveform on a straight line drawn through the origin and equidistant from it (Figure 33.9).

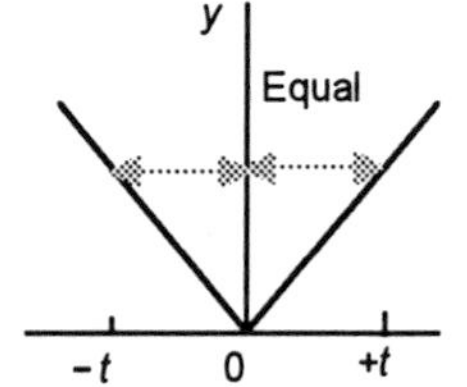

Figure 33.10 *Even symmetry*

2 *Even symmetry*
A function with even symmetry is defined as having $f(-t) = f(t)$. This means that the function value for a particular positive value of time is identical to that for the corresponding negative value of that time. Thus $y = f(x) = x^2$ is an even function since $f(-2) = 4 = f(2)$. If the y-axis was a plane mirror then the reflection of the positive time values for the waveform would give the negative time values (Figure 33.10).

In determining Fourier coefficients it is necessary to consider the odd or even nature of products of two odd or even functions.

1 *Product of two even functions*
Consider $f(x)$ and $g(x)$ and the product $F(x) = f(x)g(x)$. We can write $F(-x) = f(-x)g(-x)$. Thus if $f(x)$ and $g(x)$ are both even we must have $F(-x) = f(x)g(x)$ and so $F(-x) = F(x)$. The product of two even functions is an even function.

2 *Product of two odd functions*
Consider $f(x)$ and $g(x)$ and the product $F(x) = f(x)g(x)$. We can write $F(-x) = f(-x)g(-x)$. Thus if $f(x)$ and $g(x)$ are both odd we must have $F(-x) = \{-f(x)\}\{-g(x)\}$ and so $F(-x) = F(x)$. The product of two odd functions is an even function.

3 *Product of an odd and an even function*
Consider $f(x)$ and $g(x)$ and the product $F(x) = f(x)g(x)$. We can write $F(-x) = f(-x)g(-x)$. Thus if $f(x)$ is even and $g(x)$ is odd we must have $F(-x) = f(x)\{-g(x)\} = -f(x)g(x)$ and so $F(-x) = -F(x)$. The product of an even and an odd function is an odd function.

Example

Determine whether (a) x^2, (b) $\cos 2x$ and (c) $x^2 \cos 2x$ are even or odd functions.

(a) $y = f(x) = x^2$ is an even function since if we consider some particular value of x, say -2, we have $f(-2) = 4 = f(2)$.

(b) $y = f(x) = \cos 2x$ is an even function since if we consider some particular value of x, say $-\pi/2$ we have $f(-\pi/2) = 0 = f(\pi/2)$.

(c) Since the product of two even functions is even, $x^2 \cos 2x$ is an even function.

Revision

6 Determine whether the following are even or odd functions:

(a) $\sin x$, (b) x, (c) $x \sin x$, (d) $x \cos 2x$, (e) $x^3 \cos 2x$

33.5.1 Fourier coefficients

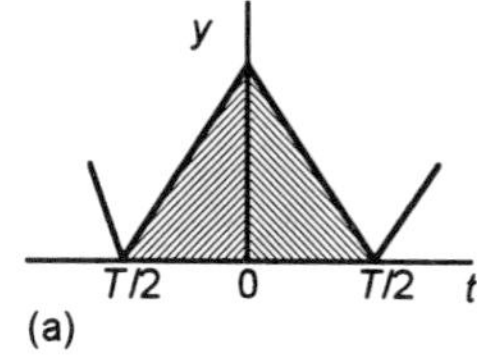

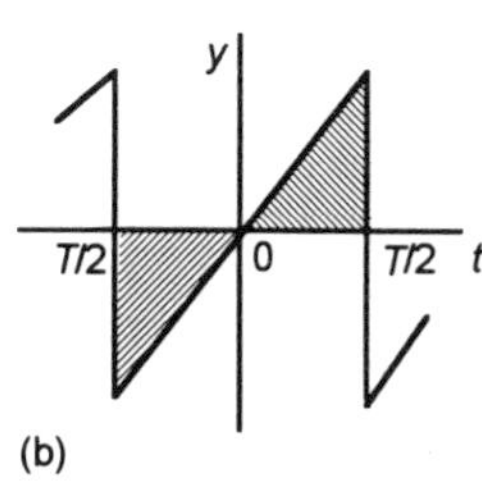

Figure 33.11 *(a) Even, (b) odd*

Consider the coefficients for a Fourier series for functions showing odd or even symmetry.

1 *a_0 coefficients*
a_0 is given by equation [20] as:

$$a_0 = \frac{2}{T}\int_0^T f(t)\,dt = \frac{2}{T}\int_{-T/2}^{T/2} f(t)\,dt$$

$$= \frac{2}{T}\int_0^{T/2} f(t)\,dt + \frac{2}{T}\int_{-T/2}^{0} f(t)\,dt$$

For a function with even symmetry we have the areas under the waveform on each side of the y-axis equal in both size and sign. Figure 33.11(a) illustrates this. A consequence of this is:

$$a_0 = 2 \times \frac{2}{T} \int_0^{T/2} f(t)\,dt \qquad [31]$$

But for an odd function (Figure 33.11(b)) the areas under the waveform on each side of the y-axis are equal in size but opposite in sign. A consequence of this is that there can be no a_0 term:

$$a_0 = \int_0^{T/2} f(t)\,dt + \int_{-T/2}^{0} f(t)\,dt = 0 \qquad [32]$$

We can look at this issue in another way. The mean value over one cycle of a waveform is given by equation [23] in Chapter 27 as:

$$\text{mean} = \frac{1}{T} \int_{-T/2}^{T/2} f(x)\,dt$$

Thus $a_0/2$ is the mean value. Thus for an odd function the mean value is 0 because the mean value is 0.

2 *a_n coefficients*
For the a_n coefficients equation [22] gives:

$$a_n = \frac{2}{T} \int_0^{T} f(t) \cos n\omega t\,dt = \frac{2}{T} \int_{-T/2}^{T/2} f(t) \cos n\omega t\,dt$$

Since cos $n\omega t$ is an even function, if $f(t)$ is even then the product is even. Hence we have, on the basis of the discussion used for a_0:

$$a_n = 2 \times \frac{2}{T} \int_0^{T/2} f(t) \cos n\omega t\,dt \qquad [33]$$

If $f(t)$ is odd then the product is odd. Thus on the basis of the discussion used for a_0:

$$a_n = \frac{2}{T} \int_{-T/2}^{T/2} f(t) \cos n\omega t\,dt = 0 \qquad [34]$$

3 *b_n coefficients*
For the b_n coefficients equation [24] gives:

$$b_n = \frac{2}{T} \int_0^{T} f(t) \sin n\omega t\,dt = \frac{2}{T} \int_{-T/2}^{T/2} f(t) \sin n\omega t\,dt$$

Since sin $n\omega t$ is an odd function, if $f(t)$ is even then the product is odd. Thus, on the basis of the discussion used for a_0:

$$b_n = \frac{2}{T} \int_{-T/2}^{T/2} f(t) \sin n\omega t\,dt = 0 \qquad [35]$$

If $f(t)$ is odd then the product is even. Thus, on the basis of the discussion used for a_0:

$$b_n = \frac{2}{T}\int_{-T/2}^{T/2} f(t)\sin n\omega t\,\mathrm{d}t = 2\times\frac{2}{T}\int_0^{T/2} f(t)\sin n\omega t\,\mathrm{d}t \qquad [36]$$

To summarise:

If f(t) is an even function then the Fourier series contains an a_0 term and only cosine terms. If f(t) is an odd function then the Fourier series contains no a_0 term and only sine terms.

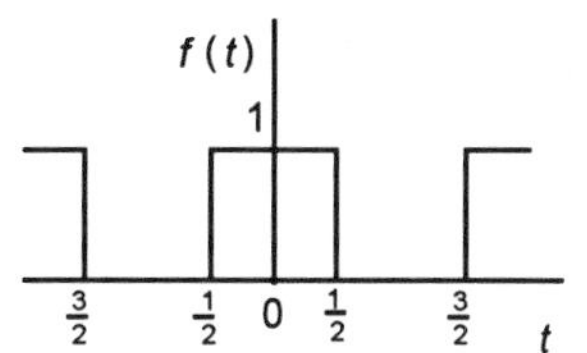

Figure 33.12 *Example*

Example

Determine the Fourier series for the function shown in Figure 33.12.

The function is an even function and so the b coefficients are all zero. The period is 2 and so $\omega = \pi$. Thus, using equation [31]:

$$a_0 = 2\times\frac{2}{T}\int_0^{T/2} f(t)\,\mathrm{d}t = 2\left(\int_0^{1/2} 1\,\mathrm{d}t + \int_{1/2}^{1} 0\,\mathrm{d}t\right) = 2[t]_0^{1/2} = 1$$

Using equation [33]:

$$a_n = 2\times\frac{2}{T}\int_0^{T/2} f(t)\cos n\omega t\,\mathrm{d}t = 2\left(\int_0^{1/2} 1\cos n\pi t\,\mathrm{d}t + \int_{1/2}^{1} 0\,\mathrm{d}t\right)$$

$$= 2\left[\frac{\sin n\pi t}{n\pi}\right]_0^{1/2} = \frac{2}{n\pi}\sin\tfrac{1}{2}n\pi$$

Thus the Fourier series is:

$$y = \tfrac{1}{2} + \frac{2}{\pi}\left(\cos\pi t - \tfrac{1}{3}\cos 3\pi t + \tfrac{1}{5}\cos 5\pi t + \dots\right)$$

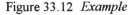

Revision

7 Determine what terms the following waveforms will contain in their Fourier series:

(a) $f(t) = 3t$ for $-\pi \le t < \pi$, period 2π,

(b) $f(t) = \cos t$ for $-\pi \le t < \pi$, period 2π,

(c) $f(t) = t^2 \cos t$ for $-\pi \le t < \pi$, period 2π

8 Determine the Fourier series for the waveform described by $f(t) = t$ for $-\pi \le t < \pi$ with a period of 2π.

33.5.2 Half-wave symmetries

By considering symmetry with respect to half-waves it is possible to determine whether odd or even harmonics will be omitted from a Fourier

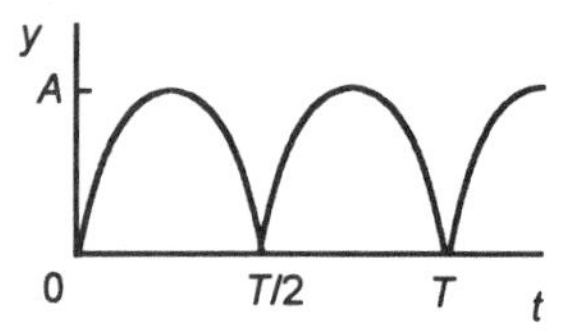

Figure 33.13 *Full-wave rectified sinusoid*

series. For some waveforms, e.g. that shown in Figure 33.13, we have successive halves of the period of a waveform identical:

$$f(t) = f(t + T/2) \qquad [37]$$

Such a form of symmetry is referred to as *half-wave repetition*. Equation [16] gives:

$$f(t) = \tfrac{1}{2}a_0 + a_1 \cos \omega t + a_2 \cos 2\omega t + \ldots + a_n \cos n\omega t$$
$$+ b_1 \sin \omega t + b_2 \sin 2\omega t + \ldots + b_n \sin n\omega t \qquad [38]$$

If we replace t by $t + T/2$ we can obtain $f(t + T/2)$. But $\omega(t + T/2) = \omega t + \pi$, and since $\cos(n\omega t + n\pi) = \cos n\omega t \cos n\pi - \sin n\omega t \sin n\pi$ and we have $\sin n\pi = 0$ for $n = 1, 2, 3, \ldots$ and even values of n give $\cos n\pi = 1$ and odd values give $\cos n\pi = -1$, then $\cos(n\omega t + n\pi) = \cos n\omega t$ for even values of n and $-\cos n\omega t$ for odd values. Similarly we can deduce that $\sin(n\omega t + n\pi) = \sin n\omega t$ for even values of n and $-\sin n\omega t$ for odd values. Thus:

$$f(t + T/2) = \tfrac{1}{2}a_0 - a_1 \cos \omega t + a_2 \cos 2\omega t - a_3 \cos 3\omega t + \ldots$$
$$- b_1 \sin \omega t + b_2 \sin 2\omega t - b_3 \sin 3\omega t + \ldots \qquad [39]$$

For a waveform with half-wave repetition, i.e. equation [37] applies, then for equation [38] to equal equation [39] we must have $a_1 = 0$, $a_3 = 0$, ... $b_1 = 0$, $b_3 = 0$, ..., etc. and so only even harmonics.

A waveform with half-wave repetition has a_0 and only even harmonics.

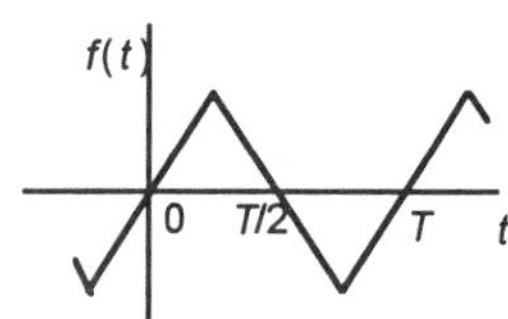

Figure 33.14 *Half-wave inversion*

Consider a waveform for which successive halves are inverted forms or previous halves, e.g. Figure 33.14:

$$-f(t) = f(t + T/2) \qquad [40]$$

Such a waveform is said to show *half-wave inversion*. Equation [16] gives:

$$-f(t) = -\tfrac{1}{2}a_0 - a_1 \cos \omega t - a_2 \cos 2\omega t - \ldots - a_n \cos n\omega t$$
$$- b_1 \sin \omega t - b_2 \sin 2\omega t - \ldots - b_n \sin n\omega t \qquad [41]$$

For equation [40] to be valid and equation [41] to equal equation [39] we must have $a_o = 0$, $a_2 = 0$, ... $b_2 = 0$, ..., etc. and so only odd harmonics.

A waveform with half-wave inversion has only odd harmonics.

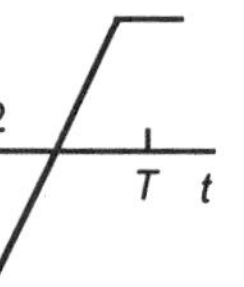

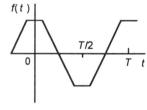

Figure 33.15 *Example*

Example

What terms will be present in the Fourier series for the waveform shown in Figure 33.15?

The waveform is even symmetrical and so only a_0 and cosine terms can be present. The waveform shows half-wave inversion and so only odd harmonics will be present. Thus the Fourier series will just consist of odd harmonic cosine terms.

Revision

9 What terms will be present in the Fourier series for the waveforms shown in Figure 33.16?

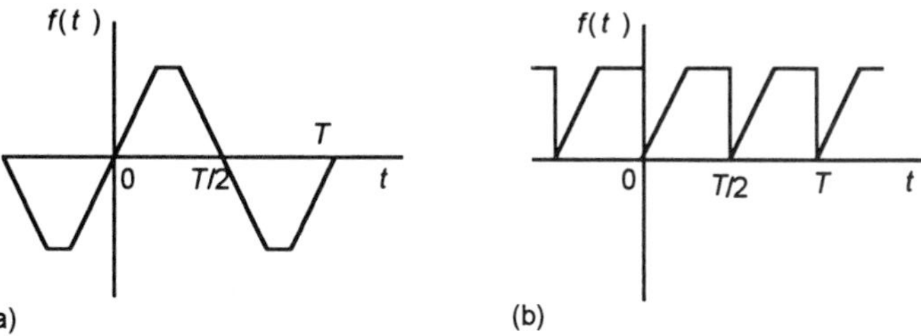

Figure 33.16 *Revision problem 9*

10 From considerations of the mean values of the waveforms in Figure 33.17, what will be the values of a_0?

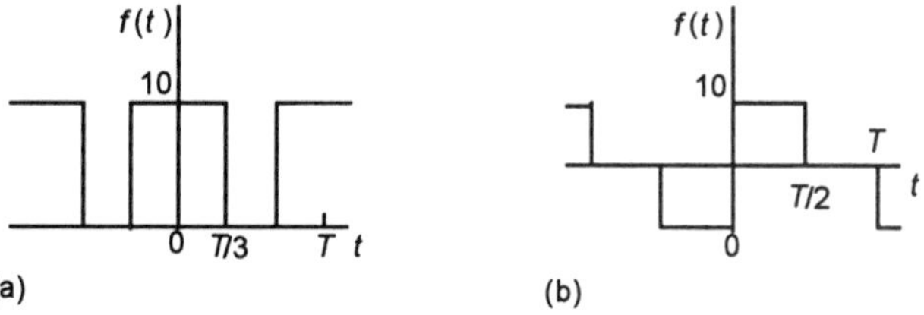

Figure 33.17 *Revision problem 10*

33.6 Frequency spectrum

The *frequency spectrum* comprises an *amplitude spectrum*, which is a graph of the amplitudes of each of the constituent sinusoidal components in the Fourier series plotted against frequency, and a *phase spectrum* which is their phases. The amplitudes are given from the Fourier coefficients by equation [19]:

$$A_n = \sqrt{a_n^2 + b_n^2} \qquad [42]$$

and the phases of sinusoidal components by equation [18] as:

$$\phi_n = \tan^{-1}\left(\frac{a_n}{b_n}\right) \qquad [43]$$

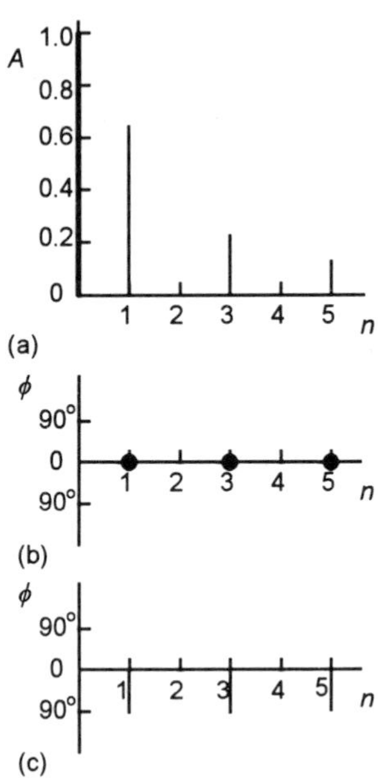

Figure 33.18 *Frequency spectrum*

This gives the Fourier series in the form shown in equation [14], namely:

$$y = A_0 + A_1 \sin(\omega t + \phi_1) + A_2 \sin(2\omega t + \phi_2) + \ldots$$

Note that if we want to refer to a series in the form:

$$y = A_0 + A_1 \cos(\omega t + \phi_1) + A_2 \cos(2\omega t + \phi_2) + \ldots$$

then, in order to take account of $\sin \omega t = \cos(\omega t - \pi/2)$, i.e. the phase difference of –90° between the cosine and sine, equation [43] becomes:

$$\phi = \tan^{-1}\left(\frac{-b_n}{a_n}\right) \qquad [44]$$

Example

Determine the frequency spectrum of the rectangular waveform which has $a_0 = 1$, $a_n = 0$ and $b_n = (1 - \cos n\pi)/n\pi$.

We have $b_1 = 2/\pi = 0.64$, $b_2 = 0$, $b_3 = 2/3\pi = 0.21$, $b_4 = 0$, $b_5 = 2/5\pi = 0.13$, etc. The A_0 term is 1. Using equation [42], the A_1 term is 0.64, the A_2 term 0, the A_3 term 0.21, the A_4 term 0, the A_5 term 0.13, etc. The phases, when referred to a sine wave, are 0° for all components. When referred to a cosine wave they are –90°. Figure 33.18(a) shows the resulting amplitude spectrum, (b) the phase spectrum when referred to sines and (c) the phase spectrum when referred to cosines.

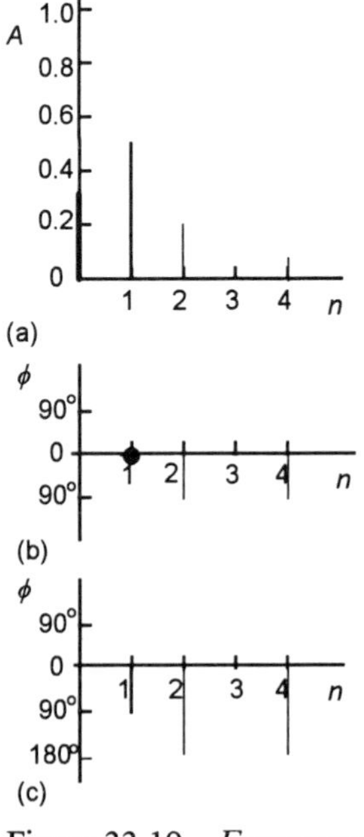

Figure 33.19 *Frequency spectrum*

Example

Determine the frequency spectrum for a half-wave rectified sinusoidal waveform if it has the Fourier series:

$$y = \frac{1}{\pi} - \frac{2}{3\pi}\cos 2\omega t - \frac{2}{15\pi}\cos 4\omega t + \ldots + \frac{1}{2}\sin \omega t$$

The A_0 term is $1/\pi = 0.32$. Using equation [42], the A_1 term is 0.5, the A_2 term $2/3\pi = 0.21$, the A_3 term 0 and the A_4 term $2/15\pi = 0.04$. The phases, when referred to a sine wave are $\phi_1 = 0$ and since $-\cos \omega t = \sin(\omega t - 90°)$, $\phi_2 = -90°$ and $\phi_4 = -90°$. When referred to a cosine they are $\phi_1 = -90°$, $\phi_2 = -180°$ and $\phi_4 = -180°$. Figure 31.19(a) shows the amplitude spectrum, (b) the phase spectrum when referred to a sine wave and (c) when referred to a cosine.

Revision

11 Determine the amplitude and phase (referred to a sine) elements for the frequency spectrum of the waveforms giving the following Fourier series:

(a) $y = \frac{1}{2} - \frac{1}{\pi}\sin \omega t - \frac{1}{2\pi}\sin 2\omega t - \frac{1}{3\pi}\sin 3\omega t - \ldots,$

(b) $a_0 = \pi/2$, $a_n = 0$ for n even and $-2/n^2\pi$ for n odd, $b_n = -(-1)^n/n$

33.7 Parseval's theorem

Situations often occur in the use of Fourier series to represent current or voltage waveforms that we require the square of the current or voltage. If we have a function represented by the Fourier series:

$$f(t) = \tfrac{1}{2}a_0 + \sum_{n=1}^{\infty}(a_n \cos n\omega t + b_n \sin n\omega t)$$

and we multiply both sides of the equation by $f(t)$:

$$[f(t)]^2 = \tfrac{1}{2}a_0 f(t) + \sum_{n=1}^{\infty}\left(a_n f(t)\cos n\omega t + b_n f(t)\sin n\omega t\right)$$

Integrating both sides of the equation with respect to t between 0 and T:

$$\int_0^T [f(t)]^2\,dt = \tfrac{1}{2}a_0 \int_0^T f(t)\,dt + \sum_{n=1}^{\infty}\left(a_n \int_0^T f(t)\cos n\omega t\,dt + b_n \int_0^T f(t)\sin n\omega t\,dt\right)$$

Using the equations for a_0, a_n and b_n ([20], [22] and [24]):

$$\frac{2}{T}\int_0^T [f(t)]^2\,dt = \tfrac{1}{2}a_0^2 + \sum_{n=1}^{\infty}(a_n^2 + b_n^2) \qquad [45]$$

Equation [45] is termed *Parseval's theorem*. Since the root-mean-square value of an alternating current or voltage is (see equation [24], Chapter 27):

$$\text{r.m.s. value} = \sqrt{\frac{1}{T}\int_0^T [f(t)]^2\,dt}$$

then equation [45] gives:

$$(\text{r.m.s. value})^2 = \tfrac{1}{4}a_0^2 + \tfrac{1}{2}\sum_{n=1}^{\infty}(a_n^2 + b_n^2) \qquad [46]$$

Example

Determine the root-mean-square value of the periodic voltage signal $v = \sin t + \frac{1}{3}\sin 3t + \frac{1}{5}\sin 5t$.

The waveform has $a_0 = 0$, $a_n = 0$, $b_1 = 1$, $b_2 = 0$, $b_3 = 1/3$, $b_4 = 0$ and $b_5 = 1/5$. Using equation [46]:

$$\text{r.m.s. value} = \sqrt{\tfrac{1}{2}\left[0 + 1^1 + 0 + \left(\tfrac{1}{3}\right)^2 + 0 + \left(\tfrac{1}{5}\right)^2\right]} = 0.76$$

Revision

12 Determine the r.m.s. value of $v = 100 \sin \omega t + 40 \sin 2\omega t + 20 \sin 3\omega t$.

Problems

1 Determine the Fourier series for the following waveforms:

(a) $y = 0$ for $0 \leq t < 5$, $y = 2$ for $5 \leq t < 10$, period = 10,

(b) $y = +1$ for $0 \leq t < 1$, $y = -1$ for $1 \leq t < 2$, period 2,

(c) $y = t$ for $0 \leq t < \pi$, $y = 0$ for $\pi \leq t < 2\pi$, period 2π,

(d) $y = t^2$ for $0 \leq t < 2\pi$, period 2π,

(e) $y = t^2$ for $-\pi \leq t < \pi$, period 2π,

(f) $y = t^2$ for $0 \leq t < \pi$, $y = 0$ for $\pi \leq t < 2\pi$, period 2π,

(g) $y = \pi + t$ for $-\pi \leq t < 0$, $y = \pi - t$ for $0 \leq t < \pi$, period 2π

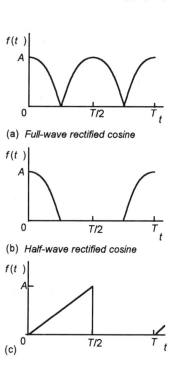

2 Determine the Fourier series for the waveforms shown in Figure 33.20.

(a) *Full-wave rectified cosine*

(b) *Half-wave rectified cosine*

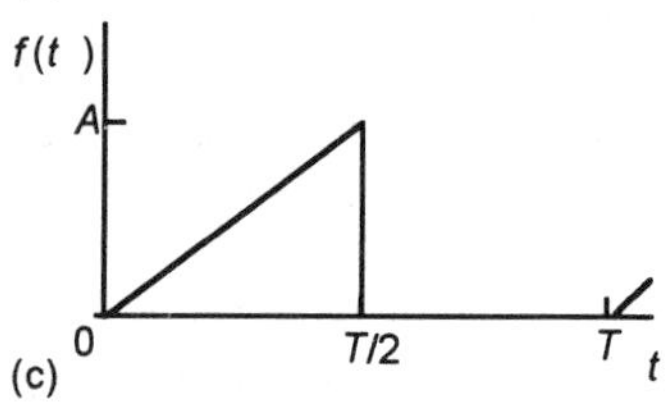

Figure 33.20 *Problem 2*

3 Determine which of the following functions will be odd and which even:

(a) $y = 2t^2$, (b) $y = 2 \cos 3t$, (c) $y = 3t$, (d) $y = t^2 \sin t$, (e) $y = \sin 2t \sin 3t$

4 Determine from consideration of symmetry, which terms will be present in the Fourier series for the waveforms shown in Figure 33.21.

5 Determine the amplitude and phase terms for a full-wave rectified sinusoidal waveform with unit amplitude, representing the phases in terms of cosines.

6 Determine the root-mean-square value of the waveform $1 + 10 \sin \omega t + 8 \sin(3\omega t + \pi/6) + 3 \sin(5\omega t + \pi/3)$.

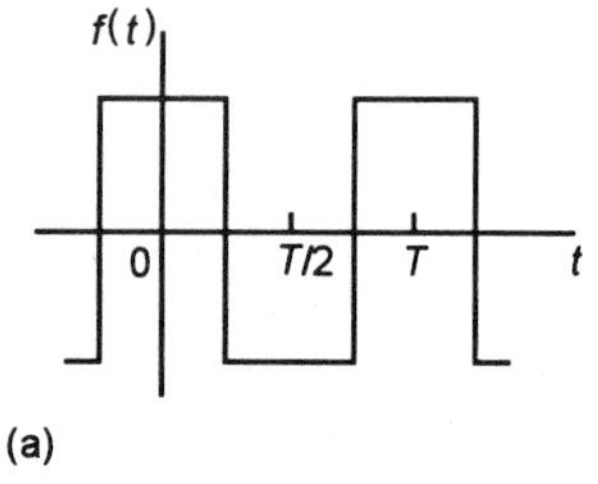

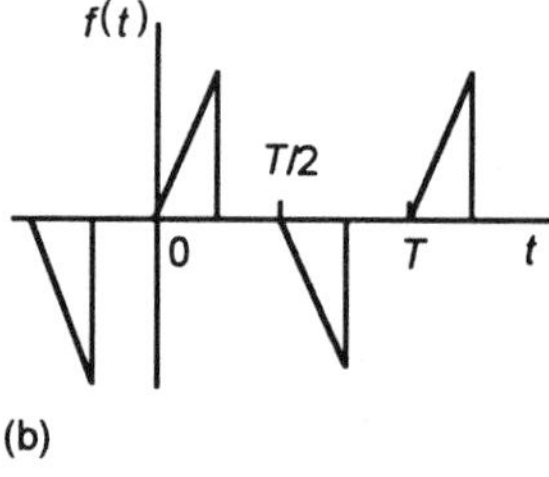

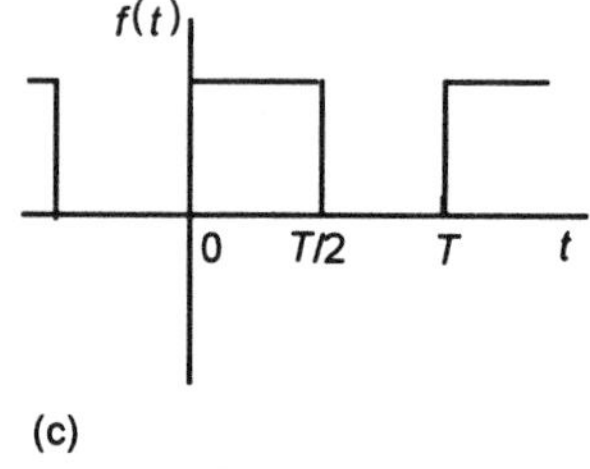

Figure 33.21 *Problem 4*

34 Non-periodic functions

34.1 Introduction

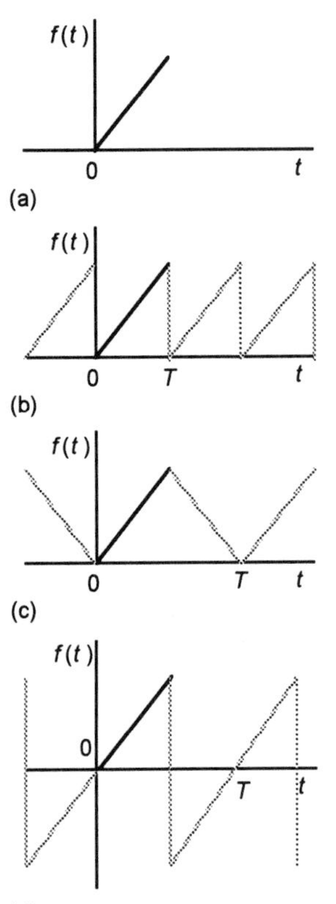

Figure 34.1 *Extensions*

Fourier series were developed in Chapter 33 for periodic functions, such functions then being represented by an infinite series of sine and cosine terms. However, situations often occur in engineering and science where a function is not periodic but only defined over a finite interval. For example, we might be considering the oscillations of a violin string which is tethered at both ends and bowed out to oscillate. The resulting waveform is restricted to a wave between the fixed ends of the string. As another example, we might be considering a single electrical pulse. We are thus interested in functions which are only defined over some finite interval and which are not periodic outside that interval. The Fourier series is, however, invariably periodic all the way to infinity. We can, however, use Fourier series to represent such functions if we pretend that the function is periodic, using the function within the specified interval but extending it outside the interval by any artificial function which is periodic.

Suppose we have the finite duration function shown in Figure 34.1(a). We might consider this to be one period of the function shown in Figure 34.1(b) or perhaps half the period of the function shown in (c) or (d). If the finite duration element is taken to be a full period of the extended function we talk of a *full-range series*, if it represents half a period the *half-range series*. We likewise could have a *quarter-range series* if the finite duration element is taken to be a quarter period of the extended function. We are entirely free to choose the form of the periodic function outside the range of the finite duration element. One of the factors that dictates the choice is the rate of convergence of the resulting series, another factor is the form of the resulting Fourier series for the periodic function. It might, for example, be useful in some situations to use a series which has perhaps just sine terms or perhaps one which has an a_0 term so that the series has a value at $t = 0$. The most used form of extension is the half-range series and the remainder of this chapter is devoted to a discussion of this form of extension.

This chapter assumes Chapter 33, using the principles and equations developed in that chapter.

34.2 Half-range series

For the half-wave series extensions shown in Figure 34.1(c) and (d), the extension in (c) has resulted in an even waveform while that in (d) an odd waveform. The even waveform would give a Fourier series containing a_0 and cosine terms. The odd waveform would give a Fourier series containing only sine terms. Thus by choosing the form of the half-range extension we can determine whether the series will be cosine or sine terms.

1 *Half-range sine series*

For the half-range sine series we have:

$$a_0 = a_n = 0 \qquad [1]$$

$$b_n = \frac{2}{T}\int_0^T f(t)\sin n\omega t\,\mathrm{d}t = \frac{4}{T}\int_T^{T/2} f(t)\sin n\omega t\,\mathrm{d}t \qquad [2]$$

2 *Half-range cosine series*
For the half-range cosine series we have:

$$a_0 = \frac{2}{T}\int_0^T f(t)\,\mathrm{d}t = \frac{4}{T}\int_0^{T/2} f(t)\,\mathrm{d}t \qquad [3]$$

$$a_n = \frac{2}{T}\int_0^T f(t)\cos n\omega t\,\mathrm{d}t = \frac{4}{T}\int_0^{T/2} f(t)\cos n\omega t\,\mathrm{d}t \qquad [4]$$

$$b_n = 0 \qquad [5]$$

Example

Determine (a) the half-range sine series and (b) the half-range cosine series for the wave pulse described by:

$$f(t) = t \text{ for } 0 \le t < \pi$$

This is the form of waveform described by Figure 34.1(a), the half-range sine series being that in Figure 34.1(d) and the half-range cosine series that in Figure 34.1(c).

(a) $T = 2\pi/\omega$ and thus as $T = 2\pi$, $\omega = 1$. As indicated by equation [1], $a_0 = a_n = 0$. Equation [2] gives:

$$b_n = \frac{4}{T}\int_T^{T/2} f(t)\sin nt\,\mathrm{d}t = \frac{4}{2\pi}\int_0^{\pi} t\sin nt\,\mathrm{d}t$$

Using integration by parts:

$$b_n = \frac{2}{\pi}\left[-\frac{t}{n}\cos nt + \frac{1}{n^2}\sin nt\right]_0^{\pi} = \frac{2}{n}(-\cos n\pi)$$

Thus the series is:

$$y = 2\left(\sin t - \tfrac{1}{2}\sin 2t + \tfrac{1}{3}\sin 3t + \dots\right)$$

(b) $T = 2\pi/\omega$ and thus as $T = 2\pi$, $\omega = 1$. As indicated by equation [3]:

$$a_0 = \frac{4}{T}\int_0^{T/2} f(t)\,\mathrm{d}t = \frac{4}{2\pi}\int_0^{\pi} t\,\mathrm{d}t = \frac{2}{\pi}\left[\frac{t^2}{2}\right]_0^{\pi} = \pi$$

Using equation [4]:

$$a_n = \frac{4}{T}\int_0^{T/2} f(t)\cos n\omega t\,\mathrm{d}t = \frac{4}{2\pi}\int_0^{\pi} t\cos nt\,\mathrm{d}t$$

Using integration by parts:

$$a_n = \frac{2}{\pi}\left[\frac{t}{n}\sin nt + \frac{1}{n^2}\cos nt\right]_0^\pi = \frac{2}{n^2\pi}(\cos n\pi - 1)$$

When n is odd then $\cos n\pi = -1$ and when n is even $\cos n\pi = 1$. Thus there are no even terms and the odd value terms have $a_n = -4/n^2\pi$. Using equation [5], $b_n = 0$. Thus the Fourier series is:

$$y = \frac{\pi}{2} - \frac{4}{\pi}\left(\cos t + \frac{1}{9}\cos 3x + \frac{1}{25}\cos 5t + \ldots\right)$$

Revision

1 Determine the half-range cosine and half-range sine series representations of the following functions:

(a) $f(t) = 1$ over the interval $0 \le t < 1$,

(b) for interval $0 \le t < \pi$, $f(t) = 0$ for $0 \le t < \pi/2$, $f(t) = 1$ for $\pi/2 \le t < \pi$,

(c) $f(t) = \sin 3t$ over the interval $0 \le t < \pi$

Problems

1 Determine the half-range cosine and half-range sine series representations of the following functions:

(a) $f(t) = 3t$ over the interval $0 \le t < \pi$,

(b) $f(t) = \cos t$ over the interval $0 \le t < \pi$,

(c) $f(t) = \pi - t$ over the interval $0 \le t < \pi$,

(d) $f(t) = \sin^2 t$ over the interval $0 \le t < \pi$

2 Determine a Fourier series involving just sines which can be used to represent the displacement x of a string which is fixed at both ends of a length L and pulled aside by (a) y_0 at its midpoint, (b) y_0 a distance x from one end.

3 Determine a Fourier series which can be used to represent the function $f(t) = \pi - t$ over a time interval $0 \le t < \pi$ if the series is to have a zero value at time $t = 0$.

4 Determine a Fourier series involving just sines which can be used to represent the temperature T along a bar of length L if $T = K(Lx - x^2)$ where x is the distance along the bar and K a constant.

35 Exponential form

35.1 Introduction

In Chapters 33 and 34 Fourier series were expressed in terms of cosines and sines. There is, however, another way that Fourier series can be expressed and that is as complex exponentials. Such a form, though appearing more complicated, is often easier to handle. It more directly gives the amplitude and phase terms in the frequency spectrum and leads into the important topic of Fourier transforms (see Chapter 36).

This chapter shows how complex exponentials can be used to represent Fourier series, assuming Chapter 33 on the Fourier series and Part 2 on complex numbers.

35.2 Complex exponential form of Fourier series

The Fourier series (equation [17], Chapter 33):

$$y = \tfrac{1}{2}a_0 + \sum_{n=1}^{\infty} a_n \cos n\omega t + \sum_{n=1}^{\infty} b_n \sin n\omega t \qquad [1]$$

can be written, using Euler's formula (equations [7] and [8], Chapter 8), as:

$$y = \tfrac{1}{2}a_0 + \sum_{n=1}^{\infty} a_n \frac{e^{j\omega t} + e^{-jn\omega t}}{2} + \sum_{n=1}^{\infty} b_n \frac{e^{j\omega t} - e^{-jn\omega t}}{2j}$$

Multiplying the b_n term by j/j and rearranging gives:

$$y = \tfrac{1}{2}a_0 + \sum_{n=1}^{\infty}\left(\frac{a_n - jb_n}{2}e^{jn\omega t} + \frac{a_n + jb_n}{2}e^{-jn\omega t}\right) \qquad [2]$$

We can simplify the equation by letting:

$$c_0 = \tfrac{1}{2}a_0 \qquad [3]$$

$$c_n = \frac{a_n - jb_n}{2} \qquad [4]$$

$$c_{-n} = \frac{a_n + jb_n}{2} \qquad [5]$$

These are termed the *complex Fourier coefficients*. c_{-n} is the complex conjugate of c_n and is thus sometimes written as c_n^*. Thus, using these complex coefficients, equation [2] becomes:

$$y = c_0 + \sum_{n=1}^{\infty}(c_n e^{jn\omega t} + c_{-n} e^{-jn\omega t}) \qquad [6]$$

Since $e^0 = 1$ we can write the c_0 term as $c_0\, e^0$. This then becomes just the $c_n\, e^{jn\omega t}$ term when we let $n = 0$ and so we can incorporate it into that summation to give:

$$y = \sum_{n=0}^{\infty} c_n\, e^{jn\omega t} + \sum_{n=1}^{\infty} c_{-n}\, e^{-jn\omega t} \qquad [7]$$

The second summation can be written in a different form. As written above it is the sum of complex conjugate terms extending from $n = 1$ to infinity. If, with the sum of c_n terms between $n = 1$ and ∞, we let $n = -1$ and sum to $-\infty$ then the exponential becomes negative and c_n becomes c_{-n}. Thus:

$$\sum_{n=1}^{\infty} c_{-n}\, e^{-jn\omega t} = \sum_{n=-1}^{-\infty} c_n\, e^{jn\omega t}$$

and so equation [7] can be written as:

$$y = \sum_{n=-\infty}^{\infty} c_n\, e^{jn\omega t} \qquad [8]$$

This is the complex exponential form of the Fourier series.

35.2.1 Determining the complex coefficients

Since (equations [22] and [24], Chapter 33):

$$a_n = \frac{2}{T}\int_0^T f(t)\cos n\omega t\, dt$$

$$b_n = \frac{2}{T}\int_0^T f(t)\sin n\omega t\, dt$$

Then equation [4] gives:

$$c_n = \frac{a_n - jb_n}{2} = \frac{1}{T}\int_0^T f(t)\cos n\omega t\, dt - j\frac{1}{T}\int_0^T f(t)\sin n\omega t\, dt$$

Writing the cosine and sine in terms of exponentials by the use of Euler's formula:

$$c_n = \frac{1}{T}\int_0^T f(t)\frac{e^{jn\omega t} + e^{-jn\omega t}}{2}\, dt - j\frac{1}{T}\int_0^T f(t)\frac{e^{jn\omega t} - e^{-jn\omega t}}{2j}\, dt$$

$$= \frac{1}{T}\int_0^T f(t)\, e^{-jn\omega t}\, dt \qquad [9]$$

Care has to be exercised in determining c_0 with the above equation since with $n = 0$ the expressions can end up invalid. In such situations c_0 can be evaluated from $a_0/2$ using equation [20] in Chapter 33:

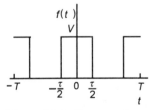

$$c_0 = \frac{a_0}{2} = \frac{1}{T}\int_0^T f(t)\,\mathrm{d}t \qquad [10]$$

Example

Determine the complex exponential form of the Fourier series for the waveform shown in Figure 35.1.

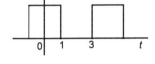

Figure 35.1 *Example*

This waveform has a period of 2 and thus $\omega = \pi$. If we take the period from $-T/2$ to $+T/2$:

$y = 1$ for $-1 \leq t < +1$
$y = 0$ otherwise

Equation [9] gives:

$$c_n = \frac{1}{T}\int_{-T/2}^{T/2} f(t)\,\mathrm{e}^{-\mathrm{j}n\omega t}\,\mathrm{d}t$$

$$= \tfrac{1}{2}\left(\int_{-1}^{1} 1\mathrm{e}^{-\mathrm{j}n\pi t}\mathrm{d}t + 0\right) = \tfrac{1}{2}\left[-\frac{1}{\mathrm{j}n\pi}\,\mathrm{e}^{-\mathrm{j}n\pi t}\right]_{-1}^{1}$$

$$= \frac{1}{\mathrm{j}2n\pi}(\mathrm{e}^{\mathrm{j}n\pi} - \mathrm{e}^{-\mathrm{j}n\pi}) = \frac{1}{n\pi}\left(\frac{\mathrm{e}^{\mathrm{j}n\pi} - \mathrm{e}^{-\mathrm{j}n\pi}}{2\mathrm{j}}\right)$$

$$= \frac{1}{n\pi}\sin n\pi$$

If we put $n = 0$ we have an invalid value. Thus for c_0 we use equation [10]:

$$c_0 = \frac{a_0}{2} = \frac{1}{T}\int_0^T f(t)\,\mathrm{d}t = \tfrac{1}{2}\int_{-1}^{1} 1\,\mathrm{d}t = \tfrac{1}{2}[t]_{-1}^{1} = 1$$

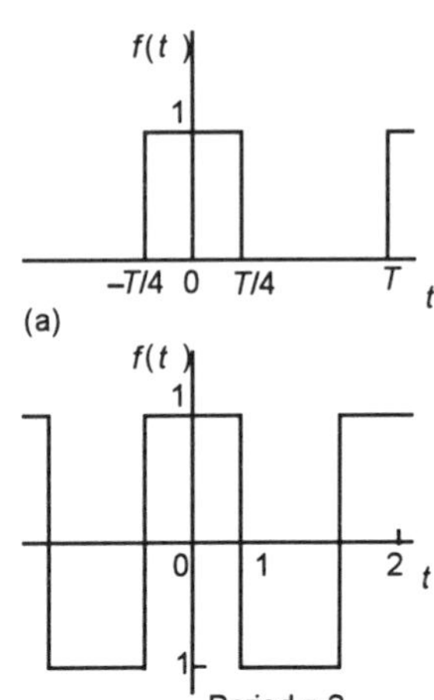

Figure 35.2 *Revision problem 1*

Revision

1 Determine the complex exponential form of the Fourier series for the waveforms shown in Figure 35.2.
In simplifying solutions note that $\dfrac{\mathrm{e}^{\mathrm{j}n\pi} - \mathrm{e}^{\mathrm{j}n\pi}}{2\mathrm{j}} = \sin n\pi = 0$.

35.2.2 Rectangular pulses

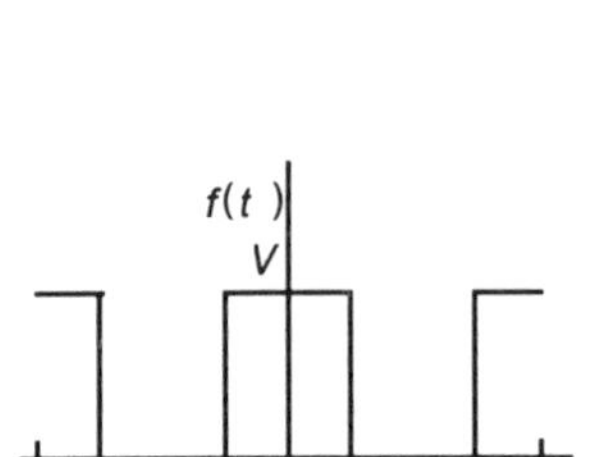

Figure 35.3 *Rectangular pulses*

A particular important waveform is that of periodic rectangular pulses of duration τ (Figure 35.3). The complex coefficient c_n is given by equation [9] as:

$$c_n = \frac{1}{T}\int_{-T/2}^{T/2} f(t)\,\mathrm{e}^{-\mathrm{j}n\omega t}\,\mathrm{d}t = \frac{1}{T}\int_{-\tau/2}^{\tau/2} V\,\mathrm{e}^{-\mathrm{j}n\omega t}\,\mathrm{d}t$$

$$= \frac{V}{T}\left[-\frac{1}{jn\omega}e^{-jn\omega t}\right]_{-\tau/2}^{\tau/2}$$

Since $\omega = 2\pi/T$:

$$c_n = -\frac{V}{j2n\pi}(e^{-jn\pi\tau/T} - e^{jn\pi\tau/T}) = \frac{V}{n\pi}\left(\frac{e^{jn\pi} - e^{-jn\pi}}{2j}\right)$$

$$= \frac{V}{n\pi}\sin\frac{n\pi\tau}{T} \qquad [11]$$

Consider what happens if we let the pulse width τ remain small and increase the period T. The spacing between the harmonics will decrease. If the periodic time tends to an infinite value, i.e. we effectively have just a single pulse, the spacing between the harmonics will become so small that they merge. We thus end up with a continuous spectrum of frequencies as the Fourier series for a pulse.

35.3 Symmetry

The effect of symmetry on the a_n and b_n coefficients was discussed in Sections 33.4.1 and 33.4.2. Since $c_n = (a_n - jb_n)/2$ then c_n is affected by symmetry.

1 *Even symmetry*

For even symmetry, $b_n = 0$ and $a_n = \frac{4}{T}\int_0^{T/2} f(t)\cos n\omega t\,dt$, thus:

$$c_n = \frac{2}{T}\int_0^{T/2} f(t)\cos n\omega t\,dt \qquad [12]$$

and is only real.

2 *Odd symmetry*

For odd symmetry, $a_n = 0$ and $b_n = \frac{4}{T}\int_0^{T/2} f(t)\sin n\omega t\,dt$, thus:

$$c_n = -j\frac{2}{T}\int_0^{T/2} f(t)\sin n\omega t\,dt \qquad [13]$$

and is only imaginary.

3 *Half-wave repetition*

With half-wave repetition there are only a_0 and even harmonics.

4 *Half-wave inversion*

With half-wave inversion there are only odd harmonics.

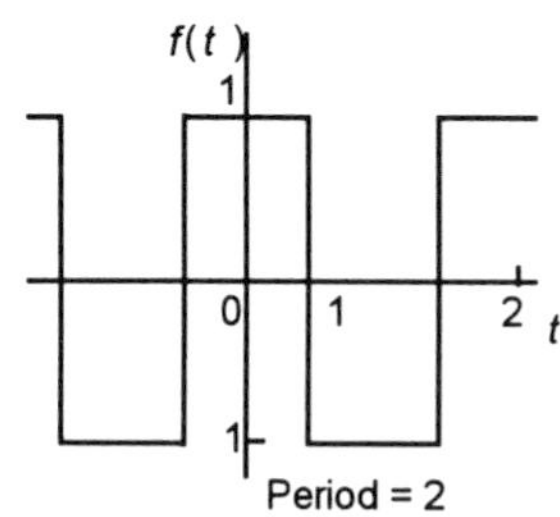

Figure 35.4 *Example*

Example

Determine the symmetries present in the waveform shown in Figure 35.4 and hence the value of the complex coefficient.

The function is an even function with half-wave inversion. Thus there will only be odd harmonics and c_n is given by equation [12] as:

$$c_n = \frac{2}{T}\int_0^{T/2} f(t)\cos n\omega t\,\mathrm{d}t$$

$$= \frac{2}{2}\left(\int_0^{1/2} 1\cos n\pi t\,\mathrm{d}t + \int_{1/2}^{1}(-1)\cos n\pi t\,\mathrm{d}t\right)$$

$$= \left[\frac{1}{n\pi}\sin n\pi t\right]_0^{1/2} - \left[\frac{1}{n\pi}\sin n\pi t\right]_{1/2}^{1} = \frac{2}{n\pi}\sin\frac{n\pi}{2}$$

When $n = 1$ we have $c_1 = 2/\pi$, $n = 2$ gives $c_2 = 0$, $n = 3$ gives $c_3 = -2/3\pi$, $n = 4$ gives $c_4 = 0$, etc.

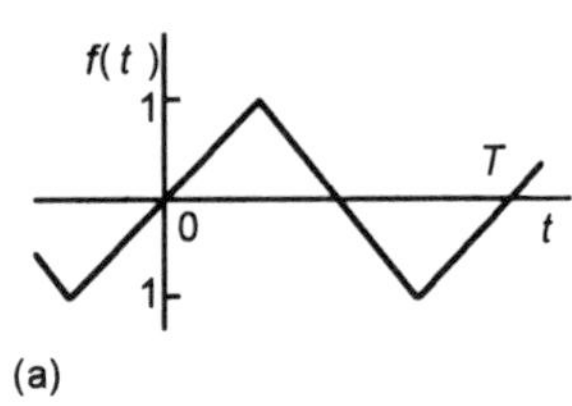

(a)

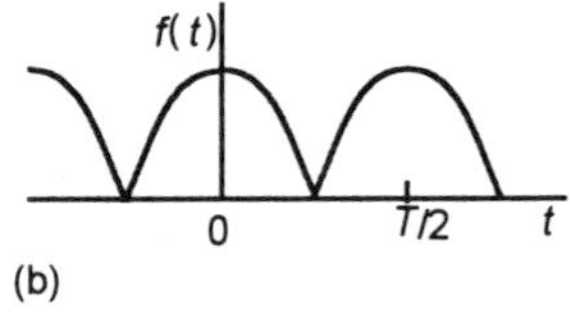

(b)

Figure 35.5 *Revision problem 2*

Revision

2 State which complex coefficients will be present in the Fourier series for the waveforms shown in Figure 35.5.

35.4 Frequency spectrum

The complex coefficient c_n in being $(a_n - \mathrm{j}b_n)/2$ (equation [4]) gives information about both the amplitude and phase of the $+n$th and $-n$th harmonics. As indicated in Section 33.2 (equations [19] and [18]), for a harmonic $A_n \sin(n\omega t + \phi_n)$ the amplitude A_n is given by:

$$A_n = \sqrt{a_n^2 + b_n^2} \qquad [14]$$

and its phase ϕ_n by:

$$\phi_n = \tan^{-1}\left(\frac{a_n}{b_n}\right) \qquad [15]$$

For the complex coefficient we can write:

$$c_n^2 = \left(\frac{a_n - \mathrm{j}b_n}{2}\right)^2 = \frac{a_n^2 + b_n^2}{4}$$

and so:

$$A_n^2 = 4c_n^2$$

$$A_n = 2|c_n| \qquad [16]$$

Thus the complex coefficient has a magnitude which is just half the amplitude. As equation [15] indicates, the phase is given by:

$$\phi_n = \tan^{-1}\left(\frac{-\text{real part of } c_n}{\text{imag. part of } c_n}\right) \qquad [17]$$

Example

A waveform gives a Fourier series with $c_n = \text{j}2/n\pi$ for all harmonics other than $n = 0$ for which $c_0 = 2$. Determine the amplitudes and phases of the constituent sinusoidal waves.

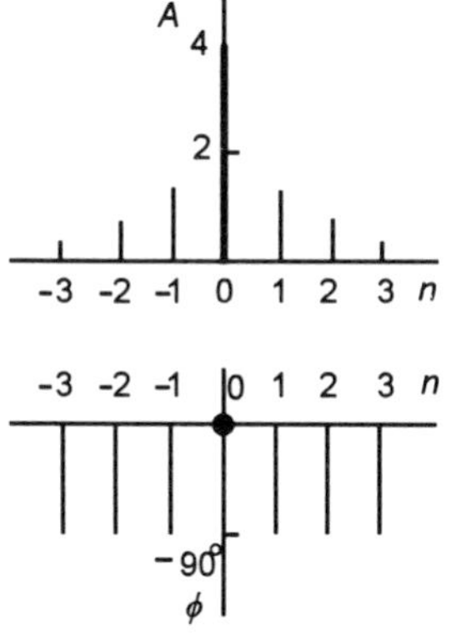

Figure 35.6 *Example*

For the d.c. component we have $A_0 = 2|c_0| = 4$. For the first harmonics we have $A_1 = 2|c_1|$ where $|c_1|$ is the square root of $|(\text{j}2/\pi)^2|$ and so $A_1 = 4/\pi$. The phases of the first harmonics are, since the real part of the complex coefficient is zero, $\tan^{-1}(-0)$ or $-90°$. For the second harmonics we have $A_2 = 2/\pi$ and phases of $\tan^{-1}(-0)$ or $-90°$. For the third harmonics we have $A_2 = 2/3\pi$ and phases of $\tan^{-1}(-0)$ or $-90°$. Figure 35.6 shows these elements of the frequency spectrum. Note that the waves occur in pairs, one for a positive value of n and one for a negative value. We thus have a *two-sided amplitude spectrum* and a *two-sided phase spectrum*. We can think of the positive n-value elements being waves produced by phasors (see Chapter 9) rotating in the conventional anticlockwise manner with the negative n-value elements being waves produced with phasors rotating clockwise. See Section 35.4.2.

Revision

3 A full-wave rectified sinusoidal waveform gives a Fourier series with:

$$c_n = -\frac{2}{(4n^2 - 1)\pi}$$

Determine the amplitudes and phases of the constituent harmonics.

35.4.1 Frequency spectrum of rectangular pulses

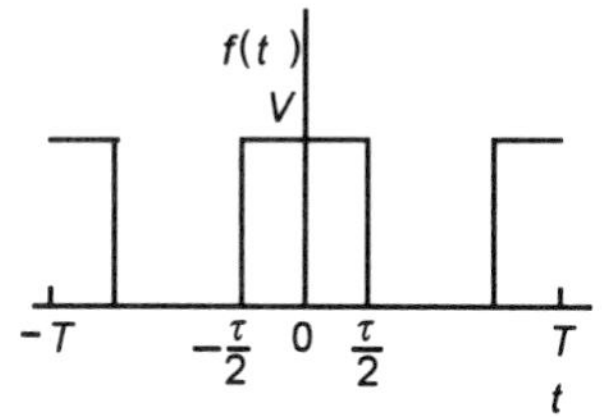

Figure 35.7 *Rectangular pulses*

A frequency spectrum which is of particular interest is that for rectangular pulses (Figure 35.7). In Section 35.2.2 the complex coefficient was derived for a rectangular pulse of width τ and period T (equation [11]) as:

$$c_n = \frac{V}{n\pi} \sin \frac{n\pi\tau}{T}$$

This is generally written as:

$$c_n = \frac{V\tau}{T} \frac{\sin \frac{n\pi\tau}{T}}{\frac{n\pi\tau}{T}} = \frac{V\tau}{T} \operatorname{sinc} \frac{n\pi\tau}{T} \qquad [18]$$

The term $(\sin x)/x$ is referred to as sinc x. At $x = 0$ we have sinc $x = 1.0$, at $x = 1$ we have sinc $x = 0$, between $x = 1.0$ and 2.0 we have negative values for sinc x, between $x = 2.0$ and $x = 3.0$ we have positive values, between $x = 3.0$ and $x = 4.0$ we have negative values, and so on.

Suppose we have the pulse height $V = 5$ and $\tau = T/5$. Then:

$$c_n = 1 \text{ sinc } \frac{n\pi}{5} \text{ or } 1\frac{\sin\frac{n\pi}{5}}{\frac{n\pi}{5}}$$

Since the amplitude $A_n = 2|c_n|$, the sinc term tells us the factor by which the amplitude of the harmonics is changed from the $n = 0$ wave. When $n = 1$ we have $c_n = 0.94$, with $n = 2$ we have $c_n = 0.76$, with $n = 3$ we have $c_n = 0.50$, with $n = 4$ we have $c_n = 0.23$, with $n = 5$ we have $c_n = 0$, with $n = 6$ we have $c_n = -0.16$, and so on. Since c_n is real with no imaginary component, we must have $\phi = \tan^{-1} \pm 0$ depending on the sign of the sine. When c_n is positive $\phi = 0°$, otherwise $\pm 180°$. Figure 35.8 shows the frequency spectrum for the first ten harmonics.

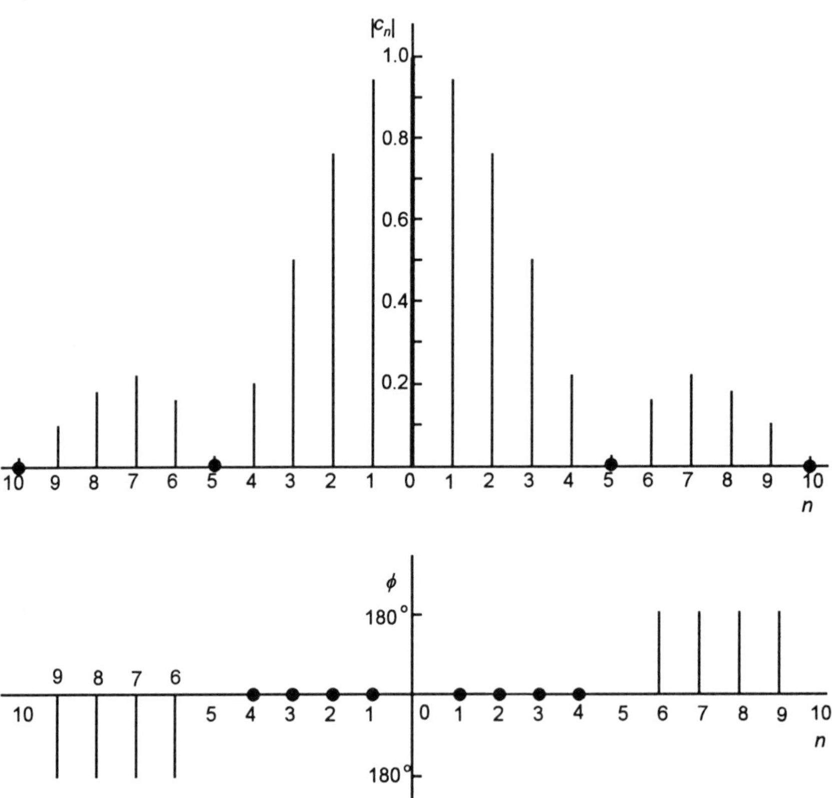

Figure 35.8 *Frequency spectrum*

35.4.2 Phasors

Consider an alternating voltage $v = V\cos(\omega t + \phi)$, where V is the maximum value and ϕ its phase. We can write this, using Euler's formula, as:

$$v = \tfrac{1}{2}V(e^{j(\omega t+\phi)} + e^{-j(\omega t+\phi)}) = \tfrac{1}{2}Ve^{j(\omega t+\phi)} + \tfrac{1}{2}Ve^{-j(\omega t+\phi)}$$

$$= \tfrac{1}{2}Ve^{j\omega t}e^{j\phi} + \tfrac{1}{2}Ve^{-j\omega t}e^{-j\phi}$$

$\tfrac{1}{2}Ve^{j\omega t}e^{j\phi}$ defines a phasor of magnitude $\tfrac{1}{2}V$ rotating anticlockwise with angular velocity ω and having a phase ϕ. Likewise $\tfrac{1}{2}Ve^{-j\omega t}e^{-j\phi}$ defines a phasor of magnitude $\tfrac{1}{2}V$ rotating with angular velocity $-\omega$, and so in a clockwise direction, and having a phase $-\phi$.

Problems

1 Determine the complex exponential form of the Fourier series for the waveforms shown in Figure 35.9

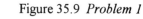

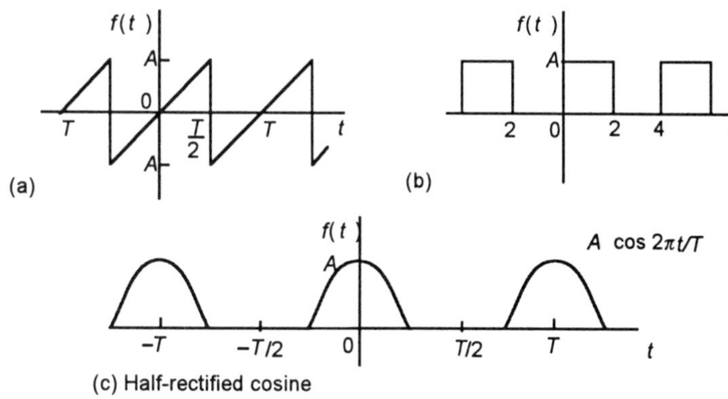

Figure 35.9 *Problem 1*

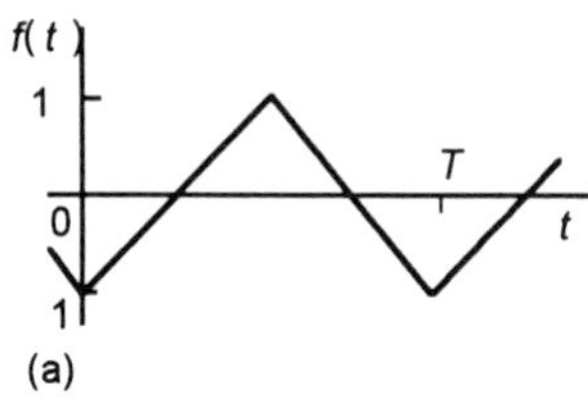

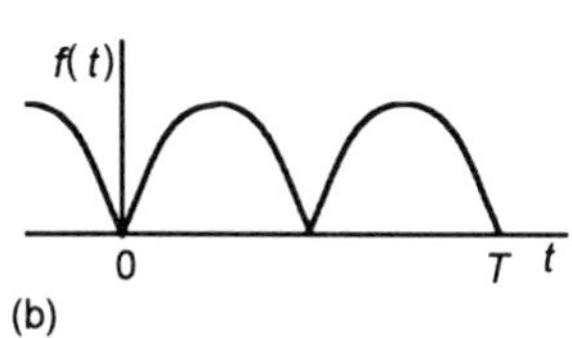

Figure 35.10 *Problem 2*

2 From a consideration of the symmetry, state which complex coefficients will be present in the Fourier series for the waveforms shown in Figure 35.10.

3 A waveform gives a Fourier series with:

$$c_n = -j\frac{3}{n\pi} \text{ for all except } n = 0,\ c_0 = -\tfrac{1}{2}$$

Determine the amplitudes and phases of the waves in the series.

4 Determine the magnitudes of c_n for the first five elements in the Fourier series for a periodic rectangular pulse train with pulses having a height of 4 and a width of quarter of the period.

36 Fourier transform

36.1 Introduction

In the previous chapters in this part, a periodic signal has been expressed as the sum of a set of sinusoidal or exponential functions, the functions being harmonically related. The frequency spectrum of such signals shows a number of discrete frequencies. When non-periodic signals were encountered, the technique adopted was to pretend that they were really were part of periodic signals (the half-range sine and half-range cosines series). In this chapter non-periodic signals are considered, but as what a periodic signal would become if the period was infinite. Thus in the entire span of time we only see one cycle of the periodic waveform and so see what appears to be a non-periodic waveform.

The Fourier series represents a waveform in terms of angular frequencies, thus we can specify a time varying signal by its Fourier series frequencies. The chapter is titled *Fourier transform*. This is because we use the techniques outlined in this chapter to transform a time varying signal into a frequency specification. The term *inverse Fourier transform* is used when we transform a signal specified by its frequencies back into a time varying signal.

This chapter assumes the earlier chapters in this part, in particular Chapter 35, and dexterity with integration and complex numbers. The chapter should only be considered an introduction to Fourier transforms.

36.2 From Fourier series to Fourier transform

Consider the Fourier series for a sequence of rectangular pulses, each pulse being of height V, duration τ and with a period of T. This was derived in Section 35.2.2, equations [11] and [18], as a series with the complex coefficient:

$$c_n = \frac{V}{n\pi} \sin \frac{n\pi\tau}{T} = \frac{V\tau}{T} \operatorname{sinc} \frac{n\pi\tau}{T} \qquad [1]$$

Figure 35.8 showed the frequency spectrum with $\tau = T/5$.

Consider the significance in general of equation [1] if have a first harmonic frequency of ω_1, Figure 36.1 showing the general form of the graph of c_n against angular frequency:

1 *Amplitudes*

c_n, and hence the amplitude, becomes zero when $\sin n\pi\tau/T = 0$. This occurs when $n\pi\tau/T = \pi$, or 2π, or 3π, etc. and so the first occurrence is when $n = T/\tau$. This gives the number of harmonics up to the first zero amplitude value. The frequency of this harmonic is $\omega = n\omega_1 = n(2\pi/T) = 2\pi/\tau$. The number of harmonics up to the second zero amplitude value is $n = 2T/\tau$ and thus its frequency will be $\omega = n\omega_1 = n(2\pi/T) = 4\pi/\tau$. Thus the frequencies at which c_n becomes zero depend only on the pulse size and are independent of the period.

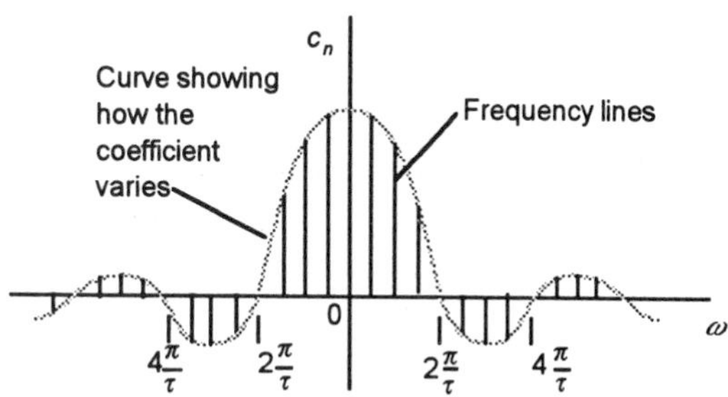

Figure 36.1 *The general form of the spectrum*

2 *Spacing of lines in frequency spectrum*
With $n = 1$ we have the first harmonic at a frequency $\omega_1 = 2\pi/T$. With $n = 2$ the second harmonic is at a frequency of $2\omega_1 = 2 \times 2\pi/T$. With $n = 3$ the third harmonic is at a frequency of $3\omega_1 = 3 \times 2\pi/T$, and so on. The frequency spacing between the harmonics is thus $2\pi/T$. Thus if the period is increased then the spacing between the frequency lines decreases. Eventually as T tends to an infinite value the lines will all merge into a continuum.

Consider what happens as the period tends to an infinite value and we end up considering just a single pulse. The frequencies at which c_n becomes zero do not change as the period changes. The envelope curve in Figure 36.1 shows how the amplitude is distributed with frequency for the lines in the spectrum and since at infinite values of T we have a continuum of frequencies, the curve shows the amplitude variation with frequency for this continuous spectrum.

36.2.1 The Fourier transform

Mathematically we can express the above general situation in the following way. The Fourier series can be expressed as (equation [8], Chapter 35):

$$f(t) = \sum_{n=-\infty}^{\infty} c_n \, \mathrm{e}^{\mathrm{j}n\omega_1 t} \qquad [2]$$

with (equation [8], Chapter 35):

$$c_n = \frac{1}{T}\int_{-T/2}^{T/2} f(t)\, \mathrm{e}^{-\mathrm{j}n\omega_1 t}\, \mathrm{d}t \qquad [3]$$

As $T \to \infty$ then equation [3] indicates that $c_n \to 0$, i.e. the amplitudes of all the frequencies tend to zero. However, if we consider c_nT we have a quantity which does not tend to 0 as $T \to \infty$. Since the amplitude spectrum becomes a continuum as $T \to \infty$ we can write ω for $n\omega_1$. Thus as $T \to \infty$ we have:

$$c_n T \to \int_{-\infty}^{+\infty} f(t)\, \mathrm{e}^{-\mathrm{j}\omega t}\, \mathrm{d}t \qquad [4]$$

The product $c_n T = c_n(2\pi/\omega)$ and is thus a function of frequency. We can therefore represent the product as $F(\omega)$ to give:

$$F(\omega) = \int_{-\infty}^{+\infty} f(t)\, \mathrm{e}^{-\mathrm{j}\omega t}\, \mathrm{d}t \qquad [5]$$

This equation transforms a function of time $f(t)$ into a function of frequency $F(\omega)$. $F(\omega)$ is said to be the *Fourier transform* of $f(t)$. The transformation can be written as:

$$F(\omega) = \mathcal{F}\{f(t)\} \qquad [6]$$

where $\mathcal{F}$ is used to indicate that the Fourier transform is being taken of $f(t)$. Note that as the integral in the defining equation [5] for the Fourier transform includes the factor $\mathrm{e}^{-\mathrm{j}\omega t}$ it may be complex.

Equation [3], explaining how c_n can be obtained from a function of time, was used to obtain the above transformation. Equation [2], explaining how a function of time can be obtained from c_n, can be used to obtain the inverse transformation. As $T \to \infty$ we can write ω for $n\omega_1$ and $c_n T \to F(\omega)$. Since $1/T = \omega_1/2\pi$ and, as $T \to \infty$ the spacing between the harmonics becomes very small and we can write $\omega_1 \to \mathrm{d}\omega$, then $1/T \to \mathrm{d}\omega/2\pi$. With such small spacing between frequencies we can replace the summation by an integral and thus obtain:

$$f(t) = \frac{1}{2\pi} \int_{-\infty}^{+\infty} F(\omega)\, \mathrm{e}^{\mathrm{j}\omega t}\, \mathrm{d}\omega \qquad [7]$$

This equation defines how a function of frequency can be converted into a function of time. Because this is the inverse Fourier transform it is written as:

$$f(t) = \mathcal{F}^{-1}\{F(\omega)\} \qquad [8]$$

$\mathcal{F}^{-1}$ is used to indicate that the inverse Fourier transform is being taken of $F(\omega)$. Equations [5] and [7] are together called the *Fourier transform pair* since they describe how the Fourier transform and the inverse Fourier transform can be obtained.

The Fourier transform is used to transform a non-periodic waveform into its frequency spectrum by using equation [5]:

$$F(\omega) = \int_{-\infty}^{+\infty} f(t)\, \mathrm{e}^{-\mathrm{j}\omega t}\, \mathrm{d}t$$

The inverse Fourier transform is used to construct a non-periodic waveform from a continuous spectrum of frequency components by using equation [7]:

$$f(t) = \frac{1}{2\pi}\int_{-\infty}^{+\infty} F(\omega)\, e^{j\omega t}\, d\omega$$

36.3 Fourier transforms

The following indicates how the Fourier transforms of some common signals can be obtained by the use of equation [5].

36.3.1 Rectangular pulse

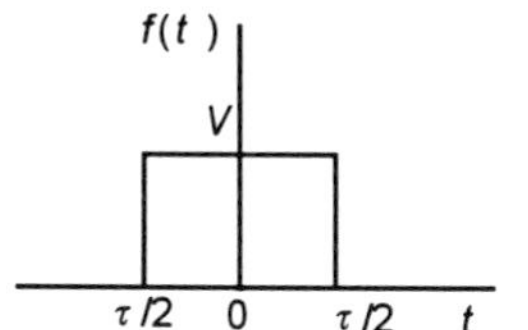

Figure 36.2 *Rectangular pulse*

Consider a rectangular pulse of height V and duration τ (Figure 36.2). At times greater than $+\tau/2$ or more negative than $-\tau/2$ the function has a value of 0. Thus in integrating from $-\infty$ to $+\infty$ we only need consider from $-\tau/2$ to $+\tau/2$. Using equation [5]:

$$F(\omega) = \int_{-\infty}^{+\infty} f(t)\, e^{-j\omega t}\, dt = \int_{-\tau/2}^{+\tau/2} V\, e^{-j\omega t}\, d\omega = V\left[-\frac{e^{-j\omega t}}{j\omega}\right]_{-\tau/2}^{+\tau/2}$$

$$= -\frac{V}{j\omega}(e^{-j\omega\tau/2} - e^{j\omega\tau/2}) = \frac{2V}{\omega}\sin\frac{\omega\tau}{2} = V\tau\frac{\sin\frac{\omega\tau}{2}}{\frac{\omega\tau}{2}} \qquad [9]$$

This equation is similar to equation [1] for c_n for a train of rectangular pulses and so will give a graph of the same form as the envelope curve in Figure 36.1. A graph of $F(\omega)$ plotted against ω is shown in Figure 36.3. If a rectangular pulse is an input to an electronic system then we can consider a continuum of frequencies to be inputted. However, since the amplitudes of the frequencies are much larger between angular frequencies of plus and minus $2\pi/\tau$, we need to have an electronic system which can transmit frequencies between these limits, i.e. between $f = \pm(2\pi/\tau)/2\pi = \pm 1/\tau$. The narrower the pulse the smaller τ and so the greater the bandwidth required of the system to keep distortion low.

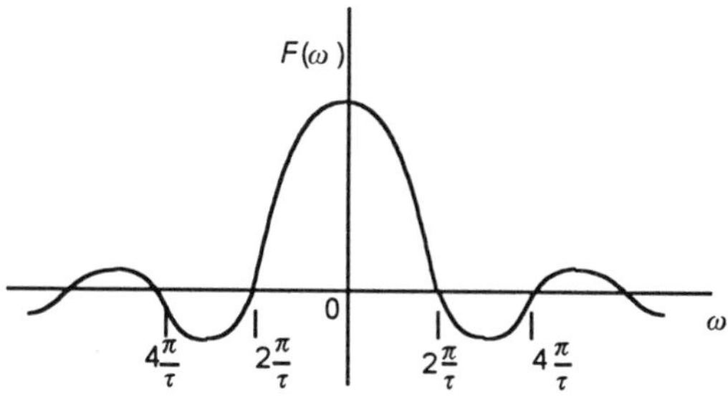

Figure 36.3 *Fourier transform for a rectangular pulse*

Revision

1 A rectangular pulse has a duration of 1 ms. Estimate the bandwidth required of an electronic system if the pulse is the input and distortion is to be kept to a minimum.

36.3.2 Positive time exponential

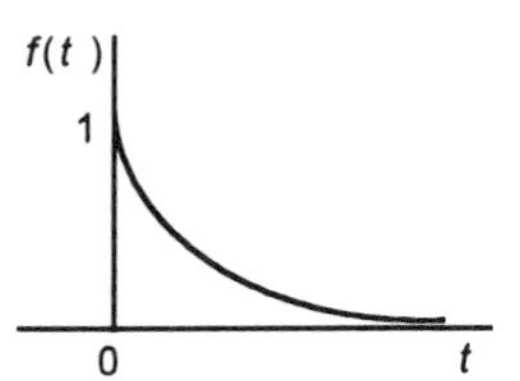

Figure 36.4 *Positive time exponential*

Consider an exponential signal of height 1 at $t = 0$ and which decays with time (Figure 36.4). We have $f(t) = e^{-at}$ for times greater than 0 and $f(t) = 0$ for times less than 0. This can be represented by $u(t)\ e^{-at}$ (see Section 6.3.2). Thus equation [5] gives:

$$F(\omega) = \int_{-\infty}^{+\infty} f(t)\ e^{-j\omega t}\ dt = \int_{0}^{+\infty} e^{-at}\ e^{-j\omega t}\ d\omega = \int_{0}^{+\infty} e^{-(a+j\omega)t}\ d\omega$$

$$= -\frac{1}{a+j\omega}[e^{-(a+j\omega)t}]_0^{\infty} = \frac{1}{a+j\omega} \qquad [10]$$

This can be written as:

$$F(\omega) = \frac{1}{a+j\omega} \times \frac{a-j\omega}{a-j\omega} = \frac{a-j\omega}{a^2+\omega^2}$$

The Fourier transform has thus a real term and an imaginary term. We can write it in the form:

$$F(\omega) = A(\omega) - jB(\omega)$$

The magnitude or modulus $|F(\omega)|$ of such a complex number (see Section 7.2.5) is:

$$|F(\omega)| = \sqrt{[A(\omega)]^2 + [B(\omega)]^2} \qquad [11]$$

and its phase $\phi(\omega)$:

$$\phi(\omega) = \tan^{-1}\frac{B(\omega)}{A(\omega)} \qquad [12]$$

Thus for the exponential we have:

$$|F(\omega)| = \sqrt{\left(\frac{a}{a^2+\omega^2}\right)^2 + \left(\frac{\omega}{a^2+\omega^2}\right)^2} = \sqrt{\frac{a^2+\omega^2}{(a^2+\omega^2)^2}} = \frac{1}{\sqrt{a^2+\omega^2}}$$

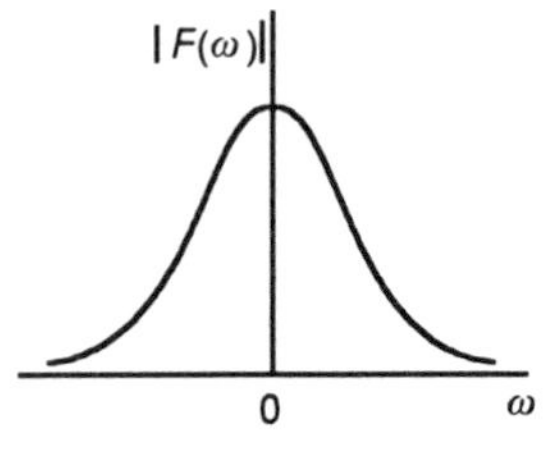

Figure 36.5 *Spectrum*

Figure 36.5 shows a graph of how the magnitude varies with angular frequency. The phase varies across the spectrum, being $\tan^{-1}(\omega/a)$.

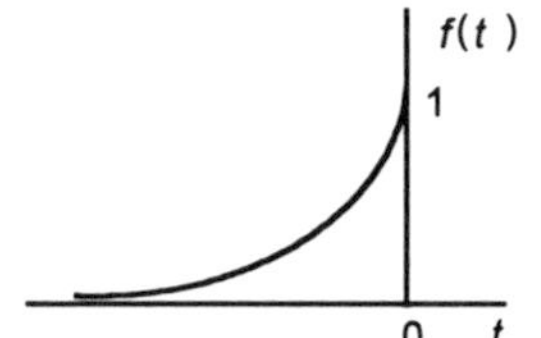

Figure 36.6 *Negative time exponential*

Revision

2 Determine the Fourier transform of the negative time exponential signal (Figure 36.6), i.e. an exponential signal of height 1 at t = 0 and for which $f(t) = e^{-at}$ for times less than 0, i.e. negative values of t, and $f(t)$ = 0 for times greater than 0.

36.3.3 Impulse

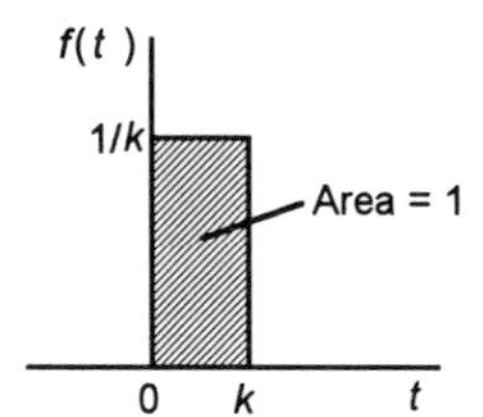

Figure 36.7 *Unit-area pulse*

We approach the consideration of an impulse by first considering a rectangular pulse (Figure 36.7) of width k and height $1/k$. This means it has an area of 1. To maintain this constant area as we reduce the pulse width, we increase the pulse height. In the limit when $k \to 0$ we have a vertical line at t = 0, the height of the line going off to infinity. Such a line is termed a unit strength impulse because the area enclosed is 1. This function is represented by $\delta(t)$ and is termed the *unit-impulse function* or the *Dirac-delta function*.

The Fourier transform for the impulse function (Figure 36.8) is thus given by equation [5], since it only exists at t = 0 and the integral limits are thus from one side of 0 to the other and indicated as 0+ and 0–, as:

$$F(\omega) = \int_{-\infty}^{+\infty} \delta(t)\, e^{-j\omega t}\, dt = \int_{0-}^{0+} \delta(t)\, e^{-j\omega t}\, dt$$

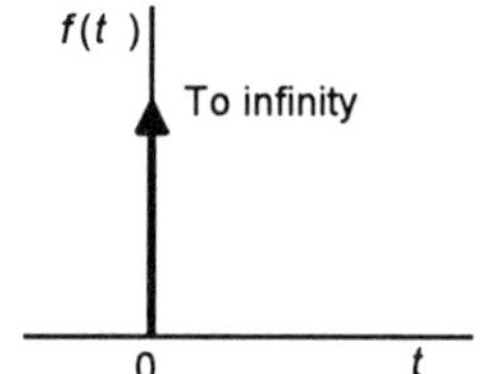

Figure 36.8 *The unit impulse*

Since we only have a signal at t = 0, $e^{-j\omega t}$ must equal e^0 = 1. Thus:

$$F(\omega) = \int_{0-}^{0+} \delta(t) 1\, d\omega$$

The integral of $\delta(t)$ over time is the area under the impulse function. Since this is 1 we have $F(\omega)$ = 1 and thus:

$$\mathcal{F}\{\delta(t)\} = 1 \qquad [13]$$

The magnitude of this transform $|F(\omega)|$ = 1. Thus the impulse function contains all frequencies with the same magnitude.

36.4 Properties of Fourier transforms

The following are some of the basic properties of Fourier transforms, such properties enabling transforms to be obtained for a range of functions from just a few standard transforms.

1 *Linearity*

Suppose that we have a function which is the sum of two other functions $f(t)$ and $g(t)$. Then:

$$\mathcal{F}\{f(t) + g(t)\} = \int_{-\infty}^{+\infty} [f(t) + g(t)]\, e^{-j\omega t}\, dt$$

$$= \int_{-\infty}^{+\infty} f(t)\,\mathrm{e}^{-\mathrm{j}\omega t}\,\mathrm{d}t + \int_{-\infty}^{+\infty} g(t)\,\mathrm{e}^{-\mathrm{j}\omega t}\,\mathrm{d}t$$

$$= \mathcal{F}\{f(t)\} + \mathcal{F}\{g(t)\} \qquad [14]$$

The Fourier transform of a sum of two functions is the sum of the Fourier transforms of the two functions considered separately.

Example

Determine the Fourier transform of the function $\mathrm{e}^{-t} + \mathrm{e}^{-2t}$ which starts at $t = 0$ and extends from $t = 0$ to $+\infty$. Note that this can be written as $u(t)\,\mathrm{e}^{-t} + u(t)\,\mathrm{e}^{-2t}$ (see Section 6.3.2).

Using equation [5]:

$$F(\omega) = \int_{-\infty}^{+\infty} f(t)\,\mathrm{e}^{-\mathrm{j}\omega t}\,\mathrm{d}t = \int_{0}^{+\infty} (\mathrm{e}^{-t} + \mathrm{e}^{-2t})\,\mathrm{e}^{-\mathrm{j}\omega t}\,\mathrm{d}t$$

$$= \int_{0}^{+\infty} \mathrm{e}^{-t}\,\mathrm{e}^{-\mathrm{j}\omega t}\,\mathrm{d}t + \int \mathrm{e}^{-2t}\,\mathrm{e}^{-\mathrm{j}\omega t}\,\mathrm{d}t$$

$$= \left[\frac{\mathrm{e}^{-(1+\mathrm{j}\omega)t}}{-(1+\mathrm{j}\omega)}\right]_{0}^{+\infty} + \left[\frac{\mathrm{e}^{-(2+\mathrm{j}\omega)t}}{-(2+\mathrm{j}\omega)}\right]_{0}^{+\infty} = \frac{1}{1+\mathrm{j}\omega} + \frac{1}{2+\mathrm{j}\omega}$$

2 *Multiplication by a constant*
Suppose we have a function which is multiplied by a constant K:

$$\mathcal{F}\{Kf(t)\} = \int_{-\infty}^{+\infty} Kf(t)\,\mathrm{e}^{-\mathrm{j}\omega t}\,\mathrm{d}t = K\int_{-\infty}^{+\infty} f(t)\,\mathrm{e}^{-\mathrm{j}\omega t}\,\mathrm{d}t$$

Thus if $\mathcal{F}\{f(t)\} = F(\omega)$ then:

$$\mathcal{F}\{Kf(t)\} = KF(\omega) \qquad [15]$$

Example

Determine the Fourier transform of an impulse of strength 2π.

This is the Fourier transform of $2\pi\ \delta(t)$ and is 2π times the Fourier transform for a unit impulse. It is thus 2π.

3 *Time scaling*
Consider the Fourier transform for a pulse when the time scale is changed, i.e. the transform for $f(at)$, where a is a real constant.

$$\mathcal{F}\{f(at)\} = \int_{-\infty}^{+\infty} f(at)\,\mathrm{e}^{-\mathrm{j}\omega t}\,\mathrm{d}t$$

If we let $x = at$ then $t = x/a$ and $\mathrm{d}t = \mathrm{d}x/a$. Thus:

$$\mathcal{F}\{f(at)\} = \frac{1}{a}\int_{-\infty}^{+\infty} f(x)\,\mathrm{e}^{-\mathrm{j}(\omega/a)x}\,\mathrm{d}x$$

This is $1/a$ times the integral we would have obtained if we had replaced ω by ω/a. Thus if $\mathcal{F}\{f(t)\} = F(\omega)$ then:

$$\mathcal{F}\{f(at)\} = \frac{1}{|a|}F\left(\frac{\omega}{a}\right) \qquad [16]$$

Thus if the time scale is expanded by some factor, i.e. $a > 1$, then the frequency scale is shortened by the same factor. If the duration of a pulse is reduced, i.e. $a < 1$, then the frequency scale is extended by the same factor. Note that the amplitude is also scaled by the same factor. If a is -1 then the above equation gives

$$\mathcal{F}\{f(-t)\} = \mathcal{F}(-\omega) \qquad [17]$$

Thus the Fourier transform of a function with negative times is just the Fourier transform of the function with negative frequencies.

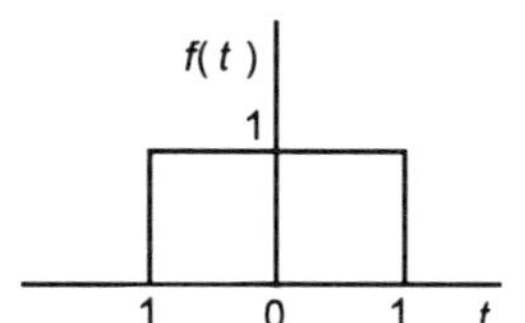

Figure 36.9 *Example*

Example

If the Fourier transform of a rectangular pulse of width 1 centred on $t = 0$ is given by $(\sin \omega/2)/(\omega/2)$, what will be the transform of a rectangular pulse of width 2 (Figure 36.9)?

A width of 2 means that if we drew a graph of the width 1 pulse we would have to divide the scale by 2 to give the width 2 pulse. Using equation [16] with $a = 1/2$, the Fourier transform is $2(\sin \omega)/\omega$.

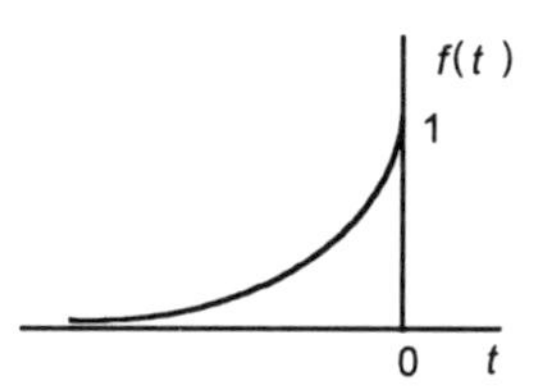

Figure 36.10 *Negative tim exponential*

Example

In Section 36.3.2 the Fourier transform for a positive time exponential $u(t)\ e^{-at}$ was derived as (equation [10]) $1/(a + j\omega)$. Derive the transform for the corresponding negative time exponential (Figure 36.10).

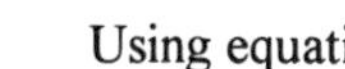

Using equation [17], the Fourier transform will be $1/(a - j\omega)$.

4 *Time shifting*
The Fourier transform when a signal is delayed by some time t_0 is:

$$\mathcal{F}\{f(t - t_0)\} = \int_{-\infty}^{+\infty} f(t - t_0)\ e^{-j\omega t}\ dt$$

We can write this as:

$$\mathcal{F}\{f(t - t_0)\} = e^{-j\omega_0 t} \int_{-\infty}^{+\infty} f(t - t_0)\ e^{-j\omega(t-t_0)}\ dt$$

The integral is just the Fourier transform we would have obtained for $f(t)$ if we had changed the time scale by t_0. Thus if $F(\omega)$ is the Fourier transform of $f(t)$:

$$\mathcal{F}\{f(t - t_0)\} = e^{-j\omega_0 t} F(\omega) \qquad [18]$$

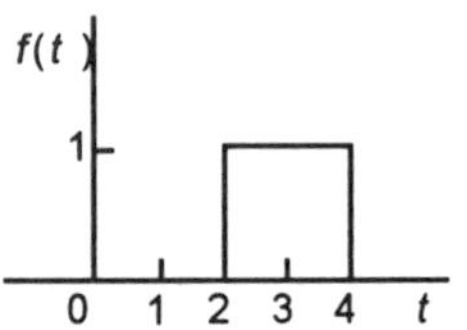

Figure 36.11 *Example*

Example

Determine the Fourier transform of a unit height rectangular pulse of width 2 which starts at $t = 2$ (Figure 36.11) if the transform for a unit height rectangular pulse of width τ centred on $t = 0$ is given by $(\tau \sin \omega\tau/2)/(\omega\tau/2)$.

For a pulse of width 2 centred on $t = 0$ we have a Fourier transform of $(2 \sin \omega)/\omega$. It is now to be centred on $t = 3$. Thus the Fourier transform of the delayed pulse is given by equation [18] as $(2\ e^{-j3\omega} \sin \omega)/\omega$.

5 *Frequency shifting*
Consider the Fourier transform of a product of $e^{j\omega_0 t}$ and a function of time $f(t)$:

$$\mathcal{F}\{e^{j\omega_0 t}f(t)\} = \int_{-\infty}^{+\infty} e^{j\omega_0 t}f(t)\ e^{-j\omega t}\ dt = \int_{-\infty}^{+\infty} f(t)\ e^{-j(\omega-\omega_0)t}\ dt$$

This is, however, just the Fourier transform with a frequency shift of ω_0. Thus:

$$\mathcal{F}\{e^{j\omega_0 t}f(t)\} = F(\omega - \omega_0) \qquad [19]$$

Example

With amplitude modulation the amplitude of a signal is used to control the amplitude of a sinusoidal carrier. If the sinusoidal carrier is $\cos \omega_c t$ and the signal is $f(t)$, the amplitude of the modulated signal is given by $f(t) \cos \omega_c t$. Determine the Fourier transform of this signal.

With the cosine written in exponential form by the use of Euler's formula:

$$\mathcal{F}\{f(t) \cos z_c t\} = \mathcal{F}\{f(t)\tfrac{1}{2}(e^{j\omega_c t} + e^{-j\omega_c t})\}$$

$$= \mathcal{F}\{f(t)\tfrac{1}{2}e^{j\omega_c t}\} + \mathcal{F}\{f(t)\tfrac{1}{2}e^{-j\omega_c t}\}$$

If $\mathcal{F}\{f(t)\} = F(\omega)$ then, using equation [19]:

$$\mathcal{F}\{f(t) \cos \omega_c t\} = \tfrac{1}{2}F(\omega - \omega_c) + \tfrac{1}{2}F(\omega + \omega_c)$$

The result of the amplitude modulation is thus the signal spectrum of $f(t)$ centred on the frequencies of $+\omega_c$ and $-\omega_c$ (Figure 36.12).

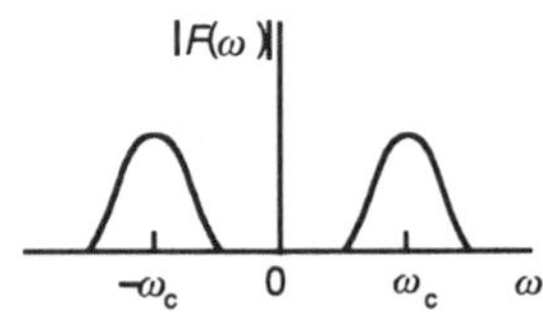

Figure 36.12 *Example*

6 *Even and odd symmetries*
Consider the effect of whether the symmetry of a function $f(t)$ is even or odd on the form of the Fourier transform of the function. The Fourier transform can be expressed, using Euler's formula, as the sum of a real part and an imaginary part:

$$F(\omega) = \int_{-\infty}^{+\infty} f(t)\, e^{-j\omega t}\, dt = \int_{-\infty}^{+\infty} f(t)\left[\cos \omega t - j \sin \omega t\right] dt$$

$$= \int_{-\infty}^{+\infty} f(t) \cos \omega t\, dt - \int_{-\infty}^{+\infty} f(t)\, j \sin \omega t\, dt$$

If the function is even, the integral with the sine is zero and thus the Fourier transform is real. Such a function gives rise to a purely cosine Fourier series. If the function is odd, the integral with the cosine has a zero value and thus the Fourier transform is imaginary. Such a function gives rise to a purely sine Fourier series. If the function is neither even or odd then the Fourier transform has both real and imaginary parts.

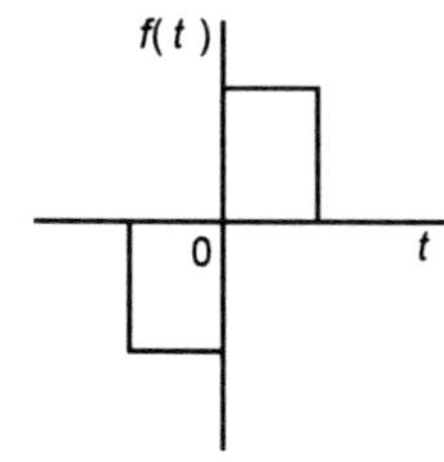

Figure 36.13 *Example*

Example

Will the function in Figure 36.13 have a real, imaginary or complex Fourier transform?

The function is odd and so the transform will be purely imaginary.

7 *Duality/symmetry property*
If you consider the equation for the Fourier transform of a function (equation [5]) and the equation for the inverse Fourier transform (equation [7]), there is obviously some symmetry of structure in the two. The following shows how we can establish the form of the symmetry. Equation [7] gives for the inverse transform:

$$f(t) = \frac{1}{2\pi} \int_{-\infty}^{+\infty} F(\omega)\, e^{j\omega t}\, d\omega \qquad [19]$$

If we replace ω by z then we can write:

$$2\pi f(t) = \int_{-\infty}^{+\infty} F(z)\, e^{jzt}\, dz$$

If we now replace t by $-\omega$ we obtain:

$$2\pi f(-\omega) = \int_{-\infty}^{+\infty} F(z)\, e^{-j\omega z}\, dz$$

The right-hand side of the above equation is the Fourier transform for $F(z)$ (equation [5]), or if we replace z by t the Fourier transform for $F(t)$. Thus:

$$\mathcal{F}\{F(t)\} = 2\pi f(-\omega) \qquad [21]$$

If $F(\omega)$ is the Fourier transform of $f(t)$ then $2\pi f(-\omega)$ is the Fourier transform of $F(t)$. This is known as the *duality* or *symmetry principle*.

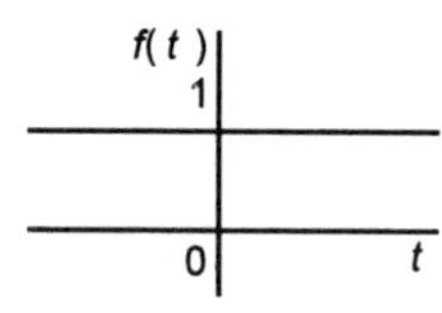

Figure 36.14 *Example*

Example

Determine the Fourier transform of a constant unit strength signal (Figure 36.14).

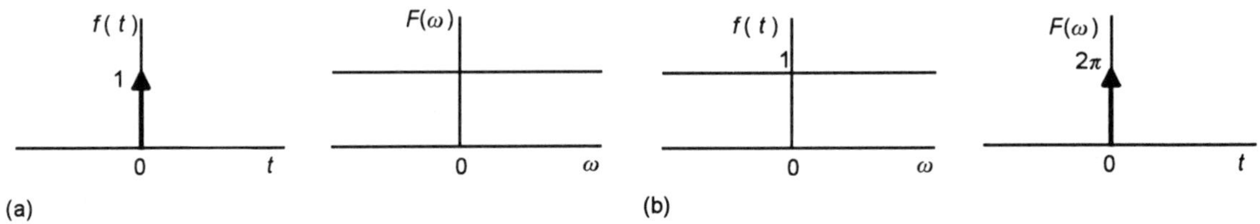

Figure 36.15 *(a) Impulse and its transform, (b) constant signal and its transform*

The Fourier transform of a unit strength impulse is 1, i.e. $\mathcal{F}\{\delta(t)\} = 1$. Then $\delta(-\omega) = (1/2\pi)\,\mathcal{F}(1)$ and so $\mathcal{F}(1) = 2\pi\,\delta(-\omega)$. Since $\delta(-\omega) = \delta(\omega)$ we can write $\mathcal{F}(1) = 2\pi\,\delta(\omega)$. It has thus a spectrum of a single impulse of strength 2π at $\omega = 0$. Figure 36.15 illustrates this duality.

Revision

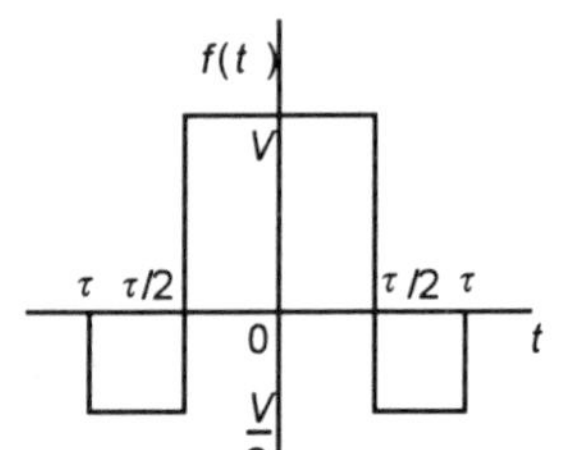

Figure 36.16 *Example*

3 Determine the Fourier transforms of the following functions using the methods indicated:

(a) The pulse shown in Figure 36.16 using scaling, time delayed pulses and linearity. Use Euler's formula to tidy the answer.

(b) The positive time exponential function which is delayed by a time of 4, i.e. $e^{-(t-4)}u(t-4)$ using time shifting.

(c) The complex exponential $e^{j\omega_0 t}$, using duality with the time delayed impulse.

(d) A function for which $f(t) = e^{j3}$ for $|t| \leq 1$ and $f(t) = 0$ for $|t| > 1$, using frequency shifting.

36.5 More Fourier transforms

There are a number of Fourier transforms which can be determined by considering the limiting values of functions. Consider the function shown in Figure 36.17, the function being termed the *signum function* and written as sgn(t). This has $f(t) = 1$ for $t > 0$ and $f(t) = -1$ for $t < 0$. It can be expressed in terms of unit step functions (see Section 6.3.2) as:

$$f(t) = u(t) - u(-t)$$

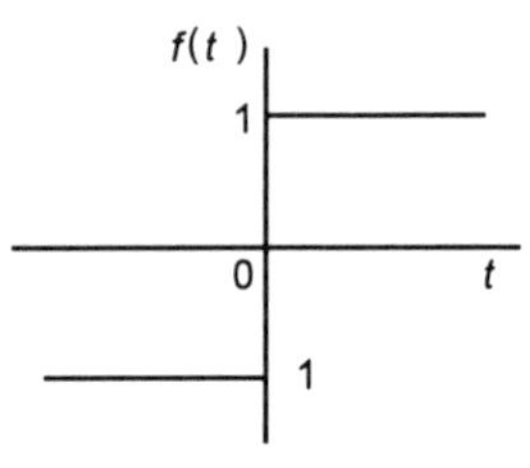

Figure 36.17 *Signum function*

To determine the Fourier transform we consider a function that approaches the signum function in the limit. This we can achieve by considering two exponentials (Figure 36.18). As a tends to 0 so the functions tend to give the signum function:

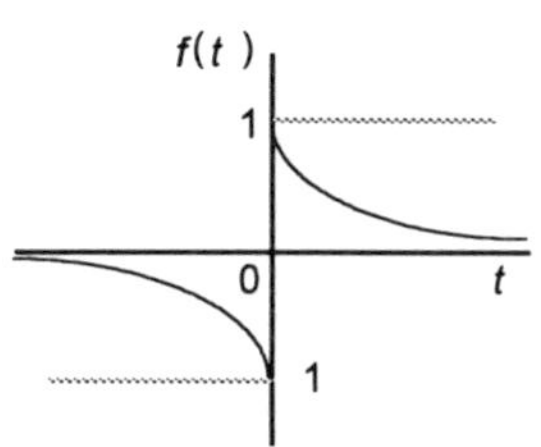

Figure 36.18 *Signum function as the limit*

$$f(t) = \lim_{a \to 0} \left[e^{-at}u(t) - e^{at}u(-t) \right]$$

The Fourier transform of the function composed of the two exponentials is:

$$\mathcal{F}\{f(t)\} = \frac{1}{a + j\omega} - \frac{1}{a - j\omega} = -\frac{2j\omega}{a^2 + \omega^2}$$

In the limit as $a \to 0$, $f(t)$ tends to the signum function and the Fourier transform tends to $-2j\omega/\omega^2 = 2/j\omega$. Thus the Fourier transform of the signum pulse is:

$$\mathcal{F}\{\text{sgn}(t)\} = \frac{2}{j\omega} \qquad [22]$$

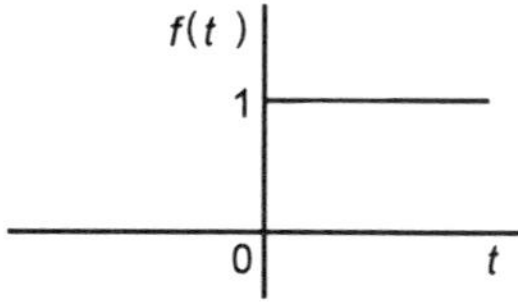

Figure 36.19 *Unit step*

We can use the above result to determine the Fourier transform of a unit step function (Figure 36.19). We can consider such a function to be the sum of two functions, one of ½ and the other ½ a signum function (Figure 36.20):

$$u(t) = \tfrac{1}{2} + \tfrac{1}{2}\text{sgn}(t)$$

Thus:

$$\mathcal{F}\{u(t)\} = \mathcal{F}\left(\tfrac{1}{2}\right) + \mathcal{F}\left(\tfrac{1}{2}\text{sgn}(t)\right)$$

$$= \pi\delta(\omega) + \frac{1}{j\omega} \qquad [23]$$

Revision

4 Show by considering the function $f(t) = A$ as the limit of two exponentials that the Fourier transform of $f(t)$ is $2\pi A\,\delta(\omega)$.

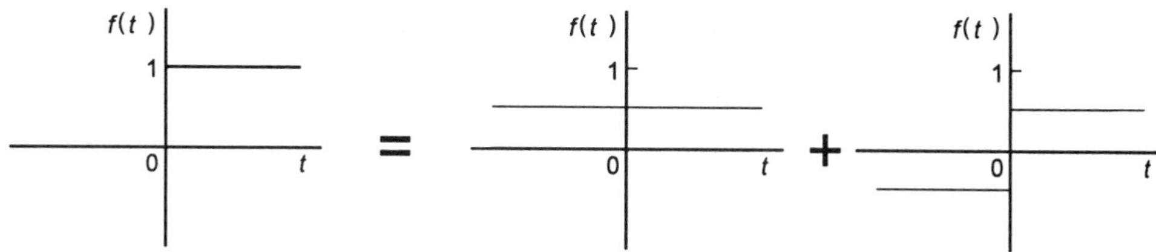

Figure 36.20 *Unit step as the sum of two functions*

36.5.1 Fourier transforms of periodic functions

Consider a periodic signal $f(t) = \cos at$. Using Euler's formula:

$$\cos at = \tfrac{1}{2}(e^{jat} + e^{-jat})$$

Hence if we take the Fourier transform we obtain:

$$\mathcal{F}\{\cos at\} = \mathcal{F}\{\tfrac{1}{2}(e^{jat} + e^{-jat})\}$$

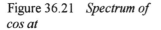

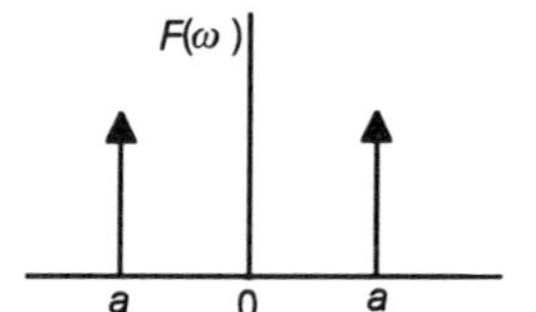

Figure 36.21 *Spectrum of cos at*

The Fourier transform of a complex exponential (see Revision problem 3(c)) is $\mathcal{F}\{e^{jat}\} = 2\pi\,\delta(\omega - a)$ and $\mathcal{F}\{e^{-jat}\} = 2\pi\,\delta(\omega + a)$. Thus:

$$\mathcal{F}\{\cos at\} = \pi\,\delta(\omega - a) + \pi\,\delta(\omega + a) \qquad [24]$$

The spectrum of cos at thus consists of single frequency lines at $\omega = +a$ and $\omega = -a$ (Figure 36.21).

Revision

5 Determine the Fourier transform of sin at.

36.5.2 Standard Fourier transforms

Table 36.1 lists commonly encountered functions and their Fourier transforms.

Table 36.1 *Fourier transforms*

	$F(t)$	$F(\omega)$
1	Impulse: $\delta(t)$	1
2	Constant: A	$2A\pi\,\delta(\omega)$
3	Pulse of width τ centred on $t = 0$: $u(t + \tau/2) - u(t - \tau/2)$	$\tau\dfrac{\sin \omega\tau/2}{\omega\tau/2}$
4	Unit step: $u(t)$	$\pi\,\delta(\omega) + \dfrac{1}{j\omega}$
5	Signum function: $u(t) - u(-t)$	$\dfrac{2}{j\omega}$
6	Positive time exponential: $e^{-at}u(t)$	$\dfrac{1}{a + j\omega}$
7	Negative time exponential: $e^{at}u(-t)$	$\dfrac{1}{a - j\omega}$
8	Complex exponential: e^{jat}	$2\pi\,\delta(\omega - a)$
9	Cosine: $\cos at$	$\pi[\delta(\omega + a) + \delta(\omega - a)]$
10	Sine: $\sin at$	$j\pi[\delta(\omega + a) - \delta(\omega - a)]$

36.6 Differentiation

In this section the formulas are developed that are needed for use in determining the Fourier transform of differential equations.

A function $f(t)$ has the Fourier transform:

$$\mathcal{F}\{f(t)\} = F(\omega) = \int_{-\infty}^{+\infty} f(t)\, e^{-j\omega t}\, dt$$

Consider the derivative of the function with respect to time. We can write for the Fourier transform of the derivative:

$$\mathcal{F}\left\{\frac{df(t)}{dt}\right\} = \int_{-\infty}^{+\infty} \frac{df(t)}{dt}\, e^{-j\omega t}\, dt$$

Using integration by parts with $u = e^{-j\omega t}$ and $dv = (df(t)/dt)\, dt$:

$$\mathcal{F}\left\{\frac{df(t)}{dt}\right\} = e^{-j\omega t}\big[f(t)\big]_{-\infty}^{+\infty} - \int_{-\infty}^{+\infty} f(t)(-j\omega\, e^{-j\omega t})\, dt$$

We can assume that $\lim_{t\to\infty} f(t) = \lim_{t\to-\infty} f(t) = 0$, thus:

$$\mathcal{F}\left\{\frac{df(t)}{dt}\right\} = j\omega \int_{-\infty}^{+\infty} f(t)\, e^{-j\omega t}\, dt = j\omega F(\omega) \qquad [25]$$

For second derivatives we can repeat the above to obtain:

$$\mathcal{F}\left\{\frac{d^2 f(t)}{dt^2}\right\} = (j\omega)^2 F(\omega) \qquad [26]$$

In general:

$$\mathcal{F}\left\{\frac{d^n f(t)}{dt^n}\right\} = (j\omega)^n F(\omega) \qquad [27]$$

The transform of a differentiated function is just the Fourier transform of that function multiplied by $j\omega$.

Example

Solve the differential equation $dy/dt - 4t = u(t)\, e^{-4t}$ when t lies between $-\infty$ and $+\infty$. You can think of the differential equation describing a first-order system to which there has been an input signal of $u(t)\, e^{-4t}$.

Taking the Fourier transform of the equation:

$$\mathcal{F}\{dy/dt\} - \mathcal{F}\{4y\} = \mathcal{F}\{u(t)\, e^{-4t}\}$$

Writing $F(\omega)$ for $\mathcal{F}\{y\}$:

$$j\omega F(\omega) - 4F(\omega) = \frac{1}{4 + j\omega}$$

Thus:

$$F(\omega) = \frac{-1}{(4 + \mathrm{j}\omega)(4 - \mathrm{j}\omega)} = -\frac{1}{4^2 + \omega^2}$$

Taking the inverse of this transform gives:

$$y = -\tfrac{1}{8}\,\mathrm{e}^{-4t}$$

Revision

6 Solve the differential equation $\mathrm{d}y/\mathrm{d}t - 4t = \delta(t)$.

36.6.1 More Fourier transforms

A useful application of differentiation is to obtain Fourier transforms of functions.

Example

Determine the Fourier transform of $f(t) = t\,\mathrm{e}^{-at}$.

Differentiating the function gives:

$$\frac{\mathrm{d}f(t)}{\mathrm{d}t} = -at\,\mathrm{e}^{-at} + \mathrm{e}^{-at} = -af(t) + \mathrm{e}^{-at}$$

Taking the Fourier transform of the equation:

$$\mathrm{j}\omega F(\omega) = --aF(\omega) + \frac{1}{a + \mathrm{j}\omega}$$

Hence:

$$F(\omega) = \frac{1}{(a + \mathrm{j}\omega)^2}$$

Revision

7 Determine, by differentiating the function, the Fourier transform of $f(t) = t^2\,\mathrm{e}^{-at}$.

36.7 Transfer function

Consider a system with an input which is a function of time $x(t)$ and an output $y(t)$ which is also a function of time and for which the input and output are related by a differential equation. For example, for a first-order system we might have:

$$a_1 \frac{dy(t)}{dt} + a_0 y(t) = bx(t)$$

where a_1, a_0 and b are constants. Taking the Fourier transform of the equation and using equation [25]:

$$j\omega a_1 Y(\omega) + a_0 Y(\omega) = bX(\omega)$$

where $Y(\omega)$ is $\mathcal{F}\{y(t)\}$ and $X(\omega)$ is $\mathcal{F}\{x(t)\}$. This equation can be written as:

$$Y(\omega) = \left(\frac{b}{j\omega a_1 + a_0}\right) X(\omega)$$

The term in the brackets is called the *transfer function* $H(\omega)$. It is the function which relates the Fourier transforms of the input and output.

$$Y(\omega) = H(\omega)X(\omega) \qquad [28]$$

Thus if we know the transfer function of a system and the transform of the input, we can compute the transform of the output. By then taking the inverse transform we can obtain the output as a function of time.

To find the response of a system to some input:

1 Find the Fourier transform of the input.

2 Multiply the Fourier transform of the input by the transfer function of the system.

3 Take the inverse Fourier transform of the result.

If the input to a system is a unit strength impulse then the Fourier transform of the input $X(\omega)$ is 1 and $Y(\omega) = H(\omega)$. Thus the transfer function is equal to the Fourier transform of the output of a system when there is a unit strength impulse input.

Example

Determine the response of a system to a unit strength impulse if it has a transfer function of $1/(1 + j\omega)$.

The input has a Fourier transform of 1. Thus the output has a transform of:

$$Y(\omega) = H(\omega)X(\omega) = \frac{1}{1 + j\omega} 1 = \frac{1}{1 + j\omega}$$

The inverse transform thus gives:

$$y(t) = e^{-t} u(t)$$

Example

Determine the response of a system to a unit step input if it has a response to an impulse input of 2 $e^{-t}u(t)$.

The unit step input, i.e. $u(t)$, has the Fourier transform of $\pi\,\delta(\omega) + 1/j\omega$. The transfer function of the system is the Fourier transform of its response to an impulse. Thus:

$$H(\omega) = \mathcal{F}\{2e^{-t}u(t)\} = \frac{2}{1 + j\omega}$$

and the Fourier transform of the output is:

$$Y(\omega) = H(\omega)X(\omega) = \frac{2}{1 + j\omega}\left[\pi\,\delta(\omega) + \frac{1}{j\omega}\right]$$

$$= \frac{2\pi}{1 + j\omega}\delta(\omega) + \frac{2}{j\omega(1 + j\omega)}$$

Since the multiplication of any function of frequency by an impulse just gives an impulse, we can write $2\pi\,\delta(\omega)$ for the first term above. The inverse transform of this is 1. The second term can be expanded by the use of partial fractions (see Section 2.6.1) to give the fractions $(2/j\omega) - 2/(1 + j\omega)$. This has the inverse transform $\text{sgn}(t) - 2\,e^{-t}u(t)$. Thus the output is:

$$y = 1 + \text{sgn}(t) - 2\,e^{-t}u(t) = 2u(t) - 2\,e^{-t}u(t) = 2(1 - e^{-t})u(t)$$

Revision

8 Determine the response of a system to an input of 20 $\text{sgn}(t)$ if it has a transfer function of $1/(4 + j\omega)$.

9 Determine the response of a system to a unit strength impulse if it has a transfer function of $2/(1 + j\omega)$.

10 Determine the response of a system to an input of $4 - 4u(t)$ if it has a transfer function of $-10/(4 + j\omega)$.

36.7.1 Convolution

With the variable t changed to τ to avoid later confusion, equation [5] gives:

$$X(\omega) = \int_{-\infty}^{+\infty} x(t)\,e^{-j\omega t}\,dt = \int_{-\infty}^{+\infty} x(\tau)\,e^{-j\omega\tau}\,d\tau$$

If we have $Y(\omega) = H(\omega)X(\omega)$ (equation [28]), then:

$$Y(\omega) = H(\omega)\int_{-\infty}^{+\infty} x(\tau)\,e^{-j\omega\tau}\,d\tau = \int_{-\infty}^{+\infty} x(\tau)\,e^{-j\omega\tau}H(\omega)\,d\tau$$

$$= \int_{-\infty}^{+\infty} x(\tau)\,e^{-j\omega\tau}\left[\int_{-\infty}^{+\infty} h(\lambda)\,e^{-j\omega\lambda}\,d\lambda\right] d\tau$$

$$= \int_{-\infty}^{+\infty} x(\tau)\left[\int_{-\infty}^{+\infty} h(\lambda)\,e^{-j\omega(\lambda+\tau)}\,d\lambda\right] d\tau$$

Let $t = \lambda + \tau$, then for the inner integral $dt = d\lambda$ and when $\lambda = \infty$ we have $t = \infty$, when $\lambda = -\infty$ we have $t = -\infty$. Thus:

$$Y(\omega) = \int_{-\infty}^{+\infty} x(\tau)\left[\int_{-\infty}^{+\infty} h(t-\lambda)\,e^{-j\omega t}\,dt\right] d\tau$$

$$= \int_{-\infty}^{+\infty}\left[\int_{-\infty}^{+\infty} x(\tau)h(t-\lambda)\,d\tau\right] e^{-j\omega t}\,dt$$

But we have:

$$Y(\omega) = \int_{-\infty}^{+\infty} y(t)\,e^{-j\omega t}\,dt$$

Thus:

$$y(t) = \int_{-\infty}^{+\infty} x(\tau)h(t-\lambda)\,d\tau \qquad [29]$$

This relationship is known as the *convolution integral*. This equation sometimes appears in a slightly different but equivalent form. We could have grouped the terms for the inner integral differently and obtained:

$$y(t) = \int_{-\infty}^{+\infty} h(\tau)x(t-\lambda)\,d\tau \qquad [30]$$

The relationships, [29] and [30], are often written as:

$$y(t) = x(t) * h(t) \qquad [31]$$

where * reads as 'convolved with'.

Convolution enables us to determine the output signal from a system when there is an input with all the calculation taking place in the time domain rather than the frequency domain.

Example

Determine the output from a system at time t when it is subject to a ramp input of $x(t) = 2t$ for $t > 0$ and the response of the system to a unit strength impulse is e^{-t} for $t > 0$.

Using equation [30]:

$$y(t) = \int_{-\infty}^{+\infty} h(\tau)x(t-\lambda)\,d\tau = \int_0^t e^{-\tau}2(t-\tau)\,d\tau$$

$$= 2t\int_0^t e^{-\tau}\,d\tau - 2\int_0^t \tau\,e^{-\tau}\,d\tau$$

Using integration by parts for the second integral:

$$y(t) = 2t[-e^{-\tau}]_0^t - 2\left([-\tau\,e^{-\tau}]_0^t - [e^{-\tau}]_0^t\right)$$

$$= -2t\,e^{-t} + 2t + 2t\,e^{-t} + 2\,e^{-t} + 2 = 2t + 2\,e^{-t} + 2$$

Revision

11 Show that the response of a system to the input $x(t) = 2\,\delta(t-4)$ if it has a response to a unit strength impulse of $3\,e^{-5t}u(t)$ is $6\,e^{-5(t-4)}u(t)$.

Problems

1 Determine the Fourier transforms for the following:

(a) $f(t) = 2$, (b) $f(t) = e^{-at}$ for $t \geq 0$, $f(t) = e^{at}$ for $t < 0$, with $a > 0$,

(c) $f(t) = u(t)\,e^{-t/\tau}$ where τ is a constant, (d) $f(t) = 5[u(t-3) - u(t-1)]$,

(e) $f(t) = 3\,e^{-4|t+2|}$, (f) $f(t) = \dfrac{5}{4 + jt}$

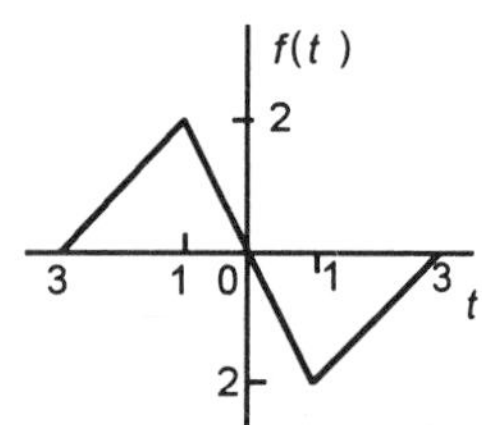

Figure 36.22 *Problem 2*

2 Determine, by differentiating the function, the Fourier transform of the function described by Figure 36.22.

3 Determine the Fourier transforms of (a) $2\sin\omega_0 t$, (b) $5\sin^2 3t$.

4 Determine the Fourier transform of $e^{-at}\sin\omega_0 t\; u(t)$. Hint: use Euler's formula for the sine.

5 Determine the response of a system to an input of a unit strength impulse if it has a transfer function of $4/(2 + j\omega)$.

6 If a system has an output of $3\,e^{-3t}u(t)$ for an input of a unit strength impulse, what is the input which would be required to give an output of $y(t) = 3(e^{-3t} - e^{-4t})u(t)$.

7 Determine using convolution the output of a system which has an input of:

(a) $e^{-t}u(t)$ if its response to a unit strength impulse is $u(t-2)$,

(b) $tu(t)$ if its response to a unit strength impulse is $e^{-t}u(t)$,

(c) $u(t) - u(t-1)$ if its response to a unit strength impulse is $e^{-t}u(t)$

37 Harmonic analysis

37.1 Introduction

In many engineering situations a waveform may not be given explicitly as a function $y(t)$ but as a series of readings of y at particular times t. Direct integration of the function to obtain the Fourier series coefficients is then not possible and a numerical method of integration has to be used (see Chapter 26 for a discussion of numerical integration). The numerical method of determining the Fourier coefficients using the trapezium rule (see Section 26.3) is termed *harmonic analysis*. This chapter illustrates how such a method is used.

This chapter assumes the trapezium rule in Chapter 26 and the Fourier series in Chapter 33.

37.2 The Trapezium rule

For one period of a periodic waveform $y = f(t)$ if we divide the area under the curve into k strips (Figure 37.1) then the trapezium rule gives for the area under the graph for one period T (equation [4], Chapter 26):

$$\int_0^T f(t)\,\mathrm{d}t \approx \frac{T}{k}\left[\tfrac{1}{2}\left(y_0 + y_k\right) + y_1 + y_2 + \ldots + y_{k-1}\right]$$

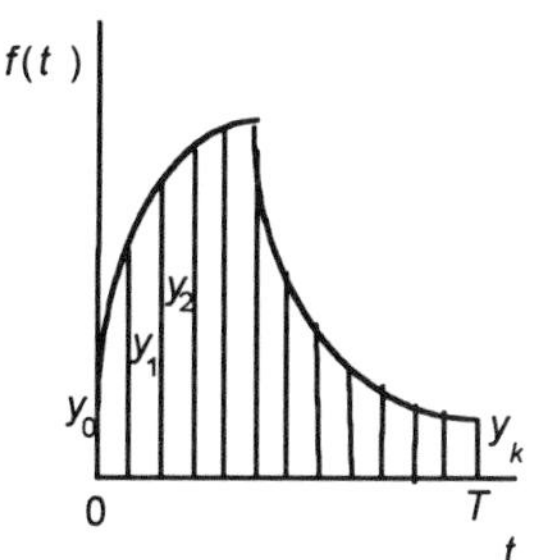

Figure 37.1 *Using the trapezium rule*

where y_0 is the height of the strip at the beginning of the period, y_1 the height at the end of the first strip, y_2 the height at the end of the second strip, etc. with y_k being the height at the end of the kth strip and the end of a period. Because the function is periodic we must have $y_0 = y_k$. Thus

$$\int_0^T f(t)\,\mathrm{d}t \approx \frac{T}{k}\left(y_0 + y_1 + y_2 + \ldots + y_{k-1}\right) \qquad [1]$$

37.2.1 The Fourier series coefficients

Using equation [1] we can derive values for the Fourier coefficients in terms of the ordinate values. Thus using equation [20], Chapter 33:

$$a_0 = \frac{2}{T}\int_0^T f(t)\,\mathrm{d}t \approx \frac{2}{k}\left(y_0 + y_1 + y_2 + \ldots + y_{k-1}\right) \qquad [2]$$

Using equation [22], Chapter 33:

$$a_n = \frac{2}{T}\int_0^T f(t)\cos n\omega t\,\mathrm{d}t$$

$$\approx \frac{2}{k}\left(y_0 \cos\frac{2\pi n}{T}t_0 + y_1 \cos\frac{2\pi n}{T}t_1 + \ldots + y_{k-1}\cos\frac{2\pi n}{T}t_{k-1}\right) \qquad [3]$$

Using equation [24], Chapter 33:

$$b_n = \frac{2}{T}\int_0^T f(t)\sin n\omega t\,\mathrm{d}t$$

$$\approx \frac{2}{k}\left(y_0 \sin\frac{2\pi n}{T}t_0 + y_1 \sin\frac{2\pi n}{T}t_1 + \ldots + y_{k-1}\sin\frac{2\pi n}{T}t_{k-1}\right) \qquad [4]$$

Generally the period is taken as being 2π, this simplifying the arithmetic.

37.2.2 Twelve-point analysis

The number of strips used to subdivide a period is a crucial question with harmonic analysis. Suppose a function has a fifth harmonic expressed as a sine. This means that every 1/10th of the fundamental period the harmonic will have a zero value. Thus if we divide the fundamental period into 10 equal size strips, we can obtain no information about the fifth harmonic since the ordinates at each strip will have a zero value. Harmonic analysis carried out up to the fifth harmonic thus requires more strips than 10. Generally only analysis up to the fifth harmonic is used and so 12 equal size strips are used to subdivide a period of 2π. Twelve also has the advantage of dividing 360° or 2π into 'nice' segments of 30° or $\pi/6$. Such analysis is referred to as *twelve-point analysis*. If analysis is required for more than the fifth harmonic, either more strips must be used or the waveform due to the sum of the first five harmonics subtracted from the function and the twelve-point analysis then repeated for the residue.

Example

Determine the Fourier series up to the fifth harmonic for the waveform described by the following data:

$t°$	0	30	60	90	120	150	180	210	240	270	300	330	360
$f(t)$	0	48	67	55	36	26	0	−48	−67	−55	−36	−26	0

Because the data indicates half-wave inversion, the second part of the period being merely the inversion of the first part, there will only be odd harmonics. Equation [3] for the first harmonic, with the period as 2π or 360°, gives:

$$a_1 = \tfrac{2}{12}(0 + 48\cos 30° + 67\cos 60° + 55\cos 90° + 36\cos 120° + 26\cos 150° + 0 - 48\cos 210° - 67\cos 240° - 55\cos 270° - 36\cos 300° - 26\cos 330°)$$

$$= \tfrac{1}{6}(0 + 41.57 + 33.5 + 0 - 18 - 22.52 + 0 + 41.57 + 33.5 + 0 - 18 - 22.52)$$

$$= 11.52$$

We can determine the other coefficients using equations [3] and [4]. Table 37.1 shows the data developed and the results.

Table 37.1 *Data for example*

$t°$	$f(t)$	$\cos t$	$y \cos t$	$\sin t$	$y \sin t$	$\cos 3t$	$y \cos 3t$	$\sin 3t$	$y \sin 3t$	$\cos 5t$	$y \cos 5t$	$\sin 5t$	$y \sin 5t$
0	0	1	0	0	0	1	0	0	0	1	0	0	0
30	48	0.866	41.57	0.5	24	0	0	1	48	−0.866	−41.57	0.5	24
60	67	0.5	33.5	0.866	58.02	−1	−67	0	0	0.5	33.5	−0.866	−58.02
90	55	0	0	1	55	0	0	−1	−55	0	0	1	55
120	36	−0.5	−18	0.866	31.18	1	36	0	0	−0.5	−18	−0.866	−31.18
150	26	−0.866	−22.52	0.5	13	0	0	1	26	0.866	22.52	0.5	13
180	0	−1	0	0	0	−1	0	0	0	−1	0	0	0
210	−48	−0.866	41.57	−0.5	24	0	0	−1	48	0.866	−41.57	−0.5	24
240	−67	−0.5	33.5	−0.866	58.02	1	−67	0	0	−0.5	33.5	0.866	−58.02
270	−55	0	0	−1	55	0	0	1	−55	0	0	−1	55
300	−36	0.5	−18	−0.866	31.18	−1	36	0	0	0.5	−18	0.866	−31.18
330	−26	0.866	−22.52	−0.5	13	0	0	−1	26	−0.866	22.52	−0.5	13
Sum			69.10		362.40		−62		38		−7.10		5.6
			$a_1 = 11.52$		$b_1 = 60.40$		$a_3 = -10.33$		$b_3 = 6.33$		$a_5 = -1.18$		$b_5 = 0.93$

Thus the Fourier series is:

$$y = f(t) = 11.52 \cos t - 60.40 \cos 3t - 1.18 \cos 5t \\ + 60.40 \sin t + 6.33 \sin 3t + 0.93 \sin 5t$$

Revision

1 In examining the data for twelve-point analysis of a periodic waveform the following patterns were discerned. What can be deduced about the coefficients that will be present in the Fourier series?

(a) The sum of all the function values over a period is 0.

(b) The function values from 180° to 360° are just the values from 0° to 180° with minus signs.

(c) The function values from 180° to 360° are just repeats of the values from 0° to 180°.

2 Using twelve-point analysis, determine the Fourier series up to the third harmonic for the following data:

$t°$	0	30	60	90	120	150	180	210	240	270	300	330	360
$f(t)$	10	12	24	26	24	20	16	14	12	8	6	8	10

Problems

1 Using twelve-point analysis, determine the Fourier series up to the third harmonic for the following data:

(a)

$t°$	0	30	60	90	120	150	180	210	240	270	300	330	360
$f(t)$	10	18	24	26	25	20	16	16	15	9	5	5	10

(b)

$t°$	0	30	60	90	120	150	180	210	240	270	300	330	360
$f(t)$	−10	−3	12	25	32	33	30	25	7	−8	−14	−15	−10

(c)

$t°$	0	30	60	90	120	150	180	210	240	270	300	330	360
$f(t)$	20	44	62	58	55	48	47	43	40	30	25	23	20

2 By the use of twelve-point analysis, determine the first three harmonics in the Fourier series for the full-wave rectified sinusoidal waveform, i.e. for the first half period $y = \sin t$ with the second half period being the same, and compare the results with those obtained analytically.

3 By the use of twelve-point analysis, determine the first three harmonics in the Fourier series for the waveform, period 2π, for which:

$$f(t) = 2 \text{ for } 0 \le t < \pi, f(t) = -2 \text{ for } \pi \le t < 2\pi$$

and compare the results with those obtained analytically.

38 Application: System response

38.1 Introduction

Often in considering electrical systems the input is not a simple d.c. or sinusoidal a.c. signal but perhaps a square wave periodic signal or a distorted sinusoidal signal or a half-wave rectified sinusoid. Such problems can be tackled by representing the waveform as a Fourier series. The basic principle used in circuit analysis with waveforms represented in this way is the *principle of superposition*. This states that we can find the overall effect of the waveform by summing the effects due to each term in the Fourier series considered alone. Thus if we have a voltage waveform:

$$v = V_0 + V_1 \sin \omega t + V_2 \sin 2\omega t + V_3 \sin 3\omega t + \dots$$

then we can consider the effects of each element taken alone. Thus we can calculate the current due to the voltage V_0 , then that due to $V_1 \sin \omega t$, then that due to $V_2 \sin 2\omega t$, then that due to $V_3 \sin 3\omega t$, and so on. Thus we might obtain currents of I_0, $I_1 \sin \omega t$, $I_2 \sin 2\omega t$, $I_3 \sin 3\omega t$, etc. The circuit current i due to the entire waveform is then:

$$i = I_0 + I_1 \sin \omega t + I_2 \sin 2\omega t + I_3 \sin 3\omega t + \dots$$

This chapter uses the trigonometrical form of the Fourier series for the representation of waveforms and so assumes Chapter 33. The section illustrating the use of the Fourier transform with electrical circuits assumes Chapter 36. A knowledge of phasors is assumed.

38.2 Waveforms represented by Fourier series

Consider the application to a pure resistance R of a voltage of:

$$v = V_0 + V_1 \sin \omega t + V_2 \sin 2\omega t + \dots + V_n \sin n\omega t \qquad [1]$$

Since $i = v/R$:

$$i = \frac{V_0}{R} + \frac{V_1}{R}\sin\omega t + \frac{V_2}{R}\sin 2\omega t + \dots + \frac{V_n}{R}\sin n\omega t \qquad [2]$$

Because the resistance is the same for each harmonic, the amplitude of each voltage harmonic is reduced by the same factor by the resistance. The phases of each harmonic are not changed.

Consider the application to a pure inductance L of a current represented by the Fourier series:

$$i = I_0 + I_1 \sin \omega t + I_2 \sin 2\omega t + \dots + I_n \sin n\omega t \qquad [3]$$

Since the voltage v across the inductor is L di/dt then:

$$v = L\frac{\mathrm{d}}{\mathrm{d}t}I_0 + L\frac{\mathrm{d}}{\mathrm{d}t}I_1 \sin\omega t + L\frac{\mathrm{d}}{\mathrm{d}t}I_2 \sin 2\omega t + \ldots + L\frac{\mathrm{d}}{\mathrm{d}t}I_n \sin n\omega t$$

Since I_0 does not change with time, the first term is zero. Thus:

$$v = \omega LI_1 \cos\omega t + 2\omega LI_2 \cos 2\omega t + \ldots + n\omega LI_3 \cos n\omega t$$

This can be written as:

$$v = \omega LI_1 \sin(\omega t + \tfrac{\pi}{2}) + 2\omega LI_2 \sin(2\omega t + \tfrac{\pi}{2}) + \ldots + n\omega L_n \sin(n\omega t + \tfrac{\pi}{2}) \quad [4]$$

Each of the current terms in [3] has had their amplitude altered by a different amount but the phase changed by the same amount. The result of this is that the shape of the voltage waveform is different to that of the current waveform.

We can represent the result of applying the current to the inductance in terms of phasors (see Chapter 9). Since the impedance Z = **V/I**, then for each of the Fourier series terms we have:

$$Z_n = \frac{\mathbf{V}_n}{\mathbf{I}_n} = \frac{V_n\angle\pi/2}{I_n} = n\omega L\angle\pi/2 \quad [5]$$

$n\omega L$ is the reactance X_L for the harmonic concerned. We can represent the above relationship in complex notation as $Z_n = \mathrm{j}X_L = \mathrm{j}n\omega L$.

Consider a pure capacitance C when a voltage:

$$v = V_0 + V_1 \sin\omega t + V_2 \sin 2\omega t + \ldots + V_n \sin n\omega t \quad [6]$$

is applied across it. Since $i = C$ dv/dt then:

$$i = C\frac{\mathrm{d}}{\mathrm{d}t}V_0 + C\frac{\mathrm{d}}{\mathrm{d}t}V_1 \sin\omega t + C\frac{\mathrm{d}}{\mathrm{d}t}V_2 \sin 2\omega t + \ldots + C\frac{\mathrm{d}}{\mathrm{d}t}V_n \sin n\omega t$$

Since V_0 does not change with time, the first term is zero. Thus:

$$i = \omega CV_1 \cos\omega t + 2\omega CV_2 \cos 2\omega t + \ldots + n\omega CV_n \cos n\omega t$$

This can be written as:

$$i = \omega CV_1 \sin(\omega t + \tfrac{\pi}{2}) + 2\omega CV_2 \sin(2\omega t + \tfrac{\pi}{2}) + \ldots + n\omega CV_n \sin(n\omega t + \tfrac{\pi}{2}) \quad [7]$$

Each of the voltage terms in [5] has had their amplitude altered by a different amount but the phase changed by the same amount. The result of this is that the shape of the current waveform is different to that of the voltage waveform.

We can represent the result of applying the current to the capacitance in terms of phasors (see Chapter 9). Since the impedance $Z = \mathbf{V}/\mathbf{I}$, then for each of the Fourier series terms we have:

$$Z_n = \frac{\mathbf{V}_n}{\mathbf{I}_n} = \frac{V_n}{I_n \angle \pi/2} = \frac{1}{n\omega C} \angle(-\pi/2) \qquad [8]$$

$1/n\omega C$ is the reactance X_C for the harmonic concerned. We can represent the above relationship in complex notation as $Z_n = -\mathrm{j}X_C = -\mathrm{j}(1/n\omega C)$ or $1/\mathrm{j}n\omega C$.

For circuit elements in series, the total impedance is the sum of the impedances of the separate elements. Thus if we have an inductance L in series with resistance R when there is an input of a voltage having harmonics, the impedance Z_n of the nth harmonic is the sum of the impedances for the nth harmonic of the two elements and is thus:

$$Z_n = R + \mathrm{j}n\omega L \qquad [9]$$

If we had a resistance R, an inductance L and a capacitance C in series then:

$$Z_n = R + \mathrm{j}n\omega L + \frac{1}{\mathrm{j}n\omega C} \qquad [10]$$

For parallel circuits, say a resistance R in parallel with an inductance L, when there is an input of a voltage having harmonics, the total impedance Z_n of the nth harmonic is given by:

$$\frac{1}{Z_n} = \frac{1}{R} + \frac{1}{\mathrm{j}n\omega L} \qquad [11]$$

If we had a resistance R, an inductance L and a capacitance C in parallel then:

$$\frac{1}{Z_n} = \frac{1}{R} + \frac{1}{\mathrm{j}n\omega L} + \frac{1}{1/\mathrm{j}n\omega C} \qquad [12]$$

Example

Determine the waveform of the current occurring when a 2μF capacitor has connected across it the half-wave rectified sinusoidal voltage $v = 0.32 + 0.5 \cos 100t + 0.21 \cos 200t$ V.

Since $i = C\,\mathrm{d}v/\mathrm{d}t$ there will be no current arising from the d.c. term. For the first harmonic we have a current of $-2 \times 10^{-6} \times 0.5 \times 100 \sin 100t$. For the second harmonic a current of $-2 \times 10^{-6} \times 0.21 \times 200 \sin 200t$. Thus the resulting current is:

$$i = -2 \times 10^{-6} \times 0.5 \times 100 \sin 100t - 2 \times 10^{-6} \times 0.21 \times 200 \sin 200t$$

$$= -0.1 \cos(100t + \pi/2) - 0.084 \cos(200t + \pi/2) \text{ mA}$$

Example

A voltage of $v = 100 \cos 314t + 50 \sin(5 \times 314t - \pi/6)$ V is applied to a series circuit consisting of a 10 Ω resistor, a 0.02 H inductor and a 50 μF capacitor. Determine the circuit current.

For the first harmonic, the resistance is 10 Ω, the inductive reactance is $\omega L = 314 \times 0.02 = 6.28$ Ω and the capacitive reactance is $1/\omega C = 1/(314 \times 50 \times 10^{-6}) = 63.8$ Ω. Thus the total impedance is:

$$Z_1 = 10 + j6.28 - j63.8 = 10 - j57.52$$

$$= \sqrt{10^2 + 57.52^2} \angle \tan^{-1}\frac{-57.52}{10} = 58.4\angle(-80.1°)\ \Omega$$

Thus the current due to the first harmonic is:

$$i_1 = \frac{100\angle 0°}{58.4\angle(-80.1°)} = 1.71\angle 80.1°\ \text{A}$$

For the fifth harmonic, the resistance is 10 Ω, the inductive reactance is $5\omega L = 5 \times 314 \times 0.02 = 31.4$ Ω and the capacitive reactance is $1/5\omega C = 1/(5 \times 314 \times 50 \times 10^{-6}) = 12.76$ Ω. Thus the total impedance is:

$$Z_5 = 10 + j31.4 - j12.76 = 10 + j18.64$$

$$= \sqrt{10^2 + 18.64^2} \angle \tan^{-1}\frac{18.64}{10} = 21.2\angle 61.8°\ \Omega$$

Thus the current due to the first harmonic is:

$$i_5 = \frac{50\angle(-30°)}{21.2\angle 61.8°} = 2.36\angle(-91.8°)\ \text{A}$$

Thus the current waveform is:

$$i = 1.71 \cos(314t + 81.1°) + 2.36 \cos(3 \times 314t - 91.8°)\ \text{A}$$

Example

A half-wave rectified sinusoidal voltage:

$$v = 0.32 + 0.5 \sin \pi t - 0.21 \cos 2\pi t - 0.04 \cos 4\pi t\ \text{V}$$

is applied to a circuit consisting of a 1 Ω resistor in series with a 1 F capacitor. Determine the waveform of the voltage output across the capacitor.

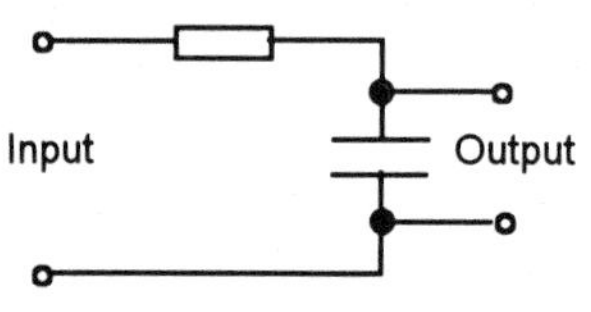

Figure 38.1 *Example*

Figure 38.1 shows the circuit. The output is the fraction of the input voltage that is across the capacitor. Thus, using phasors and the component values given:

$$\mathbf{V}_{out} = \frac{1/jn\omega C}{(1/jn\omega C) + R}\mathbf{V}_{in} = \frac{1}{1 + jn\omega CR}\mathbf{V}_{in} = \frac{1}{1 + jn\omega}\mathbf{V}_{in}$$

where $\omega = \pi$ is the fundamental frequency. For the d.c. component, with $\omega = 0$, we have $\mathbf{V}_{out\ 0} = \mathbf{V}_{in\ 0} = 0.32$ V. For the first harmonic we have $\mathbf{V}_{in\ 1} = -j0.5$ V and thus the output due to this term is:

$$\mathbf{V}_{out\ 1} = \frac{1}{1 + j\pi}(-j0.5) = \frac{1 - j\pi}{(1 + j\pi)(1 - j\pi)}(-j0.5) = \frac{-0.5\pi - j0.5}{1 + \pi^2}$$

$$= \sqrt{\frac{0.5^2\pi^2 + 0.5^2}{(1 + \pi^2)^2}} \angle \tan^{-1}\frac{-0.5}{-0.5\pi} = 0.15\angle(-162.3^\circ) \text{ V}$$

For the second harmonic we have $\mathbf{V}_{in\ 2} = -j0.21$ V and thus the output due to this term is:

$$\mathbf{V}_{out\ 2} = \frac{1}{1 + j2\pi}(-j0.21) = \frac{-0.42\pi - j0.21}{1 + 4\pi^2} = 0.033\angle(-189^\circ) \text{ V}$$

For the fourth harmonic we have $\mathbf{V}_{in\ 4} = -j0.04$ V and thus the output due to this term is:

$$\mathbf{V}_{out\ 4} = \frac{1}{1 + j4\pi}(-j0.04) = \frac{-0.16\pi - j0.04}{1 + 16\pi^2} = 0.003\,2\angle(-184.5^\circ) \text{ V}$$

Thus the output is:

$$\mathbf{V}_{out} = 0.32 + 0.15 \sin(\pi t - 162.3^\circ) + 0.033 \cos(2\pi t - 189^\circ) + 0.003\,2 \cos(4\pi t - 184.5^\circ) \text{ V}$$

Revision

1 A voltage of $2.5 + 3.2 \sin 100t + 1.6 \sin 200t$ V is applied across a 10 μF capacitor. Determine the current.

2 A voltage of $200 \cos 314t - 40 \sin 2 \times 314t$ V is applied to a circuit consisting of a 20 Ω resistor in series with a 100 μF capacitor. Determine the current in the circuit.

3 A voltage of $100 \sin 1000t + 50 \sin(3000t - \pi/6)$ V is applied to a circuit consisting of a resistance of 5 Ω in series with a parallel arrangement of a resistance of 10 Ω and an inductance having a reactance of 5 Ω at $\omega = 1000$ rad/s. Determine the total circuit current.

4 Determine the output voltage across the capacitor for the circuit shown in Figure 38.2 when the rectangular waveform voltage shown is applied to the input.

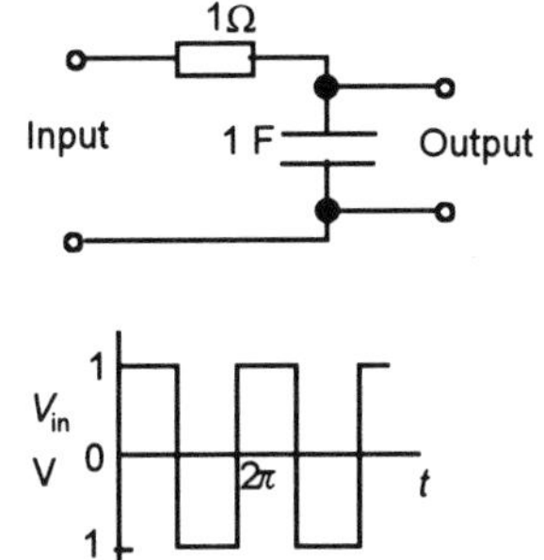

Figure 38.2 *Revision problem 4*

38.2.1 Root-mean-square values

The root-mean-square value of a periodic function $y = f(t)$ is given by (equation [24], Chapter 27):

$$\text{r.m.s. value} = \sqrt{\frac{1}{T}\int_0^T \{f(t)\}^2\,\mathrm{d}t} \qquad [13]$$

If we represent the periodic waveform by its Fourier series, then writing $\{f(t)\}^2$ as $\{f(t)\}\{f(t)\}$ and just writing one of the $f(t)$ in terms of the Fourier series:

$$\text{r.m.s. value} = \sqrt{\frac{1}{T}\int_0^T \left\{\tfrac{1}{2}a_0 + \sum_{n=1}^{\infty}(a_n \cos n\omega t + b_n \sin n\omega t)\right\} f(t)\,\mathrm{d}t} \qquad [14]$$

But (equation [20], Chapter 33):

$$a_0 = \frac{2}{T}\int_0^T f(t)\,\mathrm{d}t$$

and so for the integral of first term in equation [14]:

$$\frac{1}{T}\frac{1}{2}a_0 \int_0^T f(t)\,\mathrm{d}t = \frac{a_0^2}{4}$$

We also have (equations [22] and [24], Chapter 33):

$$a_n = \frac{2}{T}\int_0^T f(t)\cos n\omega t\,\mathrm{d}t$$

$$b_n = \frac{2}{T}\int_0^T f(t)\sin n\omega t\,\mathrm{d}t$$

and thus for the integral of the second term in equation [14] we have:

$$\sum_{n=1}^{\infty}\tfrac{1}{2}(a_n^2 + b_n^2)$$

Hence:

$$\text{r.m.s. value} = \sqrt{\tfrac{1}{2}\left[\tfrac{1}{2}a_0^2 + \sum_{n=1}^{\infty}(a_n^2 + b_n^2)\right]} \qquad [15]$$

Since we can write the Fourier series for the function in the form (equations [15], [18] and [19], Chapter 23):

$$f(t) = A_0 + \sum_{n=1}^{\infty} A_n \sin(n\omega t + \phi_n)$$

where $A_0 = a_0/2$ and $A_n = \sqrt{a_n^2 + b_n^2}$ then equation [15] can be written as:

$$\text{r.m.s. value} = \sqrt{A_0^2 + \tfrac{1}{2}\sum_{n=1}^{\infty} A_n^2} \qquad [16]$$

Thus if we have, say, a current with amplitudes for the d.c. component of I_0 and the harmonics of I_1, I_2, I_3, etc., equation [16] gives:

$$\text{r.m.s. current} = \sqrt{I_0^2 + \tfrac{1}{2}\left(I_1^2 + I_2^2 + I_3^2 + \dots\right)} \qquad [17]$$

Example

Determine the root-mean-square value of a waveform current for which we have $i = 15.0 + 27.0 \cos \omega t + 19.1 \cos 2\omega t + 9.0 \cos 3\omega t$ mA.

Using equation [17]:

$$\text{r.m.s. current} = \sqrt{15.0^2 + \tfrac{1}{2}(27.0^2 + 19.1^2 + 9.0^2)} = 28.5 \text{ mA}$$

Revision

5 Determine the root-mean-square values of the following currents:

(a) $\frac{2}{\pi}(1 - \frac{2}{3}\cos 2\omega t - \frac{2}{15}\cos 4\omega t)$ A,

(b) $\frac{4}{\pi}(\sin \omega t + \frac{1}{3}\sin 3\omega t + \frac{1}{5}\sin 5\omega t)$ A

38.3 Circuit analysis using Fourier transforms

For a resistance R the current $i(t)$ is related to the potential difference $v(t)$ across it by $v(t) = Ri(t)$. If we take Fourier transforms then $V(\omega) = RI(\omega)$. If we define impedance $Z(\omega)$ as the ratio of the voltage transform to the current transform, then for resistance:

$$Z(\omega) = \frac{V(\omega)}{I(\omega)} = R \qquad [18]$$

For a pure inductance L, since $v(t) = L\,\mathrm{d}i(t)/\mathrm{d}t$, when Fourier transforms are taken we have $V(\omega) = \mathrm{j}\omega LI(\omega)$ and so:

$$Z(\omega) = \frac{V(\omega)}{I(\omega)} = \mathrm{j}\omega L \qquad [19]$$

For a pure capacitance C, since $i(t) = C\,\mathrm{d}v(t)/\mathrm{d}t$, when Fourier transforms are taken we have $I(\omega) = \mathrm{j}\omega CV(\omega)$ and so:

$$Z(\omega) = \frac{V(\omega)}{I(\omega)} = \frac{1}{\mathrm{j}\omega C} \qquad [20]$$

We thus obtain the same expressions for the impedances as obtained with phasors in Chapter 9.

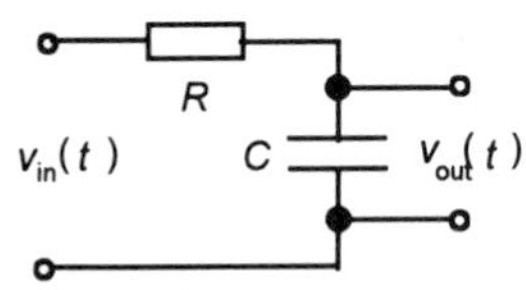

Figure 38.3 *Example*

Example

Determine the transfer function for the circuit shown in Figure 38.3 in which an input voltage is applied to a resistor in series with a capacitor and the output voltage is taken from across the capacitor. Hence determine the response when there is an input of a unit strength impulse.

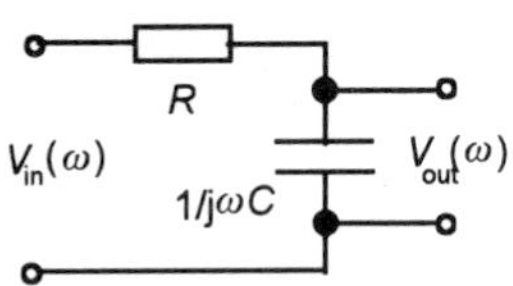

Figure 38.4 *Example*

When we use the Fourier transforms for the impedances, the circuit becomes as shown in Figure 38.4. We can now, working in this frequency domain, just consider the voltage output across the capacitor as a fraction of the input voltage applied across the series arrangement of resistance and capacitance. Thus:

$$V_{out}(\omega) = \frac{1/j\omega C}{R + 1/j\omega C} V_{in}(\omega) = \frac{1}{j\omega CR + 1} V_{in}(\omega)$$

Thus the transfer function $H(\omega)$ is:

$$H(\omega) = \frac{V_{out}(\omega)}{V_{in}(\omega)} = \frac{1}{j\omega CR + 1}$$

A unit strength impulse has a Fourier transform of 1. Thus with the input $V_{in}(\omega) = 1$:

$$V_{out}(\omega) = H(\omega)V_{in}(\omega) = \frac{1}{j\omega CR + 1} \times 1$$

In order to find the inverse and so $v_{out}(t)$ we need to find a function which has the above Fourier transform. Rearranging the transform as:

$$\frac{1}{j\omega CR + 1} = \frac{1/RC}{(1/RC) + j\omega}$$

puts it into the same form as the positive time exponential (item 6 in Table 36.1) and thus:

$$v_{out}(t) = (1/RC)\ e^{-t/RC}$$

Revision

6 Determine the transfer function for a circuit in which an input voltage is applied to a resistor of 2 kΩ in series with a capacitor of 100 μF and the output voltage is taken from across the capacitor. Hence determine the response when there is an input of a unit strength impulse.

7 A circuit consists of a 4 Ω resistor in series with an inductor of 2 H, the output voltage being taken from across the inductor. Determine the transfer function of the system and the output when the input is an exponentially decaying voltage pulse of 10 $e^{-3t}u(t)$ V.

Problems

1 A voltage of 2 sin 500*t* + 1 sin 1000*t* V is applied to a circuit consisting of a resistor of 6 Ω in series with a capacitor having a reactance of 8 Ω at ω = 500 rad/s. Determine the circuit current.

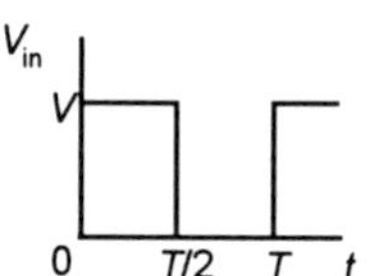

Figure 38.5 *Problem 5*

2 A voltage of 50 sin ωt + 25 sin(3ωt + π/3) V is applied to a series circuit consisting of a resistor of 8 Ω, an inductor having a reactance of 2 Ω at the fundamental frequency and a capacitor having a reactance of 8 Ω at the fundamental frequency. Determine the circuit current.

3 A voltage of 10 sin 500*t* + 5 sin 1500*t* V is applied to a circuit consisting of a resistor of 5 Ω in parallel with an inductor having a reactance of 5 Ω at ω = 500 rad/s. Determine the total circuit current.

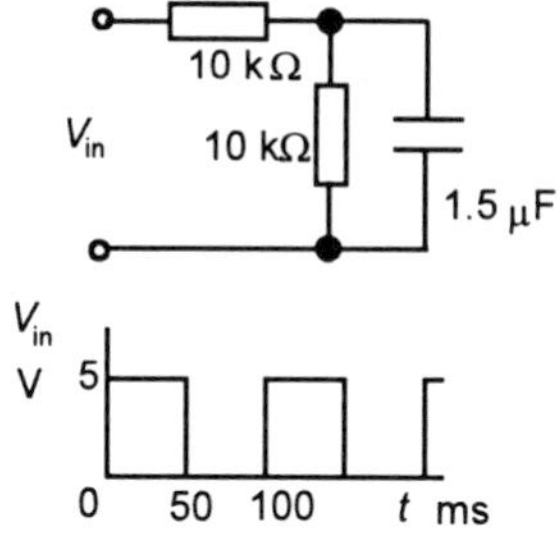

Figure 38.6 *Problem 6*

4 A voltage of 25 + 100 sin 10 000*t* + 40 sin(30 000*t* + π/6) V is applied to a circuit consisting of a resistor of 5 Ω in series with an inductor of 0.5 mH. Determine the circuit current.

5 A circuit consists of resistance *R* in series with inductance *L*. Derive an equation for the current in the circuit when the input is the rectangular waveform voltage shown in Figure 38.5.

6 Determine the total circuit current for the circuit shown in Figure 38.6 for the input voltage shown.

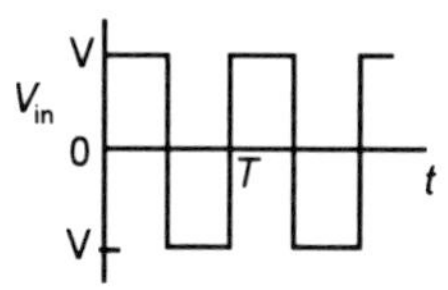

Figure 38.7 *Problem 7*

7 A circuit consists of a resistor *R* in series with a capacitance *C*. Determine the voltage output across the capacitor when the voltage input has the form shown in Figure 38.7.

8 Determine the root-mean-square voltage for the voltage described by v = 2 + 10 sin ωt + 8 sin(3ωt + π/6) + 3 sin(5ωt + π/3) – 1.5 sin 7ωt V.

9 Determine the root-mean-square value of a full-wave rectified sinusoidal waveform with amplitude *V*.

10 A voltage v = 10 + 20 sin ωt + 15 sin(2ωt + π/6) V is applied across a 10 Ω resistor. Determine the root-mean-square current.

11 Determine the root-mean-square value of the voltage 10.2 sin(ωt – 0.2) + 2 sin(3ωt + π/2) + 1 sin 5ωt V.

12 A circuit consisting of a 10 Ω resistor in series with a 15 mH inductor carries a current of 10 sin 314*t* + 5 cos 942*t* A. Determine the root-mean-square voltage across the circuit.

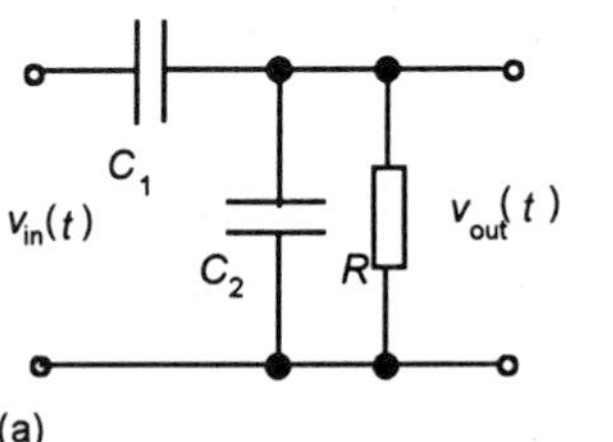

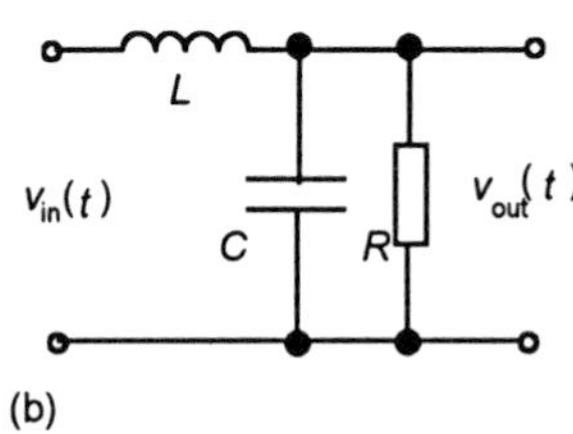

Figure 38.8 *Problem 13*

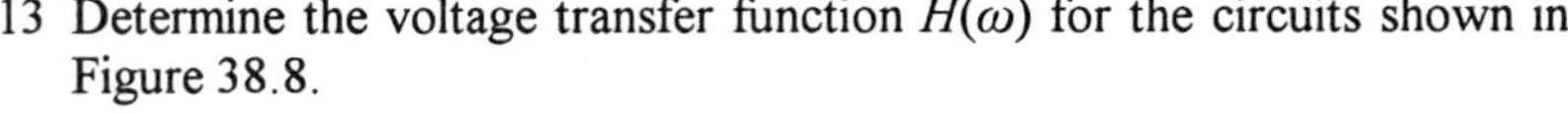

13 Determine the voltage transfer function $H(\omega)$ for the circuits shown in Figure 38.8.

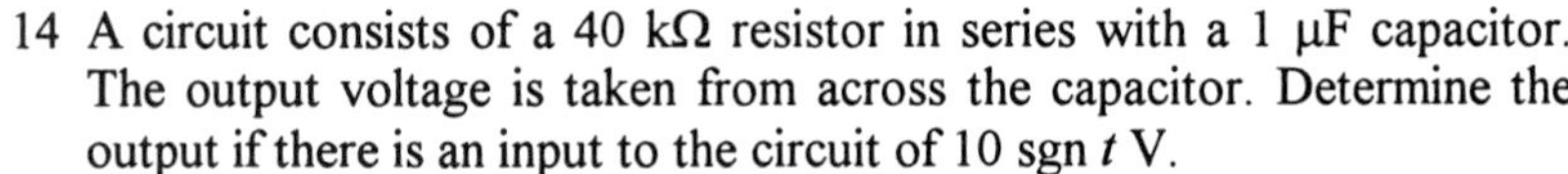

14 A circuit consists of a 40 kΩ resistor in series with a 1 μF capacitor. The output voltage is taken from across the capacitor. Determine the output if there is an input to the circuit of 10 sgn t V.

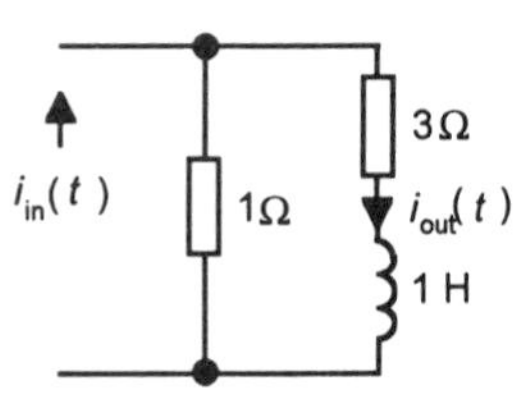

Figure 38.9 *Problem 15*

15 Determine the current transfer function for the circuit shown in Figure 38.9 and hence the output current when there is an input current of 20 sgn t A.

16 A circuit has a voltage transfer function of $10/(8 + j\omega)$. Determine the output if the input is a unit strength impulse.

17 A circuit consists of a resistor of 6 Ω in series with an inductor of 2 H. The output is the voltage across the inductor. Determine the voltage transfer function and hence the output when there is a unit strength impulse input.

Part 9 Laplace and *z*-transforms

The aims of this part are to enable the reader to:

- Use the Laplace transform to solve ordinary differential equations.
- Use the z-transform for sampled data systems.
- Determine the response of systems to various forms of input using transfer functions in the s and z domains.

This part assumes a knowledge of ordinary differential equations (Part 7) and hence the ability to use calculus (Part 6). Chapter 39 introduces the Laplace transform and the inverse Laplace transform, basic properties and how the transform can be used in the solution of differential equations for systems subject to inputs such as step, impulse and ramp. Chapter 40 shows how the Laplace transform can be used to obtain the z-transform for use with sampled data systems. Chapter 41 illustrates the use of the Laplace and z-transforms in determining the response of simple systems to various forms of input.

39 Laplace transform

39.1 Introduction

In considering the response of systems such as an electrical circuit when subject to an impulse input or perhaps a step input, we need to solve the differential equation for that system with that particular form of input. This is illustrated in Chapter 32. In this chapter a method of solving such differential equations is introduced which transforms a differential equation into an algebraic equation. This is termed the *Laplace transform*.

We can think of the Laplace transform as being rather like a function machine (Figure 39.1). As input to the machine we have some function of time $f(t)$ and as output a function we represent as $F(s)$. The input is referred to as being the *time domain* while the output is said to be in the *s-domain*. Thus we take information about a system in the time domain and use our 'machine' to transform it into information in the s-domain. Differential equations which describe the behaviour of a system in the time domain are converted into algebraic equations in the s-domain, so considerably simplifying their solution. We thus transform a differential equation into an s-domain equation, solve the equation and then use the 'machine' in inverse operation to transform the s-domain equation back into a time-domain solution.

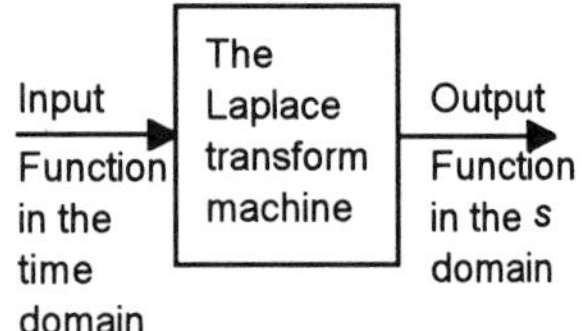

Figure 39.1 *The Laplace transform*

This chapter assumes a knowledge of ordinary differential equations (Part 7, in particular Chapter 28), and calculus, in particular Chapters 22 and 25. Aspects of Part 5 are required for the section on simultaneous differential equations.

39.2 The Laplace transform

The *Laplace transform* of some function of time is defined by:

Multiply a given function of time $f(t)$ by e^{-st} *and integrate the product between zero and infinity. The result, if it exists, is called the Laplace transform of $f(t)$ and is denoted by $\mathcal{L}\{f(t)\} = F(s)$.*

$$F(s) = \mathcal{L}\{f(t)\} = \int_0^\infty e^{-st} f(t)\, dt \qquad [1]$$

Note that the integration is between 0 and $+\infty$ and so is *one-sided* and not over the full range of time from $-\infty$ to $+\infty$.

Example

Determine the Laplace transform of $f(t) = 1$.

Using equation [1]:

$$\mathcal{L}\{f(t)\} = \int_0^\infty 1\, e^{-st}\, dt = \left[\frac{e^{-st}}{-s}\right]_0^\infty = \frac{1}{s}$$

This is provided that $s > 0$ so that $e^{-st} \to 0$ as $t \to \infty$.

Example

Determine the Laplace transform of $f(t) = e^{at}$.

Using equation [1]:

$$\mathcal{L}\{f(t)\} = \int_0^\infty e^{at}\, e^{-st}\, dt = \int_0^\infty e^{-(s-a)t}\, dt = \left[\frac{e^{-(s-a)t}}{-(s-a)}\right]_0^\infty = \frac{1}{s-a}$$

That is provided we have $(s + a) > 0$.

Example

Determine the Laplace transform of $f(t) = t$.

Using equation [1]:

$$\mathcal{L}\{f(t)\} = \int_0^\infty t\, e^{-st}\, dt$$

Using integration by parts:

$$\mathcal{L}\{f(t)\} = \left[-\frac{t}{s}\, e^{-st}\right]_0^\infty + \int_0^\infty \frac{1}{s}\, e^{-st}\, dt = \left[-\frac{1}{s^2}\, e^{-st}\right]_0^\infty = \frac{1}{s^2}$$

That is provided we have $s > 0$.

Example

Determine the Laplace transform of $f(t) = \sin at$.

Using equation [1] we can obtain:

$$\mathcal{L}\{f(t)\} = \int_0^\infty \sin at\, e^{-st}\, dt$$

and use integration by parts. However, another method is to use Euler's formula $e^{jat} = \cos at + j \sin at$. The sine is then the imaginary part of e^{jat}, written here as $\boldsymbol{I}(e^{jat})$. Thus:

$$\mathcal{L}\{\sin at\} = \mathcal{L}\{\boldsymbol{I}(e^{jat})\} = \boldsymbol{I}\int_0^\infty e^{jat}\, e^{-st}\, dt$$

$$= \boldsymbol{I}\left\{\left[\frac{e^{-(s-ja)t}}{-(s-ja)}\right]_0^\infty\right\}$$

$$= \boldsymbol{I}\left\{\frac{1}{s-ja}\right\} = \boldsymbol{I}\left\{\frac{s+ja}{s^2+a^2}\right\} = \frac{a}{s^2+a^2}$$

The same method can be used to obtain the Laplace transform of cos *at*, with the cosine being the real part of e^{jat}.

Revision

1 Determine the Laplace transforms of:

(a) $f(t) = t^2$, (b) $f(t) = t^3$, (c) $f(t) = \sinh at$. (Hint: $\sinh at = ½(e^{at} - e^{-at})$

39.2.1 Laplace transforms for step and impulse function

Consider the *unit step function* $u(t)$ shown in Figure 39.2 (see Section 6.3.2). The Laplace transform is given by equation [1] as:

$$\mathcal{L}\{u(t)\} = \int_0^{\infty} 1\ e^{-st}\ dt = \left[\frac{e^{-st}}{-s}\right]_0^{\infty} = \frac{1}{s} \qquad [2]$$

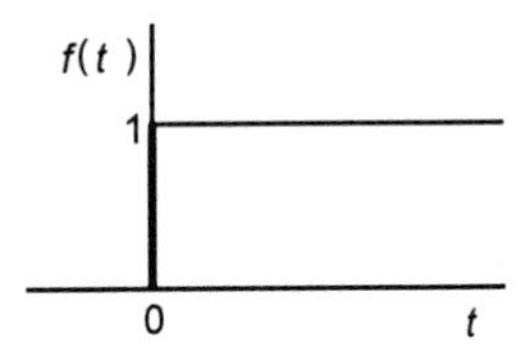

Figure 39.2 *Unit step*

Now consider obtaining the *unit impulse function* $\delta(t)$ (see Section 6.3.3). Such an impulse can be considered to be a unit area rectangular pulse which has its width k decreased to give the unit impulse in the limit when $k \to 0$. For the unit area rectangular pulse shown in Figure 39.3, the Laplace transform is:

$$\mathcal{L}\{\text{unit area pulse}\} = \int_0^{\infty} f(t)\ e^{-st}\ dt = \int_0^{k} \frac{1}{k}\ e^{-st}\ dt + \int_k^{\infty} 0\ e^{-st}\ dt$$

$$= \left[-\frac{1}{sk}\ e^{-st}\right]_0^{k} = -\frac{1}{sk}(e^{-sk} - 1)$$

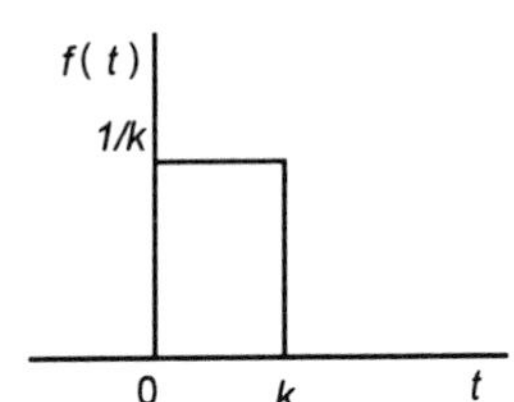

Figure 39.3 *Unit area rectangular pulse*

We can replace the exponential by a series (see Section 5.4.2), thus obtaining:

$$\mathcal{L}\{\text{unit area pulse}\} = -\frac{1}{sk}\left(1 + (-sk) + \frac{(-sk)^2}{2!} + \frac{(-sk)^3}{3!} + \ldots - 1\right)$$

Thus in the limit as $k \to 0$, the Laplace transform tends to the value 1 and so:

$$\mathcal{L}\{\delta(t)\} = 1 \qquad [3]$$

39.2.2 Standard Laplace transforms

The transforms derived above in Sections 39.2 and 39.2.1, together with others, are tabulated as a set of standard transforms so that it becomes unnecessary to derive them by the use of equation [1]. Table 39.1 gives some of the more common standard transforms. As indicated in the following section, these standard transforms can be used to derive the transforms for a wide range of functions.

Table 39.1 *Laplace transforms*

	$f(t)$	$\mathcal{L}\{f(t)\}$
1	Unit impulse $\delta(t)$	1
2	Unit step $u(t)$	$\frac{1}{s}$
3	Unit ramp t	$\frac{1}{s^2}$
4	t^n	$\frac{n!}{s^{n+1}}$
5	e^{-at}	$\frac{1}{s+a}$
6	$1 - e^{-at}$	$\frac{a}{s(s+a)}$
7	$t\,e^{-at}$	$\frac{1}{(s+a)^2}$
8	$e^{-at} - e^{-bt}$	$\frac{b-a}{(s+a)(s+b)}$
9	$(1-at)\,e^{-at}$	$\frac{s}{(s+a)^2}$
10	$1 - \frac{b}{b-a}\,e^{-at} + \frac{a}{b-a}\,e^{-bt}$	$\frac{ab}{s(s+a)(s+b)}$
11	$\frac{e^{-at}}{(b-a)(c-a)} + \frac{e^{-bt}}{(c-a)(a-b)} + \frac{e^{-ct}}{(a-c)(b-c)}$	$\frac{1}{(s+a)(s+b)(s+c)}$
12	$\sin \omega t$	$\frac{\omega}{s^2+\omega^2}$
13	$\cos \omega t$	$\frac{s}{s^2+\omega^2}$
14	$e^{-at} \sin \omega t$	$\frac{\omega}{(s+a)^2+\omega^2}$
15	$e^{-at} \cos \omega t$	$\frac{s+a}{(s+a)^2+\omega^2}$
16	$\sinh \omega t$	$\frac{\omega}{s^2 - \omega 2}$
17	$\cosh \omega t$	$\frac{s}{s^2-\omega^2}$
18	$\frac{\omega}{\sqrt{1-\zeta^2}}\, e^{-\zeta\omega t} \sin \omega \sqrt{1-\zeta^2}\, t,\ \zeta < 1$	$\frac{\omega^2}{s^2 + 2\zeta\omega s + \omega^2}$
19	$1 - \frac{1}{\sqrt{1-\zeta^2}}\, e^{-\zeta\omega t} \sin\left(\omega \sqrt{1-\zeta^2}\, t + \phi\right),\ \cos\phi = \zeta$	$\frac{\omega^2}{s(s^2 + 2\zeta\omega s + \omega^2)}$

39.2.3 Properties of Laplace transforms

The following are basic properties of Laplace transforms and can be used with the above table of standard transforms to obtain a wide range of other transforms.

1 *Linearity*
If two separate time functions $f(t)$ and $g(t)$ have Laplace transforms then the transform of the sum of the time functions, i.e. $f(t) + g(t)$, is the sum of the Laplace transforms of the two functions considered separately:

$$\mathcal{L}\{f(t) + g(t)\} = \mathcal{L}\{f(t)\} + \mathcal{L}\{g(t)\} \qquad [4]$$

This linearity property is derived by using equation [1]:

$$\mathcal{L}\{f(t) + g(t)\} = \int_0^{\infty} \{f(t) + g(t)\}\ \mathrm{e}^{-st}\ \mathrm{d}t$$

$$= \int_0^{\infty} f(t)\ \mathrm{e}^{-st}\ \mathrm{d}t + \int_0^{\infty} g(t)\ \mathrm{e}^{-st}\ \mathrm{d}t$$

$$= \mathcal{L}\{f(t)\} + \mathcal{L}\{g(t)\}$$

Since $2f(t)$ equals $f(t) + f(t)$, then the Laplace transform of $2f(t)$ will be twice the Laplace transform of $f(t)$. Thus, in general, linearity gives:

$$\mathcal{L}\{af(t)\} = a\mathcal{L}\{f(t)\} \qquad [5]$$

Example

Determine the Laplace transform of $1 + 2t$.

Using equations [4] and [5] and Table 39.1:

$$\mathcal{L}\{1 + 2t\} = \frac{1}{s} + \frac{2}{s^2}$$

Example

Determine the Laplace transform of $3 \sin 2t + \cos 2t$.

Using equations [4] and [5] and Table 39.1:

$$\mathcal{L}\{3 \sin 2t + \cos 2t\} = 3\frac{2}{s^2 + 4} + \frac{s}{s^2 + 4}$$

Example

Determine the Laplace transform of $3t^2 + 2\ \mathrm{e}^{-t}$.

Using equations [4] and [5] and Table 39.1:

$$\mathcal{L}\{3t^2 + 2\ \mathrm{e}^{-t}\} = 3\frac{2!}{s^3} + 2\frac{1}{s+1}$$

Example

Determine the Laplace transform of $\sin(\omega t + \theta)$.

We can write $\sin(\omega t + \theta)$ as $\sin \omega t \cos \theta + \cos \omega t \sin \theta$. Thus, using equations [4] and [5] and Table 39.1:

$$\mathcal{L}\{\sin(\omega t + \theta)\} = \frac{\omega}{s^2 + \omega^2} \cos\theta + \frac{s}{s^2 + \omega^2} \sin\theta$$

2 *The first shift theorem, factor* e^{-at}
This theorem states that if $\mathcal{L}\{f(t)\} = F(s)$ then $\mathcal{L}\{e^{-at}f(t)\} = F(s + a)$. Thus the substitution of $s + a$ for s corresponds to multiplying a time function by e^{-at}. This can be demonstrated if we consider equation [1] with such a function:

$$\mathcal{L}\{e^{-at}f(t)\} = \int_0^\infty e^{-at}f(t)\, e^{-st}\, dt = \int_0^\infty f(t)\, e^{-(s+a)}\, dt$$

Example

Determine the Laplace transform of $e^{-2t} \cosh 3t$.

Using the first shift theorem and the transform for $\cosh 3t$ given by Table 39.1, the transform is that of $\cosh 3t$ with the s replaced by $s + 2$:

$$\mathcal{L}\{e^{-2t} \cosh 3t\} = \frac{s+2}{(s+2)^2 - 9}$$

Example

Determine the Laplace transform of $2\, e^{-2t} \sin^2 t$.

Since $\cos 2t = 1 - 2 \sin^2 t$ we have, using the linearity property:

$$\mathcal{L}\{2 \sin^2 t\} = \mathcal{L}\{1\} - \mathcal{L}\{\cos 2t\} = \frac{1}{s} - \frac{s}{s^2 + 4}$$

Hence, using the first shift theorem and replacing the s by $s + 2$:

$$\mathcal{L}\{2\, e^{-2t} \sin^2 t\} = \frac{1}{s+2} - \frac{s+2}{(s+2)^2 + 4}$$

3 *The second shift theorem, time shifting*
The second shift theorem states that if a signal is delayed by a time T then its Laplace transform is multiplied by e^{-sT}. Thus if $F(s)$ is the Laplace transform of $f(t)$ then $\mathcal{L}\{f(t - T)u(t - T)\} = e^{-sT}F(s)$. This can be demonstrated by considering a unit step function which is delayed by a time T (Figure 39.4). Such a function is described by $u(t - T)$ (see Section 6.3.2). Equation [1] gives for such a function:

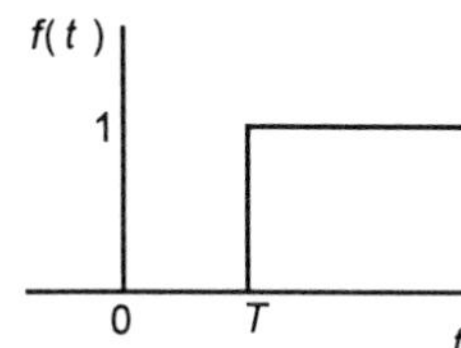

Figure 39.4 *Delayed unit step*

$$\mathcal{L}\{u(t - T)\} = \int_0^\infty u(t - T)\, e^{-st}\, dt = \int_0^T 0\, dt + \int_T^\infty 1\, e^{-st}\, dt$$

$$= \left[\frac{e^{-st}}{-s}\right]_T^\infty = e^{-sT}\frac{1}{s}$$

Example

Determine the Laplace transform for a unit impulse which occurs at a time of $t = 2$ s.

The Laplace transform for a unit impulse at $t = 0$ is 1. Thus the transform for the delayed impulse is $1\ \mathrm{e}^{-2s}$.

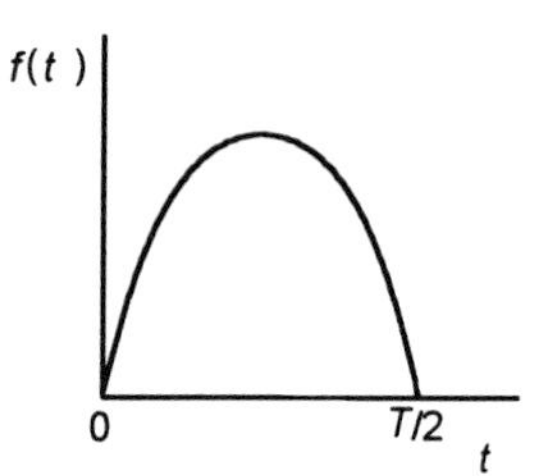

Figure 39.5 *Example*

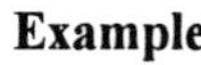

Example

Determine the Laplace transform of a single pulse consisting of just the first half of a sine wave (Figure 39.5).

We can think of such a function as being the sum of a sine function extending over an infinite number of cycles and a sine function that has had its start delayed by ½T. In this way all but the first half period waveform are cancelled out. Thus the Laplace transform is:

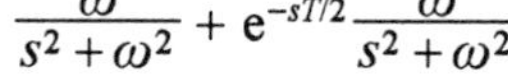

$$\frac{\omega}{s^2+\omega^2} + \mathrm{e}^{-sT/2}\frac{\omega}{s^2+\omega^2}$$

4 *Periodic functions*
A periodic function of period T has a Laplace transform of:

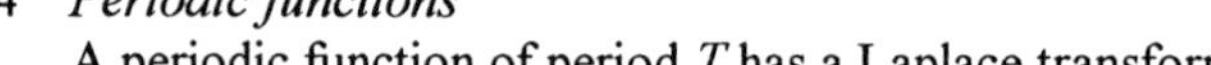

$$\frac{1}{1-\mathrm{e}^{-sT}}F_1(s)$$

where $F_1(s)$ is the Laplace transform of the function for the first period. This can be proved by considering the periodic function to be the sum of the function $f_1(t)$ describing the first period, the first period function delayed by 1 period, the first period function delayed by 2 periods, etc. The Laplace transform of the sum is thus:

$$F_1(s) + \mathrm{e}^{-sT}F_1(s) + \mathrm{e}^{-2sT}F_1(s) + \ldots = (1 + \mathrm{e}^{-sT} + \mathrm{e}^{-2sT} + \ldots)F_1(s)$$

The term in the brackets is a geometric series with the sum to infinity of $1/(1 - \mathrm{e}^{-sT})$. Thus we obtain the equation given above.

Example

Determine the Laplace transform of a full-wave rectified sine wave.

Such a wave consists of a sequence of the pulses shown in Figure 39.5. Thus the first period function has the transform:

$$\frac{\omega}{s^2+\omega^2} + \mathrm{e}^{-sT/2}\frac{\omega}{s^2+\omega^2}$$

Therefore the periodic wave has the Laplace transform:

$$\frac{1}{1-\mathrm{e}^{-sT/2}}\left(\frac{\omega}{s^2+\omega^2} + e^{-sT/2}\frac{\omega}{s^2+\omega^2}\right)$$

Revision

2 Use the linearity property to determine the Laplace transforms of the following functions:

(a) $t^2 + 3t + 2$, (b) $2 + 4 \sin 3t$, (c) $e^{4t} + \cosh 2t$, (d) $2 + 5\,e^{3t}$,

(e) $\cos 2t + \cos 3t$, (f) $t^3 + 4\,e^{-t}$

3 Use the first shift theorem to determine the Laplace transforms of the following functions:

(a) $e^{-3t} \sin 2t$, (b) $e^{4t}t^2$, (c) $e^{2t} \cos t$

4 Use the second shift theorem to determine the Laplace transform of the following functions:

(a) a unit step function which starts at $t = 5$ s,

(b) a unit impulse which occurs at $t = 4$ s,

(c) the function described by $3(t - 10)u(t - 10)$

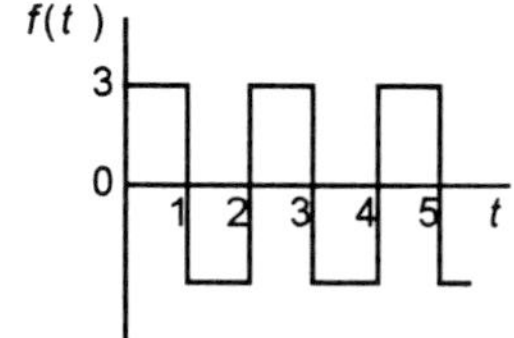

Figure 39.6 *Revision problem 5*

5 Determine the Laplace transform of the periodic function shown in Figure 39.6.

39.3 Multiplication by *t*

There are frequently situations where we require the Laplace transform of a function which is multiplied by some power of time t. For example, we might require the transform for $t \cos 2t$. The following is one way by which we can derive such a transform.

If $\mathcal{L}\{f(t)\} = F(s)$ then equation [1] gives:

$$F(s) = \int_0^\infty e^{-st} f(t)\,dt$$

Thus:

$$\frac{d}{ds}F(s) = \frac{d}{ds}\int_0^\infty e^{-st} f(t)\,dt = \int_0^\infty \frac{d(e^{-st})}{dt} f(t)\,dt$$

$$= \int_0^\infty (-t\,e^{-st}) f(t)\,dt = -\int_0^\infty e^{-st}[tf(t)]\,dt$$

$$= -\mathcal{L}\{t(f(t)\} \qquad [6]$$

Every time we differentiate $F(s)$, another t factor and another (–1) factor occur. In general:

$$\textit{If } \mathcal{L}\{f(t)\} = F(s), \textit{ then } \mathcal{L}\{t^n f(t)\} = (-1)^n \frac{d^n F(s)}{ds^n} \qquad [7]$$

Example

Determine the Laplace transform of t cos $3t$.

Table 39.1 gives:

$$\mathcal{L}\{\cos 3t\} = F(s) = \frac{s}{s^2+9}$$

Thus, using equation [6]:

$$\mathcal{L}\{t\cos 3t\} = -\frac{\mathrm{d}}{\mathrm{d}s}F(s) = -\frac{\mathrm{d}}{\mathrm{d}s}\left(\frac{s}{s^2+9}\right) = -\frac{9-s^2}{(s^2+9)^2} = \frac{s^2-9}{(s^2+9)^2}$$

Revision

6 Determine the Laplace transforms of the following functions:

(a) $t \sin 4t$, (b) $t^2 \sin 4t$, (c) $t^2\,\mathrm{e}^{at}$

39.4 Laplace transforms of derivatives

Consider the determination of the Laplace transform of the derivative of a function, i.e. $\mathcal{L}\{\mathrm{d}f(t)/\mathrm{d}t\}$. Using equation [1]:

$$\mathcal{L}\left\{\frac{\mathrm{d}}{\mathrm{d}t}f(t)\right\} = \int_0^\infty \mathrm{e}^{-st}\frac{\mathrm{d}}{\mathrm{d}t}f(t)\,\mathrm{d}t$$

Using integration by parts:

$$\mathcal{L}\left\{\frac{\mathrm{d}}{\mathrm{d}t}f(t)\right\} = -f(0) + s\int_0^\infty \mathrm{e}^{-st}f(t)\,\mathrm{d}t = -f(0) + sF(s) \qquad [7]$$

where $f(0)$ is the value of $f(t)$ when $t = 0$ and $F(s)$ is the Laplace transform of $f(t)$.

For a second derivative we can similarly obtain:

$$\mathcal{L}\left\{\frac{\mathrm{d}^2}{\mathrm{d}t^2}f(t)\right\} = \int_0^\infty \mathrm{e}^{-st}\frac{\mathrm{d}^2}{\mathrm{d}t^2}f(t)\,\mathrm{d}t = \left[\mathrm{e}^{-st}\frac{\mathrm{d}}{\mathrm{d}t}f(t)\right]_0^\infty + s\int_0^\infty \mathrm{e}^{-st}\frac{\mathrm{d}}{\mathrm{d}t}f(t)\,\mathrm{d}t$$

$$= -\frac{\mathrm{d}}{\mathrm{d}t}f(0) + s\{-f(0) + sF(s)\} = s^2F(s) - sf(0) - \frac{\mathrm{d}}{\mathrm{d}t}f(0) \qquad [8]$$

where $\mathrm{d}f(0)/\mathrm{d}t$ is the value of the first derivative when $t = 0$. Likewise for a third derivative we can obtain:

$$\mathcal{L}\left\{\frac{\mathrm{d}^3}{\mathrm{d}t^3}f(t)\right\} = s^3F(s) - s^2f(0) - s\frac{\mathrm{d}}{\mathrm{d}t}f(0) - \frac{\mathrm{d}^2}{\mathrm{d}t^2}f(0) \qquad [9]$$

where $\mathrm{d}^2f(0)/\mathrm{d}t^2$ is the value of the second derivative at $t = 0$.

Example

Given the initial condition that $x = 2$ when $t = 0$, determine the Laplace transform of $4\frac{dx}{dt}$.

Using equation [7]:

$$\mathcal{L}\left\{4\frac{dx}{dt}\right\} = 4\left[sX(s) - x(0)\right] = sX(s) - 2$$

where $X(s)$ is the Laplace transform of $x(t)$.

Example

Given the initial conditions that $x = 0$ and $dx/dt = 0$ when $t = 0$, determine the Laplace transform of $3\frac{d^2x}{dt^2} + 2\frac{dx}{dt}$.

Using equations [7] and [8]:

$$\mathcal{L}\left\{3\frac{d^2x}{dt^2} + 2\frac{dx}{dt}\right\} = 3\left[s^2X(s) - sx(0) - \frac{d}{dt}x(0)\right] + 2\left[sX(s) - x(0)\right]$$

$$= (3s^2 + 2s)X(s)$$

Revision

7 Determine the Laplace transforms of the following:

(a) $\frac{dx}{dt} + 2x$, $x = 3$ when $t = 0$,

(b) $\frac{d^2x}{dt^2} + 2\frac{dx}{dt} + 3x$, $x = 4$ and $\frac{dx}{dt} = 5$ when $t = 0$,

(c) $\frac{d^2x}{dt^2} - 5\frac{dx}{dt} + 6x$, $x = 0$ and $\frac{dx}{dt} = 1$ when $t = 0$

39.5 Laplace transforms of integrals

Consider the determination of the Laplace transform of the integral of a function, i.e.

$$\mathcal{L}\left\{\int_0^t f(t)\,dt\right\}$$

If we let $g(t) = \int_0^t f(t)\,dt$, then $\frac{d}{dt}g(t) = f(t)$. Then, using equation [7]:

$$\mathcal{L}\left\{\frac{d}{dt}g(t)\right\} = sG(s) - g(0)$$

Since $g(0) = 0$ and $G(s) = \mathcal{L}\{g(t)\}$:

$$\mathcal{L}\{f(t)\} = s\,\mathcal{L}\left\{\int_0^t f(t)\,dt\right\}$$

Thus:

$$\mathcal{L}\left\{\int_0^t f(t)\,dt\right\} = \frac{1}{s}F(s) \qquad [10]$$

Example

Determine the Laplace transform of $\int_0^t e^{-t}\,dt$.

Using equation [10]:

$$\mathcal{L}\left\{\int_0^t e^{-t}\,dt\right\} = \frac{1}{s}F(s)$$

Since:

$$F(s) = \mathcal{L}\{e^{-t}\} = \frac{1}{s+1}$$

$$\mathcal{L}\left\{\int_0^t e^{-t}\,dt\right\} = \frac{1}{s}F(s) = \frac{1}{s(s+1)}$$

Revision

8 Determine the Laplace transform of the voltage across a capacitor if it is given by:

$$\frac{1}{C}\int_0^t i(t)\,dt$$

9 Determine the Laplace transform of $\int_0^t e^{-t}\,dt$.

39.6 The inverse Laplace transform

The inverse Laplace transform is the transformation of a Laplace transform into a function of time. If $\mathcal{L}\{f(t)\} = F(s)$ then $f(t)$ is the *inverse Laplace transform* of $F(s)$, the inverse being written as:

$$f(t) = \mathcal{L}^{-1}\{F(s)\} \qquad [11]$$

The inverse can generally be obtained by using standard transforms, e.g. those in Table 39.1. The basic properties of the inverse, see Section 39.6.1, can be used with the standard transforms to obtain a wider range of transforms than just those in the table. Often $F(s)$ is the ratio of two polynomials and cannot be readily identified with a standard transform. However, the use of partial fractions (see Section 2.6) can often convert such an expression into simple fraction terms which can then be identified with standard transforms. This is illustrated in the examples given in the next section.

Example

Determine the inverse Laplace transform of $1/s^2$.

Table 39.1 indicates that the function which has the Laplace transform of $1/s^2$ is t. Thus the inverse is t.

Revision

10 Determine the inverse Laplace transforms of (a) $\frac{2}{s^2+4}$, (b) $\frac{1}{s+3}$.

39.6.1 Basic properties of the inverse transform

The *linearity* property of Laplace transforms (see Section 29.2.3) means that if we have a transform as the sum of two separate terms then we can take the inverse of each separately and the sum of the two inverse transforms is the inverse of the sum:

$$\mathcal{L}^{-1}\{F(s) + G(s)\} = \mathcal{L}^{-1}\{F(s)\} + \mathcal{L}^{-1}\{G(s)\} \qquad [12]$$

Also, linearity gives:

$$\mathcal{L}^{-1}\{aF(s)\} = a\mathcal{L}^{-1}\{F(s)\} \qquad [13]$$

where a is a constant.

The *first shift theorem* (see Section 39.2.3) can be written in inverse form as:

$$\mathcal{L}^{-1}\{F(s-a)\} = e^{at}f(t) \qquad [14]$$

where $f(t)$ is the inverse transform of $F(s)$.

The *second shift theorem* (see Section 39.2.3) can be written in inverse form as:

$$\mathcal{L}^{-1}\{e^{-sT}F(s)\} = f(t-T)u(t-T) \qquad [15]$$

Thus if the inverse transform numerator contains an e^{-sT} term, then we remove this term from the expression, determine the inverse transform of what remains and then substitute $(t - T)$ for t in the result.

Example

Determine the inverse Laplace transform of $\frac{7s}{s^2+9}$.

Table 39.1 shows the Laplace transform of $\cos \omega t$ as $s/(s^2 + \omega^2)$. Thus:

$$\mathcal{L}^{-1}\left\{\frac{s}{s^2+\omega^2}\right\} = \cos \omega t$$

Thus, using equation [13]:

$$\mathcal{L}^{-1}\left\{\frac{7s}{s^2+9}\right\} = 7\mathcal{L}^{-1}\left\{\frac{s}{s^2+9}\right\} = 7\cos 3t$$

Example

Determine the inverse Laplace transform of $\dfrac{3s-1}{s(s-1)}$.

We can write the fraction in a simpler form by the use of partial fractions. Thus:

$$\frac{3s-1}{s(s-1)} = \frac{A}{s} + \frac{B}{s-1}$$

and so we must have $3s - 1 = A(s - 1) + Bs$. Equating coefficients of s gives $3 = A + B$ and equating numerical terms gives $-1 = -A$. Hence:

$$\frac{3s-1}{s(s-1)} = \frac{1}{s} + \frac{2}{s-1}$$

The inverse transform of $1/s$ is 1 and the inverse of $1/(s - 1)$ is e^t. Thus:

$$\mathcal{L}^{-1}\left\{\frac{3s-1}{s(s-1)}\right\} = 1 + 2\,e^t$$

Example

Determine the inverse Laplace transform of $\dfrac{6}{s^2-6s+13}$.

This fraction can be rearranged as:

$$\frac{6}{s^2-6s+13} = 3\frac{2}{(s-3)^2+2^2}$$

The fraction term is now in the form $\omega/(s^2 + \omega^2)$, i.e. the transform of $\sin \omega t$ when s has been replaced by $s - 3$. This corresponds to a multiplication by e^{3t}. Thus, using equation [14]:

$$\mathcal{L}^{-1}\left\{\frac{5}{s^2-6s+13}\right\} = 3\,e^{3t}\sin 2t$$

Example

Determine the inverse Laplace transform of $\dfrac{6\,e^{-3s}}{s+2}$.

Using equation [15], extracting e^{-3s} from the expression gives $6/(s + 2)$. This has the inverse Laplace transform of $6\,e^{-2t}$. Thus the required inverse is $5(t - 3)\,e^{-2(t-3)}u(t - 3)$.

Revision

11 Determine the inverse Laplace transforms of the following:

(a) $\frac{1}{s-2}$, (b) $\frac{5}{s}$, (c) $\frac{s}{s^2+16}$, (d) $\frac{3}{s^2-9}$, (e) $\frac{5}{(s-2)^2+25}$,

(f) $\frac{1}{(s+3)^4}$, (g) $\frac{e^{-2s}}{s^2}$, (h) $\frac{e^{-3s}}{(s+2)^2}$

12 Determine, by the use of partial fractions, the inverse Laplace transforms of the following:

(a) $\frac{3s+1}{s^2-s-6}$, (b) $\frac{3s+3}{s^2+s-2}$, (c) $\frac{s-4}{(s+1)(s^2+4)}$, (d) $\frac{s+4}{s^2+4s+4}$

39.6.2 The convolution theorem

The linearity property (see Section 39.6) has the inverse transform of a sum of a number of transforms as the sum of the inverse transforms of the separate terms. But what about the inverse transform of a product? Is it the product of the inverse transforms of the separate terms? The answer is *no*, as the following example illustrates. Consider the inverse transform of $1/s^2$. As Table 39.1 indicates, this is t. But what if we consider $1/s^2$ as the product of two terms $1/s$ and $1/s$. The inverse transform of $1/s$ is 1 and thus the product of $\mathcal{L}^{-1}(1/s)$ and $\mathcal{L}^{-1}(1/s)$ is $1 \times 1 = 1$, which is certainly not the correct answer of t.

$$\mathcal{L}^{-1}\{F(s)G(s)\} \neq \mathcal{L}^{-1}\{F(s)\}\mathcal{L}^{-1}\{G(s)\} \qquad [16]$$

If $F(s)$ is the Laplace transform of $f(t)$ and $G(s)$ the Laplace transform of $g(t)$, then the product of the two Laplace transforms $F(s)$ and $G(s)$ is termed the *convolution* of the two functions $f(t)$ and $g(t)$. It is written as:

$$F(s)G(s) = \mathcal{L}\{f(t) * g(t)\} \qquad [17]$$

Thus:

$$\mathcal{L}^{-1}\{F(s)G(s)\} = f(t) * g(t) \qquad [18]$$

This can be shown to be (see Section 36.6.1 for a similar proof):

$$\mathcal{L}^{-1}\{F(s)G(s)\} = f(t) * g(t) = \int_0^t f(\tau)g(t-\tau)\,d\tau \qquad [19]$$

This is known as the *convolution theorem* and is used to determine the inverse transform of a product. Note the distinction between t and τ in equation [19], the integration being with respect to τ. As far as the integration is concerned, t is a constant.

Example

Determine the inverse Laplace transform of $\dfrac{1}{s^2(s-1)}$.

Consider it as the product of two terms $F(s) = 1/s^2$ and $G(s) = 1/(s-1)$. These give $f(t) = t$ and $g(t) = e^t$. Thus, using equation [19]:

$$\mathcal{L}^{-1}\left\{\frac{1}{s^2(s-1)}\right\} = \int_0^t \tau\, e^{t-\tau}\, d\tau = e^t \int_0^t \tau\, e^{-\tau}\, d\tau = e^t[-\tau e^{-\tau} - e^{-\tau}]_0^t$$

$$= e^t(-t e^{-t} - e^{-t} + 1) = e^t - t - 1$$

We could have obtained the same result by the use of partial fractions.

Example

Determine the inverse Laplace transform of $\dfrac{1}{s(s^2+4)}$.

Consider it as the product of two terms $F(s) = 1/s$ and $G(s) = 1/(s^2 + 4)$. These give $f(t) = 1$ and $g(t) = ½ \sin 2t$. Thus, using equation [19]:

$$\mathcal{L}^{-1}\left\{\frac{1}{s(s^2+4)}\right\} = \int_0^t 1\frac{\sin 2(t-\tau)}{2}\, d\tau = \tfrac{1}{2}\left[-\tfrac{1}{2}\cos 2(t-\tau)\right]_0^t$$

$$= \tfrac{1}{4}(1 - \cos 2t)$$

We could have obtained the same result by the use of partial fractions.

Revision

13 Use the convolution theorem to obtain the inverse Laplace transforms of the following:

(a) $\dfrac{1}{(s+1)(s+2)}$, (b) $\dfrac{1}{s^2(s+1)^2}$, (c) $\dfrac{1}{(s^2+1)^2}$

39.7 Solving differential equations

Laplace transforms offer a method of solving differential equations. The procedure adopted is:

1 Replace each term in the differential equation by its Laplace transform, inserting the given initial conditions.

2 Algebraically rearrange the equation to give the transform of the solution.

3 Invert the resulting Laplace transform to obtain the answer as a function of time.

Example

Given that $x = 0$ at $t = 0$, solve the first-order differential equation:

$$3\frac{dx}{dt} + 2x = 4$$

Taking the Laplace transform gives:

$$3[sX(s) - x(0)] + 2X(s) = \frac{4}{s}$$

Substituting the initial condition gives:

$$3sX(s) + 2X(s) = \frac{4}{s}$$

Hence:

$$X(s) = \frac{4}{s(3s+2)}$$

Simplifying by the use of partial fractions:

$$\frac{4}{s(3s+2)} = \frac{A}{s} + \frac{B}{3s+2}$$

Hence $A(3s + 2) + Bs = 4$ and so $A = 2$ and $B = -2/3$. Thus:

$$X(s) = \frac{2}{s} - \frac{2}{3(3s+2)} = \frac{2}{s} - 2\frac{\frac{2}{3}}{\frac{2}{3}\left(s+\frac{2}{3}\right)}$$

and so:

$$x(t) = 2 - 2\,e^{-2t/3}$$

Example

Given that $x = 0$ and $dx/dt = 1$ at $t = 0$, solve the second-order differential equation:

$$\frac{d^2x}{dt^2} - 5\frac{dx}{dt} + 6x = 2\,e^{-t}$$

Taking the Laplace transform gives:

$$s^2X(s) - sx(0) - \frac{d}{dt}x(0) - 5[sX(s) - x(0)] + 6X(s) = \frac{2}{s+1}$$

Substituting the initial conditions:

$$s^2X(s) - 1 - 5sX(s) + 6X(s) = \frac{2}{s+1}$$

Hence:

$$X(s) = \frac{\frac{2}{s+1} + 1}{s^2 - 5s + 6} = \frac{2}{(s+1)(s-2)(s-3)} + \frac{1}{(s-2)(s-3)}$$

We can simplify the above expression by the use of partial fractions. Thus:

$$\frac{2}{(s+1)(s-2)(s-3)} = \frac{A}{s+1} + \frac{B}{s-2} + \frac{C}{s-3}$$

Hence $A(s-2)(s-3) + B(s+1)(s-3) + C(s+1)(s-2) = 2$ and so $A = 1/6$, $B = -2/3$ and $C = ½$.

$$\frac{1}{(s-2)(s-3)} = \frac{D}{s-2} + \frac{E}{s-3}$$

Hence $D(s-3) + E(s-2) = 1$ and so $D = -1$ and $E = 1$. Thus:

$$X(s) = \frac{\frac{1}{6}}{s+1} + \frac{-\frac{2}{3}}{s-2} + \frac{\frac{1}{2}}{s-3} + \frac{-1}{s-2} + \frac{1}{s-3}$$

$$= \frac{\frac{1}{6}}{s+1} - \frac{\frac{5}{3}}{s-2} + \frac{\frac{3}{2}}{s-3}$$

The inverse transform is:

$$x(t) = \tfrac{1}{6}\,\mathrm{e}^{-t} - \tfrac{5}{3}\,\mathrm{e}^{2t} + \tfrac{3}{2}\,\mathrm{e}^{3t}$$

Revision

14 Solve the following differential equations:

(a) $2\frac{\mathrm{d}x}{\mathrm{d}t} + x = 4\,\mathrm{e}^{2t}$, with $x = 0$ when $t = 0$,

(b) $\frac{\mathrm{d}x}{\mathrm{d}t} + 5x = 2$, with $x = 0$ when $t = 0$,

(c) $\frac{\mathrm{d}^2x}{\mathrm{d}t^2} + 4x = 1$, with $x = 0$ and $\frac{\mathrm{d}x}{\mathrm{d}t} = 0$ when $t = 0$,

(d) $\frac{\mathrm{d}^2x}{\mathrm{d}t^2} + 2\frac{\mathrm{d}x}{\mathrm{d}t} + 2x = \mathrm{e}^{-t}$, with $x = 0$ and $\frac{\mathrm{d}x}{\mathrm{d}t} = 0$ when $t = 0$,

(e) $\frac{\mathrm{d}^2x}{\mathrm{d}t^2} - 2\frac{\mathrm{d}x}{\mathrm{d}t} + x = \sin t$, with $x = 1$ and $\frac{\mathrm{d}x}{\mathrm{d}t} = 0$ when $t = 0$

39.7.1 Simultaneous differential equations

A set of simultaneous differential equations can arise when Kirchhoff's laws are applied to *RLC* networks consisting of more than one mesh or node or when the oscillations are considered for a mechanical system with more than one mass. To solve such equations by the use of the Laplace transform:

1 Replace each term in the differential equations by their Laplace transforms, inserting the given initial conditions.

2 Solve the resulting simultaneous equations using any of the conventional methods for algebraic simultaneous equations.

3 Invert the resulting Laplace transforms to obtain the answers as functions of time.

Example

Solve the simultaneous equations:

$$\frac{dx}{dt} + y - x = 0, \ \frac{dy}{dt} - y - x = 0$$

with $x = 1$ and $y = 0$ at $t = 0$.

Taking the Laplace transforms of each equation:

$$sX(s) - x(0) + Y(s) - X(s) = 0, \ sY(s) - y(0) - Y(s) - X(s) = 0$$

Substituting the initial conditions and rearranging gives:

$$(s - 1)X(s) + Y(s) = 1$$

$$-X(s) + (s - 1)Y(s) = 0$$

Multiplying the second equation by $(s - 1)$ and adding it to the first equation gives:

$$Y(s) + (s - 1)^2 Y(s) = 1$$

Hence:

$$Y(s) = \frac{1}{(s-1)^2 + 1}$$

Substituting this back into one of the equations gives:

$$X(s) = \frac{s-1}{(s-1)^2 + 1}$$

Taking the inverses of these gives $y(t) = \mathrm{e}^t \sin t$ and $x(t) = \mathrm{e}^t \cos t$.

Example

Solve the following simultaneous differential equations:

$$\frac{\mathrm{d}^2x}{\mathrm{d}t^2} - 2\frac{\mathrm{d}x}{\mathrm{d}t} + 3\frac{\mathrm{d}y}{\mathrm{d}t} + 2y = 4,\ 2\frac{\mathrm{d}y}{\mathrm{d}t} - \frac{\mathrm{d}x}{\mathrm{d}t} + 3y = 0$$

with $x = 0$, $y = 0$ and $\mathrm{d}x/\mathrm{d}t = 0$ when $t = 0$.

Taking the Laplace transform of both equations:

$$s^2X(s) - sx(0) - \frac{\mathrm{d}}{\mathrm{d}t}x(0) - 2sX(s) - x(0) + 3sY(s) + 3y(0) + 2Y(s) = \frac{4}{s}$$

$$2sY(s) - y(0) - sX(s) + x(0) + 3Y(s) = 0$$

Substituting the initial conditions gives:

$$s^2X(s) - 2sX(s) + 3sY(s) + 2Y(s) = \frac{4}{s}$$

$$2sY(s) - sX(s) + 3Y(s) = 0$$

Rearranging these gives:

$$(s^2 - 2s)X(s) + (3s + 2)Y(s) = \frac{4}{s}$$

$$-sX(s) + (2s + 3)Y(s) = 0$$

We can solve these equations in a number of ways. Using Cramer's rule (see Section 18.2):

$$\frac{X(s)}{\begin{vmatrix} \frac{4}{s} & 3s+2 \\ 0 & 2s+3 \end{vmatrix}} = \frac{Y(s)}{\begin{vmatrix} s^2-2s & \frac{4}{s} \\ -s & 0 \end{vmatrix}} = \frac{1}{\begin{vmatrix} s^2-2s & 3s+2 \\ -s & 2s+3 \end{vmatrix}}$$

Hence:

$$X(s) = \frac{\frac{4}{s}(2s+3)}{(s^2-2s)(2s+3)+s(3s+2)} = \frac{2(2s+3)}{s^2(s+2)(s-1)}$$

$$Y(s) = \frac{-4}{(s^2-2s)(2s+3)+s(3s+2)} = \frac{2}{s(s+2)(s-1)}$$

We can simplify these equations by the use of partial fractions. Thus:

$$\frac{2(2s+3)}{s^2(s+2)(s-1)} = \frac{A}{s} + \frac{B}{s^2} + \frac{C}{s+2} + \frac{D}{s-1}$$

gives $2(2s + 3) = As(s + 2)(s - 1) + B(s + 2)(s - 1) + Cs^2(s + 2)(s - 1) + Ds^2(s + 2)$ and so $A = -7/2$, $B = -3$, $C = 1/6$ and $D = 10/3$. Hence:

$$X(s) = -\frac{7}{2}\frac{1}{s} - 3\frac{1}{s^2} + \frac{1}{6}\frac{1}{s+2} + \frac{10}{3}\frac{1}{s-1}$$

Likewise:

$$\frac{2}{s(s+2)(s-1)} = \frac{E}{s} + \frac{F}{s+2} + \frac{G}{s-1}$$

gives $E(s + 2)(s - 1) + Fs(s - 1) + Gs(s + 2)$ and so $E = -1$, $F = 1/3$ and $G = 2/3$. Hence:

$$Y(s) = -\frac{1}{s} + \frac{1}{3}\frac{1}{s+2} + \frac{2}{3}\frac{1}{s-1}$$

Taking the inverse Laplace transform then gives:

$$x(t) = -\tfrac{7}{2} - 3t + \tfrac{1}{6}\,\mathrm{e}^{-2t} + \tfrac{10}{3}\,\mathrm{e}^{t}$$

$$y(t) = -1 + \tfrac{1}{3}\,\mathrm{e}^{-2t} + \tfrac{2}{3}\,\mathrm{e}^{t}$$

Revision

15 Solve the following sets of simultaneous differential equations:

(a) $\frac{dx}{dt} - x - 2y = 0,\ \frac{dy}{dt} - 3x + 4y = 0,$

with $x = 0$ and $y = 1$ at $t = 0$,

(b) $\frac{dx}{dt} + 4\frac{dy}{dt} + 3y = 0,\ 3\frac{dx}{dt} + \frac{dy}{dt} + 2x = 1,$

with $x = 0$ and $y = 0$ at $t = 0$,

(c) $\frac{d^2x}{dt^2} + 2x - y = 0,\ \frac{d^2y}{dx^2} + 2y - x = 0,$

with $x = 4$, $dx/dt = 0$ and $y = 2$ at $t = 0$

39.8 Initial and final value theorems

The *initial value* of a function of time is its value at zero time, the *final value* being the value at infinite time. Often there is a need to determine the initial value and final values of systems, e.g. for an electrical circuit when there is, say, a step input. The final value in such a situation is often referred to as the *steady-state value*. The initial and final value theorems enable the initial and final values to be determined from a Laplace transform without the need to find the inverse transform.

39.8.1 The initial value theorem

The Laplace transform of $f(t)$ is given by equation [1] as:

$$\mathcal{L}\{f(t)\} = \int_0^{\infty} e^{-st} f(t)\, dt$$

and so:

$$\mathcal{L}\left\{\frac{d}{dt}f(t)\right\} = \int_0^{\infty} e^{-st} \frac{d}{dt}f(t)\, dt \qquad [20]$$

Integration by parts then gives:

$$\mathcal{L}\left\{\frac{d}{dt}f(t)\right\} = \left[e^{-st}f(t)\right]_0^{\infty} - \int_0^{\infty}(-s\, e^{-st})f(t)\, dt = -f(0) + sF(s) \qquad [21]$$

As s tends to infinity then e^{-st} tends to 0. Thus we must have, as a result of equation [20], $\mathcal{L}\{df(t)/dt)$ tending to 0 as s tends to infinity. Hence equation [21] gives:

$$\lim_{s\to\infty} [-f(0) + sF(s)] = 0$$

and so:

$$\lim_{s\to\infty} sF(s) = f(0)$$

But $f(0)$ is the initial value of the function at $t = 0$. Thus, provided a limit exists:

$$\lim_{t\to 0} f(t) = \lim_{s\to\infty} sF(s) \qquad [22]$$

This is known as the *initial value theorem*.

Example

Determine the initial value of the function $f(t)$ giving the Laplace transform $4/(s + 2)$.

Applying equation [22]:

$$\lim_{t\to 0} f(t) = \lim_{s\to\infty}\left[\frac{4s}{s+2}\right] = \lim_{s\to\infty}\left[\frac{4}{1+2/s}\right] = 4$$

Revision

16 Determine the initial values of the functions giving the following Laplace transforms:

(a) $\frac{1}{s^2+1}$, (b) $\frac{s}{s^2+1}$

39.8.2 The final value theorem

As in Section 39.8.1, for a function $f(t)$ having a Laplace transform $F(s)$ we can write (equations [20] and [21]):

$$\mathcal{L}\left\{\frac{\mathrm{d}}{\mathrm{d}t}f(t)\right\} = \int_0^\infty \mathrm{e}^{-st}\frac{\mathrm{d}}{\mathrm{d}t}f(t)\,\mathrm{d}t = -f(0) + sF(s) \qquad [23]$$

As s tends to zero then e^{-st} tends to 1 and so:

$$\lim_{s\to 0}\left[\int_0^\infty \mathrm{e}^{-st}\frac{\mathrm{d}}{\mathrm{d}t}f(t)\,\mathrm{d}t\right] = \int_0^\infty \frac{\mathrm{d}}{\mathrm{d}t}f(t)\,\mathrm{d}t$$

We can write this integral as:

$$\int_0^\infty \frac{\mathrm{d}}{\mathrm{d}t}f(t)\,\mathrm{d}t = \lim_{t\to\infty}\int_0^t \frac{\mathrm{d}}{\mathrm{d}t}f(t)\,\mathrm{d}t = \lim_{t\to\infty}\left[f(t) - f(0)\right]$$

Hence, with equation [23] we obtain:

$$\lim_{s\to 0}\left[-f(0) + sF(s)\right] = \lim_{t\to\infty}\left[f(t) - f(0)\right]$$

and so, provided a limit exists:

$$\lim_{t\to\infty} f(t) = \lim_{s\to 0} sF(s) \qquad [24]$$

This is termed the *final value theorem.*

Example

Determine the final value of the function which has the Laplace transform:

$$F(s) = \frac{2s+1}{(s+1)(s+3)}$$

Using equation [24]:

$$\lim_{t\to\infty} f(t) = \lim_{s\to\infty} sF(s) = \lim_{s\to\infty}\left[\frac{s(2s+1)}{(s+1)(s+3)}\right] = 0$$

Revision

17 Determine the final values of the functions having the following Laplace transforms:

(a) $\frac{2}{s}$, (b) $\frac{1}{s+5}$

Problems

1 Determine from the definition of the transform, the Laplace transforms of:

(a) $f(t) = \cos at$, (b) $f(t) = a$,

(c) $f(t) = \cosh at$. (Hint: $\cosh at = \frac{1}{2}(e^{at} + e^{-at})$)

2 Determine, by the use of the transforms given in Table 39.1 and the properties of Laplace transforms, the Laplace transforms of the following functions:

(a) 4, (b) $3t - 1$, (c) e^{3t}, (d) $2t + 3\,e^{t}$, (e) $t^2 + 4\,e^{-2t}$, (f) $t^2 + 2t + 1$,

(g) $2 \sin 3t$, (h) $5 \sinh 3t$, (i) $\sin 3t \cos 3t$, (j) $t\,e^{-3t}$, (k) $4 - 2\sin 3t + e^{2t}$,

(l) $t^3\,e^{-2t}$, (m) $(1 + e^{t})(1 - e^{-t})$, (n) $e^{3t}\cos t$, (o) $(1 + t)^2\,e^{-t}$, (p) $e^{-t}\sin^2 t$,

(q) $t\cosh 3t$, (r) $t^2\cosh 3t$, (s) $t^3\,e^{-3t}$

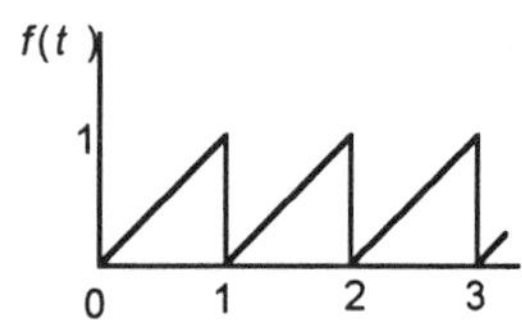

Figure 39.7 *Problem 4*

3 Determine the Laplace transforms for the following:

(a) $e^{2(t-5)}u(t - 5)$, (b) $f(t) = 0$ for $0 \le t < 2$, and $t - 2$ for $2 \le t$

4 Determine the Laplace transform for the periodic signal shown in Figure 39.7.

5 Determine the Laplace transform for the following periodic signals:

(a) $f(t) = 1$ for $0 \le t < 1$ and 0 for $1 \le t < 2$, $f(t + 2) = f(t)$,

(b) $f(t) = t$ for $0 \le t < 1$ and 0 for $1 \le t < 2$, $f(t + 2) = f(t)$,

(c) $f(t) = t$ for $0 \le t < 1$ and $2 - t$ for $1 \le t < 2$, $f(t + 2) = f(t)$

6 Determine the Laplace transforms of the following:

(a) $3\frac{dx}{dt} + x$, $x = 0$ at $t = 0$, (b) $2\frac{d^2x}{dt^2} + x$, $x = 0$ at $t = 0$,

(c) $\int_0^t t\,e^{2t}\,dt$, (d) $\int_0^t e^{t}\cos 2t\,dt$

7 Determine the inverse Laplace transforms of the following:

(a) $\dfrac{8}{s^3}$, (b) $\dfrac{s}{s^2+1}$, (c) $\dfrac{3}{2s+1}$, (d) $\dfrac{s}{s^2+1}+\dfrac{1}{s^2+1}$, (e) $\dfrac{1}{(s+2)^2}$,

(f) $\dfrac{s+1}{(s+1)^2+4}$, (g) $\dfrac{12}{(s-1)^2-9}$, (h) $\dfrac{3}{(s+2)^2}$, (i) $\dfrac{5s+1}{(s-4)(s+3)}$,

(j) $\dfrac{1}{s(s+1)}$, (k) $\dfrac{3}{(s+3)(s-2)}$, (l) $\dfrac{1}{(s+2)(s-4)(2s+1)}$,

(m) $\dfrac{2(4-s^2)}{s(s^2+2)}$, (n) $\dfrac{2s-8}{(s+1)(s^2+4)}$, (o) $\dfrac{4(s+1)}{s(s^2+4)}$, (p) $\dfrac{e^{-2(s+1)}}{(s+1)(s+2)}$

8 Use the convolution theorem to obtain the inverse Laplace transforms of the following:

(a) $\dfrac{1}{s^2(s+1)}$, (b) $\dfrac{1}{(s-1)(s-2)}$, (c) $\dfrac{s}{(s+2)(s^2+2)}$, (d) $\dfrac{1}{(s-1)(s^2+1)}$

9 Solve the following differential equations:

(a) $\dfrac{dx}{dt}-2x=3$, with $x=0$ when $t=0$,

(b) $\dfrac{dx}{dt}+4x=\cos t$, with $x=0$ when $t=0$,

(c) $\dfrac{d^2x}{dt^2}+x=3$, with $x=0$ and $\dfrac{dx}{dt}=1$ when $t=0$,

(d) $\dfrac{d^2x}{dt^2}+\dfrac{dx}{dt}-6x=5\,e^{3t}$, with $x=0$ and $\dfrac{dx}{dt}=0$ when $t=0$,

(e) $\dfrac{d^2x}{dt^2}-\dfrac{dx}{dt}-6x=\cos 2t$, with $x=0$ and $\dfrac{dx}{dt}=0$ when $t=0$,

(f) $\dfrac{d^2x}{dt^2}+2\dfrac{dy}{dx}+5x=e^{-t}\sin t$, with $x=0$ and $\dfrac{dx}{dt}=1$ when $t=0$,

(g) $\dfrac{d^2x}{dt^2}-5\dfrac{dy}{dt}-6x=t+e^{3t}$, with $x=2$ and $\dfrac{dx}{dt}=1$ when $t=0$

10 Solve the following sets of simultaneous differential equations:

(a) $\dfrac{dx}{dt}-6x+3y=8\,e^t$, $\dfrac{dy}{dx}-2x-y=4\,e^t$,

with $x=-1$ and $y=0$ when $t=0$,

(b) $\dfrac{dx}{dt}-4x+2y=2t$, $\dfrac{dy}{dt}-8x+4y=1$,

with $x = 3$ and $y = 5$ when $t = 0$,

(c) $\dfrac{dx}{dt} + 3x - y = 1, \; \dfrac{dx}{dt} + \dfrac{dy}{dt} + 3x = 0,$

with $x = 2$ and $y = 0$ when $t = 0$,

(d) $2\dfrac{dx}{dt} + \dfrac{dy}{dt} + 2x = 2, \; \dfrac{dx}{dt} + 2\dfrac{dy}{dt} + 2y = 0,$

with $x = 0$ and $y = 0$ when $t = 0$,

(e) $\dfrac{d^2x}{dt^2} + 2\dfrac{dy}{dt} + x = 0, \; \dfrac{dx}{dt} - \dfrac{dy}{dt} - 2x + 2y = \sin t,$

with $x = 0$, $dx/dt = 0$ and $y = 0$ when $t = 0$,

(f) $\dfrac{d^2x}{dt^2} + \dfrac{dx}{dt} + \dfrac{dy}{dt} + 2x - y = 0, \; \dfrac{d^2y}{dt^2} + \dfrac{dy}{dt} + \dfrac{dx}{dt} - 2y + 4x = 0,$

with $x = 1$, $dx/dt = 0$, $y = 1$ and $dy/dt = 0$ when $t = 0$

11 Determine the initial and final values, provided limits exist, of the functions giving the following Laplace transforms:

(a) $\dfrac{4}{s+1}$, (b) $\dfrac{4}{(s+1)^2}$, (c) $\dfrac{3}{s(s-1)}$, (d) $\dfrac{5s-100}{s^2+10s}$

40 z-transform

40.1 Introduction

In Chapter 39 the concern was with signals which are functions of time and which vary continuously with time, such signals being represented by $f(t)$. However, if such signals are to be processed by a microprocessor they need to be converted into digital signals. This is done by sampling the continuous time signals at regular intervals and then converting each sample into digital form, i.e. using an analogue-to-digital converter. Such signals can then be referred to as *discrete-time signals* and the system being used as a *sampled data system*. When a signal $f(t)$ has been sampled and converted to digital form we represent it by $f^*(t)$. Thus in this chapter we consider discrete-time signals as arising from the sampling of a continuous-time signal, the resulting sample being then quantised and encoded to give the digital signal. While many digital signals do not occur in this way, this approach enables us to relate digital signals to the theoretical concepts of the Laplace transform developed for continuous-time signals. The result is what is known as the *z-transform*.

This chapter assumes Chapter 39, developing the z-transform from the Laplace transform. Chapter 5 on series is also assumed.

40.2 Sampling and the z-transform

Consider a continuous-time signal $f(t)$ which is sampled periodically at times 0, T, $2T$, $3T$, $4T$, etc. Figure 40.1 shows $f(t)$ and the sampled signal $f^*(t)$. The sampling element senses the value of $f(t)$ over a short interval of time Δt. If we take Δt to be very small then the result is a series of impulses which we can represent as having sizes given by:

$$f(0), f(1T), f(2T), f(3T), \ldots f(kT)$$

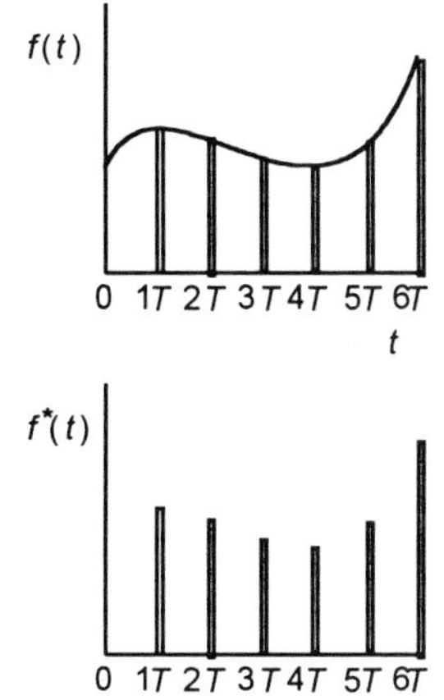

Figure 40.1 *Sampling*

where k is the number in the sequence of such impulses. The sequence can be denoted as $f[kT]$ where k can only take integer values. It is, however, common practice to denote such a sequence of digital signals as just $f[k]$ with the concept of sampling at periodic intervals of T being implied. The sampled-data signal is a series of impulses with that of $f[0]$ being one at time 0 and so can be written as $f[0]\,\delta(t)$, $f[1]$ being an impulse delayed by a time $1T$ and so written as $f[1]\,\delta(t-1T)$, $f[2]$ being an impulse delayed by a time $2T$ and so written as $f[2]\,\delta(t-2T)$, $f[3]$ being an impulse delayed by a time $3T$ and so written as $f[3]\,\delta(t-3T)$, etc. Thus:

$$f^*(t) = f[0]\,\delta(t) + f[1]\,\delta(t-1T) + f[2]\,\delta(t-2T) + f[3]\,\delta(t-3T) + \ldots \\ + f[k]\,\delta(t-kT)$$

The Laplace transform of an impulse at $t = 0$ is 1, at time $1T$ is e^{-Ts}, at time $2T$ is e^{-2Ts}, at time $3T$ is e^{-3Ts}, etc. Thus the Laplace transform of $f^*(t)$ is:

$$\mathcal{L}\{f^*(t)\} = F^*(s) = f[0]1 + f[1]\,e^{-Ts} + f[2]\,e^{-2Ts} + f[3]\,e^{-3Ts} + \ldots f[k]\,e^{-kTs}$$

$$= \sum_{k=0}^{\infty} f[k]\,e^{-kTs}$$

If we let $z = e^{Ts}$, i.e.

$$s = \frac{1}{T}\ln z \qquad [1]$$

then the above equation can be written as:

$$\mathcal{Z}\{f(k)\} = F(z) = f[0] + f[1]z^{-1} + f[2]z^{-2} + f[3]z^{-3} + \ldots + f[k]z^{-k}$$

$$= \sum_{k=1}^{\infty} f[k]z^{-k} \qquad [2]$$

When we have replaced s by z using equation [1] it is referred to as the *z-transform* of the sequence of impulses and written as $\mathcal{Z}\{f(k)\} = F(z)$.

In general, if any continuous function has a Laplace transform then the corresponding sampled function has a z-transform.

40.2.1 The z-transform

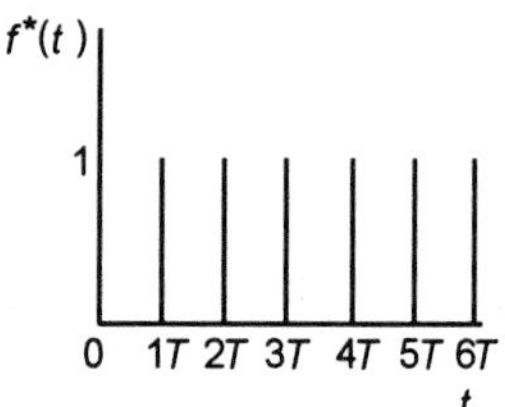

Figure 40.2 *Sampled unit step*

Consider the z-transform for a *sampled unit step* (Figure 40.2). Such a step has $f(t) = 1$ for all values of t greater than 0. Thus, using equation [2]:

$$F(z) = f[0] + f[1]z^{-1} + f[2]z^{-2} + f[3]z^{-3} + \ldots$$

$$= 1z^0 + 1z^{-1} + 1z^{-2} + 1z^{-3} + \ldots \qquad [3]$$

We can express this series in a *closed form*. Equation [3] is a geometric series (see Section 5.3.1) of the form $1 + x + x^2 + \ldots$ with the sum to infinity, provided the series converges (i.e. $|x| < 1$), of $1/(1 - x)$. Thus if we write $1/z$ for x:

$$F(z) = \frac{1}{1-(1/z)} = \frac{z}{z-1} \qquad [4]$$

provided $|z| > 1$.

Consider the z-transform of the *sampled ramp function* $f(t) = t$ (Figure 40.3). Equation [2] gives:

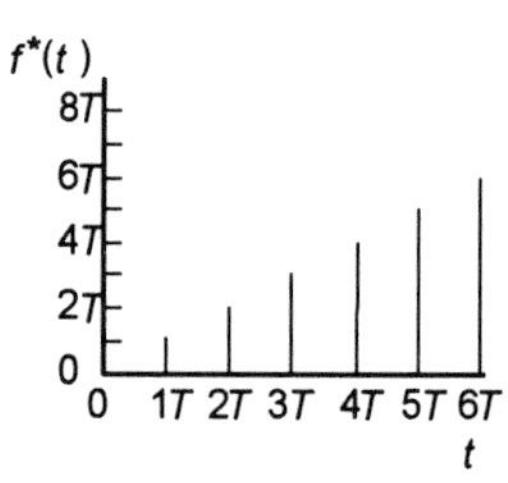

Figure 40.3 *Sampled ramp*

$$F(z) = 0 + Tz^{-1} + 2Tz^{-2} + 3Tz^{-3} + \ldots$$

This can be written as:

$$\frac{zF(z)}{T} = 1 + 2z^{-1} + 3z^{-2} + \ldots \qquad [5]$$

We can use the binomial theorem (see Section 5.5.1) for $(1 - x)^{-2}$ to obtain a series $1 + 2x + 3x^2 + \dots$. Hence, provided $|z| > 1$, equation [5] gives:

$$\frac{zF(z)}{T} = \frac{1}{(1 - 1/z)^2}$$

Thus:

$$F(z) = \frac{Tz}{(z - 1)^2} \tag{6}$$

Consider the z-transform for a *discrete-time signal* which has a regularly spaced sequence of pulses, i.e. the form 1, 1, 1, 1, 1, 1, ..., etc. This can be written as $f[k] = 1$. The z-transform is the sum of the transforms of the sequence of impulses. Thus, using equation [2]:

$$F(z) = f[0] + f[1]z^{-1} + f[2]z^{-2} + f[3]z^{-3} + \dots$$

$$= 1z^0 + 1z^{-1} + 1z^{-2} + 1z^{-3} + \dots \tag{7}$$

We can express this series in a *closed form*. Equation [7] is a geometric series (see Section 5.3.1) of the form $1 + x + x^2 + \dots$ with the sum to infinity, provided the series converges (i.e. $|x| < 1$), of $1/(1 - x)$. Thus if we write $1/z$ for x:

$$F(z) = \frac{1}{1 - (1/z)} = \frac{z}{z - 1} \tag{8}$$

provided $|z| > 1$.

Consider the z-transform of a *discrete-time signal* $f[k] = a^k$, i.e. the sequence a^0, a^1, a^2, a^3, ... etc. The z-transform is the sum of the transforms of the sequence of impulses. Thus, using equation [2]:

$$F(z) = f[0] + f[1]z^{-1} + f[2]z^{-2} + f[3]z^{-3} + \dots$$

$$= a^0 + a^1z^{-1} + a^2z^{-2} + a^3z^{-3} + \dots \tag{9}$$

We can express this series in a *closed form*. Equation [9] is a geometric series (see Section 5.3.1) of the form $1 + ax + a^2x^2 + \dots$ with the sum to infinity, provided the series converges (i.e. $|x| < 1$), of $1/(1 - ax)$. Thus if we write $1/z$ for x:

$$F(z) = \frac{1}{1 - a(1/z)} = \frac{z}{z - a} \tag{10}$$

provided $|z| > a$.

Revision

1 Determine the closed form of the z-transforms for (a) the sampled function $f(t) = e^{-at}$, (b) the discrete-time signal $x[k] = k$.

40.2.2 Standard z-transforms

It is not usually necessary to work from first principles to obtain the z-transforms of functions. Tables of commonly encountered transforms exist and these combined with basic properties of transforms enable z-transforms to be obtained for a wide range of functions. Table 40.1 gives some of the more commonly encountered z-transforms for sampled signals.

Table 40.1 *z-transforms*

	Sampled $f(t)$, sampling period T	$F(z)$
1	Unit impulse, $\delta(t)$	1
2	Unit step, $u(t)$	$\dfrac{z}{z-1}$
3	Unit ramp, t	$\dfrac{Tz}{(z-1)^2}$
4	t^2	$\dfrac{T^2z(z+1)}{(z-1)^3}$
5	e^{-at}	$\dfrac{z}{z-e^{-aT}}$
6	$1-e^{-at}$	$\dfrac{z(1-e^{-aT})}{(z-1)(z-e^{-aT})}$
7	$t\,e^{-at}$	$\dfrac{Tz\,e^{-aT}}{(z-e^{-aT})^2}$
8	$\sin \omega t$	$\dfrac{z \sin \omega T}{z^2 - 2z\cos \omega T + 1}$
9	$\cos \omega t$	$\dfrac{z(z-\cos \omega T)}{z^2 - 2z\cos \omega T + 1}$
10	$e^{-at} \sin \omega t$	$\dfrac{z\,e^{-aT} \sin \omega T}{z^2 - 2z\,e^{-aT} \cos \omega T + e^{-2aT}}$
11	$e^{-at} \cos \omega t$	$\dfrac{z(z-e^{-aT} \cos \omega T)}{z^2 - 2z\,e^{-aT} \cos \omega T + e^{-2aT}}$

The above table has given the transforms for sampled functions, they can, however, easily be adapted for discrete-time sequences. For example, if we require the z-transform of $f[k] = k$, i.e. the sequence 1, 2, 3, 4, etc., then this is the same as the transform of the sampled function $f(t) = t$ when $T = 1$ and so is $z/(z-1)^2$. If we require the transform of $f[k] = e^{-ak}$, i.e. the sequence e^{-0}, e^{-a}, e^{-2a}, etc., then this is the same as the sampled function $f(t) = e^{-at}$ when $T = 1$ and so is $z/(z - e^{-a})$. Table 40.2 gives some of the common transforms for discrete-time sequences.

Table 40.2 *z-transforms for discrete-time sequences*

	$f[k]$	$f[0], f[1], f[2], f[3], \dots$	$F(z)$
1	$1u[k]$	$1, 1, 1, 1, \dots$	$\dfrac{z}{z-1}$
2	a^k	$a^0, a^1, a^2, a^3, \dots$	$\dfrac{z}{z-a}$
3	k	$0, 1, 2, 3, \dots$	$\dfrac{z}{(z-1)^2}$
4	ka^k	$0, a^1, 2a^2, 3a^3, \dots$	$\dfrac{az}{(z-a)^2}$
5	e^{-ak}	$e^0, e^{-a}, e^{-2a}, e^{-3a}, \dots$	$\dfrac{z}{z-e^{-a}}$

Revision

2 Determine, using Table 40.1, the z-transforms for the sampled functions (a) $f(t) = e^{-2t}$, (b) $f(t) = \cos 3t$ and Table 40.2 for the discrete-time sequences (c) $f[k] = 2^k$, (d) $f[k] = k3^k$.

40.2.3 Properties of z-transforms

The following are basic properties of the z-transform. Because of the relationships between the Laplace transform and the z-transform, many of the properties of the Laplace transform are mirrored in those of the z-transform.

1 *Linearity*

The z-transform of the sum of two sequences is the sum of the z-transforms of the two sequences when considered separately:

$$\mathcal{Z}\{f[k] + g[k]\} = \mathcal{Z}\{f[k]\} + \mathcal{Z}\{g[k]\} \qquad [11]$$

This follows from the definition of the z-transform in equation [2]:

$$\mathcal{Z}\{f[k] + g[k]\} = \sum_{k=1}^{\infty}(f[k] + g[k])z^{-k} = \sum_{k=1}^{\infty} f[k]z^{-k} + \sum_{k=1}^{\infty} g[k]z^{-k}$$

The z-transform of a sequence multiplied by a constant is the same as multiplying by the constant the z-transform of the sequence:

$$\mathcal{Z}\{af[k]\} = a\mathcal{Z}\{f[k]\} \qquad [12]$$

Example

Determine the z-transform of the function $f(t) = t + e^{-t}$ when it is sampled every 1 s.

Using equation [11] and Table 40.1:

$$\mathcal{Z}\{\text{sampled }(t + e^{-t})\} = \frac{z}{(z-1)^2} + \frac{z}{z - e^{-1}}$$

Example

Determine the z-transform of the sequence $f[k] = k + e^{-k}$.

Using equation [7] and Table 40.2:

$$\mathcal{Z}\{k + e^{-k}\} = \frac{z}{(z-1)^2} + \frac{z}{z - e^{-1}}$$

2 *Shift theorems*
Consider the relationship between the z-transform of an *advanced* version of a sequence and that of the original sequence. If $f[k]$ is a sequence and $F(z)$ its z-transform, then the z-transform of the advanced sequence $f[k + n]$ is given by:

$$\mathcal{Z}\{f[k+n]\} = z^nF(z) - (z^nf[0] + z^{n-1}f[1] + z^{n-2}f[2] + \ldots + zf[n-1]) \qquad [13]$$

If $n = 1$ then equation [13] gives:

$$\mathcal{Z}\{f[k+1]\} = zF(z) - zf[0] \qquad [14]$$

If $n = 2$ then equation [13] gives:

$$\mathcal{Z}\{f[k+2]\} = z^2F(z) - z^2f[0] - zf[1] \qquad [15]$$

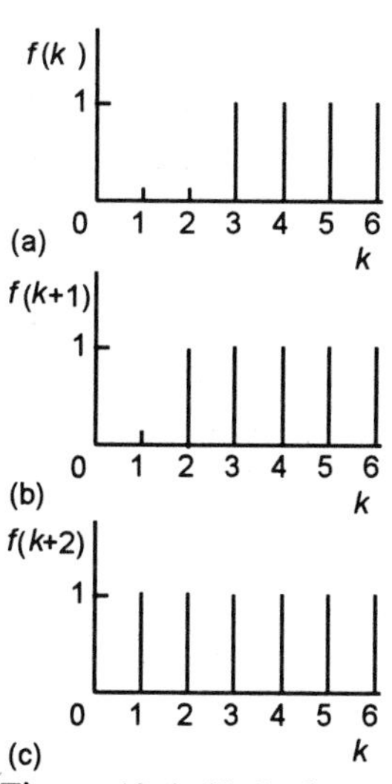

Figure 40.4 *Shifted sequences*

Note the similarity between the above equations and the Laplace transforms of derivatives. Figure 40.4 shows the sequences for a sampled unit step when the sequence has been shifted by 1 and by 2 steps. In (a) we have the sequence 0, 0, 0, 1, 1, 1, 1. In (b) we start at $k = 0$ with the signal that in (a) occurred at $k = 0 + 1$ and thus have 0, 0, 1, 1, 1, 1, 1. In (c) we start with the signal that occurred in (a) at $k = 0 + 2$ and thus have 0, 1, 1, 1, 1, 1, 1.

As an illustration of the derivation, consider that of equation [14]. If $y[k]$ is the sequence which is a single step advanced version of $x[k]$, i.e. we have $y[k] = x[k + 1]$, then equation [2] gives:

$$\mathcal{Z}\{y[k]\} = \sum_{k=0}^{\infty} y[k]z^{-k} = \sum_{k=0}^{\infty} x[k+1]z^{-k} = z\sum_{k=0}^{\infty} x[k+1]z^{-(k+1)}$$

Putting $p = k + 1$ gives:

$$Z\{y[k]\} = z\sum_{p=1}^{\infty} x[p]z^{-p} = z\left(\sum_{p=0}^{\infty} x[p] - x(0)\right) = zX(z) - zx(0)$$

A second shift theorem relates the z-transform of a *delayed sequence* to that of the original sequence. If a sampled function $f(t)u(t)$ is shifted

to the right by n sample intervals, then the z-transform of the shifted sampled function is given by:

$$\mathcal{Z}\{f[k-n]u[k-n]\} = z^{-n}F(z) \qquad [16]$$

We can derive this equation as follows. If $y[k]$ is the delayed version of the sequence $x[k]$, with a delay of n steps we have $y[k] = x[k-n]$. Using equation [2]:

$$\mathcal{Z}\{y[k]\} = \sum_{k=0}^{\infty} y[k]z^{-k} = \sum_{k=0}^{\infty} x[k-n]z^{-k}$$

If we let $p = k - n$, then:

$$\mathcal{Z}\{y[k]\} = \sum_{p=n}^{\infty} x[p]z^{-(p+n)}$$

If $x[k]$ has $x[p] = 0$ for $p < 0$ then:

$$\mathcal{Z}\{x[k-n]\} = \sum_{p=0}^{\infty} x[p]z^{-(p+n)} = z^{-n}\sum_{p=0}^{\infty} x[p]z^{-p} = z^{-n}X(z)$$

The shift theorems indicate that we can think of z as being a *time- shift operator*. Multiplication by z is equivalent to a time advance by one sampling interval, division by z being equivalent to a time delay by one sampling interval.

Example

Determine the z-transform of the sampled signal $(t-1)u(t-1)$ when sampled every 1 s.

We will consider this from first principles and then using equation [16]. The z-transform of the sampled $tu(t)$ signal, i.e. the sequence 0, 1, 2, 3, 4, etc., is:

$$F(z) = 0 + 1z^{-1} + 2z^{-2} + 3z^{-3} + 4z^{-4} + \ldots$$

If we write $zF(z) = 1 + 2z^{-1} + 3z^{-2} + 4z^{-3} + \ldots = 1/(1 - 1/z)^2$, then $F(z) = z/(z-1)^2$ (item 3 in Table 40.1). We require the transform of the sequence 0, 0, 1, 2, 3, 4, etc. This has the z-transform:

$$F(z) = 0 + 0 + 1z^{-2} + 2z^{-3} + 3z^{-4} + 4z^{-5} + \ldots$$

If we write $z^2F(z) = 1 + 2z^{-1} + 3z^{-2} + 4z^{-3} + \ldots = 1/(1 - 1/z)^2$, then $F(z) = 1/(z-1)^2$.

Using equation [16] with $n = -1$:

$$\mathcal{Z}\{\text{sampled } (t-1)u(t-1)\} = z^{-1}\frac{z}{(z-1)^2} = \frac{1}{(z-1)^2}$$

4 *Complex translation*
The z-transform of the sampled function $f(t)$ when multiplied by e^{-at} involves the substitution of $z\,e^{-aT}$ for z in the z-transform of the sampled $f(t)$, i.e.

$$\mathcal{Z}\{e^{-akT}f[k]\} = F(e^{aT}z) \qquad [17]$$

where $F(z)$ is the z-transform of $f[k]$.

Example

Determine the z-transform of the sampled function $t\,e^{-at}$ given that the z-transform of the sampled function t is $Tz/(z-1)^2$.

Using equation [17] and so substituting $z\,e^{aT}$ for z gives:

$$\frac{Tz\,e^{aT}}{(z\,e^{aT}-1)^2} = \frac{Tz\,e^{-aT}}{(z-e^{-aT})^2}$$

Revision

3 Use the linearity property and Table 40.1 to determine the z-transforms of the following functions when sampled:

(a) $f(t) = 3u(t)$, (b) $f(t) = 2t$, (c) $f(t) = t + t^2$, (d) $e^{-2t} - e^{-3t}$

4 Use the shift properties and Table 40.1 to determine the z-transforms of the following:

(a) the sequence 0, 0, 1, 0, 0, ..., (b) the sequence 0, 1, 2, 3, ...

5 Use the complex translation theorem to determine the z-transforms of the following functions when sampled:

(a) $e^{-at}\sin\omega t$, given the transform for $\sin\omega t$,

(b) $t^2\,e^{-at}$, given the transform for t^2

40.2.4 Laplace and z-transforms

Often the z-transform is required for a function which is specified by its Laplace transform. Note that the z-transform *cannot* be obtained by just replacing the s by a z. A basic approach that can be used is to determine the inverse Laplace transform, i.e. obtain $f(t)$, and then make the z-transformation.

Example

Determine the z-transform for the sampled function which has the Laplace transform of $1/(s + 2)(s + 3)$.

Using partial fractions to simplify the expression:

$$\frac{1}{(s+2)(s+3)} = \frac{A}{s+2} + \frac{B}{s+3}$$

$A(s + 3) + B(s + 2) = 1$ and so $A = 1$ and $B = -1$. Thus:

$$F(s) = \frac{1}{s+2} - \frac{1}{s+3}$$

Hence $f(t) = e^{-2t} - e^{-3t}$. Hence, using Table 40.1 and the linearity property:

$$F(z) = \frac{z}{z - e^{-2T}} - \frac{z}{z - e^{-3T}}$$

Revision

6 Determine the z-transforms for the sampled functions which have the following Laplace transforms:

(a) $\frac{1}{s}$, (b) $\frac{2}{s(s+2)}$, (c) $\frac{2}{(s+3)^2}$

40.3 The inverse *z*-transform

If $\mathcal{Z}\{f^*(t)\} = F(z)$ or $\mathcal{Z}\{f[k]\} = F(z)$ then the inverse z-transform is represented by $\mathcal{Z}^{-1}\{F(z)\}$. The inverse can be obtained in a number of ways. Here we will consider:

1 *Using tables*
Making use of the tables of transforms in order to recognise the function of time which gave rise to a particular transform.

Example

Determine the inverse transform of $4z/(z - 1)$.

This is recognisable from Table 40.1 as the sampled function $4u(t)$ or, from Table 40.2 as the discrete-time sequence $f[k] = 4u[k]$.

2 *Partial fractions*
This involves using partial fractions to simplify the z-transform and put it into a form which can be recognised in terms of standard transforms.

Example

Determine the inverse transform of $F(z) = \dfrac{z}{(z-1)(z-0.5)}$.

While we could determine the partial fractions of the above, a procedure which more often leads to standard forms is to obtain the partial fractions for $F(z)/z$. Thus:

$$\frac{F(z)}{z} = \frac{1}{(z-1)(z-0.5)} = \frac{A}{z-1} + \frac{B}{z-0.5}$$

$(z - 0.5)A + (z - 1)B = 1$ and so $A = 2$ and $B = -2$. Thus:

$$F(z) = \frac{2z}{z-1} - \frac{2z}{z-0.5}$$

Using Table 40.2:

$$f[k] = 2u[k] - 2 \times 0.5^k$$

This is the discrete-time sequence 0, 1, 1.5, 1.75, … .

3 *Expansion as a power series by long division*
This involves putting the transform into a power series by direct division of the numerator and denominator polynomials.

Example

Determine the inverse transform of $F(z) = \dfrac{z}{(z-1)(z-0.5)}$.

This is the example considered above for inversion by partial fractions. We can write it as:

$$F(z) = \frac{z}{z^2 - 1.5z + 0.5}$$

Using long division to expand this as a power series:

$$\begin{array}{r|l}
 & z^{-1} + 1.5z^{-2} + 1.75z^{-3} + \\
z^2 - 1.5z + 0.5 & z \\
 & \underline{z - 1.5z^0 + 0.5z^{-1}} \\
 & 1.5z^0 - 0.5z^{-1} \\
 & \underline{1.5z^0 - 2.25z^{-1} + 0.75z^{-2}} \\
 & 1.75z^{-1} - 0.75z^{-2} \\
 & \underline{1.75z^{-1} - 2.625z^{-2} + 0.875z^{-2}} \\
 & 1.875z^{-2} - 0.857z^{-2}
\end{array}$$

Hence $F[k] = 0z^0 + 1z^{-1} + 1.5z^{-2} + 1.75z^{-3} + \ldots$. This is the discrete-time sequence 0, 1, 1.5, 1.75, … .

Revision

7 Determine, using Table 40.1 and the properties of the transforms, the inverse transforms of the following:

(a) $\dfrac{z}{z-1}$, (b) $\dfrac{z}{z-\mathrm{e}^{-2T}}$

8 Determine, using (i) Table 40.2 and partial fractions and (ii) the power series expansion, the inverse transforms of the following, giving the answer as the first four terms in the discrete-time sequences:

(a) $\dfrac{z^2}{(z-1)(z-0.2)}$, (b) $\dfrac{0.3z}{(z-1)(z-0.7)}$

40.3.1 Initial and final value theorems

The initial and final value theorems enable the initial and final behaviour of a time sequence to be determined from the z-transform without the need to determine the inverse transform.

Using equation [2]:

$$\mathcal{Z}\{f[k]\} = F(z) = \sum_{k=0}^{\infty} f[k]z^{-k} = f[0] + \sum_{1}^{\infty} f[k]z^{-k}$$

As $z \to \infty$ then the summation tends to 0 and so:

$$f[0] = \lim_{t\to 0} f[k] = \lim_{z\to\infty} F(z) \qquad [18]$$

This is known as the *initial value theorem*.

The *final value theorem* can be stated as:

$$f[\infty] = \lim_{t\to\infty} f[k] = \lim_{z\to 1}\,(1-z^{-1})F(z) = \lim_{z\to 1}\frac{z-1}{z}F(z) \qquad [19]$$

provided the limit exists. It is given here without proof.

Example

Determine the initial and final values of the discrete-time sequence having the transform $F(z) = z^2/(z-1)(z-0.2)$.

We can write the transform as:

$$\frac{z^2}{z^2-1.2z+0.2} = \frac{1}{1-\dfrac{1.2}{z}+\dfrac{0.2}{z^2}}$$

Then, using the initial value theorem, as $z \to \infty$ we have $f[0] = 1$. The final value theorem gives:

$$f[\infty] = \lim_{z \to 1} \frac{z-1}{z} F(z) = \lim_{z \to 1} \frac{z}{z-0.2} = 1.25$$

Revision

9 Determine the initial and final values of the discrete-time sequences having the following z-transforms:

(a) $\frac{z}{(z-1)(z-0.1)}$, (b) $1 + \frac{3}{z} + \frac{1}{z^2}$

Problems

1 Use Table 40.1 or 40.2 and the properties of z-transforms to determine the z-transforms of the following sampled functions and sequences:

(a) $f(t) = \cos 4t$, (b) $f(t) = 3t$, (c) $2 - e^{-3t}$, (d) $3t\,e^{-4t}$, (e) $u(t - 4T)$,

(f) 1, 0, 0, 0, 0, ..., (g) 0, 1, 0, 0, 0, ..., (h) 0, 1, 1, 1, 1, ...,

(i) 0, 0, 1, 2, 3, 4, ...

2 Determine the z-transforms for the sampled functions which have the following Laplace transforms:

(a) $\frac{1}{s^2}$, (b) $\frac{3}{s^2+9}$, (c) $\frac{4}{(s+1)^2+16}$, (d) $\frac{4}{(s+5)(s+1)}$

3 Determine the inverse transforms of the following, expressing the results as the first four terms in the discrete-time sequence:

(a) $\frac{z}{z-3}$, (b) $\frac{z}{z-1}$, (c) $\frac{1}{z} + \frac{2}{z^3}$, (d) $\frac{2}{z^2} + \frac{1}{z} + 3$, (e) $\frac{z}{z-0.5}$,

(f) $\frac{z}{(z-1)(z-2)}$, (g) $\frac{z^4 + 2z^3 + z^2 + z}{z^4}$, (h) $\frac{z}{(z-1)(z-2)(z-3)}$,

(i) $\frac{z-3}{(z-1)(z-2)}$, (j) $\frac{2z+1}{z^2+2}$, (k) $\frac{z^3}{(z-1)^3}$

4 Determine the initial and final values of the discrete-time sequences having the following z-transforms:

(a) $\frac{z}{z-1}$, (b) $\frac{z}{z-0.5}$, (c) $\frac{2}{z^2} + \frac{1}{z} + 3$

41 Application: System response

41.1 Introduction

This follows on from Chapters 39 and 40, and shows how the Laplace transform can be used to determine the responses of systems, such as electrical circuits, to inputs such as impulses, step functions and ramps and the *z*-transform for discrete-time systems. This introduces the concept of difference equations.

Chapters 39 and 40 are assumed, together with a basic knowledge of electrical circuit analysis. The chapter can be usefully linked with Chapter 31 on the dynamic response of systems.

41.2 Circuits in the *s*-domain

While we could write differential equations to represent electrical circuits and then solve them by the use of the Laplace transform, a simpler method is to replace time-domain components by their equivalents in the *s*-domain.

Resistance R in the time domain is defined as $v(t)/i(t)$. Taking the Laplace transform of this equation gives a definition of resistance in the *s*-domain (Figure 41.1) as:

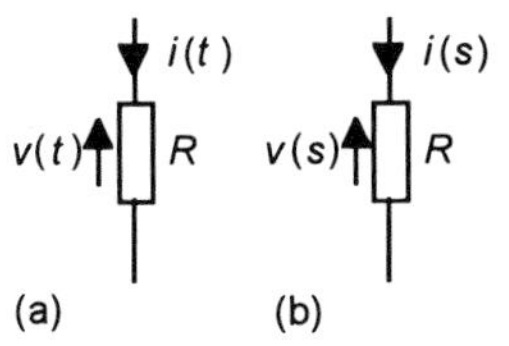

Figure 41.1 *Resistance: (a) time, (b) s-domain*

$$R = \frac{V(s)}{I(s)} \qquad [1]$$

Inductance L in the time domain (Figure 41.2(a)) is defined by:

$$v(t) = L\frac{\mathrm{d}i(t)}{\mathrm{d}t}$$

The Laplace transform of this equation is $V(s) = L[sI(s) - i(0)]$. With zero initial current then $V(s) = sLI(s)$. Impedance in the *s*-domain $Z(s)$ is defined as $V(s)/I(s)$, thus for inductance (Figure 41.2(b)):

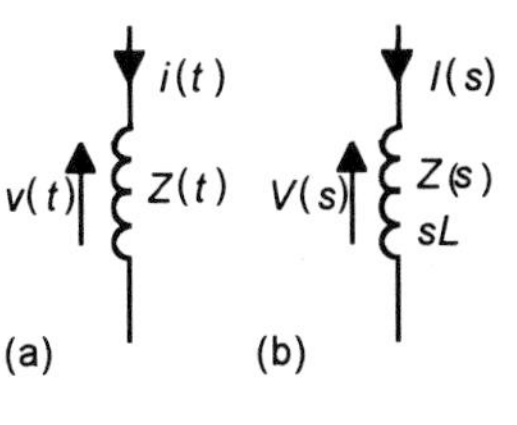

$$Z(s) = \frac{V(s)}{I(s)} = sL \qquad [2]$$

If the current was not initially zero but $i(0) = i_0$, then $V(s) = sLI(s) - Li_0$. This equation can be considered to describe two series elements (Figure 41.2(c)). The first term then represents the potential difference across the inductance L, being $Z(s)I(s)$, and the second term a voltage generator of $(-Li_0)$. Alternatively we can rearrange equation $V(s) = sLI(s) - Li_0$ in a form to represent two parallel elements (Figure 41.2(d)):

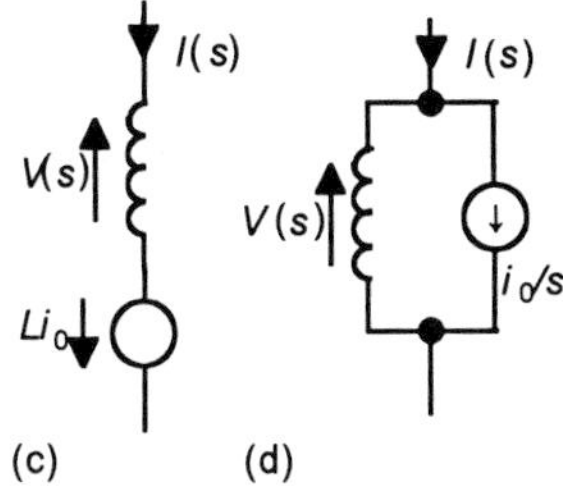

Figure 41.2 *Inductance: (a) time, (b), (c), (d) s-domain*

$$I(s) = \frac{V(s) + Li_0}{sL} = \frac{V(s)}{sL} + \frac{i_0}{s}$$

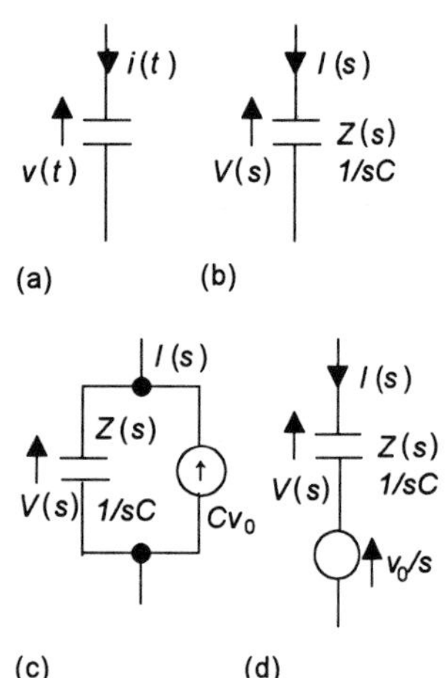

Figure 41.3 *Capacitance: (a) time domain, (b), (c), (d) s-domain*

$I(s)$ is the current into the system, $V(s)/sL = V(s)/Z(s)$ can be considered to be the current through the inductance and i_0/s a parallel current source.

Capacitance C in the time domain (Figure 41.3(a)) is defined by:

$$i(t) = C\frac{\mathrm{d}v(t)}{\mathrm{d}t}$$

The Laplace transform of this equation is $I(s) = C[sV(s) - v(0)]$. If we have $v(0) = 0$ then (Figure 41.3(b)):

$$Z(s) = \frac{V(s)}{I(s)} = \frac{1}{sC} \qquad [3]$$

If $v(0) = v_0$ then $I(s) = CsV(s) - Cv(0) = CsV(s) - Cv_0$. We can think of this representing $I(s)$ entering a parallel arrangement (Figure 41.3(c)) of a capacitor, and giving a current through it is $V(s)/Z(s) = CsV(s)$, and a current source $(-Cv_0)$. Alternatively we can rearrange the equation as:

$$V(s) = \frac{1}{sC}I(s) + \frac{v_0}{s}$$

This equation now represents a capacitor in series with a voltage source of v_0/s (Figure 41.3(d)).

Example

Determine the impedance and equivalent series circuit in the s-domain of an inductance of 50 mH if there is a current of 0.1 A at time $t = 0$.

The impedance in the s-domain is given by equation [2] as $0.050s$ Ω. Its equivalent series circuit with the initial condition $i(0) = 0.1$ A is of a voltage source of $-0.050 \times 0.1 = 0.005$ V in series with the impedance of $0.050s$ Ω.

Example

Determine the impedance in the s-domain of a capacitance of 0.1 μF and its equivalent series circuit when the capacitor has been charged to 5 V at time $t = 0$.

The impedance in the s-domain is given by equation [3] as $1/sC = 1/(0.1 \times 10^{-6}s)$ Ω, and its equivalent series circuit with the initial condition $v(0) = 5$ V is of a voltage source of $-5/s$ in series with the impedance of $10^7/s$ Ω.

Revision

1 Determine the series and parallel models in the s-domain for (a) an inductance of 10 mH when $i(0) = 0.2$ A, (b) a capacitance of 2 μF when $v(0) = 5$ V.

41.2.1 Circuit analysis

Because of the additive property of the Laplace transform, the transform of a number of time-domain functions is the sum of the transforms of each separate function. Thus with *Kirchhoff's current law*, the algebraic sum of the time-domain currents at a junction is zero and so the sum of the transformed currents is also zero. With *Kirchhoff's voltage law*, the sum of the time-domain voltages around a closed loop is zero and thus the sum of the transformed voltages is also zero. A consequence of this is that:

All the techniques developed for use in the analysis of circuits in the time domain can be used in the s-domain.

Example

Determine the impedance in the s-domain of a 10 Ω resistance in (a) series and (b) parallel with a 1 mH inductance.

(a) For impedances in series $Z(s) = Z_1(s) + Z_2(s) = 10 + 0.001s$ Ω.
(b) For impedances in parallel we have:

$$\frac{1}{Z(s)} = \frac{1}{Z_1(s)} + \frac{1}{Z_2(s)} = \frac{1}{10} + \frac{1}{0.001s} = \frac{0.001s + 10}{0.01s}$$

Hence $Z(s) = 0.01s/(0.001s + 10)$ Ω.

Example

Determine how the circuit current varies with time for a circuit having a resistance R in series with an initially uncharged capacitance C when the input to the circuit is a step voltage V at time $t = 0$.

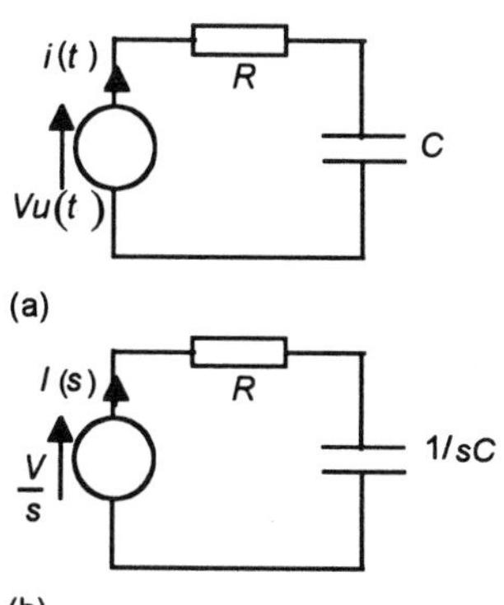

Figure 41.4 *Example*

Figure 41.4(a) shows the circuit in the time domain and Figure 41.4(b) the equivalent circuit in the s-domain. A unit step at $t = 0$ has the Laplace transform $1/s$ and thus a voltage step of V has a transform of V/s. The impedance of the capacitance is $1/sC$. Thus, applying Kirchhoff's voltage law to the circuit:

$$\frac{V}{s} = RI(s) + \frac{1}{sC}I(s)$$

and so:

$$I(s) = \frac{V}{Rs + 1/C} = \frac{V(1/R)}{s + (1/RC)}$$

This is of the form of a constant multiplied by $1/(s + a)$ and thus the inverse transform is:

$$i(t) = \frac{V}{R}\, e^{-t/RC}$$

Example

A ramp voltage of $v = kt$ is applied at time $t = 0$ to a circuit consisting of an inductance L in series with a resistance R. If initially at $t = 0$ there is no current in the circuit, determine how the circuit current varies with time.

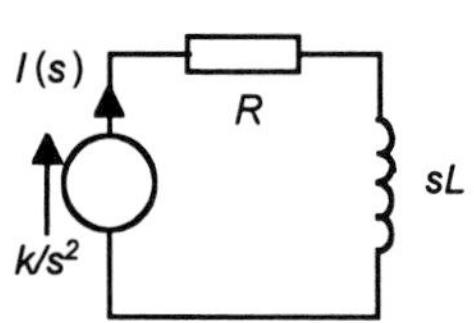

Figure 41.5 *Example*

The Laplace transform of kt is k/s^2. The inductance has an impedance in the s-domain of sL. Thus the circuit in the s-domain is as shown in Figure 41.5. Applying Kirchhoff's voltage law to the circuit gives:

$$\frac{k}{s^2} = sLI(s) + RI(s)$$

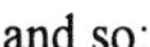
and so:

$$I(s) = \frac{k}{s^2(sL + R)} = \frac{(k/R)(R/L)}{s^2(s + R/L)}$$

This can be simplified by partial fractions, writing a for R/L:

$$\frac{a}{s^2(s+a)} = \frac{A}{s^2} + \frac{B}{s} + \frac{C}{s+a} = \frac{A(s+a) + Bs(s+a) + Cs^2}{s^2(s+a)}$$

Hence $A = 1$, $B = -1/a$ and $C = 1/a$. Thus:

$$I(s) = \frac{k}{R}\left(\frac{1}{s^2} - \frac{1}{(R/L)s} + \frac{1}{(R/L)(s + R/L)}\right)$$

Hence:

$$i(t) = \frac{k}{R}\left(t - \frac{1}{R/L} + \frac{e^{-Rt/L}}{R/L}\right)$$

Example

For the circuit shown in Figure 41.6(a), determine how the current through the 30 Ω resistor varies with time when there is an input of a step voltage of 10 V at time $t = 0$. The capacitors have no initial voltage.

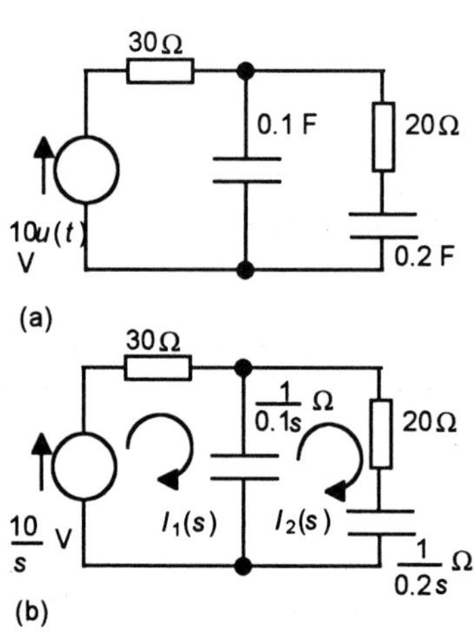

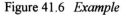

Figure 41.6 *Example*

Figure 41.6(b) shows the circuit in the s-domain. Using mesh analysis, Kirchhoff's voltage law gives for each mesh:

$$30I_1(s) + \frac{1}{0.1s}[I_1(s) - I_2(s)] = \frac{10}{s}$$

$$20I_2(s) + \frac{1}{0.2s}I_2(s) + \frac{1}{0.1s}[I_2(s) - I_1(s)] = 0$$

These can be simplified to give:

$$[3s + 1]I_1(s) - I_2(s) = 1$$

$$[4s + 3]I_2(s) - 2I_1(s) = 0$$

These two simultaneous equations can then be solved to give $I_1(s)$. Thus:

$$[3s + 1]I_1(s) - \frac{2I_1(s)}{4s + 3} = 1$$

and so:

$$I_1(s) = \frac{4s + 3}{(12s + 1)(s + 1)}$$

This can be simplified by the use of partial fractions:

$$\frac{4s + 3}{(12s + 1)(s + 1)} = \frac{A}{12s + 1} + \frac{B}{s + 1}$$

$4s + 3 = A(s + 1) + B(12s + 1)$ and thus $A = 32/11$ and $B = 1/11$.

$$I_1(s) = \frac{32/11}{12s + 1} + \frac{1/11}{s + 1} = \frac{32/11}{12(s + 1/12)} + \frac{1/11}{s + 1}$$

$$i_1(t) = \tfrac{8}{33}\, e^{-t/12} + \tfrac{1}{11}\, e^{-t} \text{ A}$$

Revision

2 Determine the impedance in the s-domain of a resistance of 10 Ω in (a) series, (b) parallel with a 2 mH inductance.

3 Determine how the current varies with time when a charged capacitor, with a potential difference of v_0, is allowed to discharge through a resistance R.

4 Determine how the current varies with time when a step voltage $Vu(t)$ is applied to a circuit consisting of a resistance R in series with an inductance L, there being no initial current in the circuit.

5 Determine how the current varies with time when a 1 V impulse is applied at time $t = 0$ to a circuit consisting of a resistance R in series with a capacitance C, there being no initial potential difference across the capacitor.

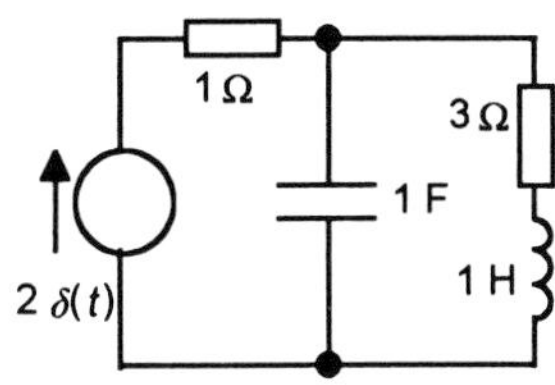

Figure 41.7 *Revision problem 6*

6 For the circuit shown in Figure 41.7, determine how the current through the 1 Ω resistor varies with time when there is an input to the circuit at time $t = 0$ of a 2 V impulse. There are no initial voltages or currents in the circuit.

41.3 Transfer function

If the input to a linear system has a Laplace transform of $X(s)$ and an output of $Y(s)$ then the *transfer function* $G(s)$ of the system is defined as:

$$G(s) = \frac{Y(s)}{X(s)} \qquad [4]$$

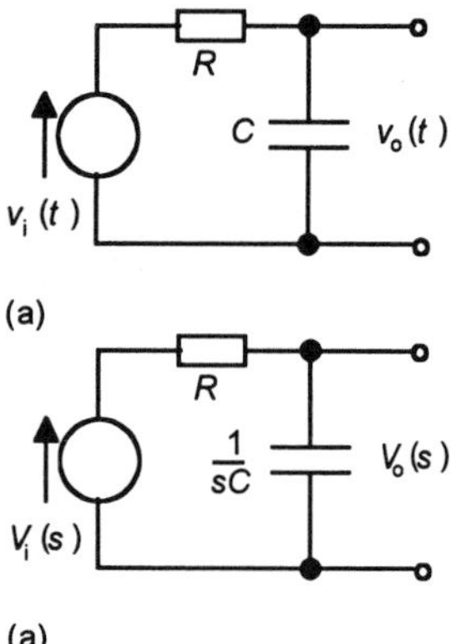

Figure 41.8 *Example*

when all the initial conditions are zero. The relation holds true for any input.

When there is a unit impulse input, i.e. $X(s) = 1$, then the output $Y(s) = G(s)$. For this reason the transfer function is sometimes referred to as the *impulse response*.

Example

Determine the transfer function relating the voltage output across the capacitor to the voltage input for the circuit shown in Figure 41.8(a).

Figure 41.8(b) shows the circuit in the *s*-domain. We have a potential divider circuit with the output $V_o(s)$ being related to the input $V_i(s)$ by:

$$\frac{V_o(s)}{V_i(s)} = \frac{1/sC}{R + 1/sC} = \frac{1}{sRC + 1}$$

This is the transfer function.

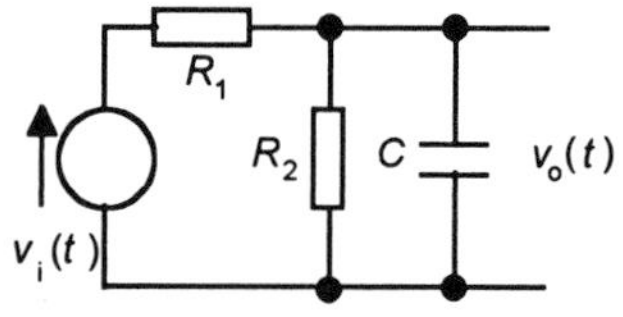

Figure 41.9 *Revision problem 7*

Revision

7 Determine the transfer function relating the voltage output across the capacitor with the input voltage for the circuit shown in Figure 41.9.

41.3.1 First-order systems

In general, a first-order system has its output $y(t)$ related to its input $x(t)$ by a differential equation of the form:

$$a_1 \frac{dy}{dt} + a_0 y = b_0 x \qquad [5]$$

Taking the Laplace transform of equation [5]:

$$a_1[sY(s) - y(0)] + a_0 Y(s) = b_0 X(s)$$

With initial conditions zero, we can write the transfer function as:

$$G(s) = \frac{Y(s)}{X(s)} = \frac{b_0}{a_1 s + a_0}$$

This is generally written in the form:

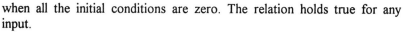

$$G(s) = \frac{b_0/a_0}{(a_1/a_0)s+1} = \frac{G}{\tau s+1} \qquad [6]$$

where G is the *gain* of the system when there are steady-state conditions, i.e. no dy/dt term, and $a_1/a_1 = \tau$ which is termed the *time constant.*

Example

Determine the response to a unit step input of a first-order system with a transfer function $G/(\tau s + 1)$.

The unit step input has a Laplace transform of $1/s$, thus equation [4] gives:

$$Y(s) = G(s)X(s) = \frac{G}{s(\tau s+1)} = G\frac{(1/\tau)}{s(s+1/\tau)}$$

Hence $y(t) = G(1 - e^{-t/\tau})$.

Revision

8 Determine the response to a unit ramp input of a first-order system with a transfer function $G/(\tau s + 1)$.

41.3.2 Second-order systems

In general, a second-order system has its output $y(t)$ related to its input $x(t)$ by a differential equation of the form:

$$a_2\frac{d^2y}{dt^2} + a_1\frac{dy}{dx} + a_0y = b_0x \qquad [7]$$

Taking the Laplace transform, with all initial conditions zero:

$$a_2s^2Y(s) + a_1sY(s) + a_0Y(s) = X(s)$$

Hence:

$$G(s) = \frac{Y(s)}{X(s)} = \frac{b_0}{a_2s^2 + a_1s + a_0} \qquad [8]$$

This is often written as:

$$G(s) = \frac{b_0}{(a_2/a_0)s^2 + (a_1/a_0)s + 1} = \frac{b_0\omega_n^2}{s^2 + 2\zeta\omega_n s + \omega_n^2} \qquad [9]$$

where $\omega_n = \sqrt{(a_0/a_2)}$, termed the *natural angular frequency*, and $\zeta = a_1/2\sqrt{(a_0a_2)}$, termed the *damping ratio*.

Example

Determine the response of a second-order system to a unit step input.

A unit step has the Laplace transform $1/s$ and thus, using the form given in equation [9] for a second-order system, the output transform $Y(s)$ is:

$$Y(s) = G(s)X(s) = \frac{b_0\omega_n^2}{s(s^2 + 2\zeta\omega_n s + \omega_n^2)}$$

This can be written in the form:

$$Y(s) = \frac{b_0\omega_n^2}{s(s + p_1)(s + p_2)}$$

where p_1 and p_2 are roots of the equation $s^2 + 2\zeta\omega_n s + \omega_n^2 = 0$.

$$p = \frac{-2\zeta\omega_n \pm \sqrt{4\zeta^2\omega_n^2 - 4\omega_n^2}}{2} = -\zeta\omega_n \pm \omega_n\sqrt{\zeta^2 - 1}$$

With $\zeta > 1$ the square root term is real and thus the inverse transform of $Y(s)$ can be obtained by simplifying the above equation using partial fractions:

$$\frac{1}{s(s + p_1)(s + p_2)} = \frac{A}{s} + \frac{B}{s + p_1} + \frac{C}{s + p_2}$$

$A(s + p_1)(s + p_2) + Bs(s + p_2) + Cs(s + p_1) = 1$ and so $A = 1/p_1p_2$, $B = -1/p_1(p_2 - p_1)$ and $C = 1/p_2(p_2 - p_1)$. Hence:

$$y(t) = \frac{b_0\omega_n^2}{p_1p_2}\left(1 - \frac{p_2}{p_2 - p_1}\,e^{-p_1t} + \frac{p_1}{p_2 - p_1}\,e^{-p_2t}\right)$$

The response is over-damped. With $\zeta = 1$ we have $p_1 = p_2 = -\omega_n$. The inverse transform then becomes:

$$Y(s) = \frac{b_0\omega_n^2}{s(s + \omega_n)^2}$$

This can be simplified by partial fractions to give:

$$Y(s) = b_0\left[\frac{1}{s} - \frac{1}{s + \omega_n} - \frac{\omega_n}{(s + \omega_n)^2}\right]$$

$$y(t) = b_0(1 - e^{-\omega_n t} - \omega_n t\,e^{-\omega_n t})$$

The response is critically damped. With $\zeta < 1$ the roots are imaginary. Hence by using Table 39.1, or simplifying using partial fractions, the inverse transform gives:

$$y(t) = b_0\left[1 - \frac{e^{-\zeta\omega_n t}}{\sqrt{1-\zeta^2}}\sin\left(\omega_n\sqrt{1-\zeta^2}\,t+\phi\right)\right]$$

where $\cos\phi = \zeta$. The response is under-damped.

Revision

9 Determine the response to a unit step input of a system having a transfer function $16/(s^2 + 8s + 16)$.

10 Determine the response to an input of $6\,e^{-2t}u(t)$ of a system having a transfer function of $s/(s^2 + 4s + 3)$.

41.3.3 Combining systems

If a system consists of a number of subsystems in series (Figure 41.10) then the overall transfer function $G(s)$ of the system is given by:

$$G(s) = \frac{Y(s)}{X(s)} = \frac{Y_1(s)}{X(s)} \times \frac{Y_2(s)}{Y_1(s)} \times \frac{Y(s)}{Y_2(s)} = G_1(s) \times G_2(s) + G_3(s) \qquad [10]$$

Thus:

The overall transfer function for a system composed of elements in series is the product of the transfer functions of the individual series elements.

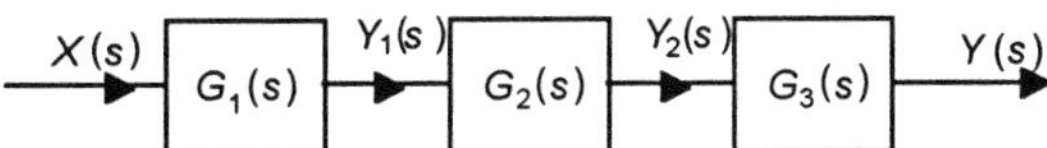

Figure 41.10 *System elements in series*

For systems with a feedback loop we can have the situation shown in Figure 41.11 where the output is fed back via a system with a transfer function $H(s)$ to subtract from the input to the system $G(s)$. This is termed *negative feedback*. The feedback signal to the input is $H(s)Y(s)$ and thus the input to the $G(s)$ system is $X(s) - H(s)Y(s)$. Hence:

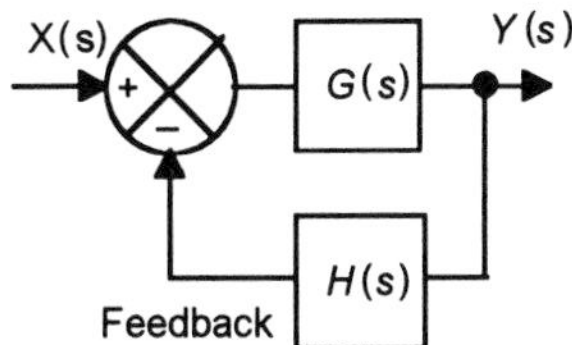

Figure 41.11 *Negative feedback system*

$$G(s) = \frac{Y(s)}{X(s) - H(s)Y(s)}$$

and so:

$$[1 + G(s)H(s)]Y(s) = G(s)X(s)$$

$$\text{transfer function of system} = \frac{Y(s)}{X(s)} = \frac{G(s)}{1 + G(s)H(s)} \qquad [11]$$

Example

Determine the overall transfer function for a system which consists of two elements in series, one having a transfer function of 1/(s + 1) and the other 1/(s + 2).

Using equation [10]:

$$\text{overall transfer function} = \frac{1}{s+1} \times \frac{1}{s+2} = \frac{1}{(s+1)(s+2)}$$

Example

Determine the overall transfer function for a control system which has a negative feedback loop with a transfer function 4 and a forward path transfer function of 2/(s + 2).

Using equation [11]:

$$\text{transfer function of system} = \frac{\frac{2}{s+2}}{1 - 4 \times \frac{2}{s+2}} = \frac{2}{s-6}$$

Revision

11 Determine the overall transfer function for a field-controlled d.c. motor if it can be considered to consist of three elements in series, a field circuit with a transfer function of $1/(Ls + R)$, an armature coil with a transfer function of k and a load with a transfer function of $1/(Is + c)$.

12 Determine the overall transfer function for a control system having a forward path transfer function of $5/(s + 3)$ and a negative feedback path with transfer function 10.

41.4 Discrete-time systems

With discrete-time systems, e.g. a microprocessor, there is an input of a sequence of pulses and an output of a sequence of pulses, the output being computed by the system by processing of the present pulse input and previous pulse inputs and possibly previous system outputs. Figure 41.12 illustrates a simple element of such a system. At the time of step k the pulse $x[k]$ enters the system as the input. This has subtracted from it the feedback signal of $ay[k]$, a being a scaling factor, and then the result, $x[k] - ay[k]$ proceeds to a unit delay block. A unit delay block takes an input of $y[k + 1]$ and gives an output of $y[k]$, i.e. it delays a sequence by one period. Thus:

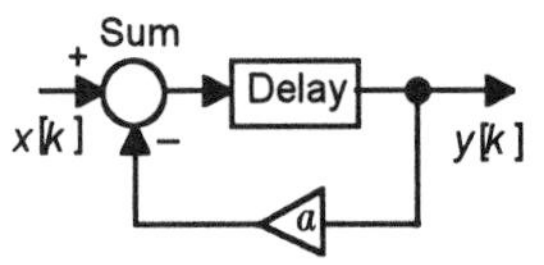

Figure 41.12 *Digital signal processor*

$$y[k+1] = x[k] - ay[k]$$

or

$$y[k+1] + ay[k] = x[k] \qquad [12]$$

Such an equation is termed a *difference equation* and is comparable to the differential equation relating the input and output of a system for analogue signals. This illustrates the types of action carried out by discrete-time signal processors.

As another illustration, consider representing the difference equation $y[k+2] - 2y[k+1] + 3y[k] = x[k]$ by a block diagram. We need a one unit delay block to convert $y[k+2]$ to $y[k+1]$ and another unit delay block to convert $y[k+1]$ to $y[k]$. The feedback of $y[k+1]$ is scaled by 2 and added to the input because it is –2, with the feedback of $y[k]$ scaled by 3. Figure 41.13 shows the resulting block diagram.

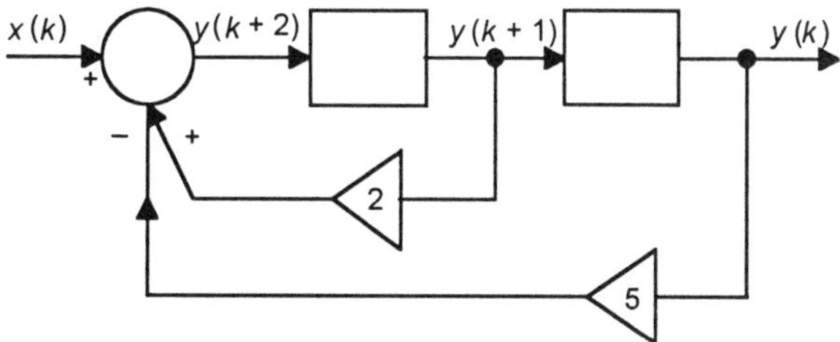

Figure 41.13 *Block diagram*

Revision

13 Draw a time-domain block diagram to represent the difference equation $y[k+1] - 3y[k] = x[k]$.

41.4.1 Solving difference equations

The analysis of a discrete-time signal processor involves solving its difference equation. This can be done by taking the z-transform of the equation. Thus with equation [12], i.e. $y[k+1] + ay[k] = x[k]$:

$$\mathcal{Z}\{y[k+1]\} + \mathcal{Z}\{ay[k]\} = \mathcal{Z}\{x[k]\} \qquad [13]$$

Suppose the input $x[k] = 1, 1, 1, 1$, etc. Then $\mathcal{Z}\{x[k]\} = z/(z-1)$ and, using the shift theorem (equation [14] in Chapter 40):

$$zF(z) - zf[0] + aF(z) = \frac{z}{z-1}$$

Thus:

$$F(z) = \frac{1}{z+a}\left(\frac{z}{z-1} + zx(0)\right) \qquad [14]$$

If, for example, we had $x(0) = 0$ and $a = 2$ then we would have:

$$F(z) = \frac{z}{(z+2)(z-1)}$$

Using long division we can write this as:

$$F(z) = z^{-1} - z^{-2} + 3z^{-2} + \dots$$

Hence:

$$y[k] = 0,\ 1,\ -1,\ 3,\ \dots$$

Revision

14 Solve the difference equations:

(a) $y[k] - 0.5y[k-1] = x[k]$, when $x[k]$ is the sequence 1, 1, 1, 1, ... and $y(0) = 0$,

(b) $y[k+2] + y[k+1] - 2y[k] = x[k]$, when $x[k]$ is the sequence 1, 1, 1, 1, ... and $y(0) = 0$, $y(1) = 1$

41.4.2 Pulse-transfer function

For a system which gives an output with a z-transform of $Y(z)$ for an input which has a z-transform of $X(z)$, we define the *pulse-transfer function* $G(z)$ as:

$$G(z) = \frac{Y(z)}{X(z)} \qquad [15]$$

when the system is at 'rest', i.e. there are no non-zero values stored on delay elements before the initial time.

Example

Determine the pulse-transfer function for a discrete-time system having the difference equation $y[k+2] + 5y[k+1] - 2y[k] = x[k]$.

Taking the z-transform with 'rest conditions':

$$z^2Y[z] + 5zY[z] - 2Y[z] = X[z]$$

Hence:

$$G(z) = \frac{Y(z)}{X(z)} = \frac{1}{z^2 + 5z - 2}$$

Example

Determine the response to a unit impulse of a system having a pulse transfer function of $z/(z + 1)(z + 2)$.

A unit impulse has a z-transform of 1. Thus:

$$Y(z) = G(z)X(z) = \frac{z}{(z+1)(z+2)}$$

This can be simplified by taking partial fractions or long division. Using long division:

$$Y(z) = z^{-1} - 3z^{-2} + 7z^{-3} - 15z^{-4} + \ldots$$

Hence:

$$y[k] = 0,\ 1,\ -3,\ 7,\ -15,\ \ldots$$

Revision

15 Determine the pulse-transfer function for a discrete-time system having the difference equation $y[k + 2] + y[k + 1] - 2y[k] = x[k]$.

16 Determine the response to a unit impulse of a system having a pulse-transfer function of $z/(z - 1)$.

41.4.3 Block diagrams in the z-domain

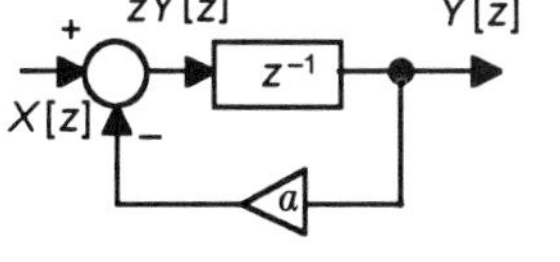

Figure 41.14 *Digital signal processor*

In the z-domain a block which produces a delay of one period has the transfer function z^{-1}. To illustrate this, consider the difference equation $y[k + 1] + ay[k] = x[k]$ (equation [12]), this being represented in the time-domain by Figure 41.12. Taking the z-transform of the equation gives $\mathcal{Z}\{y[k + 1]\} + \mathcal{Z}\{ay[k]\} = \mathcal{Z}\{x[k]\}$ (equation [13]). With zero initial conditions this becomes $zY[z] + aY[z] = X[z]$. This can be represented by the block diagram in Figure 41.14.

Revision

17 Represent the difference equation $y[k + 2] + 2y[k + 1] + 3y[k] = x[k]$ by a block diagram in the z-domain.

18 Represent a system having a pulse-transfer function $1/(z^2 + 3z + 1)$ by a block diagram in the z-domain.

Problems

1 Determine the series and parallel models in the s-domain for:

(a) an inductance of 1 mH when $i(0) = 2$ mA,

(b) a capacitance of 8 μF when $v(0) = 10$ V

2 A charged capacitor, capacitance 0.1 F and initial voltage 20 V, is allowed to discharge through a resistance of 5 Ω. Determine how the current changes with time.

3 A step voltage of 2 V is applied at time $t = 0$ to a circuit consisting of an inductance of 0.5 H in series with a resistance of 1 Ω. If there is an initial current of 1 A already in the circuit at $t = 0$, determine how the circuit current changes with time.

4 A step voltage of 10 V is applied at time $t = 0$ to a circuit consisting of an inductance of 2 H, a capacitance of 2 F and a resistance of 1 Ω in series. Determine how the circuit current will vary with time if the capacitor at $t = 0$ was uncharged.

5 A step voltage of 15 V is applied at time $t = 0$ to a circuit consisting of an inductance of 1 H, a capacitance of 0.5 F and a resistance of 3 Ω in series. Determine how the circuit current will vary with time if the capacitor at $t = 0$ had a voltage of 10 V.

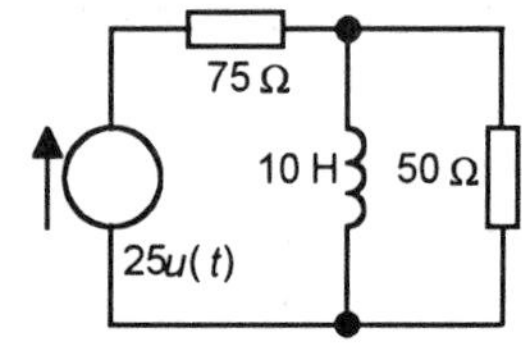

Figure 41.15 *Problem 6*

6 For the circuit shown in Figure 41.15, determine how the current through the 75 Ω resistor will vary with time when a step voltage of 25 V is applied at time $t = 0$. The initial conditions are zero.

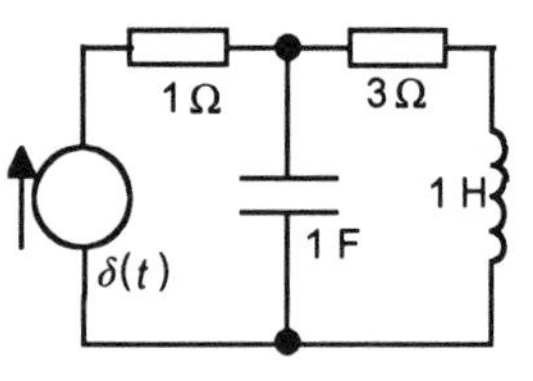

Figure 41.16 *Problem 7*

7 For the circuit shown in Figure 41.16. determine how the current through the 1 Ω resistor varies with time when there is a 1 V impulse applied to the input at time $t = 0$, the initial conditions being zero.

8 A circuit consists of a 2 Ω resistor in series with a 1 F capacitor. The voltage output is the potential difference across the capacitor. Determine the transfer function for the system.

9 Determine the response of a first-order system with a transfer function $G/(\tau s + 1)$ to a unit impulse input.

10 Determine the transfer function for a hydraulic system which has an input q and an output h related by the differential equation:

$$q = A\frac{dh}{dt} + \frac{\rho g h}{R}$$

11 Determine the transfer function for a spring-dashpot-mass system with an input F and an output x related by the differential equation:

$$m\frac{d^2x}{dt^2}+c\frac{dx}{dt}+kx=F$$

12 Determine the output, for an input of a unit step, of a system having the transfer function $s/(s+5)^2$.

13 Determine the output, for an input of a unit impulse, of a system having the transfer function $2/(s+3)(s+4)$.

14 Determine the output, for an input of $tu(t)$, from a system having a transfer function of $1/(s+1)$.

15 Determine the overall transfer function for a system if it can be considered to consist of three elements in series, the first with a transfer function of $1/(5s+1)$, the second with a transfer function of 5 and the third with a transfer function of $1/(3s+1)$.

16 Determine the overall transfer function for a control system having a forward path transfer function of $4/s(s+1)$ and a negative feedback path with transfer function $1/s$.

17 Determine the difference equation representing the discrete-time system shown by the block diagram in Figure 41.17.

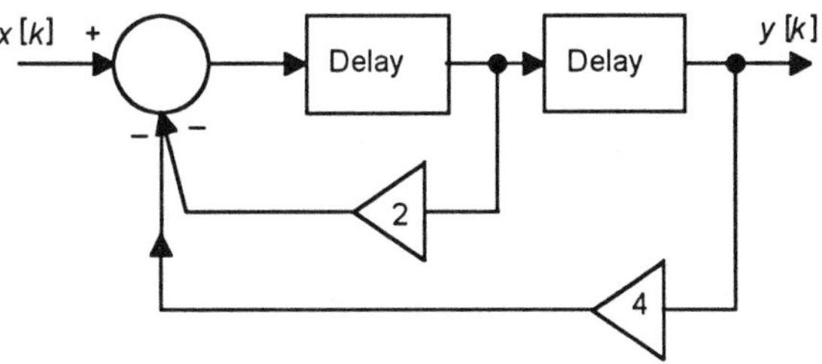

Figure 41.17 *Problem 17*

18 Solve the difference equations:

(a) $2y[k+1]-y[k]=x[k]$, with $x[k]=2^k$ and $y[0]=0$,

(b) $y[k+1]+y[k]=x[k]$, with $x[k]=1+2k$ and $y[0]=0$,

(c) $y[k+2]-4y[k]=x[k]$, with $x[k]=3k-5$ and $y[0]=0$, $y[1]=0$,

(d) $y[k+2]-5y[k+1]+6y[k]=x[k]$, with $x[k]=0$, $y[0]=0$, $y[1]=2$

19 Determine the pulse-transfer functions for discrete-time systems having the following difference equations:

(a) $y[k+1]+2y[k]=x[k]$,

(b) $y(k+2)-5y(k+1)+6y[k]=x[k]$,

(c) $y[k+2]+2y[k+1]+y[k]=x[k]$,

(d) $y[k+2)-y[k+1]+2y[k]=x[k]$

20 Represent the difference equation $y[k+3]+3y[k+1]+5y[k]=x[k]$ by a block diagram involving unit delay blocks and feedback in (a) the time domain, (b) the z-domain.

21 Determine the response to a unit impulse of systems having pulse-transfer functions of (a) $z/(z-2)$, (b) $z/(z-1)^2$.

Part 10 Probability and statistics

The aims of this part are to enable the reader to:

- Determine the probabilities of events occurring which are mutually exclusive, complementary or conditional.
- Explain and use the term probability density function.
- Define and use the measures of location and dispersion, mean and standard deviation.
- Define and use the terms permutation and combination.
- Describe the binomial, Poisson and normal distributions.
- Use the method of least squares to determine the best straight line through experimental points.
- Determine the overall errors in quantities resulting from errors in measurements used in their determination.

Probability is introduced through set notation and Venn diagrams and so Chapter 13 is assumed. Use is made of integration for the areas under curves and thus Chapter 25 is also assumed. Chapter 43 assumes Chapter 42, Chapter 44 assumes Chapter 43 and Chapter 45 assumes Chapter 43.

42 Probability

42.1 Introduction

What is the chance an engineering system will fail? What is the chance that a product emerging from a production line will be of the right quality? What is the chance that if you make a measurement in some experiment that it will be the true value of that quantity? Within what range of experimental error might you expect a measurement to be of the true value? These, and many other questions in engineering and science, involve a consideration of chances of events occurring. The term *probability* is more often used in mathematics than chance and has the same meaning in the above questions.

This chapter is about probability, its definition and determination in a number of situations. Some use is made of sets and Venn notation and so Chapter 13 is assumed. The consideration of probability leads in Chapter 43 to the concept of probability distributions and descriptions of measures of their location and dispersion, namely means and standard deviations.

42.2 Defining probability

If you flip a coin into the air, what is the chance that it will land heads uppermost? We can try such an experiment and determine the outcomes. The result of a large number of trials leads to the result that about half the time it lands heads uppermost and half the time tails uppermost. If n is the number of trials then we can define probability P as:

$$P = \lim_{n \to \infty} \frac{\text{number of times an event occurs}}{n} \qquad [1]$$

This view of probability is *the relative frequency in the long run with which an event occurs*. In the case of the coin this leads to a probability of ½ = 0.5. If an event occurs all the time then the probability is 1. If it never occurs the probability is 0.

The result of flipping the coin might seem obvious since there are just two ways a coin can land and just one of the ways leads to heads uppermost. If there is no reason to expect one way is more likely than the other then we can define probability P as *the degree of uncertainty about the way an event can occur* and as:

$$P = \frac{\text{number of ways a particular event can occur}}{\text{total number of ways events can occur}} \qquad [2]$$

In the case of the coin, this also gives a probability of 0.5. If every possible way events can occur is the required way, then the probability is 1. If none of the possible ways are the event required, then the probability is 0.

Another way the term probability is used is as *degree of belief*. Thus we might consider the probability of a particular horse winning a race as being 1 in 5 or 0.2. The probability in this case is highly subjective.

Example

In the testing of products on a production line, for every 100 tested 5 were found to contain faults. What is the probability that in selecting one item from 100 on the production line that it will be faulty?

There are 100 ways the item can be selected and 5 of the ways give faulty items. Thus, using equation [2], the probability is 5/100 = 0.05.

Revision

1 In a testing period of 1 year, 4 out of 50 of the items tested failed. What is the probability of finding one of the items failing?

2 In a pack of 52 cards there are 4 aces. What is the probability of selecting, at random, an ace from the pack?

42.2.1 Experiments, sample spaces and events

In considering probability we have to collect data on possible outcomes, this data generating process being called an *experiment.*

An experiment is any process which gives rise to data.

The term *random* or *stochastic* is used to describe an experiment where there is more than one possible outcome and the one that can occur is not known with certainty before the experiment is conducted.

An experiment, such as observing the results of flipping a coin, can result in a number of possible outcomes. A list, or set, of all possible outcomes of an experiment is called a *sample space*.

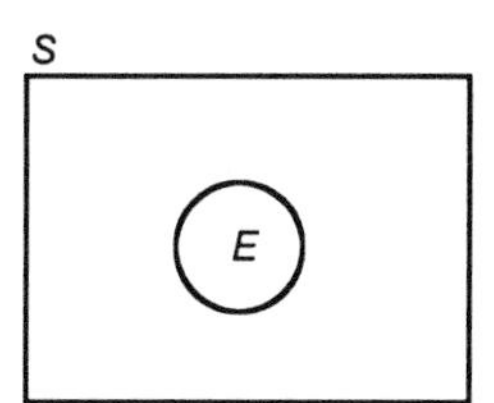

Figure 42.1 *Probability of an event*

A sample space S is a set that includes all possible outcomes for an experiment.

Thus with the flipping of a coin the sample space is the set {heads, tails}. A particular outcome, e.g. heads uppermost, is termed an *event.*

An event is any occurrence which results from the performance of an experiment and is a subset of the sample space.

The probability $P(E)$ of an event E occurring within a sample space S (Figure 42.1) is:

$$P(E) = \frac{\text{number of outcomes in } E}{\text{number of outcomes in } S} \qquad [3]$$

Consider a die-tossing experiment. A die can land in six equally likely ways, with uppermost 1, 2, 3, 4, 5, or 6. The sample space is thus {1, 2, 3, 4, 5, 6}. The Venn diagram for obtaining a 6, the required event, is shown

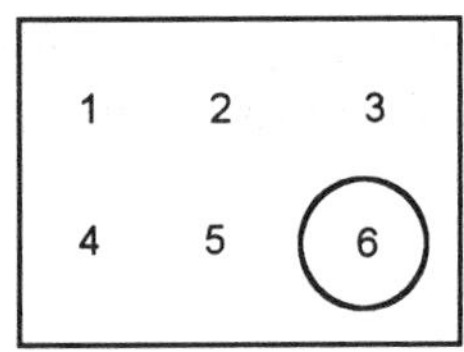

Figure 42.2 *Probability of a 6*

in Figure 42.2. Of the six possible ways the die could land, only one way is with 6 uppermost. Thus using definition [2], the probability of obtaining a 6 is 1/6. To use definition [1] we would have to carry out a large number of such experiments and determine the relative frequency with which the 6 occurred uppermost. The probability of *not* obtaining a 6 is 5/6 since there are 5 ways out of the 6 possible ways we can obtain an outcome which is not a 6. The probability of obtaining an outcome of 1, 2, 3, 4, 5 or 6, i.e. all the possible ways outcomes can occur, is 6 ways out of six and so 6/6 = 1. The probability of obtaining a particular event E plus the probability of not obtaining that event $\bar{E}$ (Figure 42.3) must thus be 1, i.e.

$$P(E) + P(\bar{E}) = 1 \qquad [4]$$

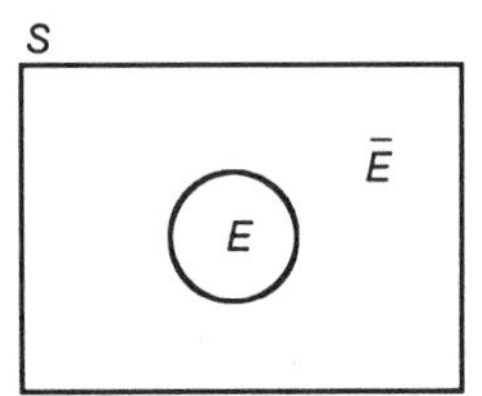

Figure 42.3 *Probability of an event*

If two events A and B are *complementary*, so that either one or the other occurs (as with an event occurring or not occurring and so we have one or the other situation), then we have the situation shown in Figure 42.3 and so:

$$P(A) + P(B) = 1$$

$$P(A \cap B) = 0$$

Complementary events are usually indicated with a bar over them, e.g. $\bar{A}$.

Suppose with the die-tossing experiment we were looking for the probability that the outcome would be an even number. The Venn diagram for obtaining an even number is shown in Figure 42.4. Of the six possible outcomes of the experiment, three ways give the required outcome. Thus, using definition [2], the probability of obtaining an even number is 3/6 = 0.5. This is the sum of the probabilities of 2 occurring, 4 occurring and 6 occurring, i.e. 1/6 + 1/6 + 1/6. The 2, the 4 and the 6 are mutually exclusive events in that if the 2 occurs then 4 or 6 cannot also be occurring.

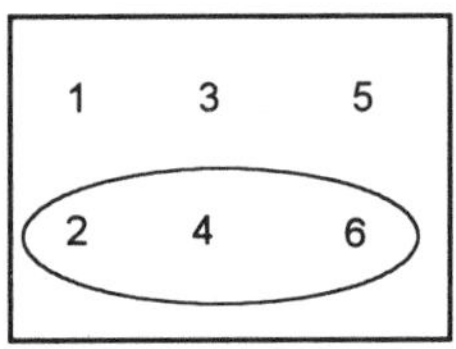

Figure 42.4 *Probability of an even number*

Mutually exclusive events are ones for which each outcome is such that one outcome excludes the occurrence of the other.

Thus if A and B are mutually exclusive then the probability of A or B occurring is the sum of the probabilities of A occurring and of B occurring (Figure 42.5), i.e.

$$P(A \cup B) = P(A) + P(B) \qquad [5]$$

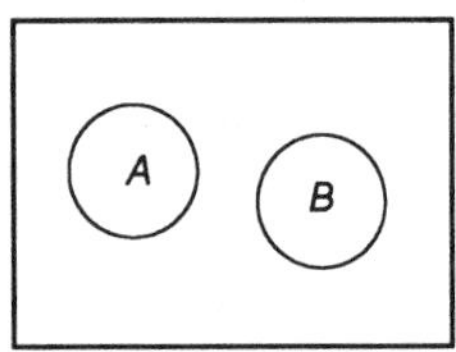

Figure 42.5 *Probability of mutually exclusive events*

Addition rule: If an event can happen in a number of different and mutually exclusive ways, the probability of its happening is the sum of the separate probabilities that each event happens.

If event A and event B cannot both occur we have:

$$A \cap B = \varnothing \qquad [6]$$

where ∅ indicates an impossible event. If A and B are *not* mutually exclusive (Figure 42.6), then the probability of A or B occurring but not both A and B is:

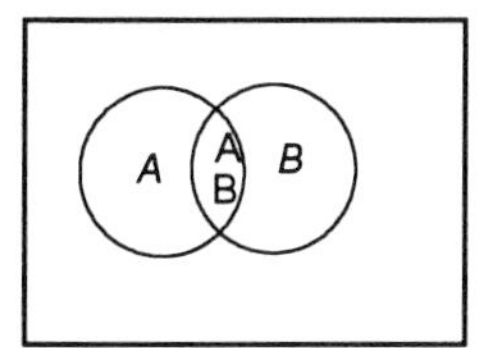

Figure 42.6 *Probability of A or B but not both*

$$P(A \cup B) = P(A) + P(B) - P(AB) \quad [7]$$

Example

The probability that a circuit will malfunction is 0.01. What is the probability that it will function?

The probability that it will function and the probability that it will not function are complementary. Thus, if $P(A)$ is the probability that it will function, equation [4] gives $P(A) + 0.01 = 1$. Hence the probability that it will function is 0.99.

Example

A company manufactures two products A and B. Market research over a month showed 30% of enquiries by potential customers resulting in product A alone being bought, 50% buying product B alone, 10% buying both A and B and 10% buying neither. Determine the probability that an enquiry will result in (a) product A alone being bought, (b) product A being bought, (c) both product A and product B being bought, (d) product B not being bought.

(a) 30% buy product A alone so the probability is 0.30.
(b) 30% buy product A alone and 10% buy A in conjunction with B. Thus the probability of A being bought is $0.30 + 0.10 = 0.40$.
(c) 10% buy products A and B so the probability is 0.10.
(d) 50% buy product B alone and 10% buy B in conjunction with A. Thus the probability of buying B is $0.50 + 0.10 = 0.60$. The probability of not buying B is thus $1 - 0.60 = 0.40$.

Revision

3 Tests of an electronic product show that 1% have defective integrated circuits alone, 2% have defective connectors alone and 1% have both defective integrated circuits and connectors. What is the probability of one of the products being found to have a (a) defective integrated circuit alone, (b) defective integrated circuit, (c) defective connector, (d) no defects?

4 Cars coming to a junction can turn to the left, to the right or go straight on. If observations indicate that all the possible outcomes are equally likely, determine the probability that a car will (a) go straight on, (b) turn from the straight-on direction.

42.3 Counting rules for numbers of ways

Suppose we flip two coins. What is the probability that we will end up with both showing heads uppermost? The ways in which the coins can land are:

HH HT TH TT

There are four possible results with just one of the ways giving HH. Thus the probability of obtaining HH is ¼ = 0.25.

There were two possible outcomes from the experiment of tossing the first coin and two possible outcomes from the experiment of tossing the second coin. For each of the outcomes from the first experiment there were two outcomes from the second experiment. Thus for the two experiments the number of possible outcomes is $2 \times 2 = 4$. This is an example of the *multiplication rule*.

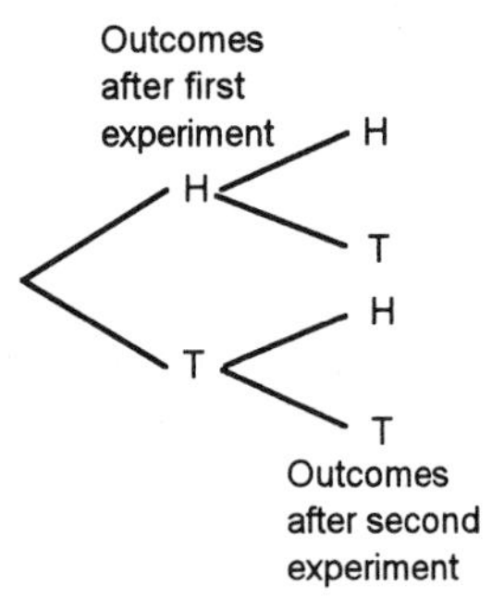

Figure 42.7 *Tree diagram*

Multiplication rule: If one experiment has n_1 possible outcomes and a second experiment n_2 possible outcomes then the compound experiment of the first experiment followed by the second has $n_1 \times n_2$ possible outcomes.

Tree diagrams can be used to visualise the outcomes in such situations, Figure 42.7 showing this for the two experiments of tossing coins.

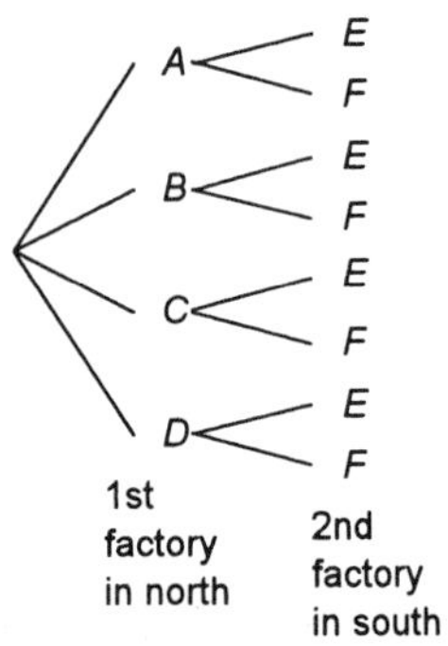

Figure 42.8 *Example*

Example

A company is deciding to build two new factories, one of them to be in the north and one in the south. There are four potential sites in the north and two potential sites in the south. Determine the number of possible outcomes.

For the first experiment there are 4 possible outcomes A, B, C and D and for the second 2 possible outcomes E and F. Thus the total number of possible outcomes is given by the multiplication rule as 8. Figure 42.8 shows the tree diagram.

Revision

5 Determine the possible number of outcomes from two throws of a die.

6 Each week a day for a safety inspection is chosen from the five working days. Determine the number of possible combinations of days in a four-week period.

42.3.1 Permutations

Suppose we had to select two items from a possible three different items A, B, C. The first item can be selected in three ways. Then, since the removal of the first item leaves just two remaining, the second item can be selected in two ways. Thus the selections we can have are:

AB AC BA BC CA CB

Each of the ordered arrangements is known as a *permutation*, each representing the way distinct objects can be arranged.

If there are n ways of selecting the first object, there will be $(n-1)$ ways of selecting the second object, $(n-2)$ ways of selecting the third object and $(n-r+1)$ ways of selecting the rth object. Thus by the multiplication rule, the total number of different permutations of selecting r objects from n distinct objects is thus:

$$n(n-1)(n-2)\ldots(n-r+1)$$

The number $n(n-1)(n-2)\ldots(3)(2)(1)$ is termed n factorial and represented by $n!$. The number of permutations of k objects chosen from n distinct objects is represented by ${}^{n}P_{r}$ or ${}_{n}P_{r}$ or $\binom{n}{r}$ and is thus:

$${}^{n}P_{r} = n(n-1)(n-2)\ldots(n-r+1) = \frac{n!}{(n-r)!} \qquad [8]$$

r taking values from 0 to n. Note that 0! is taken as having the value 1. The number of permutations of n objects chosen from n distinct objects is represented by ${}^{n}P_{n}$ or $\binom{n}{n}$ and is thus:

$${}^{n}P_{n} = n! \qquad [10]$$

Example

In the wiring up of an electronic component there are four assemblies that can be wired up in any order. In how many different ways can the component be wired?

This involves determining the number of permutations of four objects from four. Thus, using equation [8]:

$${}^{n}P_{n} = \frac{n!}{(n-n)!} = \frac{4\times3\times2\times1}{0!} = 24$$

or equation [9], ${}^{n}P_{n} = n! = 4! = 24$.

Example

How many four-digit numbers can be formed from the digits 0 to 9 if no digit is to be repeated within any one number?

This involves determining the number of permutations of 4 objects from 10. Thus, using equation [8]:

$${}^{n}P_{n} = \frac{n!}{(n-r)!} = \frac{10!}{(10-4)!}$$

$$= \frac{10\times9\times8\times7\times6\times5\times4\times3\times2\times1}{6\times5\times4\times3\times2\times1} = 5040$$

Revision

7 In how many ways can (a) 2 items be chosen from 5 distinct items, (b) 4 items be chosen from 52 distinct items?

8 In how many ways can five employees be assigned to five work shifts if there is to be only one employee working each shift.

42.3.2 Combinations

There are often situations where we want to know the number of ways r items can be selected from n objects without being concerned with the order in which the objects are selected. Suppose we had to select two items from a possible three different items A, B, C. The selections, i.e. permutations, we can have are:

AB AC BA BC CA CB

But if we are not concerned with the sequence of the letters then we only have the three ways *AB*, *AC* and *BC*. Such an unordered set is termed a *combination*.

Consider the selection of a combination of r items from n distinct objects. In the selected r items there will be $r!$ permutations (equation [10]) of distinct objects so that the permutation of r items from n contains each group of r items $r!$ times. Since there are $n!/(n-r)!$ different permutations of r items from n we must have:

$$r! \times {}^nC_r = \frac{n!}{(n-r)!}$$

where ${}^nC_r$, ${}_nC_r$ or $\binom{n}{r}$ is used to represent the combination of r items from n. Thus:

$${}^nC_r = \frac{n!}{r!(n-r)!} \qquad [11]$$

${}^nC_r$ is often termed a *binomial coefficient*. This is because numbers of this form appear in the expansion of $(x + y)^n$ by the binomial theorem (see Section 5.5.1 and Chapter 43).

When r items are selected from n distinct objects, $n - r$ items are left. The number of ways of selecting r items from n is given by equation [11] as $n!/r!(n-r)!$. The number of ways of selecting $n - r$ items from n is given by equation [11] as:

$${}^nC_{n-r} = \frac{n!}{(n-r)!(n-\{n-r\})!} = \frac{n!}{r!(n-r)!}$$

Thus we can say that there are as many ways of selecting r items from n as selecting $n - r$ objects from n:

$$ {}^nC_r = {}^nC_{n-r} \qquad [12] $$

There is just one combination of n items from n objects. Thus ${}^nC_n = 1$. If we select 0 items from n, then because equation [11] gives ${}^nC_0 = n!/0!$ and we take $1/0! = 1$, we have ${}^nC_0 = 1$. Evidently there are as many ways of selecting none of the items in a set of n as there are of choosing the n objects that are left.

Example

In how many ways can three objects be chosen from a sample of 20?

Using equation [11]:

$$ {}^{20}C_3 = \frac{20!}{3!17!} = \frac{20 \times 19 \times 18}{1 \times 2 \times 3} = 1140 $$

Example

If a batch of 20 objects contains 3 with faults and a sample of 5 is chosen, what is the probability of obtaining a sample with (a) 0, (b) 1, (c) 2 faulty items?

The number of ways we choose the sample of 5 items out of 20 is, using equation [11]:

$$ {}^{20}C_5 = \frac{20!}{5!15!} $$

(a) The number of ways we can choose a sample with 0 defective items is the number of ways we choose 5 items from 17 good items and is thus:

$$ {}^{17}C_5 = \frac{17!}{5!12!} $$

Thus the probability of choosing a sample with 0 faulty items is:

$$ \text{probability} = \frac{{}^{17}C_5}{{}^{20}C_5} = \frac{17!}{5!15!}\frac{5!15!}{20!} = \frac{91}{228} $$

(b) The number of ways we can choose a sample with 1 faulty item and 4 good items, i.e. selecting 1 faulty item from 3 faulty items and 4 good items from 17 good items, is given by the multiplication rule as ${}^3C_1 \times {}^{17}C_4$. Thus the probability of choosing a sample with 1 faulty item is:

$$\text{probability} = \frac{\frac{3!}{1!2!} \times \frac{17!}{4! \times 13!}}{\frac{20!}{5!15!}} = \frac{35}{76}$$

(c) The number of ways we can choose a sample with 2 faulty items and 3 good items, i.e. selecting 2 faulty items from 3 faulty items and 3 good items from 17 good items, is given by the multiplication rule as ${}^3C_2 \times {}^{17}C_3$. Thus the probability of choosing a sample with 2 faulty items is:

$$\text{probability} = \frac{\frac{3!}{2!1!} \times \frac{17!}{3! \times 14!}}{\frac{20!}{5!15!}} = \frac{5}{38}$$

Revision

9 In how many ways can 2 items be selected from 15 objects?

10 A sample contains 10 objects, two of them are red and 8 black. What is the probability that there is just 1 red object in a sample of 3 taken from the 10?

11 A batch of 12 components includes 1 that is defective. In how many ways can a sample of 3 components be selected with (a) none being faulty, (b) one being faulty?

12 If there are 20 red and 15 black objects, what is the probability that a sample of 5 taken at random will contain 3 red objects?

13 A batch of 100 components contains 5 faulty items. What is the probability that in a random sample of 6 there will be 1 faulty item?

42.4 Conditional probability

Suppose we have 50 objects of which 15 are faulty. What is the probability that the second object selected is faulty given that the first object selected was fault-free? This is a probability problem where the answer depends on the additional knowledge given that the first selection was fault-free. Such a problem is said to involve *conditional probability*.

To determine the conditional probability that event A occurs given that event B occurs, $P(A|B)$ representing this probability, we divide the probability that both A and B occur by the probability that B occurs, i.e.

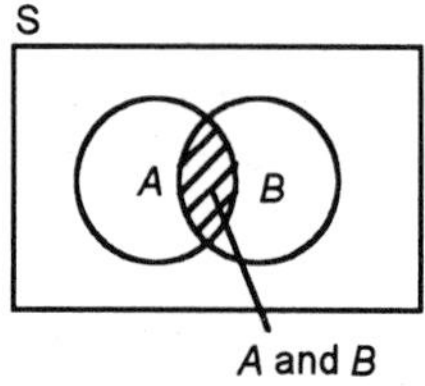

Figure 42.9 *Conditional probability*

$$P(A|B) = \frac{P(A \cap B)}{P(B)} \qquad [13]$$

assuming that $P(B) \neq 0$. Figure 42.9 illustrates the derivation of this equation for conditional probability by a Venn diagram. If we start with event B occurring it follows that its probability is B/S, with S being the entire sample space. If we now have to follow with event A then the sample space is now just B and so for an occurrence we must have a point in both A and B. Thus the probability that the event sequence $A|B$ will occur is $A \cap B$ divided by B which is the same as the probability of $A \cap B$ divided by the probability of B.

Example

Suppose we have 50 objects of which 15 are faulty. What is the probability that the second object selected is faulty given that the first object selected was fault-free?

Selecting the first object from 50 as fault-free has a probability of 35/50. Because the first object was fault-free we now have 34 fault-free and 15 faulty objects remaining. Now selecting a faulty object from 49 has a probability of 15/49. Using the multiplication rule gives the probability of the first object being fault-free followed by the second object faulty as (35/50)(15/49) = 0.21.

We could have used equation [13]. We have $P(A|B) = 15/49$ and $P(B) = 35/50$. Thus equation [13] gives:

$$P(A \cap B) = P(A|B) \times P(B) = \frac{15}{49} \times \frac{35}{50} = 0.21$$

Example

The probability that a product has no component faults is 0.8, the probability that it is assembled correctly is 0.9 and the probability that it is both free of component faults and assembled correctly is 0.7. Determine the probability that a component that is free of component faults is then assembled correctly.

Using equation [13]:

$$P(A|B) = \frac{P(A \cap B)}{P(B)} = \frac{0.7}{0.8} = 0.875$$

Revision

14 The probability that a product will have no faulty passive components is 0.8, the probability that it will have no faulty active components is 0.6 and the probability that it will have no faulty passive or active components is 0.3. Determine the probability that a product that is selected and found to have no faulty active components will have no faulty passive components.

15 A bag contains 100 balls, 40 of them black, 40 red and 20 white. If balls are drawn at random from the bag without replacement, what is the probability of drawing (a) a red ball, (b) a red ball followed by a black ball, (c) no white balls among the first three balls drawn.

16 If there is a 20% chance that a company will set up a new department and a 40% chance that you will be made head of the new department, what is the probability that you will become the head of the department?

42.4.1 Independence

Equation [13] for conditional probability $P(A|B)$ can be written as:

$$P(A \cap B) = P(B) \times P(A|B) \qquad [14]$$

Thus the probability that two events will both occur is the product of the probability $P(B)$ that one of the events will occur and the conditional probability that the other event will occur given that the first event has already occurred.

Consider the experiment of tossing a coin twice. Suppose the first toss gives heads. Does knowing that heads has occurred affect the probability that heads will occur for the second toss of the coin? The outcomes of the two experiments are: HH, HT, TH, TT, and thus the probability of obtaining heads on the second throw after obtaining heads on the first throw is ½. But this is the same as the probability of obtaining heads from a single throw of a coin. Knowing the outcome of the first throw was heads has not affected the probability of heads occurring with the second throw. Each toss of the coin is independent of previous tosses of the coin. When this occurs we say that two events are *independent*.

Events A and B are independent if the probability of the occurrence of B is not affected by the probability that A has occurred.

When A and B are two *independent* events, we can have $P(A|B) = P(A)$ and so equation [14] can be written as:

$$P(A \cap B) = P(B) \times P(A) \qquad [15]$$

This is the *multiplication rule*.

If we consider the probability of drawing two successive aces from a pack of cards, then if after drawing the first card we replace it back in the pack then the event of drawing the second card is independent of the first card. The probability of drawing an ace on both the first and second occasion is 4/52. The events are independent. However, if after drawing one ace we put it to one side and do not put it back in the pack then the probability of drawing a second ace is changed since the number of aces in the pack is now only 3 and the total number of cards 51 and so the probability is 3/51. We now have dependent events.

Example

A burglar alarm system has two sensors scanning a room. If the probability of each detecting an intruder in the room is 0.8, what is the probability that the alarm will sound when there is an intruder in the room?

The possible outcomes for the two sensors are OO, OD, DO, DD, where D stands for detected and O for not detected. The probability of a sensor detecting the intruder is 0.8, the probability of not detecting being $1 - 0.8 = 0.2$. Assuming that the detectors operate independently, equation [15] gives for both not detecting the intruder, i.e. OO, the probability $0.2 \times 0.2 = 0.04$, the probability of OD as $0.2 \times 0.8 = 0.16$, the probability of DO as $0.8 \times 0.2 = 0.16$ and the probability of DD as $0.8 \times 0.8 = 0.64$. Since at least one of the sensors must detect the intruder the probability is $1 - 0.04 = 0.96$, which is the same as that obtained by adding the probabilities of a D occurring, $0.16 + 0.16 + 0.64 = 0.96$.

Example

A manufacturer produces transistors with a reliability of 0.9 and resistors with a reliability of 0.8. If one transistor and one resistor are selected, what is the probability that both are acceptable?

Reliability is taken as meaning probability of not failing. Using equation [15]:

$$P(A \cap B) = P(B) \times P(A) = 0.9 \times 0.8 = 0.72$$

Revision

17 Resistors are available with a reliability of 0.92, capacitors with a reliability of 0.85 and inductors with a reliability of 0.95. What is the reliability if one of each is selected?

18 A production line is found to be producing products with 20% of them defective. What is the probability that two items in succession off the line are defective?

19 A box contains 5 red balls and 12 black ones. What is the possibility of drawing, at random, a red ball followed by a black ball if the red ball is (a) replaced after the first selection, (b) not replaced?

Problems

1 Testing of a particular item bought for incorporation in a product shows that of 100 items tested, 4 were found to be faulty. What is the probability that taking one item at random it will be found to be (a) faulty, (b) free from faults?

2 Resistors manufactured as 10 Ω by a company are tested and 5% are found to have values below 9.5 Ω and 10% above 10.5 Ω. What is the

probability that one resistor selected at random will have a resistance between 9.5 Ω and 10.5 Ω?

3 100 integrated circuits are tested and 3 are found to be faulty. What is the probability that one, taken at random, will result in a working circuit?

4 In how many ways can (a) 8 items be selected from 8 distinct objects, (b) 4 items be selected from 7 distinct items, (c) 2 items be selected from 6 distinct items?

5 In how many ways can (a) 2 items be selected from 7 objects, (b) 5 items be selected from 7 objects, (c) 7 items be selected from 7 objects?

6 How many samples of 4 can be taken from a batch of 25 items?

7 A batch of 24 components includes 2 that are faulty. If a sample of 2 is taken, what is the probability that it will contain (a) no faulty components, (b) 1 faulty component, (c) 2 faulty components?

8 A batch of 10 components includes 3 that are faulty. If a sample of 2 is taken, what is the probability that it will contain (a) no faulty components, (b) 2 faulty components?

9 If there are 5 red and 4 black objects and 3 of them are taken at random, what is the probability that all 3 are red?

10 Of 10 items manufactured, 2 are faulty. If a sample of 3 is taken at random, what is the probability of the sample containing (b) both the faulty items, (b) at least 1 faulty item?

11 A security alarm system is activated and deactivated by keying-in a three-digit number in the proper sequence. What is the total number of possible code combinations if digits may be used more than once?

12 When checking on the computers used in a company it was found that the probability of one having the latest microprocessor was 0.8, the probability of having the latest software 0.6 and the probability of having the latest processor and latest software 0.3. Determine the probability that a computer selected as having the latest software will also have the latest microprocessor.

13 A die is tossed. If you are told that the number obtained was 1, 2 or 3, what is the probability that it is an even number?

14 The probability that a bus leaves on time is 0.8 and the probability that a bus will leave on time and arrive on time is 0.7. What is the probability that if you catch a bus that leaves on time that it will arrive on time?

15 The probability that a student on the computer course will buy a computer is 0.5 and the probability that if he/she buys one it will be from a particular manufacturer is 0.7. What is the probability that a student on the course will have a computer from this manufacturer?

16 When a die is tossed twice, what is the probability of obtaining a 4 on the first throw and then a number less than 4 on the second throw?

17 A manufacturer finds that 90% of a product remains reliable for at least 2 years and 80% remains reliable for at least 3 years. What is the probability that a product item that has remained reliable for 2 years will remain reliable for 4 years?

18 A pair of dice are tossed twice. What is the probability of getting a sum of 7 on either of the two tosses?

19 Determine the probability of drawing two aces from a pack of cards if the first card is (a) replaced, (b) not replaced, after its selection.

20 One box contains 3 red and 5 black balls with another box containing 4 red and 2 black balls. If one ball is selected at random from each box, what is the probability that (a) one is red and the other black, (b) both are black?

21 A fail-safe system has two systems connected in parallel so that it only fails if both systems fails. If the probability that one of the systems will fail is 0.1, and the probability that the other system will fail is also 0.1, what is the probability of the entire system failing?

22 A batch of 100 items are taken for inspection. When inspected 12 are found to be oversize and 7 undersize. What is the probability that if (a) the first item selected had been acceptable and replaced in the batch, that a second item taken would also have been acceptable, (b) the first item had been oversize and replaced in the batch, that the second item would have been undersize?

43 Distributions

43.1 Introduction

All measurements are affected by random uncertainties and thus repeated measurements will give readings which fluctuate in a random manner from each other. This chapter follows on from Chapter 42 and considers the statistical approach to such variability of data, dealing with the measures of location and dispersion, i.e. mean, standard deviation and standard error, and the binomial, Poisson and Normal distributions.

In discussing variables the following terms are used:

Quantities whose variation contains an element of chance are called random variables.

Variables which can only assume a number of particular values are called discrete variables.

Variables which can assume any value in some range are called continuous variables.

Thus, for example, if we count the number of times per hour that cars pass a particular point then the result will be series of numbers such as 12, 30, 17, etc. The variable is thus discrete. However, if we repeatedly measure the time taken for 100 oscillations of a pendulum then the results will vary as a result of experimental errors and will be a series of values within a range of, say, 20.0 to 21.0 s. The variable can assume any value within that range and so is said to be a continuous variable.

43.2 Probability distributions

Consider the collection of data on the number of cars per hour passing some point and suppose we have the following results:

10, 12, 11, 13, 11, 12, 14, 12, 12, 11

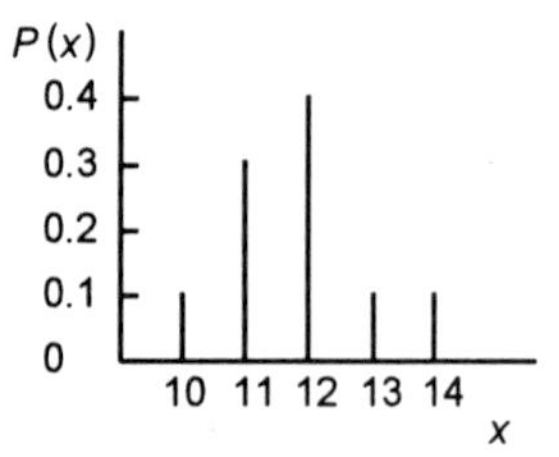

Figure 43.1 *Probability distribution*

When the discrete variable is sampled 10 times the value 10 appears once. Thus the probability of 10 appearing is 1/10. The value 11 appears three times and so its probability is 3/10, 12 has a probability of 4/10, 13 has the probability 1/10, 14 has the probability 1/10. Figure 43.1 shows how we can represent these probability values as a *probability distribution*.

Consider some experiment in which repeated measurements are made of the time taken for 100 oscillations of a simple pendulum and suppose we have the following results:

20.1, 20.3, 20.8, 20.5, 21.0, 20.8, 20.3, 20.4, 20.7, 20.6,
20.5, 20.7, 20.5, 20.1, 20.6, 20.4, 20.7, 20.5, 20.6, 20.3

With a continuous variable there are an infinite number of values that can occur within a particular range so the probability of one particular value occurring is effectively zero. However, it is meaningful to consider the probability of the variable falling within a particular subinterval. The term *frequency* is used for the number of times a measurement occurs within an interval and the term *relative frequency* or *probability* $P(x)$ for the fraction of the total number of readings in a segment. Thus if we divide the range of the above results into 0.2 intervals, we have:

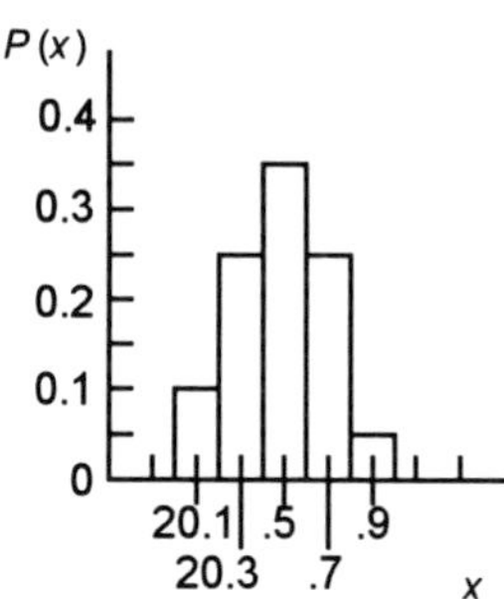

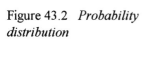

Figure 43.2 *Probability distribution*

values >20.0 and ≤20.2 come up twice, thus $P(x) = 2/20$
values >20.2 and ≤20.4 come up five times, thus $P(x) = 5/20$
values >20.4 and ≤20.6 come up seven times, thus $P(x) = 7/20$
values >20.6 and ≤20.8 come up five times, thus $P(x) = 5/20$
values >20.8 and ≤21.0 come up once, thus $P(x) = 1/20$

The probability always has a value less than 1 and the sum of all the probabilities is 1. Figure 43.2 shows how we can represent this graphically. The probability that x lies within a particular interval is thus the height of the rectangle for that strip divided by the sum of the heights of all the rectangles. Since each strip has the same width w:

$$\text{probability of } x \text{ being in an interval} = \frac{\text{area of strip}}{\text{total area}} \qquad [1]$$

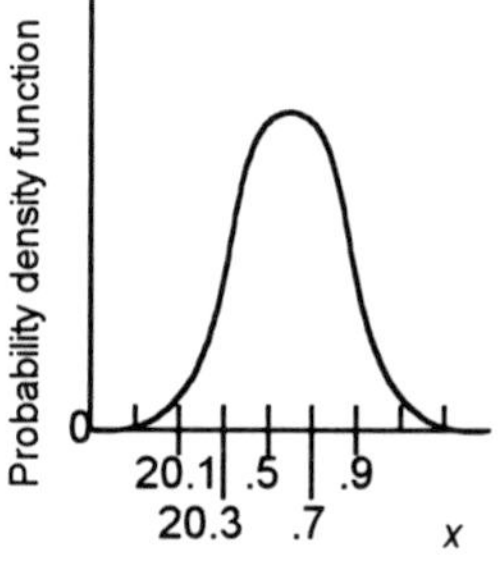

Figure 43.3 *Probability distribution function*

The histogram shown in Figure 43.2 has a jagged appearance. This is because it represents only a few values. If we had taken a very large number of readings then we could have divided the range into smaller segments and still had an appreciable number of values in each segment. The result of plotting the histogram would now be to give one with a much smoother appearance. When the probability distribution graph is a smooth curve, with the area under the curve scaled to have the value 1, then it is referred to as the *probability density function* $f(x)$ (Figure 43.3). Then equation [1] gives:

probability of x being in the interval $a < x \leq b$
= area under the function of that interval

$$= \int_a^b f(x)\,\mathrm{d}x \qquad [2]$$

The probability density function is a function that allocates probabilities to all of the range of values that the random variable can take. The probability that the variable will be in any particular interval is obtained by integrating the probability density function over that interval.

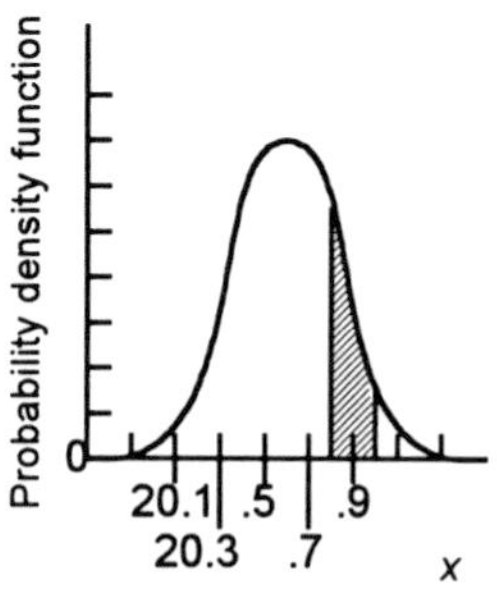

Figure 43.4 *Probability for interval 20.8 to 21.0*

Consider the probability, with a very large number of readings, of obtaining a value between 20.8 and 21.0 with the probability distribution function shown in Figure 43.4. If we take a segment 20.8 to 21.0 then the area of that segment is the probability. Suppose the result in this case is 0.30. The probability of taking a single measurement and finding it in that interval is thus 0.30, i.e. the measurement occurs on average 30 times in every 100 values taken.

Example

The following readings, in metres, were made for a measurement of the distance travelled by an object in 10 s. Plot the results as a distribution with segments of width 0.01 m.

13.478, 13.509, 13.502, 13.457, 13.492, 13.512, 13.475, 13.504, 13.473, 13.482, 13.492, 13.500, 13.493, 13.501, 13.472, 13.477

With segments of width 0.01 m we have:

Segment 13.45 to 13.46, frequency 1 and so probability 1/16
Segment 13.46 to 13.47, frequency 0 and so probability 0
Segment 13.47 to 13.48, frequency 5 and so probability 5/16
Segment 13.48 to 13.49, frequency 1 and so probability 1/16
Segment 13.49 to 13.50, frequency 4 and so probability 4/16
Segment 13.50 to 13.51, frequency 4 and so probability 4/16
Segment 13.51 to 13.52, frequency 1 and so probability 1/16

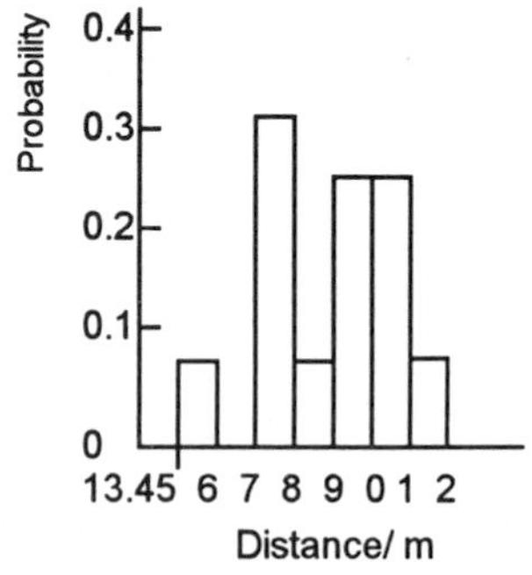

Figure 43.5 *Example*

Figure 43.5 shows the resulting distribution.

Example

A random variable x has a probability distribution function of e^{-x} for $x \geq 0$ and 0 for $x < 0$. Determine the probability of x being greater than 1. (Note that e^{-x} is a probability density function because the integral from 0 to infinity, i.e. area under the graph, has the value 1.)

The probability is given by equation [2] as:

$$P(x > 1) = \int_1^\infty e^{-x}\,dx = [-e^{-x}]_1^\infty = 0.368$$

Revision

1 A random variable x has a probability distribution function of $2\,e^{-2x}$ for $x \geq 0$ and 0 for $x < 0$. What is the probability of x being greater than 1?

2 A random variable x has a probability distribution function of $0.25/x^{1/2}$ for $0 \leq x < 4$ and 0 for all other values. What is the probability of x being greater than 1?

43.3 Measures of location and dispersion

Parameters which can be specified for distributions to give an indication of location and a measure of the dispersion or spread of the distribution about that value are the *mean* for the location and the *standard deviation* for the measure of dispersion.

43.3.1 Mean

The mean value $\bar{x}$ of a set of readings can be obtained in a number of ways, depending on the form with which the data is presented:

1 For a list of discrete readings, sum all the readings and divide by the number N of readings, i.e.:

$$\bar{x} = \frac{x_1 + x_2 + x_3 + \ldots + x_j}{N} = \frac{\Sigma x_j}{N} \qquad [3]$$

2 For a distribution of discrete readings, if we have n_1 readings with value x_1, n_2 readings with value x_2, n_3 readings with value x_3, etc., then the above equation for the mean becomes:

$$\bar{x} = \frac{n_1 x_1 + n_2 x_2 + n_3 x_3 + \ldots + n_j x_j}{N} \qquad [4]$$

But n_1/N is the relative frequency or probability of value x_1, n_2/N is the relative frequency or probability of value x_2, etc. Thus, to obtain the mean, multiply each reading by its relative frequency or probability P and sum over all the values:

$$\bar{x} = \sum_{j=1}^{n_j} P_j x_j \qquad [5]$$

3 For readings presented as a continuous distribution curve, we can consider that we have a discrete-value distribution with very large numbers of very thin segments. Thus if $f(x)$ represents the probability distribution and x the measurement values, the rule given above for discrete-value distributions translates into:

$$\bar{x} = \int_{-\infty}^{\infty} x f(x)\,\mathrm{d}x \qquad [6]$$

With a very large number of readings, the mean value is taken as being the *true value* about which the random fluctuations occur. The mean value of a probability distribution function is often termed the *expected value*.

Example

Determine the mean value of the probability density function $f(x) = 0.5\ \mathrm{e}^{-x/2}$ if $0 \le x$ and the function is 0 for $x < 0$.

Using equation [6]:

$$\bar{x} = \int_{0}^{\infty} 0.5x\ \mathrm{e}^{-x/2}\ \mathrm{d}x$$

This can be solved using integration by parts to give:

$$x = [-x\,e^{-x/2} - 2\,e^{-x/2}]_0^{\infty} = 2$$

Revision

3 Determine the mean value of a variable which can have the discrete values of 2, 3, 4 and 5 and for which 2 occurs twice, 3 occurs three times, 4 occurs three times and 5 occurs once.

4 Determine the mean value of a variable which has a probability distribution function of $2x$ for $0 \leq x \leq 1$ and 0 elsewhere.

43.3.2 Standard deviation

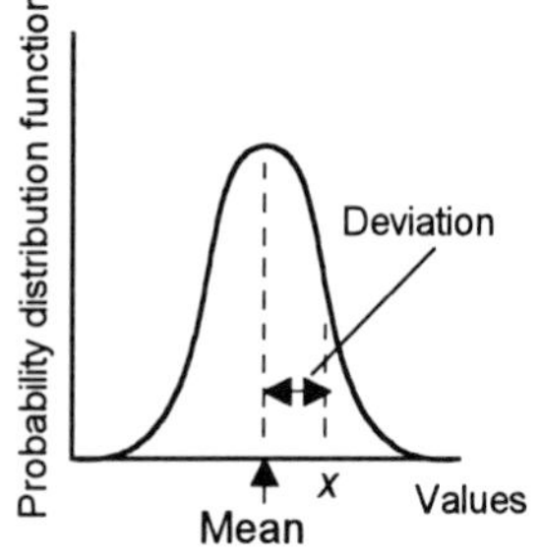

Figure 43.6 *Deviation*

Any single reading x in a distribution (Figure 43.6) will deviate from the mean of that distribution by:

$$\text{deviation} = x - \bar{x} \qquad [7]$$

With one distribution we might have a series of values which is widely scattered around the mean while another has readings closely grouped round the mean. Figure 43.7 shows the type of curves that might occur.

A measure of the spread of a distribution cannot be obtained by taking the mean deviation, since for every positive value of a deviation there will be a corresponding negative deviation and so the sum will be zero. The measure used is the *standard deviation*.

The standard deviation σ is the root-mean-square value of the deviations for all the measurements in the distribution. The quantity σ^2 is known as the variance of the distribution.

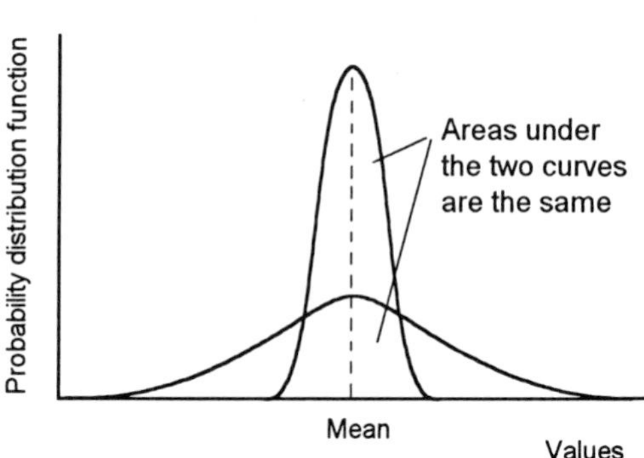

Figure 43.7 *Distributions with different spreads but the same mean*

Thus, for a number of discrete values, x_1, x_2, x_3, ..., etc., we can write for the mean value of the sum of the squares of their deviations from the mean of the set of results:

$$\text{sum of squares of deviation} = \frac{(x_1 - \bar{x})^2 + (x_2 - \bar{x})^2 + (x_3 - \bar{x})^2 + \dots}{N}$$

Hence the mean of the square root of this sum of the squares of the deviations, i.e. the standard deviation, is:

$$\sigma = \sqrt{\frac{\left((x_1 - \bar{x})^2 + (x_2 - \bar{x})^2 + (x_3 - \bar{x})^2 + \dots\right)}{N}} \qquad [8]$$

However, we need to distinguish between the standard deviation s of a sample and the standard deviation σ of the entire population of readings that are possible and from which we have only considered a sample (many statistics textbooks adopt the convention of using Greek letters when referring to the entire population and Roman for samples). When we are dealing with a sample we need to write:

$$s = \sqrt{\frac{\left((x_1 - \bar{x}_s)^2 + (x_2 - \bar{x}_s)^2 + (x_3 - \bar{x}_s)^2 + \dots\right)}{N-1}} \qquad [9]$$

with $\bar{x}_s$ being the mean of the sample. The reason for using $N - 1$ rather than N is that the root-mean-square of the deviations of the readings in a sample around the sample mean is less than around any other figure. Hence, if the true mean of the entire population were known, the estimate of the standard deviation of the sample data about it would be greater than that about the sample mean. Therefore, by using the sample mean, an underestimate of the population standard deviation is given. This bias can be corrected by using one less than the number of observations in the sample in order to give the sample mean.

A form of this equation which can simplify calculations, by not requiring each deviation to be individually calculated, is obtained in the following way. We can write the above equation for the sample standard deviation s as:

$$s^2 = \frac{1}{N-1}\sum_{j=1}^{N}\left(x_j - \bar{x}_s\right)^2$$

where j has the values 1, 2, 3, ..., etc., up to N. Hence:

$$s^2 = \frac{1}{N-1}\sum_{j=1}^{N}\left(x_j^2 - 2\bar{x}_s x_j + \bar{x}_s^2\right) = \frac{1}{N-1}\left[\sum_{j=1}^{N} x_j^2 - \sum_{j=1}^{N} 2\bar{x}_s x_j + \sum_{j=1}^{N} \bar{x}_s^2\right]$$

$$= \frac{1}{N-1}\left[\sum_{j=1}^{N} x_j^2 - 2\bar{x}_s \sum_{j=1}^{N} x_j + N\bar{x}_s^2\right]$$

$$= \frac{1}{N-1}\left[\sum_{j=1}^{N} x_j^2 - 2\bar{x}_s N\bar{x}_s + N\bar{x}_s^2\right] = \frac{1}{N-1}\left[\sum_{j=1}^{N} x_j^2 - N\bar{x}_s^2\right]$$

Hence:

$$s^2 = \frac{1}{N-1}\sum_{j=1}^{N} x_j^2 - \frac{N}{N-1}\bar{x}_s^2 \qquad [10]$$

If we were dealing with the entire population then the above equation would be:

$$\sigma^2 = \frac{1}{N}\sum_{j=1}^{N} x_j^2 - \bar{x}^2 \qquad [11]$$

i.e. the mean values of the squares minus the square of the mean.

When dealing with the entire population, if we have a number of discrete values, x_1, x_2, x_3, ..., etc., with frequencies n_1, n_2, n_3, etc., we can write:

$$\sigma^2 = \frac{(x_1-\bar{x})^2 n_1 + (x_2-\bar{x})^2 n_2 + \ldots}{N} = \frac{1}{N}\sum_{j=1}^{N} n_j\left(x_j - \bar{x}\right)^2 \qquad [12]$$

or when rearranged as:

$$\sigma^2 = \frac{1}{N-1}\left[\sum_{j=1}^{N} n_j x_j^2 - N\bar{x}^2\right] \qquad [13]$$

In the case of a sample, these equations become:

$$s^2 = \frac{1}{N-1}\sum_{j=1}^{N} n_j\left(x_j - \bar{x}\right)^2 \qquad [14]$$

$$s^2 = \frac{1}{N-1}\left[\sum_{j=1}^{N} n_j x_j^2 - N\bar{x}^2\right] \qquad [15]$$

For a continuous probability density distribution, since $(n_j/N)\ \Delta x$ is the probability for that interval Δx, i.e. $f(x)\ \Delta x$ where $f(x)$ is the probability distribution function, the above equation [12] becomes:

$$\sigma = \int_{-\infty}^{\infty} (x-\bar{x})^2 f(x)\,\mathrm{d}x \qquad [16]$$

We can write this equation in a more useful form for calculation:

$$\sigma^2 = \int_{-\infty}^{\infty} x^2 f(x)\,\mathrm{d}x - 2\bar{x}\int_{-\infty}^{\infty} x f(x)\,\mathrm{d}x + \bar{x}^2\int_{-\infty}^{\infty} f(x)\,\mathrm{d}x$$

Since the total area under the probability density function curve is 1, the third integral has the value 1 and so the third term is $\bar{x}^2$. The second integral is $\bar{x}$. The first integral is the mean value of x^2. Thus:

$$\sigma^2 = \overline{x^2} - \bar{x}^2 \qquad [17]$$

i.e. the mean value of x^2 minus the square of the mean value.

Example

Determine the mean value and the standard deviation of the sample of 10 readings 8, 6, 8, 4, 7, 5, 7, 6, 6, 4.

The mean value is (8 + 6 + 8 + 4 + 7 + 5 + 7 + 6 + 6 + 4)/10 = 6.1. The standard deviation of the sample can be calculated by considering the deviations of each reading from the mean or using equation [10]:

$$s^2 = \frac{1}{N-1}\sum_{j=1}^{N} x_j^2 - \frac{N}{N-1}\bar{x}_s^2$$

$$= \tfrac{1}{9}(8^2 + 6^2 + 8^2 + 4^2 + 7^2 + 5^2 + 7^2 + 6^2 + 6^2 + 4^2) - \tfrac{10}{9}6.1^2$$

Hence the standard deviation is 1.4.

Example

In an experiment involving the counting of the number of events that occurred in equal size time intervals the following data was obtained:

0 events 13 times, 1 event 12 times, 2 events 9 times
3 events 5 times, 4 events once

Determine the mean number of events occurring in the time interval and the standard deviation.

The total number of measurements is 13 + 12 + 9 + 5 + 1 = 40 and so the mean value is (0 × 13 + 1 × 12 + 2 × 9 + 3 × 5 + 4 × 1)/40 = 1.25. The standard deviation is given by equation [15] as:

$$s^2 = \frac{1}{N-1}\left[\sum_{j=1}^{N} n_j x_j^2 - N\bar{x}^2\right]$$

$$= \tfrac{1}{39}[(13 \times 0^2 + 12 \times 1^2 + 9 \times 2^2 + 5 \times 3^2 + 1 \times 4^2) - 40 \times 1.25^2]$$

Hence the standard deviation is 1.1.

Example

A probability density function has $f(x) = 1$ for $0 \le x < 1$ and elsewhere 0. Determine its mean and standard deviation.

The mean value is given by equation [6] as:

$$\bar{x} = \int_{-\infty}^{\infty} xf(x)\,\mathrm{d}x = \int_0^1 x\,\mathrm{d}x = \left[\frac{x^2}{2}\right]_0^1 = \frac{1}{2}$$

The standard deviation is given by equation [16]:

$$\sigma^2 = \int_{-\infty}^{\infty} (x-\bar{x})^2 f(x)\,\mathrm{d}x = \int_0^1 \left(x - \tfrac{1}{2}\right)^2 \mathrm{d}x = \int_0^1 \left(x^2 - x + \tfrac{1}{4}\right) \mathrm{d}x$$

$$= \left[\frac{x^3}{3} - \frac{x^2}{2} + \frac{x}{4}\right]_0^1 = \tfrac{1}{3} - \tfrac{1}{2} + \tfrac{1}{4} = \tfrac{1}{12} = 0.29$$

Alternatively, using equation [17] $\sigma^2 = \overline{x^2} - \bar{x}^2$, since:

$$\overline{x^2} = \int_{-\infty}^{\infty} x^2 f(x)\,\mathrm{d}x = \int_0^1 x^2\,\mathrm{d}x = \left[\frac{x^3}{3}\right]_0^1 = \tfrac{1}{3}$$

then

$$\sigma^2 = \tfrac{1}{3} - \left(\tfrac{1}{2}\right)^2 = \tfrac{1}{3} - \tfrac{1}{4} = \tfrac{1}{12} = 0.29$$

Revision

5 Determine the standard deviation of the resistance values for a sample of 12 resistors taken from a batch if the values are: 98, 95, 109, 99, 102, 99, 106, 96, 101, 108, 94, 102 Ω.

6 Determine the standard deviation of the six values: 1.3, 1.4, 0.8, 0.9, 1.2, 1,0.

7 Determine the standard deviation of the probability distribution function $f(x) = 2x$ for $0 \le x < 1$ and 0 elsewhere.

8 Determine the standard deviation of a random variable for which there are the following probabilities: for $x = 1$, probability 0.1; for $x = 2$, probability 0.2; for $x = 3$, probability 0.2; for $x = 4$, probability 0.3; for $x = 5$, probability 0.2.

43.4 Standard error of the mean

With a sample set of readings taken from a large population we can determine its mean, but what is generally required is an estimate of the error of that mean from the true value, i.e. the mean of an infinitely large number of readings. We can consider any set of readings as being just a sample taken from the very large set.

Consider a container holding six counters each having one of the numbers 3, 4, 5, 6, 7, 8. Now suppose we draw, at random, a sample of two of these counters. There are 15 different combinations of two counters we can draw. The following are the possible samples and their means:

3, 4	3, 5	3, 6	3, 7	3, 8	4, 5	4, 6	4, 7	4, 8	5, 6	5, 7	5, 8	6, 7	6, 8	7, 8
3.5	4.0	4.5	5.0	5.5	4.5	5.0	5.5	6.0	5.5	6.0	6.5	6.5	7.0	7.5

The mean values of the samples vary from 3.7 to 7.5. The mean of these means is 5.5 and the standard deviation of the means from 5.5 is about 1.1. Now suppose that, instead of samples of two, we take samples of four. There are 15 different combinations of four counters we can draw. The following are the possible samples and their means:

3, 4, 5, 6	3, 4, 5, 7	3, 4, 5, 8	3, 4, 6, 7	3, 4, 6, 8	3, 5, 6, 7	3, 5, 6, 8
4.5	4.75	5.0	5.0	5.25	5.25	5.5

3, 5, 7, 8	3, 6, 7, 8	3, 4, 7, 8	4, 5, 6, 7	4, 5, 6, 8	4, 5, 7, 8	4, 6, 7, 8
5.75	6.0	5.5	5.5	5.75	6.0	6.25

5, 6, 7, 8
6.5

The mean values of the samples vary from 4.5 to 6.5. The mean of these means is 5.5 and the standard deviation of the means from 5.5 is about 0.5. The standard deviation is smaller than that occurring with the samples of two taken from the population. Increasing the sample size by a factor of 2 has reduced the standard deviation. With the sample size of four, the means are thus much more closely bunched around the mean than with a sample size of two. The larger sample size thus reduces the chance of any one sample being too far from the mean.

The following is an algebraic analysis of the effect of sample size on the standard deviation of a mean from the true mean. Consider one sample of readings with n values: x_1, x_2, x_3, ... x_n. The deviation or error e_j in the jth reading is $e_j = x_j - X$, where X is the true value of the quantity. This true value is not known, being the value of the mean when the sample is of infinite size and so the total population. The mean of this sample is:

$$\bar{x} = \frac{1}{n} \sum x_j$$

This mean will have a deviation or error E from the true mean value X of $E = \bar{x} - X$. Hence we can write:

$$E = \left(\frac{1}{n} \sum x_j\right) - X = \frac{1}{n} \sum \left(x_j - X\right) = \frac{1}{n} \sum e_j$$

and so:

$$E = \frac{1}{n}\left(e_1 + e_2 + e_3 + \dots\ e_j\right)$$

Thus:

$$E^2 = \frac{1}{n^2}(e_1^2 + e_2^2 + e_3^2 + \ldots \text{ + products such as } e_1e_2\text{, etc.})$$

E is the error from the mean for a single sample of readings. Now, consider a large number of such samples with each set having the same number n of readings. We can write such an equation as above for each sample. If we add together the equations for all the samples and divide by the number of samples considered, we obtain an average value over all the samples of E^2. Thus E is the standard deviation of the means and is known as the *standard error of the means* e_m (more usually the symbol σ). Adding together all the error product terms will give a total value of zero, since as many of the error values will be negative as well as positive. The average of all the Σe_j^2 terms is ne_s^2, where e_s is, what can be termed, the *standard error of the sample*. Thus:

$$e_m = \frac{e_s}{\sqrt{n}}$$

But how we can we obtain a measure of the standard error of the sample? The standard error is measured from the true value X, which is not known. What we can measure is the standard deviation of the sample from its mean value. The best estimate of the standard error for a sample turns out to be the standard deviation s of a sample when we define it as:

$$s^2 = \frac{1}{n-1}\Sigma(x_j - \bar{x})^2$$

i.e. with a denominator of $N - 1$, rather than just N. Thus the best estimate of the standard error of the mean can be written as:

$$\text{standard error of the mean} = \frac{s}{\sqrt{n}} \qquad [18]$$

If random samples of size n are taken from a distribution and have a standard deviation σ then the sample means form a distribution with a standard deviation of $\sigma/\sqrt{n}$, this being termed the standard error.

Example

Measurements are to be made of the percentage of an element in a chemical by making measurements on a number of samples. The standard deviation of any one sample is found to be 2%. How many measurements must be made to give a standard error of 0.5% in the estimated percentage of the element.

If n measurements are made, then the standard error of the sample mean is given by equation [18] and so:

$$n = \frac{2^2}{0.5^2} = 16$$

Revision

9 A random sample of 25 items is taken and found to have a standard deviation of 2.0. (a) What is the standard error of the sample? (b) What sample size would have been required if a standard error of 0.5 was acceptable?

43.6 Common distributions

There are three basic forms of distribution which are found to represent many forms of distributions commonly encountered in engineering and science. These are the binomial distribution, the Poisson distribution and the normal distribution. The binomial and Poisson distributions are discrete probability distributions, the normal distribution being a continuous probability distribution.

43.6.1 Binomial distribution

In the tossing of a single coin the result is either heads or tails uppermost. We can consider this as an example of an experiment where the results might be termed as either success or failure, one result being the complement of the other. If the probability of succeeding is p then the probability of failing is $1 - p$. Such a form of experiment is termed a *Bernoulli trial*.

Suppose the trial is the throwing of a die with a 6 uppermost being success. The probability of obtaining a 6 as the result of one toss of the die is 1/6 and the probability of not obtaining a 6 is 5/6. Suppose we toss the die n times. The probability of obtaining no 6s in any of the trials is given by the product rule as $(5/6)^n$. The probability of obtaining one 6 in, say, just the first trial out of the n is $(5/6)^{n-1}$ (1/6). But we could have obtained the one 6 in any one of the n trials. Thus the probability of one 6 is $n(5/6)^{n-1}$ (1/6). The probability of obtaining two 6s in, say, just the first two trials is $(5/6)^{n-2}(1/6)^2$. But these two 6s may occur in the n trials in a number of combinations $n!/2!(n-2)$ (see Section 42.3.2). Thus the probability of two 6s in n trials is $[n!/2!(n-2)](5/6)^{n-2}(1/6)^2$. We can continue this for three 6s, 4s, etc.

In general, if we have n independent Bernoulli trials, each with a success probability p, and of those n trials k give successes, and $(n-k)$ failures, the probability of this occurring is given by the product rule as:

$$^nC_k p^k(1-p)^{n-k} = \frac{n!}{k!(n-k)!}p^k(1-p)^{n-k} \qquad [19]$$

This is termed the *binomial distribution*. This term is used because, for k = 0, 1, 2, 3, ... n, the values of the probabilities are the successive terms of the binomial expansion of $[(1-p)+p]^n$.

For a single Bernoulli trial of a random variable x with probability of success p, the mean value is p. The standard deviation is given by equation [17] as:

$$\sigma^2 = \overline{x^2} - \bar{x}^2 = p^2 - p = p(1-p)$$

For n such trials:

$$\text{mean value} = np \qquad [20]$$

$$\text{standard deviation} = \sqrt{np(1-p)} \qquad [21]$$

The characteristics of a variable that gives a binomial distribution are that the experiment consists of n identical trials, there are two possible complementary outcomes, success or failure, for each trial and the probability of a success is the same for each trial, the trials are independent and the distribution variable is the number of successes in n trials.

Example

Three identical dice are tossed. Determine the distribution of the random variable x if x is a 6 uppermost.

Using equation [29]:

$$P(0) = \frac{n!}{k!(n-k)!}p^k(1-p)^{n-k} = \frac{3!}{0!3!}\left(\tfrac{1}{6}\right)^0\left(\tfrac{5}{6}\right)^3 = 0.58$$

$$P(1) = \frac{n!}{k!(n-k)!}p^k(1-p)^{n-k} = \frac{3!}{1!2!}\left(\tfrac{1}{6}\right)^1\left(\tfrac{5}{6}\right)^2 = 0.35$$

$$P(2) = \frac{n!}{k!(n-k)!}p^k(1-p)^{n-k} = \frac{3!}{2!1!}\left(\tfrac{1}{6}\right)^2\left(\tfrac{5}{6}\right)^1 = 0.07$$

$$P(3) = \frac{n!}{k!(n-k)!}p^k(1-p)^{n-k} = \frac{3!}{3!0!}\left(\tfrac{1}{6}\right)^3\left(\tfrac{5}{6}\right)^0 = 0.005$$

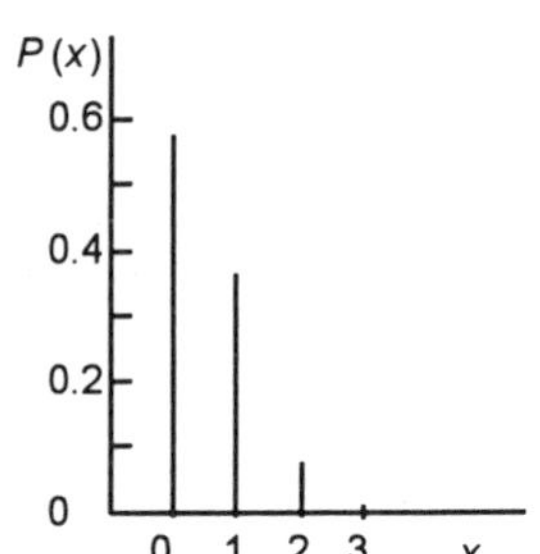

Figure 43.8 *Example*

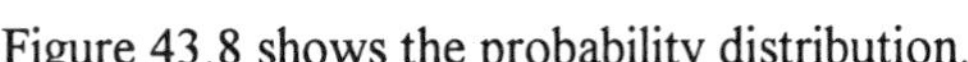

Figure 43.8 shows the probability distribution.

Example

The probability that an enquiry from a potential customer will lead to a sale is 0.30. Determine the probabilities that among six enquiries there will be 0, 1, 2, 3, 4, 5, 6 sales.

Using equation [29]:

$$P(0) = \frac{n!}{k!(n-k)!}p^k(1-p)^{n-k} = \frac{6!}{0!6!}(0.30)^0(0.70)^6 = 0.118$$

$$P(1) = \frac{n!}{k!(n-k)!}p^k(1-p)^{n-k} = \frac{6!}{1!5!}(0.30)^1(0.70)^2 = 0.303$$

$$P(2) = \frac{n!}{k!(n-k)!}p^k(1-p)^{n-k} = \frac{6!}{2!4!}(0.30)^2(0.70)^4 = 0.324$$

$$P(3) = \frac{n!}{k!(n-k)!}p^k(1-p)^{n-k} = \frac{6!}{3!3!}(0.30)^3(0.70)^3 = 0.185$$

$$P(4) = \frac{n!}{k!(n-k)!}p^k(1-p)^{n-k} = \frac{6!}{4!2!}(0.30)^4(0.70)^2 = 0.060$$

$$P(5) = \frac{n!}{k!(n-k)!}p^k(1-p)^{n-k} = \frac{6!}{5!1!}(0.30)^5(0.70)^1 = 0.010$$

$$P(6) = \frac{n!}{k!(n-k)!}p^k(1-p)^{n-k} = \frac{6!}{6!0!}(0.30)^6(0.70)^0 = 0.001$$

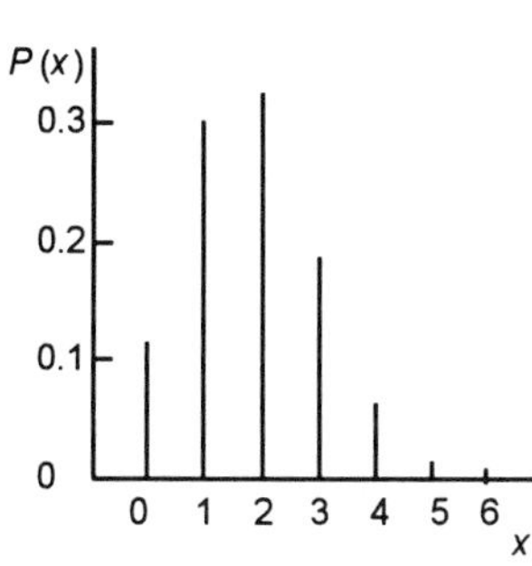

Figure 43.9 *Example*

Figure 43.9 shows the distribution.

Revision

10 The probability that any one item from a production line will be accepted is 0.70. What is the probability that when 5 items are randomly selected that there will be 2 unacceptable items?

11 Packets are filled automatically on a production line and, from past experience, 2% of them are expected to be underweight. If an inspector takes a random sample of 10, what will be the probability that (a) 0, (b) 1 of the packets will be underweight?

12 1% of the resistors produced by a factory are faulty. If a sample of 100 is randomly taken, what is the probability of the sample containing no faulty resistors?

13 A box contains 2 red balls and 5 black balls. A ball is drawn randomly from the box, its colour noted and then replaced. If 50 balls are drawn in this way, what is the expected, i.e. mean, number of red balls drawn and their standard deviation?

43.5.2 Poisson distribution

The Poisson distribution for a variable λ is:

$$P(k) = \frac{\lambda^k \mathrm{e}^{-\lambda}}{k!} \qquad [22]$$

for $k = 0, 1, 2, 3$, etc. The mean of this distribution is λ and the standard deviation is $\sqrt{\lambda}$. When the number n of trials is very large and the probability p small, e.g. $n > 25$ and $p < 0.1$, binomial probabilities are often approximated by the *Poisson distribution*. Thus, since the mean of the binomial distribution is np (equation [20]) and the standard deviation (equation [21]) approximates to $\sqrt{np}$ when p is small, we can consider λ to represent np. Thus λ can be considered to represent the average number of successes per unit time or unit length or some other parameter.

Example

2% of the output per month of a mass produced product have faults. What is the probability that of a sample of 400 taken that 5 will have faults?

Assuming we can apply the Poisson distribution, we have $\lambda = np = 400 \times 0.02 = 8$ and so equation [22] gives for $k = 5$:

$$P(5) = \frac{8^5\,e^{-8}}{5!} = 0.093$$

Example

There is a 1.5% probability that a machine will produce a faulty component. What is the probability that there will be at least 2 faulty items in a batch of 100?

Assuming the Poisson distribution can be used, we have $\lambda = np = 100 \times 0.015 = 1.5$ and so the probability of at least 2 faulty items will be:

$$P(\geq 2) = 1 - P(0) - P(1) = 1 - \frac{1.5^0}{0!}\,e^{-1.5} - \frac{1.5^1}{1!}\,e^{-1.5} = 0.442$$

Revision

14 The major accident rate per 10 000 employees for companies is, on average, 2 per year. Determine the probability that there will be at least one major accident in a year for a company with 10 000 employees.

15 The chance of a car breaking down while being driven through a tunnel has been found to be 0.006. Determine the probability that at least 2 cars will break down out of 100 passing through the tunnel.

16 The number of cars that enter a car park follows a Poisson distribution with a mean of 4. If the car park can accommodate 12 cars, determine the probability that the car park is filled up by the end of the first hour it is open.

43.5.3 Normal distribution

A particular form of distribution, known as the *normal distribution* or *Gaussian distribution*, is very widely used and works well as a model for experimental measurements when there are random errors. This form of distribution has a characteristic bell shape (Figure 43.10). It is symmetric about its mean value, having its maximum value at that point, and tends

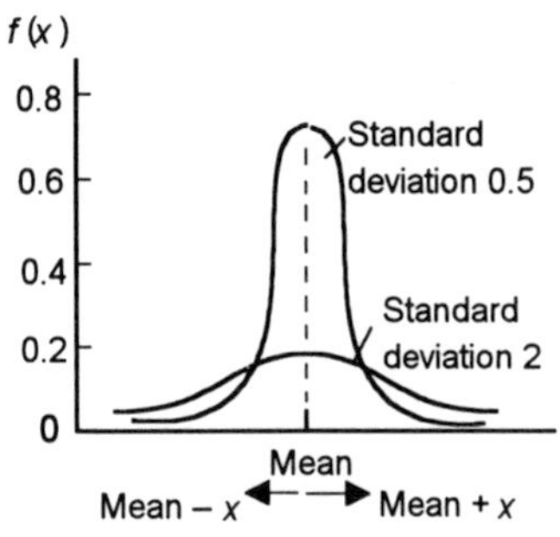

Figure 43.10 *Normal distribution*

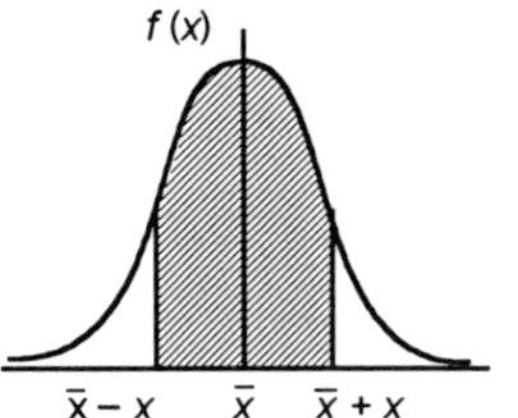

Figure 43.11 *Values within + or – x of the mean*

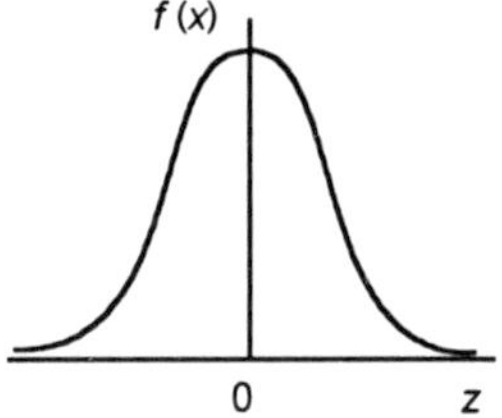

Figure 43.12 *Standard normal distribution*

rapidly to zero as x increases or decreases from the mean. It can be completely described in terms of its mean and its standard deviation. The following equation describes how the values are distributed about the mean.

$$f(x) = \frac{1}{\sigma\sqrt{2\pi}} e^{-(x-\bar{x})^2/2\sigma^2} \qquad [23]$$

The fraction of the total number of values that lies between $-x$ and $+x$ from the mean is the fraction of the total area under the curve that lies between those ordinates (Figure 43.11). We can obtain areas under the curve by integration.

To save the labour of carrying out the integration, the results have been calculated and are available in tables. As the form of the graph depends on the value of the standard deviation, as illustrated in Figure 43.10, the area depends on the value of the standard deviation σ. In order not to give tables of the areas for different values of x for each value of σ, the distribution is considered in terms of the value of $(x-\bar{x})/\sigma$, this commonly being designated by the symbol z, and areas tabulated against this quantity. z is known as the *standard normal random variable* and the distributions obtained with this as the variable are termed the *standard normal distribution* (Figure 43.12). Any other normal random variable can be obtained from the standard normal random variable by multiplying by the required standard deviation and adding the mean, i.e. $x = \sigma z + \bar{x}$. Table 43.1 shows examples of the type of data given in tables for z.

When $x - \bar{x} = 1\sigma$, then $z = 1.0$ and the area between the ordinate at the mean and the ordinate at 1σ as a fraction of the total area is 0.341 3. The area within $\pm 1\sigma$ of the mean is thus the fraction 0.681 6 of the total area under the curve, i.e. 68.16%. This means that the chance of a value being within $\pm 1\sigma$ of the mean is 68.16%, i.e. roughly two-thirds of the values.

Table 43.1 *Areas under normal curve*

z	Area from mean	z	Area from mean
0	0.000 0	1.6	0.445 2
0.2	0.079 3	1.8	0.464 1
0.4	0.155 5	2.0	0.477 2
0.6	0.225 7	2.2	0.486 1
0.8	0.288 1	2.4	0.491 8
1.0	0.341 3	2.6	0.495 3
1.2	0.384 9	2.8	0.497 4
1.4	0.419 2	3.0	0.498 7

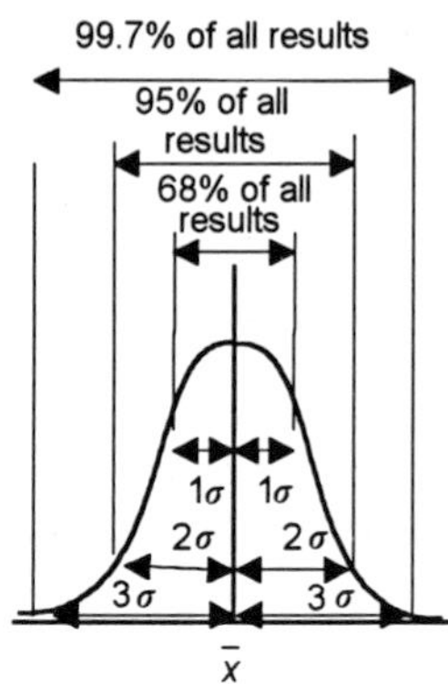

Figure 43.13 *Normal distribution*

When $x-\bar{x}=2\sigma$, then z = 2.0 and the area between the ordinate at the mean and the ordinate at 1σ as a fraction of the total area is 0.477 2. The area within $\pm 2\sigma$ of the mean is thus the fraction 0.954 4 of the total area under the curve, i.e. 95.44%. This means that the chance of a value being within $\pm 2\sigma$ of the mean is 95.44%.

When $x-\bar{x}=3\sigma$, then z = 3.0 and the area between the ordinate at the mean and the ordinate at 3σ as a fraction of the total area is 0.498 7. The area within $\pm 1\sigma$ of the mean is thus the fraction 0.997 4 of the total area, i.e. 99.74%. This means that the chance of a reading being within $\pm 3\sigma$ of the mean is 99.74%. Thus, virtually all the readings will lie within $\pm 3\sigma$ of the mean. Figure 43.13 illustrates the above.

Example

Measurements of the diameter of a rod at a number of points give a mean of 2.500 cm and a standard deviation of 0.005 cm. What will be the chance that a diameter reading will exceed 2.504 cm if five further measurements are made? Assume that the readings give a normal distribution.

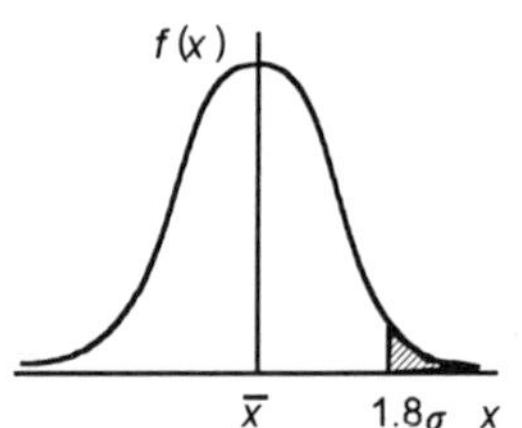

Figure 43.14 *Example*

The means can be expected to follow a normal distribution. The standard error of the mean when 5 readings are taken is thus $0.005/\sqrt{5}=0.0022$ cm. We have z = (2.504 – 2.500)/0.002 2 = 1.8 and thus Table 43.1 gives the area between the mean and this value as 0.464 1. The area greater than the 1.8 value (Figure 43.14) is thus 0.500 – 0.464 1 = 0.046 9 (the area between the mean and an extreme for the distribution is half the total area). Thus about 4.7% of the readings might be expected to have values greater than 2.504 cm. In the above analysis, it was assumed that the mean given was the true value or a good enough approximation.

Example

Measurements are made of the tensile strengths of samples taken from a batch of steel sheet. The mean value of the strength is 800 MPa and it is observed that 8% of the samples give values that are below an acceptable level of 760 MPa. What is the standard deviation of the distribution if it is assumed to be normal?

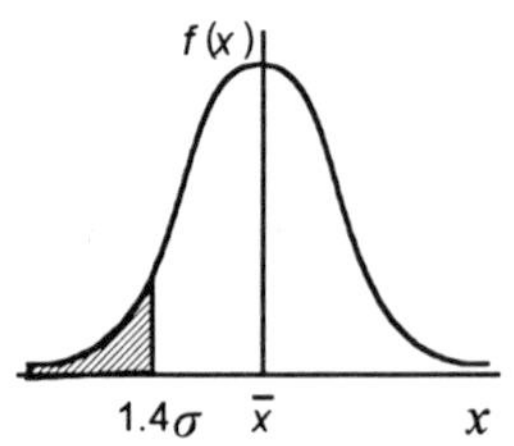

Figure 43.15 *Example*

This means that an area from the mean of 0.50 – 0.08 = 0.42. To the accuracy given in Table 43.1, this occurs when z = 1.4 (Figure 43.15). Thus, $(x-\bar{x})/\sigma=(760-800)/\sigma=1.4$ and so the standard deviation is 29 MPa. In the above analysis, it was assumed that the mean given was the true value or a good enough approximation.

Revision

17 A series of measurements was made of the periodic time of a simple pendulum and gave a mean of 1.23 s with a standard deviation of

0.01 s. What is the chance that, when a measurement is made, it will lie between 1.23 and 1.24 s?

18 The measured resistance per metre of samples of a wire have a mean resistance of 0.13 Ω and a standard deviation of 0.005 Ω. Determine the probability that a randomly selected wire will have a resistance per metre of between 0.12 and 0.14 Ω.

19 A set of measurements has a mean of 10 and a standard deviation of 5. Determine the probability that a measurement will lie between 12 and 15.

20 The time taken to complete a particular type of job is found to have a normal distribution with a mean of 20 minutes and a standard deviation of 3.33 minutes. Determine the probability that the time will be more than 23 minutes.

21 A set of measurements has a normal distribution with a mean of 10 and a standard deviation of 2.1. Determine the probability of a reading having a value (a) greater than 11 and (b) between 7.6 and 12.2.

43.5.4 The central limit theorem

This theorem asserts that:

The sum of a number n of independent random variables has a distribution that tends to a normal distribution as n tends to infinity.

This theorem explains why we can use the normal distribution for so many different situations. Physically there are many separate independent random components adding up to produce measured variables and thus, for this reason, measurements tend frequently to have a normal distribution. It also indicates that the binomial distribution for large values of n can tend to the normal distribution.

Problems

1 A random variable x has a probability distribution function of e^{-x} for $x \geq 0$ and 0 for $x < 0$. Determine the probability that x will be less than 0.5.

2 A random variable x has a probability distribution function of $0.25/x^{1/2}$ for $1 \leq x < 4$ and 0 for all other values. Determine the probability that x will be greater than 2.

3 The probability density function of a random variable x is given by $\frac{1}{2}x$ for $0 \leq x < 2$ and 0 for all other values. Determine the mean value of the variable.

4 Determine the mean and the standard deviation for the following data: 10, 20, 30, 40, 50.

5 The following are the results of 100 measurements of the times for 50 oscillations of a simple pendulum:

Between 58.5 and 61.5 s, 2 measurements
Between 61.5 and 64.5 s, 6 measurements
Between 64.5 and 67.5 s, 22 measurements
Between 67.5 and 70.5 s, 32 measurements
Between 70.5 and 73.5 s, 28 measurements
Between 73.5 and 76.5 s, 8 measurements
Between 76.5 and 79.5 s, 2 measurements

(a) Determine the relative frequencies of each segment.
(b) Determine the mean and the standard deviation.

6 The probability that a break could occur in a length of wire 2 m long is the same at all points along the length. Determine the probability that a break will occur between 0.5 and 1.5 m.

7 A random variable x has a probability distribution function of $\frac{1}{2} e^{-x/2}$ for $x \geq 0$ and 0 for $x < 0$. Determine the standard deviation.

8 It has been found that 10% of the screws produced are defective. Determine the probabilities that a random sample of 20 will contain 0, 1, 2, 3, 4, 5, 6, 7, 8, 9, 10 defectives.

9 Four coins are tossed. What are the probabilities of obtaining 0, 1, 2, 3, 4 heads?

10 The probability of a mass-produced item being faulty has been determined to be 0.10. What are the probabilities that a random sample of 50 will contain 0, 1, 2, 3, 4, 5, 6, 7, 8, 9, 10 faulty items?

11 A product is guaranteed not to contain more than 2% that are outside the specified tolerances. In a random sample of 10, what is the probability of getting 2 or more outside the specified tolerances?

12 A large consignment of resistors is known to have 1% outside the specified tolerances. What would be the expected, i.e. mean, number of resistors outside the specified tolerances in a batch of 10 000 and the standard deviation?

13 The number of telephone enquiries received by a company averages four per minute. What is the probability that at least two calls will be received in a particular one-minute period?

14 On average six of the cars coming per day off a production line have faults. What is the probability that four faulty cars will come off the line in one day?

15 The number of breakdowns per month for a machine averages 1.8. Determine the probability that the machine will function for a month with only one breakdown.

16 On the average, three accidents occur per month at a particular road junction. What is the probability that there will be at least one accident this month?

17 Measurements of the resistances of resistors in a batch gave a mean of 12 Ω with a standard deviation of 2 Ω. If the resistances can be assumed to have a normal distribution about this mean, how many from a batch of 300 resistors are likely to have resistances more than 15 Ω?

18 Measurements are made of the lengths of components as they come off the production line and a mean value of 12 mm with a standard deviation of 2 mm. If a normal distribution can be assumed, in a sample of 100 how many might be expected to have (a) lengths of 15 mm or more, (b) lengths between 13.7 and 16.1 mm?

19 Measurements of the times taken for workers to complete a particular job have a mean of 29 minutes and a standard deviation of 2.5. Assuming a normal distribution, what percentage of the times will be (a) between 31 and 32 minutes, (b) less than 26 minutes?

20 Inspection of the lengths of components yields a normal distribution with a mean of 102 mm and a standard deviation of 1.5 mm. Determine the probability that if a component is selected at random it will have a length (a) less than 100 mm, (b) more than 104 mm, (c) between 100 and 104 mm.

21 A machine makes resistors with a mean value of 50 Ω and a standard deviation of 2 Ω. Assuming a normal distribution, what limits should be used on the values if the resistance if there are to be not more than 1 reject in 1000.

44 Simple regression

44.1 Introduction

In engineering and science there is often the problem of determining the best straight line through a series of points on a graph. In this chapter we consider the *least squares method* of fitting a straight line to a plot of points and determining the slope and intercept of the line. In essence, what we have is a method of taking data for experimental points and finding the values of m and c that will give the best fit when that data is inserted into an equation of the form $y = mx + c$.

The building of models in which one variable can be written in terms of another is called the *regression* of one variable upon another. Lines fitted to data describing the mathematical relationship between variables are called *regression lines*. The term *simple regression* is used when there is only one independent variable, the term *multiple regression* being used when there are more than one independent variable. In this chapter we are just concerned with the fitting of data to the straight-line relationship and so are only concerned with simple linear regression.

44.2 The method of least squares

Consider the data points shown in the graph of Figure 44.1 and the drawing of the best straight line through them. The best straight line might be drawn by balancing the points so that those above the line balance those below the line. To simplify the discussion, we will consider that the errors are entirely in the y measurements, i.e. the dependent variable. Thus the best line is drawn by considering the deviations of the points in the y directions from the line and trying to achieve a balance. The deviation d of any one point is given by:

$$\text{deviation } d = y - \hat{y}$$

where y is the measured value and $\hat{y}$ the value given by the straight line. The value given by the straight line is:

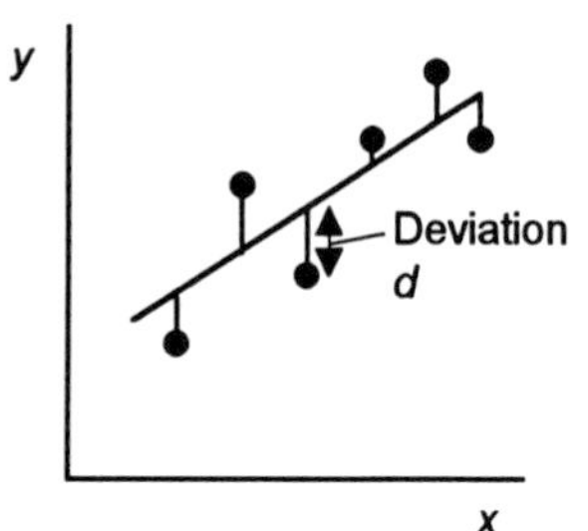

Figure 44.1 *The best straight line*

$$\hat{y} = mx + c$$

where x is the value of the independent variable, m the gradient of the graph and c its intercept with the y-axis. Thus, we can write:

$$d = y - (mx + c)$$

We can write such an equation for each data point. Thus, for point j we can write for its deviation d_j:

$$d_j = y_j - (mx_j + c)$$

with y_j and x_j being the pair of data values.

We might consider that the best line would be the one that minimises the sum of the deviations. However, a better solution is to find the line that minimises the standard errors of the points in their deviations from the straight line. This means choosing the values of m and c which minimise the sum of the squares of these deviations. This is termed the *least squares estimate* for the best straight line. Hence the best values of m and c are obtained when the following sum is minimised:

$$\text{sum of the squares } S = \sum_{j=1}^{n} d_j^2 = \sum_{j=1}^{n} [y_j - (mx_j + c)]^2$$

We can consider there to be many values of m and c, i.e. lots of straight lines that can be drawn, and we need to determine the values which result in a minimum value for the sum. If we had a simple equation such as $S = mx^2$ we could find the minimum value of the sum with respect to m by differentiating and then equating the derivative to zero, i.e. finding the value of m that gave $\mathrm{d}S/m = 0$. But we have two variables, m and c. Thus, in this case we have to determine the partial derivatives $\partial S/\partial m$ and $\partial S/\partial c$ and equate them to zero. Expanding the equation for the sum gives:

$$S = \sum_{j=1}^{n} \left[y_j^2 - 2y_j(mx_j + c) + (mx_j + c)^2 \right]$$

Hence:

$$\frac{\partial S}{\partial m} = -2 \sum_{j=1}^{n} x_j(y_j - mx_j - c)$$

$$\frac{\partial S}{\partial c} = -2 \sum_{j=1}^{n} (y_j - mx_j - c)$$

Equating both these partial derivatives to zero gives:

$$-2 \sum_{j=1}^{n} x_j(y_j - mx_j - c) = 0$$

$$-2 \sum_{j=1}^{n} (y_j - mx_j - c) = 0$$

We can rearrange these equations to give:

$$m \sum_{j=1}^{n} x_j^2 + c \sum_{j=1}^{n} x_j = \sum_{j=1}^{n} x_j y_j \qquad [1]$$

$$m \sum_{j=1}^{n} x_j + nc = \sum_{j=1}^{n} y_j \qquad [2]$$

Thus all we need to do is put the data into the above two equations and solve the pair of simultaneous equations to obtain values for m and c.

Alternatively, we can solve these two simultaneous equations [1] and [2] to give m as:

$$m = \frac{n\sum_{j=1}^{n} x_j y_j - \sum_{j=1}^{n} y_j \sum_{j=1}^{n} x_j}{n\sum_{j=1}^{n} x_j^2 - \left(\sum_{j=1}^{n} x_j\right)^2} \qquad [3]$$

This form of the equation is a useful form for using with a calculator. We can, however, put the equation in another form. Since the mean values of x and y are given by:

$$\bar{x} = \frac{1}{n}\sum_{j=1}^{n} x_j$$

$$\bar{y} = \frac{1}{n}\sum_{j=1}^{n} y_j$$

we can write m as:

$$m = \frac{\sum_{j=1}^{n} x_j(y_j - \bar{y})}{\sum_{j=1}^{n} x_j(x_j - \bar{x})}$$

and, since the sum of the deviations of y about the mean value of y must be zero, i.e. $\Sigma(y_j - \bar{y}) = 0$, then:

$$m = \frac{\sum_{j=1}^{n}(x_j - \bar{x})(y_j - \bar{y})}{\sum_{j=1}^{n}(x_j - \bar{x})^2} \qquad [4]$$

To obtain a value for c, we can substitute this value of m into equation [2] when rewritten in terms of the mean values as:

$$c = \bar{y} - m\bar{x} \qquad [5]$$

Incidentally, this equation is of the form $\bar{y} = m\bar{x} + c$ and so shows that the least squares line passes through the mean values of the data.

To obtain a value for c, an alternative to using equation [5] is to solve the simultaneous equations [1] and [2] for c to obtain:

$$c = \frac{\sum_{j=1}^{n} x_j^2 \sum_{j=1}^{n} y_j - \sum_{j=1}^{n} x_j \sum_{j=1}^{n} x_j y_j}{n\sum_{j=1}^{n} x_j^2 - \left(\sum_{j=1}^{n} x_j\right)^2} \qquad [7]$$

Example

The following data was obtained from an experiment in which the resistance R of a coil of metal wire was determined at different temperatures θ. The relationship between the resistance and temperature is envisaged as being of the form $R = m\theta + c$. Determine the values of m and c which best fit the data.

Resistance (Ω)	76.4	82.7	87.8	94.0	103.5
Temperature (°C)	20.5	32.5	52.0	73.0	96.0

To use equations [1] and [2] we need to find the values of the sum of the temperature data, the sum of the squares of the temperature data, the sum of the resistance data and the sum of the products of the two sets of data. Table 44.1 shows the calculations of these values.

Table 44.1 *Calculations for example*

R (Ω)	θ (°C)	θ^2 (°C)^2	$R\theta$ (Ω °C)
76.4	20.5	420.25	1566.20
82.7	32.5	1056.25	1687.75
87.8	52.0	2704.00	4565.60
94.0	73.0	5329.00	6862.00
103.5	96.0	9216.00	9936.00
$\Sigma R = 444.4$	$\Sigma \theta = 274$	$\Sigma \theta^2 = 18\ 725.50$	$\Sigma R\theta = 25\ 617.55$

Hence equation [1], namely:

$$m\sum_{j=1}^{n} x_j^2 + c\sum_{j=1}^{n} x_j = \sum_{j=1}^{n} x_j y_j$$

becomes:

$$18\ 725.50m + 274c = 25\ 617.55$$

Equation [2], namely:

$$m\sum_{j=1}^{n} x_j + nc = \sum_{j=1}^{n} y_j$$

becomes:

$$274m + 5c = 444.4$$

These two simultaneous equations can then be solved to give the values of m and c.

Multiplying the first equation by 5/274 gives:

$$341.71m + 5c = 467.47$$

Subtracting the second equation from the above equation gives:

$$67.71m = 23.07$$

Hence $m = 0.341$. Substituting this value into one of the equations gives $c = 70.19$. Hence the equation is:

$$R = 0.341\theta + 70.19$$

As an alternative to putting values in the simultaneous equations [1] and [2], we could have used equations [3] and [2] or equations [4] and [5]. Using equation [3], namely:

$$m = \frac{n \sum_{j=1}^{n} x_j y_j - \sum_{j=1}^{n} y_j \sum x_j}{n \sum_{j=1}^{n} x_j^2 - \left(\sum_{j=1}^{n} x_j\right)^2} = \frac{5 \times 25\,617.55 - 444.4 \times 274}{5 \times 18\,725.50 - 274^2} = 0.341$$

c can be found as before.

To use equations [4] and [5], we need to tabulate the data in a different form. Table 44.2 shows the table. Using equation [4]:

$$m = \frac{\sum_{j=1}^{n} (x_j - \bar{x})(y_j - \bar{y})}{\sum_{j=1}^{n} (x_j - \bar{x})^2} = \frac{1263.59}{3710.30} = 0.341$$

Table 44.2 *Calculations for example*

R (Ω)	θ (°C)	R – mean (Ω)	θ – mean (°C)	(R – mean)(θ – mean)	(θ – mean)^2
76.4	20.5	–12.48	–34.3	428.06	1176.49
82.7	32.5	–6.18	–22.3	137.81	497.29
87.8	52.0	–1.08	– 2.8	3.02	7.84
94.0	73.0	5.12	18.2	93.18	331.24
103.5	96.0	14.6	41.2	601.52	1697.44
Mean = 88.88	Mean = 54.8			Σ = 1263.59	Σ = 3710.30

Using equation [5], namely:

$$c = \bar{y} - m\bar{x} = 88.88 - 0.341 \times 54.8 = 70.19$$

Thus, as before:

$$R = 0.341\theta + 70.19$$

Example

The following is data obtained for the volume of a gas measured when it is measured at different volumes. The relationships between the volume v and pressure p is expected to be of the form $pv^{\gamma} = C$, where γ and C are constants. Assuming that the only significant errors are in the volume, determine a value for γ.

p (10^5 Pa)	0.5	1.0	1.5	2.0	2.5	3.0
v (dm^3)	1.62	1.00	0.75	0.62	0.52	0.46

Taking logarithms of the equation gives lg p + γ lg v = lg C. If we let X = lg p and y = lg v then, with some rearrangement, we have:

$$Y = (-1/\gamma)X + (1/\gamma)\lg C = mX + c$$

We can use the least squares method with the data and equations [3] and [2]. Table 44.3 gives intermediate calculation. Equation [4] gives

$$m = \frac{n\sum_{j=1}^{n} x_j y_j - \sum_{j=1}^{n} y_j \sum_{j=1}^{n} x_j}{n\sum_{j=1}^{n} x_j^2 - \left(\sum_{j=1}^{n} x_j\right)^2}$$

Table 44.3 *Calculations for example*

lg v, i.e. Y	lg p, i.e. X	XY	X^2
−0.301 0	0.209 5	−0.063 1	0.090 6
0	0	0	0
0.176 1	−0.124 9	−0.022 0	0.031 0
0.301 0	−0.207 6	−0.062 5	0.090 6
0.393 9	−0.284 0	−0.113 0	0.158 3
0.477 1	−0.339 2	−0.160 9	0.227 6
Σ = 1.051 1	Σ = −0.744 2	Σ = −0.421 5	Σ = 0.598 1

$$= \frac{6 \times (-0.421\,5) - 1.051\,1 \times (-0.744\,2)}{6 \times 0.598\,1 - 1.051\,1^2} = -0.703\,3$$

Thus $(-1/\gamma) = -0.703\ 3$ and so $\gamma = 1.42$.

Revision

1 Determine, by means of the least squares method and assuming that the only significant errors are in the y values, the best straight-line relationship to fit the following data:

x	65	68	71	75	77	80	84	87	93	98
y	72	72	80	82	74	78	89	91	96	95

2 The extension e of a spring is measured when different loads W are applied to stretch it and the following data obtained:

W (kg)	1.0	2.0	3.0	4.0	5.0
e (mm)	8	16	27	32	35

If the error is only in the extension values, determine the gradient and the intercept of the best straight line through the points.

Problems

1 The mass m of a compound which will dissolve in 100 g of water is measured at different temperatures θ and the following results obtained. If the error is only in the mass values, determine the gradient and the intercept of the best straight line through the points.

θ (°C)	0	10	20	30	40	50	60	70	80	90	100
m (g)	43.5	49.5	55.2	60.6	65.5	70.2	75.5	80.0	85.0	89.2	94.0

2 The depression d of the end of a cantilever as a result of weight W being added was measured and the following results obtained. If the error is only in the depression values, determine the gradient and the intercept of the best straight line through the points.

W (kg)	0.25	1.00	2.25	4.00	6.25
d (mm)	2.2	10.0	22.3	39.4	61.7

3 Determine, by means of the least squares method and assuming that the only significant errors are in the y values, the best straight-line relationship to fit the following data:

x	1	2	3	4	5	6
y	1	2	2	3	5	5

4 Determine, by means of the least squares method and assuming that the only significant errors are in the y values, the best straight-line relationship to fit the following data:

x	1	2	3	4	5	6	7	8	9	10
y	2.3	4.5	5.2	6.0	8.0	9.8	11.1	13.5	14.0	16.6

5 Measurements are made of the effort E to raise load W and the following results obtained. Determine, assuming the only significant errors are in E, the equation of the best straight line through the data.

W (N)	14	42	84	112
E (N)	6.1	14.3	27.0	36.3

6 The following data is expected to fit a relationship of the form $y = b\,e^{ax}$. By putting the relationship in a form which would give a straight-line graph, use the least squares method to determine the values of a and b. The errors are only significant for y.

x	1.00	1.25	1.50	1.75	2.00
y	5.10	5.79	6.53	7.45	8.46

7 The following data is expected to fit a relationship of the form $y = ax^n$. By putting the relationship in a form which would give a straight-line graph, use the least squares method to determine the values of a and n. The errors are only significant for y.

x	5.7	11.1	23.4	31.5	42.7
y	7.71	23.95	85.06	141.0	236.9

8 The following data is expected to fit a relationship of the form $y = a\,e^{kx}$. By putting the relationship in a form which would give a straight-line graph, use the least squares method to determine the values of a and n. The errors are only significant for y.

x	0	1	2	3
y	1.05	2.10	3.85	8.30

45 Application: Experimental errors

45.1 Introduction

Experimental *error* is defined as the difference between the result of a measurement and the true value:

$$\text{error} = \text{measured value} - \text{true value} \qquad [1]$$

Errors can arise from such causes as instrument imperfections, human imprecision in making measurements and random fluctuations in the quantity being measured. This chapter is a brief consideration of the estimation of errors and their determination in a quantity which is a function of more than one measured variable. Chapter 43 on distributions is assumed.

45.2 Errors

With measurements made with an instrument, errors can arise from fluctuations in readings of the instrument scale due to perhaps the settings not being exactly reproducible and operating errors because of human imprecision in making the observations. The term *random error* is used for those errors which vary in a random manner between successive readings of the same quantity. The term *systematic error* is used for errors which do not vary from one reading to another, e.g. those arising from a wrongly set zero. Random errors can be determined and minimised by the use of statistical analysis, systematic errors require the use of a different instrument or measurement technique to establish them.

With random errors, repeated measurements give a distribution of values. This can be generally assumed to be a normal distribution (see Section 43.4.4 for a statement of the central limit theorem). The standard error of the mean of the experimental values can be estimated from the spread of the values and it is this which is generally quoted as the error, there being a 68% probability that the mean will lie within plus or minus one standard error of the true mean. Note that the standard error does not represent the maximum possible error. Indeed there is a 32% probability of the mean being outside the plus and minus standard error interval.

With just a single measurement, say the measurement of a temperature by means of a thermometer, the error is generally quoted as being plus or minus one-half of the smallest scale division. This, termed the *reading error*, is then taken as an estimate of the standard deviation that would occur for that measurement if it had been repeated many times.

Example

A rule used for the measurement of a length has scale readings every 1 mm. Estimate the error to be quoted when the rule is used to make a single measurement of a length.

The error is quoted as ±0.5 mm.

Example

Measurements of the tensile strengths of test pieces taken from a batch of incoming material gave the following results: 40, 42, 39, 41, 45, 40, 41, 43, 45, 46 MPa. Determine the mean tensile strength and its error.

The mean is given by equation [3], Chapter 43, as (40 + 42 + 39 + 41 + 45 + 40 + 41 + 43 + 45 + 44)/10 = 42. The standard deviation can be calculated by the use of equation [10], Chapter 43. Since ($40^2 + 42^2 + 39^2 + 41^2 + 45^2 + 40^2 + 41^2 + 43^2 + 45^2 + 44^2$) = 17 682 then:

$$s^2 = \frac{1}{N-1}\sum_{j=1}^{N} x_j^2 - \frac{N}{N-1}\bar{x}^2 = \tfrac{1}{9}17\,682 - \tfrac{10}{9} \times 42^2 = 4.67$$

and the standard deviation is 2.2 MPa. The standard error is thus $2.2/\sqrt{10}$ = 0.7 MPa. Thus the result can be quoted as 42 ± 0.7 MPa.

Revision

1 Determine the mean value and standard error for the measured diameter of a wire if it is measured at a number of points and gave the following results: 2.11, 2.05, 2.15, 2.12, 2.16, 2.14, 2.16, 2.17, 2.13, 2.15 mm.

2 An ammeter has a scale with graduations at intervals of 0.1 A. Give an estimate of the standard deviation.

45.2.1 Statistical errors

In addition to measurement errors arising from the use of instruments, there are what might be termed *statistical errors*. These are not due to any errors arising from an instrument but from statistical fluctuations in the quantity being measured, e.g. the count rate of radioactive materials. The observed values here are distributed about their mean in a Poisson distribution and so the standard deviation is the square root of the mean value (see Section 43.4.2).

Example

In an experiment, the number of alpha particles emitted over a fixed period of time is measured as 4206. Determine the standard deviation of the count.

Assuming the count follows a Poisson distribution, the standard deviation will be the square root of 4206 and so 65. Thus the count can be recorded as 4206 ± 65.

Revision

3 In an experiment the number of gamma rays emitted over a fixed period of time is measured as 5210. Determine the standard deviation of the count.

45.3 Combining errors

An experiment might require several quantities to be measured and then the values inserted into an equation so that the required variable can be calculated. For example, a determination of the density of a material might involve a determination of its mass and volume, the density then being calculated from mass/volume. If the mass and volume each have errors, how do we combine these errors in order to determine the error in the density?

Consider a variable Z which is to be determined from sets of measurements of A and B and for which we have the relationship $Z = A + B$. If we have A with an error ΔA and B with an error ΔB then we might consider that $Z + \Delta Z = A + \Delta A + B + \Delta B$ and so we should have

$$\Delta Z = \Delta A + \Delta B \qquad [1]$$

i.e. the error in Z is the sum of the errors in A and B. However, this ignores the fact that the error is the standard error and so is just the value at which there is a probability of ±68% that the mean value for A or B will be within that amount of the true mean. If we consider the set of measurements that were used to obtain the mean value of A and its standard error and the set of measurements to obtain the mean value of B and its standard error and consider the adding together of individual measurements of A and B then we can write $\Delta Z = \Delta A + \Delta B$ for each pair of measurements. Squaring this gives:

$$\Delta Z^2 = (\Delta A + \Delta B)^2 = (\Delta A)^2 + (\Delta B)^2 + 2\,\Delta A\,\Delta B$$

We can write such an equation for each of the possible combinations of measurements of A and B. If we add together all the possible equations and divide by the number of such equations, we would expect the $2\,\Delta A\,\Delta B$ terms to cancel out since there will be as many situations with it having a negative value as a positive value. Thus the equation we should use to find the error in Z is:

$$\Delta Z^2 = (\Delta A)^2 + (\Delta B)^2 \qquad [2]$$

The same equation is obtained for $Z = A - B$.

Now consider the error in Z when $Z = AB$. As before, we might argue that $Z + \Delta Z = (A + \Delta A)(B + \Delta B)$ and so $\Delta Z = B\,\Delta A + A\,\Delta B$, if we ignore as insignificant the $\Delta A\,\Delta B$ term. Hence:

$$\frac{\Delta Z}{Z} = \frac{B\,\Delta A + A\,\Delta B}{AB} = \frac{\Delta A}{A} + \frac{\Delta B}{B} \qquad [3]$$

i.e. the fractional error in Z is the sum of the fractional errors in A and B or the percentage error in Z is the sum of the percentage errors in A and B. However, this ignores the fact that the error is the standard error and so is just the value at which there is a probability of ±68% that the mean value for A or B will be within that amount of the true mean. If we consider the set of measurements that were used to obtain the mean value of A and its standard error and the set of measurements to obtain the mean value of B and its standard error and use equation [3] for each such combination, then we can write:

$$\left(\frac{\Delta Z}{Z}\right)^2 = \left(\frac{\Delta A}{A} + \frac{\Delta B}{B}\right)^2 = \left(\frac{\Delta A}{A}\right)^2 + \left(\frac{\Delta B}{B}\right)^2 + 2\left(\frac{\Delta A}{A}\right)\left(\frac{\Delta B}{B}\right)$$

We can write such an equation for each of the possible combinations of measurements of A and B. If we add together all the possible equations and divide by the number of such equations, the $2(\Delta A/A)(\Delta B/B)$ terms cancel out since there will be as many situations with it having a negative value as a positive value. Thus the equation we should use to find the error in Z is:

$$\left(\frac{\Delta Z}{Z}\right)^2 = \left(\frac{\Delta A}{A}\right)^2 + \left(\frac{\Delta B}{B}\right)^2 \qquad [4]$$

The same equation is obtained for $Z = A/B$. If $Z = A^2$ then this is just the product of A and A and so equation [4] gives $(\Delta Z/Z)^2 = 2(\Delta A/A)^2$. Thus for $Z = A^n$ we have:

$$\left(\frac{\Delta Z}{Z}\right)^2 = n\left(\frac{\Delta A}{A}\right)^2 \qquad [5]$$

In all the above discussion it was assumed that the mean value of Z was given when the mean values of A and B were used in the defining equation.

Example

The resistance of a resistor is determined from measurements of the potential difference across it and the current through it. If the potential difference has been measured as 2.1 ± 0.2 V and the current as 0.25 ± 0.01 A, what is the resistance and its error?

The mean resistance is 2.1/0.25 = 8.4 Ω. The fractional error in the potential difference is 0.2/2.1 = 0.095 and the fractional error in the current is 0.01/0.25 = 0.04. Hence the fractional error in the resistance is $\sqrt{(0.095^2 + 0.04^2)} = 0.10$. Thus the resistance is 8.4 ± 0.9 Ω.

Example

If $g = 4\pi^2L/T^2$ and L has been measured as 1.000 ± 0.005 m and T as 2.0 ± 0.1 s, determine g and its error.

The mean value of g is $4\pi^2(1.000)/2.0^2 = 9.87$ m/s^2. The fractional error in L is 0.005 m and that in T is 0.05 s. Thus:

$$(\text{fractional error in } g)^2 = 0.005^2 + 2 \times 0.05^2$$

Thus the fractional error in g is 0.071 and so $g = 9.87 \pm 0.7$ m/s^2.

Revision

4 The cross-sectional area of a wire is determined from a measurement of the diameter. If the diameter measurement gives 2.5 ± 0.1 mm, determine the area of the wire and its error.

5 Determine the mean value and error for Z when (a) $Z = A + B$, (b) $Z = AB$, (c) $Z = A/B$ if $A = 100 \pm 3$ and $B = 50 \pm 2$.

45.3.1 The general case

In Section 45.3 the errors were considered for certain particular forms of function involving just two independent variables. In general we have $Z = f(A, B, ..., \text{etc.})$, where A, B, C, ...,etc. are independent variables. If we differentiate we can determine the change in Z produced by changes in the variables. Because there is more than one independent variable we use partial differentiation (see Chapter 24). Thus:

$$\Delta Z = \frac{\partial Z}{\partial A}\Delta A + \frac{\partial Z}{\partial B}\Delta B + \frac{\partial Z}{\partial C}\Delta C + \dots \qquad [6]$$

This gives the error in Z due to one combination of errors in A, B, C, etc. We can square the equation to give:

$$(\Delta Z)^2 = \left(\frac{\partial Z}{\partial A}\Delta A + \frac{\partial Z}{\partial B}\Delta B + \frac{\partial Z}{\partial C}\Delta C + \dots\right)^2$$

We can write such an equation for each of the possible combinations of measurements of the variables. If we add together all the possible equations and divide by the number of such equations, we would expect all the cross-product terms such as $\Delta A\ \Delta B$ to cancel out since there will be as many situations with them having a negative value as a positive value. Thus the equation we should use to find the error in Z is:

$$(\Delta Z)^2 = \left(\frac{\partial Z}{\partial A}\Delta A\right)^2 + \left(\frac{\partial Z}{\partial B}\Delta B\right)^2 + \left(\frac{\partial Z}{\partial C}\Delta C\right)^2 + \dots \qquad [7]$$

Example

If $Z = A\,e^B$, determine the error in Z resulting from errors in A and B.

Taking B as constant gives $\partial Z/\partial A = e^B$ and taking A as constant gives $\partial Z/\partial B = A\,e^B$. Thus, using equation [7]:

$$(\Delta Z)^2 = (e^B\,\Delta A)^2 + (A\,e^B\,\Delta B)^2$$

Example

Two resistors R_1 and R_2 are connected in parallel. Determine the error in the combination resulting from errors in each resistor.

For resistors in parallel:

$$\frac{1}{R} = \frac{1}{R_1} + \frac{1}{R_2}$$

This can be written as:

$$R = \frac{R_1 R_2}{R_1 + R_2}$$

This gives:

$$\frac{\partial R}{\partial R_1} = \frac{R_2(R_1 + R_2) - R_1 R_1}{(R_1 + R_2)^2} \text{ and } \frac{\partial R}{\partial R_2} = \frac{R_1(R_1 + R_2) - R_1 R_1}{(R_1 + R_2)^2}$$

Hence:

$$(\Delta R)^2 = \left[\frac{R_2(R_1 + R_2) - R_1 R_1}{(R_1 + R_2)^2}\right]^2 (\Delta R_1)^2 + \left[\frac{R_1(R_1 + R_2) - R_1 R_1}{(R_1 + R_2)^2}\right]^2 (\Delta R_2)^2$$

Revision

6 The refractive index n is determined from measurements of the angle of incidence i and the angle of refraction r with $n = \sin i/\sin r$. Determine the error in n due to errors in i and r.

7 Determine the error in a quantity Z if it is calculated by means of the equation $Z = A(1 + 20B)^{1/2}$ and $A = 1088 \pm 0.5$ and $B = 0.003\ 68 \pm 0.000\ 1$.

8 If $Z = A \ln B$ determine the error in Z resulting from errors in A and B.

Problems

1 Determine the mean and the standard error for the resistance of a resistor if repeated measurements gave 51.1, 51.2, 51.0, 51.4, 50.9 Ω.

2 How big a count should be made of the gamma radiation emitted from a radioactive material if the percentage error should be less than 1%?

3 Repeated measurements of the voltage necessary to cause the breakdown of a dielectric gave the results 38.9, 39.3, 38.6, 38.8, 38.8, 39.0, 38.7, 39.4, 39.7, 38.4, 39.0, 39.1, 39.1, 39.2 kV. Determine the mean value and the standard error of the mean.

4 Determine the mean value and error for Z when (a) $Z = A - B$, (b) $Z = 2AB$, (c) $Z = A^3$, (d) $Z = B/A$ if $A = 100 \pm 3$ and $B = 50 \pm 2$.

5 The resistivity of a wire is determined from measurements of the resistance R, diameter d and length L. If the resistivity is RA/L, where A is the cross-sectional area, which measurement requires determining to the greatest accuracy if it is not to contribute the most to the overall error in the resistivity?

6 The focal length f of a lens is determined from measurements of the object distance u and image distance v with $1/f = 1/u + 1/v$. Determine the error in f due to errors in u and v.

7 If $Z = \ln A$, determine the error in Z resulting from an error in A.

8 If $Z = k\,e^{cA}$, determine the error in Z resulting from an error in A, c and k being constants.

Answers

Chapter 1

Revision

1 (a) 3, (b) 5 **2** (a) 0, (b) 6 **3**(a) 2, (b) 1 **4** (a) 0, (b) 2
5 (a) 1.99 s, (b) 6.28 s **6** (a) 2 m/s, (b) 7 m/s
7 $v = 0$ for $t < 0$, $v = 2t$ for $0 \le t < 1$, $v = 0$ for $1 < t$
8 See Figure A.1 for first period
9 $v = 1$ for $0 \le t < 1$, $v = 0$ for $1 \le t < 2$, period $T = 2$
10 (a) $3x + 1$, (b) $2x + 2$, (c) $2x + 1$ **11** (a) $\frac{1}{5}(x+3)$, (b) $x - 4$, (c) $\sqrt[3]{x}$, (d) $\sqrt[3]{\frac{1}{2}(x+1)}$ **12** $x \ge 0$, $\sqrt{x}$ **13** See Figure A.2

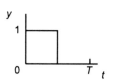

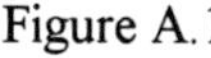

Figure A.1

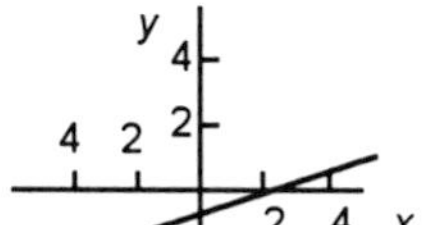

Figure A.2

Problems

1 (a) 3, 4, (b) 4, 5, (c) −2, 1 **2** (a) 2, (b) 0 **3** (a) 2, (b) 4
4 (a) 300, (b) 800 **5** $v = 0$ for $0 \le t < 2$, $v = 10$ V for $2 \le t$
6 See Figure A.3 **7** (a) and (c) **8** (a) $x^2 + 3x + 1$, (b) $9x^2 + 3$, (c) $3x^2 + 3$, (d) $x^2 - 3x - 1$, (e) $9x^2 + 12x + 5$
9 (a) $\frac{1}{2}(x-3)$, (b) $\frac{1}{4}x$, (c) $\frac{1}{3}(x+4)$, (d) $(x-2)^2$, (e) $(x+3)^{1/3} - 1$
10 See Figure A.4 **11** No. If domain restricted to $x \ge 0$ then yes.
12 (a) $x \ge 1$, $1 + \sqrt{x}$, (b) $x \ge -1$, $-1 + \sqrt{x+4}$

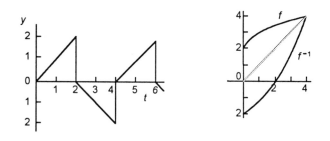

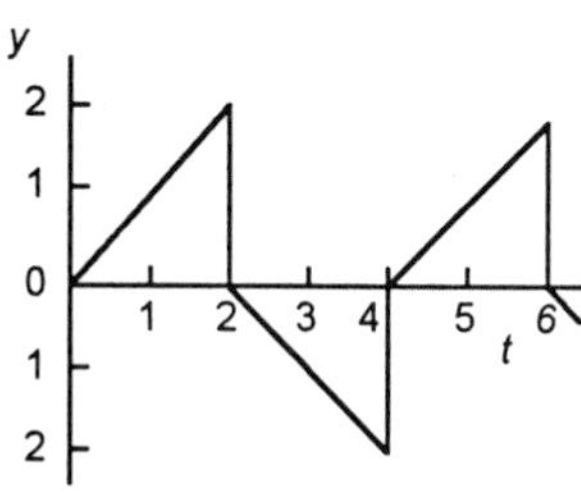

Figure A.3

Figure A.4

Chapter 2

Revision

1 (a) Straight line, yes, (b) straight line, no, (c) non-linear, (d) non-linear
2 (a) $i = t/2 + 2$, (b) $e = 1.2L$ **3** 10 mA **4** (a) 1.125 m, (b) 1.275 m
5 See Figure A.5 **6** See Figure A.6 **7** (a) $(x-4)(x+4)$, (b) $(x+3)(x+2)$, (c) $(x+1)(x+1)$, (d) $x(x+1)(x+1)$, (e) $(x+1)(x^2+2x-4)$ **8** (a) Yes, (b) yes, (c) no, (d) yes, (e) yes
9 (a) 1, 2, (b) 2, 1, (c) 2, 2 **10** (a) 3, 0, (b) 1, 1, (c) 2, 1, (d) 1, 0
11 (a) −1.6, 0.6, (b) −2.4, 0.4, (c) −1.5, (d) −5.1, (e) 0.25
12 (a) −5, −5, (b) −4, +4, (c) −3, −4, (d) −6, +2, (e) −4, −4
13 (a) −4, 2, (b) 1.46, −5.46, (c) 2.6, 0.38 **14** (a) −3.2, 1.2, (b) −2.6, −0.38, (c) 3.8, −0.41, (d) no real roots
15 38.2°C, 261.8°C **16** 0.15 s, 0.66 s

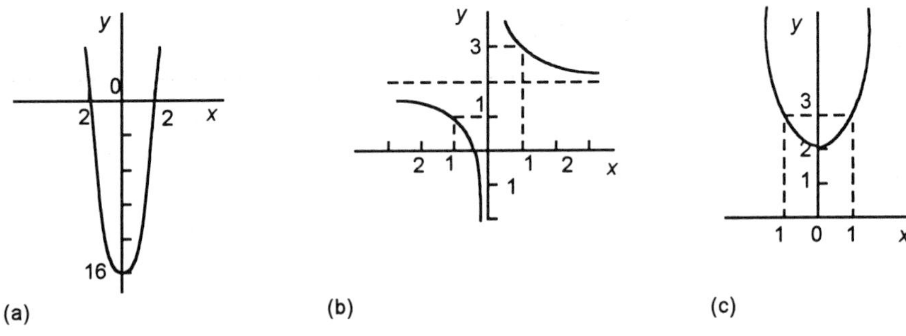

Figure A.5

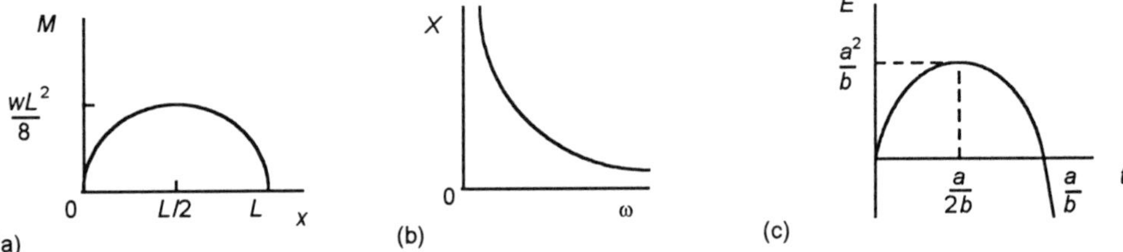

Figure A.6

17 $\frac{mgL^3}{48EI} \pm \sqrt{\left(\frac{mgL^3}{48EI}\right)^2 + 2h\left(\frac{mgL^3}{48EI}\right)}$ **18** 4.46

19 (a) $x - \frac{x}{x^2+1}$, (b) $x + 2 - \frac{x+3}{x^2-3x+2}$, (c) $3x - 7 + \frac{12}{x+2}$

20 (a) $\frac{5}{x-1} - \frac{4}{x-2}$, (b) $\frac{4}{x+1} - \frac{3}{x+2}$, (c) $\frac{3}{x+2} - \frac{2}{x-3}$

21 (a) $-\frac{1}{x-1} + \frac{1}{x-2} + \frac{4}{(x-2)^2}$, (b) $\frac{2}{(x+1)^2} - \frac{1}{x+1} + \frac{1}{x-2}$,
(c) $\frac{2}{x+1} - \frac{3}{(x+1)^2}$ **22** (a) $\frac{2}{x-1} - \frac{2x+3}{x^2+x-1}$,
(b) $\frac{1}{x-2} - \frac{x+1}{x^2+2x+2}$, (c) $\frac{7}{9(x-1)} - \frac{2}{3(x-1)^2} - \frac{7x+1}{9(x^2+2)}$

23 (a) $\frac{2}{x^2+1} - \frac{2}{(x^2+1)^2}$, (b) $\frac{2x}{x^2+2} - \frac{4x}{(x^2+2)^2}$, (c) $\frac{1}{x^2+2} + \frac{x}{(x^2+2)^2}$

24 (a) $1 - \frac{3}{x+4} + \frac{1}{x+2}$, (b) $3x - 7 + \frac{12}{x+2}$, (c) $x + 6 - \frac{8}{x-2} + \frac{37}{x-4}$

Problems

1 (a) Linear through origin, (b) non-linear, (c) linear but not through origin, (d) linear through origin **2** (a) $V = 8I$, (b) $F = 0.4N$, (c) $v = 1.2t + 5$, (d) $R = 0.15\theta + 20$ **3** 4 mA **4** See Figure A.7

5 See Figure A.8 **6** (a) $(x-1)(x+1)$, (b) $(x+13)(x-1)$, (c) $x(x+4)$, (d) $(x-1)(x+2)(x+1)$, (e) $(x+2)(x^2+x+1)$

7 (a) 5.5, 0.3, (b) 0.6, –3.6, (c) 1.7, –1.0, (d) 1, 0.37, –1.37, (e) 1, 1, –2.3 **8** (a) +1, –3, (b) +2, –1, (c) 0, +2, +1

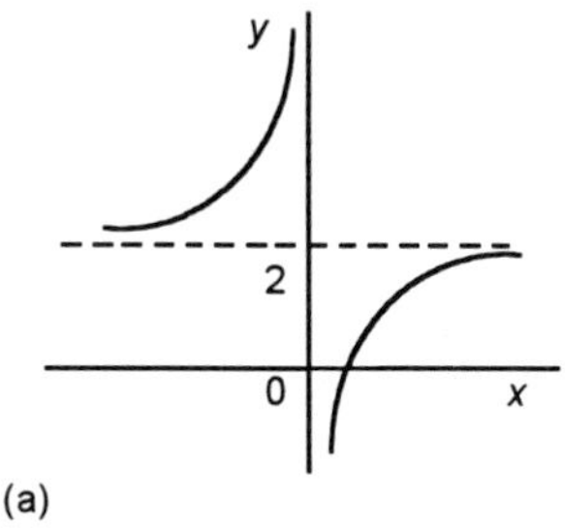

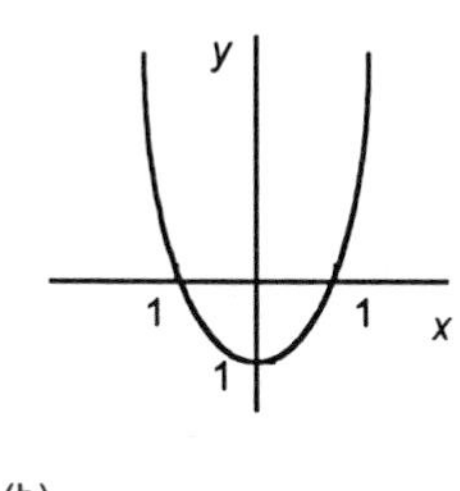

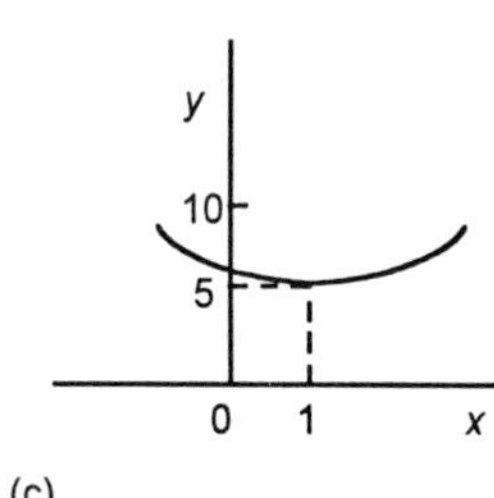

Figure A.7

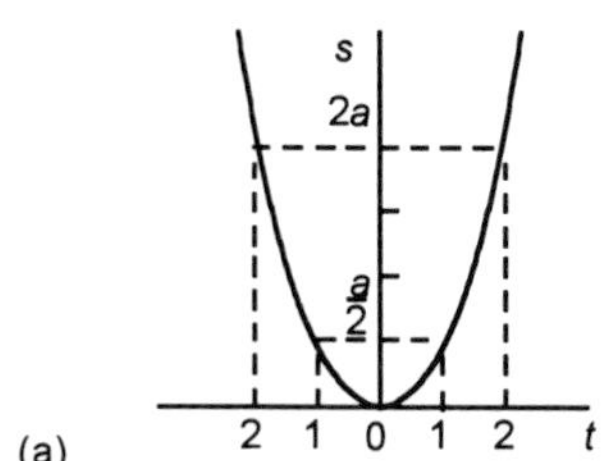

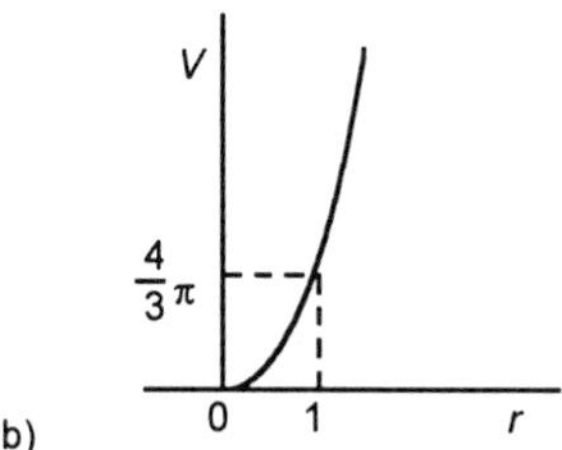

Figure A.8

9 (a) 5.9, −1.4, (b) −0.38, −2.6, (c) no real roots, (d) −1, −1
10 7.8 × 12.8 mm **11** 3.41 s, 0.59 s
12 $y=\frac{MgL}{EA}\pm\sqrt{\left(\frac{MgL}{EA}\right)^2+\frac{2Mgh}{EA}}$ **13** $h=\frac{gT^2}{8\pi^2}\pm\sqrt{\frac{g^2T^4}{64\pi^4}-k^2}$
14 (a) $\frac{2}{3(x-1)}+\frac{5}{3(2x+1)}$, (b) $\frac{8}{7(x-2)}+\frac{13}{7(x+5)}$,
(c) $\frac{5}{6(x-2)}+\frac{1}{6(x+2)}$,
(d) $\frac{1}{27(x-2)}-\frac{1}{27(x+1)}-\frac{1}{9(x+1)^2}-\frac{1}{3(x+1)^3}$,
(e) $\frac{2}{9(x-2)}+\frac{1}{3(x-2)^2}-\frac{2}{9(x+1)}$, (f) $\frac{1}{x-1}+\frac{x+1}{x^2+1}$,
(g) $\frac{3}{2x}-\frac{x}{2(x^2+2)}$, (h) $\frac{7(x+1)}{3(x^2+2x-2)}-\frac{4}{3(x+1)}$,
(i) $1-\frac{7}{4(x+3)}-\frac{1}{4(x-1)}$, (j) $x+\frac{2}{x-1}-\frac{1}{x+1}$,
(k) $x+2+\frac{1}{4(x+1)}+\frac{27}{4(x-3)}$, (l) $\frac{x}{x^2-4}+\frac{4x}{(x^2-4)^2}$,
(m) $\frac{2}{(x+1)^2}-\frac{1}{x+1}+\frac{1}{x-2}$, (n) $\frac{1}{x}-\frac{1}{x+1}$,
(o) $\frac{2x-1}{5(x^2+1)}+\frac{3}{5(x-2)}$, (p) $2x-2+\frac{3}{x-1}+\frac{4}{x+2}$,
(q) $x-\frac{1}{2(x-1)}+\frac{1}{2(x+1)}$, (r) $\frac{1}{x}+\frac{1}{2(x+1)}-\frac{1}{2(x-1)}$,
(s) $\frac{1}{x^2+1}+\frac{x-2}{(x^2+1)^2}$, (t) $\frac{1-x}{2(x^2+3)}+\frac{1}{2(x+3)}$,

(u) $\frac{1}{2x^2}-\frac{1}{4x}+\frac{1}{4(x+2)}$, (v) $x-\frac{x}{x^2+1}$,
(w) $\frac{11}{9(x-2)}+\frac{4}{3(x-2)^2}-\frac{2}{9(x+1)}$, (x) $2x+1-\frac{3}{x^2+x-1}$

Chapter 3

Revision

1 (a) a^{5x}, (b) a^{-4x}, (c) a^{7x}, (d) a^{2x} **2** (a) 0.09, (b) 0.68, (c) 5, (d) 36.94, (e) 272.99 **3** (a) 272.99, (b) 36.94, (c) 5, (d) 0.68, (e) 0.09
4 (a) 10 V, (b) 3.68 V, (c) 0 **5** (a) 2000, (b) 3264, (c) 0
6 (a) e, (b) $e^{2x}+2\,e^x$, (c) $e^{-x}+2$ **7** (a) 24.75 days, (b) 49.5 days
8 (a) x^4, (b) $\ln(x^5/2)$ **9** (a) $\lg b+0.5\lg 2-\lg a-\lg c$
10 (a) 5.19, (b) −0.593, (c) 0.419 **11** (a) $\sec\theta$, (b) 1 **12** As text.
13 30°, 90° **14** $5\sin(\theta+0.927)$ **15** (a) 14.5° + 360°n, 165.5° + 360°n, (b) 42° + 180°n **16** (a) 2, $2\pi/5$, 1, (b) 6, $2\pi/3$, 0, (c) 6, 3π, 1/3, (d) 2, 2π, −0.6 **17** (a) 0.64, (b) 0.64, (c) 0.46, (d) −0.64 **18** (a) 3.627, (b) 74.210, (c) 0.964, (d) −3.627, (e) 0.525, (f) 0.748 **19** (a) $\sinh x-\sinh y=2\sinh\frac{1}{2}(x-y)\cosh\frac{1}{2}(x+y)$, (b) $\cosh 3x=4\cosh^2 x-3\cosh x$
20 $-\coth x$ **21** (a) ±0.693, (b) 0.639 or −2.248
22 (a) 0.481, (b) 0.549, (c) 1.317, (d) −1.444

Problems

1 (a) 4^{3x}, (b) e^{6x}, (c) 2^{-2x}, (d) $1+2\,e^{2x}+e^{4x}$, (e) ¼ e^x **2** (a) 2, ∞, (b) 10, ∞, (c) ∞, 2, (d) 0, 2 **3** (a) 12.21, 14.92, (b) 8.19, 6.70, (c) 1.81, 3.30 **4** 0.95 μC **5** 0.030 **6** (a) 1.26 A, (b) 1.73 A **7** (a) 4.4×10^{-7} A, (b) 1.2 mA
8 (a) $\lg a+\lg b+\lg c$, (b) $\lg a+2\lg b+2\lg c$, (c) $\lg a-\lg b-\lg c$, (d) ½ $\lg a$ + ½ $\lg b-3\lg c$ **9** 6.49 **10** (a) 3.170, (b) 2.262, (c) 1.657, (d) −2.710, (e) −1.950, (f) 2.996, (g) 0.347, (h) −1.386
11 (a) $\tan\theta$, (b) $\sin\theta$, (c) $\sin^2\theta$ **12** As text **13** (a) 30°, 150°, 270°, (b) 90°, 270°, 45°, 225°, (c) 36.26°, 144.76°, 215.26°, 324.74°
14 (a) $\pi/6$, $5\pi/6$, $7\pi/6$, $11\pi/6$, (b) $\pi/2$, $7\pi/6$, $11\pi/6$, (c) 0, $2\pi/3$, $4\pi/3$, (d) $\pi/4$, $3\pi/4$, $5\pi/4$, $7\pi/4$, (e) 0 **15** (a) $\sqrt{41}\sin(\theta-5.608)$, (b) $\sqrt{41}\cos(\theta+5.387)$ **16** (a) 0.83, (b) 1.47, (c) 0.67, (d) −0.41
17 (a) 6, π, 1, (b) 2, $2\pi/9$, 0, (c) 6, 5π, −1/5, (d) 2, 2π, −0.2, (e) 6, $\pi/2$, $\pi/8$, (f) ½, 2π, $-\pi/6$ **18** 40, 20 Hz **19** 16.2 V
20 (a) 1.509, (b) 4.144, (c) 0.833, (d) 2.352, (e) 0.778, (f) 2.757
21 (a) 1.983, (b) 1.099, (c) 1.317, (d) 0.733 **22** $\cosh x+5\sinh x$
23 (a) $\cosh(x+y)=\cosh x\cosh y+\sinh x\sinh y$, (b) $\tanh 2x=\frac{2x}{1+\tanh^2 x}$ **24** (a) 0.21, (b) 0 or 1.39, (c) 0.916
25 As text **26** 1.622 m
27 (a) $\sqrt{\left[\left(\frac{g\lambda}{2\pi}+\frac{2\pi\gamma}{\rho\lambda}\right)\frac{2\pi h}{\lambda}\right]}$, note when surface tension neglected $\sqrt{gh}$, (b) $\sqrt{\left(\frac{g\lambda}{2\pi}+\frac{2\pi\gamma}{\rho\lambda}\right)}$

Chapter 4

Revision

1 About: (a) −1.9, 0.4, 1.5, (b) −4.5 ($-3\pi/2$), 0, +4.5 ($+3\pi/2$), (c) 0.8, −1.4, (d) 1.3 **2** (a) 5.2, (b) 1.4, (c) −1.6 **3** (a) 2.49, (b) −1.67, (c) 0.19, (d) 0.43, (e) 4.49, (f) 0.96 **4** 1.37 **5** 2.71
6 (a) 0.739 1, (b) 0.585 8, (c) 4.493 4, (d) 2.029 46, (e) −1.233 8
7 (a) 1.165 6, (b) 0, 0.256 4, (c) 1.024 99, (d) 0.619 1 **8** 2.154 4

Problems

1 (a) 0.5, (b) −1.4, 0.4, 1.0 **2** (a) 1.0, (b) 1.83, (c) 1.5, (d) −1.5, 0.5
3 (a) 0.74, (b) 0.41, (c) 0.41, (d) −0.74, (e) 1.84, (f) 0.54, (g) 2.28, (h) 4.73 **4** 2.92 **5** (a) −1.225, (b) 2.690 7, (c) 0.360 4, (d) −0.459 0, (e) 2.278 9, (f) 1.295 7, (g) 1.306 3, −3.096 0
6 (a) 0.876 7, (b) 0.426 3, (c) 1.253 4, (d) 0, 0.074 69
7 0.416 **8** 803.1 km **9** 0.87

Chapter 5

Revision

1 0, 1, 0, −1, 0, 1, ... **2** (a) 0, 1, 2, 3, 4, (b) 1, 0.37, 0.13, 0.05, 0.007
3 (a) 116, (b) 0.75 **4** (a) $(-1)^{k-1}$, (b) $5k$, (c) $2.5 - 0.5k$
5 £$0.6^k \times 10\,000$ **6** (a) 0, (b) ∞, (c) 4, (d) 0, (e) 3, (f) undefined
7 (a) 7.5, (b) 23.98, (c) 1023 **8** £103.20 **9** (a) 38, (b) 125, (c) 17
10 (a) 12, (b) 16, (c) 48 **11** (a) Convergent, (b) divergent
12 (a) Convergent, (b) divergent **13** (a) $|x| < 1$, (b) $|x| < 1$
14 (a) $1 + 2x - 4x^2 + 2x^3$, (b) $1 + (x-1) - (x-1)^2$, (c) $0 + 2(x-2) + 2(x-2)^2 + (x-2)^3$ **15** (a) $1 - x + x^2 - x^3$, (b) $\sin 1 + (x-1)\cos 1 - \frac{1}{2}(x-1)^2 \sin 1 - \frac{1}{6}(x-1)^3 \cos 1$
16 $1 - 2x^2 + \frac{2}{3}x^4$ **17** $R - R_0 = (-kC\,e^{-cT_0})(T - T_0)$
18 $E - E_0 = (T_0 + 2bT_0)(T - T_0)$ **19** $P - P_0 = (2RI_0)(I - I_0)$, 2 W, 2.1 W
20 $i_B - 4 = -8(v_A + 2)$ **21** $\delta F = \left(\frac{u}{u+v}\right)^2 \delta v + \ldots$ **22** 0.508 69
23 As given in the problem **24** (a) 1.414, (b) 1.732, (c) 2.646
25 2.095 **26** 0.265 **27** (a) $1 - x + x^2 - x^3 + \ldots$, (b) $x + x^2 + \frac{x^3}{3} - \frac{x^5}{30} + \ldots$, (c) $1 + 4x + 6x^2 + 4x^3 + x^4$, (d) $x + \frac{x^3}{3} + \frac{2x^5}{15} + \ldots$ **28** $1 + 2x + 2x^2 + \frac{8}{3}x^3 + \ldots$, 1.053 8
29 $2\left(x + \frac{x^3}{3} + \frac{x^5}{5} + \frac{x^7}{7} + \ldots\right)$, 0.693 1 **30** (a) $1 + 4x + 6x^2 + 4x^3 + x^4$, (b) $1 + \frac{3x}{2} + \frac{3x^2}{8} - \frac{x^3}{16}$, (c) $1 + \frac{5x}{2} + \frac{35x^2}{8} + \frac{105x^3}{16}$, (d) $1 - 0.25 + 0.062 - 0.015$, (e) $2 + \frac{x}{4} - \frac{x^2}{64} + \frac{x^3}{512}$
31 (a) $1 + 2x + 2x^2 + \frac{4}{3}x^3 + \ldots$, (b) $1 + x - \frac{1}{2}x^3 + \ldots$, (c) $1 - \frac{1}{2}x + \frac{1}{4}x^2 - \frac{5}{8}x^3 + \ldots$, (d) $x + \frac{x^2}{2!} + \frac{2x^3}{3!} + \ldots$, (e) $1 + \frac{1}{2}x^2 + \frac{5}{24}x^4 + \ldots$, (f) $1 - x^2 + \frac{1}{3}x^4 - \ldots$
32 As given in the problem **33** $x + \frac{x^3}{3!} + \frac{x^5}{5!} + \ldots$
34 As given in the problem **35** As given in the problem

Problems

1 1, 0.54, −0.42, −0.99, 0.28 **2** (a) 0, 1, 4, 9, 16, (b) 1, 2.72, 7.39, 20.09, 54.60, (c) 0, 2.5, 6, 10.5, 16 **3** 13 **4** 0.5 **5** (a) 0.25^k, (b) $2(-1)^k$, (c) $3 + 0.1^k$ **6** (a) 0.1, 0.01, 0.001, (b) 5.1, 5.01, 5.001, (c) −1, +1, −1 **7** (a) 8, (b) ∞, (c) 0, (d) undefined, (e) 3, (f) ∞
8 0.9^kC **9** (a) 222, (b) 9.998, (c) 28.70 **10** (a) 60, (b) 110, (c) 108
11 (a) Divergent, (b) convergent, (c) convergent, (d) convergent, (e) convergent, (f) divergent **12** $1+x-x^2-\frac{1}{3}x^3$
13 $2 + 2(x-1) + 3(x-1)^2$ **14** $1+x+\frac{1}{2}x^2+\frac{1}{6}x^3$
15 (a) $p(V) = I_s(e^{40V_a}-1)+40I_s\, e^{40V_s}(V-V_a)$,
(b) $p(V) = I_s(e^{40V_a}-1)+40I_s\, e^{40V_s}(V-V_a)+1600I_s\, e^{40V_a}\frac{1}{2}(V-V_a)^2$
16 $-1-(x+1)-(x+1)^2$ **17** $T = (Mg)\theta$, T_0 and $\theta_0 = 0$
18 $F - F_0 = (2kx_0)(x - x_0)$ **19** $I_D - 2 = \sqrt{4\beta(1+\lambda V_{DS})}\,(V_{GS}-0.5)$
20 $I - 1 = 0.2(V-5)$ **21** 0.857 17 **22** As given in the problem
23 $\delta i = k\sec^2\theta(\delta\theta) + k\sec^2\theta\tan\theta(\delta\theta)^2 + \ldots$ **24** (a) 2.449, (b) 2.828, (c) 3.606 **25** −1.552 **26** 0.347 **27** 1.895
28 (a) $1-2x^2+\frac{2}{3}x^4+\ldots$, (b) $3x+9x^2+\frac{162}{5}x^5+\ldots$,
(c) $1+2x-\frac{5}{2}x^2+\ldots$, (d) $x+x^2-\frac{x^3}{3!}+\frac{x^5}{5!}+\ldots$
29 (a) $2x+\frac{8}{3}x^3+\ldots$, (b) $1+\frac{1}{2}x+\frac{1}{8}x^2+\frac{1}{48}x^3+\ldots$,
(c) $1+\frac{1}{2}x-\frac{1}{8}x^2+\frac{1}{16}x^3+\ldots$, (d) $x-\frac{1}{2}x^2+\frac{1}{6}x^3-\ldots$, $-\pi/2 < x < \pi/2$,
(e) $1 + 3x + 6x^2 + 10x^3 + \ldots$, (f) $x+x^2+\frac{2}{3}x^3+\frac{1}{3}x^4+\ldots$,
(g) $1-\frac{1}{3}x^2-\frac{1}{45}x^4+\ldots$, (h) $1+\frac{x^2}{2}-\frac{x^4}{2!4}+\frac{x^6}{4!6}+\ldots$
30 As given in the problem **31** As given in the problem
32 $1-\frac{x^2}{2!}+\frac{x^4}{4!}+\ldots$ **33** $x+\frac{1}{3}x^3+\frac{2}{5}x^5+\ldots$
34 (a) $1 + 12x + 66x^2 + 220x^3 + \ldots$, (b) $1 + 4x + 12x^2 + 32x^3 + \ldots$,
(c) $3^{2/5}\left(1-\frac{4}{15}x-\frac{4}{75}x^2-\frac{64}{3375}x^3+\ldots\right)$, (d) $1 + x + x^2 + x^3 + \ldots$,
(e) $1-\frac{3}{2}x+\frac{27}{8}x^2-\frac{135}{16}x^3+\ldots$, (f) $1 + 2x^3 + 3x^6 + 4x^9 + \ldots$
35 (a) $16x^4 + 32x^3y + 24x^2y^2 + 8xy^3 + y^4$,
(b) $32 - 240x + 720x^2 - 1080x^3 + 810x^4 + 243x^5$ **36** 1.013 2

Chapter 6

Revision

1 (a) Even, (b) even, (c) odd, (d) odd **2** See Figure A.9
3 See Figure A.10 **4** See Figure A.11
5 (a) $\delta(t-1)$, (b) $\delta(t-1)+\delta(t-2)$, (c) $\delta[k-1]$, (d) $\delta[k-1]+\delta[k-2]$, (e) $\delta[k]+\delta[k-1]+\delta[k-2]+\delta[k-3]$

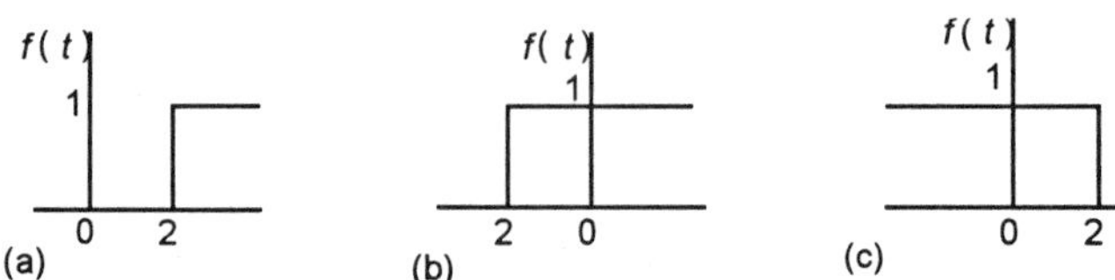

Figure A.9

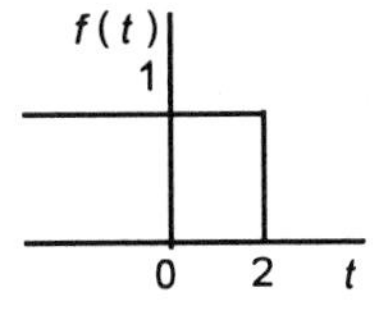

Figure A.10

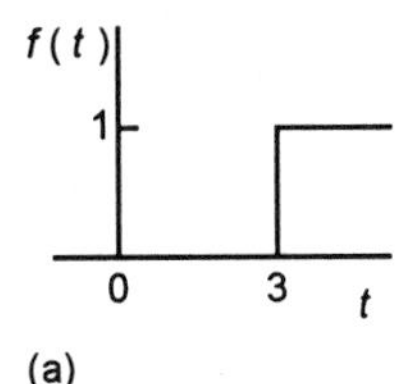

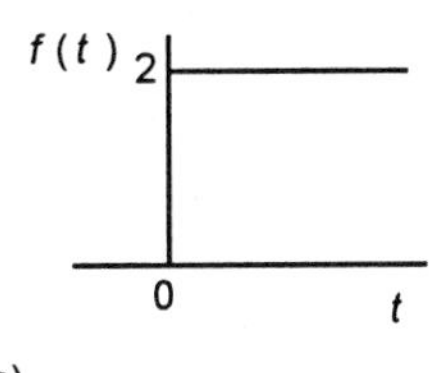

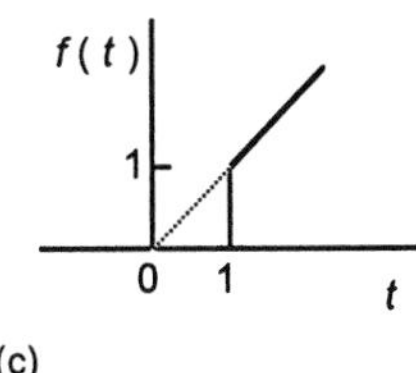

Figure A.11

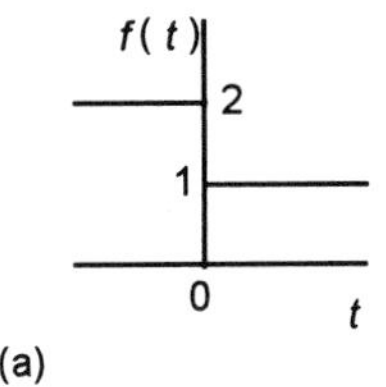

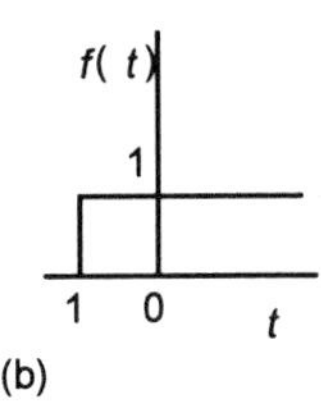

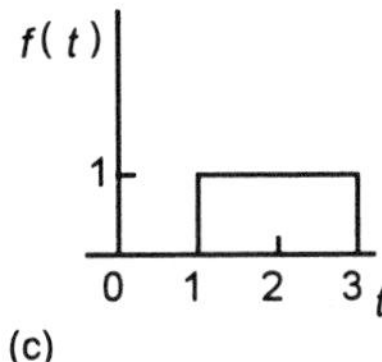

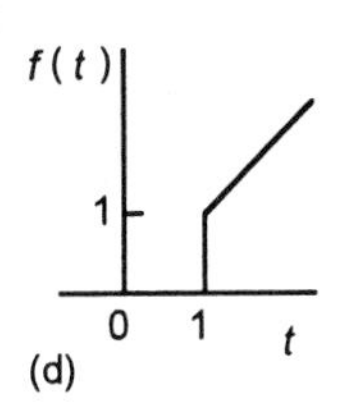

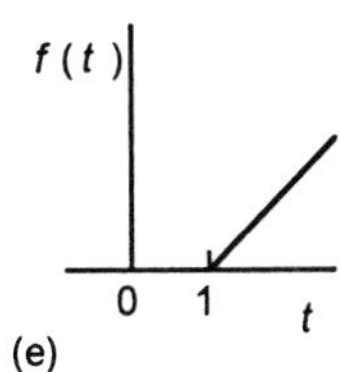

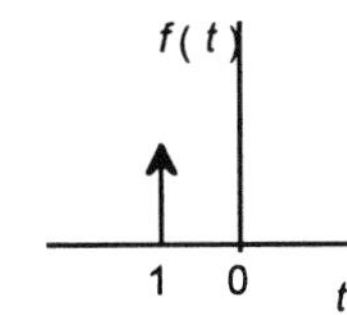

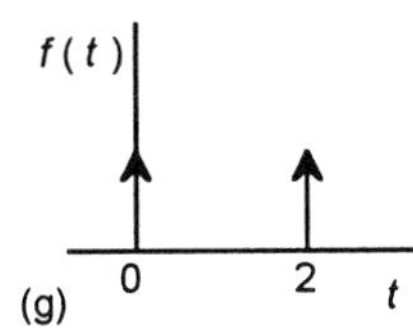

Figure A.12

Problems

1 (a) Odd, (b) even **2** See Figure A.12 **3** (a) $u(t-2)$, (b) $u(t+1)$, (c) $\frac{1}{3}(t+3).u(t+3)$, (d) $\delta(t-2)$, (e) $-\delta(t+2)$, (f) $2\delta[k-6]$

Chapter 7

Revision

1 (a) j, (b) -1 **2** (a) $\pm$j5, (b) $\pm$j9 **3** (a) $2 \pm$ j3, (b) $-3 \pm$ j2, (c) $2 \pm$ j2
4 See Figure A.13 **5** (a) $5\angle 127^\circ$, (b) $3\angle 0^\circ$, (c) $5\angle 53^\circ$, (d) $5\angle 233^\circ$, (e) $4\angle 270^\circ$ **6** (a) $1.4\angle 315^\circ$, (b) $5\angle 0^\circ$, (c) $2.8\angle 135^\circ$, (d) $2\angle 90^\circ$, (e) $5\angle 233^\circ$ **7** (a) 3 + j8, (b) 5 + j1, (c) 7 + j6, (d) 1 + j2, (e) -3 + j5
8 (a) 22 $-$ j14, (b) 11 $-$ j2, (c) 10 + j5, (d) $10\angle 40^\circ$, (e) $2\angle 40^\circ$, (f) $3\angle 120^\circ$
9 (a) 3 + j5, (b) -2 $-$ j4, (c) -4 + j6 **10** (a) -0.24 + j0.68, (b) 4 + j1, (c) 0.5 + j0.5, (d) $\frac{4}{29} + \mathrm{j}\frac{19}{29}$, (e) $2\angle 50^\circ$, (f) $2.5\angle(-50^\circ)$, (g) $0.1\angle 150^\circ$, (h) $0.5\angle 140^\circ$ **11** (a) $2\angle 40^\circ$, (b) $2\angle 160^\circ$, (c) $2\angle 280^\circ$
12 $2\angle 30^\circ$, $2\angle 150^\circ$, $2\angle 270^\circ$, 1.73 + j1, -1.73 + j1, $-$j2 **13** 2, 1 + j1.73, -1 + j1.73, -2, -1 $-$ j1.73, 1 $-$ j1.73 **14** 0.87 + j0.5, -0.5 + j0.87, -0.87 $-$ j0.5, 0.5 $-$ j0.87 **15** 0.32 + j0.78, -0.32 $-$ j0.78
16 (a) -5.66 + j5.66, (b) 117 + j44 **17** (a) $3\cos^2\theta\sin\theta - \sin^3\theta$, (b) $\cos^4\theta - 6\cos^2\theta\sin^2\theta + \sin^4\theta$, (c) $5\cos^4\theta\sin\theta - 10\cos^2\theta\sin^3\theta + \sin^5\theta$
18 (a) $\frac{1}{8}(\cos 4\theta - 4\cos 2\theta + 3)$, (b) $\frac{1}{16}(\cos 5\theta + 5\cos 3\theta + 10\cos\theta)$

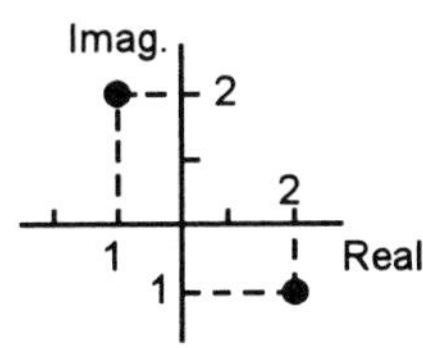

Figure A.13

Problems

1 (a) −j, (b) 1, (c) −j, (d) −1 **2** (a) ±j4, (b) −2 ± j2, (c) 0.5 ± j1.1
3 (a) $4.12\angle 166°$, (b) $5\angle 233°$, (c) $3\angle 0°$, (d) $6\angle 270°$, (e) $1.4\angle 45°$, (f) $3.61\angle 326°$ **4** (a) −2.5 + j4.3, (b) 7.07 + j7.07, (c) −6, (d) 0.68 + j2.72, (e) 1.73 + j1, (f) 1.5 − j2.6 **5** (a) 1 + j6, (b) 5 − j2, (c) −14 + j8, (d) 0.23 − j0.15, (e) 0.1 − j0.8 **6** (a) 5 − j2, (b) −2 − j1, (c) −1 + j7, (d) 1, (e) 12 + j8, (f) −10 + j6, (g) 11 − j2, (h) 13, (i) 10 + j5, (j) 0.9 + j1.2, (k) 0.23 − j0.15, (l) j1, (m) −0.3 + j1.1
7 (a) $20\angle 60°$, (b) $50\angle 80°$, (c) $0.1\angle(-20°)$, (d) $0.5\angle(-40°)$, (e) $5\angle(-20°)$, (f) $0.4\angle(-20°)$ **8** (a) 2 + j3, −2 − j3, (b) 3.46 + j2, −3.46 − j2, (c) 2.33 + j0.64, −2.33 − j0.64, (d) −0.97 + j0.26, −0.26 + j0.97, −0.97 − j0.26, 0.26 − j0.97, (e) 1, −0.5 + j0.86, −0.5 − j0.86, (f) 1.06 + j0.17, 0.17 + j1.06, −0.96 + j0.49, −0.76 − j0.76, 0.49 − j0.96
9 (a) $8\angle 60°$, 4 + j6.93, (b) −46.02 + j8.99, (c) −234 + j415, (d) −117 + j44, (e) −j32 770 **10** (a) $\cos^5\theta - 10\cos^3\theta\sin^2\theta + 5\cos\theta\sin^4\theta$, (b) $7\sin\theta - 56\sin^3\theta + 112\sin^5\theta - 64\sin^7\theta$, (c) $9\cos^8\theta\sin\theta - 84\cos^6\theta\sin^3\theta + 126\cos^4\theta\sin^5\theta - 36\cos^2\theta\sin^7\theta + \sin^9\theta$ **11** (a) $\frac{1}{32}(\cos 6\theta + 6\cos 4\theta + 15\cos 2\theta + 10)$, (b) $\frac{1}{128}(\cos 8\theta - 8\cos 6\theta + 28\cos 4\theta - 56\cos 2\theta + 35)$
12 As given in the problem **13** As given in the problem

Chapter 8

Revision

1 (a) $\sqrt{2}\ e^{j\pi/4}$, (b) $3\ e^{j\pi/2}$, (c) $\sqrt{2}\ e^{-j\pi/4}$, (d) $\sqrt{2}\ e^{j3\pi/4}$
2 (a) cos 1 + j sin 1, (b) e^{π}(cos 1 − j sin 1), (c) j5, (d) cos 3π/2 + j sin 3π/2 = −j, (e) $-1/\sqrt{2} + j1/\sqrt{2}$
3 6(cos π/3 − j sin π/3) **4** $\frac{1}{8}\cos 4\theta + \frac{1}{2}\cos 2\theta + \frac{3}{8}$
5 As given in the problem **6** $\dfrac{e^{j\theta} - e^{-j\theta}}{j(e^{j\theta} + e^{-j\theta})}$ **7** 9.152 − j0.417
8 −j sin 4 **9** j sin 1 **10** (a) 0.69 + j(π/2 + 2πn), (b) 1.15 + j(0.32 + 2πn), (c) j2πn, (d) 0.35 + j(π/4 + 2πn)
11 (a) $j\ e^{-(\pi/2 + 2\pi n)}$, (b) $8\ e^{2\pi n}$(cos 0.69 − j sin 0.69)

Problems

1 (a) $5\ e^{-j0.93}$, (b) $5.83\ e^{j0.54}$, (c) $3\ e^{j\pi/2}$ **2** (a) 22.2(cos (−3) + j sin (−3)), (b) 5.4(cos 4 + j sin 4), (c) 3(cos 3 + j sin 3), (d) e^2(cos 3 + j sin 3), (e) e^2(cos π/4 + j sin π/4) **3** 6(cos 9π/4 + j sin 9π/4)
4 $\frac{1}{16}(\sin 5\theta - 5\sin 3\theta + 10\sin\theta)$ **5** As given in the problem
6 $\frac{1}{2}(e^{j2} + e^{-j2})$ **7** 3.17 + j1.96 **8** As given in the problem
9 (a) 2.20 + j(π + 2πn), (b) 0.80 + j(1.11 + 2πn), (c) j(π/2 + 2πn), (d) 1.50 + j(1.11 + 2πn), (e) 1.61 + j(π/2 + 2πn)
10 (a) $e^{-2\pi n}$(cos 0.69 + j sin 0.69), (b) $e^{0.35 + \pi/4 - 2\pi n}$[cos(0.35 − π/4) + j sin(0.35 − π/4)] **11** 0.347 + j0.785

Chapter 9

Revision

1 Using maximum values: (a) $10\angle 0°$, 10 + j0, 10 e^0, (b) $2\angle\pi/3$, 2 cos π/3 + j2 sin π/3 = 1 + j1.73 , 2 $e^{j\pi/3}$, (c) $5\angle\pi/2$,

5 cos $\pi/2$ + j5 sin $\pi/2$ = 0 + j1, 5 $e^{j\pi/2}$ **2** (a) 3 – j2 = 3.61∠33.7° V, (b) 2 – j5 = 5.39∠68.2° V, (c) 5.5 + j2.6 = 6.1∠25.3° V, (d) 12.07 + j7.07 = 14.0∠30.4° V, (e) 8 + j1 = 8.06∠7.1° V, (f) 10 – j4 = 11.7∠21.8° V
3 (a) 36 + j2, (b) 8∠90°, (c) 8∠30° **4** (a) –0.8 + j1.4, (b) 3∠20°
5 1.45∠45.5° A, 1.45 sin(314t + 45.5°) A **6** 6.80∠38.4° V
7 12∠85° V **8** 0.4∠(–90°) = –j0.4 Ω **9** (a) 6.26 + j5.70 Ω, (b) 1 + j5 Ω, (c) 5 + j5 Ω, (d) 0.217 + j1.45 Ω **10** (a) 20 + j100 Ω, (b) 100 – j40 Ω, (c) 10 + j15 Ω, (d) 4 + j8 Ω, (e) 100 – j100 Ω
11 0.134∠(–26.6°) A

Problems

1 (a) 10∠(–30°), 8.66 – j5, (b) 10∠150°, –8.66 + j5, (c) 22∠45°, 15.6 + j15.6 **2** (a) 5.5 + j2.6, 6.1∠25.3°, (b) –2 + j7, 7.3∠105.9°, (c) 3.7 + j5.2, 6.4∠54.3° **3** (a) 25∠90°, (b) 20∠75°, (c) 44.5∠83.3°, (d) 4∠(–30°), (e) 1.25∠15°, (f) 0.164∠9.2° **4** (a) 20 + j17.32 = 26.46∠40.9° V, (b) 26.46 sin(ωt + 40.9°) V **5** 25 sin(314t – $\pi/6$) Ω
6 2∠(–36.8°) A **7** (a) 12 – j5 Ω, (b) 136.6 + j136.6 Ω, (c) 32.1 + j7.4 Ω, (d) 3.51 – j3.49 Ω, (e) 0.384 – j1.922 Ω, (f) j13.3 Ω
8 (a) 5 + j2 Ω, (b) 50 – j10 Ω, (c) 2 + j1 Ω, (d) 1.47 + j0.88 Ω, (e) –j12.5 Ω **9** 0.0844 sin(314t – 45°) A
10 170 sin(ωt ± 45°) mA **11** 2.83 sin(ωt + 45°) A
12 (a) 3.16∠(–18.4°) A, 3 – j1 A, (b) 1.41∠45°, 1 + j1 A
13 7.81∠(–38.7°) V, 9.38∠(–128.7°) V, 15.63∠51.3° V
14 70∠(–19°) V

Chapter 10

Revision

1 (a) 50 km/h, 53° N of E, (b) 28 m, N 30.3° E, (c) 21.2 N vertically
2 (a) 50 m, 53° S of E, (b) 6.4 N at 51° **3** 0.5**a** + **b** **4** (a) $\overrightarrow{AE}$, (b) 0, (c) $\overrightarrow{AB}$ **5** (a) 2.5 m/s², 4.3 m/s², (b) 173 mm, 100 mm, (c) 1.41 kN, 1.41 kN **6** (a) 5, 53.1°, (b) 2.8, 45°, (c) 13.9, 59.7°
7 (a) –4**i** + 7**j**, (b) 8**i** – 1**j**, (c) 10**i** + 11**j** **8** (a) 7**i** + 5**j**, (b) –1**i** – 1**j**, (c) –5**i** – 4**j** **9** (a) 7**i** + 4**j**, (b) –1**i** + 2**j**, (c) 1**i** + 1**j** **10** (a) 5**i** + 8**j**, (b) 9**i** – 2**j**, (c) 12**i** + 11**j**, (d) 20**i** – 9**j** **11** 3, [2/3, 1/3, 2/3]
12 9, [1/9, –4/9, 8/9] **13** 63.6° **14** As given in the problem

Problems

1 (a) 7.43 m/s, N 73° W, (b) 3.58 m/s, N 54° E, (c) 8.16 m/s, N 75° E
2 (a) 13 m, N 67° E, (b) 13 m, N 67° W, (c) 13 m, S 67° E, (d) 24.5 m, N 78° E **3** (a) $\overrightarrow{AC}$, (b) $\overrightarrow{BD}$, (c) $\overrightarrow{DB}$ **4** (a) **b** – **a**, (b) **a** + **b**, (c) **a** – 3**b**, (d) 2**b** **5** 7.8 N at 54° to AB **6** (a) $\overrightarrow{AD}$, (b) $\overrightarrow{AE}$, (c) 0, (d) $\overrightarrow{AC}$ **7** 1.36 N, 8.82 N **8** (a) 3.6, 56.3°, (b) 5.4, 21.8°, (c) 4.2, 45° **9** (a) 4**i** + 6**j**, (b) –8**i**, (c) 10**i** + 9**j**
10 (a) 7**i** + 5**j**, (b) 3**i** – 1**j**, (c) 1**i** – 4**j** **11** (a) 9**i** + 2**j**, (b) 3**i** + 4**j**, (c) –13**i** – 3**j** **12** (a) 2**i**, (b) 4**i** – 4**j** + 2**k** **13** (a) 5**i** + 4**j** + 3**k**, (b) –1**i** + 7**k**, (c) 11**i** + 8**j** –1**k**, (d) 1**i** + 2**j** + 12**k**

14 $\frac{4}{9}, \frac{4}{9}, \frac{-7}{9}; \frac{5}{\sqrt{65}}, \frac{-2}{\sqrt{65}}, \frac{6}{\sqrt{65}}; \frac{1}{\sqrt{206}}, \frac{-6}{\sqrt{206}}, \frac{13}{\sqrt{206}}$
15 (a) $\sqrt{74}$; $\frac{3}{\sqrt{74}}, \frac{7}{\sqrt{74}}, \frac{-4}{\sqrt{74}}$, (b) $\sqrt{38}$; $\frac{2}{\sqrt{38}}, \frac{3}{\sqrt{38}}, \frac{5}{\sqrt{38}}$,
(c) $\sqrt{38}$; $\frac{-3}{\sqrt{38}}, \frac{5}{\sqrt{38}}, \frac{2}{\sqrt{38}}$ **16** As given in the problem
17 76.0° **18** $\sqrt{78}$; $\frac{2}{\sqrt{78}}, \frac{-5}{\sqrt{78}}, \frac{2}{\sqrt{78}}$
19 17**i** + 11**j** + 5**k**, $\sqrt{255}$, $\sqrt{93}$, $\sqrt{21}$

Chapter 11

Revision

1 (a) 39.6, (b) 26.0, (c) 17.5 **2** (a) a^2, (b) $2a^2$, (c) $3a^2$ **3** (a) 0, (b) −22, (c) 6, (d) 24 **4** 29 **5** (a) 101.1°, (b) 43.1°
6 (a) −13**i** − 14**j** + 8**k**, (b) −5**i** − 1**j** + 6**k**, (c) −20**i** + 7**j** + 2**k** **7** 7**k**
8 4**i** + 5**j** + 7**k** **9** $\frac{1}{\sqrt{6}}$(1**i** + 2**j** + 1**k**) **10** (a) −15, (b) 4, (c) 42
11 −13 cubic units **12** 5 **13** (a) 43, (b) −43, (c) 43, (d) −43
14 As given in the problem **15** (a) 3**i** − 2**j** − 49**k**, (b) 9**i** − 6**j** − 147**k**, (c) 6**i** − 4**j** − 98**k**, (d) 12**i** − 8**j** − 196**k**

Problems

1 (a) 5, (b) 0, (c) 8, (d) −7 **2** 14 **3** (a) $2b^2$, (b) $5b^2$ **4** (a) 109.1°, (b) 72.7° **5** (a) −9, (b) 114.1° **6** (a) 1**i** − 2**j** − 4**k**, (b) 10**i** − 9**j** + 7**k**, (c) 5**i** + 7**j** + 3**k** **7** 7 square units
8 10**i** + 9**j** + 7**k** **9** 1**i** − 1**j** − 2**k** **10** −2**i** − 1**j** + 3**k**
11 (a) −7**i** − 3**j** + 8**k**, (b) −9**i** − 1**j** + 8**k**, (c) −4**i** + 4**j** − 2**k** **12** (a) −9, (b) 4, (c) 0 **13** −4 **14** As given in the problem
15 (a) 6**i** + 2**j** − 3**k**, (b) 12**i** + 4**j** − 6**k** **16** (a) 20**i** − 6**j** − 15**k**, (b) 40**i** − 12**j** −30**k**, (c) 60**i** − 18**j** − 45**k**, (d) 120**i** − 36**j** − 90**k**
17 (a) 9**i** + 8**j** − 11**k**, (b) 18**i** + 16**j** − 22**k**, (c) 27**i** + 24**j** − 33**k**, (d) 54**i** + 48**j** − 66**k**

Chapter 12

Revision

1 (a) 3.6 m/s, (b) 6.4 m/s, (c) 6.2 m/s, (d) 7.0 m/s **2** 6**i** + 11**j** + 2**k** m
3 17.4 km/h **4** 17.1 m/s, N 84.2° E **5** 6.8° N of E
6 34.9° upstream of straight across path **7** 30**i** + 20**j** km/h
8 2**i** + 14**j** m/s **9** 3**i** − 4**j** m/s **10** (a) 11.6 N, 12° to 7 N force, (b) 7.6 N, 28.4° to 5 N force, (c) 10.75 N, 9.1° to 7 N force
11 (a) 8**i** + 9**j** N, (b) −3**i** + 12**j** N, (c) −4**i** + 6**j** N **12** 3**i** − 11**j** N
13 2 N parallel to AB **14** −3**i** − 8**j** N **15** 3**i** − 3**j** m/s², 4.2 m/s² at −45° to **i** direction **16** −0.4**i** + 1**j** m/s², 1.1 m/s² at 112.8° to **i**
17 (a) 10 J, (b) 10 J, (c) 23 J, (d) 2 J **18** −12**i** + 6**j** N m
19 4**i** + 5**j** + 2**k** **20** 277.3 N m **21** 11.7 N m, 14 N m, −37.3 N m
22 −6**k** m/s **23** 5**j** m/s

Problems

1 (a) 5 m/s, (b) 5.4 m/s, (c) 7.3 m/s, (d) 3.7 m/s **2** 7**i** + 8**j** m

3 (a) $5\mathbf{i} - 4\mathbf{j}$ m/s, (b) $9\mathbf{j}$ m/s, (c) $9\mathbf{i} + 20\mathbf{j}$ m/s **4** (a) 12 m/s, N 36.9° W, (b) 4.25 km/h, N 48.3° W, (c) 15.7 m/s, S 77.9° W **5** 36.9° upstream of straight across path **6** $8\mathbf{i} + 1\mathbf{j}$ m/s **7** $1\mathbf{i} + 8\mathbf{j}$ m/s
8 $7.1\mathbf{i} + 8.9\mathbf{j}$ m/s **9** (a) $6\mathbf{i} + 7\mathbf{j}$ N, (b) $-7\mathbf{i} + 10\mathbf{j}$ N, (c) $6\mathbf{j}$ N
10 6.2 N at 56.3° to AB **11** 5 N, $\frac{3}{5\sqrt{2}}$, $\frac{4}{5\sqrt{2}}$, $\frac{1}{\sqrt{2}}$ **12** 20 N at 60° to AB **13** 17.2 N at 63° to AB **14** $-1\mathbf{i} + 1\mathbf{j}$ at 135° to $\mathbf{i}$ direction
15 (a) 27 J, (b) 12 J, (c) 6 J, (d) 12 J **16** 20 J **17** (a) 12 J, (b) −12 J, (c) 16 J **18** $-4\mathbf{i} - 8\mathbf{j} - 4\mathbf{k}$ N m **19** 137 N m, −167 N m, −255 N m **20** $35\mathbf{i} - 50\mathbf{k}$ N m **21** 282.8 N m **22** $-6\mathbf{k}$ m/s

Chapter 13

Revision

1 (a) $\{x : x \in \mathbb{R} \text{ and } 0 \le x \le 1\}$, (b) $\{x : x \in \mathbb{R} \text{ and } -3 \le x \le 3\}$, (c) $\{x : x \in \mathbb{N} \text{ and } 0 \le x \le 3\}$ **2** (a) {1, 2, 3, 4, 5, 6}, {3, 4}, (b) $\{x : x \in \mathbb{R} \text{ and } -1 \le x \le 5\}$, $\{x : x \in \mathbb{R} \text{ and } 0 \le x \le 3\}$, (c) $\{x : x \in \mathbb{R} \text{ and } 2 \le x \le 13\}$, $\{x : x \in \mathbb{R} \text{ and } 5 \le x \le 9\}$
3 (a) See Figure A.15(a), (b) see Figure A.15(b), (c) see Figure A.15(c)
4 As given in the problem **5** (a) $\overline{A \cup B \cup C}$, (b) $A \cap B \cap C$, (c) A, (d) $\bar{B}$, (e) $C \cap \bar{A}$ **6** 5

Problems

1 (a) $\{x : x \in \mathbb{R} \text{ and } 0 \le x \le 4\}\}$, (b) $\{x : x \in \mathbb{R} \text{ and } -3 \le x \le 5\}$, (c) $\{x : x \in \mathbb{N} \text{ and } 0 \le x \le 6\}$ **2** (a) $\{b, c, d, e, f, g\}$, $\{d, e\}$, (b) $\{x : x \in \mathbb{R} \text{ and } -1 \le x \le 3\}$, $\{x : x \in \mathbb{R} \text{ and } 0 \le x \le 1\}$, (c) $\{x : x \in \mathbb{N} \text{ and } 1 \le x \le 12\}$, $\{x : x \in \mathbb{N} \text{ and } 5 \le x \le 9\}$
3 (a) See Figure A.16(a), (b) see Figure A.16(b), (c) see Figure A.16(c)
4 As given in the problem **5** (a) $\{a, b, c, d, e)$, the components for A and B, (b) $\{a, b, c, d, e, f)$, the components for A, B and C **6** 4
7 67

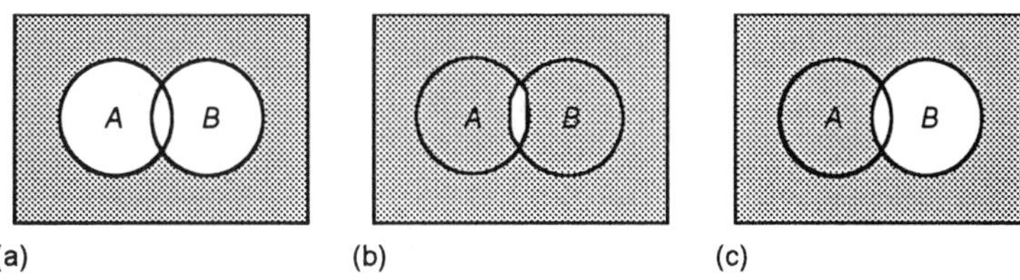

Figure A.15

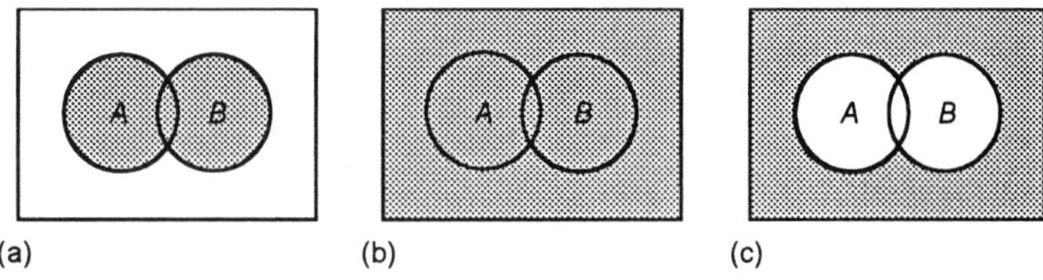

Figure A.16

Chapter 14

Revision

1 (a) 1, (b) 1, (c) 0 **2** (a) $a \cdot b$, (b) $a + b + c$, (c) b, (d) $a \cdot \bar{b}$, (e) $a + \bar{b}$

3 (a) $(b + c) \cdot d + a$, (b) $(a + \bar{a}) \cdot (b + \bar{b})$, (c) $a \cdot b + \bar{a} \cdot \bar{b}$, (d) $a \cdot \bar{c} + \bar{a} \cdot b + \bar{a} \cdot \bar{b} \cdot \bar{c}$

4

a	b	$\bar{a}$	$\bar{b}$	$a \cdot \bar{b}$	$\bar{a} \cdot b$	$(a \cdot \bar{b}) + (\bar{a} \cdot b)$
0	0	1	1	0	0	0
0	1	1	0	0	1	1
1	0	0	1	1	0	1
1	1	0	0	0	0	0

5 (a) See Figure A.17(a), (b) see Figure A.17(b), (c) see Figure A.17(c)

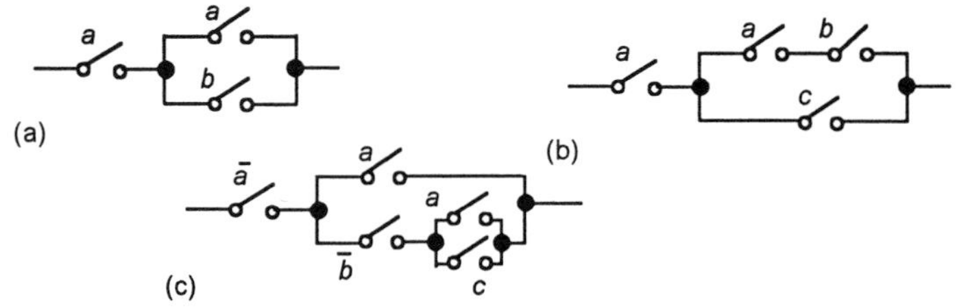

Figure A.17

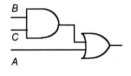

Figure A.18

Figure A.19

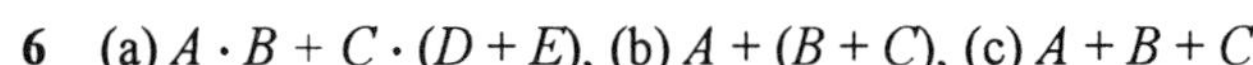

6 (a) $A \cdot B + C \cdot (D + E)$, (b) $A + (B + C)$, (c) $A + B + C$

7 (a) $(A + B) \cdot B \cdot C + A$, $B \cdot C + A$, see Figure A.18, (b) $\overline{A \cdot B} \cdot B$, $\bar{A} \cdot B$, see Figure A.19

Problems

1 (a) $a \cdot c \cdot d + b \cdot c \cdot d$, (b) $a + \bar{b} + c$, (c) $b \cdot \bar{c}$, (d) $a + \bar{b} \cdot c$, (e) $\bar{a} + b + \bar{c}$, (f) $a \cdot (b + c)$ **2** (a) $a \cdot (b + c)$, (b) $a + b + \bar{a} + \bar{b}$, (c) $\bar{a} \cdot b + a \cdot \bar{b} + a \cdot b$, (d) $c \cdot (a \cdot (b + \bar{c}) + \bar{a} \cdot b \cdot c)$

3 (a)

a	b	c	$(a + \bar{b}) + (a + \bar{c})$
0	0	0	1
0	0	1	0
0	1	0	0
0	1	1	0
1	0	0	1
1	0	1	1
1	1	0	1
1	1	1	1

(b)

a	b	$b\cdot a$	$\bar{b}$	$a\cdot b+\bar{b}$	$\bar{a}$	$\bar{a}\cdot(a\cdot b+\bar{b})\cdot\bar{b}$
0	0	0	1	1	1	1
0	1	0	0	0	1	0
1	0	0	1	1	0	0
1	1	1	0	1	0	0

4 (a) See Figure A.20(a), (b) see Figure A.20(b), (c) see Figure A.20(c), (d) see Figure A.20(d)

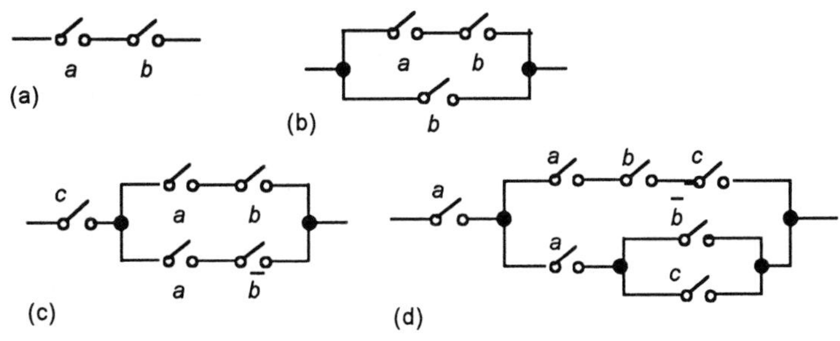

Figure A.20

5 (a) $\bar{a}\cdot\bar{b}\cdot c+a\cdot b\cdot\bar{c}$, (b) $c\cdot(a\cdot b+\bar{a}\cdot b)$

6 (a) $(A\cdot B+C\cdot D)\cdot E$, (b) $\overline{A\cdot B\cdot C}$, (c) $\bar{C}\cdot(A+\bar{B})$

7 (a) See Figure A.21(a), (b) see Figure A.21(b), (c) see Figure A.21(c), (d) see Figure A.21(d) **8** (a) $A\cdot B\cdot\overline{(\bar{A}+B\cdot C)}$, $A\cdot B\cdot\bar{C}$, Figure A.22(a), (b) $(A\cdot B+\bar{A}\cdot C)+(\bar{A}\cdot B+B\cdot C)$, $B+\bar{A}\cdot C$, Figure A.22(b)

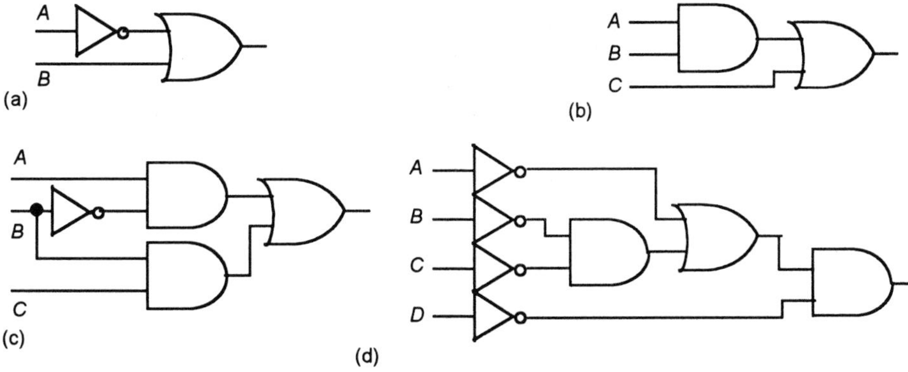

Figure A.21

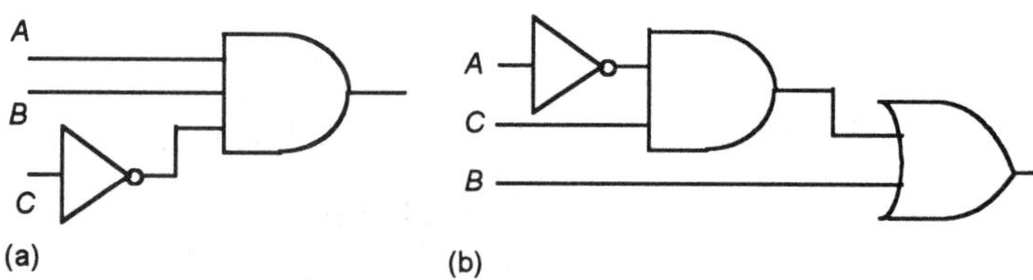

Figure A.22

Chapter 15

Revision

1 See Figure A.23 **2** See Figure A.24 **3** See Figure A.25
4 Sensor inputs high **5** See Figure A.26 **6** See Figure A.27
7 (a) 27, (b) 12, (c) 60 **8** (a) 1 100 100, (b) 1 101 111, (c) 101 010
9 (a) 159, (b) 1662, (c) 30 **10** (a) 11 101, (b) 1 110, (c) 1 11 0 111 001 **11** (a) 000 110, (b) 100110, (c) 011 011 110, (d) 010 001, (e) 010 111 101 **12** (a) 010 011, (b) 11 011 101
13 (b), (d) **14** (a) 11 111 011, (b) 11 101 001, (c) 10 000 011
15 (a) 8, (b) 15, (c) 48, (d) 14 **16** (a) Carry-out $= A \cdot B$, sum $= \bar{A} \cdot B + A \cdot \bar{B}$, (b) carry-out $= C_{in} \cdot (\bar{A} \cdot B + A \cdot \bar{B})$, sum $= \bar{C}_{in} \cdot (A + B) + \bar{C}_{in} \cdot \overline{(A + B)}$

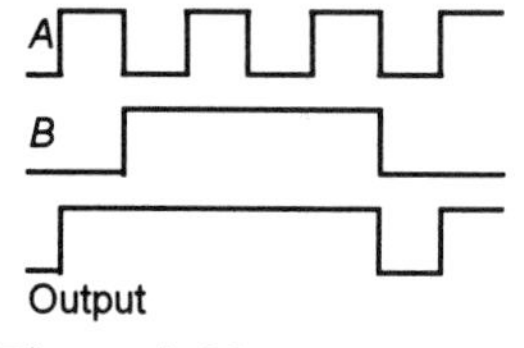

Figure A.23

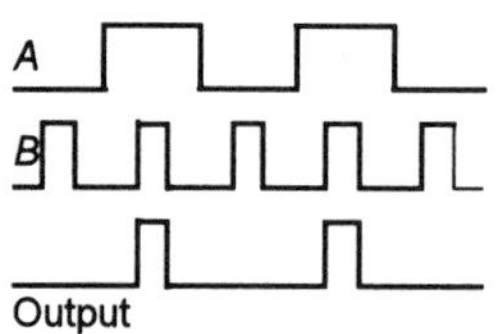

Figure A.24

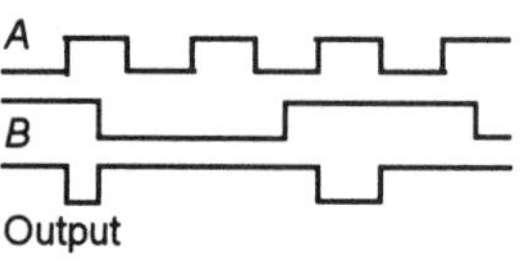

Figure A.25

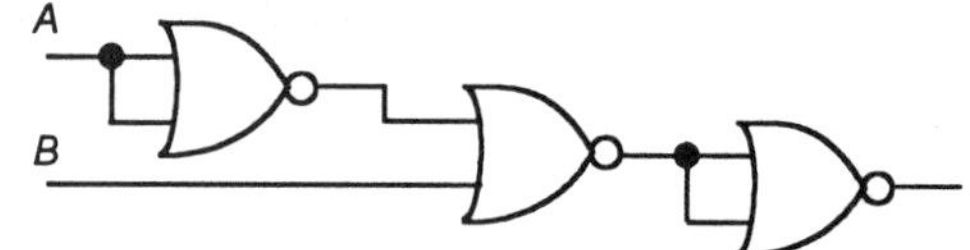

Figure A.26

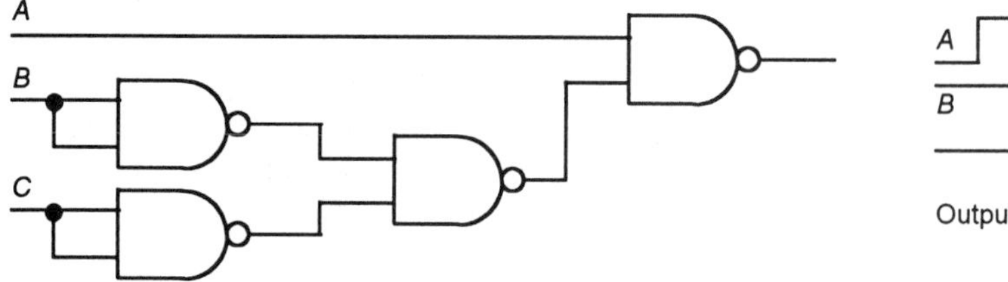

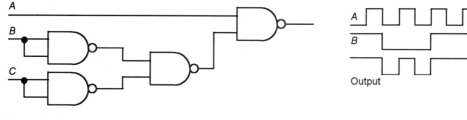

Figure A.27

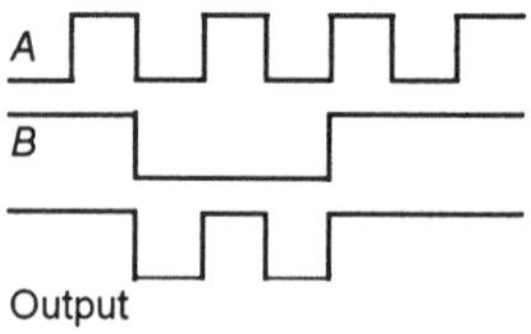

Figure A.28

Problems

1 See Figure A.28 **2** See Figure A.29 **3** See Figure A.30
4 See Figure A.31 **5** See Figure A.32 **6** See Figure A.33
7 (a) 6, (b) 21, (c) 1935 **8** (a) 011 101, (b) 111 001, (c) 111 111
9 (a) 28, (b) 159, (c) 43 981 **10** (a) 11 100, (b) 1 011, (c) 11 111

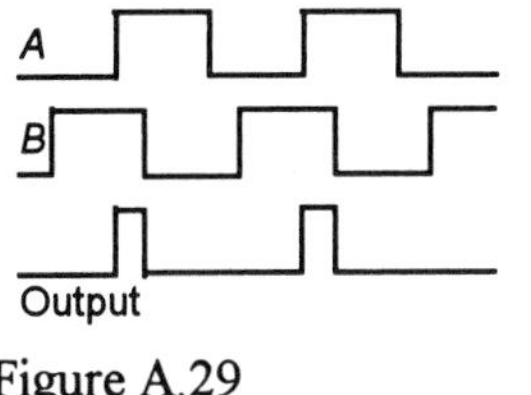

Figure A.29

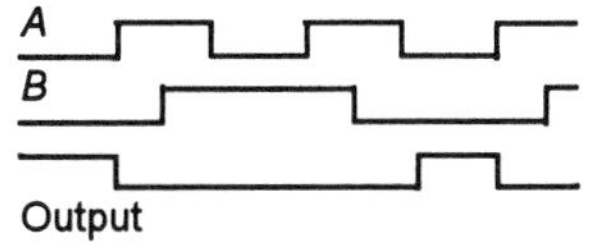

Figure A.30

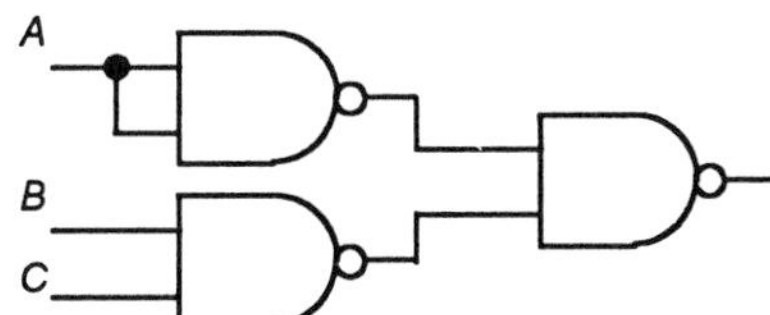

Figure A.31

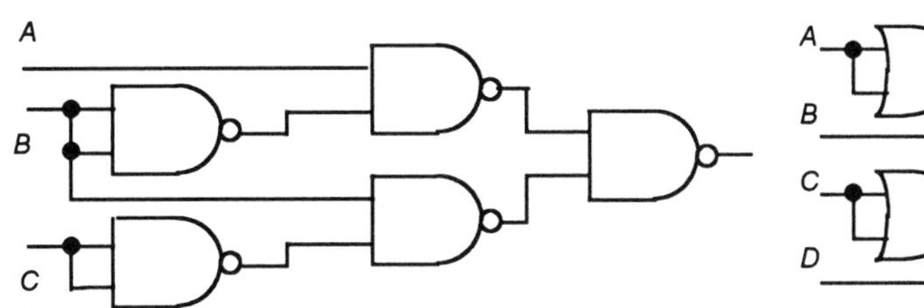

Figure A.32

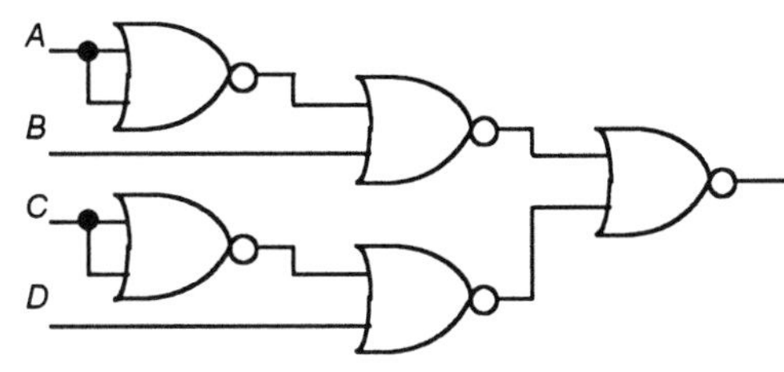

Figure A.33

11 (a) 110 001, (b) 1 100 000, (c) 000 001 000, (d) 011 011 011
12 (a) 10 011 110, (b) 11 011 101 **13** (c), (d) **14** (a) 11 111 111, (b) 11 111 101, (c) 11 111 000 **15** (a) 15, (b) 5, (c) 41
16 As given in the problem **17** As given in the problem

Chapter 16

Revision

1 (a) $x = 2, y = 1$, (b) $x = 1, y = 3$, (c) $x = -1, y = 1$, (d) $x = 2, y = -1$
2 (a) $x = 1, y = 2, z = 3$, (b) $x = 2, y = 0, z = 1$, (c) $x = 1, y = -1, z = 2$
3 $2x_1 + x_2 - x_3 = 2$, $x_1 + 2x_2 + x_3 = 4$, $3x_1 - x_2 + 3x_3 = 1$
4 $\begin{bmatrix} 1 & 4 & -1 & 4 \\ 2 & 3 & -1 & 1 \\ 3 & 1 & 2 & 5 \end{bmatrix}$ **5** $x_1 = 3, x_2 = -2, x_3 = 4$ **6** (a) $x = 1, y = 1, z = 2$, (b) $x = 2, y = -1, z = 0$, (c) $x = -1, y = 3, z = 2$, (d) $x_1 = 1, x_2 = 2, x_3 = -1, x_4 = -1$ **7** (a) Unique, (b) infinite number, (c) infinite number, (d) no solution **8** 17, 33 **9** 6 litres, 14 litres
10 15°, 45°, 120° **11** 5 litres, 20 litres, 25 litres

Problems

1 (a) $x_1 = 2, x_2 = -5, x_3 = 4$, (b) $x_1 = -2, x_2 = 6, x_3 = 2$, (c) $x_1 = 7, x_2 = -6, x_3 = -2$ **2** (a) $x = 1, y = 4, z = 2$, (b) $x = -1, y = -1, z = 3$, (c) $x = 2, y = 1, z = 4$, (d) $x = 4, y = 2, z = 1$, (e) $x_1 = 1, x_2 = 1, x_3 = 2, x_4 = 3$, (f) $x_1 = 2, x_2 = -1, x_3 = 1, x_4 = 2$ **3** (a) No solution, (b) infinite number, (c) unique **4** 42°, 138° **5** 90°, 20°, 70°
6 A 3, B 8, C 5 **7** £6, £5, £4 **8** $a \neq 1$

Chapter 17

Revision

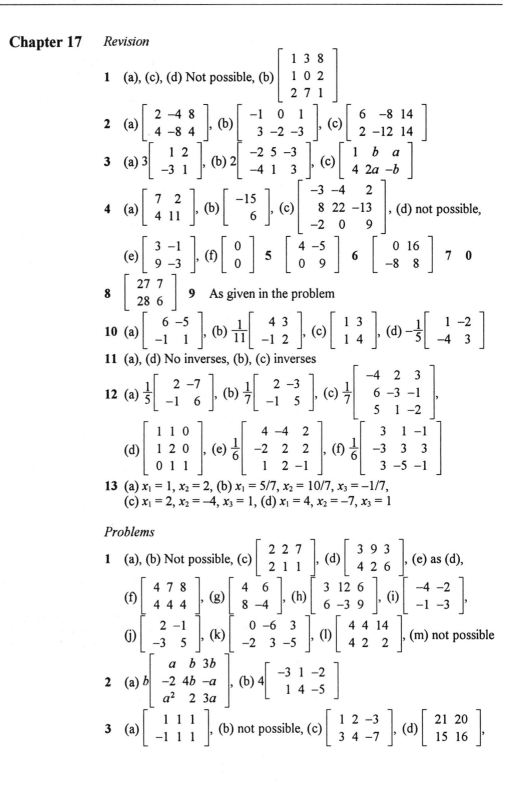

1 (a), (c), (d) Not possible, (b) $\begin{bmatrix} 1 & 3 & 8 \\ 1 & 0 & 2 \\ 2 & 7 & 1 \end{bmatrix}$

2 (a) $\begin{bmatrix} 2 & -4 & 8 \\ 4 & -8 & 4 \end{bmatrix}$, (b) $\begin{bmatrix} -1 & 0 & 1 \\ 3 & -2 & -3 \end{bmatrix}$, (c) $\begin{bmatrix} 6 & -8 & 14 \\ 2 & -12 & 14 \end{bmatrix}$

3 (a) $3\begin{bmatrix} 1 & 2 \\ -3 & 1 \end{bmatrix}$, (b) $2\begin{bmatrix} -2 & 5 & -3 \\ -4 & 1 & 3 \end{bmatrix}$, (c) $\begin{bmatrix} 1 & b & a \\ 4 & 2a & -b \end{bmatrix}$

4 (a) $\begin{bmatrix} 7 & 2 \\ 4 & 11 \end{bmatrix}$, (b) $\begin{bmatrix} -15 \\ 6 \end{bmatrix}$, (c) $\begin{bmatrix} -3 & -4 & 2 \\ 8 & 22 & -13 \\ -2 & 0 & 9 \end{bmatrix}$, (d) not possible,

(e) $\begin{bmatrix} 3 & -1 \\ 9 & -3 \end{bmatrix}$, (f) $\begin{bmatrix} 0 \\ 0 \end{bmatrix}$ **5** $\begin{bmatrix} 4 & -5 \\ 0 & 9 \end{bmatrix}$ **6** $\begin{bmatrix} 0 & 16 \\ -8 & 8 \end{bmatrix}$ **7** **0**

8 $\begin{bmatrix} 27 & 7 \\ 28 & 6 \end{bmatrix}$ **9** As given in the problem

10 (a) $\begin{bmatrix} 6 & -5 \\ -1 & 1 \end{bmatrix}$, (b) $\frac{1}{11}\begin{bmatrix} 4 & 3 \\ -1 & 2 \end{bmatrix}$, (c) $\begin{bmatrix} 1 & 3 \\ 1 & 4 \end{bmatrix}$, (d) $-\frac{1}{5}\begin{bmatrix} 1 & -2 \\ -4 & 3 \end{bmatrix}$

11 (a), (d) No inverses, (b), (c) inverses

12 (a) $\frac{1}{5}\begin{bmatrix} 2 & -7 \\ -1 & 6 \end{bmatrix}$, (b) $\frac{1}{7}\begin{bmatrix} 2 & -3 \\ -1 & 5 \end{bmatrix}$, (c) $\frac{1}{7}\begin{bmatrix} -4 & 2 & 3 \\ 6 & -3 & -1 \\ 5 & 1 & -2 \end{bmatrix}$,

(d) $\begin{bmatrix} 1 & 1 & 0 \\ 1 & 2 & 0 \\ 0 & 1 & 1 \end{bmatrix}$, (e) $\frac{1}{6}\begin{bmatrix} 4 & -4 & 2 \\ -2 & 2 & 2 \\ 1 & 2 & -1 \end{bmatrix}$, (f) $\frac{1}{6}\begin{bmatrix} 3 & 1 & -1 \\ -3 & 3 & 3 \\ 3 & -5 & -1 \end{bmatrix}$

13 (a) $x_1 = 1$, $x_2 = 2$, (b) $x_1 = 5/7$, $x_2 = 10/7$, $x_3 = -1/7$,
(c) $x_1 = 2$, $x_2 = -4$, $x_3 = 1$, (d) $x_1 = 4$, $x_2 = -7$, $x_3 = 1$

Problems

1 (a), (b) Not possible, (c) $\begin{bmatrix} 2 & 2 & 7 \\ 2 & 1 & 1 \end{bmatrix}$, (d) $\begin{bmatrix} 3 & 9 & 3 \\ 4 & 2 & 6 \end{bmatrix}$, (e) as (d),

(f) $\begin{bmatrix} 4 & 7 & 8 \\ 4 & 4 & 4 \end{bmatrix}$, (g) $\begin{bmatrix} 4 & 6 \\ 8 & -4 \end{bmatrix}$, (h) $\begin{bmatrix} 3 & 12 & 6 \\ 6 & -3 & 9 \end{bmatrix}$, (i) $\begin{bmatrix} -4 & -2 \\ -1 & -3 \end{bmatrix}$,

(j) $\begin{bmatrix} 2 & -1 \\ -3 & 5 \end{bmatrix}$, (k) $\begin{bmatrix} 0 & -6 & 3 \\ -2 & 3 & -5 \end{bmatrix}$, (l) $\begin{bmatrix} 4 & 4 & 14 \\ 4 & 2 & 2 \end{bmatrix}$, (m) not possible

2 (a) $b\begin{bmatrix} a & b & 3b \\ -2 & 4b & -a \\ a^2 & 2 & 3a \end{bmatrix}$, (b) $4\begin{bmatrix} -3 & 1 & -2 \\ 1 & 4 & -5 \end{bmatrix}$

3 (a) $\begin{bmatrix} 1 & 1 & 1 \\ -1 & 1 & 1 \end{bmatrix}$, (b) not possible, (c) $\begin{bmatrix} 1 & 2 & -3 \\ 3 & 4 & -7 \end{bmatrix}$, (d) $\begin{bmatrix} 21 & 20 \\ 15 & 16 \end{bmatrix}$,

(e) $\begin{bmatrix} 1 & 0 & 1 \\ 1 & 1 & 0 \\ 0 & 1 & 1 \end{bmatrix}$, (f) $\begin{bmatrix} 0 & 5 & -17 \\ 13 & 10 & 3 \\ 1 & -3 & 18 \end{bmatrix}$, (g) $\begin{bmatrix} 2 \\ 16 \\ -2 \end{bmatrix}$, (h) not possible

4 $\mathbf{AB} = \mathbf{BA} = \begin{bmatrix} 0 & 0 \\ 0 & 0 \end{bmatrix}$ **5** $\begin{bmatrix} -4 & 1 & -3 \\ -2 & -5 & -3 \\ -6 & -2 & -1 \end{bmatrix}$, $\begin{bmatrix} 5 & -4 & 5 \\ 7 & 5 & 6 \\ 1 & 4 & 2 \end{bmatrix}$,

$\begin{bmatrix} 2 & 5 & -1 \\ -8 & 5 & -3 \\ 16 & -2 & -1 \end{bmatrix}$, $\begin{bmatrix} -16 & -6 & 8 \\ 12 & 0 & 22 \\ 11 & -16 & 19 \end{bmatrix}$, $\begin{bmatrix} 32 & -3 & 12 \\ 36 & 29 & 24 \\ 34 & 6 & 25 \end{bmatrix}$

6 $\begin{bmatrix} 6 & 1 \\ 9 & 0 \end{bmatrix}$, $\begin{bmatrix} 6 & -2 \\ 4 & -10 \end{bmatrix}$, $\begin{bmatrix} 7 & 2 \\ -4 & 23 \end{bmatrix}$, $\begin{bmatrix} 2 & 1 \\ -29 & -21 \end{bmatrix}$, $\begin{bmatrix} 23 & 16 \\ 56 & 39 \end{bmatrix}$

7 As given in the problem

8 (a) $\frac{1}{13}\begin{bmatrix} -1 & 3 \\ 5 & 2 \end{bmatrix}$, (b) $\begin{bmatrix} 9 & -2 \\ -4 & 1 \end{bmatrix}$, (c) $-\frac{1}{10}\begin{bmatrix} -4 & 2 \\ 3 & 1 \end{bmatrix}$,

(d) $-\frac{1}{7}\begin{bmatrix} -3 & -5 \\ -2 & -1 \end{bmatrix}$ **9** (a), (c) No inverse, (b), (d) inverses

10 (a) $\frac{1}{5}\begin{bmatrix} -3 & 4 \\ 2 & -1 \end{bmatrix}$, (b) $\begin{bmatrix} 1 & 0 \\ -2 & 1 \end{bmatrix}$, (c) $\frac{1}{2}\begin{bmatrix} -8 & 6 & -8 \\ 1 & -1 & 2 \\ 12 & -8 & 12 \end{bmatrix}$,

(d) $\frac{1}{2}\begin{bmatrix} 1 & -2 & 1 \\ 1 & 0 & -1 \\ -1 & 2 & 1 \end{bmatrix}$, (e) $\frac{1}{4}\begin{bmatrix} -2 & -6 & 4 \\ -3 & -5 & 6 \\ 1 & 3 & -2 \end{bmatrix}$,

(f) $\frac{1}{25}\begin{bmatrix} 5 & 0 & -5 \\ -6 & 10 & 1 \\ 7 & 5 & 3 \end{bmatrix}$ **11** (a) $x_1 = 35/11$, $x_2 = 25/11$, (b) $x_1 = 2$,
$x_2 = -3$, (c) $x_1 = 1$, $x_2 = 3$, $x_3 = 5$, (d) $x_1 = ½$, $x_2 = ¾$, $x_3 = -½$, (e) $x_1 = 3$,
$x_2 = -5$, $x_3 = 1$ **12** As given in the problem

Chapter 18

Revision

1 (a) −2, (b) 22, (c) 23, (d) $6a$ **2** (a) $x_1 = 2$, $x_2 = -1$, (b) $x_1 = 1$, $x_2 = 3$, (c) $x_1 = 4$, $x_2 = 1$, (d) $x_1 = -1$, $x_2 = 2$ **3** (a) 18, (b) −28, (c) −8, (d) −12 **4** (a) $x_1 = 1$, $x_2 = 2$, $x_3 = -1$, (b) $x_1 = 2$, $x_2 = 1$, $x_3 = 3$, (c) $x_1 = -1$, $x_2 = 3$, $x_3 = 1$ **5** (a) 14, (b) −14, (c) −1, (d) 8, (e) 1, (f) 39, (g) 0 **6** (a) 63 000, (b) −8, (c) 0, (d) 1, (e) −1160, (f) −3

7 (a) −8, (b) 2, (c) −2

8 (a) $\begin{bmatrix} -6 & -6 & 6 \\ 3 & 5 & -7 \\ 9 & 7 & -17 \end{bmatrix}$, $\begin{bmatrix} -6 & 3 & 9 \\ -6 & 5 & 7 \\ 6 & -7 & -17 \end{bmatrix}$, $-\frac{1}{12}\begin{bmatrix} -6 & 3 & 9 \\ -6 & 5 & 7 \\ 6 & -7 & -17 \end{bmatrix}$,

(b) $\begin{bmatrix} 4 & -2 & 1 \\ 8 & -2 & -1 \\ 0 & 2 & -1 \end{bmatrix}$, $\begin{bmatrix} 4 & 8 & 0 \\ -2 & -2 & 2 \\ 1 & -1 & 1 \end{bmatrix}$, $\frac{1}{4}\begin{bmatrix} 4 & 8 & 0 \\ -2 & -2 & 2 \\ 1 & -1 & 1 \end{bmatrix}$,

(c) $\begin{bmatrix} -2 & 7 & -18 \\ -5 & -4 & -2 \\ 7 & 3 & 23 \end{bmatrix}$, $\begin{bmatrix} -2 & -5 & -7 \\ 7 & -4 & 3 \\ -18 & -2 & 23 \end{bmatrix}$, $\frac{1}{43}\begin{bmatrix} -2 & -5 & -7 \\ 7 & -4 & 3 \\ -18 & -2 & 23 \end{bmatrix}$

Problems

1 (a) 4, (b) 11, (c) 0, (d) $a - 2$, (e) 0, (f) –9, (g) 28, (h) 84, (i) 0, (j) 81, (k) –240, (l) –960 **2** (a) $x_1 = 1, x_2 = 1$, (b) $x_1 = 2, x_2 = -1$, (c) $x_1 = 1, x_2 = 3$, (d) $x_1 = 4, x_2 = 2$, (e) $x_1 = 1, x_2 = 1, x_3 = 2$, (f) $x_1 = -1, x_2 = 3, x_3 = 2$, (g) $x_1 = 3, x_2 = 1, x_3 = -2$, (h) $x_1 = 1, x_2 = 2, x_3 = -1, x_4 = 2$, (i) $x_1 = -2, x_2 = 4, x_3 = 1, x_4 = 1$ **3** (a) 4, (b) 24, (c) 4, (d) 4, (e) 4, (f) –4 **4** As given in the problem

5 (a) $\begin{bmatrix} 14 & 7 & 7 \\ -12 & -14 & -10 \\ 10 & 7 & -1 \end{bmatrix}$, $\begin{bmatrix} 14 & -12 & 10 \\ 7 & -14 & 7 \\ 7 & -10 & -1 \end{bmatrix}$, $\frac{1}{28}\begin{bmatrix} 14 & -12 & 10 \\ 7 & -14 & 7 \\ 7 & -10 & -1 \end{bmatrix}$,

(b) $\begin{bmatrix} -1 & 0 & 0 \\ 0 & 0 & -1 \\ 0 & -1 & 0 \end{bmatrix}$, $\begin{bmatrix} -1 & 0 & 0 \\ 0 & 0 & -1 \\ 0 & -1 & 0 \end{bmatrix}$, $-1\begin{bmatrix} -1 & 0 & 0 \\ 0 & 0 & -1 \\ 0 & -1 & 0 \end{bmatrix}$,

(c) $\begin{bmatrix} 1 & 1 & 0 \\ -2 & 1 & 3 \\ -1 & -1 & 3 \end{bmatrix}$, $\begin{bmatrix} 1 & -2 & -1 \\ 1 & 1 & -1 \\ 0 & 3 & 3 \end{bmatrix}$, $\frac{1}{3}\begin{bmatrix} 1 & -2 & -1 \\ 1 & 1 & -1 \\ 0 & 3 & 3 \end{bmatrix}$

Chapter 19

Revision

1 As given in the problem **2** As given in the problem

3 (a) $\lambda^2 - 5\lambda + 4 = 0$, 1, 4, (b) $\lambda^2 - 3\lambda + 2 = 0$, 1, 2, (c) $\lambda^2 - 3\lambda - 4 = 0$, –1, 4, (d) $(1 - \lambda)(\lambda - 1)(\lambda + 1) = 0$, 1, 1, –1, (e) $\lambda^2 - 8\lambda + 25 = 0$, $4 \pm j3$, (f) $\lambda^2 - 2\lambda - 5 = 0$, $1 \pm \sqrt{6}$

4 (a) 1, 4, $\begin{bmatrix} 1 \\ -1 \end{bmatrix}$, $\begin{bmatrix} 2 \\ 1 \end{bmatrix}$, (b) 3, –1, $\begin{bmatrix} \frac{1}{2} \\ 1 \end{bmatrix}$, $\begin{bmatrix} 0 \\ 1 \end{bmatrix}$,

(c) 0, $\begin{bmatrix} 1 \\ 0 \end{bmatrix}$, $\begin{bmatrix} 0 \\ 1 \end{bmatrix}$, (d) –1, 1, 4, $\begin{bmatrix} 1 \\ 1 \\ 1 \end{bmatrix}$, $\begin{bmatrix} -1 \\ -1 \\ 2 \end{bmatrix}$, $\begin{bmatrix} 1 \\ -1 \\ 0 \end{bmatrix}$

Problems

1 As given in the problem **2** (a) $\lambda^2 - 10\lambda + 9 = 0$, 1, 9, (b) $\lambda^2 - 8\lambda = 0$, 0, 8, (c) $\lambda^2 - 3\lambda - 4 = 0$, –1, 4, (d) $(1 - \lambda)(\lambda + 1)(\lambda - 2) = 0$, –1, 1, 2, (e) $-(4 - \lambda)(1 + \lambda)(1 - \lambda) = 0$, –1, 1, 4, (f) $\lambda^2 - 16\lambda + 89 = 0$, $4 \pm j5$, (g) $\lambda^2 - 3\lambda + 12 = 0$, $1.5 \pm \sqrt{8.25}$, (h) $\lambda^2 + 9\lambda + 16 = 0$, $-4.5 \pm \sqrt{4.25}$

3 (a) –4, –3, $\begin{bmatrix} 1 \\ -1 \end{bmatrix}$, $\begin{bmatrix} 2 \\ -1 \end{bmatrix}$, (b) –5, 2, $\begin{bmatrix} 7 \\ 1 \end{bmatrix}$, $\begin{bmatrix} 0 \\ 1 \end{bmatrix}$,

(c) 2, 3, $\begin{bmatrix} 1 \\ 1 \end{bmatrix}$, $\begin{bmatrix} 2 \\ 3 \end{bmatrix}$, (d) 1, $\begin{bmatrix} 1 \\ -1 \\ 1 \end{bmatrix}$, (e) 1, 1, –2, $\begin{bmatrix} 0 \\ 1 \\ 1 \end{bmatrix}$, $\begin{bmatrix} 2 \\ -1 \\ -1 \end{bmatrix}$

Chapter 20

Revision

1 (a) $x = 3, y = 1$, (b) $x = 2, y = -1$, (c) $x = 1, y = 2$, (d) $x = 3, y = 2$, (e) $x = 1, y = 2, z = 3$, (f) $x = -2, y = 2, z = -1$ **2** As in problem 1

Problems

1 (a) $x = 2, y = 1$, (b) $x = -1, y = 3$, (c) $x = 3, y = 2$, (d) $x = 5, y = -4$, (e) $x = 3, y = 2, z = 1$, (f) $x = -1, y = 2, z = 1$, (g) $x = 2, y = 4, z = 2$, (h) $x = 1, y = -3, z = 1$

Chapter 21

Revision

1 (a) 13.5 V, 6 V, (b) 9.1 V, 10.9 V, (c) 1 V, –1.5 V, –2.5 V, (d) 11 V, 17 V **2** (a) 8 A, 2 A, 6 A, (b) 3.18 A, –0.38 A, –2.39 A, 3.46 A, –2.01 A, (c) 5.6 A, 2.0 A, –0.8 A, 3.6 A, 2.8 A

Problems

1 (a) 6 V, 10 V, (b) 17.1 V, 7.3 V, (c) 8 V, 12 V, (d)1.3 V, 10 V, 3.0 V
2 (a) –0.2 A, –1.1 A, –0.9 A, (b) 0.35 A, –0.08 A, –0.40 A, 0.43 A, 0.32 A

Chapter 22

Revision

1 (a) 2, (b) 8 **2** (a) $2x + 3$, (b) $4x$ **3** (a) $4x^3$, (b) $\frac{5}{3}x^{2/3}$, (c) $-2x^{-3}$, (d) $\frac{1}{2}x^{-1/2}$ **4** (a) $5 \cos 5x$, (b) $-2 \sin 2x$, (c) $3 \sec^2 3x$
5 $314 \cos 314t$ **6** $-50 \sin 50t$ **7** (a) $4\,e^{4x}$, (b) $-3\,e^{-3x}$
8 (a) $3x^{1/2}$, (b) $-8 \sin 2x$, (c) $10\,e^{2x}$ **9** (a) $2x + 2$, (b) $2x + 3 \cos 3x$, (c) $2\pi x + 2\pi$, (d) $2 \cos x + 9 \cos 3x$, (e) $-6\,e^{-3x} - 15\,e^{-5x}$, (f) $2\,e^x$
10 (a) $4x + 2(1 + x^2)$, (b) $2x + 4$, (c) $4x\,e^{4x} + e^{4x}$, (d) $-2 \sin x \sin 2x + \cos x \cos 2x$, (e) $e^{-2x} - 2\,e^{-2x}(x + 5)$, (f) $e^x(2 \cos^2 x + \sin x \cos x - 1)$
11 (a) $\frac{6}{(x+3)^2}$, (b) 1, (c) $\frac{x-3}{(1+x)^3}$, (d) $\frac{1-\cos x}{\sin^2 x}$ **12** (a) $15(3x + 2)^4$, (b) $5 \cos(5x + 2)$, (c) $6x(x^2 + 1)^2$, (d) $5(x^2 - 6x + 1)^4(2x - 1)$, (e) $-(3/2)x^2(6 - x^3)^{-1/2}$, (f) $-4(6x^3 - x)^{-5}(18x^2 - 1)$, (g) $2x(x - 3)^3(x + 3)^{-5}$, (h) $2x^2 \cos x^2 + \sin x^2$, (i) $2(x + \sin x)(1 + \cos x)$
13 (a) $\frac{1}{2y+6}$, (b) $\frac{1}{3y^2+2}$ **14** (a) $\frac{1}{x}$, (b) $\frac{2\,e^{2x}}{1+e^{2x}}$
15 (a) $\frac{3}{\sqrt{16-9x^2}}$, (b) $\frac{1}{2\sqrt{x}\,(1+x)}$, (c) $\frac{3 \sin 3x}{\sqrt{1-9x^2}} + 3 \cos 3x \sin^{-1} 3x$, (d) $\frac{6(\tan^{-1} 2x)^2}{1+4x^2}$, (e) $\frac{-2x}{1+x^4}$, (f) $2x \cos^{-1}(x-1) - \frac{x^2}{\sqrt{2x-x^2}}$
16 (a) $\frac{1}{\sqrt{9+x^2}}$, (b) $\frac{3}{\sqrt{9x^2-1}}$, (c) $\frac{3}{1-9x^2}$ **17** (a) $-\frac{9x}{4y}$, (b) $\frac{9y-12x^5}{4y^3-9x}$, (c) $\frac{\cos(x+y)}{1-\cos(x+y)}$ **18** (a) $3t$, (b) $-\frac{4\cos\theta}{3\sin\theta}$, (c) $1 + t$, (d) $\frac{\sin\theta}{1-\cos\theta}$ **19** (a) $x^3(1 + \ln x)$, (b) $(x+3)^x\left[\ln(x+3) + \frac{x}{x+3}\right]$, (c) $3x^2\,e^{4x}(1 + x)^5 + 4x^2\,e^{4x}(1 + x)^5 + 5x^3\,e^{4x}(1 + x)^4$, (d) $\sqrt{1+x^2}\,\sin^2 x\left(\frac{x}{1+x^2} + \frac{2\cos x}{\sin x}\right)$ **20** (a) $2/x^3$, (b) $-25 \sin 5x$,

(c) $8 \sec^2 2x \tan 2x$, (d) $-(1 + 2x)^{-3/2}$ **21** (a) $4x^3 + 6x^2$, $12x^2 + 12x$, $24x + 12$, 24, (b) $2 \cos 2x$, $-4 \sin 2x$, $-8 \cos 2x$, $16 \sin 2x$
22 (a) 42 m/s, (b) 12 m/s^2 **23** $\frac{1}{2}kx^{-1/2}$, $-\frac{1}{4}kx^{-3/2}$ **24** $0.02 \cos 2t$ A/s
25 2 V **26** 100 mm^2/mm **27** $I\omega$ **28** 0.5 **29** (a) (2.5, −0.25) minimum, (b) (0, −2) minimum, (−2/3, −1.26) maximum, (c) (0, 0) maximum, (−0.71, −0.25) minimum, (0.71, −0.25) minimum, (d) (0, 0) inflexion **30** $x = y$ **31** $\frac{1}{2}(a - b)$ **32** $r = 4/3$ m **33** 6×6 cm
34 $4 \times 4 \times 2$ m **35** (a) 4.08, (b) 3.99 **36** ±12.6 cm^3 **37** 0.13 cm^2
38 1.012 5 **39** (a) $8t\mathbf{i} - (2 \sin 2t)\mathbf{j}$, $8\mathbf{i} - (4 \cos 2t)\mathbf{j}$, (b) $6t\mathbf{i}$, $6\mathbf{i}$
40 (a) $18t^2 - 6t$, (b) $(9 + 8t^3)\mathbf{k}$ **41** (a) 6, (b) $5t^4\mathbf{i} - 9t^2\mathbf{j} + 6t^2\mathbf{k}$
42 (a) $\mathbf{v} = -r\omega(\imath \sin \omega t - \mathbf{j} \cos \omega t)$, (b) $\mathbf{a} = -\omega^2\mathbf{r}$

Problems

1 (a) $5x^4$, (b) $-8x^{-5}$, (c) $-6x$, (d) ½, (e) $4\pi x$, (f) $3 \sec^2 3x$, (g) $-10 \sin 2x$, (h) $8 e^{x/2}$, (i) $-4 e^{-2x}$, (j) $9 e^{3x}$, (k) $-(5/6)x^{-3/2}$, (l) $-(12/3)x^{-3}$, (m) $-\frac{7}{2\sqrt{3}}x^{-3/2}$, (n) $-\frac{15}{8}x^{-4}$, (o) $\frac{\sqrt{3}}{2}x^{3/2}$, (p) $-24x^2 + 4x + 15$, (q) $5x \cos x + 5 \sin x$, (r) $e^{x/2} + \frac{1}{2}x e^{x/2}$, (s) $(x^2 + 1) \cos x + 2x \sin x$, (t) $\frac{-13}{(x-6)^2}$, (u) $\frac{x-1}{2x^{3/2}}$, (v) $\frac{x\cos x - \sin x}{x^2}$, (w) $\frac{2\,e^{2x}(x^2 - x + 1)}{(x^2+1)^2}$, (x) $\frac{2\cosh 2x - \cosh 3x - 3\sinh 2x \sinh 3x}{(\cosh 3x)^2}$, (y) $\frac{7}{(2-7x)^2}$, (z) $\frac{1}{2(3-x)^{3/2}}$, (aa) $3(x+5)^2$, (ab) $\frac{2x}{(x^2+1)^{3/2}(x^2-1)^{1/2}}$, (ac) $\frac{x^2+12}{3(x^2+4)^{4/3}}$, (ad) $\sin^2 x + 2x \sin x \cos x$, (ae) 3, (af) $\frac{5}{\sqrt{1-25x^2}}$, (ag) $\frac{1}{2\sqrt{x}(1+x)}$, (ah) $\frac{2}{\sqrt{e^{4x}-1}}$, (ai) $\frac{2\tan^{-1}x}{1+x^2}$, (aj) $2\sqrt{1-x^2}$, (ak) 0, (al) $\frac{1}{1-x^2}$, (am) $\frac{4}{\sqrt{16x^2-1}}$, (an) $\frac{x}{\sqrt{x^2+1}} + \sinh^{-1}x$
2 (a) $-\frac{y+2x}{2y+x}$, (b) $-\frac{4x+3y}{3x+2y-4}$, (c) $\frac{3x^2}{3y^2-1}$, (d) $\frac{2xy}{y-1}$
3 (a) $\frac{4t^3+1}{3t^2+1}$, (b) $-\frac{4\sin 4t}{2\cos 2t}$, (c) $\frac{1+e^t}{3t^2+2\pi\cos 2\pi t}$
4 (a) $(2x^2+3)^x\left[\ln(2x^2+3) + \frac{4x^2}{2x^2+3}\right]$, (b) $x^{x+1/2}\left[\left(x+\frac{1}{2}\right)\frac{1}{x} + \ln x\right]$, (c) $x^{\sqrt{x}}\left(\frac{\ln x + 2}{2\sqrt{x}}\right)$, (d) $x^{\sqrt{x+1}}\left(\frac{\ln x}{2\sqrt{x+1}} + \frac{\sqrt{x+1}}{x}\right)$
5 (a) 2, (b) $-4 \cos 2x$, (c) $6/x^4$, (d) $36x^2 - 2 - 2/x^3$, (e) $12x^2 + 12x$, (f) $\frac{3\sqrt{x} - \sqrt{x^3}}{4x^3}$ **6** 7 m/s, −4 m/s^2 **7** $6 \cos 2t - 9 \sin 3t$ m/s, $-12 \sin 2t - 27 \cos 3t$ m/s^2 **8** $0.03 \cos 5t$ A/s **9** $50 e^{-100t}$ V/s
10 πr^2 **11** $4\pi r^2$ **12** $-\frac{f_s}{c(1+v/c)^2}$ **13** 0.5 **14** −0.87
15 $L_0(a + 2bT)$ **16** (a) (2, −1) min., (b) (1, 7) max., (3, 3) min., (c) (1, −4) min., (−1, 4) max., (d) (π/2, 1) max., (3π/2, −1) min., (e) (−2, 23) max., (1, −4) min. **17** $r = h = 1$ m **18** $h = r = 4$ cm
19 ½L **20** 0, 0.32L **21** $R = r$ **22** 47.7 V **23** 3.33 m from smaller source **24** 10 mA/s **25** 0.58L **26** $-a/2b$ **27** 45°

28 4.994 **29** –4.997 6 **30** 3.07 cm^3 **31** (a) $4t\mathbf{i} + (2\cos 2t)\mathbf{j}$, $4\mathbf{i} - (4\sin 2t)\mathbf{j}$, (b) $6t\mathbf{i} + 1\mathbf{j}$, $6\mathbf{i}$ **32** (a) $6t^2 - 4t + 2$, (b) $3t^2\mathbf{i} - (4 + 4t^3)\mathbf{k}$ **34** (a) $-\frac{2}{3}\,e^{-t/5}\mathbf{i} + 1\mathbf{k}$, (b) $-\frac{2}{25}\,e^{-t/5}\mathbf{i}$

Chapter 23

Revision

1 (a) 0.21, (b) 3.5 **2** 2.7 **3** (a) 0.21, (b) 2.5 **4** 2.6 **5** 0.621
6 2.00

Problems

1 1.002 **2** 0.333 4 **3** 1.079 **4** 1.789 3 **5** –3.639
6 2.249 7

Chapter 24

Revision

1 (a) $3x^2 + 8xy^3$, $12x^2y^2 + 2y$, (b) $2x + y$, $x + 2y$, (c) $\dfrac{1}{2\sqrt{x+y}}$, $\dfrac{1}{2\sqrt{x+y}}$, (d) $2\sec^2(2x + 3y)$, $3\sec^2(2x + 3y)$, (e) e^{x+2y}, $2\,e^{x+2y}$
2 (a) 4, $-30y$, 0, (b) $8y^3 + 18x$, $24x^2y + 12$, $24xy^2$, (c) $e^{-x}\sin y$, $-e^{-x}\sin y$, $-e^{-x}\cos y$, (d) $-\dfrac{4y}{(x+y)^3}$, $\dfrac{4x}{(x+y)^3}$, $\dfrac{2(x-y)}{(x+y)^3}$, (e) $6 + y^2 e^{xy}$, $2 + x^2 e^{xy}$, $e^{xy} + xy\,e^{xy}$, (f) $2y + 9y^4\cos 3xy^2$, $6x\sin 3xy^2 + 36x^2y^2\cos 3xy^2$, $2x + 6y\sin 3xy^2 + 18xy^3\cos 3xy^2$
3 As given in the problem **4** $12t\cos(3x - y) - 2(t - 1)\cos(3x - 1)$
5 $2(e^t - e^{-t})$ **6** $(3y + 4x)\,e^{xy}$, $2(y - x)\,e^{xy}$
7 $-2t\,e^x\cos y - 2s\,e^x\sin y$, $2s\,e^x\cos y - 2t\,e^x\sin y$ **8** 3.5 cm/s
9 1200 cm^3/s **10** 4.96 units/s **11** (a) $2xy^3\,dx + 3x^2y^2\,dy$, (b) $\dfrac{3}{y}dx - \dfrac{3x}{y^2}dy$, (c) $yx^{y-1}dx + x^y\ln x\,dy$, (d) $\cos(x + y)\,dx + \cos(x + y)\,dy$ **12** 3% **13** ±7%
14 ±3.2 cm^2 **15** As in the problem **16** (a) 8, 6, (b) 4, 4, (c) 4, 3
17 (a) (–2, 3, 3) minimum, (b) (0, 0, 0) saddle, (c) (–1, 1, –4) minimum, (d) (0, 0, 2) maximum, (2, 0, –6) saddle, (e) (1, 1, 6) minimum
18 $x = 7, y = 28$ **19** 10, 10, 10 **20** (a) See Figure A.34, (0, 0, 0) saddle point, (b) see Figure A.35, (1, 2, 0) minimum

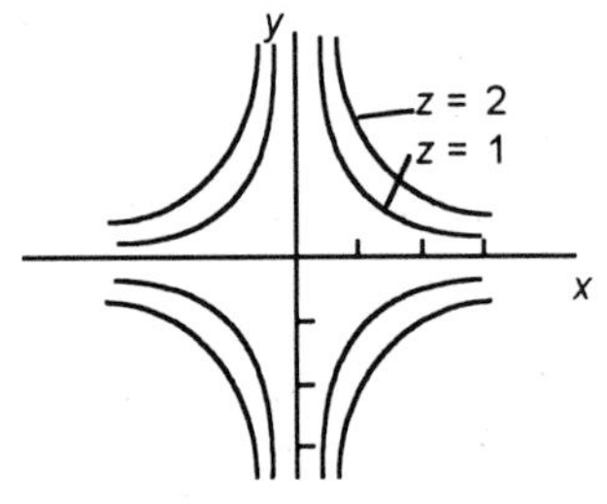

Figure A.34

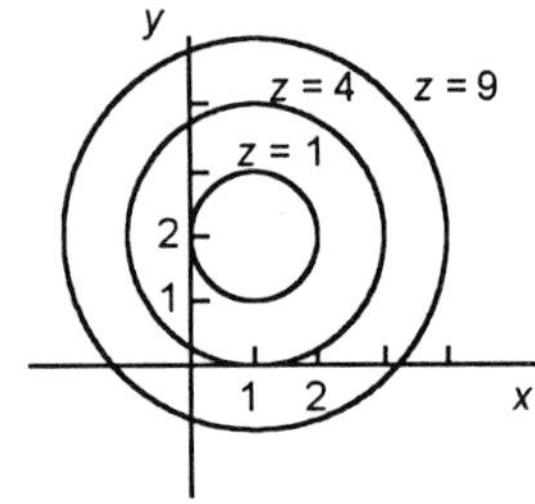

Figure A.35

Problems

1 (a) $y^2 + 6xy - 1$, $2xy + 3x^2$, (b) 3, 5, (c) $12x(2x + 5y)^2$, $30\,y(2x + 5y)^2$, (d) $\frac{2y}{(x+y)^2}$, $-\frac{2x}{(x+y)^2}$, (e) $\frac{y}{y^2+x^2}$, $-\frac{x}{y^2+x^2}$, (f) $3\,e^x + 2xy^3$, $-2\,e^y + 3x^2y^2$, (g) $2x\cos(x^2 - 3y)$, $-3\cos(x^2 - 3y)$, (h) $e^{xy}(y\cos x - \sin x)$, $e^{xy}x\cos x$, (i) $1/x$, $1/y$, (j) $-e^{-5y}[2\cos(x - 2y) + 5\sin(x - 2y)]$, $e^{-5y}\cos(x - 2y)$ **2** As given in the problem **3** (a) 4, −2, 1, (b) $-\sin x$, 0, 0, (c) 0, 0, 0, (d) $-y\sin x$, $-x\sin y$, $\cos y + \cos x$, (e) $-y^2\sin xy$, $-x^2\sin xy$, $\cos xy - xy\sin xy$
4 As given in the problem **5** As given in the problem
6 As given in the problem **7** As given in the problem
8 $8xy^3t + 6x^2y^2$ **9** $4x^3y^3(3x\cos 3t - 2y\sin 2t)$ **10** $4t$, $4s$
11 $-2s^2\sin 2t$, $2s\cos 2t$ **12** 41.97 mm²/s **13** (a) $2xy\,dx + x^2\,dy$, (b) $3x^2\,dx + 2y\,dy$, (c) $2\,dx - \sin y\,dy$, (d) $\ln y\,dx + \frac{x}{y}\,dy$
14 $dp = \frac{p}{T}\,dT - \frac{p}{V}\,dV$ **15** ±13.8 cm³ **16** ±2.8% **17** ±19%
18 $(x\sec^2\theta)\,\delta h + (\tan\theta)\,\delta x$ **19** (a) 1, −1, (b) 3.72, 2, (c) −0.25, 0.25
20 (a) (1, 2, −1) min., (b) (1, 0, 2) max., (c) (0, 0, 0) saddle, (1, 1, −1) min., (d) (0, 0, 2) max., (e) (0, 0, 0) min. **21** $x = 200$, $y = 400$
22 10, 10, 10 **23** 4 m, 4 m, 2 m **24** $x = 200/3$, $\theta = \pi/4$
25 As given in the problem **26** See Figure A.36, (0, 0, 0) minimum

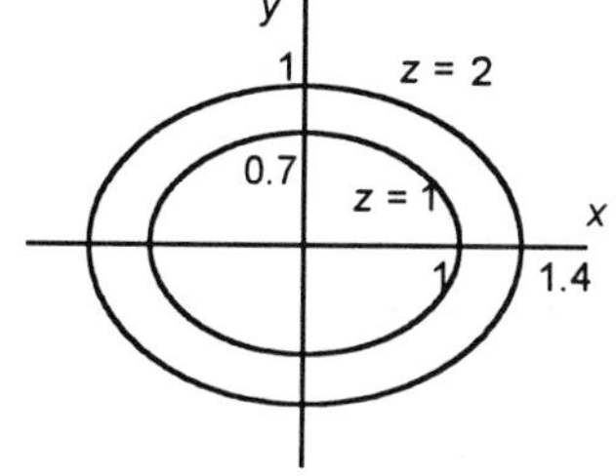

Figure A.36

Chapter 25

Revision

1 (a) ½ $e^{2x} + C$, (b) $\ln|x| + C$, (c) $x^2 + C$ **2** (a) $x^5 + x^4 + x^3 + C$, (b) $\sin 2x - 2\cos 2x$, (c) $\frac{4}{5}x^{5/2} - \frac{8}{3}x^{3/2}$ **3** (a) 2.67, (b) 7.4, (c) 22.5, (d) 9, (e) 1.39 **4** (a) 10, (b) 52/3, (c) 4, (d) 5/2
5 (a) 6, (b) 16½ **6** $20\frac{5}{6}$ **7** ½ **8** 9 **9** (a) Diverges, (b) diverges, (c) 1, (d) diverges **10** (a) $\frac{1}{6}(3x+1)^6 + C$, (b) $-\frac{1}{2}\cos 2x + C$, (c) $-\ln|\cos x| + C$, (d) $\sin^{-1}\frac{x}{2} + C$, (e) $\ln|1 + e^x| + C$, (f) $\frac{1}{3}\,e^{x^3} + C$, (g) $-\frac{2}{1+\tan\frac{x}{2}} + C$, (h) $\frac{1}{3}\sin^3 x - \frac{2}{5}\sin^5 x + \frac{1}{7}\sin^7 x + C$, (i) $\frac{1}{3}\sin^3 x - \frac{1}{5}\sin^5 x + C$
11 (a) $\frac{1}{20}$, (b) $\frac{1}{64}$, (c) $\frac{7}{48}$, (d) $\frac{61}{3}$, (e) $\sqrt{2} - 1$
12 (a) $\frac{x^2}{2}\ln|x| - \frac{x^2}{4} + C$, (b) $\frac{1}{3}x^2\,e^{3x} - \frac{2}{9}x\,e^{3x} + \frac{2}{27}\,e^{3x} + C$, (c) $x\sin x + \cos x + C$, (b) $-\frac{1}{2}x^2\cos 2x + \frac{1}{2}x\sin x + \frac{1}{4}\cos 2x + C$
13 (a) 2, (b) $\frac{\pi}{8} - \frac{1}{4}$, (c) $\frac{2}{3}\ln 2 - \frac{5}{18}$
14 (a) $\frac{1}{2}x^2 - 3x + 9\ln|x+3| + C$, (b) $\frac{1}{2}\ln|x^2 - 1| + C$, (c) $\frac{13}{2}\ln|x-6| - \frac{3}{2}\ln|x-2| + C$, (d) $4\ln|x-3| - \frac{15}{x-3} + C$, (e) $-\frac{1}{4}\ln|x| + \frac{5}{8}\ln|x-2| + \frac{5}{8}\ln|x+2| + C$, (f) $\ln|x+1| - \frac{1}{2}\ln|x^2 + 2x + 2| + C$, (g) $-\frac{1}{2}\ln|x-1| - \frac{1}{2(x-1)} + \frac{1}{4}\ln|x^2+1| + C$, (h) $\frac{\sqrt{2}}{2}\tan^{-1}\frac{x}{\sqrt{2}} - \frac{1}{2(x^2+2)} + C$ **15** (a) $t^3\mathbf{i} + 2t^2\mathbf{j} + \mathbf{c}$, (b) $\frac{1}{3}t^3\mathbf{i} + (t + t^2)\mathbf{j} + \mathbf{c}$

Problems

1 (a) $4x+C$, (b) $\frac{1}{2}x^4+C$, (c) $\frac{1}{2}x^4+\frac{5}{2}x^2+C$, (d) $\frac{3}{5}x^{5/3}-2x^{3/2}+C$, (e) $4x+\frac{1}{5}\cos 5x+C$, (f) $-\frac{2}{3}\,e^{-3x}+C$, (g) $8\,e^{x/2}+\frac{1}{3}x^3+2x+C$, (h) $4\ln|x|+C$ **2** (a) 15, (b) 8, (c) 1.10, (d) –2.25, (e) –4.5, (f) 2.67, (g) 0.83 **3** (a) 116/3, (b) 11/3 **4** ½ **5** 1/12 **6** 1 **7** 4½

8 (a) Diverges, (b) 1/18, (c) 1/3, (d) diverges

9 (a) $\frac{1}{40}(4x^2-1)(x^2+1)^4+C$, (b) $\frac{1}{2}\,e^{4x-1}+C$, (c) $\frac{1}{3}(x+2)\sqrt{2x+1}+C$, (d) $-\frac{1}{2}\cos x+C$, (e) $2\sinh^{-1}\dfrac{x}{2}+\frac{1}{2}x\sqrt{x^2+4}+C$, (f) $\frac{2}{15}(3x+2)(x-1)^{3/2}+C$, (g) $\ln\left|\dfrac{1+\tan\frac{1}{2}x}{1-\tan\frac{1}{2}x}\right|+C$, (h) $\sin^{-1}\dfrac{x}{3}+C$, (i) $\frac{1}{6}\sin^3 2x-\frac{1}{10}\sin^5 2x+C$ **10** (a) $\frac{2}{5}(x+2)^{5/2}-\frac{4}{3}(x+2)^{3/2}+C$, (b) $\dfrac{x}{\sqrt{x^2+1}}+C$, (c) $\frac{1}{3}\cos^3 x-\cos x+C$, (d) $\frac{1}{2}\sin^{-1}2x+C$, (e) $\frac{1}{3}(x^2+2)^{3/2}+C$, (f) $\frac{1}{10}\tan^{-1}\dfrac{5x}{2}+C$, (g) $\frac{1}{5}\sin^5 x-\frac{1}{7}\sin^7 x+C$, (h) $\frac{1}{4}\tan^4 x+C$, (i) $\frac{2}{3}\tan^{-1}\left(\frac{1}{3}\tan\dfrac{x}{2}\right)+C$ **11** (a) ln 2, (b) 1/90, (c) $\pi/4$, (d) $\pi/4$, (e) $\pi/8$ **12** (a) $\frac{1}{3}x^3\ln|x|-\frac{1}{9}x^3C$, (b) $\frac{1}{2}x\,e^{2x}-\frac{1}{4}\,e^{2x}+C$, (c) $x^3\sin x+3x^2\cos x-6x\sin x-6\cos x+C$, (d) $\frac{1}{25}\sin 5x-\frac{1}{5}x\cos 5x+C$, (e) $\frac{1}{2}x^2\ln|3x|-\frac{1}{4}x^2+C$, (f) $\frac{1}{2}x-\frac{1}{4}\sin 2x+C$ **13** (a) $\dfrac{\pi}{2}-1$, (b) $\frac{1}{16}\pi^2+\frac{1}{4}$, (c) 2

14 (a) $\frac{1}{4}x^2+\frac{3}{4}x+\frac{9}{8}\ln|2x-3|+C$, (b) $-\frac{1}{2}x-\frac{1}{4}\ln|1-2x|+C$, (c) $\frac{1}{2}x+\frac{1}{5}\ln|x-1|-\frac{9}{20}\ln|2x+3|+C$, (d) $\frac{1}{6}\ln|x-1|+\frac{1}{2}\ln|x+1|-\frac{1}{6}\ln|2x+1|+C$, (e) $-\frac{1}{4}\ln|x|+\frac{3}{8}\ln|x-2|-\frac{1}{8}\ln|x+2|+C$, (f) $\frac{2}{3}\ln|x-1|+\frac{5}{6}\ln|2x+1|+C$, (g) $\frac{1}{2}x^2+2x+\frac{2}{3}\ln|x-1|+\frac{5}{6}\ln|2x+1|+C$, (h) $\ln|x-3|-\ln|x-2|+C$, (i) $6\ln|x|-\ln|x+1|-\dfrac{9}{x+1}+C$, (j) $2\ln|x|-2\ln|x-1|+\ln|x^2+4|+2\tan^{-1}\frac{1}{2}x+C$, (k) $-\dfrac{1}{x}-\tan^{-1}x+C$ **15** $\left(\dfrac{t^2}{2}-\dfrac{t^3}{3}\right)\mathbf{i}+\dfrac{t^2}{2}\mathbf{j}+4t\mathbf{k}+\mathbf{c}$

Chapter 26

Revision

1 (a) 1.448, 1.460, (b) 0.791, 0.787, (c) 2.255, 2.326 **2** (a) 0.783, (b) 0.352, (c) 1.405, (d) 0 **3** (a) 0.785, (b) 3.196, (c) 1.368, (d) –0.294

Problems

1 (a) 25.25, (b) 1.108 **2** (a) 0.512, (b) 1.006, (c) 0.509, (d) 0.342

3 (a) 0.511, (b) 4.000, (c) 0.500, (d) 1.187, (e) 10.070, (f) 0.479

Chapter 27

Revision

1 3.09 square units **2** (a) 4.5 square units, (b) 4.5 square units, (c) 4.5 square units **3** (a) 10.7π cubic units, (b) 0.184 cubic units,

(c) 3.9π cubic units **4** (a) $3\pi/4$ cubic units, (b) $\pi/6$ cubic units
5 $\pi/10$ cubic units **6** 25.6π cubic units **7** (a) 1.43 units, (b) 6.04 units, (c) 755.9 units **8** 3 units **9** (a) 258.8 square units, (b) 30.85 square units, (c) 14.42 square units **10** 11.19π square units **11** 21.3π square units **12** (a) (2.31, 2.79), (b) (2, 4), (c) (0, 2.4), (d) (0.4, 0.5) **13** $(4r/3\pi, 0)$ **14** $\sqrt{3}\,\pi L^2$, $\frac{1}{4}\pi L^3$
15 33 092 mm^2 **16** (2.25, 5.4) **17** $\frac{1}{3}ML^2$ **18** $\frac{1}{3}ML^2$
19 $0.3Mh^2\tan^2\theta$ **20** $0.3Mr^2$ **21** $\frac{4}{3}ML^2$ **22** $\frac{1}{4}\pi r^4$
23 (a) 0.57 unit^4, (b) 1.07 unit^4 **24** (a) 1, (b) 7, (c) 16, (d) 0.5, (e) 3.91, (f) 17.3 **25** $2A/\pi$ **26** $0.623N_0$ **27** (a) 4.92, (b) 1.15, (c) 1.23, (d) 0.707, (e) 1.35 **28** $V/2$

Problems

1 3.62 square units **2** 20.8 square units **3** (a) 318.4π cubic units, (b) 525π cubic units, (c) $11\pi^2 + 24\pi$ cubic units **4** 17.1π cubic units
5 21.3π cubic units **6** $\frac{4}{3}\pi ab^2$ **7** As given in the problem
8 As given in the problem **9** $\frac{1}{3}\pi h^2(3r-h)$ **10** $5\pi^2a^3$
11 (a) 40 sinh 0.5, (b) 24.3, (c) 0.881, (d) 0.82, (e) 2.24 **12** 215 m
13 $8a$ **14** (a) 8.84 square units, (b) 620 square units, (c) 48π square units, (d) 1.13π square units **15** $\pi r\sqrt{r^2+h^2}$ **16** (a) $(\pi/4, \pi/8)$, (b) (4/3, 2), (c) (3/2, 6/5), (d) (0.9, 1.8), (e) (4/5, 2/7), (f) (3/2, 24/5)
17 (a) $(3\pi a/16, a/5)$, (b) $2\pi a^3/15$ **18** (a) $4\sqrt{2}\,\pi L^2$, (b) $\sqrt{2}\,\pi L^3$
19 $160\pi^2$ cubic units **20** $128\pi/3$ cubic units **21** (a) $Mh^2/18$, (b) $Mh^2/6$ **22** $5Mr^2/2$ **23** $\sqrt{\dfrac{4ab}{5}}$ **24** (a) $\frac{1}{12}ML^2$, (b) $M\left(\frac{1}{12}L^2+d^2\right)$ **25** (a) $bh^3/36$, (b) $bh^3/12$ **26** (a) 1.91, (b) 2.5, (c) 0.33, (d) 0.5, (e) 6.37 **27** 1.27 **28** (a) 5.29, (b) 0.730, (c) 3.54, (d) 1.41, (e) 2.27 **29** $I_1/\sqrt{2}$, $I_2/\sqrt{2}$, $\sqrt{(I_1^2+I_2^2)/2}$
30 $V/\sqrt{3}$ **31** V

Chapter 28

Revision

1 (a) $m\dfrac{dv}{dt}+kv^2=mg$, (b) $m\dfrac{d^2x}{dt^2}+c\dfrac{dx}{dt}+kx=0$ **2** $\dfrac{L}{R}\dfrac{di}{dt}+i=0$
3 $A\dfrac{dh}{dt}+\sqrt{2gh}=0$ **4** As given in the problem **5** (a) $y=2\,e^x$, (b) $y=4\,e^x-1$, (c) $y=3\sin x$ **6** $v_C=-V\,e^{-t/RC}+V$

Problems

1 (a) $m\dfrac{dv}{dt}+kv=0$, (b) $m\dfrac{dv}{dt}+kv^2=mg$, (c) $\dfrac{dI}{dx}+kx=0$, (d) $\dfrac{A}{\rho g}\dfrac{dp}{dt}+\sqrt{\dfrac{2p}{\rho}}=q_1$ **2** As given in the problem
3 (a) $y=x\,e^x$, (b) $y=\dfrac{1}{\omega}\sin\omega t+2\cos\omega t$, (c) $y=(2+x^2)\,e^{-x}$
4 $y=-\dfrac{w}{2EI}\left(\frac{1}{2}L^2x^2-\frac{1}{3}Lx^3+\frac{1}{12}x^4\right)$

Chapter 29

Revision

1 (a) $y = \ln x + A$, (b) $y = 2 \sin \frac{1}{2}x + A$, (c) $-1/y = x + A$, (d) $\ln y = -2x + A$ or $y = C\,e^{-2x}$, (e) $\tan^{-1} y = x^2 + A$, (f) $y = \ln(x^3 + A)$ or $e^y = x^3 + A$ **2** $y = 1/(2 - x^2)$ **3** $V = V_0\,e^{-t/RC}$ **4** $N = N_0\,e^{-kt}$

5 (a) $y = C\,e^{3x} - e^{2x}$, (b) $y = x \ln x - 1 + Ax$, (c) $y = (e^x + A)/x$, (d) $y = \sin x \cos x + A \cos x$ **6** (a) 5, (b) $3 + 2x$, (c) $5 + x^2$, (d) $\frac{1}{6}\,e^{5x}$, (e) $\frac{1}{26}\sin 5x - \frac{5}{26}\cos 5x$ **7** (a) $y = A\,e^{-2x} + x - \frac{1}{2}$, (b) $y = A\,e^{-2x} + e^x/3$, (c) $y = A\,e^{-2x} + x - 4$

Problems

1 (a) $y = 4x + x^3 + A$, (b) $-1/y = 2x + A$, (c) $e^y = x^3 + A$, (d) $y^2 = x + A$, (e) $y = A\,e^{-1/x}$, (f) $y = -2/(x^2 + A)$ **2** $i = \frac{V}{R}(1 - e^{-Rt/L})$ **3** $y = 2 - e^{-x}$

4 $v = 10 - 10\,e^{-t}$ **5** $T = T_0\,e^{\mu\theta}$ **6** 3.41 hours **7** 51.4°C

8 As given in the problem

9 (a) $y = x^4(4 \ln x + A)$, (b) $y = 2 + A(x^2 + 1)^{-3/2}$, (c) $y = \frac{x+1}{x-1}\left[\ln(x+1) + A\right]$, (d) $y = x + \cot x + A \operatorname{cosec} x$, (e) $y = \frac{1}{4}x^3 + \frac{A}{x}$, (f) $y = \left[\ln(\sec x + \tan x) + A\right]\cos^2 x$

10 $\dfrac{12(10t + t^2) + 500}{(10 + 2t)^2}$ kg/m^3 **11** (a) $y = A\,e^{-3x} + \frac{1}{3}$, (b) $y = A\,e^x + 2\,e^{3x}$, (c) $y = A\,e^{-x} + x - 1$

Chapter 30

Revision

1 (a) $y = \frac{3}{2}x^2 + 4x + 4$, (b) $y = \frac{3}{5}\,e^{2x} + \frac{2}{5}\,e^{-3x}$, (c) $y = e^{-2x} + 2x\,e^{-2x}$

2 (a) $y = A\,e^{x/2} + B\,e^{-x}$, (b) $y = (A + Bx)\,e^{3x}$, (c) $y = (A + Bx)\,e^{5x}$, (d) $y = e^{2x}(A \cos x + B \sin x)$, (e) $y = e^x(A \cos 2x + B \sin 2x)$, (f) $y = A\,e^{2x} + B\,e^{-5x}$ **3** (a) $y = -\frac{3}{4}\,e^{-5x} + \frac{3}{4}\,e^{-x}$, (b) $y = e^{3x}(3\cos 4x - 2\sin 4x)$, (c) $y = (2 + 4x)\,e^{-3x/2}$

4 (a) $y = A\,e^{2x} + B\,e^{-2x} + \frac{2}{5}\,e^{3x}$, (b) $y = e^{-3x/2}\left(\cos\sqrt{\frac{7}{2}}\,x + \sin\sqrt{\frac{7}{2}}\,x\right) + \frac{3}{4}x - \frac{1}{16}$, (c) $y = A\,e^{4x} + B\,e^{2x} + \frac{21}{85}\cos x - \frac{18}{85}\sin x$, (d) $y = A\,e^{2x} + B\,e^{3x} + \frac{1}{2}e^x$, (e) $y = A\,e^{2x} + B\,e^{3x} + e^x - \frac{1}{4}\,e^{-x}$, (f) $y = A\,e^{2x} + B\,e^{-x} + \frac{1}{4}\cos 2x - \frac{3}{4}\sin 2x$

5 (a) $y = A\,e^{4x} + B\,e^{-2x} - \frac{1}{2}x\,e^{-2x}$, (b) $y = \sin 3x + \cos 3x - \frac{5}{6}x\cos 3x$

Problems

1 (a) $y = \frac{5}{2}\,e^x - \frac{5}{2}\,e^{-x}$, (b) $y = 2\,e^x - e^{2x}$, (c) $y = 2\,e^{-x} + x\,e^{-x}$

2 (a) $y = e^{2x}(A \cos x + B \sin x)$, (b) $y = A\,e^{2x} + B\,e^{-4x}$, (c) $y = A\,e^{-4x} + B\,e^x$, (d) $y = e^{-x}(A\cos\sqrt{3}\,x + B\sin\sqrt{3}\,x)$, (e) $y = (A + Bx)\,e^{-3x}$, (f) $y = (A + Bx)\,e^{-x/3}$ **3** (a) $y = 2\,e^{3x} - e^{4x}$, (b) $y = (1 + 2x)\,e^{-x}$, (c) $y = e^{3x}(\cos 4x - \sin 4x)$ **4** (a) $y = A\,e^x + B\,e^{2x} - 2 - 3x - x^2$, (b) $y = A\,e^x + B\,e^{2x} - \frac{1}{6}\,e^{-2x}$, (c) $y = A\,e^x + B\,e^{2x} - \frac{1}{4}\sin 2x + \frac{3}{4}\cos 2x$, (d) $y = A\,e^{3x} + B\,e^{2x} + \frac{19}{108} + \frac{30}{108}x + \frac{18}{108}x^2$, (e) $y = (A + Bx)\,e^x + \frac{1}{2}x^2\,e^x$, (f) $y = A\,e^{3x} + B\,e^{-2x} + \frac{1}{15}x\,e^{3x}$, (g) $y = A\,e^x + B\,e^{-3x} - \frac{5}{4}x\,e^{-3x} - \frac{4}{455}\cos 2x - \frac{1}{65}\sin 2x$,

(h) $y = A\,e^{-x} + B\,e^{-3x} + x\,e^{-x}$, (i) $y = A\cos x + B\sin x + x\sin x$,
(j) $y = A + B\,e^{-2x} + \frac{1}{4}x^2 - \frac{1}{4}x$ **5** $y = e^{3x} - 3\,e^{2x}$
6 $y = \cos 3x - \frac{2}{15}\sin 3x + \frac{1}{5}\sin 2x$

Chapter 31

Revision

1 (a) 0, 1.00; 0.1, 1.10; 0.2, 1.22; 0.3, 1.36; 0.4, 1.53, (b) 0, 1; 0.2, 1.20; 0.4, 1.52; 0.6, 1.98; 0.8, 2.62, (c) 0, 1; 0.1, 1.001; 0.2, 1.008; 0.3, 1.027; 0.4, 1.066 **2** (a) 2.067 0, (b) 0.508 18, (c) 1.214 08
3 (a) $\frac{dy}{dx} = v$, $\frac{dv}{dy} = -y$, (b) $\frac{dy}{dx} = v$, $\frac{dv}{dx} + 3v = 2$ **4** 1.298
5 1.289 57

Problems

1 (a) 0, 1; 0.1, 1.1; 0.2, 1.22; 0.3, 1.38, 0.4, 1.57, (b) 0, 1; 0.1, 1.2; 0.2, 1.43; 0.3, 1.673; 0.4, 1.941, (c) 0, 1; 0.1, 1; 0.2, 1.01; 0.3, 1.03; 0.4, 1.06 **2** 0, 2; 0.1, 1.8; 0.2, 1.62; 0.3, 4.94; 0.4, 5.92 **3** 0, 0; 0.2, 1; 0.4, 1.90; 0.6, 2.71; 0.8, 3.44 **4** (a) 2.421 4, (b) 0.089 7, (c) 0.074 1 **5** 2.060 **6** 2.090 68 **7** 1.034 05 **8** 1.068 32

Chapter 32

Revision

1 RC **2** 10.9 V **3** $i = I(1 - e^{-Rt/L})$ **4** $x = A\,e^{-4t} + B\,e^{-t}$
5 (a) $x = 0.2\cos 3t$, (b) $x = e^{-t}(0.2\cos 2.83t + 0.070\sin 2.83t)$

Problems

1 (a) $x = 1 - e^{-t/2}$, (b) $x = 8\,e^{-t/2} + 4t - 8$ **2** $x = 5(1 - e^{-t/\tau})$
3 (a) 36.8%, (b) 13.5% **4** 12 s **5** (a) $2\frac{dx}{dt} + x = 45$, (b) $x = 45 - 25\,e^{-t/2}$ **6** $x = 1.5t - 0.75 + 0.75\,e^{-2t}$ **7** 0.44
8 (a) 5 rad/s, (b) 1.25 **9** 316 rad/s, 6.3 N s/m **10** Over damped
11 6 N s/m **12** 2.6 rad/s, 0.76 **13** $e^{-t}(0.2\cos 2.24t + 0.22\sin 2.24t)$
14 $\theta = \frac{4}{9}\pi\,e^{-t} - \frac{1}{9}\pi\,e^{-4t}$ **15** As given in the problem

Chapter 33

Revision

1 (a) $y = -10\cos(3t + \pi/2)$, (b) $y = 3.6\cos(4t - 0.98)$
2 $\frac{4A}{\pi}\left(\sin\omega t + \frac{1}{3}\sin 3\omega t + \frac{1}{5}\sin 5\omega t + \ldots\right)$
3 $\frac{A}{2} - \frac{4A}{\pi^2}\left(\sin\omega t + \frac{1}{9}\sin 3\omega t + \frac{1}{25}\sin 5\omega t + \ldots\right)$
4 $\frac{A}{2} - \frac{4A}{\pi}\left(\frac{1}{3}\cos\omega t + \frac{1}{15}\cos 2\omega t + \frac{1}{35}\cos 3\omega t + \ldots\right)$
5 $0.5 + \frac{6}{\pi}\left(\cos\omega t - \frac{1}{3}\cos 3\omega t + \ldots\right)$ **6** (a), (b), (d) odd, (c), (e) even
7 (a) sine, (b) a_0 and cosine, (c) a_0 and cosine
8 $2\left(\sin t - \frac{1}{2}\sin 2t + \frac{1}{3}\sin 3t + \ldots\right)$ **9** (a) Odd sines, (b) a_0, even sines and cosines **10** (a) 10/6, (b) 0 **11** (a) 0.5, 0.31, 0.16, 0.11, 180°, 180°, 180°, (b) 1.57, 1.19, 0.5, 0.13, –32°, 180°, –12° **12** 77.5

Problems

1 (a) $1+\frac{4}{\pi}\left(\sin\frac{\pi t}{5}+\frac{1}{3}\sin\frac{3\pi t}{5}+\frac{1}{5}\sin\frac{5\pi t}{5}+\ldots\right)$,
(b) $\frac{4}{\pi}\left(\sin\pi t+\frac{1}{3}\sin 3\pi t+\frac{1}{5}\sin 5\pi t+\ldots\right)$,
(c) $\frac{\pi}{4}-\frac{2}{\pi}\left(\cos t+\frac{1}{9}\cos 3t+\ldots\right)+\left(\sin t-\frac{1}{2}\sin 2t+\ldots\right)$,
(d) $\frac{4\pi^2}{3}+(4\cos t+\cos 2t+\ldots)-\pi(4\sin t+2\sin 2t+\ldots)$,
(e) $\frac{\pi^2}{3}-4\cos t+\cos 2t-\frac{4}{9}\cos 3t+\ldots$
(f) $\frac{\pi^2}{6}-\left(2\cos t-\frac{1}{2}\cos 2t+\ldots\right)+\left(\frac{\pi^2-4}{\pi}\sin t-\frac{\pi}{2}\sin 2t+\ldots\right)$,
(g) $\frac{\pi}{2}+\frac{4}{\pi}\left(\cos t+\frac{1}{9}\cos 3t+\frac{1}{25}\cos 5t+\ldots\right)$
2 (a) $\frac{2A}{4}+\frac{4A}{\pi}\left(\frac{1}{3}\cos\omega t-\frac{1}{15}\cos 2\omega t+\ldots\right)$,
(b) $\frac{A}{\pi}+\frac{2A}{\pi}\left(\frac{1}{3}\cos 2\omega t-\frac{1}{15}\cos 4t+\ldots\right)$,
(c) $\frac{A}{4}-\frac{2A}{\pi^2}\left(\cos\omega t+\frac{1}{9}\cos 3\omega t+\ldots\right)+\frac{A}{\pi}\left(\sin\omega t-\frac{1}{2}\sin 2\omega t+\ldots\right)$
3 (a), (b), (d), (e) even, (c) odd **4** (a) Odd cosines, (b) odd sines and cosines, (c) a_0 and odd sines **5** 0.64, 0.42, 0.08, 0.036, –180°, –180°, –180° **6** 9.33

Chapter 34

Revision

1 (a) 1, $\frac{4}{\pi}\left(\sin\pi t+\frac{1}{3}\sin 3\pi t+\frac{1}{5}\sin 5\pi t+\ldots\right)$,
(b) $\frac{1}{2}-\frac{2}{\pi}\left(\cos t-\frac{1}{3}\cos 3t+\frac{1}{5}\cos 5t+\ldots\right)$,
$\frac{2}{\pi}\left(\sin t-\sin 2t+\frac{1}{3}\sin 3t+\ldots\right)$,
(c) $\frac{2}{3\pi}+\frac{12}{\pi}\left(\frac{1}{5}\cos 2t-\frac{1}{7}\cos 4t+\ldots\right)$, $\sin 3t$

Problems

1 (a) $\frac{3\pi}{2}-\frac{12}{\pi}\left(\cos t+\frac{1}{9}\cos 3t+\frac{1}{25}\cos 5t+\ldots\right)$,
$6\left(\sin t-\frac{1}{2}\sin 2t+\frac{1}{3}\sin 3t-\frac{1}{4}\sin 4t+\ldots\right)$,
(b) $\cos t$, $\frac{8}{\pi}\left(\frac{1}{3}\sin 2t+\frac{2}{15}\sin 4t+\frac{3}{35}\sin 6t+\ldots\right)$,
(c) $\frac{\pi}{2}+\frac{4}{\pi}\left(\cos t+\frac{1}{9}\cos 3t+\frac{1}{25}\cos 5t+\ldots\right)$,
$-2\left(\sin t-\frac{1}{2}\sin 2t+\frac{1}{3}\sin 3t+\ldots\right)$,
(d) $\frac{1}{2}(1-\cos 2t)$, $\frac{8}{\pi}\left(\frac{1}{3}\sin t-\frac{1}{15}\sin 3t-\frac{1}{105}\sin 5t+\ldots\right)$
2 (a) $\frac{8y_0}{\pi^2}\left(-\sin\frac{\pi x}{L}+\frac{1}{9}\sin\frac{3\pi x}{L}+\ldots\right)$,
(b) $\frac{32y_0}{3\pi^2}\left(\sqrt{2}\sin\frac{\pi x}{L}+\frac{1}{4}\sin\frac{2\pi x}{L}+\frac{\sqrt{2}}{9}\sin\frac{3\pi x}{L}+\ldots\right)$
3 $2\sin t+\sin 2t+\frac{2}{3}\sin 3t+\ldots$ **4** $\frac{8KL^2}{\pi^3}\left(\sin\frac{\pi x}{L}+\frac{1}{27}\sin\frac{3\pi x}{L}+\ldots\right)$

Chapter 35

Revision

1 (a) $c_n=\frac{1}{n\pi}\sin\frac{n\pi}{2}$, $c_0=\frac{1}{2}$, (b) $c_n=\frac{2}{n\pi}\sin\frac{n\pi}{2}$, $c_0=0$

2 (a) Only odd values of n, imaginary, (b) only even values of n, real
3 $4/\pi$, $4/3\pi$, $4/15\pi$, ..., all phases 0°

Problems

1 (a) $c_n = \mathrm{j}\frac{A}{n\pi}(-1)^n$, $c_0 = 0$, (b) $c_n = \mathrm{j}\frac{A}{2n\pi}\{(-1)^n - 1\}$, $c_0 = \frac{A}{2}$,
(c) $c_n = \frac{A}{\pi(1-n^2)}\cos\frac{n\pi}{2}$, $n \neq \pm 1$, $\frac{A}{4}$ when $n = \pm 1$
2 (a) Only even values of n, real, (b) only even values of n, real
3 1, $6/\pi$, $3/\pi$, $2/\pi$, 0°, 90° for rest **4** 1, 0.90, 0.64, 0.30, 0

Chapter 36

Revision

1 ±1000 Hz **2** $1/(a - \mathrm{j}\omega)$
3 (a) $V\tau\left[\frac{\sin\omega\tau/2}{\omega\tau/2} - \frac{1}{2}\frac{\sin\omega\tau/4}{\omega\tau/4}\cos\frac{3\omega\tau}{4}\right]$, (b) $\frac{\mathrm{e} - \mathrm{j}4\omega}{1 + \mathrm{j}\omega}$,
(c) $2\pi\delta(\omega - \omega_0)$, (d) $\frac{2\sin(\omega - 3)}{\omega - 3}$ **4** As given in the problem
5 $\frac{\pi}{\mathrm{j}}\left[\delta(\omega - a) - \delta(\omega + a)\right]$ **6** $-\mathrm{e}^{4t}$ **7** $\frac{2}{(a + \mathrm{j}\omega)^3}$
8 $5\,\mathrm{sgn}(t) - 10\,\mathrm{e}^{-4t}u(t)$ **9** $2\,\mathrm{e}^{-t}u(t)$ **10** $-10 + 10(1 - \mathrm{e}^{-4t})u(t)$
11 As given in the problem

Problems

1 (a) $4\pi\,\delta(\omega)$, (b) $2a/(a^2 + \omega^2)$, (c) $\tau/(1 + \mathrm{j}\omega\tau)$, (d) $(10/\omega)\,\mathrm{e}^{-7\mathrm{j}\omega}\sin 4\omega$,
(e) $[24/(16 + \omega^2)]\,\mathrm{e}^{2\mathrm{j}\omega}$, (f) $10\pi\,\mathrm{e}^{4\omega}$ if $\omega \leq 0$
2 $(\mathrm{j}2/\omega^2)(3\sin\omega - \sin 3\omega)$ **3** (a) $\mathrm{j}\pi2[\delta(\omega + \omega_0) - \delta(\omega - \omega_0)]$,
(b) $2.5\pi[2\,\delta(\omega) + \delta(\omega + 6) + \delta(\omega - 6)]$ **4** $\frac{\omega_0}{(\mathrm{j}\omega + a)^2 + \omega_0^2}$
5 $4\,\mathrm{e}^{-t}u(t)$ **6** $\mathrm{e}^{-4t}u(t)$ **7** (a) $1 - \mathrm{e}^{2-t}$ for $t > 2$, 0 for $t < 2$,
(b) $t - 1 + \mathrm{e}^{-t}$, (c) $1 - \mathrm{e}^{-t}$ for $0 < t < 1$, $(\mathrm{e} - 1)\,\mathrm{e}^{-t}$ for $t > 1$

Chapter 37

Revision

1 (a) $a_0 = 0$, (b) only odd harmonics, (c) only even harmonics
2 $15 - 3.53\cos t - 2.33\cos 2t + 0 + 8.18\sin t + 0.58\sin 2t - 1.33\sin 3t$

Problems

1 (a) $15.8 - 3.79\cos t - 2.33\cos 2t + 0.83\cos 3t + 8.44\sin t + 2.60\sin 2t + 0$, (b) $10 - 21.6\cos t - 3.9\cos 2t + 1.2\cos 3t + 13.7\sin t + 1.6\sin 2t - 2.2\sin 3t$, (c) $41.3 - 8.5\cos t - 5.5\cos 2t - 16.9\cos 3t + 14.3\sin t + 5.3\sin 2t - 0.4\sin 3t$ **2** As given in the problem
3 As given in the problem

Chapter 38

Revision

1 $3.2\cos(100t + \pi/2) + 3.2\cos(200t + \pi/2)$ mA
2 $5.33\cos(314t + 57.8°) - 1.57\sin(628t + 38.6°)$ A
3 $12.4\sin(1000t - 29.7°) + 3.9\sin(3000t - 51.1°)$ A
4 $0.90\sin(t - 45°) + 0.13\sin(t - 71.6°) + 0.05\sin(t - 78.7°)$ A

5 (a) 0.71 A, (b) 0.96 A **6** $1/(1 + j\omega 0.2)$, $5\,e^{-5t}u(t)$
7 $j2\omega/(4 + j2\omega)$, $(30\,e^{-3t} - 20\,e^{-2t})u(t)$

Problems

1 $0.2\sin(500t + 53.1^\circ) + 0.14\sin(500t + 33.7^\circ)$ A
2 $5\sin(\omega t + 36.9^\circ) + 2.89\sin(3\omega t + 37.4^\circ)$ A
3 $2.9\sin(500t - 45^\circ) + 1.1\sin(1500t - 18.4^\circ)$ A
4 $5 + 14.1\sin(10\,000t - 45.3^\circ) + 2.53\sin(30\,000t - 41.8^\circ)$ A
5 $\dfrac{V}{2R} + \dfrac{2V}{\pi}\sum_{n=1,3,5}^{\infty} \dfrac{\sin(n\omega_0 t - \theta_n)}{n\sqrt{R^2 + (n\omega_0 L)2}}$, $\theta_n = \tan^{-1}\dfrac{n\omega_0 L}{R}$
6 $125 + 197.8\cos(62.8t - 71.9^\circ) + 91.9\cos(185.5t - 74.2^\circ) + 59.9\cos(314.2t - 79.0^\circ)$ A **7** $\dfrac{4V}{\pi}\sum_{n=1,3,5}^{\infty} \dfrac{\sin(n\omega_0 t - \theta_n)}{n\sqrt{1 + (n\omega_0 RC)^2}}$, $\theta_n = \tan^{-1} n\omega_0 RC$ **8** 9.57 V **9** $V/\sqrt{2}$ **10** 2.03 A **11** 7.4 V
12 99 V **13** (a) $\dfrac{j\omega RC_1}{1 + j\omega R(C_1 + C_2)}$, (b) $\dfrac{1}{1 + j\omega(L/R) - \omega^2 LC}$
14 $10\,\mathrm{sgn}\,t - 20\,e^{-25t}u(t)$ V **15** $5\,\mathrm{sgn}\,t - 10\,e^{-4t}u(t)$ A
16 $10\,e^{-8t}u(t)$ V **17** $3/(3 + j\omega)$, $3\,e^{-3t}u(t)$ V

Chapter 39

Revision

1 (a) $\dfrac{2}{s^3}$, (b) $\dfrac{6}{s^4}$, (c) $\dfrac{a}{s^2 - a^2}$ **2** (a) $\dfrac{2}{s^3} + \dfrac{3}{s^2} + \dfrac{2}{s}$, (b) $\dfrac{2}{s} + \dfrac{12}{s^2 + 9}$, (c) $\dfrac{1}{s-4} + \dfrac{s}{s^2 - 4}$, (d) $\dfrac{2}{s} + \dfrac{5}{s-3}$, (e) $\dfrac{s}{s^2 + 4} + \dfrac{s}{s^2 + 9}$, (f) $\dfrac{6}{s^4} + \dfrac{4}{s+1}$
3 (a) $\dfrac{2}{(s+3)^2 + 4}$, (b) $\dfrac{2}{(s-4)^3}$, (c) $\dfrac{s-2}{(s-2)^2 + 1}$ **4** (a) $e^{-5s}\dfrac{1}{s}$, (b) $1\,e^{-4s}$ (c) $\dfrac{3\,e^{-10s}}{s^2}$ **5** $\dfrac{3(1 - e^{-s})}{s(1 + e^{-s})}$ **6** (a) $\dfrac{8s}{(s^2 + 16)^2}$, (b) $\dfrac{8(3s^2 - 16)}{(s^2 + 16)^3}$, (c) $\dfrac{2}{(s-a)^3}$ **7** (a) $(s + 2)X(s) - 3$, (b) $(s^2 + 2s + 3)X(s) - 4s - 13$, (c) $(s^2 - 5s + 6)X(s) - 1$ **8** $\dfrac{1}{Cs}I(s)$
9 $\dfrac{1}{s(s+1)}$ **10** (a) $\sin 2t$, (b) e^{-3t} **11** (a) e^{2t}, (b) 5, (c) $\cos 4t$, (d) $\sinh 3t$, (e) $e^{2t}\sin 5t$, (f) $\frac{1}{6}t^3\,e^{-3t}$, (g) $(t-2)u(t-2)$, (h) $(t-3)\,e^{-2(t-3)}u(t-3)$ **12** (a) $e^{-2t} + 2\,e^{3t}$, (b) $2\,e^{t} + e^{-2t}$, (c) $\cos 2t - e^{-t}$, (d) $e^{-2t} + 2t\,e^{-2t}$ **13** (a) $e^{-t} - e^{-2t}$, (b) $t\,e^{-t} + 2\,e^{-t} + t - 2$, (c) $½(\sin t - t\cos t)$ **14** (a) $x = 3\,e^{2t} - 3\,e^{-t/2}$, (b) $x = \frac{2}{5} - \frac{2}{5}\,e^{-5t}$, (c) $x = \frac{1}{4} - \frac{1}{4}\cos 2t$, (d) $x = e^{-t} - e^{-t}\cos t$, (e) $x = \frac{1}{2}\,e^{t} - \frac{1}{2}t\,e^{t} + \frac{1}{2}\cos t$ **15** (a) $x = \frac{2}{7}(e^{2t} - e^{-5t})$, $y = \frac{1}{7}(e^{2t} + 6\,e^{-5t})$, (b) $x = \frac{1}{10}(5 - 3\,e^{-6t/11} - 2\,e^{-t})$, $y = \frac{1}{5}(e^{-t} - e^{-6t/11})$, (c) $x = 3\cos t + \cos\sqrt{3}\,t$, $y = 3\cos t - \cos\sqrt{3}\,t$
16 (a) 0, (b) 1 **17** (a) 2, (b) 0

Problems

1 (a) $\dfrac{s}{s^2 + a^2}$, (b) $\dfrac{a}{s}$, (c) $\dfrac{s}{s^2 - a^2}$ **2** (a) $\dfrac{4}{s}$, (b) $\dfrac{3}{s^2} - \dfrac{1}{s}$, (c) $\dfrac{1}{s-3}$, (d) $\dfrac{2}{s^2} + \dfrac{3}{s-1}$, (e) $\dfrac{2}{s^3} + \dfrac{4}{s+2}$, (f) $\dfrac{2}{s^3} + \dfrac{2}{s^2} + \dfrac{1}{s}$, (g) $\dfrac{6}{s^2 + 9}$,

(h) $\frac{15}{s^2-9}$, (i) $\frac{3}{s^2+36}$, (j) $\frac{1}{(s+3)^2}$, (k) $\frac{4}{s}-\frac{6}{s^2+9}+\frac{1}{s-2}$,
(l) $\frac{6}{(s+2)^2}$, (m) $\frac{2}{s^2-1}$, (n) $\frac{s-3}{s^2-6s+10}$, (o) $\frac{s^2+4s+5}{(s+1)^3}$,
(p) $\frac{4}{(s+1)(s^2+2s+5)}$, (q) $\frac{s^2+9}{(s^2-9)^2}$, (r) $\frac{2s(s^2+27)}{(s^2-9)^3}$, (s) $\frac{6}{(s+3)^4}$

3 (a) $\frac{e^{-5s}}{s-2}$, (b) $\frac{e^{-2s}}{s^2}$ **4** $\frac{1}{s^2}-\frac{e^{-s}}{s(1-e^{-s})}$

5 (a) $\frac{1}{s(1+e^{-s})}$, (b) $\frac{1}{s^2(1+e^{-s})}-\frac{e^{-s}}{s(1-e^{-2s})}$, (c) $\frac{1-e^{-s}}{s^2(1+e^{-s})}$

6 (a) $(3s+1)X(s)$, (b) $(2s^2+1)X(s)$, (c) $\frac{1}{s(s-2)^2}$, (d) $\frac{4(s-1)}{s[(s-1)^2+4]^2}$

7 (a) $4t^2$, (b) $\cos t$, (c) $\frac{3}{2}e^{-t/2}$, (d) $\cos t+\sin t$, (e) $t\,e^{-2t}$, (f) $e^{-t}\cos 2t$,
(g) $4\,e^t\sinh 3t$, (h) $3t\,e^{-2t}$, (i) $3\,e^{4t}+2\,e^{-3t}$, (j) $1-e^{-t}$, (k) $\frac{3}{5}e^{2t}-\frac{3}{5}e^{-3t}$,
(l) $-\frac{1}{18}e^{-2t}+\frac{1}{54}e^{4t}-\frac{27}{8}e^{-t/2}$, (m) $4-6\cos\sqrt{2}\,t$, (n) $2\cos 2t-2\,e^{-t}$,
(o) $1-\cos 2t+2\sin 2t$, (p) $(e^{-t}-e^{-2t+2})u(t-2)$ **8** (a) $t-1+e^{-t}$,
(b) $\frac{1}{3}(e^{2t}-e^{-t})$, (c) $\frac{1}{6}(2\cos\sqrt{2}\,t+\sqrt{2}\sin\sqrt{2}\,t-2\,e^{-2t})$,
(d) $\frac{1}{2}(e^t-\sin t-\cos t)$ **9** (a) $x=2\,e^t-1$,
(b) $x=\frac{4}{17}\cos t+\frac{2}{17}\sin t-\frac{4}{17}e^{-4t}$, (c) $x=3-3\cos t+\sin t$,
(d) $x=\frac{1}{5}e^{-3t}+t\,e^{2t}-\frac{1}{5}e^{2t}$,
(e) $x=\frac{3}{65}e^{3t}+\frac{1}{20}e^{-2t}-\frac{1}{52}\sin 2t-\frac{5}{52}\cos 2t$,
(f) $x=\frac{1}{3}e^{-t}(\sin t+\sin 2t)$, (g) $x=\frac{5}{36}-\frac{1}{6}t-\frac{1}{12}e^{3t}+\frac{121}{252}e^{6t}+\frac{369}{252}e^{-t}$

10 (a) $x=-2\,e^t+e^{4t}, y=-\frac{2}{3}e^t+\frac{2}{3}e^{4t}$, (b) $x=-2\,e^t+5\,e^{4t}$,
$y=-4\,e^t+4\,e^{4t}$, (c) $x=\frac{1}{2}e^{-t}+\frac{3}{2}e^{-3t}$, $y=e^{-t}-1$,
(d) $x=1-\frac{1}{2}e^{-2t/3}-\frac{1}{2}e^{-2t}$, $y=\frac{1}{2}e^{-2t}-\frac{1}{2}e^{-2t/3}$,
(e) $x=\frac{1}{9}e^{-t}+\frac{4}{45}e^{2t}-\frac{1}{5}\cos t-\frac{2}{5}\sin t+\frac{1}{3}t\,e^{-t}$,
$y=\frac{1}{9}e^{-t}-\frac{1}{9}e^{2t}+\frac{1}{3}t\,e^{-t}$, (f) $x=\frac{2}{3}+e^{-3t/2}(\cos\sqrt{3}\,t/2+\sqrt{3}\sin\sqrt{3}\,t/2)$,
$y=\frac{28}{21}-\frac{3}{7}e^t+\frac{2}{21}e^{-3t/2}(\cos\sqrt{3}\,t/2+4\sqrt{3}\sin\sqrt{3}\,t/2)$

11 (a) 4, 0, (b) 0, 0, (c) 3, 0, (d) 5, –10

Chapter 40

Revision

1 (a) $\frac{z}{z-e^{-aT}}$, (b) $\frac{z}{(z-1)^2}$ **2** (a) $\frac{z}{z-e^{-2T}}$,
(b) $\frac{z(z-\cos 3T)}{z^2-2z\cos 3T+1}$, (c) $\frac{z}{z-2}$, (d) $\frac{3z}{(z-3)^2}$ **3** (a) $\frac{3z}{z-1}$,
(b) $\frac{2Tz}{(z-1)^2}$, (c) $\frac{Tz}{(z-1)^2}+\frac{T^2z(z+1)}{(z-1)^3}$, (d) $\frac{z}{z-e^{-2T}}-\frac{z}{z-e^{-3T}}$

4 (a) $\frac{1}{z^2}$, (b) $\frac{1}{(z-1)^2}$ **5** (a) $\frac{z\,e^{-aT}\sin\omega T}{z^2-2z\,e^{-aT}\cos\omega T+e^{-2aT}}$,
(b) $\frac{T^2z\,e^{aT}(e^{aT}+1)}{(e^{aT}-1)^3}$ **6** (a) $\frac{z}{z-1}$, (b) $\frac{z(1-e^{-2T})}{(z-1)(z-e^{-2T})}$,
(c) $\frac{2Tz\,e^{-3T}}{(z-e^{-3T})^2}$ **7** (a) Sampled $u(t)$, (b) sampled e^{-2t}

8 (a) 1, 1.2, 1.24, 1.248, (b) 0, 0.3, 0.51, 0.657 **9** (a) 0, 1.11, (b) 1, 0

Problems

1 (a) $\dfrac{z(z-\cos 4T)}{z^2-2z\cos 4t+1}$, (b) $\dfrac{3Tz}{(z-1)^2}$, (c) $\dfrac{2z}{z-1}-\dfrac{z}{z-\mathrm{e}^{-3T}}$, (d) $\dfrac{3Tz\,\mathrm{e}^{-4T}}{(z-\mathrm{e}^{-4T})^2}$, (e) $\dfrac{1}{z^3(z-1)}$, (f) 1, (g) z^{-1}, (h) $\dfrac{1}{z-1}$, (i) $\dfrac{T}{z(z-1)^2}$

2 (a) $\dfrac{Tz}{(z-1)^2}$, (b) $\dfrac{z\sin 3T}{z^2-2z\cos 3T+1}$, (c) $\dfrac{z\,\mathrm{e}^{-T}\sin 4T}{z^2-2z\,\mathrm{e}^{-T}\cos 4T+\mathrm{e}^{-2T}}$, (d) $\dfrac{z(\mathrm{e}^{-5T}-\mathrm{e}^{-T})}{(z-\mathrm{e}^{-5T})(z-\mathrm{e}^{-T})}$ **3** (a) 1, 3, 9, 27, (b) 1, 1, 1, 1, (c) 0, 1, 0, 1, (d) 3, 1, 2, 0, (e) 1, 0.5, 0.25, 0.125, (f) 0, 1, 3, 7, (g) 1, 2, 1, 1, (h) 0, 1, 6, 25, (i) 0, 1, 0, –2, (j) 0, 2, –1, –4, (k) 1, 3, 6, 10

4 (a) 1, 1, (b) 1, 0, (c) 3, 0

Chapter 41

Revision

1 (a) $0.01s\ \Omega$ in series with –0.002 V, in parallel with $0.2/s$ A, (b) $0.5/s\ \mathrm{M}\Omega$ in series with $5/s$ V, in parallel with 10 μA.

2 (a) $10+0.002s\ \Omega$, (b) $\dfrac{0.02s}{10+0.002s}\ \Omega$ **3** $i(t)=\dfrac{v_0}{R}\,\mathrm{e}^{-t/RC}$

4 $i(t)=\dfrac{V}{R}-\dfrac{V}{R}\,\mathrm{e}^{-Rt/L}$ **5** $i(t)=\dfrac{1}{R}\left(\delta(t)-\dfrac{1}{RC}\,\mathrm{e}^{-t/RC}\right)$

6 $i(t)=2[1-(1+t)\,\mathrm{e}^{-2t}]$ A **7** $\dfrac{R_2}{R_1+R_2+R_1R_2Cs}$

8 $G[t-\tau(1-\mathrm{e}^{-t/\tau})]$ **9** $1-\mathrm{e}^{-4t}-4t\,\mathrm{e}^{-4t}$ **10** $12\,\mathrm{e}^{-2t}-9\,\mathrm{e}^{-3t}-3\,\mathrm{e}^{-t}$

11 $\dfrac{k}{(Ls+R)(Is+c)}$ **12** $5/(s+53)$ **13** See Figure A.37

14 (a) 0, 1, 1.5, 1.75, 1.88, ..., (b) 0, 1, 0, 3, ... **15** $\dfrac{1}{z^2+z-2}$

16 1, 1, 1, ... **17** See Figure A.38 **18** See Figure A.39

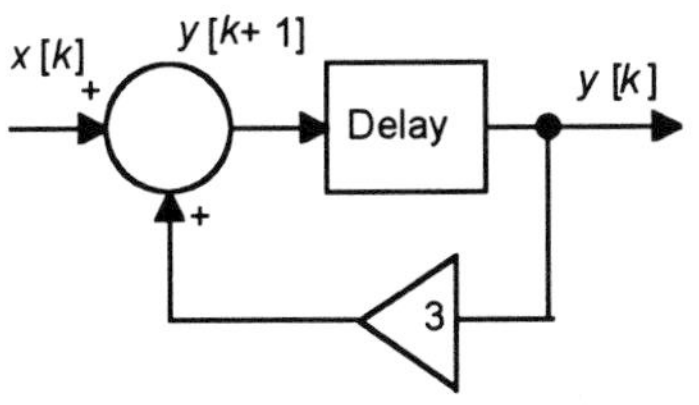

Figure A.37

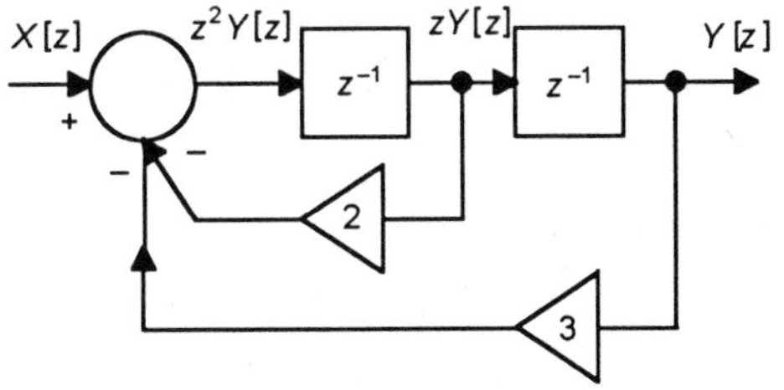

Figure A.38

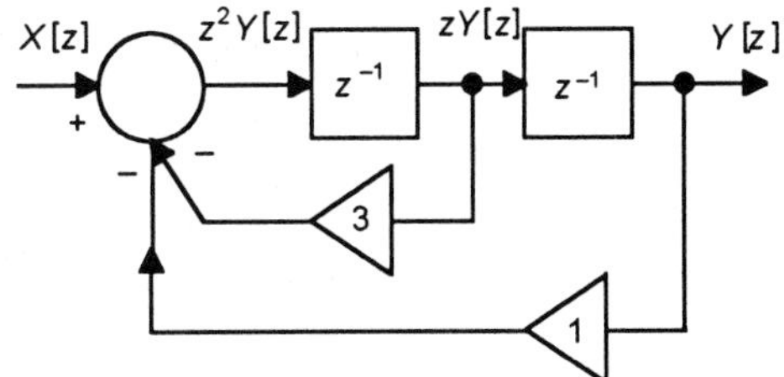

Figure A.39

Problems

1 (a) $0.001s\ \Omega$ in series with $2\ \mu V$, in parallel with $2/s$ mA, (b) $1/8s\ M\Omega$ in series with $10/s$ V, in parallel with $80\ \mu A$ **2** $4\ e^{-2t}$
3 $2 - e^{-2t}$ **4** $11.6\ e^{-0.25t} \sin 0.43t$ A **5** $5\ e^{-t} - 5\ e^{-2t}$ A
6 $\frac{1}{25}(5 - 2\ e^{-3t})$ A **7** $1 - (1 + t)\ e^{-2t}$ A **8** $0.5/(s + 0.5)$
9 $\frac{G}{\tau}\ e^{-t/\tau}$ **10** $1/(As + \rho g/R)$ **11** $1/(ms^2 + cs + k)$ **12** $t\ e^{-5t}$
13 $2\ e^{-4t} - 2\ e^{-3t}$ **14** $t - 1 + e^{-t}$ **15** $5/(5s + 1)(3s + 1)$
16 $\frac{4s}{s^2(s+1)+1}$ **17** $y[k + 2] + 2y[k + 1] + 4y[k] = x[k]$
18 (a) 8/3, 3/2, 7/4, ..., (b) 0, 1, 2, 3, .. , (c) 1, 0, –5, –2, ..., (d) 0, 2, 10, 38, ... **19** (a) $1/(z + 2)$, (b) $1/(z^2 - 5z + 6)$, (c) $1/(z^2 + 2z + 1)$, (d) $1/(z^2 - z + 2)$ **20** See Figure A.40
21 (a) 1, 2, 4, 8, ..., (b) 0, 1, 2, 3, ...

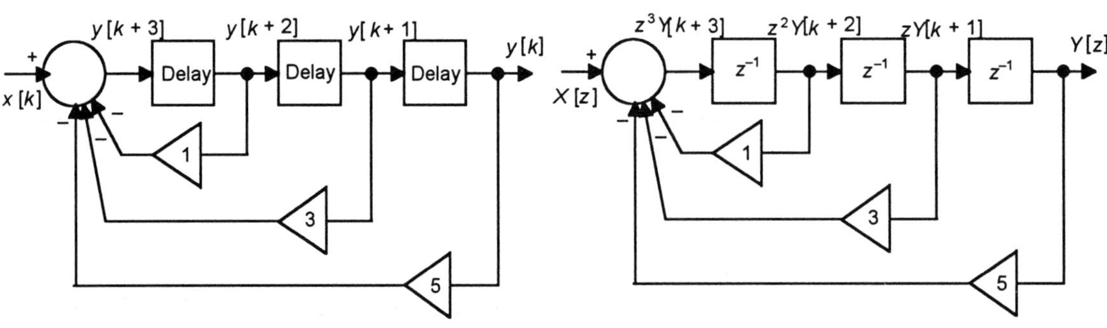

Figure A.40

Chapter 42

Revision

1 4/50 **2** 4/52 **3** (a) 0.01, (b) 0.02, (c) 0.03, (d) 0.96 **4** (a) 1/3, (b) 2/3 **5** 36 **6** 5^4 **7** (a) 20, (b) 6 497 400 **8** 120
9 105 **10** 7/15 **11** (a) 120, (b) 90 **12** 0.369 **13** 0.271
14 0.5 **15** (a) 0.4, (b) 0.162, (c) 0.508 **16** 0.08 **17** 0.74
18 0.04 **19** (a) 0.21, (b) 0.22

Problems

1 (a) 0.004, (b) 0.96 **2** 0.85 **3** 0.97 **4** (a) 40 320, (b) 840, (c) 30 **5** 1260 **6** (a) 21, (b) 21, (c) 1 **7** (a) 77/92, (b) 11/69, (c) 1/276 **8** (a) 1/15, (b) 7/15 **9** 5/42 **10** (a) 1/15, (b) 8/15
11 1000 **12** 0.5 **13** 1/3 **14** 0.875 **15** 0.35 **16** 1/12
17 0.89 **18** 1/36 **19** (a) 1/169, (b) 1/221 **20** (a) 13/24, (b) 5/24
21 0.01 **22** (a) 0.818, (b) 0.008 4

Chapter 43

Revision

1 0.135 **2** 0.586 **3** 3.0 **4** 2/3 **5** 5 **6** 0.237 **7** 0.24
8 1.4 **9** (a) 0.4, (b) 7 **10** 0.132 **11** (a) 0.817, (b) 0.016

12 0.366 **13** 14.3, 3.2 **14** 0.865 **15** 0.122 **16** 0.001 **17** 0.34 **18** 0.954 **19** 0.185 9 **20** 0.184 1 **21** (a) 0.315 6, (b) 0.726 0

Problems

1 0.393 **2** 0.5 **3** 4/3 **4** 40, 31.6 **5** (a) 0.02, 0.06, 0.22, 0.32, 0.28, 0.08, 0.02, (b) 69.3, 2.3 **6** 0.5 **7** 2 **8** 0.122, 0.270, 0.285, 0.190, 0.090, 0.032, 0.009, 0.002, 0.000 4, 0.000 1 **9** 1/16, 4/16, 6/16, 4/16, 1/16 **10** 0.005, 0.029, 0.078, 0.138, 0.181, 0.185, 0.154, 0.108, 0.064, 0.033 **11** 0.016 **12** 100, 9.9 **13** 0.908 **14** 0.135 **15** 0.297 **16** 0.950 **17** 20 **18** (a) 6.68, (b) 17.75 **19** (a) 9.68, (b) 11.5 **20** (a) 0.091 3, (b) 0.091 3, (c) 0.817 4 **21** $50 \pm 6.6\ \Omega$

Chapter 44

Revision

1 $y = 0.66x + 29.15$ **2** $e = 6.0W + 5.6$

Problems

1 $m = 0.500\theta + 44.86$ **2** $d = 9.884w + 0.062$ **3** $y = 0.857x + 0$
4 $y = 1.538x + 0.640$ **5** $E = 0.307W + 1.567$ **6** $y = 3.071\ e^{0.5056x}$
7 $y = 0.40x^{1.70}$ **8** $y = 1.044\ e^{0.682x}$

Chapter 45

Revision

1 2.134 mm, 0.011 mm **2** 0.05 A **3** 72 **4** $4.9 \pm 0.3\ \text{mm}^2$
5 (a) 150 ± 3.6, (b) 5000 ± 250, (c) 2 ± 0.1
6 $(\Delta Z)^2 = \left(\frac{\cos i}{\cos r}\right)^2 (\Delta i)^2 + \left(\frac{\sin i \cos r}{\sin^2 r}\right)^2 (\Delta r)^2$ **7** 1.2
8 $(\Delta Z)^2 = (\ln B)^2(\Delta A)^2 + (A/B)^2(\Delta B)^2$

Problems

1 $51.12\ \Omega$, $0.08\ \Omega$ **2** 10 000 **3** 39.0 kV, 0.11 kV
4 (a) 50 ± 3.6, (b) $10\,000 \pm 500$, (c) $1\,000\,000 \pm 52\,000$, (d) 2 ± 0.1
5 Diameter
6 $(\Delta f)^2 = \left(\frac{(u+v)v - uv}{(u+v)^2}\right)(\Delta u)^2 + \left(\frac{(u+v)u - uv}{(u+v)^2}\right)(\Delta v)^2$
7 $\Delta Z = \Delta A/A$ **8** $\Delta Z/Z = c\,\Delta A$

Index

86874